'This is a superlative collection, bringing together some of the finest minds in the field essays offer both range and depth, and anyone interested in the history of war from a global perspective will want to have this volume on their bookshelves.'

Richard Reid, *Professor of African History, University of Oxford*

'A Collection of great importance not only to military historians but also to those interested in global history. Conceptually alert, this volume deserves much attention.'

Jeremy Black, *Emeritus Professor, Exeter University*

'In 35 chapters written by distinguished and established historians the Handbook explores warfare from the agrarian Bronze Age through to the challenges of contemporary, 'postmodern' conflict. The perspective is uncompromisingly global, and offers remarkable scope for comparative analysis and understanding. The volume is an invaluable and indispensable work for all military historians.'

David Parrott, *Professor of Early Modern History, University of Oxford*

'This timely compendium, drawing on an impressive mix of established and emerging scholars, provides insights into the conduct and character of wars on a global scale, stretching from the earliest recorded conflicts to the postcolonial and postindustrial conflicts of today. It will quickly become a valuable platform for further scholarship.'

Douglas M. Peers, *Professor of History, University of Waterloo*

ROUTLEDGE HANDBOOK OF THE GLOBAL HISTORY OF WARFARE

This handbook examines key aspects of the development of the global history of warfare and the changing patterns of warfare over time.

Although scholarship has long eschewed a chronological narrative of the evolution of warfare that privileges the Western experience, global histories of warfare have had difficulty avoiding an overemphasis on the West. The present volume is a collection of themes rather than a history per se; it provides important perspectives on the emergence of warfare as a global historical experience from the ancient past to the present day. Drawing together numerous experts, it tells a broader, more inclusive story of the global, human experience with wars and warfare. The 35 chapters are organised in eight thematic Parts:

Part I: Origins of Warfare
Part II: Polities and Armed Forces in the Pre-Modern Era
Part III: Steppe Nomads of Eurasia
Part IV: Naval Warfare and Piracy in the Pre-Industrial World
Part V: The Impact of Gunpowder
Part VI: Transition from Industrial to Total War
Part VII: Wars of Decolonisation and Cold War
Part VIII: Postmodern/New Wars

These Parts offer an overview of the global experience of warfare to help readers understand how the wars and the militaries we see today have been shaped by historical developments across the globe.

This handbook will be of great interest to students of military history, naval history, strategic studies and world history in general.

Kaushik Roy is Guru Nanak Chair Professor in the Department of History, Jadavpur University, Kolkata, India. He specialises in warfare in Eurasia and is currently associated with the 'Warring with Machines' Project of the Peace Research Institute Oslo (PRIO), Norway.

Michael W. Charney is Professor of Asian and Military History at SOAS, University of London, where he teaches the history of violence, cities and the emergence of the contemporary world. His work mainly focuses on Southeast Asia and Africa.

ROUTLEDGE HANDBOOK OF THE GLOBAL HISTORY OF WARFARE

Edited by Kaushik Roy and Michael W. Charney

Routledge
Taylor & Francis Group

LONDON AND NEW YORK

Cover image: *Le Petit Journal* (Paris) 29 July 1893, p. 240

First published 2024
by Routledge
4 Park Square, Milton Park, Abingdon, Oxon OX14 4RN

and by Routledge
605 Third Avenue, New York, NY 10158

Routledge is an imprint of the Taylor & Francis Group, an informa business

British Library Cataloguing-in-Publication Data
A catalog record has been requested for this book
Library of Congress Cataloging-in-Publication Data
Names: Roy, Kaushik, 1971– editor. |
Charney, Michael W., editor.
Title: Routledge handbook of the global history of warfare /
edited by Kaushik Roy and Michael W. Charney.
Description: London ; New York, NY : Routledge,
Taylor & Francis Group, 2024. | Includes bibliographical
references and index.
Identifiers: LCCN 2023038527 (print) |
LCCN 2023038528 (ebook) |
ISBN 9781138345386 (hardback) |
ISBN 9781032676296 (paperback) |
ISBN 9780429437915 (ebook)
Subjects: LCSH: Military art and science—History.
Classification: LCC U27 .R68 2024 (print) |
LCC U27 (ebook) | DDC 355.0209—dc23/eng/20231115
LC record available at https://lccn.loc.gov/2023038527
LC ebook record available at https://lccn.loc.gov/2023038528

ISBN: 978-1-138-34538-6 (hbk)
ISBN: 978-1-032-67629-6 (pbk)
ISBN: 978-0-429-43791-5 (ebk)

DOI: 10.4324/9780429437915

Typeset in Galliard
by Apex CoVantage, LLC

CONTENTS

Contents

Contents

Contents

PREFACE AND ACKNOWLEDGEMENTS

The collection of essays in this handbook is intended as a guide to important aspects of the development of the global history of warfare, and the changing patterns of warfare over time, from its more local and regional multi-centred origins to the global terrain of warfare we see today. The Introduction sets out the context of the book and portrays the recent debates in the field from the point of view of warfare being worthy of study as a global phenomenon. Although scholarship has long eschewed a chronological narrative of the evolution of warfare privileging the Western experience and its export abroad, world or global histories of warfare have had difficulty avoiding an overemphasis on the West. Some scholars have achieved an effective balance regarding specific topics. But there has not yet emerged a satisfying and comprehensive approach to the global history of warfare that achieves such a balance throughout. In order to meet that need, this handbook provides important perspectives on the emergence of warfare as a global historical experience from the ancient past to the present. The contributors are drawn from varied backgrounds (historians, international relations experts, archaeologists, anthropologists and retired military officers) and from four continents (North America, Europe, Asia and Australia) in order to get the widest possible coverage and interdisciplinary approach.

This handbook comprises 35 chapters and an Introduction. The chapters are organised under eight Parts. The Parts are arranged thematically and within each, the chapters are organised chronologically, as far as possible. The chapters focus on the interplay of technology, society, culture and economy in the evolution of war in different parts of the world at various moments. Further, all the contributors also provide the relevant historiographical matrix before setting out on their subjects. This handbook gives equal attention to regular and irregular warfare, the theory and practice of warfare and deals with land, air and sea warfare. Moreover, unlike other handbooks, companions and general global histories, the *Routledge Handbook of the Global History of Warfare* does not merely concentrate on the military history of the West but gives due attention to the development of militaries in Africa, Asia, Latin America and pre-colonial North America. This handbook especially focuses on the varieties of military cultures beyond the West and the rich military history of these regions before the rise of the West in the early modern era.

The first Part deals with the origins of war and prehistoric warfare, and this periodisation depends upon the local context. Part II deals with the interlinkages between the formation of polities and the evolution of armed forces in the premodern era. Part III is concerned with the military activities of the steppe nomads of Eurasia, while Part IV turns the focus on regular and irregular warfare at sea during the preindustrial era. Part V focuses on the role played by the gunpowder weapons, and Part VI delineates the transition from industrial war to total war. Part VII deals with wars of decolonisation, the Cold War and the emergence of postcolonial states. Finally, Part VIII discusses the role of new wars in more recent times.

Owing to crises beyond our control, this handbook ran later than first hoped. There was of course the Covid-19 period. Then, our mentor Professor Dennis Showalter, who had agreed to write one of the chapters, passed away suddenly. Another shock was our close friend-cum-*guru* Bruce Vandervort, who had intended to write a different chapter and then fell victim to cancer. Another one of our contributors, Matthew Trundle, also is no longer with us. Brian Davies, another scholar-cum-friend who was to contribute to this handbook, fell seriously ill. Finally, we wish to thank our contributors for their patience and especially our editor at Routledge, Andrew Humphrys, who generously allowed us extra time.

Kaushik Roy, Kolkata, and Michael Charney, London

ABBREVIATIONS

AESA	Advanced Electronically Scanned Array
AEWC	Airborne Early Warning and Control
AFV	Armoured Fighting Vehicle
AI	Artificial Intelligence
AMLRS	Artillery Multiple-Launch Rocket Systems
AQI	Al Qaeda Iraq
AQIM	Al Qaeda in the Maghreb
AWACS	Airborne Warning and Control System
BEF	British Expeditionary Force
BIOT	British Indian Ocean Territory
BVR	Beyond Visual Range
CIA	Central Intelligence Agency
COIN	Counterinsurgency
CPA	Coalition Provisional Authority
CRPF	Central Reserve Police Force
CSBMs	Confidence and Security Building Measures
EEZ	Exclusive Economic Zone
EIC	British East India Company
EU	European Union
FOB	Forward Operating Base
FONOPs	Freedom of Navigation Operations
GDP	Gross Domestic Product
GHQ	General Headquarters
4GW	Fourth Generation Warfare
5GW	Fifth Generation Warfare
GWOT	Global War on Terror
IAF	Indian Air Force
IDF	Israeli Defence Forces
IED	Improvised Explosive Device
IJA	Imperial Japanese Army

IJN	Imperial Japanese Navy
IMF	International Monetary Fund
INA	Indian National Army
IPKF	Indian Peacekeeping Force
ISAF	International Security and Assistance Force
ISI	Inter-Services Intelligence
ISIS	Islamic State of Iraq and Syria
ISR	Intelligence, Surveillance and Reconnaissance
IT	Information Technology
IW	Information Warfare
IWM	Imperial War Museum, London
LICO	Low Intensity Conflict Operation
LOAC	Laws of Armed Conflict
LOOs	Lines of Operations
LTTE	Liberation Tigers of Tamil Eelam
MANPADS	Man Portable Air Defence Systems
MEF	Marine Expeditionary Force
MEO	Military Evacuation Organisation
MTR	Military Technical Revolution
NATO	North Atlantic Treaty Organisation
NCW	Net Centric War
NEI	Netherlands East Indies
NGO	Non-Government Organisation
NSWC	New South Wales Corps
OEF	Operation Enduring Freedom
OODA	Observe, Orient, Decide and Act
PBF	Punjab Boundary Force
PIE	Proto Indo-European
PKK	Kurdistan Workers' Party
PLO	Palestine Liberation Organization
PMC	Private Military Company/Corporation
POK	Pakistan-Occupied Kashmir
POW	Prisoner of War
PRC	People's Republic of China
PRT	Provincial Reconstruction Team
RAF	Royal Air Force
RFC	Royal Flying Corps
RIAF	Royal Indian Air Force
RMA	Revolution in Military Affairs
RPG	Rocket-Propelled Grenade
RSS	*Rashtriya Swayamsevak Sangh*
SAM	Surface to Air Missile
SEAC	South East Asia Command
SIPRI	Stockholm International Peace Research Institute
SLAF	Sri Lanka Air Force
SLBM	Submarine-launched Ballistic Missile
SLOC	Sea Lines of Communication

Abbreviations

SOF	Special Operations Forces
SOFA	State of Forces Agreement
TBA	Tactical Battle Area
TTP	Tactics, Techniques and Procedures
UAE	United Arab Emirates
UAV	Unmanned Aerial Vehicle
UN	United Nations
UNCLOS	United Nations Convention of the Law of the Sea
UNSCR	United Nations Security Council Resolution
USAAC	United States Army Air Corps
USAAF	United States Army Air Forces
USAF	United States Air Force
USN	United States Navy
VOC	Dutch East India Company
WMD	Weapons of Mass Destruction

GLOSSARY

Amir/emir The title means a royal prince or a chieftain or commander of the army

Askari African infantry soldier in employ of the colonising European power

Bey Governor of a district in the Ottoman Empire. It was an honorific Turkish title

Blitzkrieg Literal meaning lightning war. *Blitzkrieg* refers to the combined arms operation (armour-infantry-self-propelled artillery and ground attack aircraft) practised by the *Wehrmacht* between 1939 and 1942

Boyar A high-ranking feudal elite of the preindustrial Russia

Caliph Spiritual and temporal head of the Islamic *umma* (community)

Caliphate Rule or domain of the *caliph* constituting *dar al Islam* (land of light, righteousness)

Daimyo A feudal landlord of medieval Japan. Technically *daimyo* was a vassal of the *shogun*. A *daimyo* exercised administrative power and controlled a band of warriors

Dey Turkish rulers of Algiers and Tripoli in the Ottoman Empire were given this title

Emirate Domain of the *emir*

Ghulam Slave soldier of the *Caliphate*. They were mainly mounted archers and due to their specialised training and skill in horse archery were considered elite. Many were given higher education and manned the bureaucracy of the Islamic polities

Guliai gorod Mobile fortifications (field works) constructed by the Russian (Tsarist) infantry during the sixteenth and seventeenth centuries. These fortifications were constructed of wooden logs with wheels attached to them for mobility. Behind such wooden walls, the Russian infantry took position and fired at the hostile force. It was a modification of the wagon system used by the Central Eurasian steppe nomads

Helot Rural serfs of Messenia and Laconia. They provided food and provisions to the Spartan citizens

Hoplite Heavily armed drilled and disciplined Greek infantry soldier who was organised in *phalanx* formation

Intifada Popular uprising of the Palestinians against the Israeli occupation force

Janissary Disciplined elite infantry of the Ottoman Empire. From the sixteenth century onwards, they constituted the drilled gunpowder infantry of the Ottoman military machine

Jatha Sikh warbands

Jihad   Literal meaning holy war in Islam. The wars undertaken to defend Islam and to spread the religion of the prophet are known as *jihad*. It is undertaken to transform the *dar al harb* (land of darkness/heathens) into *dar al Islam* (land of light)

Jihadism   A conservative revolutionary programme geared to change the structure of both the Islamic and non-Islamic societies and to establish a new society in accordance with the golden pristine principles of early Islam. This sort of programme is followed by militant Islamic groups

Jihadist   A Muslim who advocates *jihad* or participates in the Islamic holy war

Lascar   The term refers to a coolie as well as a non-combatant Indian naval crew. During the British period, this term was also applied to an Indian coolie who helped the gunner and the loader

Luftwaffe   German Air Force under the Third Reich

Mamluk   Same as *ghulam*. In the medieval Islamic world, many *mamluks* from professional soldiers became generals, sultan makers and finally sultan themselves

Mujahideen   An armed fighter (a holy warrior) who engages in *jihad* (Islamic holy war)

Pa   Maori defensive work

Pasha   A title given by the Ottoman Court to generals and ministers

Peshwa   Prime minister of the Maratha Confederacy

Phalanx   Rectangular formation of disciplined Greek heavy infantry armed with long pikes

Pharaoh   Ruler (king) of ancient Egypt

Pinnace   A small boat with sail or oars or both

Polis   A self-governing city state of Classical Greece. Polis was a political entity with the city controlling the agricultural hinterland around it

Pomeste   Feudal land tenure similar to the *jagir* system. The *pomeste* holder enjoyed a grant of land in return for providing military service to the tsar/czar. Generally, the pomeste holder provided cavalry

Porte   Ottoman Sultan

Samurai   Hereditary military nobility of medieval Japan. The *samurai* is equivalent to medieval India's Rajput and Western Europe's knight. A *samurai* was a mounted warrior and served under a *daimyo*

Shaikh   An Arab term which refers to the head of a family or clan or a village or even a tribe

Shaman   A mystic who can control evil spirits and heal a person

Shogunate   Realm (regime or rule) of the *shogun* who happened to be the military dictator of medieval Japan. In the *shogunate*, the Japanese emperor became a puppet

Sipahi   Refers to a soldier, especially foot soldier

Tagmata   Regimental or battalion-size military formation of the Byzantine Army. Such units comprised heavy cavalry. *Tagmata* comprised standing permanent force of professional soldiers of the Byzantine emperors

Tajik   Inhabitants of Tajikistan. They speak Tajiki, a form of Iranian/Persian language. The Tajiks are probably members of Proto Indo-European language-speaking people

Tsunami   Huge waves created in the sea or ocean due to earthquake or volcanic eruption

Umma   The term refers to the Muslim community as a whole

Vizier   Minister of the Ottoman Sultan

Wehrmacht   German armed forces during the Third Reich

Wokou   Japanese pirates who operated along coastal waters of China during the medieval period

FIGURES, MAPS AND TABLES

Figures

Maps

Tables

NOTES ON CONTRIBUTORS

Robert J. Antony, before his retirement in 2019, was Distinguished Professor and Senior Researcher at Guangzhou University. Recently he has been a visiting scholar at the School of Historical Studies, Institute for Advanced Studies in Princeton. His newest books are *The Golden Age of Piracy, 1520–1810: A Short History with Documents* (2022) and *Rats, Cats, Rogues, and Heroes: Glimpses of China's Hidden Past* (2023).

Bernard S. Bachrach, Emeritus University of Minnesota, earned his PhD from the University of California, Berkeley in 1966 and was elected a Fellow of the Medieval Academy in 1986; after 1986 Professor Bachrach earned his A.B. from Queens College, NYC, in History and Classical Languages. He is the founding editor of *The Journal of Medieval Military History* and co-founding editor of *Medieval Prosopography*. His most recent book is *Medieval Military History*, co-authored with David S. Bachrach (2017).

Andrew Bamford obtained his PhD from the University of Leeds for a thesis on British manpower management during the Napoleonic Wars, subsequently published, by the University of Oklahoma Press, as *Sickness, Suffering and the Sword*. He has continued to write extensively on the British Army in the eighteenth and early nineteenth centuries, and currently edits the book series 'From Reason to Revolution, 1721–1815' for Helion and Company. He has served for a number of years on the Council of the Society for Army Historical Research, of which he is currently Hon. Secretary.

Mary Kathryn Barbier is a Professor of History at Mississippi State University. She is co-editor of *War in History*, co-director of the Second World War Research Group, North America (SWWRG, NA) and a North American Society for Intelligence History (NASIH) board member and officer. She is the author of several articles, book chapters in edited collections, and books, including *Spies, Lies, and Citizenship: The Hunt for Nazi Criminals* (2018) and *D-Day Deception: Operation Fortitude and the Normandy Invasion* (2007).

Cameron Barnes' original specialisation was medieval economic history. In recent years, he has published in different fields, from bibliometrics and university pedagogy to Hunnic history.

Ian F.W. Beckett retired as Professor of Military History from the University of Kent in 2015. A Fellow of the Royal Historical Society and also of the Society for Army Historical Research, he has held chairs in both the United States and the UK. He is internationally known for his work on the history of the British Army in the nineteenth and twentieth centuries and on the First World War. He was chairman of the Council of the UK Army Records Society from 2000 to 2014. His most recent book is *The British Army: A New Short History* (2023).

Michael W. Charney is Professor of Asian and Military History at SOAS, University of London, where he teaches the history of violence, cities, and the emergence of the contemporary world. His work mainly focuses on Southeast Asia and West Africa.

Erica Charters is Professor at the Faculty of History in the University of Oxford. Her research examines the history of war, disease and bodies, particularly in the British and French empires. More specifically, her research focuses on manpower during the eighteenth century, examining the history of bodies as well as the history of methods used to measure and enhance bodies, labour, and population as a whole, including the history of statistics.

O.R. Coates is Director of Studies in History and Politics and Bye-Fellow at St Edmund's College, Cambridge, and Associate Researcher at the Institut des mondes africains, C.N.R.S., Paris. Dr Coates' recent research has appeared in the *International Journal of African Historical Studies* and the *Journal of African Military History*.

John Connor is Senior Lecturer in History and Head of the Humanities Research Group at University of New South Wales, Canberra. He is the author of several books, including *The Australian Frontier Wars, 1788–1838* (2003), shortlisted for the Royal United Services Institute's Westminster Medal for Military Literature, and *Anzac and Empire: George Foster Pearce and the Foundations of Australian Defence* (2011).

Robert Drews is Professor of Classics Emeritus, Vanderbilt University. He received his PhD from Johns Hopkins University in 1960 and taught ancient history at Vanderbilt for 45 years, and as an adjunct professor taught occasional courses until 2014. His latest book is *Militarism and the Indo-Europeanizing of Europe* (2017). His principal interests have been the evolution of warfare and of religion in antiquity.

Mark Charles Fissel (PhD California, Berkeley 1983) has published two monographs on early modern military history. At present, he is doing research on amphibious warfare and military revolutions. Fissel is a Fellow of the Royal Historical Society.

A. James Fuller is a Professor of History at the University of Indianapolis. Among his many publications are six books, including *Chaplain to the Confederacy: Basil Manly and Baptist Life in the Old South, 1798-1868* (2000); *The Election of 1860 Reconsidered* (2013); and *Oliver P. Morton and the Politics of the Civil War and Reconstruction* (2017).

Andrew de la Garza completed his MA and PhD in History at Ohio State University, with specialties in military and South Asian history, and graduated in 2010. He is currently a

Senior Instructor of History at the University of Louisiana at Lafayette. He is the author of *The Mughal Empire at War: Babur, Akbar and the Indian Military Revolution: 1500–1605* (2016).

John Haldon is emeritus Professor of Byzantine History & Hellenic Studies at Princeton University and is currently Director of the Princeton Climate Change and History Research Initiative. A Fellow of the British Academy and a Corresponding Member of the Austrian Academy of Sciences, his research focuses on the history of the medieval eastern Roman (Byzantine) empire, in particular 7th-12thC; on pre-modern state systems and structures across the European and Islamic worlds; and on the impact of environmental stress on societal resilience in pre-modern systems.

Philip T. Hoffman is Rea A. and Lela G. Axline Professor of Business Economics and Professor of History at the California Institute of Technology.

John D. Hosler (PhD, University of Delaware) is a Professor of Military History at the Command and General Staff College at Fort Leavenworth, Kansas. A specialist on medieval warfare and the Crusades, he is the author and editor of nine books, including *Jerusalem Falls* (2022), *Seven Myths of Military History* (2022) and *The Siege of Acre, 1189–1191* (2018). He is a Past President of De Re Militari: the Society for Medieval Military History, a Trustee of the U.S. Commission for Military History and a Fellow of the Royal Historical Society.

Alan Jeffreys is Head of Equipment and Uniform at the National Army Museum, London and a visiting research fellow at the University of Greenwich. He is the author of *Approach to Battle: Training the Indian Army during the Second World War* (2016) and editor of *The Indian Army in the First World War: New Perspectives* (2018).

Robert Johnson is the Director of the Changing Character of War Research Centre at the University of Oxford and Senior Research Fellow at Pembroke College Oxford. Dr Johnson is the author of several books, including *The Afghan Way of War* (2011), co-editor of *At the End of Military Intervention* (2014) and the *True to Their Salt* (2017), as well as articles on the Iraqi and Afghan security forces. He specialises in the history of war and strategy, military thought, and on conflicts of the Eurasian region.

Mark Johnston is Head of History at Scotch College, Melbourne, Australia. He has also been an Adjunct Senior Lecturer at the Strategic and Defence Studies Centre, Canberra. He is the author of 12 books on Australian military history, including *Fighting the Enemy: Australian Soldiers and their Adversaries in World War II* (2000) and *Whispering Death: Australian Airmen in the Pacific War* (2011).

Christopher Kinsey is a Reader in Business and International Security with Defence Studies Department at King's College London. His research examines the role of the market in war. Dr Kinsey has published widely on the subject from books, book chapters and articles in leading academic journals. His present work looks at mercenary activity in Africa since the Cold War. Dr Kinsey's most recent publication is *The Mercenary: An Instrument of State Coercion* (2023).

Amarendra Kumar PhD, is Associate Professor in History at Visva-Bharati University, Santiniketan, India. Previously, he has taught at the National Defence Academy, Khadakwasla, Pune. He specialises in Maratha History. He has published articles on various aspects of Maratha history in journals and edited volumes. His research interests include military history, diplomacy and cartography.

Andrew Lambert is Laughton Professor of Naval History in the Department of War Studies at King's College, London. He has written several books and his latest one is *The British Way of War: Julian Corbett and the Battle for a National Strategy* (2021).

Timothy May, PhD, is Associate Dean of Arts & Letters and Professor Central Eurasian History. He writes on the Mongol Empire.

Rob McLaughlin is a Professor at the Australian National Centre for Ocean Resources and Security at the University of Wollongong, and a Senior Fellow at the Stockton Centre for International Law at the US Naval War College. Prior to his academic career, he was a warfare officer and a legal officer in the Royal Australian Navy, and then Head of the UN Office on Drugs and Crime's Maritime Crime Program.

Sarah C. Melville (PhD, Yale University) is Professor of Ancient History at Clarkson University in Potsdam, New York. She specialises in Neo-Assyrian history and warfare and her latest book is *The Campaigns of Sargon II, King of Assyria, 721–705 BC* (2016).

Rob Geist Pinfold is a Lecturer in International Peace and Security at Durham University's School of Government and International Affairs and a Research Fellow at the Peace Research Center Prague, Charles University. Previously, Dr Pinfold has worked and taught at King's College London, the University of Haifa and the Hebrew University of Jerusalem. He is a scholar of international security whose research intersects the study of strategy and territorial conflict, with a focus on the Middle East and the Arab–Israeli conflict.

Kaushik Roy is Guru Nanak Chair Professor in the Department of History, Jadavpur University, Kolkata, India. He specialises in warfare in Eurasia and is currently associated with 'Warring with Machines' Project of Peace Research Institute Oslo, Norway.

Andrew K. Scherer is an Associate Professor of Anthropology and Archaeology and Director of the Joukowsky Institute for Archaeology and the Ancient World at Brown University. He is an archaeologist and biological anthropologist whose research focuses on the precolonial Maya of Mexico and Guatemala. He is author of *Mortuary Landscapes of the Ancient Maya: Rituals of Body and Soul* (2015) and co-editor of *Smoke Flames, and the Human Body in Mesoamerican Ritual Practice* (with Vera Tiesler, 2018).

Air Vice Marshal **Arjun Subramaniam** (Retd) is an airpower and military historian and the author of two definitive books on contemporary Indian military history, *India's Wars: A Military History 1947–1971* (2017) and *Full Spectrum: India's Wars 1972–2020* (2020).

Kenneth M. Swope is Professor of History and Senior Fellow of the Dale Center for the study of War and Society at the University of Southern Mississippi (USA). A Board Member of the Chinese Military History Society, he is the author and editor of numerous books and articles on Late Imperial Chinese military history, including *On the Trail of the Yellow Tiger: War, Trauma, and Social Dislocation in Southwest China during the Ming-Qing Transition* (2018).

Andrew T.H. Tan (M Phil, Cambridge; PhD, Sydney) is Professor in the Department of Security Studies and Criminology, Macquarie University, Australia. He is a strategic analyst with expertise in defence studies, terrorism, insurgency and regional geopolitics, and is the author, co-author, editor or co-editor of 23 books, mostly published by Routledge, Edward Elgar and Palgrave Macmillan.

The late **Matthew Trundle** was Chair and Professor of Classics and Ancient History at the University of Auckland, New Zealand. He passed away in 2019.

INTRODUCTION
THE EVOLUTION OF WARFARE
Global Perspectives

Kaushik Roy and Michael W. Charney

Typologising the Global History of Warfare

Warfare as organised violence can be traced back to the New Stone Age. In the words of historian Azar Gat, 'For once humans had evolved agriculture, they set in train a continuous chain of developments that have taken them further and further away from their evolutionary natural way of life as hunter-gatherers' (Gat 2006: 145). We can argue that the beginning of agriculture around 10,000 BCE gave spurt to the production of surplus, demographic growth, the beginning of fortifications for the protection of agricultural wealth and the human population and the subsequent genesis of warfare. The last 12 millennia have included both regular warfare in the form of conventional conflicts between two or more polities and irregular warfare, including guerrilla war, terrorism, etc. that consists of violent confrontations in which at least one of the sides is a non-state group.

How do we study the history of warfare stretching back more than 12,000 years? One approach is cultural, which became prominent from the 1990s. The cultural approach is an updated version of the war and society studies that became dominant in the West in the late 1950s. The war and society perspective moves away from the study of battles, campaigns and great generals. It attempts to contextualise the armed forces as an institution within the broader fabric of the host society. The American historian Michael S. Neiberg aptly explains the objectives of the war and society approach in the following words: 'War and society scholars are generally not interested in explaining victory or defeat. Events on the battlefield and the performance of military leaders usually remain in the background or are completely absent' (Neiberg 2006: 42).

The cultural and linguistic turn in the 1980s further encouraged the study of the history of warfare through a cultural lens. Some have viewed this approach as the progenitor of New Military History. Matthew Hughes and William J. Philpott note the '"new" military history that approaches war through the prism of discourse analysis, culture, gender, race and memory' (Hughes and Philpott 2006: 4). Psychological experiences and the trauma of the civilians caught up in the midst of war, memory, mourning and remembrance are the themes of cultural studies of warfare (Philpott 2006: 144). One of the practitioners of New Military History, Joanna Bourke, following Dominick LaCapra, writes that texts are not merely

DOI: 10.4324/9780429437915-1

sources of information but are important events in their own light. Bourke continues that the approach of New Military historians is not merely to lay bare what happened but how what happened was imagined by the narrator, blurring the lines between fact and fiction (Bourke 2006: 258–75).

The cultural approach, as practised by military historians like Victor Davis Hanson, John Keegan, and others, makes a binary division between Western and Eastern (Oriental or Asian) cultures. The culturalists argue that while Western military culture invented the concept of decisive battles with the objective of fight to the finish, the Easterners conducted attrition-oriented harassing warfare. Western military ethos was battle-centric and focused on drilled and disciplined infantry geared to destroy the enemy force in a single afternoon. Civic ethos and communitarian feeling along with idealism (patriotism, nationalism) inspired the ancient Greeks. In contrast, the Eastern military ethos highlighted the display of force (deployment of large numbers of undisciplined soldiers), guerrilla attacks, bribery, diplomatic intrigues and perfidy. In this paradigm, non-European soldiers were slaves and mercenaries while Western soldiers were professionals imbued with the idea of freedom.[1] The culturalists draw a linear trajectory from the so-called decisive hoplite battle invented by the Classical Greeks in the sixth century BCE to the American fiasco in Vietnam during the late 1960s and the early 1970s.[2]

A binary division of the world is simplistic and somewhat frozen in time. Stephen Morillo and Michael F. Pavkovic point out that Hanson conflates modern with Western (Morillo and Pavkovic 2006: 84). What the culturalists forget is that battle-centric strategy has not been merely the monopoly of the Western armed forces throughout history. The Byzantine Empire, when faced with the Islamic threat from the eighth century onwards, avoided battles (Kaegi Jr. 2007). Even Vegetius, the military theorist of the Later Roman Empire advised his readers to avoid battles which are uncertain and costly in their result (Phillips 1989: 75–175). Shock battle was not the mainstay of the Carolingian Army (France 2008: 331). In fact, some of the greatest decisive battles in world history were waged by the Mongols. Unlike the Classical Greeks, the Mongols did not let go of their defeated enemies but pursued the defeated hostile force till the latter collapsed completely. It was probably they and not Napoleon Bonaparte who invented the operational level of war.[3] Maintaining drilled and disciplined infantry armies was not a Western exclusivity. Several ancient Chinese empires, starting from the 'terracotta soldiers,' maintained drilled and disciplined infantry. Further, the seeds of drilled and disciplined infantry engaging in pitched battle could be traced back to ancient Mesopotamia and *Pharaonic* Egypt.

An important contribution of military historians working with the cultural paradigm is that their approach has made scholars conscious about Eurocentrism, teleology and techno determinism. Culturalists deserve our praise for sensitising us to the importance of gender aspects and marginal groups in military matters. Moreover, culturalists, by posing a serious challenge to positivism, have alerted us to the limitations of the Rankean tradition of history writing.

Another way to approach global military history is to use the war and economy perspective. In this outlook, economy functions as the principal driver of changes in the nature and scope of conflict. Not all the practitioners of this line of interpretation are Marxists but all of them more or less suffer from economic determinism. They divide the evolution of warfare into the following states: Primitive Warfare in the pre-state period; Limited War in the eras of Agricultural Revolution; Total War, which flowered in the Industrial Age; and finally, the Postmodern Warfare of the Information Age.

According to this approach, in the pre-state period, hunter-gatherers conducted ambushes and raids for pillage and plunder of women and domestic animals. The Agricultural Revolution, which started around 5000 BCE, generated larger amounts of surplus and gave birth to standing armies and bureaucracy. The Bronze Age, and the subsequent Iron Age civilisations, were based on the agricultural economy. This period witnessed the evolution of interstate conflicts and the rise of pitched battles for decimating the enemy army. The industrial era started from the ca. 1750s. The First Industrial Revolution was based on the machine production of the cotton textile industry in England. The Second Industrial Revolution, kicking off in the second half of the eighteenth century in France, Germany and Russia, resulted in the emergence of iron and steel weapons along with internal combustion engines and railways which enabled the mobilisation of large numbers of soldiers. The second decade of the twentieth century witnessed the use of aerial machines for waging war in the third dimension. Conscription became the order of the day and mass armies equipped with destructive industrial weapons fought attrition-oriented campaigns on a vast geographical scope. Paul Kennedy's *The Rise and Fall of the Great Powers* (1990) deals exclusively with Western Europe's and the United States' rise to greatness during the industrial age. Kennedy establishes a simplistic connection between economic strength and military power. Moreover, the non-Western powers play a cameo role in his global drama. Further, Kennedy, like W.H. McNeill (who had written in 1982 under the shadow of the Cold War), pointed out the 'command economy' of the oriental powers (China, India, etc.) as the villain of their economic backwardness during the eighteenth and nineteenth centuries (McNeill 1982).

Some scholars have argued that the industrial age gave birth to Total War, which started emerging under Napoleon Bonaparte and reached its logical culmination under Adolf Hitler in the Third Reich. The objective of Total War was complete annihilation of not only the enemy polity's military force but also its society. Non-combatants also became legitimate targets of war (Chickering et al. 2005).[4] Roger Chickering and Stig Forster define Total War as that which

> erodes not only the limits on the size and scope of war effort: it also encourages the radicalization of warfare, the abandonment of the last restraints on combat, which were hitherto imposed by law, moral codes, or simple civility. Moreover, in order to sustain popular commitment to the war effort, governments pursue extravagant, uncompromising war aims; and they justify these goals through systematic demonization of the enemy.
>
> *(Chickering and Forster 2005a: 2)*

The Age of Total War ended with the advent of nuclear weapons which appeared to make mass armies irrelevant. The post-Second World War period experienced Limited Wars which mostly occurred in the 'Third World' under the nuclear umbrella of the superpowers. The presence of nukes in the background prevented the escalation of Limited Wars from becoming Total Wars.

Much earlier than the Total War model proposed by Chickering and Forster, in 1961, a British Army officer named Major-General J.F.C. Fuller fused the trajectory of Industrial Revolution with development of secular political ideologies for analysing the genesis of modern war in Western Europe. In Fuller's paradigm, modern war (which is equivalent to Total War) started emerging with the French Revolution and the advent of Napoleon Bonaparte, the 'God of War.' Modern war for Fuller ceased with the advent of the Nuclear Age after the Second World War (Fuller 1979).

The above model is Eurocentric in nature. Not all parts of the world experienced the same type of evolutionary stages. For instance, the Iron Age did not occur in the two Americas. Societies in the precolonial North, Central and South America remained at the Stone Age and Bronze Age levels. The coming of Western Europeans with iron weapons, along with diseases, to a great extent enabled the Western conquest of the 'New World.' Mass industrial war never occurred in precolonial and in colonial Africa. Then the Western European powers, while preparing for Total War amongst themselves, waged Small Wars in their Afro-Asian colonies for colonial conquests and consolidation. Several features of colonial Small Wars like targeting the economic resources of the enemy, partial genocide/ethnocide and the emergence of concentration camps actually first occurred in the colonies and then in Europe during the two World Wars. The Total War model mostly focuses on European interstate wars. In regions outside Europe, intrastate wars (like the Taiping Rebellion in China [1850–1864], 1857–1859 Uprising in India) were more violent and casualty prone and occurred over a wider geographical area compared with some of the industrial wars in Western Europe like the Austro-Prussian War (June–July 1866) and the Franco-Prussian War (July 1870–May 1871) (Thampi 2010; Roy 2010).

Studying the evolution of warfare through the lens of technology constitutes the third approach. Three examples could be given. J.F.C. Fuller (1945, reprint 1998) and Bernard and Fawn Brodie (1973), in their overviews, establish a simplistic direct linkage between developments of military technology and the evolving art of war. Both the works focus on conventional wars and focus mainly on Europe and the United States (from the nineteenth century onwards). Fuller claims that weapons comprise almost 99 per cent of the conflict. Brodie at least shows the interrelationship between the general development of science and the progress in military hardware. Martin van Creveld (1991) offers an insightful but linear overview of the development of military technologies from Greece to the United States starting from 2000 BCE to the 1980s. Van Creveld mirrors Fuller in asserting that technology is almost everything in war. He divides the history of warfare into four periods: the era of muscle power (both human and animal); the period of machines, which started from ca. 1500; the period of technology emerging as a system from the 1830s; and, finally, the age of automation, which began in the late 1940s. All these works turn our attention to technology as a black box but do not unpack its essentials. However, the non-technological factors and the socio-political context remain in limbo in these works. Finally, none of these authors deals with the problematic: why the West but not the 'rest'?

Some of the followers of the war and technology perspective use the heuristic device of Military Revolution. The Military Revolution concept allows the proponents of this approach to link military innovations with changes in the broader society, economy and state formation. However, the principal driver of change remains technology. Michael Roberts first used this concept in 1956 to explain the superior combat effectiveness of Gustavus Adolphus' Swedish Army during the Thirty Years' War in Europe (1618–1648). Geoffrey Parker (1988) made this concept famous by using the concept of Military Revolution to explain the rise of Western military power vis-à-vis the rest in the early modern period. Parker, in his Lees Knowles lecture in 1984, claimed: 'the key to the Westerners' success in creating the first truly global empires between 1500 and 1750 depended upon precisely those improvements in the

ability to wage war which have been termed "the military revolution."' (Parker 1988: 4).
Then this concept spread like an epidemic among historians.

Historians working on medieval and ancient European history found traces of Military
Revolutions in their respective areas. Thus, one noted historian even before Parker asserted
that a Military Revolution occurred under Alexander of Macedonia (Ferrill 1986: 149–86).
Not to be seen lagging behind, people working on Asian military history also found evidence
of Military Revolutions in precolonial China and India. Peter Lorge argues that the Military
Revolution first started in China ca. 1100 and then gunpowder weapons spread in Southeast
and South Asia. In the early modern period, writes Lorge (2008), Western Europe caught
up with the 'east'.

So many Military Revolutions in various parts of the world in different time periods seem
to neutralise each other. Further, Jeremy Black (1991) challenged the concept of Military Revolution in early modern Europe and came up with the idea of Military Evolution.
Clifford J. Rogers (1995) then tried to accommodate the criticisms of the Military Revolution concept by claiming that military developments in Western Europe between the medieval and modern eras could be explained with the aid of punctuated equilibrium theory. This
concept argues that a Military Revolution was followed by a period of Military Evolution and
then another Military Revolution occurred.[5] Strangely, this unique phenomenon occurred
only in Western Europe.

Jack S. Levy and William R. Thompson, in their book *The Arc of War* (2011), extend and
modify the approach of Rogers. Levy and Thompson paint an overview of the history of
global warfare by arguing that, for most of the time, an evolutionary approach holds water.
However, the evolutionary pattern in the genesis of warfare during the last 6,000 years was
interspersed by three accelerations (a new name for revolutions). Each acceleration resulted
in intensification of warfare: larger army size, stronger polities, extended campaigns and
more military casualties. The first military acceleration occurred in Mesopotamia during the
Bronze Age (late fourth and third millennium BCE), the second during the Early Iron Age
(last half of the first millennium BCE) in the Eastern Mediterranean and China and the third
military acceleration was confined mostly to Western Eurasia and unfolded between 1500
and 1945. Levy and Thompson try to bring in non-technological factors while discussing
the motors behind the three military accelerations. In this endeavour, they are not very
successful. Further, India, Africa and the two Americas hardly appear in their overarching
metanarrative. Finally, their data base remains shallow. However, the model is interesting and
provides food for thought.

From the other pole, scholars started claiming that the military edge of the Western Europeans in the early modern world did not matter much. Rather, diseases and collaboration
with the indigenous people were crucial factors in establishment of European supremacy
in the extra European world. The foremost recent critique of Parker's Military Revolution
concept and its timing is advanced by the American scholar J.C. Sharman. He writes that
till 1750, the Western Europeans did not enjoy military superiority (nether technical nor
organisational) over the Africans, Asians and the Amerindians. In fact, the European enclaves
in the extra European world were based on diplomacy and sufferance of the non-Western
rulers. The European technical edge over the non-Europeans was episodic and not intrinsic
and existed only for a short time (nineteenth and early twentieth centuries). From the late
twentieth century, the European military edge was again vanishing, as exemplified by the

emergence of successful insurgencies in Afro-Asia and in the rapid rise of China and India (Sharman 2020).

The point to be noted is that indigenous people collaborated with those European invaders only when the latter displayed their military superiority. True, in the 'New World,' the indigenous populace fell victims to diseases brought by the conquistadores from the 'Old World.' However, in South Asia, the British fell victim to cholera, malaria and diarrhoea in greater numbers compared with the 'natives' between ca. 1650 and 1947. The technological superiority of the Western powers over the non-Westerners was a given fact during the early modern period in many areas of the globe. However, the technological edge was not enough by itself. Intangibles like doctrine, command, tactics and strategy along with robust political institutions were other factors which enabled the 'West' to rise above the 'rest.' Some scholars have argued that superior Western organisation and the ability to adapt spelled the difference between victory and defeat during the expansion of European rule in the nineteenth century (Ness and Stahl 1977).

From the 1990s, mainly political scientists and international relations experts started using the concepts of Military Revolution and Revolution in Military Affairs (RMA) in order to understand the present state of warfare. RMA is related to the Military Technical Revolution (MTR) concept. The seeds of both these concepts could be traced back to the Soviet military writings in the 1920s and 1930s (Henninger 2006: 16). Some historians also joined this bandwagon and tried to apply the RMA concept to explain the emergence of new weapons and changes in the dynamics of warfare in the past. RMA is a narrow concept which refers to the use of a new technological system that increased the military effectiveness of an army or any other branch of the armed forces at a tactical level; for instance, the use of the composite bow by the Hittites against the ancient Egyptians. In contrast, Military Revolution is a wider concept which shows how adoption of a single or related military technology resulted in transformation of not only the armed forces but also the host society and polity. The rise of gunpowder weapons which resulted in military fiscalism or a fiscal-military state in early modern Western Europe is a prime example of Military Revolution.[6] Some scholars have found traces of several Military Revolutions and RMAs between the fourteenth century and the third millennium in Western Europe.[7]

The RMA model is technology-centric and even the Military Revolution model can be charged with being technology reductionist. As Jeremy Black notes in his history of technology (2013), we need to contextualise technological changes which occur slowly in tandem with political and cultural variables. All the above models focus on Western Europe and the United States, and mostly on the major powers. Further, intrastate conflicts on land and sea do not receive adequate attention in these approaches. However, these metanarratives have played a crucial role, rightly says historian Peter Wilson, in encouraging comparative analysis of military developments of different parts of the world especially during the early modern period (Wilson 2023).

In this handbook, we have followed an eclectic approach which attempts to bring under the scanner the extra European world and also the varieties of warfare which occurred in different regions throughout the span of history. The assortment of military cultures which were present before the nineteenth century in various parts of the globe during the period of recorded history prevents us from offering a single metanarrative about the evolution of global warfare. So, instead of any master narrative or any overarching universal global model, this collection points to the diversity in organised violence along space and time. Jeremy Black rightly warns: 'It is also necessary to address the problem of how best to develop analytical concepts that do not treat the world as an isotropic (uniform) surface, but instead make sense of different military goals and traditions, and assess their impact' (Black 2004: 19). Our aim is to give proper attention to the period before Western dominance. Technology is given due importance in this handbook. However, for us technology is not merely hardware but also includes the necessary software of social technology[8] (the tactics, techniques and procedures for utilising a weapons system for maximum effectiveness).

One might offer a critique that the bottom-up perspective highlighting the experience of killing and being killed is not followed in this handbook. After all, John Keegan, in his *The Face of Battle* (1976), introduced the sensory history of combatants engaging in battle. A history from below approach is not followed in this volume for the simple reason that adequate materials about the nameless and faceless common mass of non-Western regions in the precolonial era are not available. The point that needs to be highlighted is that two of the prominent military history books following the bottom-up perspective focus exclusively on battles fought by the Europeans. While Gregory Daly (2002) turns the limelight on the Battle of Cannae (216 BCE), Keegan deals with Agincourt (25 October 1415), Waterloo (18 June 1815) and Somme (1 July 1916). Many non-Western societies did not record history like the Greek and Roman societies. Rather, the focus of these societies was on accumulating myths. Further, during the era of Western domination, due to a lack of patronage and as a result of the aftereffects of defeats, several non-Western sources were destroyed. In addition, the problem of language further deters scholars from interpreting whatever little is available. The Subaltern history project which kicked off in India during the 1980s failed to make any interventions in the ancient and medieval periods of Indian history due to an absence of sources. Folktales collected by Westerners and ethnographers are not adequate to allow historians to reconstruct the past of the extra-European regions before the early modern period. Further, analysing the human dimension (experience of the soldiers) hardly gives us any idea about the changing dynamics of warfare: why combat occurred at a particular time in a certain place, the nature of the weapons with which they killed each other, and the impact their fighting had at the strategic and political levels. Historian John France justifiably provides a scathing criticism of Keegan's 'Face of Battle' approach in the following words: 'But the trouble is that they are essentially descriptive, with a tendency to focus on the minutiae of war . . . and actually do not tell us a great deal about war' (France 2011: 11).

Instead of following exotic methodologies, we follow Jeremy Black's cautious statement: 'More generally, in the view of this empiricist, an awareness of particular interests and specific conjunctures should take precedence over the fascination with discourse' (Black 2015: 268). Our contributors differ in their methodologies. Considering the variety of military cultures that were present in the past, we editors consider it a strength of this handbook. The handbook is organised along eight themes, which are developed below. Within each theme, the chapters dealing with various zones of the world are arranged chronologically as far as possible.

Origins of Warfare

When, why and where did warfare begin? Warfare in the world first started in Eurasia, asserts Robert Drews in the first chapter of this handbook. He argues that with the genesis of states in Mesopotamia in around 3000 BCE, organised violence came into existence. Initially, the warriors skirmished along the city-walls. However, by around 1500 BCE, with the domestication of horses, chariot warfare was born. The basics of chariot warfare involved the taming of the wild horses of the Eurasian steppes and the construction of wagons. The credit for both these military innovations goes to the Proto Indo-European (PIE) people. The chariot warriors with their bows rode to victory in the Near East and Egypt. By around 1200 BCE, the outlying regions of temperate Europe came up with infantry armed initially with rapiers and then swords and spears. The swarming tactics of the infantry overwhelmed chariots of the Bronze Age polities. Thus, by around 1000 BCE, the paradigm of chariot warfare was replaced with the paradigm of infantry warfare. The stage for the rise of hoplites in Classical Greece was thus set.

But what was the nature of warfare beyond the 'West'? In the second chapter, Andrew Scherer argues that warfare was endemic in Pre-Columbian Mesoamerica. By focusing on the Maya and the Aztecs, Scherer argues that warfare was not merely ritualised 'flower war,' but was also different from the military tradition which emerged in Greece and Rome. Pre-colonial Mesoamerican warfare aimed at capturing enemy combatants and non-combatants rather than killing them. This trend was also present in precolonial Southeast Asian warfare. Further, public torture and execution of the captives was a characteristic feature of the Pre-Columbian Mesoamerican conflicts. To a lesser extent, the practice of the taking of body parts of the fallen enemies by the victors was also present in precolonial conflicts elsewhere in the world, including Australasia and Southeast Asia.[9]

By making a case study of the Darug and the Maori, John Connor, Michael Charney and Kaushik Roy, in the third chapter, trace the changing trajectory of warfare among the tribes of Australia and New Zealand. Before the intrusion of the British, the Aborigines fought amongst themselves for avenging personal insults and women. Since women constituted economic resources for the tribes, capturing them was a technique of weakening the economic base of the hostiles. This chapter asserts that these intertribal conflicts were simultaneously both total and limited. Since all the males of a particular tribe participated, the war was total. However, the objectives of the tribes were limited. They were not to destroy the enemy tribe completely but to weaken them economically and demographically and, if possible, to absorb them within one's own fold. Rather than set piece battles, raids and skirmishes were preferred. Initially, against the colonisers, the tribes conducted their traditional corn raids geared for collecting foodgrains from their enemies. However, with the expansion of British settlers, the Maori took to firearms and strengthened their traditional defensive settlements known as *pa*. Hence, not *stasis* but change and continuity characterised warfare among the indigenous people of Australasia. One could argue that the level of warfare (in terms of scope and lethality) did not cross a certain threshold due to an absence of a state system.

Polities and Armed Forces in the Pre-modern Era

Charles Tilly's aphorism that "war made the state and the state made war" is generally applicable for the better part of human history (Tilly 1975). The Assyrians were probably the first empire builders in history. In Chapter 4, Sarah Melville argues that the Assyrians militarised themselves due to the threats posed by the neighbouring powers. Melville shows the dialectical relationship between state formation and conduct of war. The polity expanded its tentacles to manage its war machine and the operation of the military infrastructure allowed the expansion and consolidation of state power. To give an example: for increasing the operating range of the army, the Neo-Assyrians established an elaborate logistical infrastructure. As a result, this raised the staying power of the army, which in turn allowed the Assyrian centre to completely dominate the periphery. Again, the rise of a standing army increased royal control and the latter development resulted in further expansion of the permanent standing force at the cost of the militias raised by the nobility.

During the Neo-Assyrian period, the Assyrians, like the precolonial Mesoamerican polities, engaged in spectacular public displays of torture and the mutilation of their enemies. Shocking displays of violence were common in states that were administratively weak or enjoyed weak authority amongst the ruled. Such displays were used instrumentally to intimidate the general population into submission. But this approach can also be applied to wars of conquest, through the deliberate use of terror tactics to spread fear among opponents so that resistance would be weak or non-existent. Much later, the Mongols also used such terror tactics to cow their enemies into submission.

Whether there existed any connection between the political and social structure of the *polis* and hoplite fighting is the issue of a raging debate. One group of historians argues that the close order combat by heavily armoured infantry was the product of unique social and political structure of the Greek city states which emerged in the Classical period. From the opposite pole, a growing body of scholars argues that no such connection existed between the nature of combat and the social and political base of the *polis*. Historians of ancient Greece still differ on whether the hoplites actually implemented *osthismos* (push of the pikes by heavy infantry soldiers organised into tight phalanx) or they fought in loose formation once the battle started.[10] In Chapter 5, Matthew Trundle takes a middle position in this debate, showing that the hoplite ideology was important. However, even in the so-called hoplite battles, non-hoplite troops, such as the lightly armed infantry who skirmished, were vital. Further, in the case of Athens, the fleet played an important role in ensuring national security. Even in the case of Sparta, the classical model hoplite state, a significant number of hoplites were not Spartan citizens or landholders.

The fall of the Western Roman Empire is no more considered a *stunde null*. Traditionally, military historians claimed that, with the decline of the Western Roman Empire, disciplined infantry warfare ceased in Western Europe and the Germanic 'barbarians' ushered in cavalry war. However, a newer generation of historians (including Bernard Bachrach who is one of the contributors of this handbook) asserts that the Germanic invaders continued the Roman-style siege and infantry-centric war (Halsall 2020). Rather than the decline thesis, the transition thesis is more in vogue.

Bachrach's contribution to the present volume (Chapter 6) emphasises the continuity between Late Roman Gaul and Merovingian France in terms of military affairs. The so-called 'feudal' trends had already started appearing in the fourth century CE Roman Empire. The Roman Empire was facing difficulties in maintaining its mobile professional field army comprising legions due to political and economic factors. Continuous civil wars between the claimants of the Roman government resulted in repeated depletions of the ranks of professional legionaries. Taking advantage of the political chaos, big landlords did not release their tenants to join the imperial force. It was economically more fruitful for the landlords to keep the young, able-bodied men in their farms rather than releasing them to join the army of the central government. At the same time, the attacks by the Germanic tribes across the Rhine were increasing in scope and intensity. The Roman government responded by building defensive fortifications around the Gallic cities and mobilising local militia forces under the landlords who were also given imperial offices. Gradually these imperial offices became hereditary. In addition, some of the Germanic invaders were co-opted and incorporated into the Roman Army as auxiliaries and allowed to settle in Gaul. They received land from the government in return for fighting other Germanic clans which were moving from east of the Rhine. By the end of the fifth century CE, as the Roman central government struggled to survive, Gaul under the Germanic settlers continued with the Gallo-Roman military and administrative institutions, albeit in a modified form. The next chapter shifts the focus from the 'West' to the 'East.'

Kenneth Swope, in Chapter 7, discusses the shifts in our understanding of Late Imperial Chinese military history especially regarding the Ming and Qing/Manchu Empires. Traditionally, premodern China was regarded as pacific in nature, being influenced by Confucian culture. The military was subordinated to the civilian officials steeped in Confucian ethos. The martial culture was marginalised in the traditional imperial courts. This interpretation of peaceful China attracted Westerners in the immediate aftermath of the Cold War. Swope asserts that new studies of the Ming-Qing Empires reveal how military officers played a leading role in the traditional Chinese courts and military culture, and martial values were at the forefront of the polities' cultural policies. Powerful Ming and Qing emperors pursued a coherent grand strategy aimed at the aggrandisement of Chinese power in Mongolia, Korea, Vietnam and Tibet, while the Ming also pursued an ambitious hegemonic maritime strategy.

O.R. Coates provides an insightful survey of precolonial African warfare from the seventh century CE (when organised violence started in Africa) till the late nineteenth century in Chapter 8. He asserts that precolonial African wars were not the products of tribalism. Rather, economic, environmental and political factors shaped the dynamics of conflicts in the African continent. Though external factors like Islamic invasions, Indian Ocean trade and European traders were important, internal drivers (local conflicts and overland trade routes, power struggles among different factions, etc.) were also important in shaping the contours of warfare. Most of the precolonial African states did not have standing armies but relied on men in arms (citizen armies or the nation in arms). All the able-bodied males (and in some cases even females) were under the obligation to bear arms. These armies were not undisciplined rabbles but had a training architecture and command system. Hence these forces were capable of exhibiting tactical sophistication. Due to a lack of natural harbours, instead of oceangoing navies, the premodern African polities developed canoe fleets which carried out quasi 'joint' operations with the ground forces in the lakes and lagoons. Geography protected the premodern African states, unlike the Eurasian polities, from the ravages of mounted horsemen from the steppe region. This very threat shaped a different course for the evolution of warfare in Eurasia, our next theme.

Steppe Nomads of Eurasia

Clashes between the people inhabiting the steppe nomadic zone stretching from Manchuria in the east and Hungary in the west with the settled societies practising agriculture along the rimland of Eurasia were a constant throughout history at least till the eighteenth century. The horse-borne steppe nomads required cereals and iron implements from sedentary societies and in return the latter wanted animal products and mounts for transportation and military purposes. When free trade and commerce did not occur among these two distinct ecological zones, war occurred. The mounted warriors with their composite bows, due to their speed, mobility and longer operating range, were militarily superior compared with the footslogging infantry forces of the agrarian societies. To give a comparative example, while the Roman legionaries marched at the rate of 12 or a maximum of 16 miles per day, the Mongol cavalry could cover 40 miles each day for several days (Hildinger 2001: 17). However, tribal rivalries weakened the military power of the steppe nomads. When a political leader like Attila or Chingiz Khan or Timur was able to unite the nomads of Eurasia, then all hell broke loose over the sedentary states.

In Chapter 9, Cameron Barnes examines the impact of the European Huns on military history. The Huns arrived in the Volga-Don region ca. 360, bringing with them Inner Asian traditions of warfare and imperial statecraft. During the reign of Attila, Hunnic rule stretched from the Rhine to the Dnieper. They, for the first time, introduced mounted archery in Europe on a large scale. The Huns' mounted archery ended centuries of infantry dominance in warfare in the Mediterranean world, leading to the age of *the hippotoxotai* during the sixth century. With other barbarian peoples, concludes Barnes, the Huns shared a big responsibility for the collapse of Roman rule in the West.

The premodern world experienced nomadic attacks on the sedentary civilisations from two regions: Central Eurasia and Arabia. While the Huns and the Mongols came out rampaging from the steppes of Central Eurasia, the Bedouin nomads or Arabs, with the spread of Islam, started breaking out from the desert zone of Arabia. The Arabs, unlike the Central Eurasian nomads, did not initially ride on horses, but on camels. Tactics based on camels, especially in the frontier desert region and the Arabs' religious motivations enabled them to overthrow the Sassanid Empire whose main strength lay in heavy cavalry. Gradually, as the Arabs spread into Asia Minor, Syria, Egypt and Iraq–Iran, a territorial superstate in the shape of the *caliphate* came into being under the *caliph* who happened to be the spiritual and religious head of the *umma* (Muslim community). After the collapse of the Sassanid Empire, the Arabs turned their full attention towards the Byzantine Empire. The long attrition-oriented military confrontation between the Byzantine forces and the Arabs is the subject of John Haldon's chapter.

Haldon covers the period from the seventh to the eleventh century CE in Chapter 10. He focuses on how the challenge–response dynamic reconfigured the military organisation and tactical template of the Byzantines. The Arab forces, initially under the Umayyad *Caliphate* and then its successor the Abbasid *Caliphate*, comprised light cavalry armed with swords and spears and geared for raiding focused on pillage and plunder. The legionaries (heavy infantry) were not suited against such 'predatory warfare.' The Byzantine military reformers came up with a sort of combined arms tactics: heavy infantry, light infantry, light cavalry and heavy cavalry operating together. The Islamic enemy was not standing still. From the tenth century onwards, the trump card of the Arab *amirs/emirs* attacking Byzantine territories was the horse archer. Turks in the guise of military slaves (*mamluks*) were recruited and used in large numbers. These *mamluks*' horse archers gave a death blow to the Byzantine Army at the battle of Manzikert (26 August 1071). Though Haldon somewhat underrates the importance of this battle, he accepts that this defeat at the hands of the Islamic forces set off a civil war in the Eastern Roman Empire which, by the beginning of the twelfth century, had become a second-rate power.

The *mamluks* were slave soldiers. The Abbasid *Caliphate* recruited Central Asian steppe nomadic Turks through the *mamluk* institution. Gradually, as the Abbasid central government disintegrated in the tenth century, many regional sultanates emerged in the Middle East. The *mamluks*, in a piecemeal manner, transformed themselves from being military mercenaries to sultan makers and finally as sultan themselves. John Hosler's chapter shows the confrontation between the Mamluk Sultanate of Egypt with the Crusaders. Hosler, in Chapter 11, views Western and Eastern Warfare in the context of the Christian and Islamic Holy War during the crusading period of 1096–1291. The commonalities are many and significant: well matched in terms of armament and tactics, both 'Franks' and 'Turks' waged expeditionary campaigns with coalition allies that were motivated, in part, by religious impulses, and limited in time and space by human resources and environmental factors. Political relations conditioned the integrity of the respective coalitions, and competent generalship was required to sustain forces for long durations and in times of material privation. In contrast to the Islamic forces, crusading armies travelled much further and had to operate in broadly non-permissive environments. The Muslims faced enduring political and religious rivalries that belied any sense of their unity and complicated their reactions to Western, as well as Mongol, incursions. In the face of serious resistance, Western Christian forces maintained a territorial presence in the Middle East for nearly two centuries but were ultimately expelled by a combination of disastrous military policy on their part and the strategic and operational excellence of the Mamluk sultans.

In the thirteenth century, asserts Timothy May (Chapter 12), most societies which encountered the Mongols viewed them as an existential threat more on a par with natural disasters or apocalyptic doom than as invaders. With an empire that stretched from the Sea of Japan to the Mediterranean Sea and having conquered the most powerful empires of the era, there seemed no limit to the Mongols' ability to conquer. At their zenith, the Mongols possessed the capacity to mobilise armies so huge that they would rival modern superpowers. Yet, even as the Mongols reached their peak, limitations to their military power manifested. While the Mongol horse archer remained a dominant figure in the battlefield, other factors began to influence the outcome of the Mongols' military endeavours. Climatic challenges, political fractures within the Mongol Empire, and the resilience of their opponents combined to deliver the Mongols a series of defeats.

The Eurasian steppe nomads set up what Andrew de la Garza, in Chapter 13, terms 'frontier empires.' Some examples of such empires were the Ottoman, Safavid and the Mughals. He elaborates on the rise and fall of such empires and the nature of their policies. Echoing Jos Gommans, Garza claims that these were indeed hybrid states. Gommans shows that these empires, which he terms quasi-nomadic polities, were situated at the crossroads of two distinct ecological zones: dry semi-arid pastoral zone and hot and humid sedentary ecological clime. This ecological mix made the hybridisation of nomadic and sedentary techniques essential (Gommans 2018). The militaries of these frontier empires, notes Garza, comprised both sedentary drilled and disciplined gunpowder infantry plus Central Asian lightly armoured nomadic horse archers. The dominant military group had to share power with the sedentary 'civilised' ethnic communities for running the civilian administration. In addition, the rulers of these states had to depend on the ideology of enlightened despotism to buttress the legitimacy of their regimes. Faction fights within the ruling aristocracy added to invasions by a fresh group of nomads from outside caused the cyclical collapse of such empires. Gerard Chaliand writes that the steppe nomads of Eurasia started losing their dominance from circa 1500 onwards due to the repeated counteroffensives launched by the sedentary regimes of China and the rising importance of maritime power (Chaliand 2014: 1–79). Next, we turn to combat on water.

Naval Warfare and Piracy in the Pre-Industrial World

War at sea and on rivers had historically been mainly fought to shape conflict on land. Logistical support, transportation of troops from one place to another and providing fire support for a direct assault on enemy strongpoints were principal functions of the riverine and coastal navies. This sort of warfare is termed by modern commentators as littoral and joint warfare, and it has a long pedigree.

Mark Charles Fissel, in Chapter 14, asserts that amphibious warfare first emerged in ancient Egypt. Around 2500 BCE, the pharaohs used ships for transporting troops and stores in the riverine vessels. At times, rebel strongpoints and foreign enemy bases were also captured with river-borne troops and the ground forces advancing and fighting together. Thus, a sort of land–river combined arms operational matrix evolved. Gradually, the fleets conducting riverine warfare developed capabilities for waging amphibious warfare. By around 1500 BCE, the pharaohs' fleets were capable of transporting military stores and troops across the Eastern Mediterranean to Syria and Lebanon. Probably the ancient Egyptians developed specialised war boats and special forces (marines in modern terminology) to fight on board the combat vessels. One can say that the Egyptian expertise in conducting aquatic warfare gave birth to Egyptian sea power. In all probability *Pharaonic* Egypt was the first state in the ancient world to exercise sea power before the coming of Achaemenid and Classical Greek Mediterranean navies.

The preindustrial maritime world was infested with pirates. Robert J. Antony's chapter turns the focus to piracy in the South China Sea (Chapter 15). The coastal waters of China were infested with pirates from the time of the Yuan Dynasty onwards. When the Mongols were ruling China, most of the pirates were Japanese. During the Ming period, men from several coastal communities of South China participated in large-scale piracy. Under the Qing regime, several Europeans also joined the fray. Antony shows that piracy in the Far East exceeded that of the Atlantic World in terms of scope and intensity. Several times, the pirate fleets of Asia were able to decimate the Ming and Qing navies. Partly this was because most of the Chinese polities followed continental strategies rather than a full-blown maritime strategy. Then again, several Chinese emperors followed a 'closed door' policy which encouraged the mercantile communities and the people living in the coastal regions of South China to join piratical activities. Numerous bureaucrats were hand in glove with the pirates. The Qing Court, being unable to smash the pirates in naval battles, resorted to appeasement, divide and rule policies and opening up maritime trade by removing all state-initiated bans on trading activities. Several pirate leaders were won over with titles and posts and then they turned against their former compatriots. To a great extent, large-scale piracy prevented the emergence of a blue-water navy in imperial China. Ironically, as the present Chinese state created a blue water navy, the pirates were back in Southeast Asia.

The spirit of the Renaissance guided the age of geographical discoveries, as the European ships traversed the surface of oceans and seas to explore uncharted land and water masses across the globe. The prospects of lucrative commerce with the newly discovered world unleashed an era of intense rivalry among the European nations – further manifested in their attempts to establish military control over the sea lanes of the oceans. In an unprecedented move, through the Treaty of Tordesillas in 1494 CE, Spain and Portugal divided the 'New World' among them. The Portuguese subsequently established control over the sea route to India via the Cape of Good Hope. During the seventeenth century, European nations like England, the Netherlands, and—at a later stage—France, riding high on mercantilist aspirations and backed by naval power, vigorously challenged the primacy of Spain and Portugal in the waters of Europe as well as in the 'New World.' The opening up of new trade routes not only transformed the very nature of trade in the newly discovered areas but also had an indirect influence on sea power—which was subsequently used as a tool for acquiring colonies overseas.

In Chapter 16, Amarendra Kumar takes into account the ascendancy of British naval power in the Indian Ocean in the global context. He argues that the British East India Company's Marine Force, in concert with the Royal Navy, played a crucial role in establishing British naval supremacy in the Indian Ocean by transforming the Indian Ocean into a 'British lake.' Kumar deftly shows the interconnections between shifts in European power balance and the evolving British power structure east of Suez. Diplomacy, political acumen, and technological and managerial superiorities enabled the British to overcome both their European rivals as well as the Asian powers. Towards the end of this process, steam power replaced the 'Age of Sails' in the Indian Ocean. Gunpowder weapons gave the Western Europeans an edge both on land and sea during the early modern era, a theme which we discuss below.

The Impact of Gunpowder

Charcoal, sulphur and nitre; the three ingredients mixed in the right composition produce gunpowder, a commodity which in turn shaped the fate of civilisations around the world. The spread of gunpowder across the world is an example of connected history. Gunpowder was first developed in medieval China but Western Europe overtook the 'Middle Kingdom' by the sixteenth century as regards the production of sophisticated fire-spewing weapons. Warren Chin asserts that in Europe gunpowder provided the single most important spark that started a chain reaction of innovation in the military domain (Chin 2006: 85).

Philip Hoffman's chapter in this handbook poses the question of why Western Europe but not Asia was able to harness the chain reaction resulting from the use of gunpowder (Chapter 17). Hoffman claims that Western Europe, especially Britain, had an edge as regards the economy of production. He makes a controversial claim that endogenous developments (gunpowder war) within Europe and not exogenous factors (acquisition of overseas colonies) sparked the Industrial Revolution in Western Europe. He asserts that continuous innovation and mass production of gunpowder weapons was possible in Western Europe due to industrialisation and the latter occurred due to the emergence of strong states and high taxation. The last two characteristics emerged due to the rising cost of gunpowder warfare. This in turn accelerated financial innovations which sustained industrialisation. Thus, a military-industrial complex emerged in Western Europe during the early modern period.

One finds traces of Brewer's thesis in Hoffman's argument (Brewer 1989). Why is it that military-fiscal entities did not emerge in other parts of the world or especially in the regions of Asia with big agrarian bureaucratic polities like Ming-Manchu China, Mughal India, Safavid Iran and Ottoman Turkey?

Gunpowder played an important role, along with other interrelated factors, in the rise of Muscovy as a military power during the early modern period. In Chapter 18, Kaushik Roy looks at the transition of Russia under the tsars from a minor power to a major power in Eastern Europe. In the fourteenth century, Muscovy was threatened by the steppe nomads in the south and by Poland–Lithuania in the north. During the seventeenth and eighteenth centuries, the Muscovite ground force experienced a transition. From a cavalry-centric army, it became a Western-modelled infantry heavy force. The infantry's staying power was supplemented by field artillery. Nevertheless, for campaigning in South Russia, cavalry remained important. More importantly, the construction of an authoritarian militaristic polity by the successive tsars (especially Peter the 'Great') enabled the resource-poor Muscovy to maintain its ever-expanding costly military machine.

Kaushik Roy's chapter dealing with the confrontations between the Amerindians and the Western Europeans in the early modern period shows that the military systems of both sides underwent transformation during the conflict (Chapter 20). While the Amerindians integrated gunpowder weapons within their traditional military format thus creating a hybrid system, the Europeans also understood that the Western European Way of Warfare was impracticable in the New World. Hence, they modified the European military practices and also absorbed certain indigenous customs and techniques within their military structure. In the end, the lack of an industrial manufacturing foundation, a weak demographic base, and the small size of their polities sounded the death knell of the indigenous tribes of North America against the European onslaught.

The interrelated roles of gunpowder weapons and diseases (especially smallpox) imported from Western Europe in facilitating the colonisation of North America are the subject matter of Chapter 19, by Erica Charters. Charters modifies Alfred Crosby's argument by showing that not merely the spread of smallpox but the devastation wrought in the New World by the European invaders equipped with gunpowder weapons made the effects of the spread of that disease so deadly. True, unlike the people of the Old World who had developed immunity to smallpox due to the practice of living together in densely populated clusters, the American indigenous tribes were susceptible to the ravages of the disease. However, death and destruction caused by war made the American Indians more vulnerable to the virus causing smallpox. Large-scale dislocation of the population, famine, malnutrition due to the destruction of the herds of the Amerindians, and falling birth rates, all caused directly by the European colonists' invasion, made the American Indians extremely vulnerable to the dangers of smallpox. To an extent, the European soldiery was also susceptible to the disease, but, unlike the Amerindians, they were vaccinated by a weaker strain of smallpox causing the virus and thus generated accelerated immunity.

Michael Charney's chapter traces the divergent trajectory followed by the polities in maritime Southeast Asia after the introduction of gunpowder weapons between the sixteenth and nineteenth centuries (Chapter 21). Instead of centralised bureaucratic states which emerged in Western Europe after the introduction of gunpowder weapons, in maritime Southeast Asia, topography, cultural baggage and the constraints of the domestic societies often resulted in stagnation in the development of firepower weapons and their haphazard deployment in battles and sieges. Rather than technology per se, Charney concludes, management techniques and politics were the principal drivers in shaping the forces behind the rise and fall of the state system in Asia in general and maritime Southeast Asia in particular. By the early nineteenth century, the European powers replaced the paradigm of eighteenth-century 'Cabinet Wars' with Total War, our next agenda.

Transition from Industrial to Total War

To wage *der totale krieg*, mass armies are necessary. Mass armies arrived in world history with Napoleon Bonaparte. The paradigm of the Napoleonic War, as Andrew Bamford shows, in Chapter 22, was the product of military developments during the last two decades of the *ancien* regime. The mechanism of a *levy en masse* enunciated by Lazare Carnot enabled Napoleon to raise huge armies. Massive armies required a big industrial base and financial reserves. In the economic sphere, Britain had an edge while Russia's weak industrial base resulted in a large number of her soldiers being indifferently equipped. This became especially evident during the Crimean War (1853–1856). Big armies also required a sprawling logistical base. To control the mass armies, a new command infrastructure was necessary. In this case, France enjoyed an initial advantage: the multi-division corps system. Later, the continental powers like Austria and Prussia copied the corps system and the French military edge vanished. Elaborate supply and transportation facilities were necessary to keep hundreds of thousands of soldiers in the field. Napoleon encountered difficulties in this sphere when his *Grande Armee* marched into Russia in 1812. Dust, combined with an absence of good roads, and the scorched earth policy followed by the Tsarist regime, ravaged the *Grande Armee* before it reached Moscow. History would repeat itself 129 years later when the wheels of logistics disintegrated and the *Ostheer* skidded to a halt before Moscow.

Historians (mostly Americans) continue to fight, arguing about the nature of the American Civil War (1861–1865). A. James Fuller's chapter portrays the nuances of this *Historikerstreit* (Chapter 23). While one group, focusing on the scale and scope of the war and particularly on quantitative data (especially the number of deaths resulting from war, both direct and indirect), strongly asserts that the Civil War in North America marked the beginning of the Total War, the other lot of scholars challenge the data. In recent times, the 'challengers' have started using a quasi-cultural argument: the meaning of death and the shift in the concept of death during the late nineteenth and twentieth centuries. Further, the role of marginal groups (for instance non-white) is also brought under the historical scanner. Whatever stance one may take regarding the use of terminology to describe the American Civil War, it is hard to deny that several new trends emerged and were mutated during the course of this conflict. These trends flowered further during the era of the two World Wars.

The Industrial Age gave birth to Total War. Simultaneously, in the overseas colonies, the European imperial powers waged what they termed as Small Wars. Among the European states, Britain, which had the largest overseas empire, had to fight the bulk of the small wars. Taking a global view of warfare in the non-European world, Bruce Vandervort opines that irregular scouts and light infantry (indigenous soldiers employed by the imperial powers in the colonies), which, being equipped lightly, travelled fast and when required lived off the land, proved most successful in colonial warfare (Vandervort 2006: 208). Ian Beckett's chapter shows the transformation in organisational, theoretical and technological spheres of the British Army as it prepared from waging Small Wars during the late nineteenth century to Total War in 1914 (Chapter 24). Beckett argues that the British Army's strength was not theoretical innovations but adaptability and flexibility in integrating new technologies for raising its fighting power.

One of the tools of industrial war was armoured fighting vehicles (AFVs), popularly known as tanks. AFVs dominated ground war in the twentieth century. Michael Charney, in Chapter 25, considers the rise of tanks during the First World War, experimentation in the interwar period and, finally, the evolution of combined arms in armoured war which reached its logical culmination in the Second World War. Charney shows that during the Great War, it was Britain and France who pioneered the development of caterpillar-tracked vehicles. In the 1920s, Italy made a significant advance in armoured warfare. Only from the mid-1930s did Nazi Germany and the USSR dominate the field of AFVs. It was these two countries' massed armoured fleets which fought a 'steel war' on the Eastern Front during the Second World War. In the peripheries of the Asia-Pacific region, armoured vehicles played a less significant role.

Another tool of Total War is aircraft. The aerial vehicle made its first military foray during the First World War. Kathryn Barbier's chapter traces the evolution of the theory and practice of strategic airpower in the era of Total War (Chapter 28). During the First World War, the

Germans made some attempts at long-range bombing of London, first with their zeppelins and then with heavy, all-metal bombers. However, the bulk of combat in the third dimension occurred between opposing fighters for reconnaissance and to attack the hostile ground troops. Barbier asserts that the theory of strategic bombing flowered in the interwar period. The force structure of the air forces of the major powers was shaped by economic factors and geopolitics in addition to theory. Strategic air power reached its logical culmination in the Second World War with the initiation of strategic bombing of the Third Reich during day and night in 1943 by the USAAF and the RAF. The climax came with the dropping of two atomic bombs at Hiroshima and Nagasaki on 6 and 9 August 1945.

Air power also changed the 'face' of naval war. The Second World War witnessed the replacement of the battleship by the aircraft carrier as the premier sword arm of naval combat. Andrew Lambert's chapter provides an overview of the technological innovations and the evolution of naval war during the nineteenth and twentieth centuries (Chapter 27). Thanks to continuous technological advances, the tools and techniques of naval war were changing. In the nineteenth century, battleships remained the currency of power in the oceans. From the Second World War onwards, aircraft carriers stole the show. Towards the end of the twentieth century, neither big guns nor naval aircraft but rather nuclear missile-equipped submarines fitted with an array of electronic countermeasures equipment started emerging as the queen of naval war. However, elements of continuity pervaded the use of naval power by the maritime states. From the nineteenth century onwards, naval power indirectly helped the maritime states to sustain the attrition-oriented conflicts against the continental powers which employed mass armies. Moreover, naval power enabled the maritime states to project power along the shores of hostile states.

Small Wars, which occurred in the colonies during the industrial era, were characterised by a display of virulent racism on the part of the imperial soldiery. This comes out clearly in Colonel Charles Callwell's *magnum opus* titled *Small War* which was published in 1896. The motto of the European imperialists was to 'civilise the savages' of the extra-European world. Mark Johnston, in Chapter 26, shows how racial prejudices shaped the course of the Second World War in the Asia-Pacific region. Not only the Americans, Australians and the British but also the Japanese suffered from racial biases. The Allies considered the Japanese to be racially inferior hence they could not be good pilots. Similarly, the Japanese assumed that the soft-going Americans due to racial degeneration lacked the stomach for conducting a gruelling attrition-oriented land war. Race and geopolitics fused to shape the ideas and attitudes of the combatants on both sides of the frontline. The period following the end of the Second World War witnessed two interrelated processes: the beginning of the Cold War between Moscow and Washington DC and the meltdown of the Western European powers' overseas empires.

Wars of Decolonisation and Cold War

In the decades following the Second World War, large swathes of Afro-Asia witnessed decolonisation as the Western European overseas empires collapsed. The post-Second World War period witnessed the beginning of the Cold War on the one hand and 'limited hot wars' among the newly independent nations, which were hitherto under colonial rule, on the other. Due to the presence of nuclear weapons, no large-scale conventional war occurred between the 'East' and the 'West.' However, both the USSR and the United States intervened in the erstwhile 'Third World' and conducted war by proxies. Frequently, great power confrontations and indigenous rivalries whose roots could be traced back to colonialism also gave rise to interstate wars among the Third World nations with superpower support in the background. From the 1980s, ethnic assertions occurred among certain sections of the subject people of the newly decolonised states. Great power rivalries further added fuel to the fire resulting in intense intrastate conflicts among the Afro-Asian polities. Rob Geist Pinfold's chapter traces the amalgamation of these factors which gave rise to both conventional and unconventional conflicts in the Middle East in the aftermath of the Second World War (Chapter 29).

The British-Indian Army of the British Empire in India was a unique case. Most of the commissioned officers of this force were British but the non-commissioned officers and the rank and file were Indians. Only during the Second World War, as the army expanded to more than 2 million men, and due to rising casualties in the officer corps, the Raj (British government in India) faced an acute shortage of officers. The British Army itself was expanding and simultaneously its officer corps was suffering casualties. Circumstances forced the Raj to open up the commissioned ranks of the British-Indian Army to the university-educated, urban middle-class Indians who were 'tainted' with the ideas of nationalism. Alan Jeffreys' chapter looks at the tension and tortuous transformation of the British-Indian Army between 1945 and 1947 (Chapter 30). After the end of the Second World War, independence became imminent. While the British were preparing for decolonisation, Lord Louis Mountbatten's SEAC was under pressure due to shortages of British troops to deploy the Indian units as 'imperial fire fighters' in the erstwhile colonies of Britain and the Netherlands. At the same time, the British-Indian Army faced communal carnage in Punjab and Bengal and the presence of ex-Indian soldiers who had joined the Japanese-sponsored INA also known as *Azad Hind Fauj*. By mid-1947, not only had the British-Indian Army been divided into Indian and Pakistan armies but it had also emerged as 'national' forces.

The Falklands War, fought between April and June 1982 between Argentina and the UK for an obscure, small, economically unproductive and strategically irrelevant island in the South Pacific, is probably Britain's last imperial war. Or was it imperial, and how much imperialness pervaded this conflict? Rob MacLaughlin, in Chapter 31, takes a politico-cultural approach in attempting to answer this question. A sort of jingoism among certain sections of the populace in Britain was rekindled during the war. The Conservative government of the UK reaped its full advantage and tried to portray the conflict as a rerun of the late nineteenth-century Small War. However, not all the Dominions of the British Empire like Australia supported London's war effort. The military junta of Argentina provoked this conflict not for fighting British imperialism but to strengthen their regime within domestic society. Lastly, nobody asked what the original inhabitants of this island wanted. The technological advancement and the collapse of the 'Second World' in 1991 ushered in the Postmodern Age.

Postmodern/New Wars

Due to the extensive use of electronics, from the 1980s, American security experts started talking about net-centric war (NCW). The post-Cold War era witnessed proliferation of new types of irregular war. Small War or imperial counterinsurgency (COIN) of the late nineteenth century was transformed into Marxist anti-colonial insurgencies in the post-Second World War period. In the aftermath of the Cold War, insurgencies were transformed into what could be categorised as postmodern insurgencies. Postmodern insurgencies also generated a new type of COIN, which can be termed postmodern COIN. Mary Kaldor typologises these as new wars. Kaldor writes: 'the new wars involve a blurring of the distinctions between war (usually defined as violence between states or organized political groups for political motives), organized crime . . . and large scale violations of human rights' (Kaldor 2005: 2). One of the characteristics of post-Cold War irregular wars is that they are global in scope. Identity politics and religion (mainly Islam) are the principal drivers of these insurgencies.

In Chapter 32, Rob Johnson shows that in response to transcontinental *jihad*, the American COIN also became global in the form of the Global War on Terror (GWOT). Johnson states that the US-led Coalition forces waging NCW easily disposed of the regular forces both in Afghanistan (2001) and Iraq (2003). However, the Western regulars became bogged down in attrition-centric COIN. Urban insurgency (2004) proved to be extremely costly in Iraq. Winning battles is one thing and nation-building is another. NCW provides no answer to it. John P. Cann notes that the Western capital-intensive armies are gradually transforming their doctrines in response to the postmodern insurgencies: from firepower-intensive tactics, they are attempting to win the hearts and minds of the people in the disturbed zones (Cann 2006: 107–27). This trend is the polar opposite of what Callwell preached.

One of the important characteristics of the new wars of the twenty-first century is the privatisation of violence. Private military contractors (PMCs) are becoming important players in the game. Private forces existed in the medieval world but vanished after the emergence and spread of Westphalian nation-states from the early modern era. Both small and weak states of Africa, as well as major powers like the United States, use military contractors. Some have argued that the messy and fuzzy pre-Westphalian international system is making a comeback. Jakub B. Grygiel asserts in his monograph: 'Barbarians are back. Small groups, even individuals, administering little or no territory, with minimal resources but with a long reach are unfortunately on the front pages of the newspapers because of their destructive fury' (Grygiel 2018: 1). Christopher Kinsey's chapter in this volume focuses on the use of PMC in Afghanistan and Iraq and its implications for the near future (Chapter 33). The PMCs are carrying out non-combat tasks (associated with logistics) as well as low-intensity combat missions. Kinsey argues that the sphere of action of the PMCs will increase with time due to both ideological and instrumental reasons. What we need is a comprehensive legal review to keep the contractors under political supervision to prevent the erosion of state sovereignty.

Besides new types of insurgencies and deployment of private forces, the postmodern period is also witnessing the massing of the latest conventional weaponry by the 'rising' polities in various regions of Asia. One such hot spot is East Asia, which is the subject matter of Andrew Tan's contribution (Chapter 34). The extensive economic growth of China and Japan has resulted in a spurt of acquisition of the latest weapon systems by these two countries. China is emerging as the new superpower and its defence spending is next to that of the United States. Confident of its burgeoning economic and military prowess, the 'New Middle Kingdom' is flexing its muscles in the South China Sea and Sea of Japan. This has generated alarm and is triggering subsequent arms races between China, Japan and the United States. Tan claims that, if left unchecked, a war in East Asia might occur in the new millennium.

Arjun Subramaniam, in the last chapter of this handbook (Chapter 35), elaborates on the crucial role which airpower might play in the new type of warfare that is unfolding in the post-Cold War period. Influenced by Colonel Thomas Hammes' model,[11] Subramaniam asserts that in postmodern COIN operations, long precision strikes by fixed-wing manned fighter-bombers against transnational insurgents have a lot of potential. However, he warns that 4GW is probably mutating into 5GW, which will require new types of tactics, techniques and procedures (TTPs). Subramaniam hints at the use of Unmanned Aerial Vehicles (UAVs), which brings us to the next and last section of this introduction: the future of warfare.

Postscript: Digitised Battlefield and the End of Humanity?

The practitioners of the history of technology approach have identified four Industrial Revolutions. The First Industrial Revolution, which started at the end of the eighteenth century, was based on the use of animal and water power to produce cotton textiles. The Second Industrial Revolution began in the second half of the nineteenth century and was based on iron and steel, and later oil. The Third Industrial Revolution, beginning in the 1940s, saw the development of nuclear power and the electronics industry. The Fourth Industrial Revolution started in the 1990s and is continuing. Its principal product is Artificial Intelligence (AI), which is a conglomerate of various related technologies like the Internet of Things, three-dimensional printing, blockchain technology, machine learning, etc.

The most common AI-embedded weapon systems which are in use now are drones, or UAVs, Unmanned Sea Vessels and ground robots. A military robot could be defined as a cyber-physical system which possesses the capabilities of sensing, communication, decision making and action. The AI-augmented war machine could be supervised by humans and also has the capacity to act autonomously (Danet and Hanon 2014: xvii). Finally, the system is capable of improving its own performance through automatic learning. The principal fuel of AI is data. The use of AI is not only changing the character of war but also the nature of war. All the technologies which have appeared to date, before the use of AI, never encroached on the field of command (the decision-making process). However, the superior computational power of AI is resulting in the augmentation of human command with machines. Some scientists warn that a super-intelligent machine (Terminator-like) is not far away, which will result in the replacement of human command with machine command. Then the final product would be the complete dehumanisation of warfare, if not the end of humanity.

The future of warfare is beyond the remit of this handbook. But the study of the past can offer some insights. The problem with this technological deterministic approach is that politics and social fabric are left out of the loop. Further, as we have mentioned earlier, it is not merely the possession of a technology but how one uses it that is important. Many early modern Asian states had cannons but did not use them as effectively in battle as did the Europeans. Similarly, several things the robotics experts are saying are nothing new. The scientists and military officers of China, Russia and the United States are speaking of swarming robots/drones. Swarming as a tactic can be traced back to the Scythian horse archers who met Darius I's Achaemenid Army at the Pontic Steppe. For the time being, wars start with the decisions and actions of humans and the conduct of war remains a human venture with intelligent (automated and semiautonomous) machines playing second fiddle. So, human–machine interaction remains the key feature that will define warfare in the years to come.

Notes

1 John Keegan's global history of warfare (1994) is a good example of this approach. Victor Davis Hanson (2009) asserts that close quarter combat involving drilled and disciplined heavy infantry first emerged in Classical Greece. Hanson (1998) asserts that free farmers owning medium size plots in the Greek *poleis* were freedom loving peasant warriors.
2 Geoffrey Parker's edited volume (1995) portrays the triumphal march of the Western militaries from Classical Greece in around 600 BCE till it bogged down in Vietnam in the 1970s. The editors and the contributors of this volume assert that the Western Way of War was characterised by drilled

and disciplined professional infantry soldiers supported by states with robust economic and political institutions. Both Parker in the above-mentioned edited volume and Hanson in his *longue duree* overview of the evolution of warfare (2001) claim that the West pioneered the rational and scientific approach to war in contrast to the religious-magical approach of the non-Westerners.

3 For the view that Napoleonic warfare witnessed the birth of an operational level of war, see Martin van Creveld, 'Napoleon and the Dawn of Operational Warfare,' in John Andreas Olsen and Martin van Creveld (eds.), *The Evolution of Operational Art: From Napoleon to the Present* (Oxford: Oxford University Press, 2011), pp. 9–34. For an updated overview of the Mongol war machine see May (2007).

4 Chickering and Forster's team (which includes Greiner, Jorg Nagler, etc.) has produced a multi-volume set for analysing the unfolding of Total War from Napoleon to Hitler. See Boemeke et al. (1999), Chickering and Forster (2005a, 2005b, 2007, 2010), Chickering et al. (2005), Forster and Nagler (1997).

5 For the debate regarding Military Revolutions and Military Evolution in early modern Western Europe see Rogers (1995).

6 The credit for introducing the concept of military-fiscalism goes to the British historian John Brewer (1989). For him, a military-fiscal state was characterised by the hunger of cash in order to wage war. For extracting more revenues to maintain its ever-burgeoning armed forces, such a state builds an expanding bureaucracy which reaches the bottom levels of the society. Monetisation and banks, along with the techniques of deficit financing, were chief characteristics of such polities. For Brewer, such a state first emerged in Britain during the eighteenth century. The point to be noted is that Brewer is not concerned with Military Revolution. However, for the supporters of the Military Revolution thesis, the presence of a fiscal-military machinery was absolutely necessary in order to implement a Military Revolution.

7 See the edited collection by Knox and Murray (2003).

8 For discussion regarding social technology see James (2006: 28).

9 During the Second World War, in the Pacific theatre, the American soldiers collected body parts of the dead Japanese troops as a sort of totem (Cameron 1994).

10 The best overview of the debates regarding hoplite battles can be found in Kagan and Viggiano (2013).

11 Colonel Hammes (2006), an American marine officer, categorises the evolution of war into five generations. The First Generation War (1GW) involved the rise of gunpowder weapons in the sixteenth century. The 2GW involved the rise of machine guns and trenches in the American Civil War and the First World War. The 3GW witnessed manoeuvre war during the Second World War. The post-Second World War insurgencies are characterised by him as 4GW. Actually, 4GW is another name for postmodern insurgencies. The evolving 5GW for him involves biological and WMD attacks by small groups of networked insurgents (equivalent to Jakub B. Grygiel's new 'barbarians'). To an extent, Hammes, like Martin van Creveld (2008), is prophesising the end of conventional war. Both are rabid critics of the proponents of RMA who concern themselves with high-intensity, technology-centric interstate wars.

Bibliography

Black, Jeremy, *A Military Revolution? Military Change and European Society, 1500–1800* (London: Macmillan, 1991).

Black, Jeremy, *Rethinking Military History* (London/New York: Routledge, 2004).

Black, Jeremy, *War and Technology* (Bloomington and Indianapolis: Indiana University Press, 2013).

Black, Jeremy, *Clio's Battles: Historiography in Practice* (Bloomington, Indiana: Indiana University Press, 2015).

Boemeke, Manfred F., Chickering, Roger and Forster, Stig (eds.), *Anticipating Total War: The German and American Experiences, 1871–1914* (Cambridge: German Historical Institute and Cambridge University Press, 1999).

Bourke, Joanna, 'New Military History,' in Matthew Hughes and William Philpott (eds.), *Modern Military History* (Basingstoke: Palgrave Macmillan, 2006), pp. 258–80.

Brewer, John, *The Sinews of Power: War, Money and the English State, 1688–1783* (London: Unwin Hyman, 1989).

Brodie, Bernard and Fawn M., *From Crossbow to H-Bomb* (1962, reprint, Bloomington: Indiana University Press, 1973).

Cameron, Craig M., *American Samurai: Myth, Imagination, and the Conduct of Battle in the First Marine Division, 1941–1951* (Cambridge: Cambridge University Press, 1994).

Cann, John P., 'Low-Intensity Conflict, Insurgency, Terrorism and Revolutionary War,' in Matthew Hughes and William Philpott (eds.), *Modern Military History* (Basingstoke: Palgrave Macmillan, 2006), pp. 107–30.

Chaliand, Gerard, *A Global History of War: From Assyria to the Twenty-First Century*, tr. by Michele Mangin-Woods and David Woods (California: University of California Press, 2014).

Chickering, Roger and Forster, Stig, 'Are We There Yet? World War II and the Theory of Total War,' in Roger Chickering, Stig Forster and Bernd Greiner (eds.), *A World at Total War: Global Conflict and the Politics of Destruction* (Cambridge: German Historical Institute and Cambridge University Press, 2005a), pp. 1–16.

Chickering, Roger and Forster, Stig (eds.), *Great War, Total War: Combat and Mobilization on the Western Front* (2000, reprint, Cambridge: German Historical Institute and Cambridge University Press, 2005b).

Chickering, Roger and Forster, Stig (eds.), *The Shadows of Total War: Europe, East Asia, and the United States, 1919–1939* (2003, reprint, Cambridge: German Historical Institute and Cambridge University Press, 2007).

Chickering, Roger, and Forster, Stig (eds.), *War in an Age of Revolutions, 1775–1815* (Cambridge: German Historical Institute and Cambridge University Press, 2010).

Chickering, Roger, Forster, Stig and Greiner, Bernd (eds.), *A World at Total War: Global Conflict and the Politics of Destruction* (Cambridge: German Historical Institute and Cambridge University Press, 2005).

Chin, Warren, 'Land Warfare from Machiavelli to Desert Storm,' in Matthew Hughes and William Philpott (eds.), *Modern Military History* (Basingstoke: Palgrave Macmillan, 2006), pp. 84–106.

Daly, Gregory, *Cannae: The Experience of Battle in the Second Punic War* (London: Routledge, 2002).

Danet, Didier and Hanon, Jean-Paul, 'Introduction: Digitization and Robotization of the Battlefield: Evolution or Robolution?' in Ronan Doare et al. (eds.), *Robots on the Battlefield: Contemporary Issues and Implications for the Future* (Fort Leavenworth, KS: Combat Studies Institute Press, US Army Combined Arms Center, 2014), pp. xiii–xxxii.

Ferrill, Arther, *The Origins of War: From the Stone Age to Alexander the Great* (1985, reprint, London: Thames and Hudson, 1986).

Forster, Stig and Nagler, Jorg (eds.), *On the Road to Total War: The American Civil War and the German Wars of Unification, 1861–1871* (Cambridge: German Historical Institute and Cambridge University Press, 1997).

France, John, 'The Military History of the Carolingian Period,' in John France and Kelly DeVries (eds.), *Warfare in the Dark Ages* (Aldershot: Ashgate, 2008), pp. 321–39.

France, John, *Perilous Glory: The Rise of Western Military Power* (New Haven/London: Yale University Press, 2011).

Fuller, J.F.C., *The Conduct of War: 1789–1961* (1961, reprint, London: Methuen & Co., 1979).

Fuller, J.F.C., *Armament and History: The Influence of Armament on History from the Dawn of Classical Warfare to the End of the Second World War* (1945, reprint, New York: Da Capo Press, 1998).

Gat, Azar, *War in Human Civilization* (New York: Oxford University Press, 2006).

Gommans, Jos, *The Indian Frontier: Horse and Warband in the making of Empires* (New Delhi: Manohar, 2018).

Grygiel, Jakub J., *Return of the Barbarians: Confronting Non-State Actors from Ancient Rome to the Present* (Cambridge: Cambridge University Press, 2018).

Halsall, Guy, 'The Western European Kingdoms, 600–1000,' in Anne Curry and David A. Graff (eds.), *The Cambridge History of War*, vol. 2, *War and the Medieval World* (Cambridge: Cambridge University Press, 2020), pp. 50–82.

Hammes, Thomas X., *The Sling and the Stone: On War in the 21st Century* (St Paul, MN: Zenith Press, 2006).

Hanson, Victor Davis, *Warfare and Agriculture in Classical Greece* (Berkeley/Los Angeles: University of California Press, 1998).

Hanson, Victor Davis, *Carnage and Culture: Landmark Battles in the Rise of Western Power* (New York: Doubleday, 2001).

Hanson, Victor Davis, *The Western Way of War: Infantry Battle in Classical Greece* (1994, reprint, Berkeley/Los Angeles: University of California Press, 2009).

Henninger, Laurent, 'Military Revolutions and Military History,' in Matthew Hughes and William Philpott (eds.), *Modern Military History* (Basingstoke: Palgrave Macmillan, 2006), pp. 8–22.

Hildinger, Erik, *Warrior of the Steppe: A Military History of Central Asia, 500 BC to 1700 AD* (1997, reprint, Cambridge, MA: Da Capo, 2001).

Hughes, Matthew and Philpott, William, 'Introduction,' in Matthew Hughes and William Philpott (eds.), *Modern Military History* (Basingstoke: Palgrave Macmillan, 2006), pp. 1–7.

James, Alan, 'Warfare and the Rise of the State,' in Matthew Hughes and William Philpott (eds.), *Modern Military History* (Basingstoke: Palgrave Macmillan, 2006), pp. 23–41.

Kaegi, Jr., Walter Emil, 'Some Thoughts on Byzantine Military Strategy,' in John Haldon (ed.), *Byzantine Warfare* (Aldershot: Ashgate, 2007), pp. 251–68.

Kagan, Donald and Viggiano, Gregory F. (eds.), *Men of Bronze: Hoplite Warfare in Ancient Greece* (Princeton, NJ: Princeton University Press, 2013).

Kaldor, Mary, *New and Old Wars: Organized Violence in a Global Era* (1999, reprint, Dehradun: Natraj, 2005).

Keegan, John, *The Face of Battle: A Study of Agincourt, Waterloo and the Somme* (1976, reprint, Middlesex: Penguin, 1978).

Keegan, John, *A History of Warfare* (1993, reprint, New York: Vintage Books, 1994).

Kennedy, Paul, *The Rise and Fall of the Great Powers: Economic Change and Military Conflict from 1500 to 2000* (1988, reprint, London: Fontana, 1990).

Knox, MacGregor and Murray, Williamson (eds.), *The Dynamics of Military Revolution: 1300–2050* (2001, reprint, Cambridge/New York: Cambridge University Press, 2003).

Levy, Jack S. and Thompson, William R., *The Arc of War: Origins, Escalation, and Transformation* (Chicago: University of Chicago Press, 2011).

Lorge, Peter A., *The Asian Military Revolution: From Gunpowder to the Bomb* (Cambridge: Cambridge University Press, 2008).

May, Timothy, *The Mongol Art of War* (Barnsley: Pen & Sword, 2007).

McNeill, W.H., *The Pursuit of Power: Technology, Armed Forces and Society since AD 1000* (Chicago: University of Chicago Press, 1982).

Morillo, Stephen and Pavkovic, Michael F., *What Is Military History?* (Cambridge: Polity, 2006).

Neiberg, Michael S., 'War and Society,' in Matthew Hughes and William Philpott (eds.), *Modern Military History* (Basingstoke: Palgrave Macmillan, 2006), pp. 42–60.

Ness, Gayl D. and Stahl, William, 'Western Imperialist Armies in Asia,' *Comparative Studies in Society and History*, vol. 19, no. 1 (January 1977), pp. 2–29.

Parker, Geoffrey, *The Military Revolution: Military Innovation and the Rise of the West, 1500–1800* (Cambridge: Cambridge University Press, 1988).

Parker, Geoffrey (ed.), *The Cambridge Illustrated History of Warfare: The Triumph of the West* (Cambridge: Cambridge University Press, 1995).

Phillips, Brigadier-General Thomas R. (ed.), *Roots of Strategy: A Collection of Military Classics* (1940, reprint, Dehradun: Natraj, 1989).

Philpott, William J., 'Total War,' in Matthew Hughes and William Philpott (eds.), *Modern Military History* (Basingstoke: Palgrave Macmillan, 2006), pp. 131–52.

Roberts, M., *The Military Revolution 1560–1660: An Inaugural Lecture Delivered Before the Queen's University of Belfast* (M. Boyd, University of Belfast, 1956).

Rogers, Clifford J. (ed.), *The Military Revolution Debate: Readings on the Military Transformation of Early Modern Europe* (Boulder: Westview, 1995).

Roy, Kaushik, ''1857 Uprising: Colonial War or Small War or Total War?' in Kaushik Roy (ed.), *The Uprising of 1857* (New Delhi: Manohar, 2010), pp. 245–72.

Sharman, J.C., *Empires of the Weak: The Real Story of European Expansion and the Creation of the New World Order* (2019, reprint, Princeton: Princeton University Press, 2020).

Thampi, Madhavi, 'The Taiping Rebellion in China,' in Kaushik Roy (ed.), *The Uprising of 1857* (New Delhi: Manohar, 2010), pp. 229–44.

Tilly, Charles, 'Reflections on the History of European State-making,' in C. Tilly (ed.), *The Formation of National States in Western Europe* (Princeton, NJ: Princeton University Press, 1975), pp. 3–89.

Van Creveld, Martin, *Technology and War: From 2000 BC to the Present* (London: Brassey's, 1991).

Van Creveld, Martin, *The Changing Face of War: Combat from Marne to Iraq* (2007, reprint, New York: Ballantine Books, 2008).

Vandervort, Bruce, 'War in the Non-European World,' in Matthew Hughes and William Philpott (eds.), *Modern Military History* (Basingstoke: Palgrave Macmillan, 2006), pp. 195–213.

Wilson, Peter, 'Of "Master" and "Grand Narratives" and their Discontents: Early Modern European Military History,' *War & Society*, vol. 42, no. 1 (2023), pp. 90–8.

PART I

Origins of Warfare

1

THE EVOLUTION OF BRONZE AGE WARFARE, AND ITS HISTORICAL SIGNIFICANCE, IN WESTERN EURASIA AND THE NEAR EAST

Robert Drews

Warfare in the Near East and western Eurasia during the Bronze Age, or let us say in the third and second millennium BCE, is still an unsettled subject. Although tens of thousands of tablets, inscriptions and papyri come from that early period, the writing of history did not begin anywhere until the Iron Age. As a result, for a long time we assumed that Near Eastern wars during the Bronze Age were much the same as those in the Iron Age, but that assumption is evidently not correct. Through the reign of Hammurabi at Babylon (1750 BCE on the middle Mesopotamian chronology) warfare in the Near East seems to have been limited to siege warfare. As for Europe (here the name is used for Eurasia west of the steppes), it was once thought that from late in the fifth until early in the third millennium BCE, waves of warlike invaders from the steppes, on horseback and speaking Indo-European languages, conquered much of Europe and turned it from a peaceful place into a land of warriors. Although such a scenario still has some support, the evidence now available indicates that until the second half of the second millennium BCE wars were very infrequent in most of Europe, if wars were fought at all, and that mounted combat did not begin anywhere until ca. 1000 BCE. Altogether, it is beginning to appear that in the third millennium BCE, and a good part of the second, there was no battlefield combat. What follows in this chapter is my understanding of the evolution of warfare in western Eurasia and northern Africa during the Near Eastern Bronze Age and very early Iron Age. The reader must keep in mind, however, that other scholars would describe that evolution quite differently.

Before the Bronze Age

It is debated whether anything that could be called a war was fought during the Neolithic period (Carman 2013: 22–3). Osteological evidence shows that many people in Neolithic Eurasia and northern Africa died by human hand, but the violence seems usually to have been an individual homicide and occasionally a massacre of unsuspecting men, women and children. We have little evidence for a battle in the Neolithic period, if we understand 'battle'

DOI: 10.4324/9780429437915-3

to mean combat between two groups, all of whom were prepared and equipped for combat. When there was a hostile confrontation, it was probably a skirmish at long range—with the bow as the weapon—and ended after a few men had been killed or wounded. In the Egyptian language, 'bows' was the word for 'army,' evidently because in the Neolithic period group aggression had been synonymous with the bow (Darnell and Manassa 2007: 58).

Warfare from 3000 to 1700 BCE

In the Near East serious warfare began ca. 3000 BCE, with the rise of formal states. Apparently, however, for well over a millennium these wars were not fought on battlefields but were sieges of cities. A border quarrel, of course, had to be settled at the border. So, for example, in the middle of the third millennium BCE the kings of Umma and Lagash, two Mesopotamian cities, fought repeatedly at—and over the divinely ordained location of—the irrigation canal that served as the border for their adjacent kingdoms. Normally, however, when the king of one Mesopotamian city-state went to war against the king of another, the only battle of any significance was fought at the latter's city-wall (Dawson 2001: 76–117; Burke 2008: 27–8; Drews 2017: 61–75). Great effort was expended in building these mud-brick walls, which normally were at least 2 metres wide and seem to have been at least 10 metres high, and a correspondingly great effort had to be spent in breaching them. In the archive of 20,000 cuneiform tablets found at Mari on the middle Euphrates, and dating to the 1760s BCE, the Akkadian texts frequently refer to the sieges of cities, but not to a battle in which two armies clashed in the open country. Jack Sasson, who worked with the Mari texts for most of his career, observed that pitched battles were scarcely known in the Old Babylonian period: 'While they may have occurred, only one possible example can be retrieved from the vast Mari documentation. . . . Real battles—rather than skirmishes or raids—occur at the enemy's gate' (Sasson 2015: 467–8).

When a king went to besiege his enemy's city, he brought with him a huge labour force—an ÉRIN.MEŠ (a Sumerogram) or a *ṣābu* (an Akkadian word) of thousands and even tens of thousands of men, all of them conscripted[1]—and a relatively small number of warriors (archers, slingers and spearmen). The latter were usually recruited from the countryside, where men were accustomed to handling a weapon in order to protect their families' flocks and herds. Some of these men were the king's own subjects, but more were hired from tribes of nomadic pastoralists. In order to stop, or at least to slow, an approaching ÉRIN.MEŠ, the king of a threatened city might hide some of his archers in an ambush, but such ambushes rarely accomplished much. When the spearmen who accompanied the column ran toward the ambush, the archers who had hidden there would have had a long way to run before they reached the safety of their city, and eventually the siege-train would have reached the target city. Because the king of that city stationed most of his archers and slingers atop the city-wall, where they could rain down their arrows and slingstones on the attackers, the besiegers' own archers did their best to clear the wall, shielding themselves under canopy-shields (portable sheds; see Figure 1.1).

The first task for the thousands of labourers was to build—a short distance from the wall—a siege tower from which their archers could shoot arrows down upon the defenders standing atop the wall. Once the siege tower was in place and a portion of the wall was more or less denuded of defenders, the labourers began hauling dirt, sand and debris in order to pile up a siege mound against the wall. In this task some of the labourers were invariably killed and many more were wounded. When the mound finally reached the top of the wall, the aggressors' spearmen—in close formation, wearing protective leather and carrying

Figure 1.1 Drawing of a siege scene incised on a stone slab from pre-Sargonid level at Mari, dating
ca. 2400 BCE.

Source: From Fig. 1 in Yadin 1972 (after Parrot 1971, Plate XIV, no. 4). Courtesy *Israel Exploration Journal*.

shields—attempted to take over the wall and then the city. If they were successful, the city's
king and his councillors would have been seized for punishment and palace, houses and
perhaps even temples would have been plundered. Such was siege warfare in the Early and
Middle Bronze Age of the Near East (for details see Kern 1999 and Burke 2008).

Combat in the open country occurred when a king sent his hired warriors—archers and
spearmen—to punish pastoral nomads or other troublemakers who did not live in a city. In
these encounters most of the fighting was at long range and was more a skirmish than a battle.
The nomads hoped only to ward off the king's men, whose intention was only to punish the
nomads: to kill or wound some of the defenders, to destroy their camp, and to capture their
livestock and perhaps several of their women and children. Although the commander of the
king's men may have had a composite bow, which had a range of at least 200 metres, most of
the archers carried a self bow, whose range was at most half that of a composite bow. The king's
spearmen protected the archers and dealt with the opponents whom the archers had wounded.

Unlike the Near East and eventually the island of Crete (with its 'Minoan' palaces), Eu-
rope was not yet at the state or the urban level in the third and early second millennium BCE.
Long-range skirmishes must have occurred here from time to time, because arrowheads have
been found imbedded in the skeletons of some adult males who were buried in the Bell
Beaker, the Corded Ware and other archaeological cultures of Neolithic and Early Bronze
Age Europe. But evidence for hand-to-hand battles between groups of men is still lacking.
In temperate Europe, metal spearheads do not seem to have come into use until the second
quarter of the second millennium BCE (Harding 2000: 281; Leshtakov 2011: 45–8), centuries after they had become familiar in the Near East, in southern Caucasia and on the steppes.

Until that time the most lethal close-range weapons in Europe were the dagger and the so-called but misnamed 'battle-axe,' both of which were personal weapons (routinely carried by men for personal protection) rather than weapons designed for combat.

In the grassland and forest steppes, from the Carpathian Mountains eastward to the Urals and beyond, men occasionally engaged in combat, unlike their contemporaries in Europe west of the Carpathians. Most people in the steppes were pastoralists, moving with their herds and flocks from one pastureland to another. Pastoralism and combat must also have characterised the Abashevo culture to the north, along the upper Volga and Kama rivers. A collective burial of men who died a violent death, and therefore evidence of combat, was found under an Abashevo kurgan near Pepkino, in the Volga Upland. Here, in a grave dating very early in the second millennium BCE, were buried the bodies of 28 young men. Arrowheads imbedded in their skeletons indicate that the young men were killed or wounded at long range. Their enemies then used axes to crush the skulls of the men or to decapitate them (Koryakova and Epimakhov 2007: 62–3). Other early evidence for combat comes from southern Caucasia, also a land of pastoralists. A silver goblet, found under a kurgan at the Armenian village of Karashamb, was locally made shortly before 2000 BCE and was embossed with scenes of hunting and warfare. In one scene, men are wielding spears and protecting themselves with shields, and in another the victors are decapitating the opponents whom they have captured (see Kohl 2007, fig. 3.28).

Battlefield warfare and the military use of chariots apparently began at roughly the same time. The Eurasian steppe was the natural habitat for wild horses, vast numbers of them, and archaeological evidence shows that on the steppe horses were domesticated well before 4000 BCE and for a long time were raised for their meat and milk. Although they probably began to be used also as pack animals soon after domestication, it was not until ca. 2000 BCE that horses were 'tamed' for either draft or riding. The original device for enforcing commands was perhaps a nose-ring, but far more effective control came when people near the southern Ural Mountains learned to place an organic bit in the horse's mouth.

Despite the invention of the bit, riding a horse was a dangerous undertaking (saddles, to say nothing of stirrups, were still far in the future), and the speed of the horse was best utilised by hitching a team of horses to a chariot, a light cart that rolled on two spoked wheels. The earliest evidence for chariots, dating ca. 2000 BCE, has been found in the Sintashta-Petrovka archaeological culture, which lies in the forest steppe along the southern Urals (Anthony 2007: 397–405; Drews 2017: 41–6). Burials there from that period occasionally include a team of bitted horses and the chariot which they had pulled. Remains of a composite bow, other weapons, and even scales of bone armour have also been found in these chariot burials. Carrying both a driver and an archer, a chariot made it possible to hunt not only prey but also predators that were too swift and too dangerous to be hunted on foot. The skills learned for hunting were obviously soon used against other pastoralists.

Chariots were skilfully built and therefore costly, and from the Sintashta cemeteries it appears that only one man in twenty had the means to acquire a chariot (Chechushkov and Epimakhov 2018: 472). While the elite went into battle on their chariots, using the vehicle as a mobile firing platform for an archer armed with a composite bow, each chariot crew would probably have been accompanied by several commoners on foot, who were armed with spears and whose duty it would have been not only to protect their betters from a hand-to-hand attack but also to converge upon opposing chariot crews who had been immobilised. Such must have been the beginning of chariot combat. Because most of the people living in horse country seem to have spoken an Indo-European language, so also would most of the pioneers in the taming of horses and in chariot warfare.

Warfare in the Near East and the Aegean from 1700 to 1200 BCE

Although chariot combat on the Eurasian steppe and in southern Caucasia may have begun very early in the second millennium BCE, people in those lands left us no written records. Our earliest documentary evidence for chariots in battle dates ca. 1750 BCE and comes from southeastern Anatolia, where Anitta, king of Kanesh, either met or deployed 40 chariot teams in his siege of Šalatiwara. We do not know how chariots were first used militarily in the Near East, but very likely they were initially a mobile replacement for the stationary ambush as a means to harass the thousands of conscripts in an approaching ÉRIN.MEŠ as it lumbered along: a continuing stream of arrows shot from continually moving chariots (which would have stayed some distance away from the siege-train) would have killed some of the labourers and wounded many more. In order to protect his ÉRIN.MEŠ, the aggressor king would have needed chariots of his own, and so battlefield warfare in the Near East would have begun. In any case, throughout the Near East, chariots moved battles from the city-wall to the open country, and thereby initiated the battlefield warfare that is well documented in later periods. Cities continued to be walled, and sieges continued to be laid all through antiquity, but now the siege could not begin until the opponent's offensive capacity had been neutralised in battle. By the middle of the seventeenth century BCE the Hittite king, Hattushili I, had assembled an army of several hundred chariots, with which he created the Great Kingdom of Hatti. At about the same time, another chariot force conquered Lower Egypt, and put 'foreigners' (*hyksos*) in power as the Fifteenth Dynasty. At Mycenae on the Greek mainland the men who were buried in the shaft graves (ca. 1650–1500 BCE) were proud of their chariot prowess. These chariot lords in Greece, as was the case also in Mittani (a Great Kingdom stretching from western Syria to Nuzi and Arrapha, 300 miles to the east), spoke an Indo-European language and seem to have come from horse country to take over lands and populations that were ripe for conquest. The same may have happened in the Indian subcontinent, bringing an Indo-Iranian language to what is now Pakistan.

The typical chariot carried a driver and an archer, both wearing a helmet and body armour. In the Aegean world the latter was sometimes plate armour reaching down to the thighs. In the Near East it was a *sariam*: a long leather tunic covered with bronze scales. A recent calculation concluded that a *sariam*, covered with between 400 and 600 bronze scales, weighed between 37 and 57 pounds, or at least 17 kg (Howard 2011: 65–74). The head and forequarters of a chariot horse were usually protected by a leather *gurpisu*. The chariot archer was armed with a composite bow, and on the chariot was mounted a bow-case and a quiver holding 50 or 60 arrows. Hittite chariots often carried a third man: a shield bearer, whose role was much less demanding and status much lower than those of the driver and archer. In a Vedic hymn (Rig Veda 6.75; Drews 2017: 111–12), dating from the second half of the second millennium BCE, the chariot crewmen in India celebrate their armour, their bow, their team of horses and their chariot. It is not too much to say that militarism began with 'the Chariot Revolution' and that with that revolution there appeared for the first time an elite military class (Hacker 2012).

Far below this class, although their role was at least as perilous, were the chariot runners. These non-skilled warriors were men on foot who were armed with a hand-to-hand weapon and carried a small shield. Dressed for speed, they seldom wore anything more than a kilt and a leather helmet. The runners' task was to attack the crewmen of enemy chariots whose vehicles had been disabled because of a wounded horse or a broken wheel, and to deal with enemy runners. The chariot crewman himself either carried a hand-to-hand weapon at his side or kept a spear within reach. Because of the heavy *sariam* that he wore, however, he had difficulty moving and if their chariot was halted in battle the crewmen's lives depended on the runners assigned to their squadron. Although some runners came from within the kingdoms, civilised men tended to regard the runner's task as too dangerous for the modest rewards to be gained from it. Most runners were therefore hired from backward lands such as Nubia, Libya, Sicily, northern Greece, Lycia and the hill country of the Levant. Of the various 'barbarian' runners in Egyptian armies of the 19th Dynasty the Sardinians (*sherden*, or *shardana*) may have been especially efficient (Abbas 2017).

In Late Bronze Age armies, the runners may have outnumbered the chariot crewmen, but they have received little notice from historians. Even the existence of runners was unrecognised until 1963, when Alan Schulman called attention to them in an article that appeared in a low-circulation periodical (Schulman 1963, 89–90). The role of the runners was first described 20 years later: Nigel Stillman and Nigel Tallis boldly sketched it, although in a book meant for war-gamers rather than for scholars (Stillman and Tallis 1984: 56; Drews 1993: 142–47). Egyptian inscriptions mention the runner (*phrr*) but say little about him, and while Egyptian iconography celebrates the Pharaoh and his exploits on the chariot it usually ignores the runners. The best evidence for them comes from temple reliefs at Abydos, which were discovered in the nineteenth century but were scarcely known until they were photographed by the late Vronwy Hankey.[2] Ordered by Ramesses II, the reliefs show Sardinian runners in action at the Battle of Kadesh (see Figure 1.2).

Figure 1.2 Sardinian runner cutting off the hand of a Hittite charioteer. From a relief on Ramesses the Great's temple at Abydos, Egypt.

Source: Courtesy of the Vronwy Hankey estate.

Evidence for the runners has also come from a fragmentary painted papyrus found at Amarna, where artists strove for realism unprecedented in Egyptian art. This papyrus, published in 1994, depicts runners who wear boar's-tusk helmets (and so are presumably from Greece) serving alongside Egyptian runners in Akhenaton's army. The runners are apparently rushing to save a comrade whose throat is about to be cut by Libyan tribesmen (Schofield and Parkinson 1994, figs. 1 and 2; Darnell and Manassa 2007, fig. 27).

The chariot battles fought at Megiddo ca. 1468 BC and at Kadesh ca. 1286 BC (on the conventional Egyptian chronology) are by far the best known, thanks to the self-aggrandising inscriptions and reliefs ordered by the Pharaohs who led the Egyptian armies. At Megiddo Thutmose III defeated a coalition of Levantine rulers and claimed to have captured 2,041 horses and 924 chariots after the Levantine crews had abandoned them and clambered up the walls of Megiddo to safety. At Kadesh Ramesses II battled against the Hittite king Muwatalli II and claimed to have defeated an army of some 3,000 chariots. Ramesses celebrated his dubious victory on several monuments but especially on the chariot frieze of his temple at Abydos (Spalinger 2003). Although no great chariot battle is likely to have taken place in the Mycenaean world, the palaces at Knossos and Pylos each maintained a force of several hundred chariots, and employed runners to go with them.

Chariot warfare changed society in many ways. With it came militarism and a military elite, and kings glorified themselves as battlefield warriors. An Egyptian Pharaoh now had himself portrayed on a temple wall as simultaneously driving a chariot (with the reins tied around his waist) and shooting an arrow from his composite bow. The *maryannu*—charioteers and chariot warriors—were a privileged class in Near Eastern kingdoms. The runners were not in this class, were paid a subsistence wage, and in battle were given a bonus for each enemy hand or penis that they brought to the tallyman. In the Near East and in Mediterranean lands there were not yet hand-to-hand battles between heavily armed and armored warriors. The *Iliad*'s battle narratives reflect some aspects of warfare in the early Iron Age, but not in the Bronze Age (Buchholz 2010: 30).

Warfare in Temperate Europe from 1700 to 750 BCE

In temperate Europe too militarism made its first appearance toward the middle of the second millennium BCE. Surveying 'the archaeology of violence' in prehistoric Europe, Albrecht Jockenhövel showed that although violent death was not uncommon in the Neolithic period it was not until well into the Bronze Age that the prerequisites for warfare made their first appearance in Europe (Jockenhövel 2004–2005: 126). Likewise, Kristian Kristiansen, in his *Warfare in Bronze Age Society*, concluded that in central and northern Europe 'warfare became institutionalized and professionalized in the Middle Bronze Age' (Kristiansen 2018: 41). And in Europe too the arrival of militarism apparently coincided with the arrival of chariot warriors. Although in temperate Europe we have no texts to inform us, archaeological evidence shows that 'tamed' horses came to the Carpathian basin ca. 1600 BCE, and not much later to northern Italy and also (although less demonstrably) to southern Scandinavia (Drews 2017: 141–67). Horses and chariots were apparently brought to temperate Europe, as to Greece, by miliary companies from horse country to the east, where the vernacular languages were Indo-European. The takeovers therefore seem to have launched the Indo-Europeanising as well as the militarising of much of Europe.

Because warfare in Europe, including Greece, seems to have been unfamiliar before the arrival of military forces from the east, the 'conquerors' in Europe seem to have formed an elite military class that initially, at least, was seldom required to fight. The chariot armies in Greece were meant to deter the chariot armies of Egypt, Hatti and other large Near Eastern kingdoms. In temperate Europe (where the terrain was, in any case, unsuitable for chariot warfare) the chariot, not needed for combat, soon became only a status symbol, and the military class began to pride itself on the use and display of hand-to-hand weapons, and especially of bronze swords.[3] In eastern Turkey a few swords have been found in contexts from early in the third millennium BC, but these were display pieces and not weapons meant for battle. The era of swordsmanship began ca. 2000 BCE, when men of status (probably chariot crewmen) in southern Caucasia began carrying rapiers. Because the sword was the first weapon that was designed specifically for killing men, it quickly became the supreme symbol of the warrior. For several centuries it remained more a symbol than an effective weapon, but after chariot warriors had taken over populations in Greece and the Carpathian basin its utility was greatly improved. The south Caucasian swords ca. 2000 BCE were Type A rapiers, which were elegant but poorly hilted. The Type A rapier was brought to Crete before 1700 BCE and to Greece and the Carpathian basin shortly before 1600 BCE. Within a few generations (ca. 1550 BCE) the military class in those lands designed more reliable bronze weapons: the Type B rapier in Greece, and the Boiu rapier and the short Apa sword in the Carpathian basin. The Boiu soon was brought to northern Italy and the Apa to southern Scandinavia, both probably in military takeovers. Because they continued to be status symbols in addition to their service as lethal weapons, and because the men who carried them were an elite class, many of these swords were ornately decorated (see Figure 1.3).

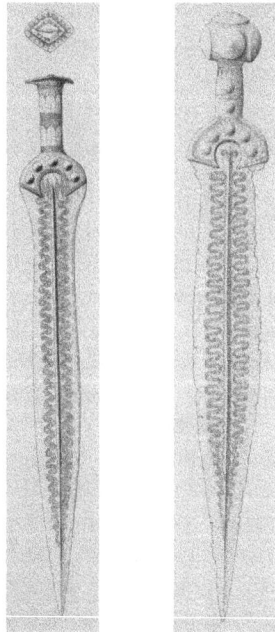

Figure 1.3 Short swords dating ca. 1500 BCE found at Hajdúsámson (Hungary) and Apa (Romania).

Source: From Kemenczei 1991, Tafel 1, no. 1, and from Bader 1991, Tafel 5, no. 25. Courtesy Franz Steiner Verlag.

As the generations passed, the military men in much of temperate Europe—from the Vistula River and the arc of the Carpathians westward to the Rhine and northern Italy—began fighting among themselves. As combat became frequent, the military men had obvious incentives to increase their numbers from below. Some 'producers,' to use Georges Dumézil's term, were enlisted as archers, and others as spearmen. By the thirteenth century BCE, many men in temperate Europe and northern Italy—probably now speaking languages ancestral to Keltic, Germanic or Italic—engaged in combat. How belligerent parts of Europe had become by its Late Bronze Age (Bz D and Hallstatt A and B) was shown recently, when archaeologists began excavating the riverbed of the Tollense river, in northeastern Germany. Thus far the bones of five horses and at least 130 men, most of them in their twenties and thirties, have been found. From what remains to be dug, the archaeologists believe that several thousand men took part in the battle, which can be dated ca. 1250 BCE (Jantzen et al. 2014: 14; see also Lidke et al. 2017).

Battles probably still began at long range, with bows, but now they ended with hand-to-hand weapons and the close-range fighting was decisive. The commander often carried a sword, sometimes decorated with repoussé or chased designs. His followers went into battle carrying shields, wearing helmets, corselets and even greaves, and wielding spears. At several sites in Europe and Britain archaeologists have found a man's skeleton with a spearhead lodged in the pelvis (Thorpe 2013: 235). Cumbersome corselets, made of plate bronze and reaching to the thighs, had been worn by chariot crews in Greece already in the fifteenth century BCE, but our earliest examples of bronze corselets from temperate Europe date to the Bz D period (ca. 1300–1200 BCE). Covering the warrior from the base of his neck to his hips, at least 30 of these bronze corselets have been found from Hungary, Slovakia and the Czech Republic westward to eastern France (Mödlinger 2012).

It was at about this time, ca. 1300 BCE, that the Urnfield culture began to spread especially in central and northern Europe. It receives its name from cremation cemeteries: large fields of buried urns, the urns containing the bones and ashes of the dead. The culture was highly militarised: bronze weapons and armour seem to have been the most prized possessions of Urnfield men, although the armour may have been meant as much for display as for protection (Harding 2000: 287–9). In the fourteenth century BCE several fortified settlements, large but short-lived, were built in central Europe. The largest known has been found at Bernstorf in Bavaria, a land where archaeologists have also found several extensive urnfields. The Bernstorf settlement extended over 12 hectares and was protected by a wood and sandy loam rampart at least 1.6 kilometres long (Röpke and Dietl 2014).

In Europe the Late Bronze Age (Reinecke's Bz D and Halstatt A and B) is thought to have continued until about 750 BCE. The Urnfield cultures, if we think in terms of various Urnfield cultures, lasted approximately as long, before becoming fully Iron Age cultures. War in Europe during this period was still on a smaller scale than in the Near East, because European society was still not at the state level. The importance of warfare, however, and the status of the warrior were apparently just as high in Europe as in the Near East. In Europe the deposition of weapons, sometimes in hoards and often in wet places (rivers, bogs, lagoons) was frequent, and often the weapons were elaborately decorated. And when a warrior died, he was often buried with the weapons and armour that had distinguished him. The weapons were swords and spears, the bow having lost much of its importance.

In what may have been a series of events parallel to the commencement of the Urnfield Culture (although urnfields have not been found here), military companies from central Europe seem to have taken over the tin-bearing mountains of southwestern Britain, northwestern France (Brittany, or Armorica) and the western coast of the Iberian Peninsula (Galicia and northern Portugal), and also the nearby bronze-working lands (such as Ireland) that had long depended on the tin deposits. What is called the Atlantic Bronze Age began ca. 1300 BCE and continued for approximately 500 years. This archaeological culture was centred on metallurgy: both the production and the use of bronze artifacts of high quality. The Atlantic Bronze Age has been described as 'the Age of the Warrior,' because battle weapons made their first appearance here ca. 1300 BCE, as did communal feasting equipment (large bronze cauldrons, and elaborately decorated spits for the roasting of meat). Swords were evidently the most important part of the warriors' panoply and a mark of their status. Flange-hilted slashing swords were pioneered in Central Europe and the quite sudden arrival of such swords (especially the Hemigkofen and Erbenheim types) in the zone of the Atlantic Bronze Age suggests that ca. 1300 BCE 'fighting and feasting' companies from lands along the upper and middle Rhine (and probably speaking the Proto-Keltic language) sailed down the river and into the Atlantic and began taking over the tin deposits and the nearby metal-working regions of southern Britain, Ireland, Brittany and the western coast of the Iberian Peninsula. From the south-central part of that peninsula well over a hundred 'warrior stelae' have been found from the area's Late Bronze Age (ca. 1250–850 BCE). Typically, the stelae depict swords, spears and shields.

The End of Chariot Warfare in the Near East and the Aegean: ca. 1250–1175 BCE

While hand-to-hand combat was characteristic of temperate Europe in the latter half of the second millennium BCE, chariot warfare remained the norm in the Near East and the Aegean through most of the Late Bronze Age. This was a highly specialised form of warfare, not for every man, and chariotries were very expensive to assemble and maintain. The construction of chariots required skilled carpenters, wheelwrights, joiners and tanners, and the keeping of horses required not only trainers, grooms, farriers and veterinarians but also extensive pastures and grain fields. When we add to all this the hiring, arming and armouring of skilled drivers and archers, we see how costly this military force was. The average kingdom could afford no more than a few hundred chariots and even a kingdom so large and rich as Egypt could afford only a few thousand. A chariot army, in most cities composed of a few hundred chariot crewmen (the elite) and a slightly larger number of chariot runners (most of them poor barbarians), was quite small when measured against the size of the population that it defended.

Barbarian runners, in contrast, were cheap and the supply was almost limitless. Late in the thirteenth century BCE it dawned on the barbarians among whom runners were traditionally recruited that if enough of these men joined together, each man with a fistful of javelins and a hand-to-hand weapon, they could defeat the chariot army of a Near Eastern or Aegean king. Until this time young men from backward lands had no higher hope than to be recruited as mercenary runners in the chariotry of a Near Eastern or Aegean king. Now their sights were set much higher: to join a host of raiders large enough to destroy the chariotry of a king, and then to sack his city and his palace. Plenty of volunteers were obviously available. In his fifth year as Pharaoh of Egypt (1208 BCE), Merneptah faced a Libyan king or chieftain who was intent on setting himself up as ruler of the western Delta. The Libyan appears to have had no chariots, but in their place he had recruited over 10,000 running warriors. Many of them were Libyans, but others came from northern Greece, Lycia, Sardinia, Sicily and Italy. This was a novelty: a host of warriors on the ground doing battle against a chariot army that was quite limited in size. Although numerous, the warriors on the ground must have been more a mob than an army, and certainly were not an infantry: the warriors apparently swarmed over the battlefield, each man on his own, depending not on a tactical unit into which he had been placed but on his own speed, agility and wits. Merneptah was able to defeat the thousands of running warriors because he had enough time to assemble his chariotry and meet the intruders in 'the fields of Perire,' which seem to have lain not far to the west of the Delta (Manassa 2003: 91–2). We have no details about the battle, but in some ways it may have resembled the Parthians' destruction of Crassus' army in 53 BCE. Merneptah's chariotry, that is, may have encircled the Libyan's runners, staying far enough away from them that it suffered few casualties, while its arrows decimated and finally destroyed the Libyan host.

More successful than the Libyan chieftain were hosts of raiders who in the late thirteenth and early twelfth centuries BCE attacked, besieged and sacked cities, and by so doing brought to an end the great civilisation of the Late Bronze Age (Drews 2020). Some of these attacks were overland, but others featured sea-borne raiders who surprised the capital cities of coastal kingdoms. Among the more than 60 cities that were sacked and burned, several to be permanently abandoned, were such places as Ashkelon, Hazor, Ugarit, Hattusha, Knossos, Troy and Mycenae. If the raiders arrived at their destination before dawn and surrounded a city as the sentries were alerting the chariot army, they could cut down the chariots as they sallied singly from the gates. After the city had been besieged for a few days, the gates would have been opened, the inhabitants herded out, and the raiders would have sacked the city and then burned much of it.

Coastal raiders owed much of their success to the capacity and speed of their ships (Emanuel 2017). The typical raiders' ship was an oared galley powered by however many raiders—40, 50, and possibly more—could be seated on its benches (the Catalogue of the Ships tells us at *Iliad* 2.510 that each of the Boiotian ships carried 120 men to Troy). The galley was also equipped with a sail, and unlike earlier sails this one was brailed. The raiders' ship therefore not only carried scores of men but also moved quickly through the water. The men at the oars guaranteed that all of the ships in a flotilla would arrive at their destination at more or less the same time. And because it had a relatively shallow draft, the raiders' galley could be easily beached and easily launched. The cities had no means of confronting the raiders' ships while still in the water, because the warship had not yet been invented: an oared galley at whose bow was an underwater naval ram, with a timber core coated with bronze, for puncturing and sinking an enemy ship. Such warships are familiar from later naval battles, such as those fought at Salamis and Actium, but ca. 1200 BCE naval rams did not yet exist (experiments with rams or cutwaters—at Greek or Phoenician coastal cities—do not seem to have begun until 1000 BCE at the earliest).

The Beginning of Infantry Warfare: ca. 1175–1000 BCE

The sacking and burning of cities in the Near East and the Aegean, which continued well into the twelfth century BCE, effectively marked the end of the chariot warfare that had begun 500 years earlier. In the Near East, chariots continued to have a role on the battlefield during the early Iron Age, but now their role was peripheral rather than central. More important than chariots was the number of men on the ground.

For defence against numerous but unorganised barbarians on the ground, those kingdoms that survived the catastrophe ca. 1200 BCE had no other choice but to turn to their own men, and to organise them in infantries. A thousand or more infantrymen, although inexperienced and not nearly so athletic or so daring as the runners, could be emboldened and effective if placed in a disciplined line. Because the running raiders swarmed over the battlefield in *mêlée* fashion, the best way to counter them was to have a long and close-order line moving steadily against them, each man in the line equipped—as in the Warrior Vase from Mycenae, probably dating from the LH IIIC period—with a spear and shield and wearing a leather helmet, corselet and greaves.

Unlike the chariot crewmen and runners upon whom a king had depended through most of the Late Bronze Age, the men in the infantry line were not professionals. They were ordinary conscripts or volunteers, strengthened by military discipline and by the *esprit de corps* that went with it. This was an historic innovation on battlefields during the transition from the Bronze Age to the Iron Age in the Near East and in the Mediterranean.

In many early infantries the men were apparently organised in three divisions, which probably should be understood as a centre, a right wing and a left wing. Each of these 'thirds' would have functioned as a tactical unit, and while the centre held its ground the wings could pivot to flank the enemy runners and finally envelop them within a triangle. In the earliest Roman infantry about which we have any information the 'thirds' (*tribūs*, from which comes our word 'tribes') were named the Ramnes, the Tities and the Luceres. Each *tribūs* was in turn divided into ten *curiae*, and every able-bodied male between the ages of 17 and 45 was assigned to one of the 30 *curiae*. Three divisions are first attested for the Greek Dorians, who early (ca. 1000 BCE) gained a reputation as spear-bearing infantrymen.[4] For early Israel we may also have an echo of a tripartite division. At 2 Samuel 23, the author speaks of King David's 'three mighty men' and of 'the thirty.' If the 'three mighty men' were the leaders of three divisions, then 'the thirty' would have been their subordinate officers.

The Beginning of Mounted Warfare: ca. 1000–850 BCE

In the Near East, chariots were still employed in the early Iron Age, although they were now ancillary to a line of infantrymen. King David of Judah and Israel, ca. 1000 BCE, evidently employed a few (40 or 50) chariots but he relied almost entirely on thousands of infantrymen to become the most powerful ruler in the Levant. In the first few centuries of the Near Eastern Iron Age, chariot-borne archers probably served to harass infantries at long range. A chariotry was very expensive, as we have seen, and a cheaper alternative was the mounted archer. When horses were bridled with jointed bronze (or iron) snaffle bits, riding horseback became much safer than it had been. In the Near East, some jointed snaffle bits were used by chariot drivers already in the Late Bronze Age, but in the horse country of the Eurasian steppe the jointed snaffle mouthpiece did not come into use until somewhat later. In Mongolia and other areas of Central Asia, for instance, it was only toward the end of the second millennium BCE, and not earlier, that mounted riders began to dominate the economy (Taylor et al. 2020). Now more confident that he could control his mount, the rider sat forward, on a saddle or a saddle pad just behind the withers (Drews 2004: 65–98).

With their new-found security, riders soon learned to use a short bow in hunting and then in combat. Mounted archers must have appeared on the Eurasian steppe and in Iran and southern Caucasia well before they showed up in the Near East or the Aegean, but the earliest iconographic evidence we have for mounted archers comes from Assyrian reliefs set up during the reign of Ashurnasirpal II (883–859 BCE). Neither these Assyrian horsemen nor the more impressive horsemen against whom they fought should be called a cavalry, because they did not yet function as a unit and do the kinds of things—especially making a concerted charge against the enemy—that cavalries did from the eighth century BCE until the nineteenth century CE. Militarily, horses in the early Iron Age were therefore less important than they had been in the Bronze Age, or than they would soon become again when the Kimmerians/Skythians pioneered the cavalry charge against infantrymen, each cavalryman wielding a hand-to-hand weapon (Drews 2004: 105–48). If the infantrymen were deployed in a single line, a cavalry charge could easily break the line and then run down the infantrymen singly as they fled. Only when some Anatolian Greeks or Carians created the phalanx, at least four ranks deep, did infantrymen find the courage or feel the pressure to stand their ground against a cavalry charge: horses could not be made to charge into a stationary formation of spear-bearing and shield-carrying infantrymen.

In this chapter I have traced what seems to me to have been the evolution of warfare in western Eurasia and northeastern Africa from the beginnings of formal states to the aftermath of the Bronze Age. A critical study of this evolution is still at an early stage, however, and much remains to be learned. Ample revision of what has been presented here can be expected in future surveys.

Notes

1 See Postgate 2013: 21: 'There is no wholly satisfactory English word to convey the meaning of *ṣābu* (ÉRIN.MEŠ): it is used as a general term for people en masse, male and female, young and old, without implying a precise social class but with the implication that they are under the control of others.'
2 Vronwy Hankey graciously sent me two of her photographs for inclusion in *End of the Bronze Age*. See Drews 1993, Plates 3–4. She had also provided them for Nancy Sandars (see Sandars 1985, Illustration 11).
3 Thousands of bronze swords found in prehistoric Eurasia have been published in the 23 volumes of *Prähistorische Bronzefunde*, Abteilung IV (Schwerter). Until 1990, *PBF* was published by C.H. Beck Verlag in Munich, and since 1990 by Franz Steiner Verlag in Stuttgart. For a brief summary of the evolution of swords in the second millennium BC see Harding 2007: 71–7.
4 On the three *phylai* ('divisions') of the Dorians on Rhodes see *Iliad* 2.653–57 and 668 (the Catalogue of Ships probably dates no later than 1000 BCE).

Bibliography

Abbas, Mohamed Raafat, 'A Survey of the Military Role of the Sherden Warriors in the Egyptian Army during the Ramesside Period,' *ENiM* 10 (2017), pp. 7–23 (ENiM is an online periodical, an acronym for *Égypte nilotique et méditerranéenne*.)
Anthony, David, *The Horse, the Wheel and Language: How Bronze Age Riders from the Eurasian Steppe Shaped the Modern World* (Princeton: Princeton University Press, 2007).
Bader, Tiberiu, *Die Schwerter in Rumänien* (Stuttgart: Franz Steiner Verlag, 1991 [*PBF* Abt. IV, Bd. 8]).
Buchholz, Hans-Günter, *Kriegswesen, Teil 3: Ergänzungen und Zusammenfassung* (Archaeologia Homerica Bd. 1, Kapitel E3) (Göttingen: Vandenhoeck & Ruprecht, 2010).
Burke, Aaron, *'Walled up to Heaven': The Evolution of Middle Bronze Age Fortification Strategies in the Levant* (Winona Lake: Eisenbrauns, 2008).
Carman, John, *Archaeologies of Conflict* (London: Bloomsbury, 2013).
Chechushkov, Igor, and Andrei Epimakhov, 'Eurasian Steppe Chariots and Social Complexity during the Bronze Age,' *Journal of World Prehistory*, vol. 31 (2018), pp. 435–83.
Cunliffe, Barry, and Koch, John (eds.), *Celtic from the West*, vols. 1–3 (Oxford: Oxbow, 2012, 2013 and 2016).

Darnell, John, and Manassa, Colleen, *Tutankhamun's Armies: Battle and Conquest during Egypt's Late Eighteenth Dynasty* (Hoboken, New Jersey: John Wiley and Sons, 2007).

Dawson, Doyne, *The First Armies* (London: Cassell, 2001).

Drews, Robert, *The End of the Bronze Age: Changes in Warfare and the Catastrophe ca. 1200 BC* (Princeton: Princeton University Press, 1993).

Drews, Robert, *Early Riders: The Beginnings of Mounted Warfare in Asia and Europe* (London: Routledge, 2004).

Drews, Robert, *Militarism and the Indo-Europeanizing of Europe* (London: Routledge, 2017).

Drews, Robert, 'Catastrophe Revisited,' in Guy Middleton (ed.), *Collapse and Transformation: The Late Bronze Age and Early Iron Age in the Aegean* (Oxford and Philadelphia: Oxbow, 2020), pp. 229–35.

Emanuel, Jeffrey, *Black Ships and Sea Raiders: The Late Bronze-Early Iron Age Context of Odysseus' Second Cretan Lie* (Lanham, Maryland: Rowman and Littlefield, 2017).

Hacker, Barton, 'Horse, Wheel, and Saddle: Recent Works on Two Ancient Military Revolutions,' *International Bibliography of Military History*, vol. 32 (2012), pp. 175–91.

Hamblin, William, *Warfare in the Ancient Near East to c. 1600 BC* (London: Routledge, 2006).

Harding, Anthony, *European Societies in the Bronze Age* (Cambridge: Cambridge University Press, 2000).

Harding, Anthony, *Warriors and Weapons in Bronze Age Europe* (Budapest: Archaeolingua, 2007).

Horn, Christian, and Kristiansen, Kristian (eds.), *Warfare in Bronze Age Society* (Cambridge: Cambridge University Press, 2018).

Howard, Dan, *Bronze Age Military Equipment* (Barnsley, South Yorkshire: Pen & Sword Books, 2011).

Jantzen, Detlef, et al., 'An Early Bronze Age Causeway in the Tollense Valley, Mecklenburg-Western Pomerania: The Starting-point of a Violent Conflict 3300 Years Ago?' *Bericht der Römisch-Germanischen Kommision*, vol. 95 (2014 [2017]), pp. 13–49.

Jockenhövel, Albrecht, 'Zur Archäologie der Gewalt: Bemerkungen zu Aggression und Krieg in der Bronzezeit Alteuropas,' *Anodos. Studies of the Ancient World*, vols. 4–5 (2004–2005), pp. 101–32.

Kemenczei, Tibor, *Die Schwerter in Ungarn*, vol. 2 (Stuttgart: Franz Steiner Verlag, 1991 [*PBF* Abt. IV, Bd. 9]).

Kern, Paul, *Ancient Siege Warfare* (Bloomington: Indiana University Press, 1999).

Kohl, Philip, *The Making of Bronze Age Eurasia* (Cambridge: Cambridge University Press, 2007).

Koryakova, Ludmila, and Epimakhov, Andrej, *The Urals and Western Siberia in the Bronze and Iron Ages* (Cambridge: Cambridge University Press, 2007).

Kristiansen, Kristian, 'Warfare and the Political Economy: Europe 1500–1100 BC,' in Christian Horn and Kristian Kristiansen (eds.), *Warfare in Bronze Age Society* (Cambridge: Cambridge University Press, 2018), pp. 23–46.

Leshtakov, Lyuben, 'Late Bronze and Early Iron Age Bronze Spear- and Javelinheads in Bulgaria in the Context of Southeastern Europe,' *Archaeologia Bulgarica*, vol. 15 (2011), pp. 25–52.

Lidke, Gundula, et al., 'The Bronze Age Battlefield in the Tollense Valley, Northeast Germany: Conflict Scenario Research,' Chapter 6 in *Conflict Archaeology: Materialities of Collective Violence from Prehistory to Late Antiquity*, edited by Manuel Fernández-Götz and Nico Roymans (London: Routledge, 2017), pp. 61–8.

Manassa, Colleen, *The Great Karnak Inscription of Merneptah: Grand Strategy in the 13th Century BC* (*Yale Egyptological Studies*, 5. New Haven: Yale Egyptological Seminar, 2003).

Mödlinger, Marianne, 'European Bronze Age Cuirasses: Aspects of Chronology, Typology, Manufacture and Usage,' *Jahrbuch des Römisch-Germanischen Zentralmuseums Mainz*, vol. 59 (2012), pp. 1–49.

Parrot, André, 'Les fouilles de Mari: Dix-neuvième campagne (printemps 1971),' *Syria*, vol. 48 (1971), pp. 253–70.

Postgate, Nicholas, *Bronze Age Bureaucracy: Writing and the Practice of Government in Assyria* (Cambridge: Cambridge University Press, 2013).

Röpke, Astrid, and Carlo Dietl, 'The Vitrified Bronze Age Fortification of Bernstorf (Bavaria, Germany): an Integrative Geoarchaeological Approach,' *European Geologist*, vol. 38 (2014), pp. 24–32.

Sandars, Nancy, *The Sea Peoples: Warriors of the Ancient Mediterranean 1250–1150 BC*, revised edition (London: Thames and Hudson, 1985).

Sasson, Jack, 'Siege Mentality: Fighting at the City Gate in the Mari Archives,' in S. Yona et al. (eds.), *Marbeh Ḥokmah: Studies in the Bible and the Ancient Near East in Loving Memory of Victor Avigdor Hurowitz* (Winona Lake, Indiana: Eisenbrauns, 2015), pp. 465–78.

Schofield, Louise, and R.B. Parkinson, 'Of Helmets and Heretics: A Possible Egyptian Representation of Mycenaean Warriors on a Papyrus from El-Amarna,' *ABSA*, vol. 89 (1994), pp. 157–70.

Schulman, Alan, 'Egyptian Chariotry: A Re-examination,' *Journal of the American Research Center in Egypt*, vol. 2 (1963), pp. 75–98.

Spalinger, Anthony, 'The Battle of Kadesh: The Chariot Frieze at Abydos,' *Ägypten und Levante*, vol. 13 (2003), pp. 163–99.

Stillman, Nigel, and Nigel Tallis, *Armies of the Ancient Near East, 3000 B.C. to 539 B.C.* A Wargames Research Group Publication (Worthing, Sussex: Flexiprint, 1984).

Taylor, William, et al., 'Early Pastoral Economies and Herding Transitions in Eastern Eurasia,' *Scientific Reports* 10, 1001 (2020). doi:10.1038/s41598–020–57735-y

Thorpe, Nick, 'Warfare in the European Bronze Age,' in Harry Fokkens and Anthony Harding (eds.), *The Oxford Handbook of the European Bronze Age* (Oxford: Oxford University Press, 2013), pp. 234–47.

Yadin, Yigael, 'The Earliest Representation of a Siege Scene and a "Scythian" Bow from Mari,' *Israel Exploration Journal*, vol. 22 (1972), pp. 89–94.

2

WAR, CAPTIVE-TAKING AND WARRIORS IN PRE-COLUMBIAN MESOAMERICA

Andrew K. Scherer

There are many excellent book-length studies of war in Pre-Columbian Mesoamerica. These include Ross Hassig's (1992) classic overview, and numerous edited volumes such as the one by Kathryn Brown and Travis Stanton (2003), Richard Chacon and Rubén Mendoza's collection of essays on indigenous warfare in Latin America (2007), Axel Nielsen and William Walker's (2009) edited volume on the study of ancient warfare in the Americas through the lens of practice theory, and a volume I co-edited with John Verano (Scherer and Verano 2014a) focusing on war, place and the human body in ancient Mesoamerica and the Andes. Due to the limited space, it would be impossible to synthesise what is presented in these fine works while also providing a new synthesis on Mesoamerican war. Instead, this chapter begins with a brief overview of Mesoamerica, followed by broad comment on the nature of war and violent conflict in Pre-Columbian Mesoamerica, highlighting the commonalities and key differences among its most well-known and best studied societies. By doing so, the chapter will touch on the central question, why did indigenous Mesoamericans go to war? It will then focus on the role of captive-taking in causes and consequences of Mesoamerican warfare and conclude with an overview of the warriors who fought in ancient Mesoamerican conflicts.

Pre-Columbian Mesoamerica

Mesoamerica encompasses the current nation-states of Mexico, Guatemala, Belize and portions of Honduras and El Salvador. In defining Mesoamerica, scholars frequently cite Paul Kirchhoff (1952), who highlights key traits that Mesoamerican societies had in common. Among them are the construction of temple pyramids, the use of both a 365-day solar calendar and a 260-ritual calendar, the playing of the ball game, and a subsistence system where maize, beans and squash were of central importance. There is, of course, a certain arbitrariness to these criteria; not all Mesoamerican societies manifest these attributes in the same fashion, and many of these phenomena are also found historically among indigenous societies of the southwestern United States and in southern Central America.

Chronologically, archaeologists divide Mesoamerica's Pre-Columbian history into a Paleoindian period (pre-8000 BCE), an Archaic period (8000–2000 BCE), a Preclassic or Formative period (2000 BCE–250 CE), a Classic period (250–1000 CE) and the Postclassic period

DOI: 10.4324/9780429437915-4

(1000–1519 CE) (Evans 2013). These are best understood as loose temporal divisions that imperfectly organise time for scholars of Mesoamerica. Complex societies arose in the Pre-classic period and persisted till the Spanish conquest. It is the practice of war among those complex societies, particularly of the Classic and Postclassic periods, that is the focus here. Among the most intensively studied are the Maya, Zapotec, Teotihuacan, Mixtec, Toltec and the Aztec. For the most part, these are not names that were used by the people themselves but are instead cultural categories created by scholars. Moreover, these names do not refer to analogous social units and they are not all contemporary with one another. For example, 'Maya' refers to the indigenous peoples of Eastern Mexico (Yucatan, Quintana Roo, Campeche, Tabasco, Chiapas), Guatemala and Belize who today speak 30 closely related languages (Houston and Inomata 2009). Most of the archaeological sites throughout the same geographic area (with some exception for the Pacific Coast) are understood as ancestral to the Maya, a connection that is reinforced by shared religious practice (evident in ancient art) and language, as reconstructed from hieroglyphic texts from the Classic period (250–900 CE).

'Aztec' refers to a political alliance of Nahuatl speakers from three city-states in Central Mexico, all of whom traced their origins to the mythological Aztlan (Smith 2012). The most powerful city in the alliance was Tenochtitlan, home to the Mexica people. Nahuatl and other related Nahua languages are still widely spoken in Central Mexico. The Aztec saw themselves as inheritors of the Toltec tradition, though did not trace their origins to the Toltec capital at Tula, which flourished from 900–1150 CE. The language of the Toltec is unknown, though it possibly was a form of Nahua, which may also have been related to the language spoken at the powerful yet enigmatic superpower of Teotihuacan, a massive archaeological site located just north of modern Mexico City (Cowgill 1997, 2008). On one hand, the political power of Teotihuacan is evident by the far reach of its rulers' influence, apparently affecting politics in the Maya area during the Early Classic period (250–600 CE) (Stuart 2000). Yet we still do not know how the city was governed, what was the spatial extent of its rulers' dominion, or even how and why it waged war.

War and Polity

The basic political unit throughout Pre-Columbian Mesoamerica for much of its history was the community—a village, town or city (Yaeger and Canuto 2000). Within and between those communities, lineages also served as an important axis by which social belonging and identity were defined, although the degree to which lineal descent served as a principal organising structure was situationally contingent (Watanabe 2004). Although form and structure varied considerably across time and space, in the most general sense, as Pre-Columbian societies grew in size and complexity in the Middle Preclassic (ca. 900 BCE–300 CE) and Late Preclassic periods (300 BCE–300 CE), inequalities in wealth and power increased. At the time of the conquest, elite status was closely linked to lineal descent and this appears to have been the case at least as far back as the Classic period.

By the Late Preclassic period (and in some places as early as the Middle Preclassic period), the largest Mesoamerican communities had populations that numbered in the thousands. Governed by an elite class, we can speak of such places as polities (Charlton and Nichols 1997; Webster 1997). In these earliest polities, a lord's authority may have encompassed only the residents of the immediate community he ruled. In that sense, many Mesoamerican polities were city-states. This is especially true of the Maya during the Early Classic period. However, as populations grew and political hierarchies became more complex over the course of the Classic period, Mesoamerican city-states in some places morphed into polities with regional dominion over lesser communities (Golden and Scherer 2013). Again, this process is especially well understood for the Maya, thanks to their rich corpus of hieroglyphic texts (Houston and Inomata 2009; Martin and Grube 2008; Tokovinine 2013). Over time, some polities sought alliances with one another, others competed and even warred, and in some cases hegemonies were established. A good example is the long-standing political rivalry between the Maya polities of Tikal and Calakmul, the two most powerful political centres of the Classic period who, in their great rivalry, established brief over-lordship over lesser Maya polities (Martin and Grube 1995). Other such hegemonies are known to have existed in Pre-Columbian Mesoamerica; the most well-known being the Aztec Triple Alliance (which took on the character of empire) of the Postclassic period (Berdan et al. 1996) and presumably Teotihuacan during the Classic period (Cowgill 1997, 2008). By the eighth and ninth centuries, most Maya polities of Southern Mexico, Northern Guatemala and Belize collapsed, and these settlements were abandoned just as new polities emerged, especially in the Northern Yucatan and the Chiapas and Guatemalan Highlands. Cycles of polity coalescence, fluorescence and collapse likely began in the Mesoamerican Preclassic period, and continued up to the time of the Spanish conquest, and are especially evident at Teotihuacan, Monte Alban (Zapotec) and Tula (Toltec) (Cowgill 2012; Webster 2012).

For the first half of the twentieth century, scholars mused that the western half of Mesoamerica had a deep history of war-making. The military prowess of the Aztec was well known from colonial-era sources and most scholars accepted that this history of bellicosity could be traced at least as far back as the ninth and tenth centuries, when Classic period centres collapsed and Postclassic period communities were fortified (Armillas 1951). In contrast, the eastern portion of Mesoamerica (i.e., the Maya area) was thought to be largely peaceful and without war (Morley 1946: 70; Thompson 1943: 128). However, advances in archaeology, art history and epigraphy proved that to be an erroneous assumption, with evidence for warfare in all parts of Mesoamerica as least as early as the Preclassic period (Joyce 2014; Webster 1993, 1999, 2000).

Initially, scholars were reluctant to fully concede that the Pre-Columbian peoples of Mesoamerica waged war on a large scale with real implications for economic gain and political change, instead emphasising combat as a form of ritual practice, fought primarily by elites to capture high-status enemies for sacrifice (Freidel 1986). Today, most researchers recognise that war was both ritual practice and yet had real socio-political and economic implications, and likely involved all sectors of Mesoamerican society, from nobility to commoners (Scherer and Verano 2014b).

By 'war' I am referring to violent conflict between polities. In most cases, Mesoamerican wars were waged among polities that we would define as ethnically the same (i.e., common language, shared traditions and beliefs). Bronze Age Greece is a useful analogy for thinking about Mesoamerican war in that much violence occurred among neighbouring city-states, but also involved, at times, conflict with further flung political entities. This was true of some of the wars waged by the Aztecs and perhaps also Teotihuacan (Hassig 1992; Sears 2019). There are, however, no known episodes of Pre-Columbian wars that could be described as genocide, where one group attempted to wipe out another or remove an entire populace from its territory. Nevertheless, there is a deep history of population-wide migration in Mesoamerica and we cannot rule out the possibility that some of those populations movements were triggered by conflict (Cowgill 2013). Since evidence for territorial control and border-making are quite ephemeral in Pre-Columbian Mesoamerica, wars of absolute conquest are equally obscure. My colleagues and I have documented at least one instance of territorial border-making at the Classic period Maya kingdom of Yaxchilan, which developed a line of fortifications against its northern rival Piedras Negras (Golden et al. 2008; Scherer and Golden 2009, 2014). Yet that border was short-lived, confined for the most part to the eighth century AD. That border never appears to have shifted and it ultimately dissolved upon the collapse of the polity.

Why then did Pre-Columbian people go to war? Although evidence for maintaining territory is ephemeral, control over routes of travel was likely an important prerogative in Mesoamerican warfare, although difficult to reconstruct from archaeological and textual sources (Webster 2000: 81). Another stimulus for violent conflict was revenge and personal vendetta. Classic Maya hieroglyphic texts record centuries-long conflicts that show deep memory for vengeance against longtime foes. One such example was the Maya kingdom of Piedras Negras' defeat of their rival at Pomona in 795 CE, a victory that is expressly described in the hieroglyphic texts on Piedras Negras Stela 12 as an act of retribution for a defeat suffered 250 years earlier (Golden 2010: 377; Houston et al. 2000: 101).

Yet the most important motivators for Mesoamerican warfare were likely those highlighted by Frey Diego de Landa in his account of war among the Yucatec Maya at the time of the Spanish conquest:

> The reason they waged war with one another was to take their property from them and capture their children and wives, and because it was the custom among them to pledge what they possessed to each other; upon collection and payment they began to quarrel and attacked each other, and then the lord of that Pueblo armed his people against the other, and for that reason they waged war upon each other.
>
> *(Tozzer 1941: 41)*

Thus, one principal reason Mesoamerican people waged war was to raid neighbouring polities for food and other spoils. A second major cause for warfare was to establish tributary relationships; payment was demanded at defeat and served as insurance against future attacks. Such tribute took a variety of forms and our best evidence comes from the Aztec codices where tribute lists include mostly items that were easily quantified, transported and storable, including foodstuffs (maize, chili peppers, salt, cacao), bolts of cotton cloth, warrior costumes, precious materials (jade, gold) and even slaves who worked as labourers or were offered or sold for sacrifice (Smith 2012: 169–71). A third major reason for warfare was to take captives, an act that warrants more detailed consideration in light of the many misconceptions that surround it.

Captive-Taking

Captive-taking dominates war-related imagery in Mesoamerican art (especially for the Maya) and the practice is widely reported in both indigenous and Spanish documents that date to the time of the conquest. Much of this imagery relates to the taking of captives in battle but there are also images that indicate non-combatants were also taken. One such example from an unprovenanced Classic period Maya vase, pointed out to me by Stephen Houston, seems to show warriors capturing women and children (Figure 2.1). The purpose of captive-taking

Figure 2.1 Late Classic period Maya painted vase depicting battle and captive-taking. Warriors are painted with black body paint, armed with spears, and equipped with hats and backpack-bundles for long-distance travel. Two men (left) defend themselves by throwing stones. Women and children are possible spoils of war and the two women with white exposed breasts have black cords around their necks, suggesting they may be slaves.

Source: Photograph © Justin Kerr, K5451.

in Mesoamerican war has a history of misconception, with its connection to rituals of sacrificial violence overemphasised at the expense of the political and economic implications of captives. Current evidence indicates three primary uses for captives: as slaves, as victims of human sacrifice, and for ransom-tribute and similar economic exchanges.

The practice of keeping slaves is evident in numerous contact era sources from Mesoamerica. The Italian chronicler Peter Martyr d'Anghiera ([1516] 1912: 317) reports that on Columbus' fourth voyage he witnessed two (presumably Maya) canoes off the coast of Honduras that were being drawn by slaves with ropes around their necks. Mary Miller (personal communication, 2018) calls attention to a corpus of Pre-Columbian Maya figurines from Jaina, Mexico that depict men and women with ropes around their necks as likely representations of slaves (see the two possible slaves in Figure 2.1). For the Maya, Landa *Relación* describes an extensive slave trade network that brought traders from the Yucatan Peninsula to Tabasco and Veracruz (Tozzer 1941: 94). Many Mesoamerican slaves entered bondage through non-violent means, including as punishment for criminals and a 'voluntary' choice by impoverished people (e.g., Aztec *tlacotin*) (Aguilar-Moreno 2006: 75). And yet one of the major commodities for Yucatec Maya lords at the time of the conquest was slaves taken in battle from neighbouring provinces (Clendinnen 1987: 148–9; Roys 1943: 65–7; Scholes and Roys 1968: 59). The antiquity of slave-raiding in Mesoamerica is murky, although if Miller is correct in her interpretation of the Jaina figurines as am I in my reading of the scene in Figure 2.1, the keeping of slaves dates at least to the Late Classic period in the Maya area.

Far less ambiguous is the evidence for sacrificial offering of captives taken in war. Numerous excellent studies of Mesoamerican sacrifice exist and the details of the practice are beyond the scope of this chapter (see especially López Luján and Olivier 2010). Nevertheless, an eyewitness account of the sacrifice of Spaniards captured during the siege of Tenochtitlan clarifies the fate of some Mesoamerican war captives (Diaz del Castillo [1632] 1912b: 149–50):

> [A]nd we all looked towards the lofty Cue where they [drums] were being sounded, and saw that our comrades whom they had captured when they defeated Cortes were being carried by force up the steps, and they were taking them to be sacrificed. When they got them up to a small square in [front of] the oratory, where their accursed idols are kept, we saw them place plumes on the heads of many of them and with things like fans [in their hands?] they forced them to dance before Huichilobos [an image of the Aztec patron deity], and after they had danced they immediately placed them on their backs on some rather narrow stones which had been prepared as [places for] sacrifice, and with stone knives they sawed open their chests and drew out their palpitating hearts and offered them to the idols that were there, and they kicked the bodies down the steps, and Indian butchers who were waiting below cut off the arms and feet and flayed [the skin off] the faces, and prepared it afterwards like glove leather with the beards on, and kept those for the festivals when they celebrated drunken orgies, and the flesh they ate in chilmole.

Although the scope and scale of such sacrifices varied from society to society (the Aztecs seemed to have engaged in ritual killing in numbers generally not seen in other Mesoamerican societies), the basic principles were generally consistent throughout Mesoamerica. As evident in the passage above, the Aztecs sustained their patron deity on human flesh, and the need to provide offerings of flesh and blood was widespread throughout Mesoamerica. The practice followed a basic logic: supernatural beings sacrificed of themselves to create humanity and that act needed to be reciprocated in kind (Scherer and Houston 2018). Although nobility and spiritual leaders were required to give of themselves through acts of bloodletting, obligations of human life were also paid with a substitute, ideally a foreigner. Heart sacrifice was a favoured mode of sacrifice among the Aztecs and is evident in other Mesoamerican societies. Nevertheless, other forms of killing are known, including decapitation by axe, piercing by arrows, and immolation. Such spectacles of violence and the threat of future captive-taking episodes would have also served to guarantee the flow of tribute and as a deterrent for revolt.

Although evidence for human sacrifice is abundant, the art and discourse from throughout Mesoamerica does not focus on the *killing* of captives but the *taking* of captives. This is especially clear from the imagery and text from the Maya area where the emphasis is on the taking of captives in battle, the delivery of captives by war leaders to their kings, and the subjugation of captives, often quite literally beneath the feet of a victorious king (Figure 2.2). Indeed, the hieroglyphic texts are generally mute about the ultimate fate of these captives. We can presume that those who were never heard from again were killed. And yet some famous captives, such as a king from the Maya Kingdom of Palenque, survived their capture and even continued to rule, although presumably after oaths of fealty were sworn and promises of tribute were made (McAnany 2010: 278–83; Stuart and Stuart 2008: 217).

Figure 2.2 Shield Jaguar IV of Yaxchilan (left) takes a captive in battle on Bonampak Lintel 2 and (right) sits on a throne and receives captives as tribute from a war captain on Laxtunich Lintel 1.

Source: Left: photograph by A.K. Scherer; right: drawing by A. Safronov.

Elite captives were thus a form of political and economic currency, and their capture was a key motivation for Mesoamerican warfare. One of the oldest surviving (and still performed) dramas in the Americas, the *Rabinal Achi*, a fifteenth-century Achi Maya dance performance, highlights the many dimensions of Mesoamerican warfare: it was ritual, it had political and economic consequences and it was fuelled by revenge at a perceived wrongdoing. The *Rabinal Achi* tells the story of Cawek, a captive K'iche' Maya nobleman. Cawek, a former ally, led raids to capture captives from Rabinal in retribution of a perceived slight. He is in turn captured and, after a defiant exchange of words between the captor and captive, Cawek is sacrificed by decapitation. Shortly before he is killed, Cawek muses that his body (specifically his skull) will be ransomed for tribute: 'Then it will be sent down there to my mountain, my valley, ending up as an even trade for five score seeds of pataxte [a type of cacao], five score seeds of cacao, *paid* by my children, my sons, at my mountain, at my valley' (emphasis added, Tedlock 2003: 105).

Warriors

Mesoamerican combatants went to war armed with weapons of stone and wood and without the aid of draft animals or siege engines. Combatants used both missile and melee weapons, although specific tactics varied among Mesoamerican societies, influenced by environment and the organisation of fighting forces. The most basic missile weapon was a thrown stone (see Figure 2.1), something the Spanish noted throughout their accounts of the conquest (e.g., Diaz del Castillo [1632] 1912a: 2). Caches of stones have been found archaeologically, presumably preparations for war. Some of these caches include rocks that were hammered into spheres (Alcover Firpi et al. 2019). These were likely used with slings, a weapon that continues to be used to this day for hunting. Blowguns are show in the imagery of the ancient Maya, though it is unclear the extent to which they were used in warfare. The atl-atl (spear thrower) was widely used in Central Mexico from the Classic period forward. It may have been less extensively utilised by the Maya, perhaps owing to the forested environment in which they lived. Bows and arrows were adopted in the Terminal Classic period and were widespread throughout Mesoamerica in the Postclassic period (Aoyama 2005; Cervera Obregón 2011: 101).

Perhaps the single most important weapon of war was the spear—a wooden shaft mounted with a chipped stone biface made from chert (flint) or obsidian (see Figures 2.1 and 2.2). Similar blades mounted on shorter shafts served as knives, both carried by combatants and used for heart sacrifice throughout Mesoamerica. The Aztec made use of a wooden club lined with obsidian blades, the *macuahuitl*, which Bernal Diaz (Diaz del Castillo [1632] 1912a: 66) claimed, 'cut better than our [steel] swords did' (Figure 2.3). Stone axes and maces were also wielded throughout Mesoamerica and are among the earliest weapons known (Figure 2.4). Although difficult to detect archaeologically, clubs are shown in the art of Mesoamerica. Mesoamerican warriors protected themselves with armour that was made from cotton and other padded materials. They bore shields of hide and cloth which were likely used to both protect against projectile weapons and deflect blows from spear and clubs (see Figures 2.1 and 2.3).

Figure 2.3 Aztec jaguar warrior faces a tethered captive in sacrificial combat, each man wields a *macuahuitl*.

Source: Tovar Codex, sixteenth century.

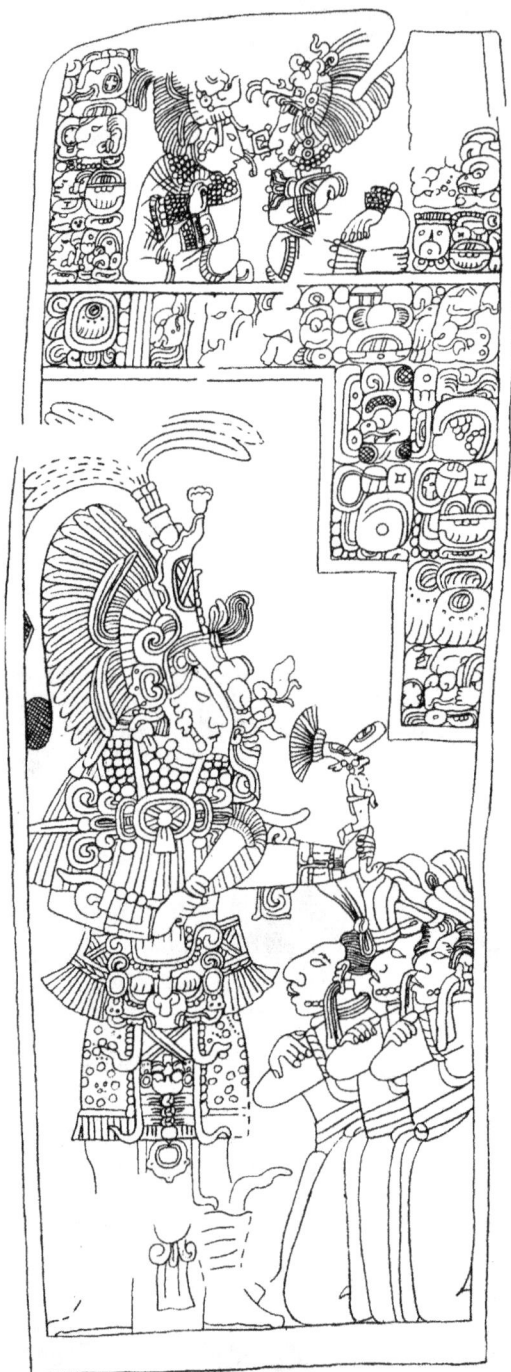

Figure 2.4 Shield Jaguar III of Yaxchilan embodies the violent rain deity Chahk by donning a mask in his likeness and wielding his ceremonial axe, K'awiil, in his left hand.

Source: Drawing by John Montgomery.

Although arranged battles fought in open terrain (i.e., 'battlefields') are known to have occurred, as in the case of the Aztec 'Flower Wars' (*xochiyaoyotl*), most Mesoamerican conflicts were likely surprise raids—attacks against unsuspecting travellers or against settlements. That is, effectively the entire landscape was a potential 'battleground' (Scherer and Verano 2014b: 9). Conflict likely began with an initial volley of missile weapons as combatants then closed for melee combat. As the objective in many battles was the taking of captives, battleground casualties were likely relatively low even if the fighting itself was quite bloody and dangerous, judging by the weaponry involved and the incidence of antemortem trauma observed in human remains (Serafin et al. 2014; Tiesler and Cucina 2012). Archaeological evidence for hasty fortifications raised at communities that were later burned indicates that towns were targeted for attack (Inomata 1997, 2003, 2008). The Aztec glyph for conquest of a polity is a burning temple with an askew roof, underscoring the connection between conflict and the built environment (specifically the destruction of a patron deity's shrine) (Boone 2000: 34). Fire too was an important element in Mesoamerican war. In Classic period texts, the Maya distinguished between helpful fire (e.g., the hearth) and the vengeful fire employed in violent conflict (Scherer and Houston 2018: 115–16), and the burning of both agricultural fields and communities was an essential tactic in Pre-Columbian war.

The question of who served as warrior has long plagued the study of Pre-Columbian Mesoamerican warfare (Webster 2000: 99). At least from the Classic period onwards, combatants among the well-studied Mesoamerican societies included elite noble warriors. Even rulers are known to have gone to battle, if not during their reign then as young warriors prior to their ascendancy to govern. For example, the Late Classic period rulers of the kingdom of Yaxchilan are all reported to have taken a captive prior to their ascendancy to the throne, a captive they then celebrated throughout their reign (Stuart 1985). One of these kings, Shield Jaguar IV, is depicted taking a captive in combat on the lintel of a building at the vassal community of Bonampak, implying he fought in battle (Figure 2.2, left). This contrasts with other sculptures, such as a lintel in the Kimbell Art Museum in Dallas, where the same king is shown receiving captives as tribute from his nobles (Figure 2.2, right). The challenge in using art to determine who fought is that elites were the makers and thus the subjects of much of the art that preserves and non-elite subjects are largely missing. At least in the case of the Aztecs, we know the bulk of the fighting force was composed of commoners (Hassig 1988: 28). Among the Classic period Maya, at least some battles seem to have been timed to correspond to lulls in the agricultural cycle (Webster 2000: 101), suggesting that farmers were expected to take up arms on occasion.

Psychological studies of violence coupled with data collected from modern battlegrounds show that humans are not predisposed to killing but instead must be conditioned to do so (Grossman 2009; Sapolsky 2017: 470–5, 644–7). This is especially true of close-quarters killing with hand-to-hand weapons like knives and bayonets (i.e., the recent equivalent of spears). A variety of techniques have been shown to be effective for increasing a person's killing capacity, including excessive training to condition combatants, dehumanisation of the foe and diffused responsibility through group participation in killing and anonymity (i.e., masking one's identity). For Mesoamerican warriors, the capacity to kill was enabled through the painting of bodies and the donning of elaborate costumes and headdresses, both of which deindividuated and dehumanised the combatants and their foes (see Figure 2.1). Such costuming may have involved transformation into a powerful predatory animal spirit, as in the eagle and jaguar warriors of Aztec society (Figure 2.3), or the embodiment of Death Gods and patron deities among the Maya, accomplished through adornment with skulls, bones and the placement of god masks in headdresses (Figure 2.4). In part, this may be understood as the channelling of these powerful beings to gain their prowess in battle. However, costuming and masking in Mesoamerican worldview are not so much a process of *impersonation* as *co-existing* or *becoming*. The warrior was no longer himself but the being he embodied. A similar process of displaced killing is evident in human sacrifice: lords, ladies and priests who conducted the killing did so while embodying supernatural beings. These same nobles are never shown in acts of violence (even in self-inflicted acts of bloodletting) garbed in their quotidian courtly wear (Scherer and Houston 2018). That these acts of ritual violence were performed publicly further diffused the killing to the collective action of the community.

Conclusion

Warfare in Mesoamerica was enmeshed in symbolism and ritual, and yet had very real social, political and economic consequences. Mesoamerican wars generally consisted of battles between neighbouring polities, among social groups that we would perceive as ethnically the same. Combat was waged with implements of wood and stone. The taking of captives was a primary objective: to serve as slaves, as victims of sacrifice or to leverage for tribute. Through the process of embodying supernatural beings and through acts of public sacrifice, the violence and killing of war were both attributed to non-human beings while being displaced to the collective whole.

Bibliography

Aguilar-Moreno, Manuel, *Handbook of Life in the Aztec World* (Oxford: Oxford University Press, 2006).

Alcover Firpi, Omar, Ricardo Rodas, Urquizú, Monica and Scherer, Andrew K., 'La guerra en el Usumacinta preclásico: tres temporadas de investigaciones en el sitio Macabilero,' in *XXXII Simposio de Investigaciones Arqueológicos en Guatemala, 2018. Museo Nacional de Arqueología y Etnología, Guatemala* (Guatemala, 2019).

Aoyama, Kazuo, 'Classic Maya Warfare and Weapons: Spear, Dart, and Arrow Points from Aguateca and Copan,' *Ancient Mesoamerica*, vol. 16 (2005), pp. 291–304.

Armillas, Pedro, 'Mesoamerican Fortifications,' *Antiquity*, vol. 25, no. 98 (1951), pp. 77–86.

Berdan, Frances F., Blanton, Richard E., Boone, Hill, Elizabeth, Hodge, Mary G., Smith, Michael E. and Umberger, Emily, *Aztec Imperial Strategies* (Washington DC: Dumbarton Oaks Research Library and Collection, 1996).

Boone, Elizabeth Hill, *Stories in Red and Black: Pictorial Histories of the Aztecs and Mixtec* (Austin: University of Texas Press, 2000).

Brown, M. Kathryn and Stanton, Travis W. (eds.), *Ancient Mesoamerican Warfare* (Walnut Creek, CA: AltaMira Press, 2003).

Cervera Obregón, Marco A., *Guerreros Aztecas: Armas, Técnicas de Combate e Historia Militar del Implacable Ejército Conquistó Mesoamérica* (Madrid: Ediciones Nowtilus, 2011).

Chacon, Richard J. and Mendoza, Rubén (eds.), *Latin American Indigenous Warfare and Ritual Violence* (Tucson: The University of Arizona Press, 2007).

Charlton, Thomas H. and Nichols, Deborah, 'Diachronic Studies of City-States: Permutations on a Theme: Central Mexico from 1700 B.C. to A.D. 1600,' in D. Nichols and T.H. Charlton (eds.), *The Archaeology of City-States: Cross-Cultural Approaches* (Washington, DC: Smithsonian Institution Press, 1997), pp. 169–208.

Clendinnen, Inga, *Ambivalent Conquests: Maya and Spaniard in Yucatan, 1517–1570* (Cambridge: Cambridge University Press, 1987).

Cowgill, George L., 'State and Society at Teotihuacan, Mexico,' *Annual Review of Anthropology*, 26 (1997), pp. 129–61.

Cowgill, George L., 'An Update on Teotihuacan,' *Antiquity*, vol. 82, no. 318 (2008), pp. 962–75.

Cowgill, George L., 'Concepts of Collapse and Regeneration in Human History,' in D.L. Nichols and C. Pool (eds.), *The Oxford Handbook of Mesoamerican Archaeology* (Oxford: Oxford University Press, 2012), pp. 301–8.

Cowgill, George L., 'Possible Migrations and Shiftinging Identities in the Central Mexican Epiclassic,' *Ancient Mesoamerica*, vol. 24, Issue 1 (2013), pp. 131–49.

Diaz del Castillo, Bernal, *The True History of the Conquest of New Spain*, vol. 1, 1632, tr. by A. Maudslay (London: M.A. Hakluyt Society, 1912a).

Diaz del Castillo, Bernal, *The True History of the Conquest of New Spain*, vol. 4, 1632, tr. by A. Maudslay (London: M.A. Hakluyt Society, 1912b).

Evans, Susan Toby, *Ancient Mexico and Central America: Achaeology and Culture History* (New York: Thames & Hudson, 2013).

Freidel, David, 'Maya Warfare: An Example of Peer Polity Interaction,' in C. Renfrew and J. Cherry (eds.), *Peer Polity Interaction and Socio-Political Change* (Cambridge: Cambridge University Press, 1986), pp. 93–108.

Golden, Charles, 'Frayed at the Edges: The Re-Creation of Histories and Memories on the Frontiers of Classic Period Maya Polities,' *Ancient Mesoamerica* vol. 21, no. 2 (2010), pp. 373–84.

Golden, Charles and Scherer, Andrew K., 'Territory, Trust, Growth and Collapse in Classic Period Maya Kingdoms,' *Current Anthropology*, vol. 54, no. 4 (2013), pp. 397–417.

Golden, Charles, Scherer, Andrew K., Muñoz, A. René and Vásquez, Rosaura, 'Piedras Negras and Yaxchilan: Divergent Political Trajectories in Adjacent Maya Polities,' *Latin American Antiquity*, vol. 19, no. 3 (2008), pp. 249–74.

Grossman, Dave, *On Killing: The Psychological Cost of Learning to Kill in War and Society* (n.d., reprint, New York: Back Bay Books, 2009).

Hassig, Ross, *Aztec Warfare: Imperial Expansion and Political Control* (Norman, OK: University of Oklahoma Press, 1988).

Hassig, Ross, *War and Society in Ancient Mesoamerica* (Berkeley: University of California Press, 1992).

Houston, Stephen D. and Inomata, Takeshi, *The Classic Maya* (Cambridge: Cambridge University Press, 2009).

Houston, Stephen D., Escobedo, Héctor, Child, Mark, Golden, Charles, Terry, Richard and Webster, David L., 'In the Land of the Turtle Lords: Archaeological Investigations at Piedras Negras, Guatemala, 2000,' *Mexicon*, vol. 22, no. 97 (2000), pp. 97–110.

Inomata, Takeshi, 'The Last Day of a Fortified Classic Maya Center: Archaeological Investigations at Aguateca, Guatemala,' *Ancient Mesoamerica*, vol. 8 (1997), pp. 337–51.

Inomata, Takeshi, 'War, Destruction, and Abandonment: The Fall of the Classic Maya Center Aguateca, Guatemala,' in T. Inomata and R.W. Webb (eds.), *The Archaeology of Settlement Abandonment in Middle America* (Salt Lake City: The University of Utah Press, 2003), pp. 43–60.

Inomata, Takeshi, *Warfare and the Fall of a Fortified Center: Archaeological Investigations at Aguateca* (Nashville, TN: Vanderbilt Institute of Mesoamerican Archaeology, Vanderbilt University Press, 2008).

Joyce, Arthur A., 'Warfare in Late/Terminal Formative-Period Oaxaca,' in A.K. Scherer and J.W. Verano (eds.), *Embattled Places, Embattled Bodies: War In Pre-Columbian Mesoamerica and the Andes* (Washington, DC: Dumbarton Oaks Research Library and Collection, 2014), pp. 117–42.

Kirchhoff, Paul, 'Meso-America,' in S. Tax (ed.), *Heritage of Conquest* (Glencoe, IL: Free Press, 1952), pp. 17–30.

López Luján, Leonardo and Olivier, Guilhem (eds.), *El sacrificio humano en la tradición religiosa mesoamericana* (Mexico City: Instituto Nacional de Antropología e Historia and the Universidad Nacional Autónoma de México, 2010).

Martin, Simon and Grube, Nikolai, 'Maya Superstates,' *Archaeology*, vol. 48, no. 6 (1995), pp. 41–6.

Martin, Simon and Grube, Nikolai, *Chronicle of the Maya Kings and Queens: Deciphering the Dynasties of the Ancient Maya* (2nd edition, New York: Thames and Hudson, 2008).

Martyr d'Anghiera, Peter, *De Orbe Novo*, vol. 1, 1516 (New York: MacNutt, 1912).

McAnany, Patricia A., *Ancestral Maya Economies in Archaeological Perspective* (Cambridge: Cambridge University Press, 2010).

Morley, Sylvanus G., *The Ancient Maya* (Stanford: Stanford University Press, CA., 1946).

Nielsen, Axel E. and Walker, William (eds.), *Warfare in Cultural Context: Practice, Agency, and The Archaeology of Violence* (Tucson: The University of Arizona Press, 2009).

Roys, Ralph L., *The Indian Background of Colonial Yucatán*. (Washington: Carnegie Institute of Washington Publication, 1943).

Sapolsky, Robert M., *Behave: The Biology of Humans at Our Best and Worst* (New York: Penguin, 2017).

Scherer, Andrew K. and Golden, Charles, 'Tecolote, Guatemala: Archaeological Evidence for a Fortified Late Classic Maya Political Border,' *Journal of Field Archaeology*, vol. 34, no. 3 (2009), pp. 285–304.

Scherer, Andrew K. and Golden, Charles, 'War in the West: History, Landscape, and Classic Maya Conflict,' in A.K. Scherer and J.W. Verano (eds.), *Embattled Places, Embattled Bodies: War In Pre-Columbian Mesoamerica and the Andes* (Washington, DC: Dumbarton Oaks Research Library and Collection, 2014), pp. 57–92.

Scherer, Andrew K. and Houston, Stephen D., 'Blood, Fire, Death: Covenants and Crisis among the Classic Maya,' in V. Tiesler and A.K. Scherer (eds.), *Smoke, Flame, and the Human Body in Mesoamerican Ritual Practice* (Washington, DC: Dumbarton Oaks Research Library and Collection, 2018), pp. 109–50.

Scherer, Andrew K. and Verano, John W. (eds.), *Embattled Places, Embattled Bodies: War In Pre-Columbian Mesoamerica and the Andes* (Washington, DC: Dumbarton Oaks Research Library and Collection, 2014a).

Scherer, Andrew K. and Verano, John W., 'Introducing War in Precolumbian Mesoamerica and the Andes,' in A.K. Scherer and J.W. Verano (eds.), *Embattled Places, Embattled Bodies: War In Pre-Columbian Mesoamerica and the Andes* (Washington, DC: Dumbarton Oaks Research Library and Collection, 2014b), pp. 1–24.

Scholes, France V. and Roys, Ralph L., *The Maya Chontal Indians of Acalan-Tixchel* (2nd edition, Norman: University of Oklahoma Press, 1968).

Sears, Matthew A., *Understanding Greek Warfare* (New York: Routledge, 2019).

Serafin, Stanley, Lope, Carlos Peraza and González, Eunice Uc, 'Bioarchaeological Investigation of Ancient Maya Violence and Warfare in Inland Northwest Yucatan, Mexico,' *American Journal of Physical Anthropology*, vol. 154, 1 (2014), pp. 140–51.

Smith, Michael E., *The Aztecs* (3rd edition, New York: Wiley-Blackwell, 2012).

Stuart, David, 'The "Count of Captives" Epithet in Classic Maya Writing,' in V.M. Fields (ed.), *Fifth Palenque Round Table, 1983* (Monterey, CA: Pre-Columbian Art Research Institute, 1985), pp. 97–101.

Stuart, David, 'The Arrival of Strangers: Teotihuacan and Tollan in Classic Maya History,' in D. Carrasco, L. Jones and J.S. Sessions (eds.), *Mesoamerica's Classic Heritage: From Teotihuacan to the Aztecs* (Boulder: University Press of Colorado, 2000), pp. 465–513.

Stuart, David and Stuart, George, *Palenque: Eternal City of the Maya* (London Thames and Hudson, 2008).

Tedlock, Dennis, *Rabinal Achi: A Mayan Drama of War and Sacrifice* (Oxford: Oxford University Press, 2003).

Thompson, J. Eric S., 'A Trial Survey of the Southern Maya Area,' *American Antiquity*, vol. 9, Issue 1 (1943), pp. 106–34.

Tiesler, Vera and Cucina, Andrea, 'Where Are the Warriors? Cranial Trauma Patterns and Conflict among the Ancient Maya,' in D.L. Martin, R.P. Harrod and Ventura R. Perez (eds.), *The Bioarchaeology of Violence* (Gainesville: University Press of Florida, 2012), pp. 160–79.

Tokovinine, Alexandre, *Place and Identity in Classic Maya Narratives* (Washington, DC: Dumbarton Oaks Research Library and Collection, 2013).

Tozzer, Alfred M., *Landa's Relación de las Cosas de Yucatan: A Translation*. Papers of the Peabody Museum of American Archaeology and Ethnology no. 18 (Cambridge, MA: Harvard University Press, 1941).

Watanabe, John M., 'Some Models in a Muddle: Lineage and House in Classic Maya Social Organization,' *Ancient Mesoamerica*, vol. 15, Issue 1 (2004), pp. 159–66.

Webster, David, 'The Study of Maya Warfare: What It Tells Us about the Maya and about Maya Archaeology,' in J.A. Sabloff and J.S. Henderson (eds.), *Lowland Maya Civilization in the Eighth Century A.D.* (Washington, DC: Dumbarton Oaks, 1993), pp. 415–44.

Webster, David, 'City-States of the Maya,' in D. Nichols and T. Charlton (eds.), *The Archaeology of City-States: Cross-Cultural Approaches* (Washington, DC: Smithsonian Institution Press, 1997), pp. 135–54.

Webster, David, 'Ancient Maya Warfare,' in K. Raaflaub and N. Rosenstein (eds.), *War and Society in the Ancient and Medieval Worlds: Asia, the Mediterranean, Europe, and Mesoamerica* (Washington, DC: Center for Hellenic Studies, Trustees for Harvard University, 1999), pp. 333–60.

Webster, David, 'The Not So Peaceful Civilization: A Review of Maya War,' *Journal of World Prehistory* vol. 14, no. 1 (2000), pp. 65–119.

Webster, David, 'The Classic Maya Collapse,' in D.L. Nichols and C. Pool (eds.), *The Oxford Handbook of Mesoamerican Archaeology* (Oxford: Oxford University Press, 2012), pp. 324–34.

Yaeger, Jason and Canuto, Marcello A., 'Introducing an Archaeology of Communities,' in M.A. Canuto and J. Yaeger (eds.), *The Archaeology of Communities* (New York: Routledge, 2000), pp. 1–15.

3

INDIGENOUS WARFARE IN AUSTRALIA AND NEW ZEALAND

John Connor, Michael W. Charney and Kaushik Roy

Historiography on Aboriginal warfare in Australia and New Zealand has generally focused on the periods for which there are substantial written records. This has meant most attention is on the period after the arrival of the Europeans and has been framed with significant European biases, including the denial of Maori tactical skill and military prowess (Belich 1986). Although archaeological and anthropological work has shed some light on precolonial Aboriginal life and warfare, much of the work of reconstruction of aboriginal (in Australia) and Maori (in New Zealand) warfare relies on interpretation of colonial documents and indigenous oral histories. The wars by which Australia and New Zealand became part of the British Empire commenced near Sydney on the banks of the Hawkesbury River in 1795 and would range across the entire continent of Australia and the islands of New Zealand for over 130 years until they were concluded in 1928 with the attacks on Aborigines at Coniston Station in the deserts of central Australia. The British had the advantage of firearms, but it was the mobility of horses that was more important in gaining victory.

Warfare in Continental Isolation

The lands that would become Australia and New Zealand separated completely about 70 million years ago, and although Australia would include the islands of Tasmania and New Guinea during the last Ice Age, this was followed by the submersion of lands between them due to a rise in the level of the ocean. As a result, the prehistoric migrations of humans into Australia represented an earlier and different wave from those of the peoples who would settle Southeast Asia, including most of the Philippine and Indonesian Archipelago, and much of Oceania. This early migration probably occurred sometime between the known human presence around 60,000 years and 120,000 years ago (Arthur and Morphy 2005). Over time, these settlers spread across the continent and the population grew so that, by the late 1780s, there were an estimated 750,000–900,000 Aboriginal Australians. Nevertheless, the geographical expanse of Australia ensured that communities remained small and localised, speaking about 250 different languages (Walsh 1993).

Traditional aboriginal warfare was both universal and limited. It was universal because each group continually attempted to prove their superiority over their neighbours, and

DOI: 10.4324/9780429437915-5

because all members of society took part in it. Boys learnt to fight by playing with toy spears, clubs and boomerangs, and every initiated male became a warrior (Collins [1798] 2003). Girls and women participated in warfare, sometimes as combatants, but more often as victims. It does not appear that wars were fought to gain territory: Aboriginal culture stresses the strong connection to one's own 'country.' The duration of fighting was limited because warriors always had to stop campaigning to resume food gathering (Collins [1798] 2003: 466, 485; Barrallier Diary, *Historical Records of New South Wales*, V, 767).

There were four types of traditional warfare: formal battles, raids for women, revenge attacks and ritual trials. Formal battles, in which two groups fought each other and ended hostilities after a few participants had been killed or wounded, had a practical purpose and were not simply a quaint cultural practice. It was impossible to control casualties in impromptu raids and ambushes, but it was possible to limit losses in formal battles and ensure group survival (Keeley 1996: 90–1). As one British settler wrote in 1796, Darug formal battles were not attended with those fatal consequences which result generally from the battles of those nations who were civilised and Christian. Raids for women were considered warfare because in traditional Aboriginal societies, women's food gathering and childbearing abilities were economic resources which were fundamental to the groups' survival. As the elder men held property rights over the women in the group, these raids were aimed at transferring property from one group to another and must be considered warfare in agricultural societies. Aboriginal society was organised in small, non-hierarchical kinship groups. To prevent inbreeding, Aboriginals recognised the necessity for men to marry women from outside their small kinship group (Connor 2013: 11, 13). Sometimes women were 'abducted' only after they had given their consent, but in other cases the raids were violent and in earnest.

Revenge attacks arose from the traditional belief held by many Aboriginal groups that deaths were caused by another's action, either by direct violence or sorcery. The Darug funeral ceremony included a ritual where the corpse was 'asked' who had caused the death and a person would be named as responsible. That person, or someone from that group, would then be killed in revenge. The final type of traditional warfare was the ritual trial. These were a form of socially condoned punishment for males in which a man was required to stand his ground and accept any wound he might receive inflicted, according to local custom, by spear, club or boomerang. The Darug carried out their trials on a hill above the small British settlement (near what is now the intersection of Sydney's George and Liverpool Streets), hurling spears at the individual using a *woomera* spear thrower that increased the spear's velocity to 160 kilometres per hour and its accurate range to 50 metres. The Darug man receiving punishment was allowed to carry a light bark *eleemong* shield to parry away the incoming spears. When the Maori Te Pahi watched one of these trials during a visit to Sydney in December 1805, the *Sydney Gazette* reported his reaction:

> Tip pa-he, who with several of his sons was present, regarded with contempt their warfare; he frequently discovered much impatience at the length of interval between the fights; and by signs exhorted them to dispatch; he considered the heel-a-man an unnecessary appendage, as the hand was sufficient to put aside and alter the direction of any number of spears; he nonetheless praised highly the wammera, or throwing stick, as from its elasticity he acknowledged the weapon to receive much additional velocity.
>
> *(Uniake, 1823)*

In the case of New Zealand, the same peoples who settled Polynesia also arrived in New Zealand in the twelfth or thirteenth centuries and eventually evolved the Maori civilisation. They were also divided into localised groups called *iwi*, which closely approximates tribe, and these were made up of different *hapu* or clans, which in turn themselves were made up of one or more *whanau* or extended families, although all of these terms overlapped in different usages and were sometimes reconstructed retroactively to interpret the past (Ka'ai 2004, 71). Nevertheless, social particularism provided sufficient terrain for endemic warfare and the demonstration of both status and warrior skills in rivalries between competing groups. Moreover, the Polynesian roots of Maori warfare would endure well into the nineteenth century, leading to comparisons of Maori combat, tactics and defences with those of other islands in the larger Polynesian context (D'Arcy 2000).

Maori traditional warfare was a significant cultural activity. Winter was the period for war, as this was the season during which crops were harvested and stored during the cooler weather when conflict could occur. Maori traditional weapons consisted of a range of war clubs and a variety of spears. These were used in hand-to-hand combat. The rationale for conflict was often a breach of personal or tribal mana or prestige. Success in warfare would result in gaining land and the restoration of mana (Haami 2000, 303–7).

It is possible that the ancestors of the Maori may have arrived with experience of constructing fortifications prior to their migration to New Zealand. The earliest identified *pa* in New Zealand was probably completed around 1500 CE. Over 6,813 fortified *pa* have been identified in New Zealand, with the majority constructed in the North Island (Haami 2000: 303–7). Pre-colonial Maori fortifications were most probably settlements in which a community lived, and converged when the *pa* was threatened. Fortifications in the early 1800s appear to have been constructed using small enclosures, each most probably being allocated to a family group ('Maori Defence Works' 2000: 299).

Darug Resistance in Australia

In different periods, Aboriginal collective violence has been presented variously as conflict and crime. Rather than depicting colonial conflict between Aborigines and the British as war between different political groups, successive British governments would assert that their claim of possession over the continent had automatically turned all indigenous Australians into British subjects. The Colonial Office and colonial governors thus did not recognise Aboriginal sovereignty over their lands or define frontier conflict as 'war.' The legal fiction that there had been no 'war' in Australia meant that the conflicts by which the continent was conquered 'country by country' have only recently been recognised (Broome 2010). This section will focus mainly on the case of the Darug resistance to British expansion as an example of the larger Aboriginal experience.

The British landed at Kamay (today Botany Bay) to establish their colony of New South Wales in 1788, but this did not automatically lead to warfare between the indigenous and non-indigenous arrivals. The small size of the British settlement and the ability of both groups to share the resources of the harbour minimised conflict. Nevertheless, the presence of the Europeans exposed the Aborigines to the spread of smallpox and other diseases for which they had no resistance, weakening the Aboriginal population and making them more vulnerable to British expansion.

The British were opposed by the indigenous leader, holy man and warrior, Pemulwuy, of the Eora group of clans, who waged a 12-year-long guerrilla war against them, bringing together the various clans in the area, including the Darug group, for this struggle. In 1790, Pemulwuy had been appointed by local elders to exact a retribution killing against a British hunter charged with providing game-meat to the colony but who was known to have killed Aborigines for sport. Pemulwuy was put on a wanted list and the governor ordered six of his kinsmen to be arrested as a deterrent. British expeditions sent against him failed, and Pemulwuy remained at large. From 1792 he began raiding farms, burning crops and killing settlers throughout the colony.

The conflict widened in 1795 when Lieutenant-Governor Francis Grose established farms at Green Hills (now Windsor) on the Hawkesbury River. Conflict arose because the land could either be used by the Darug to collect yams or by the British to grow maize. It could not be shared. In May 1795, Captain David Collins wrote that 'open war' had commenced between the natives and the settlers. Collins had been in the colony since its foundation, but this was the first time he had described violence between the British and Darug as 'war.' It was this conflict over land which made the Hawkesbury River the location of the first frontier war in Australia (Collins [1798] 2003: 323, 348).

After raids on the British farms in the first five months of 1795 had killed five settlers Grose's successor, Lieutenant-Governor William Paterson, began to fear that the farmers would abandon the Hawkesbury settlement. Unwilling to lose this fertile agricultural land, he deployed a detachment of the New South Wales Corps (NSWC)—the British Army regiment raised specially for service in the colony—to the Hawkesbury. Paterson ordered his troops to kill any Darug they found and hang their bodies from gibbets as a warning to the rest. While gruesome, this was a common British practice, and the bodies of executed convicts were placed on gibbets at Pinchgut (now Fort Denison) in Sydney Harbour. On 7 June 1795, the detachment of two officers, 66 other ranks and three drummer boys, opened fire on a Darug group in the forest not far from the farms, killing or wounding seven or eight individuals. The Darug attacks continued once the NSWC detachment withdrew, so Paterson was forced to establish a permanent garrison on the Hawkesbury. With the passage of time, the strength of NSWC continued to increase. In 1791, its ranks numbered merely 192, but in 1810 rose to 794 (including 57 convicts) (Grey 1990: 11).

Under Pemulwuy, who would be shot and killed in 1802, the Darug were the first Aboriginal people to develop new tactics for use specifically in frontier warfare. In these frontier warfare tactics, the focus of attacks shifted from settlers—sensibly avoided as they often carried firearms. Instead, the two main Darug tactics were raids on settlers' crops, particularly corn, and raids on farmhouses. Corn was easy to take and unlike other introduced grains, such as wheat and barley, could be eaten without the need for husking and grinding. Corn raids were a form of food gathering, but were also a type of warfare in the same way that cattle raids had been a form of warfare in Ireland in the medieval era. An effective corn raid required large numbers—sometimes 150 people—and reflected a high degree of coordination between the small Darug groups. When the late Pemulwuy's son, the warrior Tedbury, was captured at Pennant Hills during the harvest of 1805, it was claimed that few if any of the out-farms had escaped pillage (Collins [1798] 2003: 348; *Sydney Gazette*, 18 May and 24 June 1804).

Farmhouse raids required fewer people than did corn raids—the average raiding party was between 15 and 20—but needed the careful collection of intelligence on the number of people living in the farmhouse, how they were armed, what they owned and the best times to attack. Once the Darug learnt English, it increased their ability to gather intelligence and carry out raids. The *Sydney Gazette* complained in 1804 that the Darug acquired the language only to 'apply the talent in mischief and deception.' Three attacks in 1805 at Portland Head on the Hawkesbury were carried out using intelligence gathered from the farms by an English-speaking 13-year-old Darug girl (*Sydney Gazette*, 7 July 1805).

The Darug's frontier warfare tactics found their greatest success on the lower Hawkesbury downstream from Portland Head. Here the river meanders through a narrow valley confined by sandstone cliffs with small river flats of rich alluvial soil deposited on the bends of the river. Settlers had tried to farm the flats in 1795, but they were constantly attacked by Darug coming down from the heavily wooded ridges and abandoned the district in 1796 (Collins [1802] 2003: 11). The British made another attempt to occupy the lower Hawkesbury during the corn harvest of 1804 but were driven out again. Some Darug who met with Governor Philip Gidley King told him that they had been driven out from the upstream area. As a result, they were trying to hold on to some land with access to the river. King told the British Colonial Secretary that 'the Darug's request appear[ed] to be so just and so equitable that I assured them no more settlements should be made lower down the river.' Nevertheless, settlers did return to the lower Hawkesbury, but King negotiated a 'reconciliation' with the Daruq in July 1805 that ended the raids (Hall 1926: 555–93).

The second major phase of frontier warfare on the Hawkesbury commenced in 1814 on the southern part of the watercourse, known as the Nepean River. Here the Darug were joined in their attacks on the British corn crops and farmhouses by the Darawal and Gandangara peoples, who began to move into these districts as the Darug population declined due to disease and warfare. When the Aboriginal raids began, soldiers of the Veteran Company were sent to defend the farms. The Veteran Company consisted of men who wished to remain in the colony when the regiment left in 1810, but were no longer fit for military service. Three privates of the Veteran Company were guarding Milehouse's Farm at Appin on 7 May 1814 when they confronted an Aboriginal raiding party taking corn. The soldiers fired their muskets, but the warriors, who knew how long it took to reload these weapons, rushed them before they could fire again, killing Private Isaac Eustace and forcing the other two to flee (Campbell 2002; *Sydney Gazette*, 14 May 1814).

By April 1816, Aboriginal raids caused settlers to flee the frontier, and Governor Macquarie was forced to augment the Veteran Company with regular soldiers from the 46th Regiment to be 'Guards of Protection for those Farms which are Most exposed to the incursions of the Natives.' These small frontier garrisons provided the settlers with some sense of security, but, as Macquarie admitted, it was impossible to protect every farm. The British soldiers killed 15 Aborigines, but their foes could move faster through the Australian bush. Captain W.J.B. Shaw's party pursued a party of 15 Darug warriors around Windsor for five hours but were unable to catch up with them. The troops returned to Sydney when their supplies ran out at the end of April. Aboriginal raids resumed, and Macquarie sent out another expedition. However, by the end of 1816, Aboriginal leaders agreed to end the war of the Hawkesbury (Memo Governor Lachlan Macquarie to Captain W.J.B. Schaw, 9 April 1816, 4/1735, AONSW, CSO, National Library of Australia, Canberra). This sort of coercion formed only one part of Macquarie's Aboriginal policy; the second part was an attempt to conciliate the Darug and their neighbours with the establishment of a school for Aborigines and the encouragement of Aboriginal men to become farmers and the granting of land to individuals in the area of western Sydney that became known as Blacktown (Brook and Kohen 1991). Nevertheless, in the decades ahead, British settlers would consider the Aborigines to be like animals and call for their extermination. Settlers raped Aborigine women, emasculated the men, and they were hunted down and killed, as part of a genocide that has only recently been acknowledged in Australia (Adhikari 2022).

The Musket Wars in New Zealand (1807–1836)

The conflicts in New Zealand in the nineteenth century are controversial as European documents and historical accounts long depicted indigenous society as violent 'savages' who were thus 'saved' by colonial rule. Revisionist historiography would claim present Maori society as unified and peaceful before British rule and claim that the Maori had been tricked into signing their sovereignty away to the British in 1843. This prompted counter-revisionist historiography that would challenge these notions. The latter trend led to new interest in the so-called 'musket wars' in which there was an intensification of indigenous internecine fighting that involved about 3,000 violent encounters of various scales from 1807 to 1836. As these conflicts were waged between Maori before British rule they provide an attractive counter-example to revisionist claims (Hunt 1999).

As the title 'musket wars' suggests, these wars saw the introduction of firearms. In New Zealand, as elsewhere in Polynesia, firearms and cannons were mainly significant at first as a means of striking terror into the enemy. But once introduced, psychological fear was overcome by a clear awareness of what they could and could not do (D'Arcy 2000: 127), leading to a rapidly enhanced scale of killing, more retaliatory attacks by relatives of the victims and the desire to acquire more firearms. Ultimately, these wars cost the lives of one-third of the overall Maori population. Europeans were also involved in the fighting, both as suppliers and sometimes as participants (Crosby 2017). According to one calculation, between 1810 and 1840, warfare resulted in the death of 20,000 people (Fenton 2013: 119).

As Polynesian warfare relied on mobility, the introduction of the cannon slowed the Maori and others down. They had no animals, so they had to carry the cannons around themselves, and the cannons were slow to load and fire. They were more effectively deployed in defence works known as *pa* (D'Arcy 2000: 128).

Research on the Musket Wars in New Zealand, which has examined them from a broader Polynesian perspective, notes that eventually firearms became less effectual due to the continuity of indigenous considerations of status which the introduction of firearms, levelling the playing field, threatened to throw into disorder. In response, the Maori developed modes of combat using firearms that compromised their efficacy, such as defence works in New Zealand and the emphasis on duels from a great distance. Similar developments occurred in Tonga in the same period and after into the 1860s (D'Arcy 2000: 127–8). Even so, the scale of killing became so intense that chiefs in northern New Zealand began to entertain the idea, encouraged by Western missionaries, that British rule, bringing in an outsider and a single sovereign authority that could rule over competing tribes, could help to stop the internecine fighting that was destabilising Maori society and economy (Crosby 2017).

The New Zealand Wars, 1843–1872

As with the Musket Wars, the New Zealand Wars, from 1843–1872, were also problematically depicted in colonial historiography. The Victorian worldview could not entertain the idea that indigenous warriors could defeat European armies and, of course, the Zulu Wars and the Battle of Isandlwana would not occur until seven years after the New Zealand Wars were over. Revisionist historiography by James Belich (1986) demonstrated that instead of merely being brave warriors who resisted the British in sporadic and disorganised ways, the Maori were well organised and exhibited excellent tactical knowledge and military prowess. Moreover, this was not the story of repeated Maori defeats. Instead, Belich argues, this was at best a limited victory for the British overall and along the way they suffered repeated defeats at the hands of the Maori. Belich initially argued that during these wars the Maori developed a new technique, the forerunner to trench warfare, in the form of the 'modern *pa*,' a special defence works developed in response to the relentless of British attacks and their numerical superiority. After criticism of these claims, Belich would in later work posit that there were actually three stages of the *pa*, one of which, the intermediary 'musket *Pa*' phase, was actually developed during the aforementioned Musket Wars (D'Arcy 2000: 118). Regardless, in the end, British numbers, not miliary superiority, were the deciding factor (Belich 1986).

In the New Zealand Wars, various Maori tribes in the North Island fought against British regular soldiers, local colonial forces and Britain's Maori allies (Haami 2000: 303–7). In June 1843, British settlers travelled north, to what is now the town of Blenheim, to challenge two Maori men, Te Ruparaha and Te Rangihaeata of Ngati Toa, regarding disputed land sales in the nearby Wairau Valley. As the British were crossing the Tua Marina Stream, a musket was fired. In the ensuing confrontation, 22 Pakeha and 2 Maori were killed. The British garrison in New Zealand was increased. In 1845, a punitive military expedition was sent north to occupy Maori territory (Cowan 1922: 34–5).

The New Zealand Wars commenced with conflict between Ngapuhi Maori and British soldiers in 1845 in the Bay of Islands in the North Island. Here, Hone Heke Pokai and his comrades had chopped down the flagpole at Kororaeka four times to demonstrate how the shift of the colony's capital from the Bay of Islands to Auckland had affected the local economy, and that the British Crown was exceeding its authority. One the other side, Tamati Waka Nene had aligned themselves with the British. Three British infantry regiments, the 58th, 96th and 99th, were deployed to the Bay of Islands in 1845, augmented by Royal Navy sailors (Keenan 2012a).

George Grey arrived in New Zealand towards the end of 1845 to take up the role of Colonial Governor. Here, he was

> escorted by the troops and the numerous persons assembled together, with many chiefs and natives of the surrounding districts. Our loyal ally, Nene, having arrived from the Bay in the *North Star*, was present, in the British naval captain's uniform, and wore the sword given to him by Governor Hobson.
> (New Zealand Spectator and Cook's Strait Guardian, *3 May 1845;*
> Nelson Examiner and New Zealand Chronicle, 6 December *1845)*

With the end of the Northern War, Grey sent most of his troops to Wellington, where local Maori were contesting the incursion of European settlers in the Hutt Valley. The Wesleyan Minister, Reverend Thomas Buddle, presented two lectures on the Maori at the Auckland Mechanics' Institute in March and May 1851. Buddle compared traditional Maori warfare to that of the Ancient Greeks:

> The New Zealander had no weapons by which warfare could be carried on at a distance for any length of time. It soon became necessary to enter into close combat. A mode

more adapted to the tempers and feelings with which they came into the field, than any which would have kept them at a distance from each other.

('Two Lectures,' 1851)

Conflict resumed in Taranaki in May 1863, which would become the precursor for the Waikato War. Governor Grey had prepared for this campaign and had constructed the Great South Road, which ran from Auckland to the border with the Maori King Tawhaio on the Mangatawiri River. The Kingites lost the Waikato War, though it was a narrow defeat.

During 1864, the Taranaki prophet Te Ua Haumene established a new religion known as Pai Marire or the Hauhau movement. Major Lambert, having landed his troops and weapons from the *St Kilda*, issued arms to 'the loyal natives,' including Maori leaders such as Kopu, Paora and Wairoa. The presence of the steamship *St Kilda* was reported to 'have a salutary effect on the native mind' (Keenan 2012b).

In June 1868, a new prophet, Riwha Titokowaru, emerged in Taranaki to oppose the New Zealand government's policy of confiscating Maori land. As settlers moved into South Taranaki, and began land confiscations, a dispute over the access to timber felling led to a violent dispute, resulting in the death of three sawyers. Titokowaru provided sanctuary to the Maori and refused to hand them over to the government authorities. The government recruited 400 armed constabulary who were sent to Taranaki, where they were joined by 100 Whanganui Maori (Keenan 2012b).

Titokowaru's campaign against the British commenced in June 1868 with a force of fewer than 100 warriors. On 12 July, the Maori leader commenced hostilities. In six months, Titokowaru defeated several colonial forces and occupied around 130 kilometres of territory between the Wanganui and Waingongoro rivers. On 7 September 1868, a battle was fought between 350 colonial troops and Titokowaru's 60 warriors. After the failed assault at Mororua, Titokowaru moved towards the extensive earthworks of the Taurangaika *pa*. Titokowaru retreated northwards and was sheltered in Ngati Maru lands in Taranaki. Titokowaru continued to support peace and non-violent resistance until his death in 1888. Te Kooti continued to attack his enemies until early 1872, when, with his six remaining followers, he found sanctuary in New Zealand's King Country. The following year, Te Kooti renounced violence and died in 1893.

Conclusion

Aboriginal and Maori warfare evolved independently of each other, due to geographical separation, although the decentralised nature of warfare in both resulted from comparable clan-based warrior traditions. The expansion of British exploration and soon after settlement and conquest in the late eighteenth and nineteenth centuries meant that despite their different timelines, Aboriginal warfare and Maori warfare were impacted at roughly the same time by the introduction of European firearms and soldiers. This resulted in some changes in the ways in which both societies waged war, against the British instead of the historical pattern of internecine conflict. In Australia, horses, by providing the Europeans mobility in conducting raids and securing communications, gave them an advantage over the Aborigines (Richards 2013: 30). Aborigines' warfare was effective in delaying British conquest initially. But, the steady pressure of British arms and demographic resources resulted in defeat, significant indigenous deaths, cultural dislocation and arguably genocide of the indigenous people. The

imperial wars in Australia, Tasmania, New Zealand and in the New World could be categorised as a sort of global genocide (Sousa 2004: 193–209).

Bibliography

Adhikari, Mohamed, *Destroying to Replace: Settler Genocides of Indigenous Peoples* (Indianapolis: Hackett, 2022).

Arthur, Bill and Morphy, Frances, *Macquarie Atlas of Indigenous Australia* (Stroud: Macquarie Library, 2005).

Barrallier, Ensign Francis, 'New South Wales Corps,' in *Historical Records of New South Wales*, vol. 5 (7 vols., Sydney: NSW Government Printer, 1892–1901).

Barrallier Diary, in *Historical Records of New South Wales*, V, 767 (Sydney: NSW Government Printer, 1892–1901).

Belich, James, *The New Zealand Wars and the Victorian Interpretation of Racial Conflict* (Auckland: Auckland University Press, 1986).

Binney, J., *Redemption Songs: A Life of Te Kooti Arikirange Te Turuki* (Auckland: Auckland University Press, 1995).

Brook, J. and Kohen, J.L., *The Parramatta Native Institution and the Black Town* (Kensington, New South Wales: New South Wales University Press, 1991).

Broome, Richard, *Aboriginal Australians: A History since 1788*, 4th edition (Sydney: Allen & Unwin, 2010).

Butlin, N.G., *Our Original Aggression: Aboriginal Populations of Southeastern Australia, 1788–1850* (Sydney: Allen & Unwin, 1983).

Campbell, Judy, *Invisible Invaders: Smallpox and Other Diseases in Aboriginal Australia* (Melbourne: Melbourne University Press, 2002).

Collins, David, *An Account of the English Colony of New South Wales, with Remarks on the Dispositions, Customs, Manners, etc. of the Native Inhabitants of that Country*, 2 vols [1798, 1802] (Sydney: University of Sydney Library, 2003).

Connor, John, 'Traditional Indigenous Warfare,' in Craig Stockings and John Connor (eds.), *Before the Anzac Dawn: A Military History of Australia to 1915* (Sydney: University of New South Wales Press, 2013), pp. 8–20.

Cowan, James, *The New Zealand Wars: A History of the Maori Campaigning and Pioneering Period* (Wellington: Government Printer, 1922).

Crosby, Ron D., *The Musket Wars: A History of Inter-Iwi Conflict, 1806–1845* (Auckland: Oratia Media, 2017).

D'Arcy, Paul, 'Maori and Muskets from a Pan-Polynesian Perspective,' *New Zealand Journal of History*, vol. 34, no. 1 (2000), pp. 117–32.

Fenton, Damien, 'Australians in New Zealand Wars,' in Craig Stockings and John Connor (eds.), *Before the Anzac Dawn: A Military History of Australia to 1915* (Sydney: University of New South Wales Press, 2013), pp. 118–47.

Grey, Jeffrey, *A Military History of Australia* (Cambridge: Cambridge University Press, 1990).

Haami, Bradford, 'Maori Traditional Warfare,' in Ian McGibbon (ed.), *The Oxford Companion to New Zealand Military History* (Wellington: Oxford University Press, 2000), pp. 303–7.

Hale, Horatio. *United States Exploring Expedition During the Years 1838, 1839, 1840, 1841, 1842. Under the command of Charles Wilkes, U.S.N. Ethnography and Philology* (Lea and Blanchard: Philadelphia, 1846).

Hall, Lesley D., 'Physiography & Geography of the Hawkesbury River between Windsor and Wiseman's Ferry,' *Proceedings of the Linnean Society of New South Wales* (1926), pp. 555–93.

Hunt, G., 'Study Destroys Picture of Peace and Harmony Before the Treaty,' *National Business Review (New Zealand)* (6 August 1999), n.p.

Ka'ai, Tania, *Ki Te Whalao: An Introduction to Maori Culture and Society* (Auckland: Pearson Longman, 2004).

Keeley, Lawrence, *War Before Civilisation: The Myth of the Peaceful Savage* (New York: Oxford University Press, 1996).

Keenan, Danny, 'Story: New Zealand Wars.' Te Ara, The Encyclopedia of New Zealand, https://teara.
    govt.nz/en/new-zealand-wars/page-2 (2012a), accessed 18 July 2020.
Keenan, Danny, 'Story: New Zealand Wars.' Te Ara, The Encyclopedia of New Zealand, https://
    TeAra.govt.nz.govt.nz.end/new-zealand-wars/page-9 (2012b), accessed 18 July 2020.
'Maori Defence Works,' in Ian McGibbon (ed.), *The Oxford Companion to New Zealand Military History* (Wellington: Oxford University Press, 2000).
Paine, Daniel, *The Journal of Daniel Paine 1794–1797 Together with Documents Illustrating the Beginning of Government Boat-building and Timber-gathering in New South Wales, 1795–1805*, eds.,
    R.J.B. Knight and Alan Frost (Sydney: Library of Australian History, 1983).
Perkins, Rachel. *First Australians* [film], Blackfella Films, 2008.
Richards, Jonathan, 'Frontier Warfare in Australia,' in Craig Stockings and John Connor (eds.), *Before the Anzac Dawn: A Military History of Australia to 1915* (Sydney: University of New South Wales Press, 2013), pp. 21–38.
Sousa, Ashley Riley, '"They Will Be Hunted Down Like Wild Beasts and Destroyed!" A Comparative Study of Genocide in California and Tasmania,' *Journal of Genocide Research*, vol. 6, no. 2 (2004), pp. 193–209.
Statham, Pamela (ed.), *Ins and Outs: The Composition and Disposal of the NSW Corps, 1790–1810* (Canberra: ANU Central Printery, 1892).
Titokowaru, Riwha, Belich, James, '*I Shall Not Die*': *Titowaru's War, New Zealand, 1868–9*, (Wellington: Allen & Unwin/Port Nicholson Press, 1989).
'Two Lectures Delivered by the Rev. Thos. Buddle, Welseyan Minister at the Auckland Mechanics Institute, on the Evenings of 25th March, and 12 May 1851,' in *The Aborigines of New Zealand. Maori Messenger: Te Kaeri Maori* (Library, Victoria University of Wellington, 1851).
Uniake, John, *Narrative of Mr Oxley's Expedition to survey Port Curtis and Moreton Bay, with a view to form Convict Establishments There in pursuance of the Recommendation of the Commissioner of Inquiry* (London: John Murray, [1823] 1825).
Walsh, Michael, 'Languages and Their Status in Aboriginal Australia,' in: Michael Walsh & Colin Yallop (eds.*), Language and Culture in Aboriginal Australia* (Canberra: Aboriginal Studies Press, 1993), pp. 1–13.

PART II

Polities and Armed Forces in the Pre-Modern Era

4

ASSYRIAN WARFARE

1550–609 BCE

Sarah C. Melville

In the ancient Near East, from as early as the fourth millennium BCE, urban-agrarian societies and nomadic tribes fought amongst themselves over scarce resources, territory and power. The first recorded wars were local affairs between adjacent city-states, but as economic growth allowed armies to campaign farther afield, the theatre of war expanded. The continuing cycle of war and power shifts eventually gave rise to the Neo-Assyrian Empire (1000–609 BCE), one of the largest and most enduring of the ancient world. At its height in the seventh century BCE, the empire stretched from the Persian Gulf to the Zagros and Taurus Mountains, through Syria and Palestine and into Egypt. In terms of scale and longevity, it greatly exceeded all predecessors, and its successors, the empires of Babylonia and Persia, modelled themselves on the Assyrian example.

For all their success, the Assyrians did not diverge much from the traditional ways and means of war. All competitors faced the same challenges which they met according to their abilities and resources. The agricultural calendar, dry climate and logistical limitations restricted armies to seasonal campaigns along well-known marching routes that offered potable water and food. Temporary levies made up the bulk of all armies; only in the first millennium BCE could wealthy, centralised states maintain sizeable standing forces. Armies were equipped with hand-held weapons, armour and shields. They used the same tactics, built comparable fortifications and siege machinery, and claimed to follow long-standing, internationally recognised codes of conduct. Changes occurred incrementally, and the adoption of new technology depended on the material resources available to each state.

Since military outcomes served as a barometer of divine approval and a measure of royal legitimacy, most of the kings led their armies in person. As a result, warfare became inextricably linked to ideologies of kingship and to religious beliefs (Pongratz-Leisten 2015; Gaspa 2020). Priests and seers accompanied the army on campaign to perform rituals and interpret omens (Rivaroli 2013; Jean 2013; Ulanowski 2021). Post-war victory celebrations validated the king's military decisions and tacitly acknowledged soldiers' sacrifice. People understood that war offered opportunity; at the very least it was a means to defend life and property and at best a path to wealth, power and security. Even so, the horrors of war demanded explanation, so men developed the principle of just war and created ways to find meaning in their violence.

DOI: 10.4324/9780429437915-7

A gradual accumulation of resources enabled the Assyrians to develop military and administrative systems robust enough to conquer and hold a vast territorial empire. This chapter traces the evolution of Assyrian warfare from the sixteenth century BCE to the end of the seventh century BCE, in order to demonstrate how the state's economic capacity and degree of centralised authority shaped its armies and determined how they waged war. Sources for the period under consideration include royal inscriptions, administrative documents, letters, monuments and artifacts such as arms and armour. Due to the uneven distribution of evidence across time and geographical space, much of the following discussion is provisional.

Warfare during the Middle Assyrian Period (ca. 1550–1050 BCE)

Located in what is now northern Iraq, the cities Assur, Nineveh, Calah (Nimrud) and Arbela defined Assyria's traditional heartland (Radner 2011; Pedde 2012). The Assyrians traced their royal line in an unbroken sequence back to 'kings who lived in tents,' nomadic ancestors in the distant past whose descendants ruled the city Assur on the Tigris River (Glassner 2004: 136–7). This continuity helped them maintain a sense of identity in a precarious, ever-changing world. Surrounded by enemies, they learned early that survival depended on military strength and savvy political leadership. Though Assyria was a prosperous trading centre during the early second millennium BCE, it did not evolve into an expansionist state until the Middle Assyrian period (ca. 1550–1050 BCE), when pressure from more powerful kingdoms—Hatti, Kassite Babylonia and Mitanni—forced it to militarise.

Initially, Middle Assyrian kings struggled to assert their independence from Mitanni in Syria and the Kassites in Babylonia. When the power of Mitanni declined, the Assyrian King Ashur-uballit I (ca. 1363–1328 BCE) seized the opportunity to establish his kingdom as an independent power. Over the next century, Adad-nirari I (ca. 1295–1264 BCE),

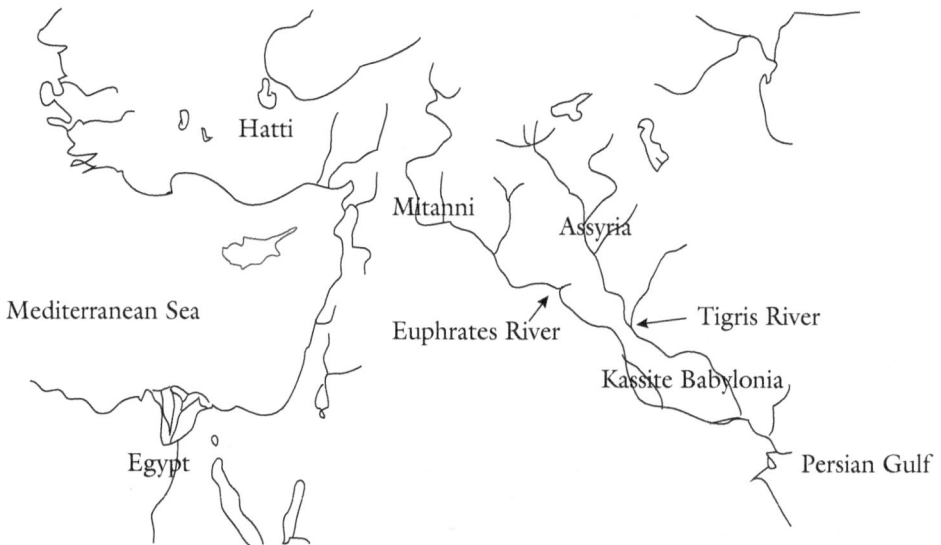

Map 4.1 The Near East in the fifteenth century BCE.

Shalmaneser I (ca. 1263–1234 BCE) and Tukulti-Ninurta I (ca. 1233–1197 BCE) managed to elevate Assyria to the lead position in the Near East. Although royal authority increased substantially during this period, the king did not wield the type of power that his Neo-Assyrian successors would enjoy. The nobility controlled large territories, technically on the king's behalf, but in fact exercising considerable autonomy (Faist 2010: 17; Düring 2015a; Jakob, 2016). During the reign of Tukulti-Ninurta I, for example, the grand vizier (*sukkallu rabiu*) styled himself 'King of Hanigalbat,' Hanigalbat being the district he governed for the crown (Fales 2012: 112). Nevertheless, this period produced the foundations of royal ideology and the pattern for expansion upon which the Neo-Assyrian kings built their empire (Düring 2018; Düring 2020).

Middle Assyrian documentary sources are not as abundant as those from the Neo-Assyrian period. Recent finds at sites such as Dur-Katlimmu, Harbe and Tell Sabi Abyad suggest that the Assyrians borrowed some administrative practices from Mitanni, though specifics of the military recruitment system remain elusive (Llop 2012; Düring, 2015a). The Assyrian king required free adult males, including conquered people forcibly removed to new locations, to perform service (*ilku*) as soldiers or labourers in return for land allotments. Officials raised troops from their own large agricultural estates (*dunnu*) and the provinces they governed on the king's behalf (Llop 2011; Llop-Raduà 2012; Düring 2015b). Some troops (*ṣābū ḫurādātu* or *ḫurādu*) served on a more permanent basis, though not in large enough numbers to constitute a regular, standing army (Jakob 2003: 202–8). Wealthier men sometimes paid others to take their place, but the nature and extent of the practice remain unclear (Postgate 2008: 89–90).

The crown supplied infantry levies with woollen capes (to double as blankets), jerkins, weapons, helmets and shields. Two shield types predominated: small bucklers made of leather or wood with metal reinforcements for use in close combat, and larger, woven wicker or leather tower-shields for protecting archers or for defence of the infantry during siege operations, especially while launching assaults. Offensive weapons included bronze-tipped spears, composite bows, swords, daggers, whips, slings and a variety of axes, staves, clubs, maces and battle staffs (Llop 2016). The bronze sickle sword, named for its distinctive curved shape, was popular throughout the Near East and Egypt at this time, though evidence of Assyrian usage is rare. One particularly fine example bears the name and lineage of the Assyrian King Adad-nirari I (1307–1275 BCE) (Maxwell-Hyslop 2002: 210–13, 216, figs. 1–3). As metal technology improved and bronze weapons became cheaper to produce, the short stabbing sword gradually superseded the sickle sword. Spears remained the favoured close-combat weapon.

During the Late Bronze Age (1600–1050 BCE), as the period is broadly known, chariots rose to prominence in Near Eastern armies. The expense of procuring and training chariot horses increased their status value, and since international standing depended on wealth and military success, large chariot forces predominated. The military's funding structure is not fully understood since it is difficult to determine what provincial officials actually owned and what belonged to the palace. It appears, however, that some officials provided their own horses, cabs and harnessing equipment (Feldman and Sauvage 2010: 89). Beyond its primary function to resolve interstate disputes, warfare reflected the organization of power within the regime. Increased campaigning in terrain unsuitable for chariots may have diminished the power of elite owners, which in turn allowed the king to take sole credit for campaign success. Few Middle Assyrian kings mention chariots in their campaign accounts, a fact that reflects both the king's struggle to wrest power from his nobles and the changing nature of warfare.

Chariots held two men, a driver (*ša mugerre*) and a warrior (*māru damqu*) armed with a composite bow made of strips of wood, sinew and horn glued together. Cabs also carried spears for use if the fight turned hand-to-hand. The charioteers' armour included bronze helmets and metal scales sewn onto leather jerkins (Llop 2016). A lightweight cab, spoked wheels and an improved yoking system increased mobility and speed, thus allowing chariots to participate more actively in battle than they had in earlier periods. Since no source details how charioteers fought—whether they worked cooperatively or singly, charged or rode parallel to the enemy line—battle tactics remain a matter of conjecture (De Backer 2009–10; Feldman and Sauvage 2010; Genz 2013). The need to acquire horses influenced both military and economic strategies, while the high cost of chariot warfare motivated the kings to engage in battle only after diplomatic options had been exhausted.

Characteristically short on details, Middle Assyrian battle descriptions omit tactics and troop arrangements in favour of royal heroism. Nevertheless, rulers had a sophisticated grasp of the military arts. *The Epic of Tukulti-Ninurta I*, for example, recounts his war with the Kassite King of Babylonia, Kashtiliash IV (ca. 1232–1225 BCE). Each king used deception, spies and political grandstanding to gain the military advantage over his enemy. Hostilities began when Tukulti-Ninurta goaded Kashtiliash into a confrontation by arresting Babylonian merchants for spying. When in response Kashtiliash detained Assyrian messengers, Tukulti-Ninurta invaded Babylonia, taking the Kassites by surprise and forcing them to resort to ambush and raid, tactics considered dishonourable at the time, at least by those on the receiving end. According to the epic, Kashtiliash bought time by deliberately dragging out negotiations, 'as a ruse until he could draw up his warriors, until he had made ready his battle plan' (Foster 2005: 309). When the two armies finally engaged, the Assyrians defeated the Babylonians, capturing Kashtiliash in the process and then looting Babylon which they controlled for another seven years until the Kassites seized power once again. Deception, treachery, spying and political chicanery were no strangers to the Assyrians who had already developed the arts of war, diplomacy and gamesmanship that would serve them so well in the new millennium. Their enemies, though equally familiar with these methods, proved less adept at implementing them—or so the Assyrian sources would have us believe.

Warfare in the Neo-Assyrian Period (1000–609 BCE)

The Assyrians entered a period of decline at the end of the second millennium BCE, as the Near East experienced widespread political upheaval and economic depression. The Hittites, Kassites and various city-states in Syria-Palestine succumbed to foreign invaders and never recovered. Egypt survived but lost most of its empire, while Assyrian territory shrank to its smallest footprint. The memory of their former glory and the experience of loss proved transformative for the Assyrians, who responded aggressively to new challenges, including rising powers Urartu, Elam and Phrygia.

Resurgence began in the tenth century under the leadership of Adad-nirari II (911–891 BCE) and his son Tukulti-Ninurta II (890–884 BCE). In the drive to recapture their western territories, these kings campaigned almost yearly, in general following the model of their predecessors, but also actively seeking new advantages. For example, Adad-nirari II

Phrygia

Taurus Mountains

Urartu

.Harran

.Nineveh

Assyria

.Arbela

Zagros Mountains

Tigris River

Mediterranean Sea

Euphrates River

.Babylon

Babylonia

Elam

Judah

Egypt

Persian Gulf

Map 4.2   The Near East 1000–609 BCE.

claimed a notable military innovation, the circumvallation of a besieged town with strategically placed earthen redoubts:

> I mustered my chariotry (and) troops (and) marched to the city Gidara, which the Arameans call Raqammatu (and) which the Arameans had taken away by force . . . . In my wisdom I placed redoubts all around it (the city), (a tactic) that did not exist among the kings my fathers.
>
> *(Grayson 1991: A.0.99.2.54–55)*

The same king created forward supply depots to facilitate the army's passage (Grayson 1991: 142). In enemy territory the army had to fend for itself, so improved logistical planning allowed it to travel farther and for longer periods.

Building on their predecessor's success but still short on central power and resources, Assurnasirpal II (883–859 BCE) and his son Shalmaneser III (858–824 BCE) used terror tactics to achieve their goals. Summary executions, head-taking and mutilation had always played a role in war, but while other Near Eastern cultures rarely celebrated such acts, the Assyrians recognised the potential for political payoff and capitalised on it. Royal inscriptions and narrative art of this period emphasise the king's overwhelming might and the dire consequences of resistance. A typical passage from Assurnasirpal's annals describes his methods:

> Within the mighty mountain I massacred them. With their blood I dyed the mountain red like red wool (and) the remainder the gorges (and) mountain torrents

swallowed. I carried off captives (and) possessions from them, I cut off the heads of their warriors (and) built a tower (of them) in front of their city. I burned their adolescent boys (and) girls.

(Grayson 1991: A.0.101.17.ii.55–62)

Lacking the means to garrison and administer many of the lands he invaded, Assurnasirpal found it expedient to kill, destroy and loot. Those who escaped the onslaught and stayed in their homelands would no longer pose a threat, whereas prisoners and goods removed to Assyria enabled massive building projects, including a complete renovation of Calah and refurbishment of the temple of Mamu at Imgur-Enlil (modern Balawat). Aiming to increase his power at the expense of the hereditary nobility, Assurnasirpal moved the capital from Assur to Calah. Soon after, Shalmaneser added a military palace (*ekal mašarti*) there for the storage of armaments and tribute, and to function as the army's muster place (Oates and Oates 2001: 144–54; Kertai, 2015: 58–74, 117–20, 147–54, 159–66). Such buildings became standard in royal cities thereafter.

Although earlier kings understood the political value of public spectacle, the Neo-Assyrians were the first to harness symbolic capital programmatically as a force multiplier. Successful campaigns, expensive building programmes, public oath-taking, gruesome executions, gift exchange, the reception of tribute and banquets all projected royal wealth and power. Kings used the prestige won through such displays to mask institutional weaknesses that emerged as the empire expanded. Terror tactics not only enhanced Assyria's fearsome reputation but motivated its enemies to submit without fighting. Therefore, calculated violence helped preserve Assyria's most important resource, its army. Though such tactics seem abhorrent now, Assurnasirpal and Shalmaneser did what they could with the limited means available to them. Extreme violence, such as flaying and impalement, demonstrated the king's fulfillment of ideological expectations, for he was duty-bound to punish treaty-breakers, avenge subjects' deaths and suppress rebellions. In the ancient Near East, kings often used brutality to communicate political messages and project power (Radner 2015). What made the Assyrians stand out from other Near Eastern societies was not their political violence, but their open acknowledgement of it.

The army continued to be the primary instrument of royal power. Recognising the need to adapt to changing circumstance, ninth-century BCE rulers focused on improving equitation. Though chariot cabs retained their Middle Assyrian size and shape, wheels became heavier, now sporting six spokes and a thicker rim in order to increase stability without sacrificing mobility. Unable to use shields, charioteers and fighters wore helmets and ankle-length robes covered in metal scales (e.g., Schachner 2007: pl. 8). Still a high-status military division, chariots remained vital to combat operations in the open terrain of Mesopotamia and Syria-Palestine, but their inability to function in rough terrain hindered conquest of mountainous regions.

Cavalry provided an ingenious solution to the problem of mobile warfare in rough terrain, though initially its effectiveness was limited since inexperienced riders sat forward on the horse's withers. Without saddles, stirrups or a martingale (rein holder), cavalry troops had to work cooperatively in pairs, one to fight, usually as an archer, while the other controlled the horse. In effect, the rider pairs acted as rough terrain, cab-less chariots (Archer 2010: 70–1). Horse soldiers wore no armour other than a helmet (e.g., North West Palace Calah, Room B, panel 9) (Drews 2004: 58). Though useful as messengers, scouts, foragers and fighters, cavalry could not ride in formation or execute complex tactical manoeuvres as a unit.

Despite these advancements, constant campaigning and costly construction overburdened the economy. Shalmaneser's reign ended in a bloody civil war that introduced a period of weakness and internal discord. Assyria did not resume imperial expansion in earnest until roughly a century later, after a rebellion in Calah propelled Tiglath-pileser III (744–727 BCE) to the throne. A talented general and political leader, Tiglath-pileser reorganised the administration and created a large standing army to conquer and occupy new territories and oversee deportations.

Although deportation was not a new practice, the Assyrians were the first to realise its full potential (Oded 1979). The deportation of conquered peoples and the recruitment of elite units from defeated armies not only improved Assyria's military capability but encouraged assimilation and reduced the likelihood of rebellion. The fact that deportees and foreign soldiers relied on the crown for their livelihood increased the king's power. To establish royal authority over all aspects of government, Sargon II (721–705 BCE) appointed eunuchs to many high-ranking positions, for without families to support, and owing their livelihood to the king, they proved more trustworthy than hereditary elites. Eunuchs also served throughout the standing army, not in segregated units but mixed in with other soldiers, usually as officers. These personnel adjustments tipped the balance of power toward the crown and away from the nobility. Moreover, centralised collection of tribute, plunder and taxes gave the king control of the resources needed to maintain his army and expand his territory.

Over the course of the imperial period, the army became more ethnically diverse, as large numbers of non-Assyrians were enlisted. From at least the mid-eighth century BCE, certain Aramaean tribes, such as the Gurreans and Itu'eans, served the king as auxiliaries. Light-armed spearmen from Anatolia, Urartu or North Syria, the Gurreans wore distinctive crested helmets and cardio-phylax type chest protectors (Dezső 2012a: 38–9, 49, 93). The Itu'eans, who wore no protective gear and fought as archers, originated in the mid-Euphrates area (Luukko 2019: 94–5). Itu'eans served in various capacities and were often seconded to the provinces on behalf of the crown. Several letters from the governors mention Itu'ean contingents as peacekeepers, intelligence gatherers and border patrol (see, e.g., Parpola 1987: nos. 32, 176). In addition to Gurreans and Itu'eans, other groups including Ru'eans, Chaldeans, Urartians, Medes, Israelites and Elamites also served in the Assyrian military (Dalley 1985).

During the empire's heyday, the armed forces consisted of temporary levies, deportee units, Assyrian standing divisions, the royal guard and auxiliaries. Each of these included infantry and equestrian units. As commander-in-chief, the king travelled on campaign and sometimes participated in combat, at least until Sargon II's death in battle in 705 BCE put an end to the practice. After that point, kings usually commanded from a safe distance or delegated leadership to one of the magnates. By the end of the eighth century BCE, a sizeable standing army occupied strategic points around the empire and formed the core of the campaign forces. It is not clear how long regular soldiers served or whether they earned compensation upon retirement (if indeed age brought an end to military service), but on active duty the crown allotted them land and sometimes a portion of the loot (Dezső 2016: 105–24).

The standing army, known as the king's cohort (*kiṣir šarri*), included chariotry, cavalry and infantry organised into decimal units of varying size, the largest numbering in the thousands and the smallest around ten. After the king, the highest military officials were the chief eunuch (*rab ša reši*), who administered the standing army and managed horse-acquisition, and the field marshal (*turtan*) who sometimes campaigned on the king's behalf. The other magnates and governors led their own household troops and the men levied in their districts. The complete chain of command is not well understood since ranks proliferated and sometimes overlapped. Basic ranks referred to the size or type of contingent; e.g., commander of 50 (*rab ḫanši*), cohort commander (*rab kiṣri*) or chariot team commander (*rab urâte*) (Dezsö 2012a, 2012b). It seems that kings and high officials arranged the army according to their own preferences 'on the basis of personal/political/tactical standards and mechanisms' (Fales 2009: 76). Sources do not suggest the type of standardisation that exists in modern armies.

Administrators continued to use the *ilku*-system to raise temporary troops, called king's men (*ṣab šarri*) for campaigns. As in earlier periods, unless the king granted a special exemption (*kidinnu*), every city, town and village in the empire had to conscript men for annual service (Holloway 2002: 293–302; Dezsö 2016: 39–58). Annual campaigns did not require empire-wide mobilisation. Rather, recruiters drew soldiers from the catchment area closest to the objective, and appointed reservists to wait at the ready in their home districts. Naturally, some troops stayed on provincial garrison duty.

Heavy infantry of the standing army wore conical helmets and scale armour or solid breastplates, and carried spears approximately nine feet long, tipped with tanged or socketed iron or bronze blades. Shield types varied by task: spearmen with tower shields protected shield-less archers and slingers; those with smaller shields joined in close-quarter combat (Barron 2010: 113–42: De Backer 2016). Most conscripts probably served as infantry archers or slingers because weapons releasing missiles in volleys required less expertise than those used in hand to hand, close quarter combat. By the same principle, the crown outfitted levies with helmets and weapons but rarely body armour. Other, probably seasonal, soldiers known as *kallapu*-troops served on horseback and on foot as messengers, scouts, the vanguard or rearguard, and as a kind of military police to keep the infantry levies in line (Scurlock 2014).

The adoption of the martingale sometime in the mid-eighth century BCE allowed riders to control their mounts and fight at the same time. The first individual horsemen appear as lancers on the reliefs of Tiglath-pileser III (745–727 BCE). Later, during the reign of Sargon II (720–705 BCE), cavalrymen are shown carrying bows, but they are not depicted shooting at the gallop until Ashurbanipal's reign (668–627 BCE) (e.g., Bereton 2018: fig. 57). Several types of cavalry served in the Assyrian military: the king's personal mounted guard (*pirru*); the horse guards (*ša qurbuti*); the regular Assyrian cavalry (*ša petḫalli*) including contingents from defeated armies, and *kallapu*-troops (Dezsö 2012b). As far as we know, none of these units rode in formation or carried out the type of manoeuvres typical of later Greek and Roman cavalry.

Cavalry did not entirely displace chariots, which by the eighth century BCE had become larger and heavier. Teams of three or even four horses pulled cabs with iron-hobnail-reinforced wheels, and a third and sometimes fourth man joined the crew as shield bearer or second fighter. After the reign of Sargon II, chariots no longer appear in battle scenes, although royal inscriptions mention recruitment of foreign chariot teams (Frame 2021: 1.75–6). The cumbersome heavy chariots of seventh-century BCE, the wheels of which appear man-height on some reliefs, suggests that they had lost battlefield function. The chariot continued as a symbol of royalty; the king used a chariot for royal hunts and the review of booty and prisoners.

As the importance of chariots waned, that of cavalry increased. Communication depended on messengers riding in relay, while the army's equestrian units played a vital role in battle and as foragers, scouts and skirmishers. The need to acquire horses fuelled Assyrian expansion into eastern Anatolia and western Iran, areas known for their excellent breeding grounds. So great was the military demand for horses that kings used any means available—conquest, taxation, trade, tribute and breeding schemes—to secure an adequate supply. Mules, donkeys and perhaps camels served as military pack animals. A large part of the Neo-Assyrian military apparatus was devoted to animal husbandry, but without accurate knowledge of the enemy, the equestrian units could not operate effectively.

Every aspect of a campaign depended on intelligence. Kings gathered information to determine targets of opportunity as well as enemy intentions. Without it they would have been unaware of trouble brewing elsewhere while they marched blind into unfamiliar foreign lands. Although the Assyrians did not have a dedicated intelligence organisation in the modern sense, they carefully maintained spies and communication networks, apparently with greater success than their competitors. Royal decision-making relied on a constant stream of information gleaned from multiple sources. Corroboration from different sources mitigated incidents of false reporting and misinformation (Dubovský 2006; Dezsö 2014). The Assyrians were well aware that success required reliable intelligence.

Once planning was completed, the army prepared for war. After mustering troops and equipment and completing the appropriate religious rites, the army set out in marching order behind the standards of major gods, including Assur, Ishtar, Adad and Nergal (Pongratz-Leisten, Bleibtreu and Deller 1992; Melville 2016b). The size of the army, which could number in the tens of thousands, and the complexity of its operations required specialised personnel: engineers to build siege machinery and oversee the construction of fortified camps; spies; grooms; scouts; cooks; doctors; diviners; priests; scribes and body servants to attend the king and other elites. Some soldiers were assigned to manage prisoners, deportees and the transportation of plunder, including horses and small cattle (sheep and goats). Camps ranged in size from impermanent marching camps to fully fortified ones built to protect the army during lengthy sieges. Kings and high-ranking officers lived in fully furnished tents, while common soldiers made do with simpler accommodation and basic food (grain and legumes) (Fales and Rigo 2014). Standard campaign operations included devastation of agricultural land and settlements, battles and sieges, usually in that order.

As the empire grew and the army penetrated more distant territories, logistics became progressively more difficult, for consumption quickly surpassed transport capacity (Fales 1990; Ponchia 2014; Fales and Rigo 2014; Marriott and Radner 2015). Mustering could take weeks and required a huge outlay of provisions. On the march, as soldiers travelled through Assyrian or client territory, grain depots supplied them, but in enemy lands they had to forage for food. Looting secured the food supply and costly prizes, while depriving the enemy of valuable resources and forcing them to seek an immediate resolution through battle or surrender. Though effective in theory, that tactic did not always work in practice. Making a subtle nod to the difficulty involved in feeding the army, royal inscriptions often declare, 'I let (my) troops eat the grain (and) dates in their gardens (and) their crops in the countryside,' or 'I opened their storage granaries and fe[d] my troops on grain rations beyond measure' (Grayson and Novotny 2012–14: 213.50; Foster 2005: 806). Soldiers carried reserve rations and kings timed campaigns to coincide with food harvests, yet starvation and disease always threatened to derail plans.

Battles entailed special risk, for a catastrophic defeat could topple a throne and bring a state to ruin. Full-scale, pitched battles were relatively rare during this period because they were too costly for evenly matched armies and a poor option for weaker ones. The sources are characteristically terse about battle tactics but indicate that the Neo-Assyrian Army could carry out basic manoeuvres such as flanking, envelopment or penetration of the centre (Scurlock 1997; De Backer 2007). Armies arrayed in full view of each other, but occasionally, as at the Battle of Mt. Uaush where Sargon II defeated the Urartians in 714 BCE, one side would break protocol and attack before preparations had been completed:

> I (Sargon II) did not ease their (the soldiers') weariness, I did not give water to wet their thirst, I did not establish a base or build a walled camp, I did not send out my warriors or rally my troops, I did not turn my right and left wings to my side, (and) I disregarded my rear . . . . With only my single chariot and the horsemen who ride at my side, who never leave me in hostile or friendly territory, the cavalry troop of (my brother) Sin-ah-usur, I fell upon him (Rusa of Urartu) like a furious arrow; I caused his defeat and turned back his attack.
>
> *(Frame 2021: 65.129–39)*

Sargon's list of negatives reveals standard military procedure. Normally, he would have set up camp, rested and fed his men, and positioned a rear guard before deploying his battle line (Fagan 2010). This account also demonstrates the soldiers' flexibility and discipline, for (apparently) they did not hesitate to follow the king's order.

Most of the battlefield killing occurred when one side broke and ran. In the event of a draw or Pyrrhic victory, both sides would withdraw, sometimes to negotiate but more often to maintain an uneasy truce until one of them could renew the fight. Sargon II fought Elamite forces (allied to Babylonia) to a draw at the battle of Der in 720 BCE and then waited ten years before invading the south. Likewise, after fighting an indecisive battle against a Babylonian–Elamite coalition at Halule in 691 BCE, Sennacherib spent a year organising a full-scale invasion of Babylonia. In both cases, strategic delay led eventually to victory. That the Assyrian elites and regular soldiers accepted their rulers' decisions to withdraw attests to the strength of royal authority, the discipline of the army and solid strategic planning.

Often, rather than face the Assyrians, opposing forces refused battle, hoping instead to survive a siege. If time and supplies were plentiful, the Assyrians would blockade the city and starve it into submission. Where circumstances invited action, they applied siege methods that had been in use for millennia (Eph'al 2009; De Backer 2012, Siddall 2019; Nadali 2019). In the account of his campaign through Judah, for example, Sennacherib described standard techniques:

> I surrounded (and) conquered forty-six of his fortified cities, fortresses, and small settlements in their environs, which were without number, by having ramps stamped down and battering rams brought up, (and by) the assault of foot soldiers, sapping, breaching, and siege engines.
>
> *(Grayson and Novotny 2012: 16.iii.74–iv.2)*

Occasionally, the attackers won entry through treachery as someone inside the city let them in (e.g., Fuchs and Parpola 2001: no. 131). Relief forces could pose a threat but rarely dislodged the invaders. In the mountainous terrain when the use of siege machinery proved impossible, soldiers had to assault by escalade, no doubt taking heavy casualties in the process. After enduring terrible living and combat conditions, soldiers often took out their frustrations on the inhabitants when a city fell. In ancient times, as in all periods since, sieges were among the most brutal of military encounters.

Once the fighting was over, the Assyrians often treated the vanquished with clemency, leaving local rulers in place as clients and establishing tribute and military obligations through treaties. Sometimes they made an example of the ringleaders and deported the general population, not as slaves but as people subject to the same laws and duties as Assyrians. But when the king needed to make a political statement, suppress a rebellion or mollify his troops, he could be ruthless and deliberately cruel: Sargon II had Ilu-bi'di, the rebel who led a coalition against him in 720 BCE, flayed alive; Esarhaddon (680–669 BCE) made captured nobles wear the severed heads of their kings around their necks while parading through Nineveh, and Ashurbanipal (668–627 BCE) had his Elamite prisoners grind up the bones of their ancestors at the gates of Nineveh prior to execution (Richardson 2007; Radner 2015). Post-campaign celebrations reinforced the impression of Assyria's invincibility, demonstrated the king's success, and invited civilian onlookers to accept the imperial mission.

By the end of the eighth century BCE, few if any existential threats to Assyria persisted: Urartu had become weak, Babylonia seemed under control and Elam could not match Assyria's resource base. Smaller states such as Judah had little hope of breaking free of the imperial juggernaut. Impressed by Assyria's military capability, the biblical prophet Isaiah warned:

> Here they come, swiftly, speedily! None of them is weary, none stumbles, none slumbers or sleeps, not a loincloth is loose, not a sandal-thong broken. Their arrows are sharp, all their bows bent; their horses' hoofs seem like flint, and their wheels like the whirlwind.
>
> *(Isaiah 5: 26–8 NRSV)*

Notably, what he found so threatening was not Assyria's fighting prowess but its ability to mobilise human and economic resources and put them into operation. Indeed, Assyria's economic capacity and manpower reserves gave it the advantage in any war.

Despite the outward signs of success, rebellions broke out at the slightest opportunity. Unable to take on Assyria by themselves, subject states often formed coalitions, but even the most well-resourced alliances rarely survived contact with the Assyrian Army. Nevertheless, officials frequently reported difficulty imposing their authority on resentful subjects (Radner 2016; Melville 2016a). One provincial governor complained of simmering resistance: 'There is permanent hostility (here) in the desert but I am not neglecting (my) guard' (Parpola 1987: no. 176). In another case, a priest and mayor teamed up to complain, 'Bibiya, the prefect of the Itu'eans, and Tarditu-Aššur, the prefect of the Itu'eans, his deputy, sit outside the Inner City (Assur), in front of the gate, eating together, drinking wine and squandering the exit-dues of the Inner City' (Cole and Machinist 1998: no. 33). Reports of desertion, slacking and incompetence were as common in the royal correspondence as denunciations and pleas for help.

It would be wrong to read too much into these missives because the royal archives stored only those letters deemed particularly important. Even so, they exposed imperial vulnerability. Assyria's control over its peripheral provinces was always contingent and no amount of political play or costly signalling could alter that fact. Toward the end of Ashurbanipal's reign, the empire began to weaken until a host of new challenges—climate change, demographic shifts, economic depression and internecine strife—left it exposed to outside invaders. After a lengthy struggle (ca. 626–612 BCE), the combined armies of the Babylonians and Medes were able to destroy it. That both the Neo-Babylonian (626–539 BCE) and Persian Empires (539–330 BCE) adopted many Assyrian administrative and military practices attests to the latter's achievement.

Conclusion

In the second millennium BCE, when Assyrian kings ruled a small-city state with limited re-sources, they could not resist the intervention of the stronger states Mitanni and Babylonia. When the opportunity arose, the Assyrians counterattacked, winning new lands and wealth. Over the following centuries, kings' fortunes waxed and waned; sometimes they could only raid and loot, but gradually they accrued the power and organisational ability to rule what they conquered. As the crown drew power to itself away from the nobility and controlled more resources, it gained military capability. By the end of the eighth century BCE, the Assyrian king had become an absolute ruler, the living link between human sphere and the divine one, a mystical warrior capable of imposing order on chaos.

Apart from a few specific innovations that competitors quickly adopted, the Assyrians did not revolutionise warfare; they simply waged it more efficiently than their competitors. Rec-ognising that the side with the most human and material resources usually won, they focused on procurement, organisation and operations. By always going on the offensive, they pro-tected their own resources and deprived the enemy of his. As the empire grew, Assyrian kings had to develop ways to project power without using force, for the army could not campaign on multiple fronts simultaneously. Expansion into new lands, development of agriculture and control of trade sustained the economy; deportations pacified conquered peoples and the threat of reprisal backed by a strong military kept enemies at bay. A coherent imperial ideology that reflected deep-seated beliefs as much as a calculated image helped to create an Assyrian identity. In effect, Assyria was the first state fully to realise the imperial potential and, like all empires, its military provided the foundation for its power.

Bibliography

Archer, R., 'Chariotry to Cavalry: Developments in the Early First Millennium,' in G.M. Fagan and M. Trundle (eds.), *New Perspectives on Ancient Warfare* (Leiden: Brill, 2010), pp. 57–80.

Barron, A.E., *Late Assyrian Arms and Armour: Art Versus Artifact*. PhD dissertation, University of Toronto, 2010.

Bereton, G. (ed.), *I am Ashurbanipal King of Assyria, King of the World* (London: Thames and Hud-son, 2018).

Cole, S. and Machinist, P., *Letters from Priests to the Kings Esarhaddon and Ashurbanipal*. State Archives of Assyria, vol. 13 (Helsinki: University of Helsinki Press, 1998).

Dalley, S., 'Foreign Chariotry and Cavalry in the Armies of Tiglath-pileser III and Sargon II,' *Iraq*, vol. 47 (1985), pp. 31–48.

De Backer F., 'Some Basic Tactics of Neo-Assyrian Warfare,' *Ugarit Forschungen*, vol. 39 (2007), pp. 69–116.

De Backer, F., 'Evolution of War Chariot Tactics in the Ancient Near East,' *Ugarit Forschungen*, vol. 41 (2009–10), pp. 29–46.

De Backer, F., *L'art du siege néo-assyrien* (Leiden: Brill, 2012).

De Backer, F., *The Neo-Assyrian Shield: Evolution, Heraldry, and Associated Tactics* (Atlanta: Lockwood Press, 2016).

Dezsö, T., *The Assyrian Army I: The Structure of the Neo-Assyrian Army, 1: Infantry* (Budapest: Eötvös University Press, 2012a).

Dezső, T., *The Assyrian Army I: The Structure of the Neo-Assyrian Army, 2: Cavalry and Chariotry* (Budapest: Eötvös University Press, 2012b).

Dezső, T., 'Neo-Assyrian Military Intelligence,' in H. Neumann, et al. (eds.), *Krieg und Frieden im Alten Vorderasien, 52e Rencontre Assyriologique Internationale, Münster, 17.-21. Juli 2006* (Münster: Ugarit Verlag, 2014), pp. 221–35.

Dezső, T., *The Assyrian Army II: Recruitment and Logistics* (Budapest: Eötvös University Press, 2016).

Drews, R., *Early Riders: The Beginnings of Mounted Warfare in Asia and Europe* (New York: Routledge, 2004).

Dubovský, P., *Hezekiah and the Assyrian Spies: Reconstruction of the Neo-Assyrian Intelligence Services and Its Significance for 2 Kings 18–19* (Rome: Pontificio Istituto Biblico, 2006).

Düring, B.S., 'The Hegemonic Practices of the Middle Assyrian Empire in Context,' in B.S. Düring (ed.), *Understanding Hegemonic Practices of the Early Assyrian Empire* (Leiden: NINO, 2015a), pp. 289–311.

Düring, B.S., 'Reassessing the *Dunnu* Institution in the Context of the Middle Assyrian Empire,' *Ancient Near Eastern Studies*, vol. 52 (2015b), pp. 47–68.

Düring, B.S., 'At the Root of the Matter: The Middle Assyrian Prelude to Empire,' in C. Tyson and V. Herrmann (eds.), *Imperial Peripheries in the Neo-Assyrian Period* (Boulder: University of Colorado Press, 2018), pp. 47–79.

Düring, B.S. *The Imperialisation of Assyria: An Archaeological Approach* (Cambridge: Cambridge University Press, 2020).

Eph'al, I., *The City Besieged: Siege and its Manifestations in the Ancient Near East* (Leiden: Brill, 2009).

Fagan, G., '"I Fell Upon Him Like a Furious Arrow": The Neo-Assyrian Tactical System,' in G.M. Fagan and M. Trundle (eds.), *New Perspectives on Ancient Warfare* (Leiden: Brill, 2010), pp. 81–100.

Faist, B., 'Kingship and Institutional Development in the Middle Assyrian Period,' in G.B. Lafranchi and R. Rollinger (eds.), *Concepts of Kingship in Antiquity: Proceedings of the European Science Foundation Exploratory Workshop* (Padova: S.A.R.G.O.N., 2010), pp. 15–24.

Fales, F.M. 'Grain Reserves, Daily Rations, and the Size of the Assyrian Army: A Quantitative Study,' *State Archives of Assyria Bulletin*, vol. 4, no. 1 (1990), pp. 23–34.

Fales, F.M., 'The Assyrian Word for "(Foot) Soldier,"' in G. Galil, M. Geller and A. Millard (eds.), *Homeland and Exile: Biblical and Ancient Near Eastern Studies in Honour of Bustenay Oded* (Leiden: Brill, 2009), pp. 71–94.

Fales, F.M., '"Ḫanigalbat" in Early Neo-Assyrian Royal Inscriptions: A Retrospective View,' in G. Galil, A. Gilboa, A.M. Maeir and D. Kahn (eds.), *The Ancient Near East in the 12th–10th Centuries BCE: Culture and History: Proceedings of the International Conference Held at the University of Haifa, 2–5 May 2010* (Münster: Ugarit-Verlag, 2012), pp. 99–119.

Fales, F.M. and Rigo, M., 'Everyday Life and Food Practices in Assyrian Military Encampments,' in L. Milano (ed.), *Paleonutrition and Food Practices in the Ancient Near East: Toward a Multidisciplinary Approach* (Padova: S.A.R.G.O.N., 2014), pp. 413–36.

Feldman, M.H. and Sauvage, C., 'Objects of Prestige? Chariots in the Late Bronze Age Eastern Mediterranean and Near East,' *Ägypten und Levante/Egypt and the Levant*, vol. 20 (2010), pp. 67–181.

Foster, B.R., *Before the Muses: An Anthology of Akkadian Literature* (Bethesda: CDL Press, 2005).

Frame, G. *The Royal Inscriptions of Sargon II, King of Assyria (721–705 BC)*. The Royal Inscriptions of the Neo-Assyrian Period Volume 2 (University Park: Eisenbrauns, 2021).

Fuchs, A. and Parpola, S., *The Correspondence of Sargon II, Part III*. State Archives of Assyria vol. 15 (Helsinki: Helsinki University Press, 2001).

Gaspa, S. 'The Assyrian King as Warrior: Legitimacy through War as a Religious and Political Issue from Middle Assyrian to Neo-Assyrian Times,' in K. Ruffing, K. Dross-Krüpe, S. Fink and R. Rollinger (eds.), *Proceedings of the tenth Symposium of the Melammu Project Held in Kassel September 26–28 2016 & Proceedings of the Eighth Symposium of the Melammu Project Held in Kiel November 11–15 2014* (Wien: Verlag der Österreichischen Akademie der Wissenschaften, 2020), pp. 113–56.

Genz, H., 'The Introduction of the Light, Horse-Drawn Chariot and the Role of Archery in the Near East at the Transition from the Middle to the Late Bronze Ages: Is there a Connection?' in A.J. Velmdmeijer and S. Ikram (eds.), *Chasing Chariots: Proceedings of the First International Chariot Conference (Cairo 2012)* (Dingwall: Sandstone Press, 2013), pp. 95–106.

Glassner, J.-J., Foster, B.R. (trans.), *Mesopotamian Chronicles* (Leiden: Brill, 2004).

Grayson, A.K., *Assyrian Rulers of the Early First Millennium BC I (1114–859 BC)*. Royal Inscriptions of Mesopotamia Assyrian Period vol. 2 (Toronto: University of Toronto Press, 1991).

Grayson, A.K. and Novotny, J., *The Royal Inscriptions of Sennacherib, King of Assyria (704–681 BC), Parts 1 and 2*. Royal Inscriptions of the Neo-Assyrian Period vol. 3 (Winona Lake, IN: Eisenbrauns, 2012–14).

Holloway, S.W., *Aššur Is King! Aššur Is King! Religion in the Exercise of Power in the Neo-Assyrian Empire* (Leiden: Brill, 2002).

The Holy Bible: New Revised Standard Version (New York: Oxford University Press, 1989).

Jakob, S., *Mittelassyrische Verwaltung und Sozialstruktur* [Middle Assyrian administration and social structure] (Leiden-Boston: Brill, 2003).

Jakob, S., 'Middle Assyrian Period (14th to 11th Century BCE),' in E. Frahm (ed.), *A Companion to Assyria* (Chichester: Wiley & Sons, 2016), pp. 117–42.

Jean, C., 'Magie et Histoire: les rituels en temps de guerre' [Magic and history: rituals in the times of war], in L. Feliu, J. Llop, A. Millet Alba and J. Sanmartín (eds.), *Time and History in the Ancient Near East: Proceedings of the 56th Rencontre Assyriologique Internationale, Barcelona 26–30 July 2010* (Winona Lake, IN: Eisenbrauns, 2013), pp. 107–12.

Kertai, D., *The Architecture of Late Assyrian Royal Palaces* (Oxford: Oxford University Press, 2015).

Llop, J., 'The Creation of the Middle Assyrian Provinces,' *Journal of the American Oriental Society*, vol. 131, no. 4 (2011), pp. 591–603.

Llop, J., 'Middle Assyrian Letters: A New Survey,' *Aula Orientalis*, vol. 30, no. 2 (2012), pp. 289–306.

Llop, J., 'The Weaponry of the Middle Assyrian Army According to the Written Sources,' *Histoire militaire ancienne. Revue Internationale d'Histoire Militaire Ancienne*, no. 3 (2016), pp. 199–222.

Llop-Raduà, J., 'The Development of the Middle Assyrian Provinces,' *Altorientalische Forschungen*, vol. 39, no. 1 (2012), pp. 87–111.

Luukko, M., 'Gurraeans and Itu'aeans in the Service of the Assyrian Empire,' in J. Dušek and J. Mynářová (eds.), *Aramaean Borders: Defining Aramaean Territories in the 10th–8th Centuries B.C.E* (Leiden: Brill, 2019), pp. 92–124.

Marriott, J. and Radner, K., 'Sustaining the Assyrian Army among Friends and Enemies in 714 BCE,' *Journal of Cuneiform Studies*, vol. 67 (2015), pp. 127–43.

Maxwell-Hyslop. K.R., 'Curved Sickle-Swords and Scimitars,' in L. al-Gailani Werr, J. Curtis, H. Martin, A. McMahon, J. Oates and J. Reade (eds.), *Of Pots and Plans: Papers on the Archaeology and History of Mesopotamia and Syria Presented to David Oates in Honour of His 75th Birthday* (London: Nabu Publications, 2002), pp. 210–17.

Melville, S.C., 'Insurgency and Counterinsurgency in the Assyrian Empire during the 8th Century BC,' in Timothy Howe and Lee L. Brice (eds.), *Brill's Companion to Insurgency and Terrorism in the Ancient Mediterranean* (Leiden: Brill, 2016a), pp. 62–92.

Melville, S.C., 'The Role of Rituals in Warfare during the Neo-Assyrian Period,' *Religion Compass*, vol. 10, no. 9 (2016b), pp. 219–29.

Nadali, D. 'Images of Assyrian Sieges: What They Show, What We Know, What Can We Say,' in J. Armstrong and M. Trundle (eds.), *Brill's Companion to Siege Warfare in the Ancient Mediterranean* (Leiden: Brill, 2019), pp. 53–68.

Oates, J. and Oates, D., *Nimrud: An Assyrian Imperial City Revealed* (London: British School of Archaeology in Iraq, 2001).

Oded, B., *Mass Deportations and Deportees in the Neo-Assyrian Empire* (Wiesbaden: Dr. Ludwig Reichert Verlag, 1979).

Parpola, S., *The Correspondence of Sargon II*, Part I. State Archives of Assyria vol. 1 (Helsinki: Helsinki University Press, 1987).

Pedde, F., 'The Assyrian Heartland,' in D.T. Potts (ed.), *A Companion to the Archaeology of the Ancient Near East* (Chichester: Wiley and Sons, 2012), pp. 851–66.

Ponchia, S., *Management of Food Resources in the Neo-Assyrian Empire: Data and Problems* (Padova: S.A.R.G.O.N, 2014).

Pongratz-Leisten, B., *Religion and Ideology in Assyria* (Berlin: De Grutyer, 2015).

Pongratz-Leisten, B., Bleibtreu, E. and Deller, K., 'Götterstreitwagen und Götterstandarten: Götter auf dem Feldzug und ihr Kult im Feldlager, *Bagdhader Mitteilungen*,' 23 (1992), pp. 291–356.

Postgate, J.N., 'The Organization of the Middle Assyrian Army: Some Fresh Evidence,' in P. Abrahami and L. Battini (eds.), *Les armées du Proche-Orient ancient (IIIe- Ier mill. Av. J.-C)* (Oxford: BAR, 2008), pp. 83–92.

Radner, K., 'The Assur-Nineveh-Arbela Triangle: Central Assyria in the Neo-Assyrian Period,' in P. Miglus and S. Mühl (eds.), *Between the Cultures: The Central Tigris Region in Mesopotamia from the 3rd to the 1st Millennium BC* (Heidelberg: Heidelberger Orient-Verlag, 2011), pp. 321–9.

Radner, K., 'High Visibility Punishment and Deterrent: Impalement in Assyrian Warfare and Legal Practice,' Zeitschrift für Altorientalische und Biblische Rechtsgeschichte/Journal for Ancient Near Eastern and Biblical Law, vol. 21 (2015), pp. 103–28.

Radner, K., 'Revolts in the Assyrian Empire: Succession Wars, Rebellions Against a False King and Independence Movements,' in J.J. Collins and J.G. Manning (eds.), Revolt and Resistance in the Ancient Classical World and the Near East: in the Crucible of Empire (Leiden: Brill, 2016), pp. 41–54.

Richardson, S.F.C., 'Death and Dismemberment in Mesopotamia: Discorporation between the Body and Body Politic,' in N. Laneri (ed.), *Performing Death: Social Analyses of Funerary Traditions in the Ancient Near East and Mediterranean* (Chicago: Oriental Institute of the University of Chicago, 2007), pp. 189–208.

Rivaroli, M., 'The Ritualization of War: The Phases of *Bellum* and Their Sacred Implications,' in C. Ambos and L. Verderame (eds.), *Approaching Rituals in Ancient Cultures. Questioni di Rito: Rituali come Fonte di Conoscenza delle Religioni e delle Concenzioni del Mondo nelle Culture Antiche* (Pisa: Fabrizio Serra, 2013), pp. 261–86.

Schachner, A., *Bilder eines Weltreichs. Kunst- und Kulturgeschichtliche Untersuchungen zu den Verzierungen eines Tores aus Balawat (Imgur-Enlil) aus der Zeit von Salmanassar III, König von Assyrien* (Turnhout: Brepols, 2007).

Scurlock, J., 'Neo-Assyrian Battle Tactics,' in G. Young, M. Chavalas, and R. Averbeck, K. Danti (eds.), *Crossing Boundaries and Linking Horizons: Studies in Honor of Michael C. Astour on His 80th Birthday* (Bethesda: CDL Press, 1997), pp. 491–515.

Scurlock, J., '*Kallāpu*: A New Proposal for a Neo-Assyrian Military Term,' in H. Neumann, R. Dittmann, S. Paulus, G. Neumann, and A Schuster-Brandis (eds.), *Krieg und Frieden im Alten Vorderasien 52e Rencontre Assyriologique Internationale International Congress of Assyriology and Near Eastern Archaeology, Münster, 17.–21. Juli 2006* (Münster: Ugarit-Verlag, 2014), pp. 725–34.

Siddall, L.R. 'The Nature of Siege Warfare in the Neo-Assyrian Period,' in J. Armstrong and M. Trundle (eds.), *Brill's Companion to Siege Warfare in the Ancient Mediterranean* (Leiden: Brill, 2019), pp. 35–52.

Ulanowski, K. *Neo-Assyrian and Greek Divination in War* (Leiden: Brill, 2021).

5

THE *POLIS* AND THE EMERGENCE OF *HOPLITE* WARFARE

Matthew Trundle

The Nature of the Question

Aristotle (*Politics* 1297b: 21–6; see also 1289b: 36–50) considered that a state's military organisation affected its political structure. Thus, he thought that in parts of Greece, where cavalry dominated, tight oligarchies ruled exclusively, as only wealthy aristocrats could own horses. Where heavy infantry dominated, states embraced new forms of looser oligarchic governments, open as they were to those who fought with heavy arms, which were cheaper than horses, but still relatively exclusive. In another passage, Aristotle (*Politics* 1321a: 5–14) notes how states with large numbers of lightly armed troops, and especially powerful navies, employed more democratic forms of government, relying on the participation of the poorer men who rowed in the fleet and voted in the assembly. These perceived links between war and politics have played a significant role in modern thinking about Greek historical development, especially in regard to the creation and survival of the *polis*—sometimes translated as city-state—as a political community.

For modern historians, the link between the rise of the *polis*, with its broad citizen base rather than exclusive aristocratic control, and the rise of the *hoplite* (heavy infantrymen), and *hoplite* warfare in particular, has become one of the most vexed and well-discussed problems of ancient Greek history.[1] Nothing can be taken for granted regarding the timing and nature of the emergence of both phenomena, let alone the establishment of a link between the two. What follows explores the problems concerning questions of their connection and discusses both sides of what is an ongoing debate. This chapter hopes to offer some conclusions concerning the emergence and predominance of the *hoplite* and *hoplite* ideology and any correlation with citizenship.

The reasons for the emergence of both the Greek *hoplite phalanx* and its relationship to the Greek *polis* remain separate problems in their own right. The connection (if any) between these two developments adds to the conundrum of their development. Indeed, the Archaic Period saw numerous developments that all appear, if we choose to associate them, interconnected and interrelated. Thus, the rise of the *polis* theoretically saw a broader base of citizens holding and farming their lands, attending political assemblies as well as serving in *hoplite* armies. The citizen appears to combine simultaneously the attributes of the farmer

DOI: 10.4324/9780429437915-8

(landholder), politician (at least as a member of a citizen assembly) and soldier. The *polis* made such associations possible. Equally viable, however, is that the appearance of these new farmer, citizen-soldiers made the *polis* possible in turn. But there was a host of other phenomena contemporary with this debate that need consideration and that may themselves have created the *polis* and its unique society independent of farming and fighting.

The period of the rise of the *polis*, therefore, saw Greeks create colonies (new *poleis*) across the Mediterranean, facilitating trade that in turn created greater wealth, increased the use of slaves (or in the case of Sparta and Thessaly Helots and Penestae often associated with 'serfdom') and greater use of metals that in turn facilitated citizens to be both leisured politicians in the case of slaves and heavily armed soldiers in the case of metals. In a cycle of intertwined phenomena, each new development on the face of it supported, and was supported by, the other new developments in the Greek states. The problem remains that the *polis* itself may have developed independent of any social, political or military conditions and similarly other changes might equally have emerged themselves independent of one or all of the noticeable developments of the Archaic Age. In short, then, questions remain about the relationship between the birth and consolidation of the *polis* with the emergence of a new type of soldier, the *hoplite*, and a new style of fighting, the *hoplite*-based *phalanx*.

What is a *Polis?*

To understand the problem of the relationship between *polis* and *hoplite* properly, we need to define the meaning of these terms. The *polis* that emerged in the Archaic Age differed from the cities and communities found in Homer and the so-called Dark Age in numerous ways. Homer's communities were run exclusively by chieftains (sometimes styled as heroes) and their families who oversaw a number of households and justified their status through birth-right and military leadership. The political community did not exist in a legally defined way. As the *polis* emerged, the old aristocracy of chieftains lost power to a broader group of landholders as the legal, political and economic Greek State came into being. This new state became the *polis* of the Archaic Age. Essentially, the *polis* was a self-governing, independent community based on a specific territory of a city with a defined legally based communal structure.[2] The term *polis* itself, however, is not entirely useful nor exclusive to what we might style as *polis* communities. *Polis* appears regularly in Homer's poems in what are pre-*polis* contexts. In Homer, therefore, the term *polis* seems to refer to a central place and one perhaps separate from the surrounding countryside. His *polis* conjures up images of a fortified site, a citadel or even a palace. The well-known depiction of an early community on the shield of Achilles shows two Homeric *poleis* with surrounding agricultural land, growing grains, grapes and pasture, while its people harvest and joined in sacrifice and ritual together. There is an assembly and judges hear lawsuits. This appears to represent a very early form of a *polis* state, despite its Homeric context. Victor Davis Hanson draws our attention to the Homeric *polis* early in his work *The Other Greeks* as he highlights Laertes' farm (*agros*), which is far from the *polis* (Homer, *Odyssey* 24: 205–12; Hanson 1997: 47).

The city and the farm in this instance are exclusive unrelated domains. Laertes appears as an outcast from the *polis* itself. The *polis* in this instance represented the centre politically where Odysseus' palace lay, but a territory separate from the surrounding fields, farms and pastureland. In the Late Archaic and Classical *polis* this exclusivity no longer existed. The farms and farm owners were an integral part of the *polis* and the farm-holders themselves were theoretically equal members of the community of the *polis*—that is, they were citizens—full-members of the *polis* called *politai*.[3]

The heart of the *polis* was simply a part of the whole, indivisible from the surrounding territory (*chora*). Thus, in Athens, the acropolis (literally 'high-city') served as a central physical and symbolic space within the *polis* as a whole, which provided a spiritual (via temples, altars and spaces for worship), military (a place of refuge) and a political (assembly places like *pnyx*) focal point for the state. The acropolis-citadel represents the most well-known example of the kind of Homeric 'city' away from, but dominating, the countryside around. The *polis* of course was more than the physical location and sites of city and country. Its essence lay in the community—the *politeia*—a word that came to embody the laws or constitution of the state and indeed the state itself. Hence its significance to those who participated in the political, economic, social and military life of the community.

What is a *Hoplite*?

The word *hoplitês* signifies a Greek heavy infantryman. There was nothing unique about heavy infantry to the Archaic Greeks. Men armed with shields, spears and swords, and often protected with helmets and armour, fighting in mass bands had dominated the battlefields of the ancient world since the first nations emerged in the river valleys of the Euphrates and Tigris in the latter centuries of the fourth millennium BCE.[4] In a recent article, Matthew Lloyd identified a common Mediterranean style of warfare that transcended regional variations in the period just prior to the emergence of the Greek *hoplite* (Lloyd 2017: 231–52).

Indeed, even in the Greek world prior to the *polis*, we can identify men armed with spears, swords, large shields, bronze armour, greaves (both metal and leather) and a variety of helmets. In around 700 BCE a new kind of soldier began to appear in the Greek world, heavily armed, but with some differences from their Mycenaean, Dark Age and Homeric forebears. The Greek *hoplite* (*hoplitês*) derived his name from his arms (*hopla*) and armour (see Hans van Wees 2004: 47–51). This new soldier's defining feature was his shield (*aspis*).[5] The word *hoplitês* appears in literature quite late, that is in the fifth century BCE, despite the fact that the *hoplite* shield with its double grip appeared in ca. 700 BCE alongside soldiers who look to us exactly like *hoplitai*. The shield weighed about 7.5 kg, had a heavy rim emphasising its convex shape, and most importantly a double grip that included a central armband (the *porpax*) and a handgrip (*antelabe*) just inside the rim. The *porpax* appears genuinely to be both unique to the Greek world and a new invention that enabled the user to 'push' with his body on the shield against his opponent. Scholars argue over whether the shield could be used in individual one on one combat.[6] These discussions significantly impact whether the shield could be used in individual combats or if it only functioned successfully in a massed band of soldiers armed with similar shields each protecting one another. This group formation was called the *phalanx*. Despite the absence of the term *hoplitês* in Archaic literature, *phalanxes* (and words associated with them) are common in Homer's *Iliad* (and then throughout Greek writing) alongside a number of terms denoting lines of men marshalled in groups on the battlefield (*stiches*, *lochoi* etc.).[7]

Homeric warfare presents soldiers massed together on the battlefield, despite the poet's emphasis on the heroes who led these contingents. Some suggest that Homeric warfare represents only a poetic construct with no bearing on a real-historical Greek moment (for example, Cartledge 2013: 78). Others see Homer's epics as a window into the past, even if they dispute which moment of the past exactly.[8] Thus, both the *Iliad* and the *Odyssey* still provide a frame of reference to understand the mindset of poets and listeners.

Other early poets like Tyrtaeus and Callinus similarly present images of men massing together to fight collectively. Not until the fifth century BCE, however, do we see uniformity in arms and armour in the *phalanxes*. Latacz illustrated the importance of *phalanx*-style warfare to Homeric battle descriptions. Homer's *phalanxes* contain diverse types of soldiers, including archers and slingers, clustered round the great heroes of the epic. These men often hide behind the shields of the great heroes. Tyrtaeus also has well-armed men protecting missile troops and unarmed men (*gymnetes*, literally naked) in the line. But, as Viggiano notes, Tyrtaeus has passages (i.e., Fragment 11) that suggest a more fluid battlefield in which *promachoi* or forward fighters led bands of supporters running forward and retreating, depending on their bravery and the situation (Viggiano 2013: 47–8). Geometric vases of the eighth century BCE show a preponderance of missile troops in their images of early Greek battles, and even those with large shields and armour also carry javelins, suggestive of a more fluid battlefield even if one of massed bands (Snodgrass 1965, 1967, 2013: 85–94).

When did the *Polis* Emerge?

If there is no *communis opinio* regarding the date and nature of the emergence of the *hoplite phalanx*, then there is equally none concerning the appearance of the Greek *polis*. Homer's poems are chronologically amorphous at best, and if they do represent any historical realities then these range from the Bronze Age through to the later Archaic Period. There are glimpses of the political and social architecture of early *poleis* in Homer's poems. When describing an early city engraved into the shield of Achilles in Book 20, the poet has it walled and surrounded by fields with mixed farming of grapes and grains, alongside farm animals (Homer *Iliad* 18.478–610). The people enjoy a communal festival and ritual, both important features of later *polis* life, despite the presence of a king over the city, who owns land worked for him by the people, but not perhaps all the land. In one scene judges dispense laws in a trial. This in itself suggests roots at the very least of the political and legal basis of the *polis* to come, if not yet the true *polis*. We have already noted that Homer also presents images that are distinctly un-*polis* like where the city (*polis*) itself stands aloof from the countryside and its people and their farms (*agros*).

Naturally, the relationship of the *polis* to *hoplite* warfare depends greatly on the timing of the appearance of both, and at the very least on a symbiotic chronology of their respective appearances before we delve more deeply into other socio-political connections. Most scholars look to the eight century BCE for the emergence of the *polis* (as for so much else in Greek culture) as a political, social, economic, religious and military institution. It was this century after all that saw the end of the so-called Dark Age and the birth of the Archaic Age. It saw the colonisation of the Mediterranean by the Greeks founding up to a thousand new cities (*poleis*). That stated, the full-grown *polis* cannot have appeared immediately and out of nothing. Processes that led states towards law codes, politics and genuine community and citizenship must have taken decades to ferment. Similarly, not all *poleis* were identical. Indeed, fifth-century BCE Athens, the city we know most about, may well have been atypical in its incorporation of so many *polis* institutions into the life of every citizen and the city-community so completely. Classical Athens had an all-embracing democratic society that fostered *polis* life to a high degree politically and socially. It was certainly unique in its allowance of citizenship to those who held no land within the community that was part of that democratic ethos. In states like Sparta (and indeed in Greek states everywhere), landholding was a prerequisite of citizenship and only a relatively small elite group held full citizenship.

The evidence prevents us from asserting absolutely that the *polis* proper existed in the eighth century BCE. Greek colonisation created new cities across the Mediterranean. These new communities needed 'instant' constitutions and laws. They created notions of citizenships and rules of landholding. It is hard not to see the roots of the true *polis* in this process. These cities also needed defending from those whose territories they occupied, reinforcing the landholding soldiers of these early communities. Evidence shows that the earliest codes and laws emerged a little later. The first written law codes, according to Greek traditions, appear in the seventh century BCE (i.e., that of Draco in 621 BCE). The first inscribed laws of governance we have (rather than a code per se) also date to the seventh century (that of Dreros ca. 620 BCE). Perhaps it is to this period we should look closely for the first *poleis*. Some suggest that the *polis* does not strictly emerge until later still. The law codes of both Draco and Solon suggest moves towards political, legal and social integrity.

When (and Where) did the *Hoplite* Emerge?

If the roots of the *polis* appear in the eighth century BCE, when do we see the first *hoplites* in Greece? Fernando Echeverria argues, from the etymological evidence available, that the term *hoplite* does not appear until sources dated to the fifth century BCE (Echeverria 2012: 291–318). Such etymological evidence is irrefutable in itself for a later naming of the phenomenon. A Greek heavy armoured infantryman that we would associate with *hoplite* warfare carried a number of key arms and armour.[9] That stated, the component parts of the *hoplite* panoply appeared from around 700 BCE: as we have noted, the rimmed circular shield (*aspis*) around 700 BCE, the bronze Corinthian helmet and bell cuirass at about the same time. Greaves for the legs were common already in Homer, as were the thrusting spear and sword. The *hoplite* then could well have been a part of Greek warfare from the beginning of the seventh century BCE.

Vases depicting *hoplites* in action, fighting as individuals surrounded by lightly armed troops, and those armed with Boeotian shields, appear at this time too. By around 670 BCE, *hoplites* fighting *hoplites* in lines or *phalanxes* also appear clearly and without supporting lighter infantrymen on proto-Corinthian vases. The famous Chigi Vase, perhaps 640 BCE, graphically illustrates lines of bronze armoured *hoplites* at the point at which they came together in battle (Hurwitt 2002). The regional location of such *hoplite* warfare seems to have been the north-eastern Peloponnese around Argos and Corinth. Thus, the iconic *hoplite*'s arms of the shield and helmet took the name 'Argive' shield and 'Corinthian' helmet. From this region, *hoplites* spread into the Arcadian Peloponnese, Sparta (see below) and north to Thebes and Athens.

John Hale has argued that early Greek *hoplites*, armed with their heavy shields and other bronze arms and armour, found service with Assyrian and Babylonian kings in the early Archaic Age (Hale 2013: 176–93). These were high-status warriors able to afford the best equipment and fight as ritualised friends of eastern rulers. Not only did these early *hoplites* fight as individuals, they also, according to Hale, learned a great deal about eastern military systems from their hosts and employers. They brought this military knowledge of heavy infantry fighting back with them to the Greek world. Of some significance, these early protagonists of *hoplite* warfare were neither citizens nor farmers of the communities for which they fought, nor were they protagonists necessarily of massed band warfare. Clearly, such men challenged the connection between citizenship, farmer landholding and the *hoplite* in the early days of *hoplite* warfare.

Polites and *Hoplites*: Connecting the *Hoplite* as Citizen?

Victor Davis Hanson has led a school of thought that connects closely the rise of the *polis* and a new kind of farmer landholder with the emergence of the *hoplite phalanx*.[10] The *polis* was a product of the unification of new farmer landholders into a single political and social entity. In turn, these new farmers developed a new kind of warfare that suited their socio-economic situation and mode of agricultural production. They grew grapes, olives and grains, which required attention throughout the year. They could afford the arms and armour required for heavy infantrymen. This armour protected them and fighting in the *phalanx* as a unit required morale and bravery but little skill as each man became a part of a whole protective unit standing together but averse to trained manoeuvre and tactics. They did not have much time to devote to warfare, but still needed to defend their land and the new emerging *polis* communities.

According to this thesis, *hoplite* warfare provided a way of devoting little time but maximum energy to fighting short, sharp wars in the form of decisive and one-off pitched battles against neighbouring states that settled disputes over marginal land. The *hoplite* battle consumed only a few days of time, and required little training. The fight itself was over very quickly. Once one line had collapsed, those who survived the initial breakthrough fled. The compromised line was exposed to the victorious enemy. The fleeing *hoplites* threw down their shields, allowing them to escape easily from their well-armed opponents. Losses on both sides were limited. Such battles symbolically provided a demonstration of victory without being devastating and enabled these amateur soldiers to return to their farms without too much damage to either side. The absence of many light troops and cavalry prevented pursuit and large-scale carnage after one of the two *phalanxes* collapsed. This has led some to see these encounters as ritualised, almost conspiratorially designed to uphold a socio-economic and political system dominated by a farming landholding elite that transcended the specific interests of the *poleis* themselves, to the exclusion of professional specialists and lightly armoured poorer infantrymen.[11]

A number of scholars have questioned this model of early *hoplite* warfare's symbiosis with the *polis*, new types of farmers and related phenomena. As we have seen, initial questions concern chronology. While there is no doubt that many aspects of the *hoplite*'s equipment appeared early enough to influence or reflect the first *poleis* in the north-eastern Peloponnese, the work of Hans van Wees and Peter Krentz in particular has shown that the stand-alone *hoplite phalanx* probably did not appear until the early fifth century BCE (Krentz 1982: 23–39; van Wees 2004). Hans van Wees even devoted a chapter in the important edited volume *Men of Bronze* to challenging Hanson's thesis regarding farming and fighting (Wees 2013: 222–55). Echeverria's important discussion supports this view, showing the very late appearance of the term *hoplite* into the Greek lexica (Echeverria 2012: 291–318). John Hale's ideas can be added to this group of dissenters with his arguments concerning the individual amateur aristocratic Greek heavy infantryman as a feature of the seventh-century BCE eastern kingdoms (Hale 2013: 176–93). He styles these men as neither patriots nor farmers—and we perhaps might add certainly not citizens of *poleis*. Archaeology has done little to help Hanson's cause in addition. The idea of farms spread out across the political landscape of early *poleis* does not seem to reflect the archaic landscape of these early states (Foxhall 2013: 194–221).

Political developments within the emerging Greek *poleis* have also been seen as potentially born from the new emerging *hoplite* citizen soldier. Scholars thus saw the decline of the old Homeric-style aristocracy linked to the rise of the *hoplite* soldier. No better was this seen than in the challenge posed in many early Greek states by the emergence of the strong men styled as tyrants in cities associated with early *hoplite* warfare like Argos and Corinth in particular. The first strong man of Argos, Pheidon, usually dated to the 670s–660s BCE, led a new kind of army across the Peloponnese, defeated the Spartans at Hysiae (ca. 669 BCE) and reorganised the Olympic Games. Pheidon remains a shadowy figure. He may have been a war leader and a tyrant, and he may have led an early form of *hoplite* army. The evidence for him makes it very hard to say anything much tangible about his career. The same issues with evidence can be said of the first tyrants of Corinth. The first of these men, Cypselus, perhaps led a popular revolution against the traditional aristocratic families of the state and established a three-generation tyranny over the city. Later tyrannies at Sicyon, Naxos and Athens may well have followed similar models of revolution. John Salmon thought that the new *hoplite* class of farmer citizen soldiers aided the rise of these tyrannies at least latently and explained the war leader qualities of these early tyrants (Salmon 1977: 84–101). But as we have seen above, chronology does little to help the connection between a sizable group of like-minded new citizen *hoplites* with the new *polis* communities and the rise of early tyrannies, especially in places like Argos and Corinth.

Sparta as the *Hoplite* State

Sparta has often been regarded as the paradigm of *hoplite* states. The full Spartan citizen, the *Spartiatês*, had to fulfil several interconnected criteria of landholding and *hoplite* status in order to attend the assembly. Each *Spartiate* progressed in his lifetime through a series of tests that began at birth. This process was called the *agôgê*. Failure at any stage resulted in social death and the removal of status, land and privilege. Membership in the *hoplite* army was critical to a Spartan's status and the '*hoplite*' ethos resonated deeply throughout Spartan society.[12]

Spartans collectively styled themselves as *homoioi* or equals (from which earlier English commentators sometimes referred to the *Spartiatai* as 'Peers'). This alikeness reflected their supposed political or *hoplite* status. *Hoplites* were armed in very similar fashion to each other. The *hoplite* shield was a paradigm of this status. Spartan traditions reflected the importance of the shield. Spartan mothers told their sons to return with or lying on their shields. As we have noted, the first thing thrown away in combat when the *phalanx* line broke was the shield. Discarding one's shield in the battle-line was a crime in Sparta and resulted in expulsion from the Spartiate-status group. Such men were styled as *tressantes* (or tremblers/cowards) and suffered innumerable social slurs. For this reason, those who fell at Thermopylae against overwhelming odds were venerated as heroes. They died obeying Sparta's laws to stand and die with their shields intact, rather than flee from the enemy.[13]

Sparta itself challenges many of Victor Davis Hanson's ideas about the relationship between *polis* and *hoplite*, despite the close relationship between *hoplite* status, ideology and Spartiate status.[14] Thus, Sparta remained highly exclusive and elitist and the state never had a tyrant, despite its associations with *hoplite* warfare. Spartan society remained intrinsically aristocratic despite ideals of supposed equality, community and identity. *Spartiatae* were landholders and owners by their status, but were not themselves farmers of the land. The connection between Spartan fighting and farming remained tenuous, despite the claims of such connections promulgated by, for example, Hanson for a link between farming and fighting. Additionally, even at Sparta, however, plenty of non-elite group *Spartiatae* fought in the *hoplite phalanx*. These men were excluded from full political status within the state. They, among other things, challenge the idea that *hoplite* status and full citizenship were intrinsically connected within emerging *poleis*.

Indeed, it seems probable that over time the numbers of full Spartan citizens declined and from the mid-fifth century BCE onwards declined dramatically. If Plutarch's figures are to be believed, there were originally 9,000 *Spartiatae* when Lycurgus (whoever he was and whenever he lived) drew up the original constitution. At the battle of Plataea, 5,000 *Spartiatae* fought of a total of 10,000 men all styled as Lacedaemonian *hoplites* accompanied by 35,000 helots (acting as retainers, baggage carriers and perhaps foragers). Herodotus notes that the total *Spartiate* population in 479 BCE was 8,000 men (Herodotus 7.234.2). These numbers had fallen to only a few thousand in the Peloponnesian War period and a few hundred by the middle of the third century BCE after the defeats at Leuctra (371 BCE) and Mantinea (362 BCE). The important take away from this is that in the Classical Period at the very least *hoplite* status and citizenship in Sparta had no direct relationship. Indeed, the exclusive place of citizenship and of landholding, both of which became more exclusive as time progressed, made it impractical from a military perspective to enforce such a relationship by the later fifth and fourth centuries BCE. The necessity of maintaining decent numbers of *hoplites* required the Spartans to utilise as many non-*Spartiatae* as possible in *hoplite* armies. This saw the recruitment of so-called *Neodamodeis* ('new men') probably from amongst trusted and elevated helots with limited political status, in the later fifth century BCE alongside the increasing use of *Perioeci* (those from non-*Spartiate* villages within Spartan territory) and allied and mercenary *hoplites* in the fourth century.

Hoplites at Athens—Draco, Solon, Peisistratus and Cleisthenes

Finally, turning to Athens we can see in various ways all the challenges concerning *hoplite* and *polis* in Athenian political development. Athenian constitutional reforms from the overthrow of traditional aristocratic rule of the families styled as Eupatridae (so-called 'Noble Born') perhaps with the emergence of the first annually appointed archons (ca. 683–682 BCE) to the earliest written constitutions of Draco (621 BCE) and Solon (594 BCE), the tyranny of the Peisistratidae (546–510 BCE) and the more democratic reforms of Cleisthenes (507–506 BCE) each demonstrate aspects of the influence of broader social and economic interest groups within the Athenian state. The emergence of three annually appointed archons from 683–682 BCE represented an important stage in Athenian political development away from the lifelong hereditary chieftainship more associable with Homer's world. Among these archons was the polemarch, or war archon, who may also reflect a new kind of 'national' army for want of a better term in what we might style as the nascent Athenian state.[15]

In the century after the establishment of the annually selected archonships, Solon in particular altered the power structure in Athens to favour wealth rather than birth, and by delineating Athenian society into four wealth classes turned an aristocracy into a plutocracy. There were military ramifications for Solon's new political apparatus. Military leadership came from amongst the most wealthy, those producing 500 bushels of wet or dry measures per year, followed by the *Hippeis* (Cavalrymen, sometimes translated as knights) who produced 300 bushels and the Zeugitae (Yoke-men able to afford perhaps a yoke of oxen to plough their land) producing over 200 bushels annually. Some have seen this group as representing the *hoplite*-class, able to hold enough land to farm well and afford a panoply of arms and armour, and have even gone so far as to suggest that such men were yoked into military service as the meaning of their title (Rosivach 2002: 33–43). The group below these were the thetes, able to hold citizenship with fewer than 200 bushels of produce. Importantly, Solon made citizenship independent from landholding. Citizenship at Athens was status based and not solely or even economic despite the economic markers of his status groups. Holding property was not a prerequisite of citizenship. The poor thetes in the fifth century BCE would make up many of those who rowed in the Athenian Fleet in the following century. This latter point reflects another aspect of the connection between military service and Greek politics as some have seen in this the basis of Athenian radical democracy (Hale 2009).

The debate between *hoplite* and *polis* remains vexed and likely to be too simplistic in and of itself to explain changes in Athenian political life connected to the make-up of oarsmen in the fleet and Athenian imperial interests. Despite some support for this position from our largely aristocratic written ancient sources, many modern scholars no longer agree that the radical democracy and the fleet's oarsmen have a direct connection. They also no longer see a direct connection between *hoplite* and *polis*.[16]

Like the earlier tyrannies of other states, that of Peisistratus from 546 BCE onwards might well have enjoyed the latent support of a new class of farming *hoplites*. His final years of tyranny, however, began with their defeat in a land battle and Pallene, supported as he was by hired Thracians and allied troops sent from Naxos. If there was a unified class of *hoplite* farmers, then they do not appear to have been politically active in this era. They do seem, however, to have played more of a role in the events after the fall of the Athenian tyrant of Peisistratus' son Hippias in 510 BCE. Hippias was driven from Athens by the Spartans, but the Spartans and their aristocratic Athenian friends were themselves ousted by a popular rising of Athenians, and perhaps this rising involved Athenian *hoplites* who now realised their political potential within the state. These same *hoplites* then repelled invasions from Thebes and Chalcis to secure the position of another aristocratic Athenian named Cleisthenes. Cleisthenes then enacted sweeping constitutional reforms that laid the foundations for the Athenian democracy. These reforms were both political and military in nature. Like Peisistratus he maintained the socio-economic architecture of Solon's system, but he made the *Deme* (village) the central platform for citizenship and military service dividing Athenians into ten *phylae* (tribes) spread throughout Attica. These tribes became the basis of political and military service. Athenians voted within the tribes at state level and served together in the army in tribal units. By 493–492 BCE, each tribe had an annually elected tribal general as its commander who worked with the traditional annually selected archon who had once commanded the whole Athenian Army. This new model army won the decisive victory at Marathon in 490 BCE over a Persian invasion army perhaps three times its size to enshrine its position in the state.

Conclusion

It remains a matter of perspective whether we can find a tangible connection between the emergence of the *polis* and the rise of the *hoplite*. We need to arbitrate between optimists like Hanson and Viggiano who do see a clear connection and pessimists like van Wees and Krentz who remain less convinced of one. There is little doubt that there was a very strong ideology that promoted the *hoplite* as the premier type of soldier within Greek *poleis* above even aristocratic cavalry, and certainly above lightly armoured missile troops and naval personnel.

This was even the case in fifth-century BCE Athens, despite its all-powerful and impressive navy. But this was only ideology. Light troops were essential to Greek warfare in the Classical Period (if not the Later Archaic Age) and no modern scholar would question the importance of the fleet to the power and security of the Athenian *polis* in this period. Even in Hanson's conception of military history, he styles *hoplites* as dinosaurs by the later fifth and fourth century BCE, as they were now joined on the battlefields of Greece by other troop types fighting long-distance and drawn-out campaigns. To him their era of centrality and dominance lay in the past. Spartan *hoplites*, by the later fifth century BCE, came from a variety of social backgrounds and increasingly non-Spartan communities. That stated, however, it remains hard for us to know the extent to which war had changed from the latter decades of the sixth century BCE down to the Classical Period. As we have seen in this discussion, the early evidence of Homer and the poets is far from conclusive concerning anything to do with Greek warfare and Greek society.

Notes

1 Many works discuss the link between the *polis* and the *hoplite* as well as the broader social, political and economic changes that occurred in the Archaic Period. Generally, see Lorimer 1947 and 1950; Snodgrass 1964, 1980 and 1993; Cartledge 1977; Salmon 1977; Latacz 1977; Ridley 1979; Murray 1980: 124–36; Hanson 1983, 1991 and 1997; Bowden 1993; Schwartz 2009. Several articles in the edited volume of Kagan and Viggiano 2013 cover this subject thoroughly.
2 Definitions abound, but for example see Murray 1980: 62 and note 4.
3 Most recently, for discussion of the definition of the *polis*, see Perris 2017. For a good definition of the *polis* itself see Hansen 2006. And for discussion, see Blok 2005: 7–40, especially 7. On the *polis* as a society, see Ober 1993 and most recently see Blok 2017.
4 As noted by Snodgrass 1965: 113. Generally, for heavy infantry shields in non-Greek contexts see Borchhardt 1977: 36–9; more recent is Raaflaub 2013: 99. For heavy infantry in Assyrian armies, see Dezső 2012: 53–81, 99–114.
5 For the etymology of the term *aspis* (shield) and *hopla* (arms) see Lazenby 1996.
6 For those who argue for the shield as only useful in a *phalanx* see Hanson 1995 and Schwartz 2009, Viggiano 2013, especially 116, and Hanson 2013. For those who see more independence of those wielding the shield, see the work of van Wees generally, e.g., 1986, 1988, 1992, 1994, 1996, 1997, 2000, 2001, 2004, 2008, 2013; Krentz 1985 and Rawlings 2000.
7 For massed band warfare in the *Iliad*, see Latacz 1977; for Iliadic *hoplites*, see Bowden 1993.
8 Raaflaub 1998, 2008; van Wees 2000.
9 Cartledge 1977; Snodgrass 1964, 1967; Viggiano and van Wees 2013.
10 Hanson 1983, 1989, 1995, 2013.
11 For support of the Hanson thesis, see Schwartz 2009; Viggiano 2013.
12 On early Spartan developments see Hammond 1950; Chrimes 1949: 397–428; Murray 1980: 159–80; Trundle 2000; Forrest 1980: 58; Jones 1968: 31–3; Hodkinson 1997: 87.
13 On Thermopylae, see Cartledge 2006; Trundle 2013, 2017.
14 The best work on Sparta's socio-economic situation remains Hodkinson 2000.
15 For Athenian military and political developments, see Pritchard 2010: 7–27. For Athenian military ideology, see Loraux 1986.
16 For support of the non-Athenian presence amongst oarsmen including slaves, see Graham 1992, 1998; Jordan 2000; Krentz 2007, especially 150; van Wees 2004: 210–12, 2008, especially 295–8.

Bibliography

Blok, J., 'Becoming Citizens: Some Notes on the Semantics of "Citizen"' in Archaic Greece and Classical Athens,' *Klio*, vol. 87, no. 1 (2005), pp. 7–40.
Blok, J., *Citizenship in Classical Athens* (Cambridge: Cambridge University Press, 2017).

Borchhardt, H., 'Fruhe griechische shildformen,' in H.-G. Bucholz and J. Wiesner (eds.), *Kiregswesen I. Archaeologia Homerica, pt. E* (Gottingen: Vandenhoeck & Ruprecht, 1977), pp. 1–56.

Bowden, H., 'Hoplites and Homer,' in J. Rich and G. Shipley (eds.), *War and Society in the Greek World* (London: Routledge, 1993), pp. 45–63.

Cartledge, P., 'Hoplites and Heroes,' *Journal of Hellenic Studies*, vol. 97 (1977), pp. 11–28.

Cartledge, P., 'The Birth of the Hoplite,' in P. Cartledge (ed.), *Spartan Reflections* (Los Angeles: University of California Press, 2001), pp. 153–66.

Cartledge, P., *Thermopylae: The Battle that Changed the World* (New York: Vintage Books, 2006).

Cartledge, P., '*Hoplitai/Politai*: Refighting Ancient Battles,' in D. Kagan and G.F. Viggiano (eds.), *Men of Bronze: Hoplite Warfare in Ancient Greece* (Princeton NJ: Princeton University Press, 2013), pp. 74–84.

Cawkwell, G.L., 'Orthodoxy and Hoplites,' *Classical Quarterly*, vol. 39 (1989), pp. 375–89.

Chrimes, K.M.T., *Ancient Sparta: A Reexamination of the Evidence* (Manchester: Manchester University Press, 1949).

Dezső, T., *The Assyrian Army I: The Structure of the Assyrian Army: Infantry, 1.* (Budapest: Monographs of the Institute of Ancient Studies, 2012).

Echeverria, Fernando, 'Hoplite and Phalanx in Archaic and Classical Greece: A Reassessment,' *Classical Philology*, vol. 107, no. 4 (2012), pp. 291–318.

Forrest, W.G., *A History of Sparta: 950–192 BC* (London: Duckworth, 1980).

Foxhall, L. 'Can We See the "Hoplite Revolution" on the Ground? Archaeological Landscapes, Material Culture, and Social Status in Early Greece,' in D. Kagan and G.F. Viggiano (eds.), *Men of Bronze: Hoplite Warfare in Ancient Greece* (Princeton NJ: Princeton University Press, 2013), pp. 194–221.

Graham, A.J. 'Abdera and Teos,' *The Journal of Hellenic Studies*, vol. 112 (1992), pp. 44–73.

Graham, A.J., 'Thucydides 7.13.2 and the Crews of Athenian Triremes: An Addendum,' *Transactions of the American Philological Association*, vol. 128 (1998), pp. 89–114.

Greenhalgh, P., 'The Homeric *Therapon* and *Opaon*,' *BICS*, vol. 29 (1982), pp. 81–90.

Hale, J., *Lords of the Sea: The Epic Story of the Athenian Navy and the Birth of Democracy* (London: Viking, 2009).

Hale, J., 'Not Patriots, Not Farmers, Not Amateurs: Greek Soldiers of Fortune and the Origins of Hoplite Warfare,' in D. Kagan and G.F. Viggiano (eds.), *Men of Bronze: Hoplite Warfare in Ancient Greece* (Princeton NJ: Princeton University Press, 2013), pp. 176–93.

Hammond, N.G.L., 'The Lycurgean Reforms at Sparta,' *Journal of Hellenic Studies*, vol. 70 (1950), pp. 42–64.

Hansen, M.H., *Polis: An Introduction to the Ancient Greek City-State* (Oxford: Oxford University Press, 2006).

Hanson, V.D., *Warfare and Agriculture in Classical Greece* (Pisa: Giardini, 1983).

Hanson, V.D., *The Western Way of War: Infantry Battle in Classical Greece* (New York: Oxford University Press, 1989).

Hanson, V.D. (ed.), *Hoplites: The Classical Greek Battle Experience* (London: Routledge, 1991).

Hanson, V.D., *The Other Greeks: The Family Farm and the Roots of Western Civilisation* (New York: Free Press, 1995).

Hanson, V.D., *Fields Without Dreams: Defending the Agrarian Ideal* (New York: Free Press, 1997).

Hanson, V.D., 'The Hoplite Narrative,' in D. Kagan and G.F. Viggiano (eds.), *Men of Bronze: Hoplite Warfare in Ancient Greece* (Princeton NJ: Princeton University Press, 2013), pp. 256–75.

Hodkinson, S., 'The Development of Early Spartan Society,' in L. Mitchell and P.J. Rhodes (eds.), *The Development of the Polis in Archaic Greece* (New York: Routledge, 1997), pp. 83–102.

Hodkinson, S., 'An Agonistic Culture?: Athletic Competition in Archaic and Classical Spartan Society,' in S. Hodkinson and A. Powell (eds.), *Sparta: New Perspectives* (London: Duckworth, 1999), pp. 147–87.

Hodkinson, S., *Property and Wealth in Classical Sparta* (London, Duckworth 2000).

Hodkinson, S., 'Was Classical Sparta a Military Society?,' in S. Hodkinson and A. Powell (eds.), *Sparta & War* (Swansea: Duckworth, 2006), pp. 111–62.

Holladay, A.J. 'Hoplites and Heresies', *Journal of Hellenic Studies*, vol. 102 (1982), pp. 94–103.

Hurwit, J.M., 'Reading the Chigi Vase,' *Hesperia*, vol. 71 (2002), pp. 1–22.

Jones, A.H.M., *Sparta* (Oxford: Blackwell, 1968).

Jordan, B., 'The Crews of Athenian Triremes,' *L'Antiquite Classique*, vol. 69 (2000), pp. 81–101.

Kagan, D. and Viggiano, G.F. (eds.), *Men of Bronze: Hoplite Warfare in Ancient Greece* (Princeton NJ: Princeton University Press, 2013).

Krentz, P., 'Fighting by the Rules: The Invention of the Hoplite Agôn,' *Hesperia*, vol. 71 (1982), pp. 23–39.

Krentz, P., 'The Nature of Hoplite Battle,' *Classical Antiquity*, vol. 4 (1985), pp. 50–61.

Krentz, P. 'War,' in P. Sabin, H. van Wees and M. Whitby (eds.), *The Cambridge History of Greek and Roman Warfare*, vol. 1, *Greece and the Hellenistic World and the Rise of Rome* (Cambridge: Cambridge University Press, 2007), pp. 147–85.

Krentz, P., 'A Cup by the Douris Painter,' in G. Fagan and M. Trundle (eds.), *New Perspectives in Ancient Warfare* (Leiden, Brill, 2010), pp. 183–204.

Latacz, J., *Kampfparänese, Kampfdarstellung und Kampfwirklichkeit* (Munich: Zetemata, 1977).

Lazenby, J., 'The Myth of the Hoplite's Hoplon,' *Classical Quarterly*, vol. 46 (1996), pp. 27–33.

Lendon, J.E., *Soldiers and Ghosts: A History of Battle in Classical Antiquity* (New Haven/London: Yale University Press, 2005).

Lloyd, M., 'Unorthodox Warfare? Variety and Change in Archaic Greek Warfare (ca. 700–ca. 480 BCE),' in B. Hughes and F. Robson (eds.), *Unconventional Warfare from Antiquity to the Present Day* (London: Palgrave Macmillan, 2017), pp. 231–52.

Loraux, N., *The Invention of Athens* (Cambridge: Cambridge University Press, 1986).

Lorimer, H.L., 'The Hoplite Phalanx with Special Reference to the Poems of Archilochus and Tyrtaeus,' *Annual of the British School at Athens*, vol. 42 (1947), pp. 76–138.

Lorimer, H.L, *Homer and the Monuments* (London: Macmillan, 1950).

Morris, I., *Burial and Ancient Society: The Rise of the Greek City State* (Cambridge: Cambridge University Press, 1987).

Murray, O., *Early Greece* (London: Fontana Press, 1980).

Ober, J., 'The "Polis" as a Society: Aristotle, John Rawls and the Athenian Social Contract,' in M.H. Hansen (ed.), *The Ancient Greek City-State* (Copenhagen: Royal Danish Academy of Sciences and Letters, 1993), pp. 129–60.

Perris, S., 'Is There a *Polis* in Euripides' *Medea*?' *Polis, The Journal for Ancient Greek Political Thought*, vol. 34 (2017), pp. 318–35.

Pritchard, D., 'Introduction,' in D. Pritchard (ed.), *War, Culture and Democracy in Classical Athens* (Cambridge: Cambridge University Press, 2010), pp. 1–65.

Pritchett, W.K., *The Greek State at War*, vol. 4 (Los Angeles: University of California Press, 1985).

Pritchett, W.K., 'A Recent Theory on Homeric Warfare,' *Studies in Ancient Greek Topography*, Part VII (Amsterdam, 1991), pp. 181–90.

Raaflaub, K., 'Soldiers, Citizens, and the Evolution of the Early Greek Polis,' in L. Mitchell and P. Rhodes (eds.), *The Development of the Polis in Archaic Greece* (New York: Routledge, 1997), pp. 49–59.

Raaflaub, K., 'A Historian's Headache: How to Read Homeric Society,' in N.R.E. Fisher and H. van Wees (eds.), *Archaic Greece: New Evidence and New Approaches* (Swansea: Duckworth, 1998), pp. 169–93.

Raaflaub, K., 'Homeric Warriors and Battles: Trying to Resolve Old Battles,' *Classical World*, vol. 101, no. 4 (2008), pp. 469–83.

Raaflaub, K., 'Early Greek Infantry Fighting in a Mediterranean Context,' in D. Kagan and G.F. Viggiano (eds.), *Men of Bronze: Hoplite Warfare in Ancient Greece* (Princeton NJ: Princeton University Press, 2013), pp. 95–111.

Rawlings, L., 'Alternative Agonies. Hoplite Martial and Combat Experiences Beyond the Phalanx,' in H. van Wees (ed.), *War and Violence in Ancient Greece* (London: Duckworth, 2000), pp. 233–50.

Ridley, R.T., 'The Hoplite as Citizen,' *Antiquité Classique*, vol. 48 (1979), pp. 508–48.

Rosivach, V., '*Zeugitai* and Hoplites,' *Ancient History Bulletin*, vol. 16 (2002), pp. 33–43.

Runciman, W.G., 'Greek Hoplites, Warrior Culture, and Indirect Bias,' *The Journal of the Royal Anthropological Institute* vol. 4 (1998), pp. 731–51.

Salmon, J., 'Political Hoplites,' *Journal of Hellenic Studies*, vol. 97 (1977), pp. 84–101.

Schwartz, A., *Reinstating the Hoplite: Arms, Armour and Phalanx Fighting in Archaic and Classical Greece* (Stuttgart: Franz Steiner Verlag, 2009).

Singor, H.W., '*Eni protoisi machesthai*: Some Remarks on the Iliadic Image of the Battlefield,' in J.P. Crielaard (ed.), *Homeric Questions* (Amsterdam: Brill, 1995), pp. 183–200.

Snodgrass, A.M., *Early Greek Armour and Weapons from the End of the Bronze Age to 600* BC (Edinburgh: Edinburgh University Press, 1964).

Snodgrass, A.M., 'The Hoplite Reform and History,' *Journal of Hellenic Studies*, vol. 85 (1965), pp. 110–22.

Snodgrass, A.M., *Arms and Armour of the Greeks* (London: Thames and Hudson, 1967).

Snodgrass, A.M., *Archaic Greece: The Age of Experiment* (Berkeley and Los Angeles: University of California Press, 1980).

Snodgrass, A.M., 'The 'Hoplite Reform' Revisited,' *Dialogues d'Histoire Ancienne*, vol. 19 (1993), pp. 47–61.

Snodgrass, A.M., 'Setting the Frame Chronologically,' in D. Kagan and G.F. Viggiano (eds.), *Men of Bronze: Hoplite Warfare in Ancient Greece* (Princeton NJ: Princeton University Press, 2013), pp. 85–94.

Trundle, M., 'The Spartan Revolution,' *War and Society*, vol. 18 (2000), pp. 1–16.

Trundle, M., 'Thermopylae,' in C. Matthew and M. Trundle (eds.), *Beyond the Gates of Fire* (Bradford: Pen and Sword, 2013), pp. 27–38.

Trundle, M., 'Spartan Responses to Defeat: From a Mythical Hysiae to a Very Real Sallassia,' in J.H. Clark and B. Turner (eds.), *Brill's Companion to Military Defeat in Ancient Mediterranean Society* (Leiden: Brill, 2017), pp. 144–61.

Vernant, J.-P., 'City-State Warfare,' in J.-P. Vernant (ed.), *Myth and Society in Ancient Greece* (Brighton: Harvester Press, 1980), pp. 19–44.

Viggiano G.F., 'The Hoplite Revolution and the Rise of the Polis,' in D. Kagan and G.F. Viggiano (eds.), *Men of Bronze: Hoplite Warfare in Ancient Greece* (Princeton NJ: Princeton University Press, 2013), pp. 112–33.

Viggiano, G.F. and Wees, H. van, 'The Arms, Armor and Iconography of Early Greek Hoplite Warfare,' in D. Kagan, and G.F. Viggiano (eds.), *Men of Bronze: Hoplite Warfare in Ancient Greece* (Princeton NJ: Princeton University Press, 2013), pp. 57–73.

Wees, H. van, 'Leaders of Men? Military Organisation in the *Iliad*,' *Classical Quarterly*, vol. 36 (1986), pp. 285–303.

Wees, H. van, 'Kings in Combat,' *Classical Quarterly*, vol. 38 (1988), pp. 1–24.

Wees, H. van., *Status Warriors: War, Violence and Society in Homer and History* (Amsterdam: J. Giebern, 1992).

Wees, H. van, 'The Homeric Way of War and the Hoplite Phalanx,' *Greece and Rome*, vol. 41 (1994), pp. 1–18.

Wees, H. van, 'Heroes, Knights and Nutters', in A. Lloyd (ed.), *Battle in Antiquity* (Swansea: Classical Press of Wales, 1996), pp. 1–86.

Wees, H. van, 'Homeric Warfare,' in I. Morris and B. Powell (eds.), *A New Companion to Homer* (Leiden: Brill, 1997), pp. 668–93.

Wees, H. van, 'The Development of the Hoplite Phalanx,' in H. van Wees (ed.), *War and Violence in Classical Greece* (Swansea: Duckworth, 2000), pp. 125–66.

Wees, H. van, 'The Myth of the Middle-Class Army,' in T. Bekker-Nielsen and L. Hannestad (eds.), *War as a Cultural and Social Force: Essays on Warfare* (Copenhagen: Kongelige Danske Videnskabernes Selskab, 2001), pp. 45–71.

Wees, H. van, *Greek Warfare: Myths and Realities* (London: Duckworth, 2004).

Wees, H. van, 'War and Society,' in P. Sabin, H. van Wees, and M. Whitby (eds.), *The Cambridge History of Greek and Roman Warfare*, vol. 1, *Greece, the Hellenistic World and the Rise of Rome* (Cambridge: Cambridge University Press, 2008), pp. 273–99.

Wees, H. van, 'Farmers and Hoplites: Models of Historical Development,' in D. Kagan and G.F. Viggiano (eds.), *Men of Bronze: Hoplite Warfare in Ancient Greece* (Princeton NJ: Princeton University Press, 2013), pp. 222–55.

6

TRANSITION FROM ROMAN TO MEROVINGIAN WARFARE

The Long Fifth Century

Bernard S. Bachrach

Introduction

A half-century ago I first addressed the subject of the transition from imperial military institutions in Gaul to the organisations utilised during the Merovingian era. At that time, I suggested that the Merovingian military was formed from many different groups, its tactics were flexible, and its strategy was variable. The most important conclusion that can be drawn from this old study, especially in light of the previous scholarship with which it dealt, was that the Merovingian military, like almost everything else in Gaul, was overwhelmingly influenced by the institutions of the Roman Empire in the West and its Latin speaking population. By contrast, comparatively little was owed institutionally or culturally to the Franks as an ethnic group, who were only a small minority of the population as a whole and a small part of the fighting forces. I concluded: 'As with many aspects of Merovingian life, the military organization recalls *Romania* and not *Germania*' (Bachrach 1972: 28).[1]

For several hundred years historians claimed that the so-called barbarian invasions in the fifth century brought an end to Roman civilisation in the West and ushered in the 'Dark Ages' (Lyon 1972).[2] However, during the decades following the Second World War, scholars have established that the so called 'Dark Ages' were a highly misleading myth purposefully developed through the misuse of narrative texts, especially by clerics touting the 'end of days,' to explain the decline and fall of the Roman Empire. The view that trumpeted the victory of barbarism and the ruin of civilisation has been replaced during the past several decades by the 'Late Antique.' According to current scholarship, Later Roman culture, broadly understood, prospered in the West well into the Early Middle Ages.[3] Now, for example, in some quarters, Charlemagne (768–814) has come to be seen as a Late Antique ruler, who not only continued to preserve and further to develop aspects of many late Roman institutions, especially the church, but also revived the imperial office in the West to which the government in Constantinople acquiesced (Brown 1971, 2003).

Gaul, as the subject of this study, is to be equated geographically with the Hexagon bounded on the east by the Rhine, on the south by the Alps, Mediterranean and Pyrenees, on the west by the Atlantic Ocean, and in the north by the English Channel. The population of Gallo-Romans in this region during the fifth century is estimated to have been in the range

DOI: 10.4324/9780429437915-9

of 10 to 12 million Latin speakers, i.e., about half the size of the population of the French Kingdom in the reign of Louis XIV (1643–1715). It is speculated that during this period there were perhaps between 150,000 and 200,000 men, women and children who were speakers of a Germanic language or the near descendants of those who had one or another Germanic language as their mother tongue.[4] Most of these settlers initially spoke a Frankish dialect and by and large over time they assimilated linguistically into the vastly larger Latin-speaking Christian population (Wright 1982; Banniard 1992).

As was the case throughout most of the Later Roman Empire, late antique Gaul was well Christianised with flourishing episcopal administrative centres in its fortress cities and thousands of churches distributed throughout the region as a whole. Also of considerable importance were numerous monastic foundations for both men and women where concerted efforts were undertaken to preserve Latin culture and educate future generations regardless of their initial ethnic origins.[5] Christian institutions flourished and grew in number and in influence as the great wealth once held by pagan religious institutions was transferred to the church. The educational system preserved late antique Latin culture and religious life was based on the use of Latin texts as well as in orally performed rituals. Practical literacy was of essential importance for administrative purposes at all levels of government, central and local, as well as in the church, and also for business purposes (Banniard 1992).

Society, as a whole, was thoroughly numerate as even farmers had to calculate quantities of seed grain available and inventory crops, while shepherds and herders had to keep track of their flocks and herds (Alföldi-Rosenbaum 1971: 1–9). The legal system was based on what modern scholars describe as 'vulgar' Roman law and even the so-called 'barbarian' legal texts, e.g., the Salic Law, were published in Latin. These texts had been thoroughly penetrated by Roman legal principles (Murray 1983). This pattern prevailed even in the Bavarian laws published in the remote southeast region of the Merovingian kingdom (Esders 2016: 1–24). The system of land tenure and property law was fundamentally Roman throughout the Hexagon. Late Roman agricultural organisation saw the introduction of a bi-partite structure for large estates that had been developed in Roman Syria during the fourth century and replaced the traditional *latafundia* system (Sarris 2004: 280–311).

The Military Background

Of fundamental importance, not only to the military infrastructure of Gaul during the fifth century but for the region as a whole throughout the Middle Ages, was the massive system of fortifications constructed by the Roman government largely in the course of the fourth century CE.[6] The strategic decision to fortify the urban centres of Gaul was initiated by the Emperor Probus (276–282) and subsequently adopted by many of his successors. The overall aim of this long-term strategy was to provide protection from piracy and banditry that was rampant during the Later Empire as a result of the exceptionally strong economy that flourished throughout the Hexagon (Bachrach 2010: 38–64, 2017: 3–34). This strategy resulted in a great many previously unwalled or poorly defended *urbes* being converted at great expense and with considerable social disruption into fortress cities (Johnson 1983; Reddé et al: 2006). Readers familiar with Viking operations a half-millennium later can easily understand imperial thinking with regard to protecting population and economic centres.

By the early fifth century there were more than 100 massively fortified urban fortresses distributed strategically throughout Gaul (Bachrach 2010: 38–64, 2017: 3–34). These fortifications dominated the coasts, river valleys, mountain passes and the major roads throughout Gaul (Bachrach 2017: 3–34). In addition, there were many scores of lesser fortifications, *castra* and *castella*, constructed by the government and strategically placed in areas that were regarded as vulnerable to illegal exploitation. This combination of fortress cities and lesser fortifications provided a system that is to be considered as providing a strategic system of defence in depth throughout Gaul. Many of the lesser fortifications, i.e., *castella*, were distributed along the Rhine Frontier and in the Alps (Reddé et al. 2006). Also noteworthy were well-designed privately built strongholds that were sited, financed and constructed by wealthy senatorial magnates with the intention of protecting their large and flourishing estates from seaborne pirates and land-based bandits (Samson 1987: 287–315).

These fortress cities, *urbes*, served as the capitals of the *civitates*, which were the basic units of local imperial administration. Each of these circumscriptions was under the jurisdiction of a count (*comes civitatis*), who made his headquarters in the fortified *urbs*. He was appointed by the imperial government and served at the pleasure of the emperor. Most fortress cities were located either on the coast or on the banks of one of the many rivers that provided a well-developed and extensive communication and transportation network throughout Gaul. In military perspective, the river systems facilitated the rapid movement of large numbers of troops and their logistical support, when necessary. The well-developed imperial road system, which was more expensive to use, also was of considerable importance especially in areas that were not well served by waterways. Both the rivers and the roads, along with numerous blue and brown water ports as well as canals, such as the Vecht, played key roles in making possible the effective defence of Hexagon. The great economic success of Gaul, which in turn helped to provide the wealth to fortify its cities and lesser population centres, was sustained, as well, by this complex system of transportation and communication (Bachrach 2012a: 27–50, 2012b: 38–64).

The Roman Army in Gaul

The hardened defences established by the imperial government were defended effectively by the imperial field army from various threats launched from beyond the frontiers. Those people living beyond the frontiers had not mastered the art of siege warfare, which required sophisticated technology and a well-developed logistical system (Bachrach 1995: 7–13). Throughout the third and fourth centuries the forces of the imperial military proved to be exceptionally effective in defending the Hexagon from enemies living beyond the frontiers (Elton 1992: 167–76; Bachrach 2014: 1–37, 2017: 3–34). A reasonably accurate sense of the size and structure of the Roman military establishment throughout the empire during the early fifth century is provided by the *Notitia Dignitatum*, which was drawn up ca. 395 and was maintained more or less up to date in Gaul at least into the later 420s. The sections regarding the Hexagon make clear that 'on paper' there were in the neighbourhood of 100,000 fighting men of various types based throughout Gaul early in the fifth century (Jones, 1964, I: 607–86, II: 1417–450; Whitby 2004: 156–86).

The regular forces of the imperial army in Gaul were divided into two large groups: the *comitatenses* and the *limitanei*. In addition to regular troops, numerous other units were based in military colonies of various kinds. To provide material support for these fighting men there were government-run factories staffed largely by slaves but administered by military personnel for the production of arms and armour to support this very substantial military establishment (Jones 1964, I: 608–10, 612–13, 649–54). Finally, on the frontiers of the Hexagon, the Roman government established various groups of so-called 'barbarians' who were designated as allies under various rubrics. By and large the fighting men in these groups were not trained in the same manner as Roman troops although the government often provided them with some of their arms, especially short swords. The strategic assignment of these allies, broadly conceived, was to act as a buffer between imperial territory and would be raiders or invaders from further afield whose aim, in general, was to seek booty within the empire rather than conquest. The most important of these allies at the beginning of the fifth century were Franks, but notice should be taken also of Alamanni, Burgundians and Bavarians (Bachrach 2014: 1–37).

The *comitatenses* constituted a field army divided into units of about 1,000 effectives that were referred to as *legiones*. These forces were exceptionally mobile as a relatively large percentage of effectives within each unit were mounted troops. The military command in Gaul tended to base the units of *comitatenses* in the environs of Gaul's fortress cities. As a result, the existing supply systems that fed Gaul's populous *urbes* also provided logistical support for the legions. Their encampments contributed to the economic health of the *urbs* near which they were based as consumers of locally produced items and especially as patrons of taverns and houses of prostitution. By their very presence in any particular location, the *comitatenses* had the potential to deter any group of raiders that might seek to undertake a plundering raid in the region in which they were based (MacMullen 1967).

By contrast with the *comitatenses*, whose deployments constituted a defence in depth throughout Gaul, the *limitanei* were established on the frontiers with special attention to the Rhine and the Alpine regions. These troops constituted the first line of defence within the imperial borders as, in principle, they were deployed to stop or slow down raiders or invaders crossing into the empire. However, as the border area was porous, the major tactical task of the *limitanei* was to ambush raiders when they tried to return to their homes with whatever loot they had acquired. The *limitanei*, in general, held the lands on which they lived and even the fortifications, often termed *burgi* and *castelli*, that they garrisoned as military lands in restricted hereditary tenure. The Roman government provided the *limitanei* with slaves to work these lands while the soldiers carried out their military duties (Bachrach 1997a: 95–122).

Military Effectiveness

Although we must regard the numbers provided by the *Notitia Dignitatum* as 'paper' figures, which may perhaps be too large but also could be too small, it is very clear that the imperial forces available to defend Gaul in combination with the material defences constructed in the course of the fourth century proved exceptionally effective early in the fifth century. The most serious test of Gaul's military during the fifth century was a massive barbarian 'invasion' of the Hexagon by some 200,000 or more men, women and children, who began crossing the frozen Rhine on 31 December 406. This was a coalition of five groups: two of Alans, two of Vandals and one of Suevi. Roman intelligence assets had been tracking the movements of this coalition from the time that it formed on the Lower Danube and began moving toward the Rhine. Following traditional diplomatic efforts, the Romans secured the defection of one group of Alans and these were settled in Gaul by the Roman government as allies soon after crossing the Rhine as were some Suevi who were enlisted in the Roman Army (Bachrach 2014: 1–37).

On the whole, the invaders were slaughtered in large numbers during the early winter of 407 by the Roman Army and their Frankish allies. Finally, it became obvious to the coalition leaders that without imperial support they would starve before the end of what was a very cold winter. Thus, the various survivors surrendered and made pacts that saw the Roman government provide the erstwhile invaders with food and land on which temporarily to encamp for as long as they caused no trouble. This situation lasted about two years when the remains of the coalition took advantage of conflict among various Roman commanders within Gaul, who supported or opposed the usurper Constantine III (407–409), and in the late summer and autumn of 409 they fled *en masse* into Spain (Bachrach 2014: 1–37).

The ability of the imperial field army in Gaul to bring under control an enemy force that initially included in the neighborhood perhaps as many as 50,000 fighting men must be regarded as an impressive tactical achievement by any standard of reckoning. The question, therefore, must be posed, as to how it came about that less than a half-century later, when Aetius (b. 391–d. 454) mobilised his forces to defeat Attila at Châlons, the imperial army of some 40,000 fighting men included comparatively few *comitatenses* who were based in Gaul. Rather, Aetius relied heavily on the largely assimilated descendants of various so-called barbarians, e.g., Franks, Visigoths and Alans, whose fathers and grandfathers had been settled in Gaul some three decades earlier, and who, of course, were now citizens of the empire (Bachrach 1994: 59–67). In short, what had happened to the Roman field army in Gaul?

It is obvious that the *comitatenses*, who had served in 406, if still alive in 451 had been retired for some two decades. In fact, the field army that had been so successful against the Coalition of 406 not only had suffered losses at that time but had been depleted during the next decade. Casualties are to be associated with the brief usurpations by Constantine III (407–409) and Jovinus (411–413) as well as the failed efforts of at least a half-dozen generals, such as Justinian, Nebiogast and Gerontus. This is made clear by the fact that in 413, Constantius, the newly appointed *magister militum* for Gaul, found it necessary to transfer units from the *limitanei* into the field army. He replaced these on the frontier with allies, e.g., Franks, Alamanni and Burgundians (O'Flynn 1983: 63–73).

The weakening of the imperial field army, however, was not primarily the result of losses in the field. Rather, the traditional system of recruitment ceased to work effectively. In effect, a great many of the local magnates, who were responsible for the recruitment of soldiers for the army, abandoned their obligations to the empire, which were costly in economic terms. Rather, these magnates took advantage of the unrest throughout the Hexagon and to a significant degree stifled the recruitment process. In addition to saving the costs attendant upon the recruitment process, local magnates worked to keep able-bodied men at home in the *civitates* in which they lived so that they could be used for the local defence. Many such potential recruits were, in fact, signed up to serve in the military households, *obsequia*, of rich and powerful men at the local level (Guilhiermoz 1902; Mathisen 2019: 137–67).

Despite the gradual breakdown of the imperial system of recruitment, Constantius commanded a field army of sufficient strength to control the Visigoths who wandered into Gaul after the sack Rome. He used these Visigoths, who served essentially as imperial mercenaries in order to obtain food, to crush the usurper Jovinus. As a result, these Goths were permitted by Constantius to encamp in the environs of Bordeaux where they were fed. The efforts by these Visigoths to break free of imperial control failed, again because of a lack of food. Then Constantius used them to defeat various Vandal and Alan groups in Spain, who were remnants of the coalition of 406, and as a reward he established them in encampments throughout the southwest of Gaul as mercenaries paid with tax receipts to protect the Iberian Frontier. In subsequent decades these Goths, who in earlier generations while living beyond the frontiers of the empire had been farmers, obtained agricultural land to support themselves (Goffart 2006).

In addition to transferring *limitanei* into the field army and deploying Visigothic mercenaries in Southern Aquitaine, the imperial authorities further recognised the need to increase the military manpower lost through the ongoing failure of the recruitment system. It had been the case throughout the history of the empire that citizens, despite the immensely effective professional army they supported with high levels of taxation, were, in general, very well armed. Thus, in the face of the decline in numbers of regular military personnel it became the norm, in a formal legal sense, that the civilian population was formed into militia levies for the local defence to defend the area in which they dwelled. Special attention was given to men, both free and even slaves, living within the fortress cities and in their immediate environs. They were required to defend the fortress cities in which they dwelled should these come under attack. Each militia man not only was required to possess a bow and two quivers of arrows but also to practise on a regular basis so that he could make a good account of himself in defence of his city. More wealthy citizens were required to undertake military operations beyond the borders of the *civitas* in which they lived. These developments were to be of exceptional importance not only during the Later Roman Empire but throughout the Middle Ages (Bachrach 1993a: 55–63, 1997b: 25–31, 1997c: 689–703, 1999: 271–307; Bachrach and Bowlus 2000: 14, 122–36).

While the local magnates in the *civitates* undermined the system of recruitment in order to assure the defence of the area in which they lived, the government tried to take advantage of the breakdown in the traditional process. Not only were ordinary citizens grouped into militia forces both for the local defence and for expeditionary purposes by imperial edict, but the magnates and their military households also were now liable to serve the government as more or less regular fighting men. As long as one or another government official was able to muster these troops the entire process of replacing regular troops with militia men prospered and, in addition, was exceptionally inexpensive. Not only was it unnecessary for the government to pay these fighting men but the militia men were required for all intents and purposes to support themselves. When government officials were in a position effectively to mobilise the militia forces for one or another deployment, the military situation could be considered to be under control. However, when government officials were unable to mobilise these forces the military situation reverted to local defence.

Following the assassination of Aetius in 454, Emperor Majorian (457–461) commanded a sufficient element of the imperial field army as well as militia elements and mercenaries to go to war against the Vandals. However, he was defeated and subsequently assassinated. Aegidius (ca. 455–462), Majorian's *magister militum per Gallias*, maintained a modicum of control north of the Loire with a surviving remnant of the field army (MacGeorge 2002: 152–8). He was succeeded by his son Syagrius (462–486), whose *regnum* likely was smaller than that of his father. In Aquitaine and Burgundy, the southwestern and southeastern quadrants of the Hexagon, respectively, various *civitates* strove to maintain independence as made clear, for example, by Sidonius Appolinaris (d. 480) in regard to the defence of Clermont. In Aquitaine, for example, Euric (466–484), who is styled *rex Visigothorum*, pursued a policy similar to that of Aegidius, i.e., dominating numerous *civitates*. However, his expansionist efforts were thwarted as he was stopped from gaining control of territory north of the Loire or east of the Rhône (Delaplace 2016: 271–81).

The process by which a military transition took place in Gaul during the later fifth century is to be identified with the gradual disappearance of the imperial field army and the fragmentation of political power throughout the Hexagon following the assassination of Majorian in 461. Primarily, political authority and military power were located in each *civitas*, which was governed from its fortress city by a count. He oversaw the administration of the territory, e.g., tax collection and judicial processes, through the expertise of traditional local government officials. This was done generally with the support of bishop of the diocese, the territory of which was, in general, co-terminus with the borders of the *civitas*. The bishop also provided a modicum of religious legitimacy for the local ruler among the overwhelmingly Christian populous under his jurisdiction.

The count's rule was supported in the main by his military household, which is to be seen as the largest and most effective *obsequium* in the region. However, in order to maintain peace and prosperity it was necessary for the count to garner the support of a sufficient number of local magnates, who supported their own military households, so that potential conflict was diminished and economic life was not negatively impacted. The military households of these magnates when combined with that of the count constituted the main force of professional soldiers based in each *civitas*. The members of these military households should be characterised as mercenaries, who, in general, were supported by the men whom they served rather than by tax receipts (Bachrach 2008: 167–92). At least some of these mercenaries are noted by contemporaries or near contemporaries as being equipped in a manner similar to imperial soldiers (MacGeorge 2002: 121–2).

Any one local magnate over the course of the latter part of the fifth century might try to expand his power over several or even many *civitates*, and, as a result, either influence or replace local rulers in the course of creating a substantially larger polity. This process was followed, for example, by several Visigothic rulers and especially by Euric, discussed above, who failed in his aim to dominate the Hexagon despite a modicum of support from imperial officials in Italy. However, he did enjoy support from many Gallo-Roman magnates, who served as administrators, whose command of the government resources in Aquitaine made possible his military adventures by providing the sinews for war.[7] Perhaps the most interesting among these men, from a military perspective, was Namatius, a friend of Sidonius Apollincus, who was admiral of Euric's fleet that protected the Atlantic Coast from pirates and invaders who sought to attack the merchant vessels that sailed over the important trade route that ran from the Mediterranean to the British Isles.[8]

Euric's failed effort to establish his rule over much more than the southwestern quadrant of Gaul may be compared with the success enjoyed by Clovis (481–511) (Rouche 1996). The latter subjected about 80 per cent of the Hexagon to his *regnum*. In his will, he divided the territory he controlled among his four sons in direct imitation of the tradition established by Constantine the Great (d. 337). In this process of conquest, Clovis gathered under his leadership the military institutions of the region and brought about much of what would be the military organisation within the Hexagon and its environs for much of the Middle Ages. These institutions, which ostensibly had been developed by the imperial government were tri-partite in nature. All able-bodied males, whether free or slave, were required to serve for the local defence in the area in which they dwelled. Men of means, who could afford to support themselves on campaign were required to serve *in expeditione* beyond the borders of their home *civitas* and to provide additional fighting men according to their wealth. Finally, all magnates who maintained an *obsequium* or military household were responsible also for expeditionary service (Bachrach 1997b: 25–31, 1997c; Bachrach and Bowlus 2000: 122–36).

Clovis' career began as the heir of his father Childeric (460–481), who was the ruler of Toxandria with its capital in the fortress city of Tournai.[9] Following the death of Majorian, who likely had appointed him, Childeric co-operated with Aegidius, who commanded what may be thought to have been the remains of the imperial field army and was the dominant political and military figure north of the Loire and as far east as Toxandria. Childeric was among those magnates who worked with Aegidius even as far west as operations in the valley of the Loire.[10] When Aegidius died, he was succeeded by his son Syagrius (462–486), who inherited control of much of what had been his father's *regnum* and established his capital at Soissons.[11]

Syagrius maintained good relations with both Childeric and Clovis until about 486, when it is generally believed that the latter, at about 20 years of age, decided that he wanted to replace Syagrius as the dominant figure in Northern Gaul. However, scholars have been unable to explain why at this time Clovis attempted to overthrow Syagrius other than that he was an exceptionally ambitious young man, who sought to be more than the ruler of Toxandria.[12] The aggressive posture putatively taken by Clovis rests upon the thoroughly biased work of Gregory of Tours, which was written during the early 590s. Gregory projects Clovis as the secular hero of his *Ten Books of History* in the tradition of King David and Constantine the Great and, if for no other reason, this account is to be regarded as suspect.[13] Rather, it seems more likely to have been the case that Syagrius at this time was interested in redressing the losses in the east suffered by Aegidius ca. 461 and was seeking to do so at Clovis' expense.[14]

It is clear, however, that Syagrius, whether on the offence or the defence, had miscalculated in regard to Clovis. The military resources potentially available to Clovis, as had been the case with his father as well likely with previous rulers of Toxandria, were considerably greater than those commanded, in general, with regard to the ruler of almost any single *civitas* in the northeast. No later than 298 CE, Constantius Chlorus, the father of the Constantine the Great, had settled a substantial number of Salian Franks in Toxandria and additional numbers are known to have been settled there by Emperor Julian (d. 363). Constantius had established the regulations for this settlement in a manner similar to the inheritance rules that governed lands possessed by *limitanei*. Thus, these lands, which became embedded in Salian custom and law, recorded in Latin as *terra Salaci*, could only be inherited by males capable of performing military service. These tenurial structures played an obvious role in assuring the availability of free allodial landowners for military service that lasted well into the eighth century.[15]

In addition to an expeditionary levy composed, in part, of long-assimilated Franks, who were descended in some cases from families that had lived under imperial administration in Toxandria for some two centuries, Clovis was able to recruit free men of sufficient economic means dwelling throughout the region to serve *in expeditione* for at least six months each campaigning season when mobilised for service beyond the borders of the *civitas* in which they lived.[16] Personally, of course, Clovis commanded his own military household, and as early as 486 had made arrangements with the leaders of other *civitates*, such as Ragnachar, who ruled at Cambrai, to provide support for him against Syagrius (Bachrach 1972: 3–17). Of considerable importance in garnering the support of various Gallo-Roman magnates following his victory over Syagrius was the work of Archbishop Remigius of Rheims (437–533), who gave sound diplomatic advice to Clovis, which he took, and which had a rallying effect on episcopal leadership throughout Gaul (Rouche 1996: 201–5, 278–80, 440–53; Daly 1994: 619–64).After defeating Syagrius in battle at Soissons, Clovis systematically secured control of the northern half of the Hexagon as well as some parts of the southeast quadrant in the course of the next quarter-century (Bachrach 1970: 21–31). Of great importance, in this context, was Clovis' ability not only to mobilise large armies but forces with which he could effectively establish and successfully maintain sieges of fortress cities. Among those that fell to Clovis were Albi, Angoulême, Avignon, Arles, Bordeaux, Clermont, Paris, Rodez, Toulouse and Verdun (Bachrach 1972: 3–17). Command of the technology and maintenance of a logistical system that made possible these successful efforts distinguishes his campaigns in a significant manner from those of Euric and indicates the integration of important imperial structures into Clovis' military forces.[17] One of the key ways in which imperial military observers, e.g., the general Merobaudes (fl. 455), had distinguished 'barbarian' warfare from Roman warfare was the inability of the former to execute successful sieges because they lacked the training, equipment and institutional elements needed to maintain logistical support (Bachrach 1995: 7–13).

An examination of Clovis' conquest of Aquitaine in 509 provides an opportunity to glimpse in some detail his role in the transition. With support from Constantinople, Clovis, whom the Emperor Anastasius (491–518) appointed both an imperial general and honorary consul, secured control of the greater part of the southwestern quadrant of the Hexagon at the expense of Alaric II (484–509), Euric's son and successor. Anastasius' envoys provided Clovis with the uniform of an imperial general, *codicili* affirming his appointment as an honorary consul as well as a *patricius*. The latter appointment made him the supreme imperial appointee in Gaul. Anastasius also provided a substantial supply of gold coins bearing the emperor's likeness to help support Clovis' army on campaign as well as to symbolise imperial control of the project (McCormick 1989: 155–80; Mathisen 2012: 79–110). In 509, Clovis' forces won a decisive battle at Vouillé about 10 miles north of Poitiers, and proceeded to take either by siege or the threat of a siege the major fortress cities of Aquitaine (Bachrach 2012c: 11–42).

Clovis not only mobilised the expeditionary levies from the *civitates* of northern Gaul but with the aid of several bishops in Aquitaine was able to integrate levies from various regions that at least nominally had been under the control of Alaric II (Bachrach 1997b: 25–31, 1997c: 689–703; Bachrach and Bowlus 2000: 122–36). Clovis assured these bishops in a formal decree that they would be secure in their possessions and issued orders to his troops to avoid looting areas that fell under his control. In fact, Clovis instituted the traditional imperial orders that only grass and water were to be taken from the countryside without compensation to the local population. All other necessities for the advance of his army, even hay, were to be paid for by the troops of the invading army. In fact, Clovis is recorded to have enforced a violation of these orders by executing two of his soldiers (Bachrach 2012c: 11–42).

Conclusion

The most significant aspect of the transition from traditional imperial military strategy in Gaul took place largely during the fourth century with the construction of massive fortress cities throughout the Hexagon. These fortifications not only determined the nature of warfare, as territory could neither be defended nor conquered without taking control of these great stone fortress cities. Locally based militia forces were developed to defend the *urbes* and, as a result, large armies were required to lay siege to them, as a ratio of about five or six attackers to each defender was required to threaten the capture of a fortress city. Essentially, the large forces required to attack such fortresses resulted in the development of expeditionary levies, who were mobilised on the basis of their wealth and thus could sustain a lengthy siege. The need to lay siege to a fortress city that might last for months required the maintenance of an effective logistical system that could support large armies.

Notes

1 When this study (Bachrach 1972) was in the process of being published, I observed to my friend and Department Chair, R.S. Hoyt, to whose memory I dedicate this chapter, that the written sources were few and severely biased while the archaeological evidence was underdeveloped in relation to matters of military organisation and warfare. At that time Hoyt observed: 'If your treatment of the subject is sound, and I believe it is, whatever new material that is to be found will support your views.'
2 Lyon 1972 provides a useful survey of the primitivist views that supported the notion of the 'Dark Ages.'
3 Wallace-Hadrill 1952 chose the periodisation 400–1000 for this study because, he argued, the period had 'unity.' For helpful introductions to 'The Late Antique,' see Brown 1971, 2003. Bowersock 2000, at p. 196, observed, 'Now, in 1995, it is probably fair to say that no responsible historian of the ancient or medieval world would want to address or acknowledge the fall of Rome as either fact or paradigm.' It is to be noted that the 'Dark Age' topos was rejected in print as early as 1931 by Laistner, who in the 'Preface' observed that 'no reasonably informed person any longer believes in the "Dark Ages"—a prolonged period of hopeless barbarism succeeding on the fall of the Western Empire.'
4 Cf. Geary 1988: 15, who underestimates the population of Gaul, in general, but is more realistic in terms of estimating the numbers of people of Frankish descent.
5 Of specific interest is McLaughlin 2019; Bailey 2016; and Hen 1995.
6 The best introduction remains Johnson 1983, and regarding costs see Bachrach 2010. The long-term impact of these fortress cities is treated by Bachrach 2000; Bachrach 2005: 61–83, 460–2. Efforts to undermine the importance of these fortress cities are critiqued by Bachrach (2002).
7 Wolfram 1988: 181–90, who rather exaggerates Euric's effectiveness in military matters.
8 Wolfram: 1988: 445. See especially note 14.
9 For an interesting discussion of the career of Childeric, see MacGeorge 2002: 95–102. Unfortunately, this work is somewhat out of date in regard to the material culture.
10 For a useful sketch of Aegidius' career, see MacGeorge 2002: 77–110.
11 Concerning Syagrius, see MacGeorge 2002: 111–36. There is an extensive debate regarding the size and strength of Syagrius' *regnum*. See, for example James 1988: 67–77, who attacks the traditional view that Syagrius dominated Gaul north of the Loire and east to the Rhine. James argues for a much smaller and weaker *regnum*, while MacGeorge argues for a more traditional view. Neither position is well documented.
12 MacGeorge 2002: 122–33 reviews the literature but can find no convincing rationale for Clovis' supposed initiative in seeking to conquer Syagrius' kingdom.
13 Regarding Gregory's bias, see Walter Goffart 1988: 219.
14 Concerning Aegidius' losses in the east of the Hexagon, cf. MacGeorge 2002: 107.
15 See Bachrach 1997a: 98–102, where these matters are discussed in considerable detail. For additional observations on these matters regarding free landowners and their military service, see Bachrach 2016.
16 For the lengthy history of select levies in Gaul, see Bachrach 2016, and especially with regard to various *antiquae consuetudines*, which were of Roman imperial origin and continued to be maintained by the Carolingians.
17 Regarding the dominance of siege warfare from the late Roman period to the end of the Middle Ages, see Bradbury 1992; and the review article by Bachrach 1993b; Petersen 2013.

Bibliography

Alföldi-Rosenbaum, E., 'The Finger Calculus in Antiquity and in the Middle Ages: Studies on Roman Game Counters I,' *Frühmittelalterliche Studien*, vol. 5 (1971), pp. 1–9.
Bachrach, Bernard S., 'Procopius and the Chronology of Clovis's Reign,' *Viator*, vol. 1 (1970), pp. 21–31.

Bachrach, Bernard S., *Merovingian Military Organization: 481–751* (Minneapolis: University of Minnesota Press, 1972).

Bachrach, Bernard S., 'Grand Strategy in the Germanic Kingdoms: Recruitment of the Rank and File,' in Françoise Vallet and Michel Kazanski (eds.), *L'Armée romaine et les barbares du IIIe au VIIe siècle* (Paris: Association française d'archéologie mérovingienne et Musée des antiquités nationales, 1993a), pp. 55–63.

Bachrach, Bernard S., 'Medieval Siege Warfare: A Reconnaissance,' *The Journal of Military History*, vol. 58 (1993b), pp. 119–33.

Bachrach, Bernard S., 'The Hun Army at the Battle of Chalons (451): An Essay in Military Demography,' in Karl Brunner and Brigitte Merta (eds.), *Ethnogenese und Übrerlieferung: Angewandte Methoden der Frühmittelterforschung* (Vienne-Munich: Ethnogenese und Übrerlieferung, 1994), pp. 59–67.

Bachrach, Bernard S., 'The Education of the "Officer Corps" in the Fifth and Sixth Centuries,' in Françoise Vallet and Michel Kazanski (eds.), *La noblesse romaine et les chefs barbares du IIIe au VIII siècle*, in quarto (Paris: Association française d'archéologie mérovingienne et Musée des antiquités nationales, 1995), pp. 7–13.

Bachrach, Bernard S., 'Military Lands in Historical Perspective,' *Journal of the Haskins Society*, vol. 9 (1997a), pp. 95–122.

Bachrach, Bernard, 'The Imperial Roots of Merovingian Military Organization,' in Anne Norgard Jorgensen and Birthe L. Clausen (eds.), *Military Aspects of Scandinavian Society in a European Perspective, AD. 1–1300* (Copenhagen: Military Aspects of Scandinavian Society, 1997b), pp. 25–31.

Bachrach, Bernard, 'Quelques observations sur la composition et les caractéristiques des armées de Clovis,' in Michel Rouche (ed.), *Clovis: Histoire et Mémoire*, 2 vols. (Paris: Presses de l'Université de Paris-Sorbonne, 1997c), pp. 689–703.

Bachrach, Bernard, 'Early Medieval Europe,' in Kurt Raaflaub and Nathan Rosenstein (eds.), *War and Society in the Ancient and Medieval Worlds: Asia, The Mediterranean, Europe, and Mesoamerica* (Cambridge-MA: Harvard University Press, 1999), pp. 271–307.

Bachrach, Bernard S., 'Imperial Walled Cities in the West: An Examination of their Early Medieval *Nachleben*,' in James T. Tracy (ed.), *City Walls: The Urban Enceinte in Global Perspective* (Cambridge: Cambridge University Press, 2000), pp. 192–218.

Bachrach, Bernard S., 'Fifth Century Metz: Later Roman Christian *Urbs* or Ghost Town?,' *Antiquité Tardive*, vol. 10 (2002), pp. 363–81.

Bachrach, Bernard S., 'On Roman Ramparts, 300–1300,' in Geoffrey Parker (ed.), *The Cambridge Illustrated History of Warfare: The Triumph of the West*, 2nd ed (Cambridge: Cambridge University Press, 2005), pp. 64–91.

Bachrach, Bernard S., 'Merovingian Mercenaries and Paid Soldiers in Imperial Perspective,' in John France (ed.), *Mercenaries and Paid Men in the Middle Ages: Proceedings of a Conference held at University of Wales, Swansea, 7th-9th July 2005* (Leiden/Boston: Brill, 2008), pp. 167–92.

Bachrach, Bernard S., 'The Fortification of Gaul and the Economy of the Third and Fourth Centuries,' *Journal of Late Antiquity*, vol. 3, no. 1 (2010), pp. 38–64.

Bachrach, Bernard S., 'Continuity in Late Antique Gaul: A Demographic and Economic Perspective,' in David M. Nicholas, Bernard S. Bachrach and James Murray (eds.), *Comparative Perspectives on History and Historians: Essays in Memory of Bryce D. Lyon (1920–2007)* (Kalamazoo-Michigan: Western Michigan University, 2012a), pp. 27–50.

Bachrach, Bernard S., 'The Fortification of Gaul,' in David M. Nicholas, Bernard S. Bachrach and James Murray (eds.), *Comparative Perspectives on History and Historians: Essays in Memory of Bryce D. Lyon (1920–2007)* (Kalamazoo-Michigan: Western Michigan University, 2012b), pp. 38–64.

Bachrach, Bernard S., 'Vouillé in the Context of the Decisive Battle Phenomenon,' in Denuta Shanzer and Ralph W. Mathisen (eds.), *The Battle of Vouillé 507 CE: Where France Began* (Berlin: De Gruyter, 2012c), pp. 11–42.

Bachrach, Bernard S., 'Some Observations Regarding Barbarian Military Demography: Geiseric's Census of 429 and Its Implications,' *Journal of Medieval Military History*, vol. 12 (2014), pp. 1–37.

Bachrach, Bernard S., 'Charlemagne's Expeditionary Levy: Observations Regarding *Liberi Homines*,' *Studies in Medieval and Renaissance History*, 3rd. ser., vol. 12 (2016), pp. 1–65.

Bachrach, Bernard S., 'Late Roman Grand Strategy: Fortification of the urbes of Gaul,' *Journal of Medieval Military History*, vol. 15 (2017), pp. 3–34.

Bachrach, Bernard and Bowlus, Charles R., 'Heerwesen,' in Heinrich Beck, et al. (eds.), *Reallexikon der Germanischen Altertumskunde* (Berlin-New York: De Gruyter, 2000), pp. 14, 122–36.

Bailey, L.K., *The Religious Worlds of Laity in Late Antique Gaul* (London/New York: Bloomsbury, 2016).

Banniard, Michel, *Viva Voce: Communication écrit et communication orale du IVe au IXe siècle en occident Latin* (Paris: Institut d'études augustiniennes, 1992).

Bowersock, Glen W., 'The Vanishing Paradigm of the Fall of Rome,' in Glen W. Bowersock, *Selected Papers on Late Antiquity* (Bari: Edipuglia, 2000), pp. 187–97.

Bradbury, Jim, *The Medieval Siege* (Woodbridge: Boydell, 1992).

Brown, Peter, *The World of Late Antiquity: A.D. 150–750* (New York: Harcourt Brace Jovanovich, 1971).

Brown, Peter, *The Rise of Western Christendom: Triumph and Diversity: A.D. 200–1000*, 2nd ed. (Malden-Mass.: Blackwell, 2003).

Daly, William, 'Clovis: How Barbaric, How Pagan?,' *Speculum*, vol. 69 (1994), pp. 619–64.

Delaplace, Christine, 'The So-Called "Conquest of the Auvergne," (469–75) in the History of the Visigothic Kingdom. Relations between the Roman Elites of Southern Gaul, the Central Imperial Power in Rome and the Military Authority of the Federates on the Periphery,' in David Brakke and Deborah Deliyannis (eds.), *Shifting Cultural Frontiers in Late Antiquity* (London: Routledge, 2016), pp. 271–81.

Elton, Hugh, 'Defence in Fifth Century Gaul,' in John Drinkwater and Hugh Elton (eds.), *Fifth Century Gaul: A Crisis or Identity?* (New York/Cambridge: Cambridge University Press, 1992), pp. 167–76.

Esders, Stefan, 'Late Roman Military Law in the Bavarian Code,' *Clio Themis*, vol. 10 (2016), pp. 1–24.

Geary, Patrick J., *Before France and Germany: The Creation and Transformation of the Merovingian World* (Oxford: Oxford University Press, 1988).

Goffart, Walter, *The Narrators of Barbarian History AD 550–800: Jordanes, Gregory of Tours, Bede, and Paul the Deacon* (Princeton: Princeton University Press, 1988).

Goffart, Walter, *Barbarian Tides: The Migration Age and the Later Roman Empire* (Philadelphia: University of Pennsylvania Press, 2006).

Guilhiermoz, Paul, *Essai sur l'origine de la noblesse en France au Moyen Age* (Paris: A. Picard et fils, 1902).

Hen, Yitzhak, *Culture and Religion in Merovingian Gaul: A.D. 481–750* (Leiden/New York: E.J. Brill, 1995).

James, Edward, *The Franks* (Oxford: Basil Blackwell, 1988).

Johnson, Stephen, *Late Roman Fortifications* (Totowa-New Jersey: Batsford, 1983).

Jones, A.H.M., *The Later Roman Empire 284–602. A Social, Economic and Administrative Survey*, 2 vols. (Norman: University of Oklahoma Press, 1964).

Laistner, M.L.W., *Thought and Letters in Western Europe: 500–900* (New York: Dial Press, 1931).

Lyon, Bryce D., *The Origins of the Middle Ages: Pirenne's Challenge to Gibbon* (New York: Norton, 1972).

MacGeorge, Penny, *Late Roman Warlords* (Oxford: Oxford University Press, 2002).

MacMullen, Ramsay, *Soldier and Civilian in the Later Roman Empire* (Cambridge, Mass.: Harvard University Press, 1967).

Mathisen, Ralph W., 'Clovis, Anastasius, and Political Status in 508 C.E.: The Frankish Aftermath of the Battle of Vouillé,' in Denuta Shanzer and Ralph W. Mathisen (eds.), *The Battle of Vouillé 507 CE: Where France Began* (Berlin: De Gruyter, 2012), pp. 79–110.

Mathisen, Ralph W., 'The Fifth Century: Age of Transformation,' in Jan Willem Drijvers and Noel Lenski with the assistance from Kevin Feeney and Sean Northrup (eds.), *Proceedings of the 12th Biennial Shifting Frontiers in Late Antiquity Conference* (Bari: Edipuglia, 2019), pp. 137–67.

McCormick, Michael, 'Clovis at Tours, Byzantine Public Ritual and the Origins of Medieval Ruler Symbolism,' in E. K. Chrysos et A. Schwarcz (eds.), *Das Reich und die Barbaren* (Vienna-Cologne: Wien, Bohlau, 1989), pp. 155–80.

McLaughlin, A.E. Tiggy, 'Ordinary Christians and the Fifth-Century Reform of the Church in Gaul,' in Jan Willem Drijvers and Noel Lenski (eds.), *The Fifth Century: Age of Transformation. Proceedings of the 12th Biennial Shifting Frontiers in Late Antiquity Conference* (Bari: Edipuglia, 2019), pp. 262–72.

Murray, Alexander C., *Germanic Kinship Structure: Studies in law and Society in Antiquity and the Early Middle Ages* (Toronto: Pontifical Institute of Mediaeval Studies, 1983).

O'Flynn, John Michael, *Generalissimos of the Western Roman Empire* (Edmonton: University of Alberta Press, 1983).

Petersen, Leif Inge Ree, *Siege Warfare and Military Organization in the Successor States (400–800 AD): Byzantium, the West and Islam* (Leiden/Boston: Brill, 2013).

Reddé, Michel with the aid of Brulet, Raymond, Fellmann, Rudolf, Haalebos, Jan-Kees, and Schnurbein, Siegmar von (eds.), *L'Architecture de la Gaule romaine. Les fortifications militaires*, Documents d'archéologie française 100, directed by (Paris-Bordeaux: Éditions de la Maison des sciences de l'homme, 2006).

Rouche, Michel, *Clovis* (Paris: Fayard, 1996).

Samson, Ross, 'The Mervingian Nobleman's House. Castle or Villa?' *Journal of Medieval Archaeology*, vol. 13 (1987), pp. 287–315.

Sarris, Peter, 'The Origins of the Manorial Economy: New Insights from Late Antiquity,' *English Historical Review*, vol. 119 (2004), pp. 280–311.

Wallace-Hadrill, J.M., *The Barbarian West: 400–1000* (London: Hutchinson, 1952).

Whitby, Michael, 'Emperors and Armies, AD 235–395,' in Simon Swain and Mark Edwards (eds.), *Approaching Late Antiquity: The Transformation from the Early to Late Empire* (Oxford: Oxford University Press, 2004), pp. 156–86.

Wolfram, Herwig, *History of the Goths*, tr. Thomas J. Dunlap (Berkeley, Los Angeles/London: University of Los Angeles, 1988).

Wright, Roger, *Late Latin and Early Romance in Spain and Carolingian France* (Liverpool: Francis Cairns, 1982).

7

NEW APPROACHES TO LATE IMPERIAL CHINESE WARFARE

Kenneth M. Swope

The serious academic study of China's military past can be dated from around 50 years ago when Frank A. Kierman Jr. and John King Fairbank published their edited collection *Chinese Ways in Warfare* (Kierman, Jr. and Fairbank 1974). Though the editors themselves admitted that the essays contained therein were highly preliminary and designed to stimulate further investigation into China's rich military past, their tentative conclusions were often accepted as authoritative. Amazingly, it would be more than two decades before a new generation of scholars began seriously re-opening the book(s) on China's long and storied military past, largely by drawing on theretofore little used but incredibly rich primary source materials. Less encumbered by the messy political correctness and pervasive anti-military biases of their predecessors who were approaching Asian history against the backdrop of the wars in Korea and Vietnam, not to mention the Second World War and its attendant wars accompanying de-colonisation, this new generation of researchers has literally created a dynamic field of study from scratch. In this process, they have created an international scholarly organisation, the Chinese Military History Society, which recently began the publication of the first scholarly journal in English exclusively dedicated to the study of China's military past and institutions.[1] While all of China's military past has enjoyed greater examination in the West in the past two decades, partly as the result of China's return to global economic and military prominence, the field of late imperial history, here broadly defined as encompassing the Ming and Qing (1368–1911) dynasties, has been a particularly fruitful area of study. Interestingly enough, many of these studies have drawn parallels to the present and/or purported to offer insights into contemporary China's military goals, strategic culture and diplomatic behaviour. This chapter shall offer an overview of how various authors have theorised Late Imperial Chinese warfare, also considering the broader implications of their studies for both China scholars and comparative military historians.

Returning to the aforementioned seminal volume by Fairbank and Kierman, its particular interpretation of Chinese military institutions and their place in Chinese culture is worth mentioning because it remained influential for such a long time and continues to have resonance today in less informed circles. While admitting that he was by no means an expert in Chinese military history, Fairbank offered a number of sweeping generalisations. First off, he characterised the Chinese style of warfare as 'somewhat less expansive

DOI: 10.4324/9780429437915-10

than our own' (Kierman, Jr. and Fairbank 1974: 1). Expanding on this observation, he drew attention to the long history of civilian control over the military, thereby establishing the supposed primacy of civil (*wen*) virtues over military (*wu*) ones in Chinese society, again drawing explicit comparisons with the Western experience, opining that the 'germ of Chinese defensiveness, her primary concern for social order at home instead of expansion abroad, came from her landlocked situation in North China remote from other centers of civilization and from sea routes communicating with them' (Kierman, Jr. and Fairbank 1974: 3). He doubled down on this point by arguing that the triumph of civil over military culture was 'no mere fiction implanted in the record by the civilian chroniclers who monopolized it' but rather 'another Chinese achievement in the ordering of society' (Kierman, Jr. and Fairbank 1974: 4). He also focused extensively on Sunzi's famous *Art of War* as the paradigmatic example of China's emphasis on winning without fighting, ignoring the myriad examples in both that work and other military classics that stress the use of overwhelming force as both a deterrent and an example.[2] As will be seen below, these characterisations were in fact rather flawed, as revealed by studies produced all over the world in the ensuing decades, many of which draw upon historical sources that were theretofore relatively obscure or unplumbed.

After spending considerable effort in extolling the virtues of the Chinese civil bureaucracy and its supposedly near total control of the military, Fairbank waxes poetic about the achievements of bureaucratic government in China, contending that the Chinese expanded 'only gradually' because 'their energies were more devoted to the organization of their society within than to its growth outward, to implosion rather than explosion' (Kierman, Jr. and Fairbank 1974: 9). Again, while this all sounds quite idyllic, people such as the Tibetans and the Uighurs might argue strongly against such characterisations of China's expansion or lack thereof. Moreover, Fairbank then proceeds to discuss Chinese relations with the nomads of Inner Asia, disingenuously arguing that these should not be viewed as foreign relations in the Western style 'because Inner Asia was from early times for military purposes a constituent sector of the Chinese military scene, even if peripheral; that the same hierarchy of means for maintaining the social order within China could be used to sustain orderly relations with the nomads' (Kierman, Jr. and Fairbank 1974: 13). In fact, as recent studies have amply demonstrated, such was not the case at all. It was not until the Manchu Qing dynasty (1644–1911) that the so-called nomad problem was solved by the astute melding of steppe and Chinese military and bureaucratic practices by a series of exceptionally gifted emperors, a point to which I shall return below.

In Fairbank's defence, his interpretation of China's military tradition was heavily coloured by recent events, as well as his own areas of research expertise, which focused on the creation of treaty ports and Sino–Western relations from the Opium War (1839–1842) through the founding of the People's Republic. This period is still known in China as the 'Century of Humiliation,' a nadir of military strength when all pride in China's considerable cultural achievements was concomitantly eroded in the face of Western challenges and the rise of China's neighbour, Japan.[3] In the face of these developments, perhaps understandably, even Chinese people embraced the notion of China as inherently pacifist, prompting scholars such as Lei Haizong to characterise traditional Chinese culture as 'a-military' during the Second World War (Lei and Lin 1989). Such views were widely embraced by Western scholars of Fairbank's generation and their students, many of whom sought in China an alternative to the hegemonic militarism of the United States during the Vietnam War era.[4] A pacifistic China gradually returning to international prominence at the time seemed to fit the bill nicely.

Thus, for the next two decades or so after the publication of *Chinese Ways in Warfare*, similar interpretations of China's military past and traditions continued to hold sway. Texts such as Geoffrey Parker's highly influential *The Military Revolution: Military Innovation and the Rise of the West, 1500–1800* and *The Cambridge Illustrated History of Warfare: The Triumph of the West* continued to contrast an explicitly dynamic, innovative and aggressive 'West' with the rest, notably China, since it was the non-Western state with the best documented military tradition even though most of these scholars ignored the vast majority of that tradition in favour of their essentialist explanations for Western ascendency in the modern era.[5] Scholars such as Victor Davis Hanson, incidentally a contributor to the latter volume, expanded upon Parker's more circumscribed findings, positing a 'Western Way of War' that extended all the way back to Classical Greece and was rooted in unique Western traditions of democracy that made it unparalleled in its lethality, thereby accounting for the domination of Western Europe and its satellite civilisations in the modern era (Hanson 2000, 2002).

Interestingly enough, most of the characteristics identified by Parker, Hanson and others to demonstrate the supposed superiority of Western warfare, such as the presence of military training manuals, the emphasis upon drill and training and a willingness to adapt new military technologies, could all be found in imperial China. But to that point no one had really bothered to look for them. Furthermore, some Chinese scholars also continued to adhere to such facile dichotomies. The noted Ming Dynasty historian Ray Huang, for example, characterised Ming China as 'introverted and lacking in competitive character,' which he attributed to an inherent Chinese conservatism and desire for homogeneity and uniformity (Huang 1999: 142–3).

Happily, this state of affairs changed significantly starting in the 1990s as a new generation of China scholars, exposed to Asian popular culture that often emphasised the martial arts and military heroes and unencumbered by the intellectual baggage of the Vietnam War era, began to more seriously explore China's military past.[6] The results have been impressive indeed. In addition to a number of translations of Chinese military classics and works of strategy such as those cited above, monographs and articles on virtually every era of Chinese history have been published, many of them profoundly altering our understanding of the place of the military in Chinese society throughout history while also delineating elements of a distinctly Chinese military culture that share certain aspects with other military cultures while also embodying unique characteristics.[7] Other enterprising scholars, such as Tonio Andrade, have actively engaged the aforementioned studies trumpeting Western superiority, presenting the Chinese side of the story while also making use of primary sources from the Western side (Andrade 2011, 2016).

The most fruitful area of study has been the Late Imperial Era, roughly defined here as encompassing the Ming and Qing dynasties (1368–1911). This is due to both the much greater availability of primary sources for this period and the fact that the era in question encompasses the timeframe in which the so-called Rise of the West took place. In other words, there is an audience for such studies beyond the norm for topics in Chinese history. Finally, while the present study is focusing primarily on English language scholarship on Late Imperial Chinese military history, there has also been an efflorescence of Chinese scholarship on military history in the past two decades, both in response to these Western investigations and as a result of a renewed interest in China's military past due to popular culture in China (i.e., movies and television programmes), and connected to China's rise as a military power in the twenty-first century. Interestingly enough, however, Chinese authorities continue to adhere to the traditional pacifist interpretation of China's military past, contending, for example, that China has never been an aggressive imperialist power and that its current aims remain purely defence, in marked contrast to those of the United States.[8] For their part, mainland Chinese scholars, with a few notable exceptions, have tended to toe the party line while simultaneously casting light on more positive aspects of China's military past in terms of technological developments in particular, emphasising military achievements within the context of achieving national unity and harmony.

While there were efforts to recover China's military past across all historical periods, it was the dynamic expansion of Qing studies starting in the 1980s that served as a springboard for the discussion and analysis of Late Imperial military history more broadly. Previously, owing to the aforementioned emphasis upon the impact–response model of modern Chinese historical analysis, the Qing was largely cast as heavily Sinicised, xenophobic and conservative, a moribund empire whose pathetic flailing in the face of Western and Japanese challenges in the nineteenth century was almost comical. That picture began to change with the publication of Frederic Wakeman Jr.'s magisterial *The Great Enterprise* in 1985 (Wakeman, Jr. 1985). True, there had been previous studies of individual Qing emperors, most notably Kangxi (r. 1662–1722) and his grandson Qianlong (r. 1736–1795), but Wakeman's book was seminal for its emphasis upon Manchu state building and the relationship between these efforts and military power and, to a lesser extent, Qing martial culture. This work was followed by a series of books by scholars such as Pamela Crossley, Evelyn Rawski and Mark Elliott, many of whom turned to Manchu sources, which had theretofore been largely ignored in Western language scholarship on the Qing (Crossley 1999; Elliott 2001; Rawski 1996). These studies revealed a Qing imperium that was consciously multi-ethnic, militarily expansive and administratively innovative, both building upon pre-existing Chinese traditions and developing its own institutions for the management of the empire and its many subject peoples. Among other things, the Qing revived traditions of imperial touring, embraced their role as Buddhist monarchs, tapping into the cultural charisma first cultivated by Mongol emperors and available to them by virtue of military victories over certain Mongol chieftains, and cast themselves as the inheritors of the Han Chinese civilising mission, bringing culture to other 'savage' minorities of the frontiers (Chang 2007; Elverskog 2006; Hostetler 2001).

Thus was born the field known as the New Qing History (Waley-Cohen 2004; Wu 2016). These studies were not without their critics, both in China and elsewhere. Some argued that the new generation had gone too far, downplaying the important contributions played by Chinese officials and Chinese cultural forms and institutions in the formation and expansion of Qing power (Ho 1998). Others, particularly in mainland China, decried the portrayal of the Qing as an expansionistic, imperialist power, arguing that China, as a prime victim of Western imperialism, could not have been an imperialist power itself. Instead, they pointed to the natural process of assimilation that culminated in the creation of today's multi-ethnic People's Republic, conveniently glossing over the ethnic unrest that has often marred communist rule in China.[9]

Returning to the works under consideration, significantly not all of this work directly pertained to traditional military history per se, but rather engaged other aspects of Qing history. Yet it was unmistakable that the threat or reality of Qing military force underpinned many of its political and institutional endeavours. In fact, the emerging image of the Qing began to much more approximate its early modern contemporaries such as the Russians and the Ottomans, both of which the Qing engaged, and even warred with, either directly or via proxies.

Thus, the reality that emerged established that the Qing was in fact a successful empire, one that, like all other empires, grounded its power in its military. Such realisations opened the door for a number of excellent studies of the Qing military. These included both narrative campaign histories and specialised studies highlighting distinctive aspects of Qing military culture, such as the erection of monumental steles commemorating military triumphs or the creation of victory paintings (Perdue 2005; Waley-Cohen 2006).

At the same time, readier access to Chinese primary sources and an expanding interest in local and regional histories resulted in a proliferation of studies that focused on processes and events on a smaller scale, thereby shedding important light on the relationship between war and local society. These sources include not only the Chinese local gazetteers and private histories, but also folk narratives and ethnographic field research, something that was difficult to conduct prior to the 1980s. Especially significant in this respect are studies focusing on Qing expansion into the southwest and into Central Asia. In the former area, the Qing continued a Ming dynasty policy known as *gaitu guiliu*, which translates as 'returning the lands to circulation,' meaning bringing these formally aboriginal or unclassified lands under regular government jurisdiction.[10] In fact, the Ming were simply building upon earlier processes.[11] Impelled in part by the widespread destruction that took place over the four decades of the Qing conquest and consolidation of power in the seventeenth century, the Qing encouraged Han settlers in particular to move into these formerly marginal lands, prompting a variety of local responses, many of which were highly contentious and often resulted in bloody, generally failed, uprisings (Weinstein 2014). Such studies are invaluable in uncovering the gritty dimensions of Qing expansion and the costs borne by local peoples. They could serve as useful comparative studies for those interested in overland expansion in the United States, Russia or Canada, though to my knowledge no one in those fields has yet appreciated this potential avenue of exploration.

Continuing the theme of comparative imperial studies, recent work on the Ming and Qing has also highlighted the presence of grand strategy in the formulations of these monarchs, something else which had long been ignored by scholars of Late Imperial China. In addition to Peter Perdue's work, cited above, Yingcong Dai has analysed the evolution of Qing grand strategy with regards to Sichuan and Tibet, noting that the conquest of these regions became imperative in the larger Qing quest for control of Central Asia (Dai 2009). My own work has examined the importance of grand strategy for the Ming rulers. For example, when the Japanese under their warlord Toyotomi Hideyoshi (1536–1598) launched their invasion of the Asian mainland, the Ming determined that it was vital that they retain their hegemony in East Asia, so they decided to send an expeditionary force to dislodge the Japanese from Korea. This strategic concern was augmented by their traditional obligations as tributary overlords to Korea, the military dimensions of which are only now being considered by scholars (Swope 2009, 2015b: 163–96).

The decision to aid Korea was made against the backdrop of a fairly coherent grand strategy under the Ming Emperor Wanli (r. 1573–1620). Advised by a capable Grand Secretary named Zhang Juzheng, Wanli's overarching goal was to make the Ming borders strong and the army fearsome once again. Zhang had grown up amidst an atmosphere of fear regarding the Mongols, who had raided the outskirts of Beijing when Zhang was a young official. Determined to prevent that from happening again, Zhang cultivated networks with capable military officials such as the famous Qi Jiguang (1528–1588), streamlined the taxation system, expanded the use of mercenaries, improved frontier defences and authorised destabilising military strikes against Mongol encampments on the steppe. After Zhang's death in 1582, Wanli continued to prioritise military affairs and consistently sought to protect and patronise his leading military commanders, a practice emulated by his Qing successor, Kangxi.[12] Unfortunately, after Wanli died, his successors proved unable to articulate a coherent grand strategy and Ming officialdom degenerated into unproductive sniping and paralysing factionalism.[13]

Returning to the Qing, while Perdue has ably chronicled the purely military dimensions of the Qing expansion across Central Asia, others have examined the religious and commercial aspects of the Qing imperial presence in Central Asia (Hodong Kim 2010; Kwangmin Kim 2016; Millward 2016). Still others have made use of the meticulous Qing military records to examine Qing logistics and military finance, again highlighting the sophistication of the Qing military machine.[14] All these studies point to the important ways in which the study of military history, processes and institutions is being used to shed greater light on late imperial history as a whole. Most importantly, such studies shatter the picture of a backwards, ossified, xenophobic late imperial state, unable or simply unwilling to compete with its neighbours or alter its diplomatic, commercial or military practices. On the contrary, we see a state that was constantly challenged by a multiplicity of threats on all sides, but consistently sought ways to maintain its strategic edge, whether through straightforward military means or via cultural assimilation programmes or even proxy rulers, as in the case of Central Asia. That not all these policies were completely successful, particularly in the face of the general Qing decline in the mid-nineteenth century should not obscure the fact that the Ming and Qing were remarkably stable for some five centuries, a feat which has virtually no parallels anywhere in the world and one which can be at least partly credited to their superior military power and organisation.[15]

While the primary aim of Qing historians was to reopen the question of the martial/ steppe influences on imperial culture and explore its implications for Qing rule, it opened the door for historians of other dynasties to reexamine the significance of *wu* in Chinese culture throughout the ages. They jumped at this opportunity and the results of their research have been fruitful indeed. Previously, studies of the Ming military tended to focus on the negative, examining such topics as sixteenth-century piracy, the late Ming peasant rebellions and the apparent weakness of the Ming in the face of the Manchu challenge in the seventeenth century (Kwan-wai So 1975; Chan 1982; Parsons 1993). While these studies were generally grounded in certain primary source materials, such interpretations were extrapolated from falsities perpetrated by the Qing conquerors in the seventeenth century to justify their invasion and highlight their own military prowess. Sources that offered alternative views of Ming military competence were glossed over as mere boasting or ignored entirely by scholars apparently eager to perpetuate the notion of absolute civilian control over the military.

This picture began to erode with the emergence of several notable Ming scholars in the 1990s. David Robinson highlighted the importance of 'men of force' in the Ming political landscape and demonstrated that military men no less than their civilian counterparts often had extensive networks that extended through all ranks of society from common bandits all the way up to the emperor himself (Robinson 2001). Robinson then turned his attention to Ming Court culture and found many fascinating cultural and institutional links to the preceding Mongol Yuan (1279–1368) Dynasty. Previous scholarship had acknowledged Ming debts to the basic Mongol military system, but had largely ignored the continuities in court culture, which were marked by the performance of martial spectacles such as royal hunts and archery contests that prove how martial culture permeated Ming society at all levels (Robinson 2013). In doing so, Robinson revealed continuities between Yuan-Ming-Qing court cultures that better contextualised the aforementioned findings of Qing historians.

My own work on the decline and fall of the Ming started with an examination of the Three Great Campaigns of the Wanli Emperor (*Wanli san da zheng*), which occurred between 1592 and 1600 (Swope 2001). These campaigns, which included the successful ouster of the Japanese from the Korean Peninsula in 1598, were traditionally portrayed as manifestations of Ming military weakness and incompetence wherein the Ming prevailed in spite of itself. But a close reading of theretofore largely ignored primary sources offered a much more nuanced understanding of the geo-political situation and the significance of these campaigns within the context of Ming grand strategy.[16] It also offered a glimpse into an imperial state that was, for all its flaws, dynamic and creative in its approach to military problems, willing to adopt new tactics and technologies to seek solutions to the constantly changing military environment. In short, Ming success in the Three Great Campaigns can be viewed as a culmination of the military reforms launched by Zhang Juzheng in the 1570s and mark the high point of the late Ming military revival presided over by the Wanli Emperor. More recent studies by other scholars have reinforced these conclusions and further drawn our attention to the importance of military history for understanding other aspects of Late Imperial society including state–society relations, the position of eunuchs, and the impact of warfare upon local populations forced to provide corvee labor or other services to the armies (Miller 2009; Hasegawa 2013).

Additionally, the ongoing debate about the significance of the Military Revolution in world history has ignited a spate of publications about the development and deployment of gunpowder weapons in China. Peter Lorge suggests that, in order to make proper use of gunpowder weapons, European states actually had to become 'more Chinese'; in other words, bureaucratised (Lorge 2008). Sun Laichen has argued convincingly that the Ming should be considered the world's first gunpowder empire and highlighted the importance of the dissemination of Ming military technology for state consolidation in early modern Southeast Asia (Sun 2003: 495–517). Much like Tonio Andrade's excellent work comparing the development of gunpowder weapons in Europe and China respectively, Sun has also traced lines of technological diffusion across both Central Asia and the Indian Ocean, thereby demonstrating the receptivity of Ming rulers, officials and soldiers to outside ideas and innovations and shattering hoary notions of Ming isolationism (Sun 2018: 119–50).

Augmenting and building upon these findings, other scholars have reopened the book on the famous naval expeditions of the early Ming, led by the eunuch Admiral Zheng He. While these were most definitely not missions of exploration, a strong case has been made for considering them examples of force projection or even proto-imperialism, in the words of Geoffrey Wade (Wade 2005: 37–58; Dreyer 2007). Indeed, these expeditions accord perfectly with Ming grand strategy goals, particularly the desire to 'manifest awe' in order to deter potential enemies and protect their trade interests. The fact that the Ming engaged in a costly failed occupation of Vietnam in this same period attests to their interest in expanding their global footprint under the ambitious usurper Emperor Yongle (r. 1403–1424). Moreover, such studies more firmly situate the Ming at the forefront of global economic and political processes in the fifteenth century, again drawing our attention to the centrality of military power in enabling the Ming to assume such a position. Not coincidentally, it was also during Yongle's reign that the Ming established the first dedicated firearms training divisions in the world (in Beijing), a fact conveniently ignored by proponents of the Western military superiority thesis.

In a nutshell, studies of the Ming military, not unlike those of its Qing counterparts, have transformed from generally negative appraisals grounded in a narrative of civil pacifism and xenophobia, to far more nuanced, and generally positive assessments that search for reasons for the long success and prosperity of the Ming rather than its downfall. For if an empire was truly as passive and incompetent as suggested by the likes of Ray Huang, how could it have not only survived, but prospered for nearly three centuries? Thus, rather than the relatively negative portrayal of Qi Jiguang as a 'lonely general' that one finds in Huang's highly influential *1587: A Year of No Significance*, recent studies (and a film) highlight Qi as a military pioneer and innovator, whose organisational structure and tactics were later adopted by Qing military leaders in their campaigns to crush the Taiping Rebellion (1851–1866), the largest civil war in human history.[17]

Among the major reasons for the expansion of interest in Late Imperial military topics is the increasing availability of primary sources, not to mention the ability and willingness to use them in creative ways. In addition to the fact that basic sources and collections such as local gazetteers, the dynastic histories, the Veritable Records (*shilu*) of the Ming and Qing dynasties and the massive compilation of historical and literary materials known as the Four Treasuries (*siku quanshu*) created under the Qianlong emperor, which takes up an entire floor of the National Library of China in hardcover, are now available in various online or electronic formats, publishers in China are continuing to produce compilations of primary sources and collected works, many of which were never widely available before or have long been out of print. These include exceptionally rich compilations of the collected works of prominent Ming and Qing dynasty military officials.[18] These compilations contain everything from private letters and poems to detailed battle reports, plans, recommendations and official correspondence between the military commanders in question and counterparts at all levels of the administrative hierarchy. Thus, they provide tremendous insight into contemporary debates and thinking and allow us to evaluate Chinese military commanders, many of whom of course did hold civil degrees, in a much more nuanced fashion. They also shed light on factional politics and personal relationships and networks, making them of great interest to social historians as well.

In fact, social historians have been profitably making forays into the realm of military history using sources such as military guard registers, family household records, genealogies, and funerary inscriptions and epitaphs. Michael Szonyi has recently offered an analysis of the ways Ming military families in southeast China manipulated government systems of registration for their own advantage and how, contrary to what earlier historians generally assumed, the status of being a military family in the Ming could in fact be desirable for a variety of reasons (Szonyi 2017). His study also reinforces points made above about the inherent dynamism of Late Imperial Chinese society. Likewise, Ivy Maria Lim has examined the way in which lineages in Southeast China responded to local defence initiatives to enhance their status within the tax and household registration systems in order to improve their local status and influence over time (Lim 2019: 258–78). In other words, the study of Late Imperial military institutions and their impact on all aspects of life has profoundly reshaped our understanding of society more generally (Dai 2017: 329–52; Robinson 2017: 297–327).

One might also reference a particular type of source that is unique to the Qing Dynasty, the *fanglue*, or campaign history. In 1749 the Qianlong Emperor ordered the creation of the Office of Military Archives (*fanglueguan*) 'for the recording and narration in approved form of all the Qing imperial wars' as a means of providing the official spin, if you will, on all Qing military campaigns.[19] Whatever their biases, such campaign histories are treasure troves for military historians as they contain a wealth of specific details on strategy, tactics and the narrative course of events that simply cannot be gleaned anywhere else. Thus, they are massive works, generally running into the hundreds of pages per campaign covered and the verbiage throughout provides a wonderful sense of the ways in which the Qing perceived military prowess as a concrete manifestation of the legitimacy of the monarch.[20] Notably, this rhetorical stance remains even in the campaign histories from the late Qing; indeed it was especially important for the Qing rulers to seek validation by this point in their existence and victory over the empire's multifarious internal foes, even when the Qing seemed less capable of matching up against the Europeans, was evidence that not all was lost.

Hopefully this brief survey has demonstrated that the field of Late Imperial Chinese military history has proven to be incredibly fertile over the past two decades and continues to be an area of growth. There remain many rich sources and a dizzying array of topics inviting further investigation. As Guo Wu (2016: 48) notes, with respect to the New Qing History, 'It has contributed to contemporary thinking about Chinese nationalism, imperialism, and ethnicity, as well as the relationship between scholarship and politics.' The same generalisations are certainly applicable to the findings of those working in Late Imperial Chinese military history. Moreover, military historians of China, perhaps even more so than their counterparts in other sub-fields, have been especially active in engaging and contributing to broader debates in the field of military history, as highlighted herein. This in turn provides ample opportunity for comparative military historians and specialists working in other geographic areas to consider the implications of the findings of specialists in the China field. That many of their conclusions might be applicable to other imperial contexts such as war is, after all, a sadly universal human experience in one form or another.

Notes

1 *The Journal of Chinese Military History* is published by Brill. It can be accessed here: https://brill.com/view/journals/jcmh/jcmh-overview.xml?lang=en
2 See the discussion in Johnston 1995. On Sunzi's *Art of War*, see Sawyer 1996. For a translation and detailed analysis of all seven of the military classics of early China, see Sawyer 1993. For a rejection of the pacifist thesis with regards to China, see Sawyer 2004: ix. Also see Swope 2015a: 597–634.
3 On this topic see Wang 2014.
4 On the evolution of China studies and their connection to broader political and intellectual trends in American society, see Cohen 2010.
5 See Parker 1996 and Parker 1995, especially pp. 2–9.
6 On the influence of Chinese martial arts films, in particular in the realm of popular culture in the West, see Hunt 2003.
7 See, for example, the fine essays in Cosmo 2009, especially the introductory essay on military culture, pp. 1–22.
8 See Holmes and Yoshihara 2008: 122–33. On the refutation of China as an imperialist power and its connection to the so-called New Qing History School, see Wu 2016, especially pp. 51–3.
9 Professor Li Zhiting, a prominent historian of the Qing working at the Chinese Academy of Social Sciences in Beijing, has been a particularly virulent critic of the Qing imperialism thesis, as he calls it. See the discussion here: https://www.jeremiahjenne.com/the-archives/2018/4/8/chinese-academy-of-social-sciences-throwing-shade-at-the-new-qing-history. This story includes links to the original Chinese article, which criticises Mark Elliott and other proponents of the New Qing History. Also see Li Zhiting, 'Xin Qingshi: Xin diguo zhuyi shixue biaoben' [The New Qing History:

The Root of Neo-Imperialism in Historical Studies] *Zhongguo shehui kexue xuebao* 4 (April 2015), http://sscp.cssn.cn/xkpd/zm_20150/201504/t201520_1592234.html

10 On this process as it evolved from the Ming to the Qing, see Shin 2006: 184–92. On the rationale for it in the early Qing, see Swope 2018: 256–71.

11 For the long view of expansion to the south-west, see Herman 2007.

12 On Wanli's patronage of military officials, see Swope 2004: 34–81. On Kangxi's practices, see Dai 2000: 71–92.

13 On the fall of the Ming and the problem of grand strategy at the end of the Ming, see the discussion in Swope 2014.

14 Theobald 2013. More recently, Yingcong Dai has also examined military finance, particularly corruption, with respect to the government's campaigns against the White Lotus rebels in 1796–1804. See Dai 2019.

15 On the remarkable nature of Late Imperial Chinese stability, see Kang 2010: 1–16.

16 Some of the most important primary sources include Song Yingchang, *Jinglue fuguo yaobian* [Important Documents Related to the Restoration of a Vassal State] 2 vols. (Taibei: Taiwan xuesheng shuju, 1986); Zhuge Yuansheng, *Liangchao pingrang lu* [Pacification Campaigns of Two Reigns] (Taibei: Taiwan xuesheng shuju, 1969); Qu Jiusi, *Zuben Wanli wugong lu* [Record of the Military Accomplishments of the Wanli Era] (Taibei: Yiwen shuguan, 1980); Mao Ruizheng, *Wanli san da zheng kao* [An Examination of the Three Great Campaigns of the Wanli Emperor] (Taibei: Wenhai chubanshe, 1971); and Qian Yiben, comp., *Wanli dichao* [Capital Gazette of the Wanli Reign] 3 vols. (Taibei: Qiming shuju, 1959).

17 Huang 1981: 156–88. For the recent reappraisal of Qi and a collection of essays addressing his broader impact, see Sim 2017. The film in question is Gordon Chan, dir., *God of War*, 2017.

18 A few recent examples include Zuo Zongtang, *Zuo Zongtang quanji* [Collected Works of Zuo Zongtang] 15 vols. (Changsha: Yuelu shushe, 2014); Sun Chuanting, *Sun Chuanting ji* [Collected Works of Sun Chuanting] 2 vols. (Taiyuan: Shanxi renmin chubanshe, 2018); Sun Chengzong, *Sun Chengzong ji* [Collected Works of Sun Chengzong] 3 vols. (Beijing: Xueyuan chubanshe, 2014); Yuan Chonghuan, *Yuan Chonghuan ji* [Collected Works of Yuan Chonghuan] (Shanghai: Shanghai guji chubanshe, 2014); Xiong Tingbi, *Xiong Tingbi ji* [Collected Works of Xiong Tingbi] (Beijing: Xueyuan chubanshe, 2010); and Yang Sichang, *Yang Sichang ji* [Collected Works of Yang Sichang] 2 vols. (Changsha: Yuelu shushe, 2008).

19 Waley-Cohen 2006: 20. Qianlong also commissioned a complete set of his own military writings and took to calling himself 'The Old Man of the Ten Complete Victories.'

20 For two examples, see Zhu Xueqin, comp., *Qin ding jiaoping nian (fei) fanglue* [Campaign History of the Pacification of the Nian Bandits] in *Zhongguo fanglue congshu* [Collected Campaign Histories of China] 32 vols., vol. 1 (Taibei: Chengwen chubanshe, 1968); Beijing Zhongguo shudian, comp., *(Qinding) Pinding qi sheng fanglue* vol. 21: *Pingding Shaan-Gan Xinjiang huifei fanglue* [Imperial Pacification Campaigns of Seven Provinces, vol. 21: The Pacification of the Muslim Bandits of Shaanxi, Gansu and Xinjiang] (Beijing: Zhongguo shudian, 1985).

Bibliography

Andrade, Tonio, *Lost Colony: The Untold Story of China's First Great Victory over the West* (Princeton: Princeton University Press, 2011).

Andrade, Tonio, *The Gunpowder Age: China, Military Innovation and the Rise of the West in World History* (Princeton: Princeton University Press, 2016).

Chan, Albert, *The Glory and Fall of the Ming Dynasty* (Norman: University of Oklahoma Press, 1982).

Chang, Michael G., *A Court on Horseback: Imperial Touring and the Construction of Qing Rule, 1680–1785* (Cambridge, MA: Harvard University Press, 2007).

Cohen, Paul A., *Discovering History in China: American Historical Writing on the Recent Chinese Past* (1984, reprint, New York: Columbia University Press, 2010).

Cosmo, Nicola Di (ed.), *Military Culture in Imperial China* (Cambridge, MA: Harvard University Press, 2009).

Crossley, Pamela Kyle, *A Translucent Mirror: History and Identity in Qing Imperial Ideology* (Berkeley: University of California Press, 1999).

Dai, Yingcong, 'To Nourish a Strong Military: Kangxi's Preferential Treatment of his Military Officials,' *War and Society*, vol. 18, no. 2 (October 2000), pp. 71–92.

Dai, Yingcong, *The Sichuan Frontier and Tibet: Imperial Strategy in the Early Qing* (Seattle: University of Washington Press, 2009).

Dai, Yingcong, 'Qing Military Institutions and their Effects on Government, Economy, and Society, 1640–1800,' *Journal of Chinese History*, vol. 1 (2017), pp. 329–52.

Dai, Yingcong, *The White Lotus War: Rebellion and Suppression in Late Imperial China* (Seattle: University of Washington Press, 2019).

Dreyer, Edward L., *Zheng He: China and the Oceans in the Early Ming Dynasty, 1405–1433* (New York: Pearson, 2007).

Elliott, Mark C., *The Manchu Way: The Eight Banners and Ethnic Identity in Late Imperial China* (Stanford: Stanford University Press, 2001).

Elverskog, Johan, *Our Great Qing: The Mongols, Buddhism and the State in Late Imperial China* (Honolulu: University of Hawaii Press, 2006).

Guo Wu, 'New Qing History: Dispute, Dialog, and Influence,' *The Chinese Historical Review*, vol. 23, no. 1 (2016), pp. 47–69.

Hanson, Victor Davis, *The Western Way of War: Infantry Battle in Classical Greece*, 2nd ed. (Berkeley: University of California Press, 2000).

Hanson, Victor Davis, *Carnage and Culture: Landmark Battles in the Rise of Western Power* (New York: Anchor Books, 2002).

Hasegawa, Masato, 'Provisions and Profits in a Wartime Borderland: Supply Lines and Society in the Border Region Between China and Korea, 1592–1644' (PhD Diss., Yale University, 2013).

Herman, John E., *Amid the Clouds and Mist: China's Colonization of Guizhou, 1200–1700* (Cambridge, MA: Harvard University Press, 2007).

Ho, Ping-ti, 'In Defense of Sinicization: A Rebuttal of Evelyn Rawski's Reenvisioning the Qing,' *The Journal of Asian Studies*, vol. 57, no. 1 (February 1998), pp. 123–55.

Holmes, James R. and Yoshihara, Toshi, 'China's Naval Ambitions in the Indian Ocean,' in Gabriel B. Collins, et al. (eds.), *China's Energy Strategy: The Impact on Beijing's Maritime Policies* (Annapolis: Naval Institute Press, 2008).

Hostetler, Laura, *Qing Colonial Enterprise: Ethnography and Cartography in Early Modern China* (Chicago: University of Chicago Press, 2001).

Huang, Ray, *1587: A Year of No Significance* (New Haven: Yale University Press, 1981).

Huang, Ray, *Broadening the Horizons of Chinese History* (Armonk, NY: M.E. Sharpe, 1999).

Hunt, Leon, *Kung Fu Cult Masters: From Bruce Lee to Crouching Tiger* (London: Wallflower Press, 2003).

Johnston, Alastair Iain, *Cultural Realism: Grand Strategy and Strategic Culture in Chinese History* (Princeton: Princeton University Press, 1995).

Kang, David C., *East Asia Before the West: Five Centuries of Trade and Tribute* (New York: Columbia University Press, 2010).

Kierman, Jr., Frank A. and Fairbank, John King (eds.), *Chinese Ways in Warfare* (Cambridge, MA: Harvard University Press, 1974).

Kim, Hodong, *Holy War in China: The Muslim Rebellion and the State in Chinese Central Asia, 1864–1877* (Stanford: Stanford University Press, 2010).

Kim, Kwangmin, *Borderland Capitalism: Turkestan Produce, Qing Silver, and the Birth of an Eastern Market* (Stanford: Stanford University Press, 2016).

Lei Haizong and Lin Tongqi, *Zhongguo wenhua yu Zhongguo di bing* [Chinese Culture and the Chinese Military] (1940, reprint), Changsha: Yuelu shushe, 1989.

Lim, Ivy Maria, 'The Lineage Organization in Ming China: A Case Study of Haining in the Sixteenth Century,' in Kenneth M. Swope (ed.), *The Ming World* (London: Routledge, 2019), pp. 258–78.

Lorge, Peter, *The Asian Military Revolution: From Gunpowder to the Bomb* (London: Routledge, 2008).

Miller, Harry S., *State Versus Gentry in Late Ming China, 1572–1644* (New York: Palgrave Macmillan, 2009).

Millward, James A., *Beyond the Pass: Economy, Ethnicity, and Empire in Qing Central Asia, 1759–1864* (Stanford: Stanford University Press, 2016).

Parker, Geoffrey (ed.), *The Cambridge Illustrated History of Warfare: The Triumph of the West* (Cambridge: Cambridge University Press, 1995).

Parker, Geoffrey, *The Military Revolution: Military Innovation and the Rise of the West, 1500–1800*, 2nd ed. (Cambridge: Cambridge University Press, 1996).

Parsons, James B., *Peasant Rebellions of the Late Ming Dynasty* (reprint, Ann Arbor: Association for Asian Studies, 1993).

Perdue, Peter, *China Marches West: The Qing Conquest of Central Asia* (Cambridge, MA: Harvard University Press, 2005).

Rawski, Evelyn S., 'Reenvisioning the Qing: The Significance of the Qing Period in Chinese History,' *The Journal of Asian Studies*, vol. 55, no. 4 (November 1996), pp. 829–50.

Robinson, David M., *Bandits, Eunuchs, and the Son of Heaven: The Economy of Violence in Ming China* (Honolulu: University of Hawaii Press, 2001).

Robinson, David M., *Martial Spectacles of the Ming Court* (Cambridge, MA: Harvard University Press, 2013).

Robinson, David, 'Why Military Institutions Matter for Ming History,' *Journal of Chinese History*, vol. 1 (2017), pp. 297–327.

Sawyer, Ralph D., *The Seven Military Classics of Ancient China* (Boulder: Westview Press, 1993).

Sawyer, Ralph D., tr., *The Complete Art of War* (Boulder: Westview Press, 1996).

Sawyer, Ralph D., *Fire and Water: The Art of Incendiary and Aquatic Warfare in China* (Boulder: Westview Press, 2004).

Shin, Leo K., *The Making of the Chinese State: Ethnicity and Expansion on the Ming Borderlands* (Cambridge: Cambridge University Press, 2006).

Sim, Y.H. Teddy (ed.), *The Maritime Defence of China: Ming General Qi Jiguang and Beyond* (Singapore: Springer, 2017).

So, Kwan-wai, *Japanese Piracy in Ming China during the Sixteenth Century* (Lansing: Michigan State University Press, 1975).

Sun Laichen, 'Military Technology Transfers from Ming China and the Emergence of Northern Mainland Southeast Asia (c. 1390–1527),' *Journal of Southeast Asian Studies*, vol. 34, no. 3 (October 2003), pp. 495–517.

Sun Laichen, 'The Military Implication of Zhu Wan's Costal Campaigns in Southeastern China: Focusing on the Matchlock Gun,' in Kenneth M. Swope and Tonio Andrade (eds.), *Early Modern East Asia: War, Commerce and Cultural Exchange* (London: Routledge, 2018), pp. 119–50.

Swope, Kenneth M., 'The Three Great Campaigns of the Wanli Emperor, 1592–1600: Court, Military, and Society in Late Sixteenth-century China' (PhD Diss., University of Michigan, 2001).

Swope, Kenneth M., 'A Few Good Men: The Li Family and China's Northern Frontier in the Late Ming,' *Ming Studies*, vol. 49 (Spring 2004), pp. 34–81.

Swope, Kenneth M., *A Dragon's Head and a Serpent's Tail: Ming China and the First Great East Asian War, 1592–1598* (Norman: University of Oklahoma Press, 2009).

Swope, Kenneth M., *The Military Collapse of China's Ming Dynasty, 1618–44* (London: Routledge, 2014).

Swope, Kenneth M., 'Manifesting Awe: Grand Strategy and Imperial Leadership in the Ming Dynasty,' *The Journal of Military History*, vol. 79, no. 3 (July 2015a), pp. 597–634.

Swope, Kenneth M., 'Ming Grand Strategy and the Intervention in Korea,' in James B. Palais (ed.), *The East Asian War, 1592–1598: International Relations, Violence, and Memory* (London: Routledge, 2015b), pp. 163–96.

Swope, Kenneth M., *On the Trail of the Yellow Tiger: War, Trauma, and Social Dislocation in Southwest China during the Ming-Qing Transition* (Lincoln: University of Nebraska Press, 2018).

Szonyi, Michael, *The Art of Being Governed: Everyday Politics in Late Imperial China* (Princeton: Princeton University Press, 2017).

Theobald, Ulrich, *War Finance and Logistics in Late Imperial China: A Study of the Second Jinchuan Campaign, 1771–1776* (Leiden: Brill, 2013).

Wade, Geoffrey, 'The Zheng He Voyages: A Reassessment,' *Journal of the Malaysian Branch of the Royal Asiatic Society*, vol. 78, part 1, no. 228 (2005), pp. 37–58.

Wakeman, Jr., Frederic, *The Great Enterprise: The Manchu Reconstruction of Imperial Order in Seventeenth-Century China*, 2 vols (Berkeley: University of California Press, 1985).

Waley-Cohen, Joanna, 'The New Qing History,' *Radical History Review*, vol. 88 (Winter 2004), pp. 193–206.

Waley-Cohen, Joanna, *The Culture of War in China: Empire and the Military Under the Qing Dynasty* (London: I.B. Tauris, 2006).

Wang, Zheng, *Never Forget National Humiliation: Historical Memory in Chinese Politics and Foreign Relations* (New York: Columbia University Press, 2014).

Weinstein, Jodi L., *Empire and Identity in Guizhou: Local Resistance to Qing Expansion* (Seattle: University of Washington Press, 2014).

8

WARFARE IN PRECOLONIAL AFRICA

O.R. Coates

'If the blaze of the fire I kindled does not consume them,' the Somali *sheikh* and warrior Sayid Muhammad 'Abdille Hasan threatened his English opponents, then 'I am not a true Muslim.' He chose the medium of Somali oral poetry to make his threat, and legitimised his claim in terms of Islamic notions of virtue and bravery (Samatar 1982: 190). Hasan's verse challenges a still pervasive notion of interminable and atavistic conflict that has long plagued the historiography of warfare in pre-colonial Africa. His call to resistance frames militarism within locally comprehensible cultural frameworks, in order to appeal to his Somali brethren. This is important to appreciate because the characterisation of unending, supposedly 'tribal' warfare persists in writing about conflict on the continent, even though it is chiefly informed by colonial racial stereotypes. Notions of 'primitive warfare,' whether in the colloquial or anthropological use of the term, continue to frame explanations of African warfare that obscure the agency of individual Africans and the range of African political, social and military institutions (Turney-High 1949: vii; Thornton 1999; Reida 2012a; Uzoigwe 1975: 469).

We shall see that warfare in pre-colonial Africa varied enormously in terms of strategy, weaponry, motivation and combatants. Considering the entirety of the continent, my survey will not consider warfare prior to 630 CE; it will draw to a close in the middle of the nineteenth century. 'Pre-colonial' here is taken to mean those periods prior to the consolidation of European power in Africa following the Berlin Conference of 1884–1885 (Falola and Heaton 2008: 86). This chapter will include discussion of Abbasid, Ottoman and Portuguese influence on the continent, as well as expansionism by African powers. Following a brief exploration of overarching questions in understanding pre-colonial warfare in Africa, my account will proceed geographically through north, west, east, central-west and south Africa. Where possible I will try to indicate the linguistic and cultural terms in which African societies conceptualised military organisation and warfare.

Precolonial African military history remains underdeveloped; while the early historiography of precolonial warfare focused on 'great men,' subsequent scholarship explored weaponry and technology, while a slimmer body of more recent research has examined the cultural and social contexts surrounding warfare (Uzoigwe 1970: 3; Uzoigwe 1977; Ade Ajayi and Smith 1964; Ogot 1972: 2; Crowder 1971; Goody 1971; Alpers 1975; Guy 1979; Reid

DOI: 10.4324/9780429437915-11

2007; Macola 2016). To labour the point, warfare in Africa cannot be understood through the racially charged notions of 'tribe,' or assertions of allegedly indeterminate and perennial violence; it was inextricable from its economic, social and cultural settings just as conflict elsewhere in the world.

Understanding Warfare in Precolonial Africa

Several major themes shape our understanding of conflict in precolonial Africa, particularly since 1500: the influence of the Atlantic and Indian Ocean slave trades, the role of horses and firearms, and more recent linguistic and environmental approaches. The first two of these remain better understood than the third. Horses played a major role in conflict in the Sahel, North Africa and the Cape; the potential of cavalry in the south of the Sahel but north of the Zambezi was severely limited by the presence of sleeping sickness, and the tsetse fly that served as a vector for the disease (Giblin 1990: 79; Ade Ajayi and Crowder 1985: 10–11). But rarely did these themes represent decisive causes in isolation. Even in the 'horseman's world' of the Sahel, low-status infantrymen still 'bore the brunt' of warfare; in any case, the cavalry were far from homogenous, and their techniques varied enormously over time and place (Iliffe 2004: 21).

Firearms undoubtedly played a major role in securing military supremacy. Introduced by European coastal traders from the 1500s, they became relatively common in the region in the second half of the eighteenth century, when West Africa received between 283,000 and 394,000 firearms from English suppliers alone, while gunpowder imports in the same period from Britain totalled some 50 million pounds (Aderinto 2018: 32). But they hardly represented an unqualified advantage. Weapons had to be well maintained and used with skill, and, in any case, they were of gravely limited use in certain environments such as on the forest paths of coastal West Africa and the Congo where visibility was poor. Despite these caveats, we should not assume that African soldiers necessarily had dilapidated weapons. Historians have been hasty to generalise across the continent, yet Ethiopia produced its own arms during the nineteenth century and boasted a sophisticated park of artillery. Guns also had an importance beyond warfare, becoming associated with gendered identity and social status in certain African societies (Macola 2016: 21–2).

Etymology and historical linguistics provide clues as to the longevity of conflict; one word for raiding in the Great Lakes region is at least 2,500–3,000 years old (Stilwell 2014: 46). Physical geography and the environment also played a key role in shaping conflict; I have already noted the significance of the tsetse fly in shaping conflict in tropical Africa. Coastlines posed similar constraints; beyond the Mediterranean, natural harbours were rare on the continent's coasts, limiting the bases for oceangoing naval forces. So, conflict instead tended to focus on lagoons, lakes and navigable rivers (Reid 2012a: 2).

African cultures boasted developed ideas of martial identity. Sometimes these were based on the prowess of the hunter; in 1659, the earliest known Hausa poet, Dan Marina praised the King of Katsina partly for his abilities as an elephant slayer (Iliffe 2004: 20–32). Lists of warriors and kings can be found dating back to the twelfth and thirteenth centuries in the nineteenth-century Kano Chronicle (Bobboyi and Hunwick 1991: 125; Hunwick and Lovejoy 1993: 95). Codes of martial honour determined warriors' behaviour in battle. For instance, when Songhai horsemen were routed, the enemy were sufficiently impressed by the bravery of the survivors that they tended to their wounds (Hunwick 1999: 147; Iliffe

2004: 17). In the Funj Sultanate, the *faris* or knight was praised in heroic poetry; the hero Hamaj Sinnar (1762–1821) was mythologised as a model of nobility and temperance (Spaulding 1985: 300). The figure of Sunjata, the reputed founder of the Mali Empire, mentioned by Ibn Khaldun and Ibn Battuta, forms the centrepiece of the complex heroic epic of Sunjata (Jansen 2011: 105–6). The Somali *sheikh* and Dervish warrior Sayyid Muhammad 'Abdille Hasan, who harried British, Ethiopian and Italian forces alike for two decades from 1900, was a talented political orator in his native Somali language (Samatar 1982: 189–90). The nineteenth-century resistance fighter 'Abd al-Kadir became a national icon who appeared in literature, biography and iconography (Achrati 2007: 146).

North Africa

The Islamic conquest of North Africa brought an Arab army of 4,000 men under 'Amr b. al-'Ās into Egypt in February 640; advancing on Alexandria, the Arabs were able to exploit local resentment at Byzantine exploitation. The Arabs moved inland and established Cairo as their capital at the head of the Nile Delta; their expansion continued in 642 CE, when a force of 20,000 under 'Ukba b. Nafi' reached the first cataract and Nubia. Here, Nubian archers formed a formidable force and their cavalry was used to devastating effect in hit and run attacks; while 'Ukba was forced to withdraw, one bruising engagement saw 250 Arabs lose their eyes to Nubian arrows. Ultimately, the Nubians were subdued following a treaty known as the *baqt* and agreed to send 360 slaves annually to Egypt. Nonetheless, further rebellions followed between 750 and 950 CE with Nubian states now joined under the kingdom of Dongola, and their armies raiding Upper Egypt (Stapleton 2013: 19–20).

Beyond Egypt, Arab attention soon turned to the Maghrib. 'Amr ibn al-'As undertook an unauthorised campaign between 642 and 645 CE, bringing Cyrenaica and Tripolitania under his control. In the Libyan interior, 'Ukba b. Nafi' led an Arab force against the Zawila oasis; he was to lead his troops to the 'Farthest Maghrib,' a term for modern-day Morocco, as well as to the 'Sea of Darkness' (the Atlantic) and the 'Sea of Sand' (or the Sahara). Participation in the *jihad* necessary to extend the caliphal territory brought rewards; Mu'awiya, founder of the Umayyad Dynasty, awarded pay, land and tax exemptions to warriors engaged in these conflicts. Arab rulers in North Africa cultivated new connections with sub-Saharan neighbours. 'Uqba also likely ventured far into the Libyan Sahara, while the governor of the Maghrib, 'Abd al-Rahman, ordered wells to be dug along routes leading from southern Morocco into the Sudan. Sub-Saharan Africans were also present in the Maghrib during this period; the Arabs and the Fatimids had corps of *'Abid* or black slaves (Levtzion 1979: 637–8, 641). On the shores of the Atlantic, the Arab conquest of the furthest Maghrib had reached its logical conclusion; as 'Uqba rode his steed into the surf, he cried: 'Oh God! If the sea had not prevented me, I would have coursed on forever like Alexander the Great' (Brett and Fentress 1998: 82).

The Berbers played a major role in resisting the Arab armies, although later they were incorporated into the ranks of their former adversaries as skilled warriors. A prophetess credited with second-sight, the Berber warrior al-Kahina, led a force to defeat the Arabs at Nini River and Gabes, before Hassan ibn Numan overcame the Berbers at the Second Battle of Gabes in 697, and killed Kahina outside Tubna in eastern Algeria (Brett and Fentress 1998: 85; Stapleton 2013: 21). Underlying the eighth-century revolts was the Berbers' status as simultaneously tribute-paying subjects of the Arabs and as warriors within their armies. Berbers had entered into military service as the Arabs gradually conquered their territories; Musa b. Nusayr's Arab forces, based at the *misr* or garrison city of Kairouan, traversed the Maghrib between 704 and 710, establishing bases at Tlemcen, Tangier and Tafilalet, a key staging post for the Saharan trade (Ajayi and Crowder 1985: 62–3). Berber warriors were integral to the Arab army from early as 711. The use of Berber *mawali* or clients provided the Arabs with a valuable source of manpower (Brett and Fentress 1998: 84). Tarik b. Ziyad led around 7,000 Berber horsemen to rout the Visigoths at the Barbate River in Spain, before a largely Berber force reached the outskirts of Toulouse in 721. For all their accomplishments, Berbers did not reach the status enjoyed by Arab warriors. Sporadic rebellions broke out in 739 and 740 influenced by Kharidjite factions, especially the Ibadi migrants who had formed communities at Tahert, Zawila and Sijilmasa. The Berbers resisted the authority of the Abbāsid *Caliphate* of Baghdad (Levtzion 1979: 641; Love 2018: 28).

The high numbers of Berbers within the Arab army drove the rulers of Egypt's Fatimid Dynasty to import Turkish slaves, known as *mamluks,* in an attempt to reduce their numbers. The Fatimid infantry was chiefly composed of sub-Saharan and Greek slaves. This contrasted with early Islamic Egypt, where the dominant South Arabian families relied on local militia controlled by the *sahib al-shurta* or chief of police. The Fatimid Dynasty rejected direct allegiance to the Abbasid *caliphs,* but maintained the infrastructure developed by their Tulunid and Ikhshidid predecessors, especially in terms of reliance on slave and mercenary armies recruited outside Egypt itself (Kennedy 2015: 264–5, 269). By the mid 1100s, taxation collapsed due to corruption and the Fatimid rulers could no longer afford to pay their soldiers; violent tensions inside the army emerged along ethnic lines as *mamluk,* Berber and Sudanese soldiers fought and undermined royal authority. A wholesale re-organisation of the army under Salah al-Din in the 1160s, with the aim of repulsing Christian crusaders in Palestine, did not prevent an unsuccessful rebellion of black Sudanese troops allied to the Fatimids in 1169 (Stapleton 2013: 22).

With the demise of the final Fatimid ruler in 1171, Salah al-Din established the Ayyubid Dynasty, a move facilitated partly by Salah's policy of incorporating landed estates into state ownership and administering them with *mamluk* officers. The *mamluks* became a landed elite in Egypt; ultimately taking power, and importing military slaves from the Turkey and Southern Russia, while expanding into Syria and Palestine, defeating the Mongols and European forces. In Africa, they turned Nubia into a vassal state, selecting a Muslim prince in 1315, and converting the cathedral at Dongola into a mosque. Ultimately, the era of *mamluk* domination was fatally challenged by the rise of the Portuguese in the Indian Ocean. A Venetian-sponsored *mamluk* fleet was defeated at Diu off the coast of India in 1509, precipitating the decline of *mamluk* influence in Arabia and the Indian Ocean.

Further west, in the Maghrib, Berber forces united during the 1040s under the reformist movement known as the Almoravids. Concentrating in the first decade of their existence to 1053, on forging an Islamic community among the Sanhaja, the Almoravids embarked on a campaign of military conquest, conquering Sijilmasa (1053), Awdaghust (1054), Sus and Dra'a (1056–1059) and Fez (1069), before expanding eastwards into modern-day Algeria in the 1070s (Abun-Nasr 1987: 81–2). Based around the new capital of Marrakesh, the Almoravid state was founded by Yusuf b. Tashfin. Adorned in the woollen clothes of the Bedouin, and leading an abstemious life reliant on barley and camel milk, Yusuf commanded his forces into his eighties, such as at the battle of Zallaqa in 1086 (Abun-Nasr 1987: 83). His army was reliant upon infantry using javelins for protection, although he also possessed cavalry and camels. Within this force, the Sanhaja constituted a dominant warrior class, distinguishing themselves with their trademark face mufflers (Stapleton 2013: 24).

The Almoravid supremacy in the Maghrib was challenged by another reform movement under Abu 'Abdulla Muhammad b. Tumart, from the Masmuda group of Berbers. Originally known as the *al-Muwahiidun*, or unitarians, Ibn Tumart's followers have come to be known by the Spanish corruption of 'Almohads' (Abun-Nasr 1987: 87). Ibn Tumart was a man of learning who detested what he regarded as the corruption of his faith by the Almoravid jurists. During his travels to the *mashriq* in search of learning, he was alleged to have met al-Ghazzali while in Baghdad. Bitter at the Almoravid prohibition of his books, the philosopher reportedly assured Ibn Tumart he would be responsible for ending Almoravid rule in the Maghrib by his own hand. In 1130, Ibn Tumart sent expeditions into the High Atlas with the aim of bringing the tribal groups there under his control. In the same year, a spring expedition against the Almoravid capital Marrakesh ended with defeat. In August 1130, Ibn Tumart died, but the event was reportedly kept secret by his inner circle of confidents for three years. 'Abdul-Mu'min took military control of the Almohad forces, commanding them for 33 years (Abun-Nasr 1987: 87, 90).

By 1140 'Abdul Mu'min had defeated Tashfin near Tilimsan. The defeated *caliph* fled to Oran, where he was killed in March 1145 as he attempted to flee the citadel on horseback. With Almoravid rule effectively removed, it was only a matter of time before the western Maghrib came under Almohad control. Fez fell in 1146, while the following year Marrakesh was conquered. On the site of the newly destroyed Almoravid palace, 'Abdul Mu'min erected the Kutubiyya mosque (Abun-Nasr 1987: 91). King Roger II's expansion from his Sicilian kingdom drove Almohad attentions to the eastern Maghrib. Mu'min's 1151 expedition changed destination from Spain to Algiers. Knowledge and military conquest were intertwined. A defender of Arab intellectual life, Roger II commissioned al-Idrisi's geographical survey of Africa *Nuzhat al-Mustaq* at the time of his plans for invasion.

The Ottomans occupied Cairo in January 1517. The new Egyptian Army included *mamluks*, who periodically rebelled and clashed with the Ottoman *janissaries*, as in 1523. Further reforms initiated by Ibrahim Pasha from 1525 ensured that the army was prohibited from extorting peasants, while separating *mamluks* and *janissaries* in most cases, and reorganising the army to include the *sipahis* (Ottoman cavalry armed with firearms), as well as an Anatolian infantry. The army was funded from tax and customs duties. Beyond Egypt, Ottoman influence in the Maghrib was experienced in fundamentally distinctive ways that require separate discussion. In Morocco, the Saadians prevented wholesale Ottoman domination. Sultan Abd Al-Malik raised a significant army to repel the Portuguese who had invaded in 1578. While the Ottomans provided political support, and Al-Malik's army was structured on Ottoman lines, with Turkish officers, though their influence remained limited. Al-Malik's successor, Ahmad al-Mansur, also used Turkish advisors to consolidate his army and adopted Turkish rank titles such as *pasha*, as well as the designation *sipahi* for his cavalry (Stapleton 2013: 28, 31). In 1591, al-Mansur conquered the Niger Valley Kingdom of Songhay, timing his invasion to coincide with the defeat of the Spanish Armada in the English Channel. The rise of the 'Alawid Dynasty in the 1660s, under Mawlay al-Rashid, created new alliances within Arab tribal groups, and saw an ethnic diversification of the military; the Wadaya Army, a largely Arab force, was based around Marrakesh, while *'abid* regiments concentrated near Miknasa consisted of sub-Saharan soldiers. These *'abid* soldiers were reorganised under Mawlay Sidi Muhammad III from 1757 and constituted a significant force in Moroccan politics from the 1770s when they were dispersed in an attempt to destroy their power. In 1798, Berbers and 'Abid soldiers helped Mawlay Sulayman establish order after a period of conflict (El Hamel 2012: 209, 230–7). The nineteenth century saw a period of decline in the Moroccan Army, and the French occupation of Algiers isolated the region from the rest of North Africa. The Moroccan Atlas would provide a power base for the Algerian rebel leader Abd al-Qadir in the 1840s.

In 1525, Khayr al-Din, a bandit leader of Greco-Turkish origin, captured Algiers from the religious leader Ahmad bin al-Qadi. The port became the lynchpin of Ottoman authority in the Maghrib. It was at the centre of a coastal province ranging from Mostaghanem to Jijel. Maritime exploits were key to the port's power, and sailors from around the Mediterranean mixed with Christian captives in the territory's state-managed piracy campaigns. Individual captains gained such distinction as to become popular legends, including Murad in the 1580s and Hamidu in the early 1800s (Abun-Nasr 1987: 153). The relationship between sailors and the powerful *janissaries* was not always harmonious. For instance, in 1671, a pirate rebellion overthrew local military leaders (Stapleton 2013: 37). Algiers continued to pose a major threat to Spanish shipping in the Mediterranean, claiming victims from the Mediterranean until the Second Barbary War saw the *dey* forced to agree to refrain from taking such captives.

Increasing numbers of Anatolian troops served in the army at Algiers. While local Algerians were excluded from service, those who were the product of Turkish soldiers' unions with local women, also known as *kulughlis*, were later recruited back into the *janissary* corps. Although the *janissaries* were proficient in firearms, disciplined and well-trained, their numbers declined markedly from 12,000 in the 1600s to 7,000 a century later, and a mere 4,000 in the 1800s. The military council, once a powerful force in Algerian politics, ceased to have real power from the 1700s. By the first decades of the nineteenth century, Algeria had become unstable and violent. To give an example, in 1805, Turkish troops massacred 200 Jews over wheat disputes, while between that year and 1816, seven *deys* were assassinated (Stapleton 2013: 37). The June 1830 French expedition to Algiers initiated French military domination in the region.

Ottoman control of Tunisia originated in Sinan Pasha's seizure of the Spanish fort at Halq al-Wadi in August 1574. The new administration focused on the military commander or *bey* and a *pasha* nominated by the Porte, and these officers were supported by some 4,000 *janissaries* under junior officers known as *deys*. By the late 1600s, the Murad family had come to dominate the politics of the interior, represented by a *bey* who stood increasingly opposed to the *deys* of Tunis. During the 1700s, the *deys'* office in Tunis was eroded and the status of Turkish soldiers challenged. The reorganisation schemes of Husayn bin Ali, founder of the Husaynid Dynasty, amassed a large local arm to defeat a tentative invasion from Algiers. Later Hasaynid rulers, such as Hammuda Bey, imported Turkish troops but attempted to prevent sustained interaction between the men and local people, often by insisting on the use of Turkish as the language of military command. Ahmed's rise to power in 1837 saw the founding of the military academy at Bardo and the formation of a large professional force of 26,000 men primarily loyal to Tunis. Despite an ambitious programme of military spending and building projects, the country amassed significant debts, prompting European powers to recognise Istanbul's long-standing claims to Tunis in 1871. Tripoli was taken by the Ottomans in 1551, and became the base for a powerful *janissary* contingent. Ottoman efforts to control the Saharan trade intensified with a 1577 expedition to the Southern Fezzan region. Some Turkish musketeers travelled to Borno in the Chad basin to aid its ruler

Idris Aloma, despite the fact that his requests for assistance had been rebuffed by Tripoli. In 1711, Ahmad Qaramanli, a *kulughli* cavalry leader, staged a coup in which 300 *janissary* officers were killed and the Ottomans were subsequently forced to recognise the Qaramanli Dynasty's authority. As in Algiers, a flourishing pirate trade operated out of Tripoli; ethnically mixed band of captains, including one Scotsman, became celebrated as heroes. Under Yusuf Pasha (1795–1832), a restored Libyan Fleet used the threat of piracy to extort money from European countries; Tripoli's lucrative piracy was only conclusively destroyed with the 1818 Congress of Aix-la-Chapelle (Stapleton 2013: 41, 43–4, 46). The decline of revenue from piracy drove Yusuf Pasha to plan the conquest of Borno, when in 1817, Muhammad al-Kanemi sought Tripoli's help in his campaign neighbouring Baghirmi, Muhammad al-Mukni, the *Bey* of Fezzan, led a force across the Sahara. The commitment was short lived as al-Mukni returned from one expedition to Borno with many slaves, but any long-term Libyan aid was quashed by British opposition (Abun-Nasr 1987: 200).

Following the evacuation of Napoleon's forces in 1801, a period of instability saw the British-backed *mamluks* vie for power with Ottoman *janissary* troops. A Turkish attempt to consolidate authority resulted in the rise of the Albanian Muhammad Ali, who cannily outmanoeuvred the Porte to take control of Egypt as *wali* at 35. In his early military exploits in Arabia, Ali had captured Mecca and Medina in 1803 and 1804 respectively. During the 1810s he consolidated his power, ordering the massacre of over 450 *mamluk amirs* in March 1811, and killing several thousand more in Upper Egypt the following year. A threefold expansion of annual revenue between 1805 and 1812 paved the way for military expansion from 1813 to 1829. Ali's campaigns of the 1810s yielded mixed results; he charged his eldest son Ibrahim Pasha with operations against the Wahhabis in Arabia from 1813, where he captured the Wahhabi capital of Dar'iyya in September 1818. This Arabian campaign proved a success and resulted in the Wahhabi leader 'Abdallah ibn Sa'ud being sent to Istanbul. By contrast, Ali's African expeditions south to Dongola and Kurfufan in 1820 were nothing short of a disaster. His Turkish army became sick due to the unfamiliar climate, lacked adequate transport and suffered a high rate of desertions (Fahmy 1998: 143, 145, 153).

The experience of the Sudan campaign showed that Turkish troops could not be relied upon for operation in Saharan Africa, and led Ali to restructure his army using conscripts from Egypt instead. He acquired 130,000 men in ten years (Fahmy 1998: 154). While Ali's officer cadre was reserved for Turkish speakers, the rank and file troops were Arabic-speaking peasants. He used his new army to dominate Sudan and to extend Egypt's influence in Arabia into the Yemen and consolidate his power within Egypt. During the 1820s, Ali's relation with the Porte vacillated; he refused help to Istanbul in June 1826, but then committed and lost his fleet at the Battle of Navarino the following year (Fahmy 1998: 155). Further expensive reforms saw seven cavalry regiments created around a new school at Giza, a fleet built inside Egypt and the creation of the *liman* of Alexandria, a feared prison and arsenal relying on the forced labour of convicts (Fahmy 1997: 72). In a final era of military expansion from 1829 to 1841, the *pasha* sent troops to Syria, a possible mustering point for any Ottoman invasion of Egypt, and his son entered Anatolia in 1832. While Egyptian forces briefly threatened Istanbul, the May 1833 Peace of Kutahia saw the Porte recognise Ali's authority in Egypt, the Hijaz and Crete, with Ibrahim receiving Syria (Al-Sayyid Marsot 1984: 231). Further conflict emerged when Ibrahim defeated Ottoman forces north of Aleppo in 1839, but a combination of European fears over Egyptian expansion in Arabia, and increasing economic problems inside Egypt, saw the Ottoman sultan limit Ali to 18,000 troops in peacetime in June 1841. Although he was named governor of Egypt for life, the *pasha*'s military power was spent (Fahmy 1997: 264).

West Africa

West Africa was home to a series of powerful Sahelian empires dating back to 1000 CE. The coastal regions became important from the mid-sixteenth century, as European traders and slavers operated along the coasts of modern-day Ghana, Nigeria and Benin. From 700 CE, the Kingdom of Ghana developed between the Senegal and Niger rivers (McIntosh and McIntosh 1988: 146). Banding together in self-defence, the Soninke farmers produced swords, spears and arrowheads, and they became middlemen in the trade across the Sahara (Stapleton 2013: 70). Their ruler, based at his capital of Kumbi-Saleh, acquired wealth from taxation of gold mining and commerce in salt and gold. He brought Arabic-speaking merchants and scholars to live in Ghana, who built mosques and resided in a dedicated quarter (Levtzion 1979: 668; 1980: 17). By the mid-eleventh century, Ghana had become the major kingdom of the Sahel; the conquest of the Tuareg trading centre of Awdaghust in 1050 marked its furthest expansion north (Levtzion 1979: 670). The Soninke Kingdom's predominant route of expansion was southwards, with migrants settling along the Niger Valley. The size of Ghana's Army remains obscure as it did not keep a standing military, and while there is no evidence of a cavalry, gold décor for horses was produced in the kingdom (Stapleton 2013: 71). Ultimately, a combination of ecological change, and new trading routes that bypassed Awdaghust to the east, precipitated the collapse of Ghana. The southern margins of the Sahara around 1600 were 250–300 km north of their equivalent position by 1850 (McIntosh 2008: 352; Stapleton 2013: 71).

By the mid-1230s, the Malinke had organised around Sundjata. His cavalry-based army began a campaign climaxing in the Battle of Kirina in 1235, where Sundjata defeated the Sosso and established authority of the Soninke people (Stapleton 2013: 73). Sundjata became the *mansa* or leader of the Mali Empire. Following his death in 1255, Sundjata's successors Mansa Ouali and Mansa Sakura extended the authority of Mali over Gao, the Takrur Kingdom (in modern-day Senegal and Mauritania), and also conquered key Tuareg copper towns in modern-day Niger (Levtzion 1977: 331–5). Mansa Musa's 1325 observance of the *hajj* attracted Arab scholars to the region, turning Jenne, Timbuktu and Gao into influential intellectual centres. Mali continued as a major power into the fifteenth century, but by the 1430s began to lose control of the Saharan trade. The Mali Empire successfully repelled European seaborne invasion. In the 1440s and 1450s, Portuguese slave raiders along the Atlantic Coast were repelled by client warriors of Mali using poison arrows and shallow boats (Stapleton 2013: 74).

Songhay, under the Sonni Dynasty, based along the Niger bend, rose to fill the vacuum left by an increasingly frail Mali. Under Sonni Ali, their armies captured former Malian cities such as Timbuktu, then occupied by the Tuareg, in 1468, and Jenne in 1480. A general of Sonni Ali, Muhammad Turi, took power after Ali's death and further strengthened Songhay, rejuvenated the Saharan trade, and came to be known as Askia the Great. Songhay expanded northwards with campaigns against the Tuareg, and east to subjugate Gobir, Katsina and Kano. Like Ghana before it, Songhay had no standing army and the individual governors were tasked with recruiting warriors. The sixteenth century saw 'Alawid intervention in Songhay, with Ahmad al-Mansur 1590 invasion, with 1,500 lancers, 500 mounted gunmen with arquebuses, as well as 2,000 infantry. The Moroccan force faced 10,000–30,000 Songhay infantry and 12,500–18,000 cavalry (Stapleton 2013: 75). 'Alawid victory on 12 March 1591 paved the way for the seizure of Gao and Timbuktu. By the time Morocco abandoned direct rule in 1610, Songhay had splintered into autonomous towns.

The Sahel of the eighteenth and nineteenth centuries witnessed a succession of conflicts leading to the creation of new Muslim states. Each *jihad* had different expressions and aims, resulting in a differing form of political and military organisation. While the Sokoto *Caliphate*, and the empires of Samori Touré and al-Hajj 'Umar Taal, are relatively well known, shorter-term *imamates* were also established by Ma Ba Jaxu, Cheikhou Amadou Ba and al-Hajj Mamadu Lamin Dramé. Many early *jihad* leaders were linked to the Qadiriyya Sufi brotherhood or *tariqa*, whereas later ones tended to be Tijani. Except for Samori Touré, instances of *jihad* were not interrelated, but instead had largely separate motivations, such as opposition to slave-raiding, exploitation and injustice, and economic grievances. Instances of *jihad* included Nasar al Din's insurrection in the Senegal Valley in 1673/4, Bundu (1680s), Fuuta Jalon (1725), Hausaland (1804), Masina (1818), the wars of al Hajj Umar Taal (1850s) and Senegambian uprisings (1860s), as well as Samori Touré's *jihad* of 1875. The *mujahideen* did not necessarily experience victory, as the wars of Nasir al-Din and the Bornu *jihad* of the nineteenth century failed (Loimeier 2010: 270–2).

'Usman dan Fodio (1754–1817) was one of the preeminent military leaders of the era and he continues to inspire political, social and military activism in the present. Born in Maratta in the powerful Hausa state of Gobir, 'Usman studied with his father at Degel, moving between teachers, before embarking on a peripatetic teaching career across Hausaland in the later 1770s and 1780s (Stapleton 2013: 275). 'Usman maintained links with the ruling class of Gobir and taught some members of the royal family (Robinson 2007: 144). He wrote critiques of the un-Islamic innovation (*bid'a*) in a series of texts that engaged with the work of Malikite jurist 'Abd al-Karim al-Maghili (Loimeier 2010: 275). Uthman and his son Muhammad Bello had to justify waging *jihad* against other Muslims, and deployed the idea of apostasy (*irtidad* or *ridda*), arguing that their enemies had engaged in friendship (*muwalat*) with unbelievers. These textual arguments enabled them to justify the waging of war against other Muslims, particularly neighbouring Borno (Kariya 2018: 181; Last and Al-Hajii 1965: 231). By the early 1800s, reformers including the *shaikh*'s family became increasingly aggrieved with Gobir, emigrated from the perceived land of unbelief (*hijra*) and, from their new vantage point in the countryside, declared war on their erstwhile hosts. They declared all preceding Hausa states as vestiges of the age of *Jahiliyya* or ignorance. By 1808, 'Usman subdued the Sultan of Gobir, and in 1809 established Sokoto and the adjacent sultanates such as Katsina and Kano were overthrown by military commanders who had once been students of 'Usman (Robinson 2007: 144). After 'Usman's death in 1817, Muhammad Bellow took the title of *khalifa* and ruled over a grouping of emirates that sent tribute to Sokoto, collectively known as the Sokoto *Caliphate*.

'Usman's activities in Hausaland inspired the 1805 rebellion of Fulbe (Fulani), against the ruling Sayfawa in Bornu, to the east in the Central Sudan (Hiribarren 2017: 1). The Sayfawa Dynasty had evolved from a cluster of social groups in the area of Kanem to the north-east of Lake Chad sometime around 900. They founded a capital at Gazargamu in the latter half of the fifteenth century but by 1800 the Borno had gone into decline (Aminu 1981: 31). The Fulani *jihadists* took advantage of this state of affairs and destroyed Ngazargamu in 1808, but their activities were halted by Muhammad al-Kanemi, a religious and military leader. He became the *de facto* ruler of Bornu in 1809 (Abun-Nasr 1987: 200). Military activities had long played a key role in Borno expansionism, although privileged traders and other could be granted a *mahram* or exemption from military service (Aminu 1981: 32). Kanemi reorganised Borno's forces, placing an emphasis on the nomadic Kanembu infantry and Shuwa Arab light cavalry. In the 1820s, a British explorer estimated Kanemi's army as consisting of some 30,000 cavalry and 9,000 Kanembu spearmen (Stapleton 2013: 81). Kanemi's correspondence with 'Uthman challenged the very legitimacy of waging *jihad* against Borno and, in particular, Sokoto's willingness to label Muslim rulers who helped those who attacked specific groups of Muslims, such as other Hausa Sultans, as 'unbelievers' (Abdullah 2009: 341; Last and Al-Haji 1965: 234–5). Further south, the so-called Slave Coast (consisting of modern-day Benin and South-Western Nigeria), as well as the Gold Coast (Ghana) became major centres for slave exports during the Atlantic slave trade from 1500. On the Gold Coast, Dutch, English, French and Danish traders competed along a 400-mile coastline. Due to its increasing hunger for human beings, the coastal trade stimulated the growth of a succession of empires in the interior and these polities engaged in warfare in order to obtain slaves to trade for cloth, metal goods and cowries and these commodities were in turn used to purchase horses and guns (Reid 2012a: 79).

From the seventeenth century, Oyo expanded north of the forest belt thanks to a developed cavalry. Although it had attacked the 'Benin Gap,' a sparsely forested patch of Southern Benin, Oyo lacked a direct link to coastal merchants until 1730, when Dahomey was given tributary status (Reid 2012a: 79). From the eighteenth century, Oyo became a key slave producer. Although it lacked a standing army, Oyo had a force of around 100,000 men, including a cavalry using lances and javelins, as well archers. Despite this impressive retinue, firearms remained relatively peripheral to Oyo's Army. In one charge against the Dahomians in 1726, the Oyo Army was rumoured to have suffered because their horses were unaccustomed to the sound of musket fire (Law 1977: 187–8). The army was organised under the authority of the *alafin* and the *basorun* who was commander of the metropolitan army, as well as a senior political chief, followed by the *kakamfo* or commander of the provincial army, and 70 *Eso* or war chiefs who resided in the capital. The officer corps was controlled by the *oyo mesi* or central council of chiefs (Law 1977: 188; Reid 2012a: 58). By the late eighteenth century, the army had become a key element in Oyo politics. When the *alafin* Awole sent units against his *kakamfo* Afonja, the latter marched on to Oyo, forcing the suicide of the *alafin*, while the behaviour of *alafin* Karan caused a mutiny (Reid 2012a: 81).

The Kingdom of Dahomey expanded with the slave trade to become a formidably militarised state. Military campaigns often became self-perpetuating because war captives were vital to sustaining the Dahomian state itself. During the 1720s, the kingdom conquered the coastal regions of Whydah and Allada in order to gain access to trade, and from 1730 the kingdom became a tributary of Oyo. Political and military affairs were often inseparable in the kingdom. During the eighteenth century, young boys were trained for the Dahomian Army, while by the nineteenth century the kingdom's famous female regiments, dubbed 'Amazons' by European observers, engaged in battle (Reid 2012a: 82–3; Alpern 1998: 9; Vandervort 1998: 14). The Dahomian Army was organised around the *agau*, assisted by his deputies the *zohenu*m and the *fosupo* (Reid 2012a: 83).

In modern-day Ghana, the Asante Kingdom emerged from a federation of Akan states during the seventeenth century, in response to competition for gold and land, as well as the opportunities posed by growing coastal trade (Reid 2012a: 83). In the 1670s, Osei Tutu, the Asantehene assumed control over Kumasi, creating a network of tributary chiefdoms, the 'golden stool' at Kumasi as a symbol of national unity, and a national army. The Ashanti Army was not a standing army, but a small group based in the capital served as the nucleus around which men from across the Empire could be raised, and at full strength the Ashanti forces consisted of about 100,000 men. Actual forces were far smaller; when the Ashanti Army fought the British Expeditionary Force in the Anglo-Ashanti Wars, it numbered around 40,000. The army boasted a command structure based around the king, privy council, and a general staff and the military police enforced heavy penalties for desertion, cowardice or failure to report for service. Entirely infantry-based, the Ashanti forces made use of 'Long Danes' or trade muskets that, while unwieldy, weighing nearly 20 lb, were used effectively against African and British opponents. A further unique feature among African militaries was the fact that, in the Asante Army, the wounded soldiers were treated by medical orderlies (Vandervort 1998: 15–16). For all its strength, the Asante Army was reliant upon slaves. In the late eighteenth century most of the slave recruits were obtained as tribute from provinces north of Kumasi (Reid 2012a: 86).

After the decline of Oyo during the 1830s, a power vacuum was created in Southern Nigeria. Violent conflict engulfed the major nineteenth-century Yoruba states of Ijaye, Ibadan, Abeokuta and Ijebu, and at the same time, the British presence in the region was consolidated following the annexation of Lagos in 1861. The Yoruba Wars can be divided into two halves, from the 1830s to 1878, and from that year until 1893. In the first stage, Ibadan emerged triumphant after wars with Ijaye and the Abeokuta based Egba, while in the second, several Yoruba states collaborated to block Ibadan's power. Before this latter conflict reached a definitive conclusion, the British seized the chance to intervene in 1892–93 to impose peace (Reid 2012b: 36). The Yoruba generally relied on foot soldiers from the 1830s, because fighting had moved into the tsetse fly zone, but they still fought with cavalry in the Ijaye War of the 1860s. A rich vocabulary described the Yoruba arsenal; the *ida* was a straight, narrow-bladed sword, whereas the *agedengbe* was a backwards-curving iron blade, and the *ogbo* was like a scimitar (Smith 1967: 90–2, 97–8, 102). Spears formed the inevitable complement to these swords, as well as the *orun* or bow, and *ofa* or arrow, along with clubs; in the latter part of the nineteenth century, the *ibon ilowo* or hand-gun was introduced, largely for horsemen.

East Africa

Between 1500 and 1880 the role of violence in many sub-Saharan societies escalated due to a variety of factors, such as external trading, the development of a slave export trade outside of the continent and the availability of firearms. In the 1880s, between 80,000 and 100,000 firearms entered East African coastal towns bound for the interior. Many of these coastal towns with ports had developed before 1400, with the Swahili serving as intermediaries in the trade between Arab sailors and the peoples of the interior (Beach 1962: 453; Stilwell 2014: 42). Portuguese and Ottoman expansion increased in the 1500s. Fort Jesus at Mombasa was completed in 1593–6. Lisbon's influence declined in the face of successive Omani raids in the seventeenth century, leading to the Omani seizure of Zanzibar in 1698 (Stapleton 2013: 113). By the nineteenth century, the Sultanate of Oman had established itself as a major power participating in the Indian Ocean trade (Coquery-Vidrovitch 2009: 86–7).

In the 1630s, Ethiopia enjoyed relative stability after 100 years of upheaval; Gondar was a major urban centre (Reid 2012a: 91). The imperial army played a key role; units made up of the powerful Oromo were co-opted into the force and the leaders could raise forces of between 50,000 and 100,000 men, and significant groups of attached non-combatants (Reid 2012a: 91). The Ethiopian Army was frequently deployed against Oromo groups on the empire's restive eastern frontiers. By the late eighteenth century, the country entered a period of extended warfare and the imperial palace at Gondar existed only as a shadow of its former self by the 1770s. The nineteenth century witnessed an attempt to revive the historic Solomonic Empire under Emperor Tewodros II (1855–1868). Tewodros's attempts to forge a national army under his own authority led to civil strife inside Ethiopia when faced with an Anglo-Indian expeditionary force in 1867. After being defeated at the battle of Aroge on 10 April 1868, he unsuccessfully attempted to broker peace with the British and subsequently killed himself (Stapleton 2013: 124–5; Vandervort 1998: 23). Under Yohannes IV, the Ethiopians fought successful campaigns against the Western-style Egyptian armies, obtaining Remington rifles and artillery from the vanquished forces, especially after having defeated the Italians at Dogali in 1887. The Emperor Yohannes IV died leading his men against Mahdist forces the following year. His successor, Menelik, led the victorious Ethiopian force of around 100,000 men against the Italians at Adowa in 1896; 70,000 of whom had repeating rifles superior to some of their opponent's weaponry (Vandervort 1998: 24). The only sub-Saharan army with its own extensive artillery in the nineteenth century, the Ethiopians had developed their own domestic armaments industry and placed riflemen as the core of their infantry.

In the Great Lakes region, to the south of Ethiopia, the centralised states of Bunyoro and Buganda had well-developed military traditions. During the eighteenth and nineteenth centuries, violence escalated, largely as a result of growing competition for resources and the expansion of Indian Ocean trade in the nineteenth century, and the spoils of war necessitated an ever-expanding cycle of militarism (Reid 2012a: 115; Stapleton 2013: 111; Cohen 1989: 270–1). On the shores of Lake Victoria, Buganda's growth was driven by the existence of banana plantations and a relatively dense population that could provide well-nourished troops for the overlords. Thanks to its canoe fleet, Buganda dominated trade on the northern stretches of Lake Victoria and it came into direct conflict with its older neighbour Bunyoro. During the reign of Kabaka Mawanda, between 1700 and 1750, Buganda entered a period of military expansion. The military appointments came under central control, and a new system of command allowed rapid mobilisation. Warfare intensified, as armies increasingly raided for cattle and slaves, or to gain possession of new provinces. The Ganda eschewed their bows in favour of wooden or iron-tipped spears. Under Mutesa (1857–1884), the Ganda Army expanded in a corridor to the west of Lake Victoria, while their canoe fleet grew into one of Africa's largest. In Bunyoro, military re-organisation led to the emergence of new rifle regiments, known as *barusura* in the 1870s. The Ganda were weaker with firearms than their Nyoro neighbours, with some fighters even viewing them as indicative of cowardice (Reid 2012a: 92, 116). Adjacent to the lacustrine region, Rwanda and Burundi also boasted long-standing traditions of militarism. By the eighteenth century, the Tutsi had developed a distinctive ethnic identity intertwined with their self-professed social status over the Hutu and used this to perpetuate social dominance.

On the East African grasslands in the area of modern-day Kenya and Tanzania, new warring groups and states emerged; the influence of the Indian Ocean trade in ivory and slaves played a key role by the 1800s in accelerating conflict. In this region, Maasai pastoralists constituted a feared and efficient military force. Organising themselves according to age sets, the Maasai were able to mobilise over large distances and defend their far-flung herds

(Reid 2012a: 95; Stapleton 2013: 138). This mobility was perfected by units of spearmen who were able to stage raids on other inhabitants of the pastureland and they used large, socketed blades to overwhelm the lighter spears of local competitors. These attacks proved integral to Masaai expansionism as they fought with the Kikuyu and Kamba. Unabashed in the acquisition of as much land as possible, the Masaai boldly claimed possession of all the world's cattle. Swayed by the Masaai's fearsome reputation, Arab and Swahili traders avoided taking the most direct routes between Zanzibar and Lake Victoria. Their concerns were well-founded as in one 1857 attack, according to Richard Burton, 800 Maasai destroyed a force of 148 Arab and Baluchi gunmen (Stapleton 2013: 138, 140). Although armed Arab-Swahili traders were beginning to venture into the interior by the 1840s, there was an important role for some groups in the region, particularly the Kamba, in acting as middlemen on the trade routes to the coast (Reid 2012b: 43).

To the south, around Lake Tanganyika, the dispersal of the Ngoni peoples from south of the Zambezi caused considerable disruption. The exact influence of these peoples in Eastern Africa, many of whom were not ethnically Ngoni, but had instead absorbed the latter's military techniques, remains controversial. Nonetheless, they played a key role in revolutionising warfare in the region (Reid 2012a: 111). By the middle of the nineteenth century, these groups raided throughout the region. They had a fearsome military reputation based on their adoption of Zulu military techniques, including the use of the short-stabbing spear known as the *assegai* and their agile mobility during raids (Reid 2012a: 96). The proliferation of lethal violence drove some local groups such as the HeHe of the Tanzanian Highlands to become major military forces in their own right so as to better repel the Nguni, while adopting many of their tactics, such as techniques of 'total warfare,' in which some Nyamwezi, Kimbu and Sukuma warriors of the 1880s all killed women and children in conquered settlements (Fage 1996: 320). A paradigmatic product of this unstable era, the Nyamwezi leader Mirambo adopted close-combat tactics and aggressive expansionism, while repelling the last of the Ngoni invaders, and thus dominated tracts of territory in modern-day Tanzania in the 1870s and 1880s (Reid 2012a: 117). Armed bands of young men with muskets, known as the *ruga ruga* were central to Mirambo's state. They continued to play an important role throughout East Africa in multiple arenas, from armies to trading caravans (Iliffe 1979: 87). Despite his dramatic rise to power, Mirambo's death in 1884 presaged the demise of his kingdom. Mpandoshalo, his successor, could not maintain stability, and the kingdom disintegrated by the late 1880s (Reid 2012a: 117; Reid 2012b: 56). Mirambo's state was not unique in the region; another Nyamwezi leader, named Msiri, built Garenganze during the 1860s and 1870s (Reid 2012b: 56).

Central Western Africa

The Congo Basin and the region of modern Angola experienced severe pressure from the Atlantic and the Indian Ocean trade for slaves and ivory. Integration into both trading networks prompted an intensification of violence from the 1500s. From 1000, the Atlantic Coast was home to a patchwork of small independent kingdoms united by common cultural links (Nziem 1998: 80). The Kingdom of Kongo, with its royal capital and administrative centre of Mbanza Kongo boasting some 10,000 to 15,000 inhabitants, emerged as pre-eminent among these coastal states (Stapleton 2013: 163). Its history can be divided into two periods; the first, from 1483 to 1540, was marked by its exposure to new influences such as coastal trade and Christianity, with the second moving from Kongo's height to its division following the 1665 Battle of Ambuila (Nziem 1998: 85). From 1512, Kongo undertook a campaign against the Mbundu people to the south. Kongo's infantry included heavy infantry with shields, using spears and bows, and often dodging enemy projectiles. Each provincial governor in Kongo drew on a professional force consisting of lower nobility, as well as conscripts; the Congo River served as the major transport artery, with a fleet of 800 dugout canoes each carrying 150 people. Kongo's southward expansion generated enormous local resistance, leading to the deaths of Bernardo I in 1567 and Henrique I in 1568 in battle (Stapleton 2013: 149, 163). The 1560s were marked by a series of attacks on the Kongo Kingdom by the Jaga and other local opponents who seized the capital and sent the King into hiding. The kingdom was rescued when, in 1571, King Sebastian of Portugal sent an expeditionary force of 600 men, and defeated the invaders (Nziem 1998: 90). Ultimately, the Portuguese presence in Kongo provoked a fearful Ndo Luvwalu I to seek protection from Rome. Successive monarchs approached non-Catholic European powers such as Holland in an effort to block Portuguese influence. In 1622, the governor of Angola attacked a Kongo chief whom he alleged to be harbouring Angolan fugitives; although the Mani Mbamba, the Kongolese military, intervened, and Portugal retracted its governor. This incident heightened tensions between the two kingdoms (Nziem 1998: 91; Thornton and Mosterman 2010: 235). Despite this respite from hostilities, tensions later flared when a future governor of Angola attempted to prospect for mines inside the kingdom. Luanda promoted rebellion against Kongo in the kingdom's southern territories, and the resulting battle at Ambuila on 29 October 1665 saw the destruction of the Mani Mbamba (Nziem 1998: 92). A period of fragmentation persisted in the Kongo area until the nineteenth century.

The nineteenth century witnessed increasing European and Arab influence in the Congo Basin (Nziem 1998: 267). The Western Congo Basin became increasingly integrated into Atlantic commercial networks, with local peoples using large dugout canoes to transport ivory and other goods. Ivory and slaves were also key commodities in the eastern tracts of the Congo Basin (Mumbanza 1997: 275–6). The Swahili-Arabs, originally from Zanzibar, were influential traders and slave raiders in this economy. Prominent among these 'Congo Arabs' was the trader and warlord Hamid bin Muhammad al-Murdjibi, better known as Tippu Tip (1837–1905), who led expeditions around Lake Tanganyika, and connected the Manyema region (Eastern Congo) to the Zanzibari trading network (Sicard n.d.). Although he cut the figure of a dandy, the Afro-Arab slave raider and merchant was also capable of commanding considerable amounts of organised violence, particularly from the bands of *askari* who searched the villages of Manyema for slaves (Page 1974: 70). Tippu Tip's apparent reluctance to cease slaving, prohibited in the Congolese Free State area since the Treaty of Zanzibar in 1887, combined with Belgian paranoia about Tip's supposed links to his Mahdist co-religionists in the Sudan, formed the justification for the 1890 *Force Publique* offensive against the Swahili-Arabs known as the Congo Arab War. In this war, tens of thousands died, although Swahili-Arabs continued to play an important role in the Belgian Congo into the 1920s (Vandervort 1998: 140–1; Draper 2019: 1021; Bontinck 1999: 379).

Southern Africa

Peoples of the South African interior had long been instrumental in trading with Dutch visitors to the Cape, as well as intermediaries of the Portuguese in Angola. Khoekhoe traders brought beads and other goods into the interior to trade with the Tswana and other peoples. At Delagoa Bay on the Eastern Coast of Southern Africa, traders arrived to obtain ivory from local chiefdoms. The evidence is patchy, but societies in the interior of the late eighteenth and early nineteenth century appear to have been become increasingly unstable and militarised. Settlers were hardly exempt from this process. Dutch colonists on the Cape sent Commando raids to seize San women and children for their farms, whilst raiders such as Coenraad de Buys operated outside of any state oversight. It appears that many polities were becoming larger and more centralised in the same period, although any ready connection between foreign trade and political development remains unproven. For example, a ruler like Makaba (ca. 1790–1824) of the Bangwaketse presided over a stratified society with a hierarchy consisting of the ruling family and a tier of subordinate groups, and a low status grouping of clients (Wright 2009: 214, 218).

A particular centre of conflict emerged between the Drakensberg Mountains and the Indian Ocean. Here, Shaka ka Senzangakhona came to power in the Zulu chiefdom. The Zulu already had a system of *amabutho* or regiments, originally for males, but later for females too. These units originally undertook hunting duties but became increasingly militarised over time. Even marriage became attached to the *amabutho*, as Shaka gave men permission to seek wives from a designated women's regiment. Zulu society operated according a tripartite division between the aristocrats, including the *izikhulu*, an intermediary class of chiefs, and low-status individuals relegated to geographically peripheral regions, sometimes designated as *amalala* or menials. Shaka's key victory over the abakwaNdwandwe at Izindololwane in 1826 marked a turning point in South African history as it ensured that the AmaZulu and the abakwaGaza Kingdoms were the dominant polities of the eastern interior. Despite this triumph, Shaka's defeat of his principal external opponent simply cleared the way for greater challenge within the Zulu royal house itself. The death of Shaka's mother Nandi in August 1827 was followed by a succession of killings (Wright 2009: 229–31, 236).

Significant population displacement occurred in the South African interior in the first decades of the nineteenth century. Some of this was due to the campaigns of Shaka and others east of the Drakensberg, but this was not such an important factor in population displacement as had been imagined by some earlier twentieth-century historians (Etherington 2004: 203–4; Hamilton 1992: 37; Wright 2009, 214). The *Mfecane* or Shaka's wars, once solely credited with driving population displacement as far as the grasslands of Eastern Africa, were overexaggerated by too great a reliance on Zulucentric sources, and little attention to the role of other Southern African peoples, or the activities of settlers and environmental factors. Displacement appears to have been the result of multiple conflicts that involved African and European actors, as well as potential factors such as overpopulation (Worden 2000: 14–15). Nonetheless, significant numbers of Africans were displaced both southwards into the Lower Orange Valley, and into the northern Veld and modern-day Zimbabwe and Mozambique. The Ndebele, led by Mzilikazi, a former lieutenant of Shaka's, moved through Mozambique, Eastern Zambia and Zimbabwe in the 1820s and 1830s. Ngoni groups crossed into the regions of Zambia, Malawi and even Tanzania. A further driver of militarisation in Southern Africa from the 1830s was the Great Trek or Boer settlers across the Orange and Vaal rivers. Possession of firearms initially allowed the Boers to broker trading agreements with Africans, but their expansion into the interior and the foundation of the two Boer 'republics' of the Transvaal and Orange Free State caused inevitable conflict with those groups dwelling in these areas. In the 1850s and 1860s, conflict between the Boers of the Orange Free State and Moshoeshoe's Sotho forced the latter to seek British protection (Reid 2012a: 221, 252). The Zulu Army experienced some important victories in the Anglo-Zulu Wars. For instance, in 1879 they inflicted a major defeat on British forces at Isandlwana, before their final defeat at Ulundi (Worden 2000: 26).

Conclusion

Standing armies were rare in Africa until the late nineteenth century. Three types of military organisation were present: bands of locally raised forces were most common (as in Ethiopia and the Ashanti Empire), amalgamations of soldiers into regiments (such as among the Ngoni peoples, the Zulu, the Ndebele), and 'citizen armies' consisting of adult males capable of bearing arms (Samori's forces, the Madhist state in the Sudan, 'Abd al-Kadir's Algeria). Within these forces, the cavalry often consisted of an aristocratic elite, while the low-status individuals served in the infantry; however, exceptions complicate this assessment, such as the Muslim empire of Masina, which allowed all males to serve in its cavalry (Vandervort 1998: 4).

The social context of warfare in Africa changed markedly between 1500 and ca. 1880. The impact of the Atlantic and Indian Ocean trading networks, particularly the demand for slaves, played a major role, but should not obscure internal drivers of change. Reid has argued that levels of violence accelerated in the eighteenth and the nineteenth centuries. This in turn led to increased militarism and a 'swell of manpower' (Reid 2012a: 145). Such an analysis might be applied in the cases of the Ashanti, Oyo or Kongo armies, or the Swahili-Arab raiders of the eastern Congo Basin, or warfare in Buganda. More generally, this spiral of aggression gave the military a central role in political power; a state of affairs only perpetuated by the wars of colonial conquest.

This trajectory should not obscure the distinctive cultures of militarism that existed prior to this period. The impact of the trans-Saharan trade among peoples of the Sahel, or the processes of Arab expansionism and Berber self-assertion in the Maghrib, represent important dynamics in the period from 640 CE to 1500 CE. Although exposed to the Indian Ocean trade via the Red Sea, Egypt remained a largely distinct case in African military history well into the nineteenth century. To give an example, Muhammad Ali's army focused on Ottoman strategic objectives, such as suppression of the Wahabbiya in Arabia, establishing an Egyptian Empire in the Sudan, and contesting the power of the Porte in the Mediterranean.

Bibliography

Abdullah, A., 'Intertextuality and West African Arabic Poetry: Reading Nigerian Arabic Poetry of the 19th and 20th Centuries,' *Journal of Arabic Literature*, vol. 40, no. 3 (2009), pp. 335–61.

Abun-Nasr, J., *A History of the Maghrib in the Islamic Period* (Cambridge: Cambridge University Press, 1987).

Achrati, N., 'Following the Leader: A History and Evolution of the Amir 'Abd al-Qadir al-Jazairi as Symbol,' *Journal of North African Studies*, vol. 12, no. 2 (2007), pp. 139–52.

Ade Ajayi, J. and Smith, R., *Yoruba Warfare in the Nineteenth Century* (Cambridge: Cambridge University Press, 1964).

Ade Ajayi, J. and Crowder M. (eds.), *Historical Atlas of Africa* (Harlow: Longman, 1985).

Aderinto, S., *Guns and Society in Colonial Nigeria: Firearms, Culture, and Public Order* (Indiana: Indiana University Press, 2018).

Al-Sayyid Marsot, A., *Egypt in the Reign of Muhammad Ali* (Cambridge: Cambridge University Press, 1984).

Alpern, S., 'On the Origins of the Amazons of Dahomey,' *History in Africa*, vol. 25 (1998), pp. 9–25.

Alpers, E., *Ivory and Slaves in East Central Africa to the Late Nineteenth Century* (Berkeley: University of California Press, 1975).

Aminu, M., 'The Place of Mahrams in the History of Kanem-Borno,' *Journal of the Historical Society of Nigeria*, vol. 10, no. 4 (1981), pp. 31–8.

Beach, R., 'The Arms Trade in East Africa in the Late Nineteenth Century,' *Journal of African History*, vol. 3, no. 3 (1962), pp. 451–67.

Bobboyi, H. and Hunwick, J., 'Falkeiana I: A Poem by Ibn Al-Sabbāgh (Dan Marina). In Praise of the Amīr Al-Mu Minīn Kariyagiwa,' *Sudanic Africa*, vol. 2 (1991), pp. 125–38.

Bontinck, F., 'Le Vertiable Sens des Surnoms Africains de Tippo Tip,' *Annales Aequatoria*, vol. 20 (1999), pp. 379–85.

Brett, M. and Fentress, E., *The Berbers* (Oxford: Blackwell, 1998).

Cohen, D., 'Peoples and States of the Great Lakes Region,' in. J. Ajayi (ed.), *UNESCO General History of Africa*, vol. 6, *Africa in the Nineteenth Century until the 1880s* (Oxford: Heinemann, 1989), pp. 270–1.

Coquery-Vidrovitch, C., *Africa and the Africans in the Nineteenth Century: A Turbulent History* (Armonk: M.E. Sharpe, 2009).

Crowder, M. (ed.), *West African Resistance: The Military Response to Colonial Occupation* (London: Hutchinson, 1971).

Draper, M., 'The *Force Publique*'s Campaigns in the Congo-Arab War, 1892–1894,' *Small Wars & Insurgencies*, vol. 30 (2019), pp. 1020–39.

El-Hamel, Chouki, *Black Morocco: A History of Slavery, Race, and Islam* (Cambridge: Cambridge University Press, 2012).

Etherington, N., 'A Tempest in A Teapot? Nineteenth-Century Contests for Land in South Africa's Caledon Valley and the Invention of the Mfecane,' *Journal of African History*, vol. 45 (2004), pp. 203–19.

Fage, J., *A History of Africa* (London: Routledge, 1996).

Fahmy, K., *All the Pasha's Men: Mehmed Ali, His Army, and the Making of Modern Egypt* (Cambridge: Cambridge University Press, 1997).

Fahmy, K., 'The Era of Muhammad "Ali Pasha,"' in M. Daly and C. Petry (eds.), *The Cambridge History of Egypt*, vol. 2, *Modern Egypt, from 1517 to the End of the Twentieth Century* (Cambridge: Cambridge University Press, 1998), pp. 139–79.

Falola, T. and Heaton, M., *A History of Nigeria* (Cambridge: Cambridge University Press, 2008).

Giblin, J., 'Trypanosomiasis Control in African History: An Evaded Issue?' *Journal of African History*, vol. 31, no. 1 (1990), pp. 59–80.

Goody, J., *Technology, Tradition, and the State in Africa* (Oxford: Oxford University Press, 1971).

Guy, J., *The Destruction of the Zulu Kingdom: The Civil War in Zululand, 1879–1884* (London: Longman, 1979).

Hamilton, C., '"The Character and Objects of Chaka": A Reconsideration of the Making of Shaka as "Mfecane" Motor,' *Journal of African History*, vol. 33 (1992), pp. 37–63.

Hiribarren, V., *A History of Borno: Trans-Saharan African Empire to failing Nigerian State* (London: Hurst, 2017).

Hunwick, J., *Timbuktu and the Songhay Empire: Al-Sa'di's Ta'rikh al-Sūdan Down to 1613, and Other Contemporary Documents* (Leiden: Brill, 1999).

Hunwick, J., and Lovejoy, P., 'Not Yet the Kano Chronicle: King-Lists with and without Narrative Elaboration from Nineteenth-century Kano,' *Sudanic Africa*, vol. 4 (1993), pp. 95–130.

Iliffe, J., *A Modern History of Tanganyika* (Cambridge: Cambridge University Press, 1979).

Iliffe, J., *Honour in African History* (Cambridge: Cambridge University Press, 2004).

Jansen, J., 'The Intimacy of Belonging: Literacy and the Experience of Sunjata in Mali,' *History in Africa*, vol. 38 (2011), pp. 103–22.

Kariya, K., 'Muwalat and Apostasy in the Early Sokoto Caliphate,' *Islamic Africa*, vol. 9 (2018), pp. 179–208.

Kennedy, H., *The Prophet and the Age of the Caliphates: The Islamic Near East from the Sixth to the Eleventh Century* (London: Routledge, 2015).

Last, D., and Al-Hajii, M., 'Attempts at Defining a Muslim in 19th Century Hausaland and Bornu,' *Journal of the Historical Society of Nigeria*, vol. 3, no. 2 (1965), pp. 231–40.

Law, R., *The Ọyọ Empire, c.1600-c.1836: A West African Imperialism in the era of the Atlantic Slave Trade* (Oxford: Clarendon Press, 1977).

Levtzion, N., 'The Western Maghrib and Sudan,' in R. Oliver and J. Fage (eds.), *The Cambridge History of Africa*, vol. 3, *From c.1050 to c.1600* (Cambridge: Cambridge University Press, 1977), pp. 331–462.

Levtzion, N., 'The Sahara and the Sudan from the Arab Conquest of the Maghrib to the Rise of the Almoravids,' in. R. Oliver and J. Fage (eds.), *The Cambridge History of Africa*, vol. 2, *From c.500 BC to AD 1050* (Cambridge: Cambridge University Press, 1979), pp. 637–84.

Levtzion, N., *Ancient Ghana and Mali* (New York: Africana, 1980).

Loimeier, R., 'Africa South of the Sahara to the First World War,' in. M. Cook and F. Robinson (eds.), *The New Cambridge History of Islam*, vol. 5, *The Islamic World in the Age of Western Dominance* (Cambridge: Cambridge University Press, 2010), pp. 269–98.

Love, P., *Ibadi Muslims of North Africa: Manuscripts, Mobilization, and the Making of a Written Tradition* (Cambridge: Cambridge University Press, 2018).

Macola, G., *The Gun in Central Africa: A History of Technology and Politics* (Athens: Ohio University Press, 2016).

McIntosh, R., and McIntosh, S., 'From Siecles Obscurs to Revolutionary Centuries on the Middle Niger,' *World Archaeology*, vol. 20, no. 1 (1988), pp. 141–65.

McIntosh, S. 'Reconceptualizing Early Ghana,' *Canadian Journal of African Studies*, vol. 42, nos. 2/3 (2008), pp. 347–73.

Mumbanza, B., 'La pirogue dans l'ouest du bassin du Congo au milieu du 19ème siècle,' *Annales Aequatoria*, vol. 18 (1997), pp. 239–98.

Nziem, I., *Histoire Générale du Congo* (Brussels: De Boeck, 1998).

Ogot, B. (ed.), *War and Society in Africa: Ten Studies* (London: Frank Cass, 1972).

Page, M., 'The Manyema Hordes of Tippu Tip: A Case Study in Social Stratification and the Slave Trade in Eastern Africa,' *International Journal of African Historical Studies*, vol. 7, no. 1 (1974), pp. 69–84.

Reid, R., *War in Pre-colonial Eastern Africa* (Oxford: James Currey, 2007).

Reid, R., *A History of Modern Africa: 1800 to the Present* (Chichester: Wiley-Blackwell, 2012a).

Reid, R., *Warfare in African History* (Cambridge: Cambridge University Press, 2012b).

Robinson, D., *Muslims Societies in African History* (Cambridge: Cambridge University Press, 2007).

Samatar, S., *Oral Poetry and Somali Nationalism: The Case of Sayid Mahammad 'Abdille Hasan'* (Cambridge: Cambridge University Press, 1982).

Sicard, S., Online. '"al-Murdjibī",' in P. Bearman, T. Bianquis, C.E. Bosworth, E. Donzel, and W.P. Heinrichs (eds.), *Encyclopaedia of Islam*, 2nd ed. Accessed online 5 December 2019, http://dx.doi.org.ezp.lib.cam.ac.uk/10.1163/1573-3912_islam_SIM_5540

Smith, R., 'Yoruba Armament,' *Journal of African History*, vol. 8, no. 1 (1967), pp. 87–106.

Spaulding, J., *The Heroic Age in Sinnar* (East Lansing: African Studies Centre of Michigan State University, 1985).

Stapleton, T., *A Military History of Africa*, vol. 1, *The Pre-colonial Period from Ancient Egypt to the Zulu Kingdom* (Santa Barbara: Praeger, 2013).

Stilwell, S., *Slavery and Slaving in African History* (Cambridge: Cambridge University Press, 2014).

Thornton, J., *Warfare in Atlantic Africa* (London: UCL Press, 1999).

Thornton, J., and Mosterman, A., 'A Re-Interpretation of the Kongo-Portuguese War of 1622 According to New Documentary Evidence,' *Journal of African History*, vol. 51 (2010), pp. 235–48.

Turney-High, H., *Primitive War: Its Practice and Concepts* (Columbia: University of South Carolina Press, 1949).

Uzoigwe, G.N., 'Kabalega and the Making of a New Kitara,' *Tarikh*, vol. 3, no. 2 (1970), pp. 3–21.

Uzoigwe, G.N., 'Pre-Colonial Military Studies in Africa,' *Journal of Modern African Studies*, vol. 13, no. 3 (1975), pp. 469–81.

Uzoigwe, G.N., 'The Warrior and the State; in A. Mazrui (ed.), *The Warrior Tradition in Modern Africa* (Leiden, Brill, 1977).

Vandervort, B., *Wars of Imperial Conquest in Africa: 1830–1944* (London: UCL Press, 1998).

Worden, N., *The Making of Modern South Africa* (Oxford: Blackwell, 2000).

Wright, J., 2009. 'Turbulent Times: Political Transformations in the North and East, 1760s–1830s,' in C. Hamilton, B. Mbenga, and R. Ross (eds.), *The Cambridge History of South Africa*, vol. 1, *From Early Times to 1885* (Cambridge: Cambridge University Press, 2009), pp. 211–52.

PART III

Steppe Nomads of Eurasia

9

THE EUROPEAN HUNS

Cameron Barnes

Introduction

The European Huns were nomads who arrived in the Volga-Don region ca. 360 (de la Vaissière 2014). To many contemporary observers, they seemed to be a people without a history. The Huns were in fact descendants of the imperial Xiongnu, a confederacy which had once dominated much of the Eurasian steppes (de la Vaissière 2005; Atwood 2012). The nomads were also heirs to Inner Asian traditions of mounted warfare and statecraft: a combination which made possible the creation of a powerful state within a few generations (Kim 2013).

The Coming of the Huns

In the third century CE, Gothic invasions had shaken the Roman Empire. By the early 360s, however, the Goths were seen as a spent force (Kulikowski 2007). The Huns drove the Goths and Romans into renewed collusion. After subduing the Alans on the Dnieper, the Huns overwhelmed the Gothic confederations of the Greuthungi and Thervingi. Although some Goths submitted to Hunnic rule, most fled. In their flight, the Goths dislodged other peoples (including the Sarmatians and Taifali), driving them towards the Roman borders. The great age of migrations, the *Völkerwanderung*, had begun (Heather 1995).

By 376, the Goths were pressing against the Danube frontier, seeking refuge. Admitted by the Emperor Valens, the Goths revolted soon after due to Roman mistreatment. In the wake of the crushing Gothic victory at Adrianople (378), the Roman Empire descended into long years of chaos and civil war (Williams and Friell 1994). Although Theodosius I restored unity in late 394, his death only a few months later led to the division of the Roman Empire between his two sons, a split which proved permanent.

Most Huns were still far to the east (Heather 2006). Writing in ca. 400, Marcian of Heraclea locates the Hunnic homeland on the lower Dnieper (Alemany 2000). However, there were also Huns south of the Danube. Hunnic mercenaries were used to good effect by Theodosius I against the usurpers Maximus (388) and Eugenius (394) (Maenchen-Helfen, 1973). The nomads' military reputation was such that both Stilicho and Rufinus—the most

DOI: 10.4324/9780429437915-13

powerful generals in the divided Roman Empire after Theodosius' death—relied on Hunnic bodyguards in the mid-390s (Janniard 2015).

In 395, Hunnic raiders from the Pontic steppes crossed the Caucasus through the Darial Pass, and descended upon Armenia, Cappadocia, Cilicia and Syria. The Huns reached Edessa and threatened Antioch (Downey 1961). Another group of Huns moved east along the Euphrates, approaching Ctesiphon. The Mesopotamian raid resulted in disaster for the Huns. A devastating Sasanian counter-attack sent the survivors fleeing through the Darband pass (Greatrex and Greatrex 1999). Despite this setback, the Huns returned again a few years later, raiding Phrygia and Cappadocia (Martindale 1980).

The Hunnic raids had a lasting impact. Already under pressure from the Huns' eastern cousins (Kidarites and Chionites), the Sasanians established an entente with the Eastern Roman Empire (Potts 2018). With Roman financial support, work began on the formidable defences of the Dariali Gorge soon after 395 (Sauer and Pitskhelauri 2018). The nomadic threat north of the Caucasus remained a moderating factor in relations between the East Romans and Iran well into the sixth century (Drijvers 2009).

By ca. 400, a Hunnic state had emerged on the Danube, ruled by a certain Uldin. Although Uldin raided Thrace in 404–405, he led a Hunnic army against Radagaisus on behalf of the Romans in 406 (Wijnendaele 2016). Uldin was soon hostile again, invading Thrace for the second time in 408. After the Romans bribed his chieftains to desert, Uldin's offensive collapsed. Although he drops out of history, the debacle of 408 did not mean the end of Hunnic power on the Danube. In ca. 412, the Huns there had a new paramount king, Charaton (Treadgold 2004).

By the 420s, the Danube Huns were under the rule of two brothers: Octar and Rua, members of the same royal clan as the Basich and Kursich who had led the raid into Mesopotamia in 395 (Blockley 1983). In 422, Rua invaded Thrace and threatened Constantinople. In response, the Eastern Roman Court agreed to pay the Huns an annual tribute of 350 pounds of gold (Croke 1977). It was probably Octar—ruler of the western part of the Hunnic realm—who provided the future patrician Aetius with an army in support of the Western Roman usurper Joannes in 425. Following Octar's death, Rua lent Aetius a second Hunnic army in 434, allowing the latter to become the most powerful man in the Western Roman Empire (Maenchen-Helfen 1973).

During the Rua's reign, the West developed a fatal dependence on Hunnic military support (Moss 1973). Without the Huns, it is doubtful if Aetius would have been able to restore Roman authority in Gaul through his victories over Franks, Burgundians and Visigoths (Heather 2006). The price was the surrender of part of Pannonia to the Huns, a transfer which occurred sometime before 449 (Martindale 1980). This decision provided the nomads with rich grazing lands as well as a strategic foothold south of the Danube (Barnes 2015). The consequences of the Western Roman Empire's dependence on the Huns were brought home in the 450s, when the end of the Hunnic alliance led to the rapid collapse of Roman power in Gaul (Heather 2016).

The Age of Attila

After Rua's death in the mid-430s, he was succeeded by his nephews, Bleda and Attila. Although the brothers secured an increase in the annual tribute, the Huns invaded the Eastern Roman Empire in 441. After seizing the border fortress of Constantia, Hunnic armies took a string of cities—Magus, Viminacium, Singidunum and Sirmium—before advancing up the Morava valley to take the strategic city of Naissus (Whitby 2017). By the autumn of 442, the war was over, after a stunning Hunnic victory (Maenchen-Helfen 1973).

Worse was yet to come for the Romans. In 445, Attila assassinated Bleda, assuming sole power. The next year, he began pressing Theodosius II for arrears of the tribute. When an earthquake levelled the walls of Constantinople in January 447, Attila crossed the Danube (Croke 1978). Near the river Vit, the Huns defeated an East Roman Army, before thrusting south-east to take cities such as Marcianopolis, Philippopolis and Arcadiopolis. Attila's host came to the outskirts of Constantinople. Theodosius II agreed to hand over 6,000 pounds of gold as arrears of tribute, and promised to pay 2,100 pounds annually (Blockley 1983).

As a result of the Hunnic raids of the 440s, Roman urbanism in the Balkans disintegrated. Many cities were abandoned following deliberate destruction. Although some recovered—Philippopolis flourished again later in the century (Topalilov 2014)—more typical was the fate of Nicopolis ad Istrum. Post-Hunnic reconstruction here was limited to a greatly reduced circuit of walls enclosing a handful of ecclesiastical and military buildings (Liebeschuetz 2001).

Although Attila's power was at its height, the new Eastern Roman Emperor, Marcian, took the risk of withholding the annual tribute in 450 (Hohlfelder 1984). Rather than strike against the devastated Balkans, Attila cut a swathe through northern Gaul in 451. The fall of Aureliani on the Loire was averted only by the approach of a Visigothic-Roman relief force under Aetius. Although the allies caught up with the retreating Huns at the Battle of the Catalaunian Fields, Aetius failed to cut off the Hunnic withdrawal. Although the battle is traditionally seen as a Hunnic defeat, the encounter was very likely a Hunnic tactical victory (Kim 2013). After the encounter, Attila's army—laden with booty and captives—recrossed the Rhine in safety.

In any case, when Attila invaded Northern Italy in 452, he faced little opposition (Burgess 1988). The Huns swept across the Julian Alps, sacking Aquileia and moving west along the Po Valley. Attila took major cities such as Mediolanum, Patavium and Ticinum. Fears that the Hunnic ruler was intent on sacking Rome seem misplaced (Linn 2019). After meeting Roman ambassadors on the river Minico, Attila again retired with his plunder and slaves.

Attila's campaigns in Gaul and Italy had little long-term impact. Merovingian traditions regarding Attila's destruction of Metz seems exaggerated: most of the area within the city walls had already been abandoned long before 451 (Halsall 2002). Despite later accounts (still repeated in many secondary sources), the Huns did not totally erase Aquileia from the earth. The city retained its administrative and ecclesiastical importance until the Lombard invasion (Sotinel 2005; Christie 2012; Marano 2012). At Mediolanum, signs of recovery were even swifter. Soon after Attila's departure, the population began reconstruction of their *basilica nova*, burnt during the Hunnic occupation (Maenchen-Helfen 1973). This picture is repeated at other sites in the West—such as Augusta Treverorum and Ticinum—where urban life continued to flourish despite the brief Hunnic occupation (Bullough 1966; Handley 2001).

The Italian campaign was Attila's last. He died suddenly in 453. When his sons quarrelled, their subjects rose in revolt, led by the Gepid King Ardaric. After the Battle of Nedao (ca. 454), Hunnic rule collapsed. At least two of Attila's sons, Dengizich and Ernach, succeeded in temporarily carving out separate kingdoms. Dengizich was dead by 469. The fate of Ernach's kingdom is unclear, although it may have possibly survived to form the nucleus of the later Bulgar state (Golden 1992). The Huns did not vanish entirely from the historic record after Attila. The Romans continued to recruit Hunnic mercenaries for decades, although it becomes increasingly difficult to separate Huns from the other Turkic-speaking nomads.

Tactics and Weapons

The overwhelming majority of Huns probably fought as lightly armoured mounted archers. Their tactics conformed to nomadic traditions. Zosimus refers to the Huns riding around the Goths 'with timely sallies and retreats, shooting down at them from their horses' and causing 'untold slaughter' (Buchanan and Davis 1967: 153). However, the bow was not their only weapon. Hunnic warriors probably switched between the bow and the lance or sword as required.

The evidence is largely indirect. Writing in the reign of Justinian, Procopius refers to Hunnic-style Roman cavalry of his day as 'shooting with bows or thrusting with spears or wielding swords' as the need required (Kaldellis 2014: 534). This picture is consistent with hints in the fifth-century sources. Sidonius claims that Aetius learnt the art of war from the Huns. When praised by the historian Frigeridus for his martial skills, Aetius is described as possessing the same attributes as the elite cavalry of Procopius' day: horsemanship, archery and skill with the lance (Martindale 1980).

Roman observers such as Procopius regarded the bows of the Huns as wonder weapons. Hunnic bows were asymmetric and constructed of wood, horn, bone and sinew (Bóna 2002). The nomads' bows were longer (typically 130–160 cm) and heavier in construction than Roman types, with multiple reinforcing bone laths. This design gave a heavier draw, increased range and greater penetrative power (Coulston 2016; Riesch 2017).

The size and design of Hunnic lances and spears is unknown, as grave finds are rare. Archaeologists are better informed regarding the Huns' swords. The typical Hunnic sword was long and narrow: a double-edged weapon with a large iron cross guard (Lebedynsky 2008). The sword was often worn in combination with a narrow seax: a single-edged cutting weapon 30–50 cm long with a 2–3 cm wide blade (Kiss 2014). Swords were probably suspended from a waist belt by means of a strap (Lebedynsky 2008), whereas the seax seems to have hung separately from the belt at a 30–45-degree angle by means of a two-point suspension (Csiky 2015).

Although most Huns were lightly equipped horse archers, literary sources refer to Huns wearing armour (Maenchen-Helfen 1973). This is consistent with evidence for the use of armour by the Huns' distant ancestors on the steppes: the ancient Xiongnu (Giele 2011; Tishkin 2018). Even before Huns, other nomads on the Pontic steppes (Scythians, Sarmatians and Alans) had fielded contingents of heavy cavalry made up of tribal aristocrats (Perevalov 2002; Groncharovski 2006; Nefedkin 2006; Simpson and Pankova 2017). It would have been surprising, therefore, if the Huns had not also done so at the height of their empire. Evidence for Hunnic horse armour is non-existent, although the possibility cannot be ruled out.

The Huns' military success depended in part on their horses. Literary sources are unstinting in their praise of the Huns' mounts. Vegetius rates the Hunnic horse as the most useful for war 'by reason of their endurance of hard work, cold and hunger' (Mezzabotta 2000: 61). He also praises the same breed for its disposition, which 'is moderate, sensible and tolerant of wounds' (Mezzabotta 2000: 62).

The Huns were probably the first to bring the high-treed wooden saddle to the Mediterranean world. Hunnic saddles (like the rest of the Huns' horse tack) were radically different from their Roman counterparts, and derived from earlier Inner Asian types (Bóna 2002). Similar to modern stock saddles, the Huns' saddles had a high front connected to the rear cantle by two longitudinal wooden stretchers. As a result, the rider was raised above the horse's back, and the weight burden was distributed on either side of the spine.

Hunnic Strategy

Autumn was the traditional raiding season in Inner Asia. In contrast, the Huns launched their raids on sedentary opponents in spring or summer, due to the relative shortage of fodder beyond the steppes. Dated events in Hunnic raids cluster around mid-year. The Huns reached the province of Osroene in July 395 (Maenchen-Helfen 1973). Metz is said to have fallen on 7 April 451, while Attila abandoned the siege of Aureliani on 14 June (Lebedynsky 2011). Aquileia fell on 18 July 452, at the start of Attila's Italian campaign (Linn 2019).

Despite the Hun's reputation for lightning raids, Hunnic armies probably moved no more rapidly than those of their sedentary opponents during major campaigns. In this respect they were similar to the later Mongols (Sinor 2007). Part of the reason was the presence of non-Hunnic levies. Jordanes notes that Attila commanded a host of 'innumerable peoples of diverse tribes' (Mierow 1915: 107), many of whom would have been foot soldiers. The speed of Hunnic armies was also limited by the Huns' horse-drawn baggage train, as well as the flocks and herds driven to provide meat and milk (Moss 1932; Maenchen-Helfen 1973).

It is often assumed that the non-Hunnic element of Attila's armies was simply expendable manpower (Rance 2015). However, many of the Huns' subject peoples had their own traditions of mounted combat. Like the Huns, the Sarmatians and Alans were horsemen armed with the bow, lance and sword (Nefedkin 2006). Both peoples were famous for their heavy cavalry in antiquity (Perevalov 2002; Nefedkin 2006). Most of the riders among the Huns' Germanic subjects were probably lance-armed shock cavalry (Elton 1996; Nefedkin 2008; Halsall 2016), although members of the elite seem to have emulated Hunnic mounted archery (Kazanski 2018). These contingents added significantly to the size and striking power of the Huns' mounted forces.

Hunnic strategy seems to have involved both opportunism and long-range planning. Many of the Hunnic raids—395, 408, 422, 441–442 and 447—coincided with military or political crises within the Eastern Roman Empire (Lee 1993). Jordanes describes Attila as 'a subtle man' who 'fought with craft before he went to war' (Mierow 1915: 103). In the months before his Gallic Campaign, Attila engaged in complex diplomatic manoeuvres designed to divide his potential opponents (Clover 1973).

An important element in Hun strategy was the ability to take large, well-fortified cities. At Naissus in 441, the Huns employed scaling ladders, rams and counter-weighted assault platforms (Blockley 1983). Attila is said to have 'built machines and brought up all kinds of artillery' at Aquileia in 452. Hunnic skill may reflect an earlier exposure to Iranian and Inner Asian poliorcetics (Petersen 2013).

It has been claimed that the 'Huns never conquered Roman territory,' and that they aimed 'at neither conquest nor settlement' (Kelly 2015: 195). This is an oversimplification. The motivations behind Hunnic campaigns were far more complex than mere plunder. In the 440s, Attila and Bleda sought not only booty, but also to open the frontiers to trade and to extort tribute, as well as territorial concessions. At the height of his rule, Attila controlled an empire stretching from the Rhine to the Dnieper: one which included former Roman territories south of the Danube, such as parts of Pannonia (Millar 2006). If Attila was not more ambitious in his territorial demands after 447, part of the reason was that he already controlled the desirable pasturelands from the Middle Danube to the Pontic Steppes (Williams and Friell 1994).

Attila's spectacular success in empire-building reflected his ability to harness traditional expectations. Hunnic values were those of a heroic society. Drinking and feasting at Attila's court were opportunities for poets to sing of past victories, and for the Hunnic audience to be moved with memories of past campaigns (Blockley 1983). Warfare was also a means by which Hunnic rulers could demonstrate their fitness for power according to steppe tradition (Fletcher 1979), as well as to accumulate the wealth required to buy support. As Uldin discovered in 408, a Hunnic leader's power could collapse overnight if he was outbid.

Attila presented himself as moderate in his own needs, but generous and able to reward loyalty (Blockley 1983). Jordanes describes him as 'gracious to supplicants and lenient to those who were once received into his protection' (Mierow 1915: 102). Archaeological evidence in part confirms this picture. Rich finds from aristocratic graves indicate the elites among the Huns' subjects shared in the profits of Hunnic aggression (Heather 2006).

Military Influence

Hunnic mounted archery ended centuries of infantry dominance in Mediterranean warfare, beginning what has been termed the age of the *hippotoxotai* in the Eastern Empire (Syvänne 2004). From ca. 400, mounted archery became part of the essential skillset of the ideal Roman warrior (Rance 2007). By the sixth century, Roman horse archers rode into battle with a nomadic panoply, including a rigid wooden saddle and Inner Asian archery equipment (Stephenson and Dixon 2003). Although contact with the Huns began this transformation, only the adoption of the Hunnic bow can be confidently dated to the fifth century CE (Coulston 1985). An early date for the replacement of the traditional Roman-Celtic horned saddle by Hunnic types is equally plausible.

Overall Hunnic influence on Germanic weapons and tactics was probably minimal. Grave finds indicate the adoption of Hunnic-style military equipment (swords, seaxes, bows, saddles and horse tack) by members of the Germanic elite as a result of Hunnic prestige (Kazanski 1991, 2013). However, the impetus behind this fashion faded after Attila's death. For example, the Hunnic swords which had been widely diffused up until the middle third of the fifth century had little influence on later Osterburken-Kamathen types (Miks 2015). The major exception was the permanent replacement of earlier saddle designs by the Hunnic wooden saddle. Despite this innovation, Germanic military equipment remained remarkably conservative until the sixth century (Elton 1996).

Hunnic influence on Sasanian weapons and tactics was equally limited. This was perhaps unsurprising. In 395, Hunnic raiders had fled before the Iranians' formidable shower shooting (Blockley 1983). Sasanian archery equipment seems to have hardly changed throughout the fifth and sixth centuries (Bálint 1978; Mode 2006). Procopius draws a sharp distinction between Hunnic-inspired Roman archery and Iranian practice (Bivar 1972). Although the Sasanian horned saddle fell into disuse ca. 400, newer Iranian patterns lacked the high cantle typical of Hunnic types (Herrmann 1989).

The Hunnic Legacy

The Huns cast a long shadow over the imagination of later peoples. Their cruelty remained proverbial in the post-Roman West, just as the shock of the Hunnic raids in the 390s endured for centuries in the Syriac and Armenian tradition, acquiring an apocalyptic colouring (van Donzel and Schmidt 2009; Mathisen 2011). The Hunnic destruction of the Gothic Greuthungi in the mid-370s formed the basis for an extraordinarily long-lived epic cycle across Germanic-speaking Europe and Scandinavia (Tolkien 1953–1957). Attila's name was also attached to another complex series of epics which took as their theme the downfall of the Burgundian Kingdom at the hands of the Huns in 437 (Thomas 1991).

The Huns' most lasting legacy was the walls of Constantinople. Without the Hunnic threat, the Byzantine capital may have never gained the formidable defences which underpinned its survival for the next millennium (Heather 2016). Work on the Theodosian Walls began around the time of Uldin's first invasion of Thrace. Completed in 413, the walls were a complex, multi-layered system of defences, the greatest urban fortifications ever constructed in Mediterranean antiquity (Crow 2008: 249–85). During the next few decades, Constantinople was provided with a system of open-air reservoirs and vaulted cisterns, ensuring that the capital did not run short of drinking water during a Hunnic siege (Magdalino 2010). The final element in the city's defensive belt was the Long Walls. These stretched 56 kilometres across the Thracian Peninsula, incorporating pentagonal towers with torsion artillery, forts, gateways and roadworks (Crow 2008: 249–85). The Long Walls were probably constructed after 447, when Attila's armies came close to the capital (Whitby 1985).

The reign of Theodosius II also saw a systematic reconstruction of fortifications and defence lines across Greece and the Southern Balkans (Croke 1983; Cherf 1984; Fowden 1988). As at Constantinople, the Hunnic threat drove Roman military engineers to experiment with ambitious new designs. At Thessalonica, capital of the Illyrian Prefecture after Attila's conquest of Sirmium in 441, a revolutionary line of triangular towers on the western edge of city eliminated the dead fire zones close to the walls, while strengthening the fabric against battering rams. This experiment influenced military architecture across the Eastern Roman Empire over the next two centuries (Rizos 2011).

The achievements of Theodosius II's military architects should not disguise the overall failure of the East Roman response to the Hunnic threat. The Hun raids in the 440s were a sharp break in the military, economic and social history of the Balkans. Any lasting reconstruction of the Danube defences was probably delayed until the reign of Justinian. Late Roman urbanism in the Balkans never recovered from the impact of the Huns. For the next few centuries, the Balkan pattern was a scatter of walled administrative centres, with a shrunken rural population clustered in small fortified towns and defended refuges (Wilkes 2013).

The impact of the Huns on the western half of the Roman Empire was no less significant. In the longer term, their actions ensured the long, slow death agony of the Western Empire. Although Hunnic mercenaries were essential to Aetius' efforts to sustain Roman rule in Gaul, the invasions of 441–442 were ultimately fatal to the West's survival. As a result of the invasions, the Romans were forced to abandon their attempt to recapture North Africa from the Vandals, a step which plunged the Western Roman Empire into a financial crisis from which it never recovered (Heather 2006). It was perhaps fitting that it was a Hun, Odovacer—the son of one of Attila's advisors—who deposed the last Western emperor in 476 (Kim 2013).

Bibliography

Alemany, A., *Sources on the Alans: A Critical Compilation* (Leiden: Brill, 2000).

Atwood, C.P., 'Huns and Xiongnu: New Thoughts on an Old Problem,' in B.J. Boeck, R.E. Martin and D. Rowland (eds.), *Dubitando: Studies in History and Culture in Honor of Donald Ostrowski* (Bloomington, IN: Slavica Publishers, 2012), pp. 27–52.

Bálint, C., 'Vestiges archéologiques de l'époque tardive des sassanides et leurs relations avec les peuples des steppes,' *Acta Archaeologica Hungarica*, vol. 30, no. 1–2 (1978), pp. 173–212.

Barnes, C., 'Rehorsing the Huns,' *War & Society*, vol. 34, no. 1 (2015), pp. 1–22.

Bivar, A.D.H., 'Cavalry Equipment and Tactics on the Euphrates Frontier,' *Dumbarton Oaks Papers*, no. 26 (1972), pp. 271–91.

Blockley, R.C., *The Fragmentary Classicising Historians of the Later Roman Empire: Eunapius, Olympiodorus, Priscus and Malchus* (Liverpool: Francis Cairns, 1983).

Bóna, I., Les Huns: *Le grand empire barbare d'Europe IVe-Ve siècle* (Paris: Errance, 2002).

Buchanan, J.J. and Davis, H.T., *Zosimus: Historia Nova* (San Antonio, TX: Trinity University Press, 1967).

Bullough, D.A., 'Urban Change in Early Medieval Italy: The Example of Pavia,' *Papers of the British School at Rome*, no. 34 (1966), pp. 82–130.

Burgess, R.W., 'A New Reading for Hydatius Chronicle 177 and the Defeat of the Huns in Italy,' *Phoenix*, vol. 42, no. 4 (1988), pp. 357–63.

Cherf, W.J., 'Procopius, Lime-mortar C^{14} Dating and the Late Roman Fortifications of Thermopylai,' *American Journal of Archaeology*, vol. 88, no. 4 (1984), pp. 594–8.

Christie, N., 'Vrbes Extinctae: Archaeologies of and Approaches to Abandoned Classical Cities,' in N. Christie, and A. Augenti (eds.), *Vrbes Extinctae: Archaeologies of Abandoned Classical Towns* (Farnham: Ashgate, 2012), pp. 1–44.

Clover, F.M., 'Geiseric and Attila,' *Historia: Zeitschrift für Alte Geschichte*, vol. 22, no. 1 (1973), pp. 104–17.

Coulston, J.C.N., 'Roman Archery Equipment,' in M.C. Bishop (ed.), *The Production and Distribution of Roman Military Equipment: Proceedings of the Second Roman Military Equipment Seminar* (Oxford: British Archaeological Reports, 1985), pp. 220–366.

Coulston, J.C.N., 'Imitation and Inspiration in "Roman" Archery,' *Journal of Roman Military Equipment Studies*, vol. 17 (2016), pp. 203–14.

Croke, B., 'Evidence for the Hun Invasion of Thrace in A.D. 422,' Greek, *Roman and Byzantine Studies*, vol. 18, no. 4 (1977), pp. 347–67.

Croke, B., 'Two Early Byzantine Earthquakes and their Liturgical Commemoration,' *Byzantion*, vol. 51, no. 1 (1978), pp. 122–47.

Croke, B., 'The Context and Date of Priscus Fragment 6,' *Classical Philology*, vol. 78, no. 4 (1983), pp. 297–308.

Crow, J., 'The Infrastructure of a Great City: Earth, Walls and Water in Late Antique Constantinople,' *Late Antique Archaeology*, vol. 4, no. 1 (2008), pp. 249–85.

Csiky, G., *Avar-age Polearms and Edged Weapons: Classification, Typology, Chronology and Technology* (Leiden: Brill, 2015).

de la Vaissière, È., 'Huns et Xiongnu,' *Central Asiatic Journal*, vol. 49, no. 1 (2005), pp. 3–26.

de la Vaissière, È., 'The Steppe World and the Rise of the Huns,' in M. Maas (ed.), *The Cambridge Companion to the Age of Attila* (Cambrige: Cambridge University Press, 2014), pp. 175–92.

Downey, G., *A History of Antioch in Syria: From Seleucus to the Arab Conquest* (Princeton, NJ: Princeton University Press, 1961).

Drijvers, J.W., 'Rome and the Sasanid Empire: Confrontation and Coexistence,' in P. Rousseau (ed.), *A Companion to Late Antiquity* (Malden, MA: Wiley-Blackwell, 2009), pp. 441–54.

Elton, H., *Warfare in Roman Europe AD 350–425* (Oxford: Clarendon Press, 1996).

Fletcher, J., 'Turco-Mongolian Monarchic Tradition in the Ottoman Empire,' *Harvard Ukrainian Studies*, vols. 3–4, no. 1 (1979), pp. 236–51.

Fowden, G., 'City and Mountain in late Roman Attica,' *Journal of Hellenic Studies*, vol. 108 (1988), pp. 48–59.

Giele, E., 'Evidence for the Xiongnu in Chinese Wooden Documents from the Han Period,' in U. Brosseder, and B.K. Miller (eds.), *Xiongnu Archaeology: Multidisciplinary Perspectives of the*

First Steppe Empire in Inner Asia (Bonn: Rheinische Friedrich-Wilhelms-Universitat, 2011), pp. 49–75.

Golden, P.B., *An Introduction to the History of the Turkic Peoples: Ethnogenesis and State Formation in Medieval and Early Modern Eurasia and the Middle East* (Wiesbaden: Otto Harrassowitz, 1992).

Greatrex, G. and Greatrex, M., 'The Hunnic Invasion of the East of 395 and the Fortress of Ziatha,' *Byzantion*, vol. 69, no. 1 (1999), pp. 65–75.

Groncharovski, V.A., 'Some Notes on Defensive Armament of the Bosporan Cavalry in the First Centuries AD,' in M. Mode, and J. Tubach (eds.), *Arms and Armour as Indicators of Cultural Transfer: The Steppes and the Ancient World from Hellenistic Times to the Early Middle Ages* (Wiesbaden: Reichert, 2006), pp. 445–51.

Halsall, G., *Settlement and Social Organization: The Merovingian Region of Metz* (Cambridge: Cambridge University Press, 2002).

Halsall, G., 'The Ostrogothic Military,' in J. Arnold, S. Bjornlie and K. Sessa (eds.), *A Companion to Ostrogothic Italy* (Leiden: Brill, 2016), pp. 173–99.

Handley, M., 'Beyond Hagiography: Epigraphic Commemoration and the Cult of Saints in Late Antique Trier,' in R.W. Mathisen and D. Shanzer (eds.), *Society and Culture in Late Antique Gaul: Revisiting the Sources* (Aldershot: Ashgate, 2001), pp. 187–200.

Heather, P., 'The Huns and the End of the Roman Empire in Western Europe,' *English Historical Review*, vol. 110, no. 435 (1995), pp. 4–41.

Heather, P., *The Fall of the Roman Empire: A New History of Rome and the Barbarians* (New York, NY: Oxford University Press, 2006).

Heather, P., 'East and West in the Fifth Century,' in U. Roberto and L. Mecalla (eds.), *Governare e riformare l'Impero al momento della sua divisione: Oriente, Occidente, Illirico* (Rome: Publications de l'École française de Rome, 2016), pp. 199–224.

Herrmann, G., 'Parthian and Sasanian Saddlery. New Light from the Roman West,' in L. De Meyer and E. Haerinck (eds.), *Archaeologia Iranica et Orientalis. Miscellanea in honorem Louis Vanden Berghe* (Gent: Peeters, 1989), pp. 757–809.

Hohlfelder, R.L., 'Marcian's Gamble: A Reassessment of Eastern Imperial Policy Toward Attila AD 450–453,' *American Journal of Ancient History*, vol. 9, no. 1 (1984), pp. 4–69.

Janniard, S., 'Les adaptations de l'armée romaine aux modes de combat des peuples des steppes (fin IVe–début VIe siècle apr. J.-C.),' in R. Umberto, and L. Mecella (eds.), *Governare e riformare l'impero al momento della sua divisione: Oriente, Occidente, Illirico* (Rome: Publications de l'École française de Rome, 2015), pp. 231–69.

Kaldellis, A., *Prokopios: The Wars of Justinian* (Indianapolis, IN: Hackett, 2014).

Kazanski, M., 'A propos des armes et des éléments de harnachement "orientaux" en Occident à l'époque des Grandes Migrations (IVe-Ve s.),' *Journal of Roman Archaeology*, vol. 4 (1991), pp. 123–39.

Kazanski, M., 'Barbarian Military Equipment and its Evolution in the Late Roman and Great Migration periods (3rd–5th c. A.D.),' in A.C. Sarantis and N. Christie (eds.), *War and Warfare in Late Antiquity: Current Perspectives* (Leiden: Brill, 2013), pp. 493–522.

Kazanski, M., 'Bowmen's Graves from the Hunnic Period in Northern Illyricum,' in M.L. Nagy and K.L. Szőlősi (eds.), *"To Make a Fairy's Whistle From a Briar Rose": Studies Presented to Eszter Istvánovits on Her Sixtieth Birthday* (Nyíregyháza: Jósa András Múzeum, 2018), pp. 407–17.

Kelly, C., 'Neither Conquest nor Settlement: Attila's Empire and its Impact,' in M. Maas (ed.), *The Cambridge Companion to the Age of Attila* (Cambridge: Cambridge University Press, 2015), pp. 193–208.

Kim, H.J., *The Huns, Rome and the Birth of Europe* (Cambridge: Cambridge University Press, 2013).

Kiss, A.P., 'Huns, Germans, Byzantines? The Origins of the Narrow Bladed Long Seaxes,' *Acta Archaeologica Carpathia*, vol. 49 (2014), pp. 111–44.

Kulikowski, M., *Rome's Gothic Wars: From the Third Century to Alaric* (Cambridge: Cambridge University Press, 2007).

Lebedynsky, I., *De l' épée scythe au sabre mongol: les armes blanches des nomes de la steppe, IXe siècle av. J.-C.–XIXe siècle apr. J.-C.* (Paris: Errance, 2008).

Lebedynsky, I., *La campagne d'Attila en Gaule. 451 apr. J.-C.* (Clermont-Ferrand: Lemme, 2011).

Lee, A.D., *Information and Frontiers: Roman Foreign Relations in Late Antiquity* (Cambridge: Cambridge University Press, 1993).

Liebeschuetz, J.H.W.G., *Decline and Fall of the Roman City* (Oxford: Oxford University Press, 2001).

Linn, J., 'Attila's Appetite: The Logistics of Attila the Hun's Invasion of Italy in 452,' *Journal of Military History*, vol. 83, no. 2 (2019), pp. 325–46.

Maenchen-Helfen, O.J., *The World of the Huns: Studies in Their History and Culture* (Berkeley, CA: University of California Press, 1973).

Magdalino, P., 'Byzantium = Constantinople,' in L. James (ed.), *A Companion to Byzantium* (Chichester: Wiley-Blackwell, 2010), pp. 43–54.

Marano, Y.A., 'Dopo Attila: Urbanesimo e storia ad Aquileia tra V e VI secolo d.C,' in J. Bonetto, and M. Salvadori (eds.), *L'architettura privata ad Aquileia in età romana. Atti del Convegno di Studio (Padova, 21–22 febbraio 2011)* (Padova: Padova University Press, 2012), pp. 571–89.

Martindale, J.R., *The Prosopography of the Later Roman Empire* (Cambridge: Cambridge University Press, 1980).

Mathisen, R.W., 'Catalogues of Barbarians in Late Antiquity,' in R.W. Mathisen, and D. Shanzer (eds.), *Romans, Barbarians, and the Transformation of the Roman world: Cultural Interaction and the Creation of Identity in Late Antiquity* (Farnham: Ashgate, 2011), pp. 17–32.

Mezzabotta, M.R., 'Aspects of Multiculturalism in the Mulomedicina of Vegetius,' *Akroterion*, vol. 45, no. 1 (2000), pp. 52–64.

Mierow, C.C., *The Gothic History of Jordanes* (Princeton, NJ: Princeton University Press, 1915).

Miks, C., 'Sword, *gladius*,' in Y. Le Bohec (ed.), *The Encyclopedia of the Roman Army* (Oxford: Wiley-Blackwell, 2015), pp. 948–70.

Millar, F., *A Greek Roman Empire: Power and Belief under Theodosius II 408–450* (Berkeley, CA: University of California Press, 2006).

Mode, M., 'Art and Ideology at Taq-i Bustan: The Armoured Equestrian,' in M. Mode and J. Tubach (eds.), *Arms and Armours as Indicators of Cultural Transfer: The Steppes and the Ancient World from Hellenistic Times to the Early Middle Ages* (Wiesbaden: Reichert, 2006), pp. 391–413.

Moss, C., 'Isaac of Antioch. Homily on the Royal City,' *Zeitschrift für Semitistik und verwandte Gebiete*, vol. 8 (1932), pp. 61–72.

Moss, J.R., 'The Effects of the Policies of Aetius on the History of Western Europe,' *Historia: Zeitschrift für Alte Geschichte*, vol. 22, no. 4 (1973), pp. 711–31.

Nefedkin, A.K., 'Sarmatian Armour According to Narrative and Archaeological Data,' in M. Mode and J. Tubach (eds.), *Arms and Armour as Indicators of Cultural Transfer: The Steppes and the Ancient World from Hellenistic Times to the Early Middle Ages* (Wiesbaden: Reichert, 2006), pp. 433–44.

Nefedkin, A.K., 'Weaponry of the Goths of the mid-3rd to the 7th Century AD,' *Journal of Roman Military Equipment Studies*, vol. 16 (2008), pp. 147–55.

Perevalov, S.M., 'The Sarmatian Lance and the Sarmatian Horse-Riding Posture,' *Anthropology and Archaeology of Eurasia*, vol. 40, no. 4 (2002), pp. 7–21.

Petersen, L.I.R., *Siege Warfare and Military Organization in the Successor States (400–800 AD): Byzantium, the West and Islam* (Leiden: Brill, 2013).

Potts, D.T., 'Sasanian Iran and Its Northeastern Frontier,' in N. DiCosmo, and M. Maas (eds.), *Empires and Exchanges in Eurasian Late Antiquity: Rome, China, Iran, and the Steppe, ca. 250–750* (Cambridge: Cambridge University Press, 2018), pp. 287–301.

Rance, P., 'Battle,' in P. Sabin, H. van Wees and M. Whitby (eds.), *The Cambridge History of Greek and Roman Warfare*, vol. 2, *Rome from the Late Republic to the Late Empire* (Cambridge: Cambridge University Press, 2007), pp. 342–78.

Rance, P., 'Siege Warfare: Late Empire,' in Y. Le Bohec (ed.), *The Encyclopedia of the Roman Army* (Oxford: Wiley-Blackwell, 2015), pp. 893–900.

Riesch, Holger, *Pfeil und Bogen in der römischen Kaiserzeit* (Ludwigshafen: Angelika Hörnig, 2017).

Rizos, E., 'The Late-Antique walls of Thessalonica and Their Place in the Development of Eastern Military Architecture,' *Journal of Roman Archaeology*, vol. 24, no. 1 (2011), pp. 450–68.

Sauer, E. and Pitskhelauri, K., 'Securing the Caucasus through Intelligence and Hard Power: Rome and Persia at the 'Caspian Gates' (Dariali Gorge, Georgia),' in C.S. Sommer and S. Matešic (eds.), *Limes XXIII: Proceedings of the 23rd International Congress of Roman Frontier Studies Ingolstadt 2015* (Mainz: Nünnerich-Asmus, 2018), pp. 273–9.

Simpson, S.J. and Pankova, S.V., *Scythians: Warriors of Ancient Siberia* (London: Thames & Hudson, 2017).

Sinor, D., 'The Mongol Art of War: Chinggis Khan and the Mongol Military System,' *Journal of Military History* vol. 71, no. 4 (2007), pp. 1223–4.

Sotinel, C., *Identité civique et christianisme: Aquilée du IIIe au VIe siècle* (Rome: École Française de Rome, 2005).

Stephenson, I.P. and Dixon, K.R., *Roman Cavalry Equipment.* (Stroud: Tempus, 2003).

Syvänne, I., *The Age of the Hippotoxotai: Art of War in Roman Military Revival and Disaster (491–636)* (Tampere: Tampere University Press, 2004).

Thomas, N., 'The Testimony of Saxo Grammaticus and the Interpretation of the Nibelungenlied,' *Oxford German Studies*, vol. 20, no. 1 (1991), pp. 7–17.

Tishkin, A.A., 'Altai During the Period of the Xiongnu Empire,' in G. Krist, and L. Zhang (eds.), *Archaeology and Conservation Along the Silk Road* (Vienna: Böhlau, 2018), pp. 157–89.

Tolkien, C., 'The Battle of the Goths and Huns,' *Saga-Book of the Viking Society*, vol. 14 (1953–57), pp. 141–63.

Topalilov, I., 'The Barbarians and the City: Comparative Study of the Impact of the Barbarian Invasions in 376–378 and 442–447 on the Urbanism of Philippopolis, Thrace,' in D. Dzino and K. Parry (eds.), *Byzantium, Its Neighbours and Its Cultures* (Brisbane: Australian Association for Byzantine Studies, 2014), pp. 223–36.

Treadgold, W., 'The Diplomatic Career and Historical Work of Olympiodorus of Thebes,' *International History Review*, vol. 26, no. 4 (2004), pp. 709–33.

van Donzel, E. and Schmidt, A., *Gog and Magog in Early Syriac and Islamic Sources: Sallam's Quest for Alexander's Wall* (Leiden: Brill, 2009).

Whitby, M., 'The Long Walls of Constantinople,' *Byzantion*, vol. 55, no. 2 (1985), pp. 560–83.

Whitby, M., 'Attila and the Huns, 442–455,' in M. Whitby, and H. Sidebottom (eds.), *Encyclopedia of Ancient Battles* (Chichester: Wiley-Blackwell, 2017), pp. 1166–73.

Wijnendaele, J.W., 'Stilicho, Radagaisus, and the So-called "Battle of Faesulae" (406 CE),' *Journal of Late Antiquity*, vol. 9, no. 1 (2016), pp. 267–84.

Wilkes, J.J., 'Homeland Security in the South-West Balkans (3rd–6th c. AD),' in A. Sarantis, and N. Christie (eds.), *War and Warfare in Late Antiquity* (Leiden: Brill, 2013), pp. 735–57.

Williams, S. and Friell, G., *Theodosius: The Empire at Bay* (London: Routledge, 1994).

10

ARAB–BYZANTINE WARFARE

660 CE–1040 CE

John Haldon

The Muslims have no enemy more dangerous than the Romans, a fact that is confirmed by the holy Qur'ān.

Tenth-century Arab geographer, Abū'l-Faraj al-Kātib al-Bagdādī Kudāma ibn Ja'far, Kitāb al-Harāj *(De Goeje 1870ff. VI: 192)*

Context

The frontier warfare that took place from the later seventh into the tenth century between East Roman, or Byzantine, forces and those of the Umayyad and Abbasid *Caliphates* dominates the political and to some degree the cultural history of Byzantium across these centuries. The Islamic conquests of the 630s and 640s CE radically transformed the later Roman state in the east (see Maps 10.1 and 10.3). By 642, all of Egypt and the Middle Eastern provinces had been lost, Arab forces had penetrated deep into Asia Minor and Libya, and imperial forces had been withdrawn into Asia Minor, to be settled across the provinces of the region as the only available means of supporting them. The East Roman Empire was left with a rump of its former lands: Central and Northern Asia Minor, the Southern Balkan coastal regions, the Aegean and, in the West, parts of Italy, Sardinia and the Central and Western North African provinces. North Africa was lost by 700, while imperial territory in Italy was constantly shrinking. Thus by 642 the empire had lost somewhere over half its area and three-quarters of its resources—a drastic loss for an imperial state which still had to maintain and equip a considerable army and an effective administrative bureaucracy if it were to survive at all.[1] The changes which accompanied the developments of the seventh century affected all areas of social, cultural and economic life. There occurred a 'ruralisation' of society, a result of the devastation, abandonment, shrinkage or displacement of many cities in Asia Minor as a result of invasions and raids. The defensive properties of 'urban' sites, their direct relevance to military, administrative or ecclesiastical needs, and so on, played the key role in whether a city survived or not.[2] Constantinople became the pre-eminent city of the empire (Mango 1986: 118–36).

Until the extinction of the Sassanid Empire by the Islamic armies in the 630s and early 640s, the Persian state had been the main rival of the Roman Empire in the East. Thereafter,

DOI: 10.4324/9780429437915-14

the Umayyad (661–750) and then Abbasid (751–1258) *Caliphates* posed a constant threat.[3] The warfare falls into several phases: from the 650s to the 720s, when Arab–Islamic invasions were a regular phenomenon aimed at the destruction of the East Roman state, entailing two major sieges of Constantinople, in 668–669 and 717–718; the 720s to 750s, when a modus vivendi had been established, but in which Muslim attacks remained a constant source of economic and political dislocation;[4] and thereafter a period of over a century punctuated by a number of large-scale campaigns into Roman territory, often led by the *Caliph* himself, members of his immediate family and entourage, or a local borderland *emir*, until the middle of the eleventh century. During the tenth and early eleventh century, the collapse and fragmentation of Abbasid authority made it possible for the empire to re-establish a military and political pre-eminence in the region. The increasingly important role of Turkish slave and mercenary soldiers in the Abbasid *Caliphate* from the 840s, but more especially the eventual arrival of the Seljuk Turks in the 1050s, drastically altered this picture after the 1050s and 1060s. A combination of internal political dissension and a relatively minor military defeat at the hands of the Seljuk Sultan Alp Aslan in Eastern Anatolia in 1071 (battle of Mantzikert, modern name Malazgirt) ushered in a period of civil war and resulted in the loss of Eastern and Central Asia Minor, which henceforth became dominated by groups of Turkish nomadic pastoralists (known as Türkmen) who presented a constant threat to all forms of sedentary occupation.

But it was the Arab Islamic conquests of the seventh century that first radically altered the strategic and political geography of the whole East Mediterranean region. The complete failure of attempts to meet and drive back the invaders in open battle induced a major shift in strategy whereby open confrontations with the Muslim armies were avoided. As a result of the disastrous defeat at the battle of the Yarmuk River in 636, the armies which had operated in Syria, Palestine and Mesopotamia were withdrawn into Anatolia, and re-established within a greatly changed economic and strategic framework. The regions across which they were based were determined by the ability of these districts to provide for the soldiers in terms of supplies and other requirements. By the last decades of the seventh century, the Anatolian provinces across which these armies were garrisoned were known collectively by the name of the army based there. The distribution of the various units of the field armies across the provinces in this way was connected with logistical demands, but the strategic implications of this meant that Roman counterattacks were relatively slow to organise, and defence was often fragmented and localised (Haldon 1997). Yet defeats reflected, on the whole, short-term difficulties and problems, while the longer-term recovery that took place from the middle of the eighth century reflected an increasingly stable strategic situation, a growing economy and, just as importantly, the increasing weakness of the empire's most dangerous and persistent enemy during the seventh and most of the eighth century, the Abbasid *Caliphate* (Whittow 1996).

East Roman Responses

From the 640s, East Roman armies attempted to control the major Anatolian passes through the mountain barrier running from the sea north-eastwards towards the Armenian highlands. The Arab conquest of Cilicia and its abandonment by the empire at the end of the seventh century forced changes to this approach. Rather than attempting to maintain a clearly defined frontier, the field armies followed a strategy of attempting to throw invading armies back following a pitched battle and, from the later part of the seventh century after this approach had met with virtually no success, harassing and dogging invaders once they had entered imperial territory and completed their attack. This strategy meant, however, that the economic hinterland of the frontier incurred substantial damage, subject as it was to regular devastation. The result was the appearance, by the first half of the eighth century, of what was in effect a 'no-man's land' between the settled and economically safer regions on both sides. The new arrangements nevertheless appear to have prevented the establishment by the Arabs on imperial territory behind the Taurus–Anti-Taurus Mountain zone, which by the early eighth century had come to represent the real frontier region between the two states (see Map 10.2). While these developments were taking place on land, a naval command, the Kibyrrhaiotai (named after the coastal town of Kibyrrha in Caria) had come into existence, probably by the 690s, in western and south-western coastal Anatolia, establishing thereby a sea-borne defence for the Western Anatolian Coast and Aegean region.

The military commands, known as *strategides* or 'generalcies,' were at first merely groupings of provinces across which different armies were based (refer to Map 10.3). By 730 or thereabouts they had acquired a clear geographical identity; and by the later eighth century some elements of fiscal as well as military administration were based on these territorial divisions, which from the early ninth century became known as 'themes' (*themata*), a term that reflected local fiscal structures introduced to support soldiers through their own communities at a provincial level.[5]

The localisation encouraged through the distribution of the field armies across Asia Minor generated a distinction between the regular elements—full-time soldiers—and the militia-like conscripts in each division. Emperor Constantine V (741–775) established a small elite force, known as the *tagmata* ('the regiments'), which soon became the elite field division of campaign armies. While never very large, it had better pay and discipline than regular and part-time provincial units, and served as a valuable nucleus of picked troops on campaign. This was, in fact, the first step in a rapidly growing tendency to recruit mercenary forces, both foreign and indigenous, to form special units and to serve for the duration of a particular campaign or group of campaigns. As imperial might recovered and grew, so the empire was able to re-assert its military strength in the East during the ninth century, and the role and the proportion of such full-time units became ever more important.

A number of key elements contributed to Byzantine strategy. Initially, local troops attempted to hold raiding forces and turn them back at the Taurus and Anti-Taurus passes. On occasion this policy worked quite well. Where it did not work (which was frequently the case), local forces generally tried to harass the invading forces and follow their movement so that the location of each party or group was known. This strategy was successful partly due to the garrisoning of numerous small forts and fortresses along the major routes, on cross-roads and locations where supplies might be stored, and above and behind the frontier passes through which enemy forces had to pass to gain access to the Byzantine hinterland. Such posts were a constant threat to any invading force, although for the raiding party to lay siege to them was more trouble than it was worth. So long as such outposts were occupied by the imperial forces, the enemy raiders could be tracked and eventually brought to battle or ambushed.

From the later eighth and early ninth century, a series of small frontier pass commands, called kleisourarchies (*kleisourarchiai*) were erected behind the frontier. These complemented the imperial strategy of concentrating on highly localised defences, employing harassment and ambushes of enemy raiders intended to limit the damage inflicted by all but the largest forces. The origins and pace of development of this system remains obscure, and the first clear evidence for such a policy appears at the end of the eighth century, even if it is likely that it had been practised in some frontier regions for some time before this (Haldon and Kennedy 1980: 79–87, 95–106; Métivier 2008).

Arab–Islamic Strategy

While the early Islamic world was immeasurably wealthier and more populous, organising its resources in such a way as to overwhelm the empire proved to be an impossible task, even if on occasion it came close, in part because its strategy also reflected ideological motives. The Roman Empire soon became the hereditary enemy of the Islamic world, offering a rival world view and political system, and warfare against it took on a particular cultural symbolic value—fighting the Romans carried much greater ideological significance than fighting the various barbarians found along other frontiers.

Once the frontier had begun to stabilise, around the 750s, and with Byzantine frontier outposts along the Taurus–Anti-Taurus chains posing a significant challenge to the peace and security of the Islamic lands behind the frontier zones, a sophisticated system of frontier posts and fortresses began to be established (see Map 10.2). The frontier lands (known as *al-thughūr*) were divided into two main zones, the Syrian frontier stretching from the Cilician Plain in the west towards the mountains of Northern Syria in the east, where it abutted the upland and mountainous regions known as al-Jazīrah, which stretched eastwards as far as the Armenian Highlands (Bonner 1994). In the latter were the fortress-cities of Mar'aš (Germanikeia), al-Hadath and Malatya (Melitēnē), each fronted by a series of smaller forts, referred to as *ḥiṣn* (as opposed to *madīnah*, city) covering major and minor routes from the Byzantine lands southwards and eastwards. In the Cilician region al-Maṣṣiṣah (Mopsouestia), Adhanah (Adana) and Tarsus fulfilled similar functions, again with a number of smaller advanced forts to protect the major routes from the north. The fortress-cities thus lay in the plains and protected their rural hinterlands as well as functioning as major foci for offensive raiding into Roman lands. Each city had a substantial garrison of regular troops, usually drawn from the local provinces in Syria or Jazīrah, or from more distant Khurāsān, and complemented by large numbers of volunteers from across the Islamic world during the annual campaigns. By the middle of the ninth century, Tarsus had become a major focus for raids into Roman territory and its army posed a substantial threat to the Byzantine provinces behind the Taurus Mountains.[6]

Already from the 660s and 670s frontier raiding took on an annual rhythm aimed both at economic disruption as well as the seizure of booty in people, livestock and other movable wealth. Palaeo-environmental data from Cappadocia, for example, coincides with historical information to show that a period of continuous raiding and enemy occupation between 664 and 678 appears to have led to a near-complete collapse of agriculture in the area. This may not have been very common, but that it could occur is made clear by this example (Eastwood et al. 2009; Haldon 2007; Ahrweiler 1962). By the later eighth century, and probably long before, there were two or three sets of raids each year, the spring, summer and winter raids.[7] The spring and summer raids were much more usual, largely because of the adverse climatic conditions of the Anatolian winter, and generally entered Byzantine territory through the main passes—the Cilician Gates north of Tarsus, the Darb al-Hadath north of Mar'aš and the Darb Malatyah (see Map 10.2). Their frequency and size varied with the political situation, both in the *caliphate* as a whole as well as at local level, especially as, from the middle or later ninth century, the central authority exercised from Baghdad was weakened and local *emirates* came to dominate the region.[8]

Jihād and Holy War

Probably the best-known feature of the Byzantine–Arab conflict is the Islamic concept of *jihād*, or religious struggle against unbelievers. Christian theologians had debated the rights and wrongs of waging war from the beginning, but once the empire had become officially Christian by the end of the fourth century it became easier to justify fighting. There remained a substantial body of opinion that condemned all forms of warfare and violence, but the imperial Church justified warfare as a defensive measure against those who threatened the Christian community—the empire—and there developed in elite military circles a strong element of militarism during the eighth century and afterwards. But no theory of 'holy war' in the later Western, Crusading sense, or indeed in the sense that evolved with Islam. Indeed, such a theory was unnecessary, since all warfare was to defend the Christian Roman empire, that was in itself adequate justification.[9]

In Islam, *jihād* was itself an idea which crystallised quite slowly, emerging in its established form only in texts of the later eighth century CE. It was seen in particular as one aspect of the ideology of the frontier and as an element of 'good' Islamic rulership, but took a little while to mature and receive the sort of exegetical attention required to develop it into a fully-fledged theory (Bonner 1992, 1996; Mottahedeh and al-Sayyid 2001). On the other hand, it played an important role in the continuous warfare along the Byzantine–Islamic frontier, and was one of the factors that determined the presence of the Abbasid *Caliphs* themselves during the annual campaign, the only front on which the *caliphs* themselves were regularly involved (Haldon and Kennedy 1980: 106). Indeed, it has been cogently argued that it was in the frontier regions of the *thughūr* during the eighth century that the cultural and political discourse of border warfare helped to consolidate a theory of holy war expressed in religious terms. Divine reward, pious companionship and communal responsibility (rather than subordination to the centralised patrimonial state) were its key elements, and its protagonists were scholars and theologians rather than soldiers and warriors.

As early as the middle of the eighth century, Christians were aware of this idea and commented upon it (Krausmüller 2004; Sypiański 2013; Lee 2013), and in the later ninth and tenth centuries Byzantine writers even attempted to elaborate a version of religiously motivated warfare within the existing framework of Christian theology. Importantly, it was only at this point—later ninth century—that East Roman writers appear to have seen Islam as an existential threat, a challenge to the very existence of Christianity, rather than simply the erroneous faith (often understood as an aberrant form of Christian heresy) of a dangerous enemy.[10]

Tactics in the Field: Changing Roman Responses

In responding to the threat from the victorious Islamic armies after the initial conquests, Byzantine tactics and strategy from the 640s had to adapt quickly to a new situation. Light cavalry now came to dominate the border warfare, with skirmishing and hit-and-run raids characterising much of the fighting from the middle of the eighth century onwards. Although substantial numbers of mounted soldiers were always resent, the bulk of the Arab forces were mostly composed of infantry, and their extensive use of camels and horses to improve their mobility enabled them to travel far more quickly than most of their enemies, and the highly mobile nature of the warfare thereafter gave the Arab mounted infantry an advantage over traditionally outfitted Roman infantry. While infantry continued to be needed by the Eastern Romans, therefore, playing an important part in several campaigns, as well as in the guerrilla warfare along the eastern frontier in the later ninth and tenth centuries, it seems that their value slowly declined, perhaps because they were drawn largely from the poorest and least well-equipped of the provincial soldiery. Hardly any of the descriptive sources for the wars along the eastern frontiers during the campaigns of the eighth and ninth centuries give any details. The increasingly seasonal campaigning and localised recruitment in the theme armies, combined with a lack of professional training, the physical dispersal of the soldiers and the type of warfare waged, meant that while infantry soldiers could fulfil garrison duties and irregular, skirmishing warfare in broken country, or lie in wait for hostile forces, they would not be reliable under regular battlefield conditions. The development of infantry tactics after the period of the first Islamic conquests as well as the higher profile of mounted warfare thus reflected quite closely the general strategic situation in which the empire found itself. In general, there took place a levelling out of the different arms into simply light cavalry and infantry. Only the *tagmata* at Constantinople appear to have provided a heavy cavalry force. It seems to have been the responsibility of local officers in the provinces to establish field units and to arm them as each specific occasion required.

East Roman provincial armies were organised from at least the middle of the seventh century into *tourmai*, *drouggoi* and *banda*: very roughly, divisions, brigades and regiments. The first and last of these came to have a territorial identity, and each *tourma* had a headquarters or base, a fortified town or fortress. Each *bandon* was identified with a clearly defined locality, from which it drew its soldiers. The *drouggos* or brigade, the middle level of this structure, always remained a tactical unit, without any territorial identity. Each commander of a *tourma*, the *tourmarchēs*, had his base in a fortress town, and was an important figure in the military administration, responsible for the fortresses and strongpoints in his district, for the safety of the local population and their goods and chattels, as well as for dealing with raids into his territory and for informing his superior of enemy movements. The sizes of units on the battlefield varied according to tactical need, and there was no neat equivalence between the territorial and administrative districts and the size of the units drawn from them. Calculating the size of Byzantine armies from this sort of information is, in consequence, rather tricky. The use of archery seems to have declined considerably among Eastern Roman soldiers as a weapon with real tactical importance after the sixth and early seventh centuries. The use of the bow did not disappear entirely, since there always seem to have been some mounted archers, and provincial infantry troops probably included some soldiers armed with bows. And, in fact, the Early Islamic enemies of the Roman Empire relied upon a very similar panoply; while there were significant differences in points of detail, style, and in some cases of fabrication, differences in the fundamental types of defensive equipment and weaponry were minimal, to a large extent a result of both familiarity and constant warfare (Zaki 1979; Nicolle 1976, 1991).

From the 830s and 840s the *caliphate* began to employ Turkish mounted archers, a development that caused the Byzantines great difficulties. The composite reflex bow used by these troops was not native to the empire's lands, and military and tactical organisation was not up to enforcing regular training and practice of the sort possible in the context of Late Roman military structures. In the sixth and seventh centuries, the Byzantine forces used Hunnic mercenaries and allies, and from this basis trained their own mounted archers accordingly.

Thereafter, the possibilities for this sort of training declined. In the later ninth and early tenth century, such soldiers were once again recruited from the steppes—Chazars, Magyars and others, for example—and when the Emperor Leo VI (886–912) wrote his military handbook called the *Tactica*, he too noted the decline in Roman archery, the many defeats which were a result, and commends that all Roman recruits practise with the bow.[11]

Raiding and Guerrilla Warfare

The failure of the great siege of Constantinople of 717–718, the second and final attempt to knock Byzantium out at a single blow, resulted in a gradual stabilization of the frontier regions. Thereafter, and up to the middle of the tenth century, warfare along the frontier zone took the form of raid, counter-raid and skirmishing, of guerrilla tactics and ambushes. A good deal is known about this style of fighting both from historians' accounts of campaigns and battles, as well as from a number of military handbooks, some of them written by serving soldiers. The early-tenth-century *Taktika* of the emperor Leo VI, for example, shows that warfare along the eastern front followed a well-established pattern. Substantial Arab raids deep into East Roman territory, successfully capturing large amounts of booty, are recorded in some of the sources. By the 950s and 960s, this was changing, as imperial successes in pushing forward the frontier rendered the traditional system of defensive warfare more or less redundant.[12]

The Roman eastern frontier was guarded by a chain of look-out posts, with small units of irregulars acting as scouts and informants along the frontier, particularly covering the various points of ingress into imperial territory. The frontier was a broad band of territory, and the location of such look-out posts seems to have changed according to the situation, while raids and counter-raids intended to destroy enemy outposts or more important local fortresses and bases frequently altered the pattern of local strategy.

An anonymous treatise written in the 960s sets out the key aspects of this type of warfare.[13] First, the local commanders should make sure that the networks of watch-posts and look-outs are in order. Scouts should be recruited from among the local population, men with experience, a good knowledge of local routes and the different qualities they possess. They should work on a 15-day rotation, and to be despatched in small groups to watch the roads and routes that might be used by the enemy. Local commanders should make extensive use of spies, including merchants and others on genuine business in the enemy's land—a long tradition in Byzantine strategic thinking. The call-up of registered soldiers should be strictly observed, and the scouting parties should be checked by an officer from time to time. They should also change their location in order to avoid capture. There were pre-planned schemes for evacuating the non-military population of the regions through which an enemy raiding party would pass, once its route had been ascertained, in order to preserve the local population, and at the same time to deprive the enemy of the chance to collect provisions and easy booty.

The most important aspect of this frontier defensive strategy was 'shadowing.' Following and harassing the enemy by exploiting one's own knowledge of the local terrain was one aspect; keeping a close watch on his column and especially his encampment, in order to attempt ambushes on forage parties and other isolated groups, was another. Crucial to all operations was the idea of bringing together several smaller forces, leading eventually either to a full-scale confrontation, but with the imperial forces at a numerical advantage, or to a pincer movement, designed to encourage the enemy force to give up and return home.

In this case, it was usually planned for imperial troops to have occupied the passes or exit routes which the enemy commander would follow. The subsequent surprise attack or ambush, which could result in the recovery of all or most of the booty, and certainly with the destruction and rout of the enemy army, was the ultimate aim. The possibility that his own forces might themselves become the victims of shadowing and ambush was ever present, however, and the Byzantine commander was urged to use scouts and outriders in order to prevent this from happening.

One of the distinguishing features of this treatise is the focus on the judgement and independence of the local commanders. Not only should they themselves organise regular, small-scale raids over the border (unless the empire had made a formal truce with the Arab *emirs* or the *caliphate* itself); they should be prepared to attack an invading force whenever an appropriate opportunity arose, and not necessarily wait for the arrival of reinforcements or the local senior commander. The author of the treatise distinguishes three types of enemy raid, differentiated by size or by timing. Small, rapid raiding parties of cavalry, which might invade Roman territory at any time, and whose entry should be communicated to the local commanders as quickly as possible by the border scouts and watch-posts, should be shadowed, met, ambushed or hemmed in, and turned back, and if possible without any substantial gains in booty. Second, there were major raids, generally in August and September, consisting of substantial forces made up of volunteers for the *jihād* as well as regular troops from the Arab borderlands—Malatya, Aleppo, Tarsos, Antioch. Such raids had both an economic and an ideological function, first in terms of the desire for booty, and to damage the Roman economy, and second in respect of the desire of many Muslims to participate in the *jihād*. The local commander was enjoined to use every means at his disposal to find out when such raids would begin, by which route, and how numerous the enemy host would be. The invading force should be shadowed, along with any accompanying raiding parties which were sent out once the main force had reached Roman territory. The invaders' logistical difficulties should be maximised by the removal of livestock and crops, or even their destruction in extreme cases. The enemy force should be subject to constant harassment as it moved, foraged for supplies, set up camp or attempted to collect booty. The passes through which it would return should be occupied and ambushes laid; the water supplies should be held by Byzantine forces. The enemy should be attacked as they returned, laden with booty. Naturally, the Romans were not always able to respond successfully to such attacks, and there are many examples where Roman preparations failed to produce the desired results, or where the Roman commanders were unable to outwit and out-general their adversary. Third, the local commander also had to be on his guard against surprise raids, launched before the local population has been evacuated or any sort of ambush or shadowing-party sent out. In an effort to delay the enemy, various measures could be applied, such as a feint attack to distract the enemy from pillaging the villages while they are being hastily evacuated. Once the local troops were in the field, the strategy of harassment and ambush, by day and by night, came into play. While he was one of the empire's most successful antagonists, the *Emir* of Aleppo was ambushed on at least three occasions using this strategy, barely escaping with his life on one occasion (Dagron and Mihaescu 1986; McGeer 1995).

This sort of warfare could also be offensive. Local commanders were advised to maintain bands of raiders, whose task it was to raid deep into enemy territory in order to foment insecurity and uncertainty. One of their most important tasks was to take prisoners, so that Byzantine commanders might learn of enemy troop movements and intentions.

John Haldon

The Maritime Frontier

Until the middle of the seventh century, the East Romans faced no serious challenges in the Mediterranean, either in the eastern or the western halves. East Roman naval squadrons, directed primarily against pirates and aimed at controlling maritime commerce, appear in the sixth century to have been limited in number, and based around the Mediterranean coastline at key ports. There is very little reliable information, and it is clear that larger forces could be drafted into service for major seaborne expeditions such as that directed against the Vandal Kingdom in North Africa in the late 520s. From the early 650s, however, the Arabs began to build warships in various naval bases around the coasts of the Levant, an enterprise that met with considerable success. A major Roman Fleet commanded by the Emperor Constans II (641–668) was defeated (the emperor narrowly escaped) in 654 (Howard-Johnston 2010: 479–80; O'Sullivan 2004; Christides 1985). The Arabs were able to mount a large seaborne attack on Constantinople shortly thereafter, again in 668–669, and once again in 717–718. All were defeated, the attacks of the 660s and 717–718 with the help of a new weapon, 'liquid fire,' a light crude oil mixed with pine resin and projected through a pump, effectively a sort of medieval flame-thrower. It clearly had devastating effects, although that it changed the course of history, as often suggested, seems too strong a claim (Haldon, Hewes and Lacey 2006).

Thereafter there evolved a degree of equilibrium, although by the later ninth century the Byzantines seem to have achieved the upper hand. This was managed through the establishment of a permanent war fleet of heavily armed vessels, *dromones*, based at Constantinople, complemented by a number of provincial fleets manned and maintained on the same fiscal basis as their terrestrial counterparts.[14] The maritime forces of the Kibyrrhaiotai, mentioned already, of the Aegean Sea and of Samos, as well as land divisions with a maritime force attached to them, such as those of Kephallenia, the Peloponnese or Hellas, all contributed substantially to this, and certainly by the time of the Emperor Leo VI the empire was well on the way to naval superiority, in spite of occasional substantial setbacks (Haldon 2014: 109; Christides 2011). Successful combined land and sea operations were also mounted, as in the 880s in Italy and the Peloponnese, for example, but there were some serious defeats—the expedition against North Syria in 911 came to grief off Crete when it was met by a substantial Arab Fleet, as did the expedition against Crete itself in 949. The island eventually fell in 961 to a smaller but better-led expedition (Pryor and Jeffreys 2006: 385; Eickhoff 1966: 65–7, 173–5, 235–7; Ahrweiler 1966: 35–44, 93–7, 111–13). While there remained substantial maritime opposition at times, notably from the rising power of the Fatimid *Caliphate*, based from the 960s in Egypt, East Roman dominance at sea was maintained well into the middle years of the eleventh century.[15]

Offensive Strategy and Tactical Change

As a result of the increasingly aggressive warfare on the eastern front carried on by the empire from the second quarter of the tenth century, the need to recruit more professional soldiers, and the need to operate effectively on campaigns which demanded more than the seasonally available forces provided by the traditional provincial or thematic armies, a number of important changes appeared in the tactical structure and in the arms and armour of Byzantine troops. A series of important technical treatises on strategy and tactics, written in the middle and later tenth century, and the narrative accounts of contemporaries, both

164

Byzantines and Arabs, corroborate much of what they say (McGeer 2008; Sullivan 2010). The changes involved the revival of a corps of disciplined, effective heavy infantry, able to stand firm in the line of battle, confront enemy infantry and cavalry, support their own cavalry, march long distances and function as garrison troops away from their home territory on a permanent basis. They also entailed the introduction of a corps of heavy cavalry armed with lances and maces, which could operate effectively alongside infantry. This would add weight to the Byzantine attack, and thus substantially enhance the aggressive power of the Byzantine cavalry. Along with these developments, field tactics were developed in which these arms operated in a complementary way, offering the commanding officer a flexible yet hard-hitting force which could respond appropriately to a range of different situations (McGeer 1988, 1992).

Evidence for these changes comes partly from the contemporary sources, especially the military handbooks already referred to, but also from the startling successes marked up by Byzantine armies in the process of reconquest and expansion from the 950s onwards (Garrood 2008). In a tract known as the 'Recapitulation of Tactics' *Syllogē taktikōn*, a new formation of infantry soldiers is described, consisting of troops wielding thick-stocked, long-necked javelins or pikes, probably similar in form to the Roman legionary *pilum*. Their task was to confront and beat back enemy heavy cavalry attacks. According to the 'Recapitulation,' there should have been about 300 soldiers equipped in this manner, arrayed in the intervals between the infantry units making up the main battle line. They were deployed in either line or wedge formation to break up an enemy attack. In a slightly later treatise known as the '*Military Precepts*' compiled some 20 years later, the tactic had evolved further, so that there were in each major infantry unit of 1,000 men 100 soldiers so equipped, integrated with 400 ordinary spearmen, 300 archers and 200 light infantry (with slings and javelins).[16]

This important change in the role of infantry was reflected in the changed political and military situation of the tenth century. Whereas the sixth-century military treatise known as the *Stratēgikon* presents its sections on infantry drill and formations after those (more detailed) dealing with cavalry, the tenth-century texts give infantry formations equal or even preferential treatment. Infantry had now become a key element of the army both numerically and tactically, outnumbering cavalry by 2:1 or more, in contrast to the normal situation in the preceding centuries. Contemporaries note the greatly improved discipline and training which such troops displayed. The importance of infantry is demonstrated in the fact that a special commander for the infantry division in each army was appointed, the hoplitarch (*hoplitarchēs*), in charge of training, discipline and fighting skill. The new tactics were embodied in a new formation, in which infantry and cavalry worked together, essentially a hollow square or rectangle, depending on the terrain, designed to cope with encircling movements from hostile cavalry, as a refuge for Byzantine mounted units when forced to retreat, and as a means of strengthening infantry cohesiveness and morale.

These new formations mark a significant change in the role of infantry, no longer drawn up in a deep line with only a limited offensive role, but actively integrated into the offensive heavy cavalry tactics of the period. Infantry units now represented a sort of mobile marching camp, with a traditionally rather unreliable force given new strength as a defensive field formation on the one hand, to provide security in defence and on the march, a mobile base and refuge for lighter troops and cavalry, and on the other as a formation which could be transformed into a solid attacking formation at a few simple commands. The cavalry also evolved at this time. A new formation of very heavy cavalry appears, called *klibanophoroi*, heavy cavalry troopers armed from head to foot in lamellar, mail and quilting, whose horse was likewise protected—face, neck, flanks and forequarters were all to be covered with armour to prevent enemy missiles and blows from injuring the cavalryman's mount. While very few in number (as they were so expensive to maintain), these became the elite strike force within a field army.[17]

Both Byzantine and Arab writers of the time comment on the impressive effects of this formation on their foes. One Arab writer notes that the horse-armour of the heavy cavalry mounts made them appear to be advancing without legs. Some of these changes were probably due to the general, later emperor, Nikephoros Phokas, who became Commander-in-Chief in the East in the 950s, and immediately embarked upon a programme of training and drilling the soldiers in an attempt to re-establish good discipline, fighting spirit and good battlefield skills. His success is evident in the effective warfare waged by Byzantine forces over the following 50 years or more.[18]

In contrast, the forces of the *emirs*, against whom the East Romans fought in the tenth and eleventh centuries, remained largely more lightly armed, although they too possessed heavy cavalry soldiers. This was in part a reflection of the fragmented nature of the empire's competitors along the eastern frontier, which meant that the extensive resources needed to equip and maintain such troops were less readily available to the *emirs* and warlords of Northern Syria and Iraq. It also reflected the reliance built into traditional Islamic frontier warfare upon speed and lightness of equipment and armament, traditionally facets which had given the Arabs and others a considerable advantage—in the early tenth century the Emperor Leo VI advised his generals to adopt the tactic of mounting their infantry, a standard practice in the armies of Tarsus, in order the better to deal with their attacks.[19]

The successes of the period from about 960 to 1025 are impressive, but they were by no means uniform. East Roman armies had achieved a powerful reputation, so much so that by the 1030s the mere threat of an imperial army marching into Northern Syria was sufficient to cow the local *emirs*. Yet, while these successes were the result of a combination of good organisation and logistics, intelligent tactics, well-armed, trained and disciplined soldiers, and good morale, the key still remained the competence and effectiveness of the commanders. While tactical order and uniformity of equipment and training certainly gave Byzantine armies at most times of the empire's history a distinct advantage, even under the most effective rulers incompetent officers led their troops to disaster. This dependence on the charisma and intelligence of its leaders, typical of medieval warfare more broadly, was one of the most significant in-built weaknesses of the imperial military system at a tactical level. Combined with short-sighted strategic planning and internal political conflict, this was to lead during the middle of the eleventh century to serious problems and to the erosion of the effectiveness of the field armies as well as the provincial defences (Haldon 2003).

The successes of the generals who led the expansionist strategy of the second half of the tenth century had several unforeseen consequences. First, the provincialised structures which had developed in the period from the later seventh to ninth centuries were substantially eroded. New strategic commands known as *ducates* and *katepanates*, much smaller territorially than the traditional *themata*/themes, established a protective curtain between the inner regions of the empire and the frontier zones, while the traditional thematic militias were neglected. The logic behind the newly established strategic units was to address local threats or respond to the need to mobilise for larger expeditionary offensives, rather than major hostile incursions aimed at the heart of the empire. While the new structures did represent a defence in depth of sorts, the fragmentation of command around the periphery had obvious disadvantages when a major threat appeared. It was essentially up to the emperor, or one of the two commanders-in-chief, rather than a local commander, to assemble an appropriate force and march to deal with it. Much, therefore, depended on the competence of the leader in question. The empire still fielded an effective and feared campaign army, however, and so long as it remained well led and successful, the empire's enemies were held at bay.

During the middle years of the eleventh century this strategic model broke down, largely a result of an imbalance between diplomacy and military strength introduced through civil conflict and provincial or military rebellion. The thematic or provincial militias had in any case long been neglected in favour of full-time, regionally recruited *tagmata* maintained on a permanent basis, and their demise cannot be blamed on the emperors of the eleventh century. Reductions in military expenditure induced a heavier dependence on non-Byzantine mercenaries than had been the case until the end of Basil II's reign, so that foreign mercenary troops, especially of Western knights—Franks, Germans and Normans—played an increasingly prominent role, usually under their own leaders. The defeat at Manzikert in 1071 at the hands of the Seljuk Turks was not in itself such a great disaster from a purely military perspective, but the ensuing civil war and internal disruption gave the invading Turks a free hand in Central Asia Minor, and produced a very different political-strategic map of the region (Haldon 2003).

Conclusion

The long period of frontier warfare, interspersed with major expeditions, the centrality of Constantinople, and the role played by each side in the apocalyptic literature and ideas of the other, could not but have an effect on both cultures, and in many different ways. An astrological literature evolved, for example, which claimed to foretell the results of wars and which suggested that both sides were locked into a cycle of victory and defeat (Dagron 1997). Indeed, this material was influential enough to encourage a certain fatalism, since some believed that when, on the basis of the cycle, it was the turn of the other side to be victorious, there was no real point in resisting. The Italian ambassador Liudprand of Cremona reported from a visit to Constantinople in 968:

> The Greeks and Saracens have books . . . where one finds written the number of years of each emperor's life . . . and whether he will have success or failure against the Saracens . . . for this . . . reason, the Greeks, full of courage, attack, and the Saracens, without hope, offer them no resistance, awaiting the moment when it will be their turn to attack and that of the Greeks to make no resistance.[20]

Horoscopes and other forms of predictions were particularly important when fighting and warfare were at issue, since not only the ordinary soldiers but the senior officers were just as interested in trying to predict the outcome of a conflict.[21] On the Byzantine side, a number of contemporary apocalyptic texts look forward to a future great victory of the Roman emperor, in spite of the chastisements and catastrophes which have befallen the Roman (i.e., the Christians). A tradition also developed concerning the siege of Constantinople, part apocalypse, part hagiography, in which the role of the Virgin Mary as the city's protector was key.[22] By the same token, there developed a parallel early Islamic apocalyptic

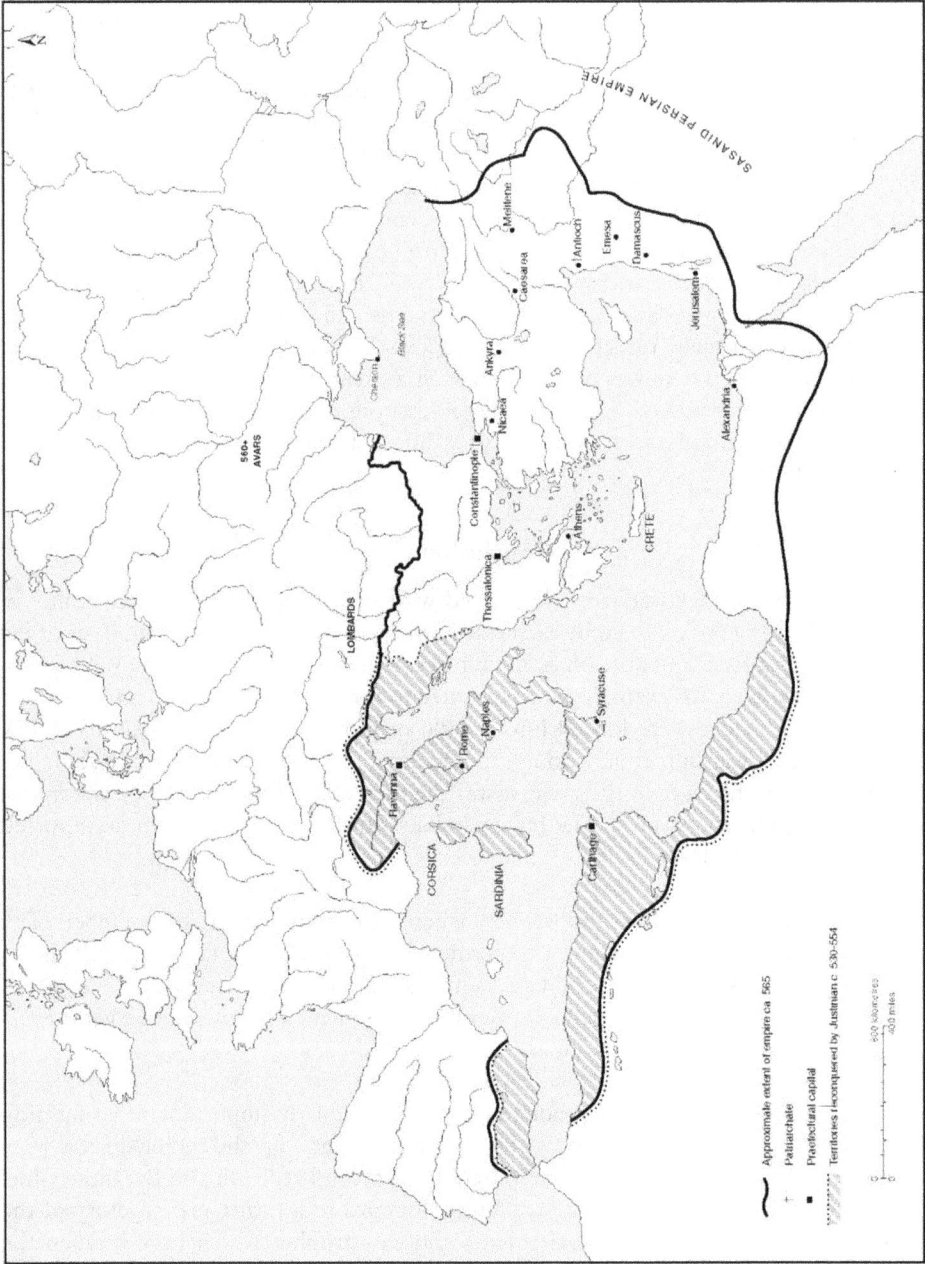

Map 10.1 The Roman empire ca. 570. Shaded areas represent the reconquests of Justinian (527–565).

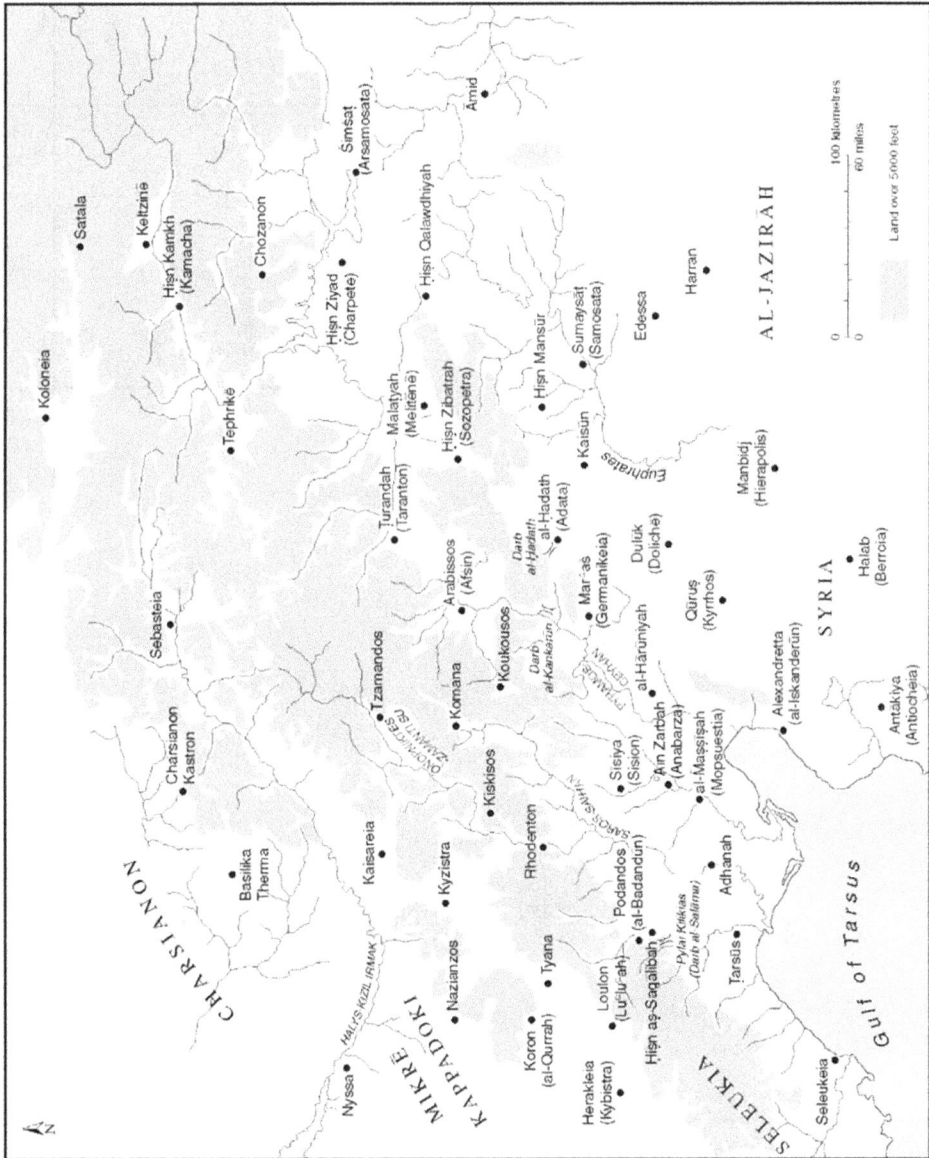

Map 10.2   The Taurus—Anti-Taurus frontier regions ca. 700–960.

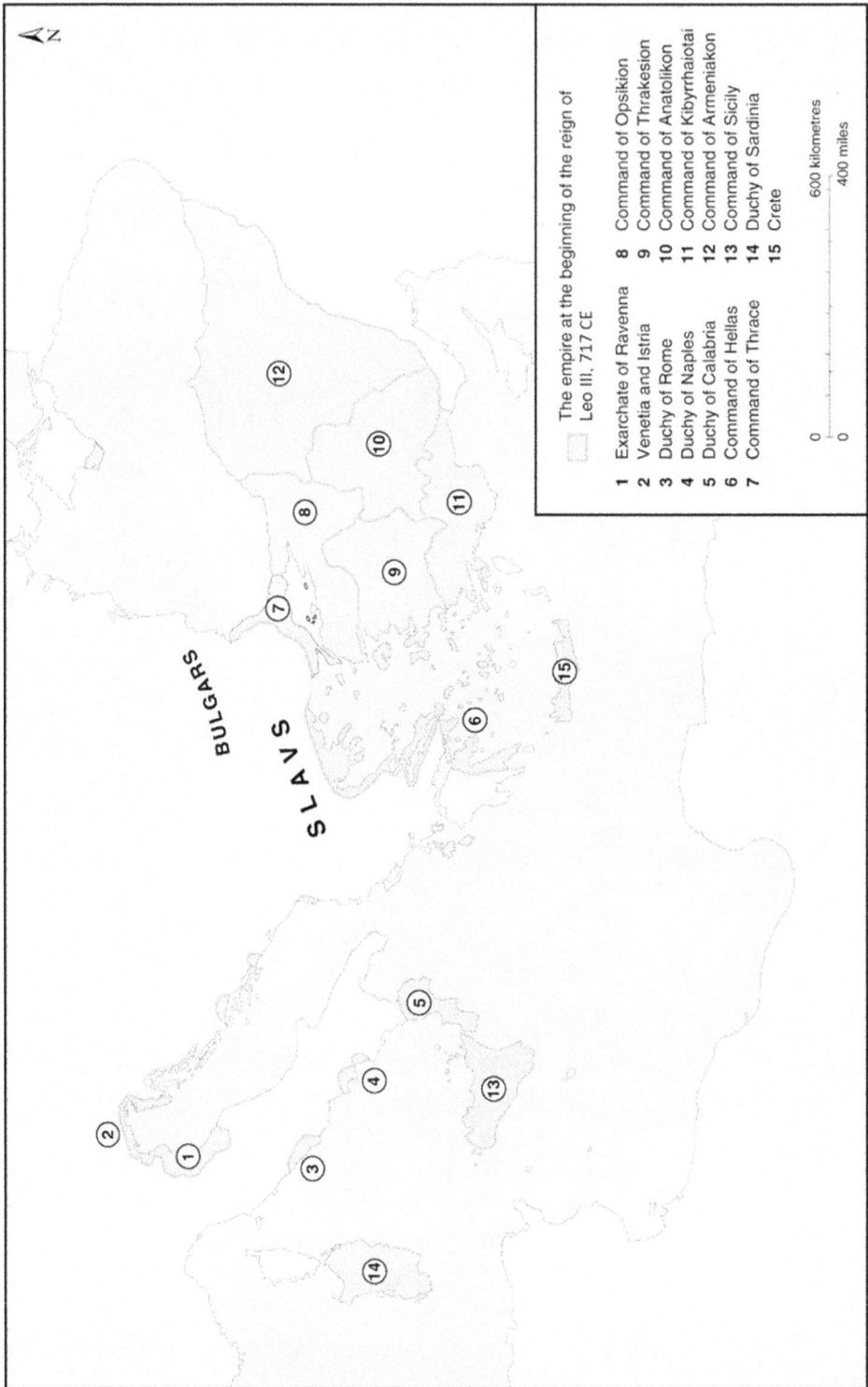

Map 10.3 East Roman military commands ca. 717.

The empire at the beginning of the reign of Leo III, 717 CE

1 Exarchate of Ravenna
2 Venetia and Istria
3 Duchy of Rome
4 Duchy of Naples
5 Duchy of Calabria
6 Command of Hellas
7 Command of Thrace
8 Command of Opsikion
9 Command of Thrakesion
10 Command of Anatolikon
11 Command of Kibyrrhaiotai
12 Command of Armeniakon
13 Command of Sicily
14 Duchy of Sardinia
15 Crete

BULGARS

SLAVS

0   600 kilometres
0   400 miles

tradition devoted to the theme of the conquest of Constantinople, underlining just how important this became for the invaders, both ideologically as well as in practical terms. One such text notes:

> You will invade Constantinople three times. In the first one you will face affliction and hardship. In the second there will be a peace treaty between you and them, so that the Muslims will build mosques in it, join them in a raid beyond Constantinople and return to it. And in the third invasion God will open it for you.
>
> *(Bashear 1991: 177)*

That tradition had a long history, well into the early Ottoman period, and it took until 29 May 1453 for the prophecies finally to be realised. Arab–Byzantine warfare was about a great deal more than waging war.

Notes

1 For more detailed accounts: Haldon 1997, 2016a.
2 Brubaker and Haldon 2011, especially chapters 6 and 7; Haldon 2006; Brandes 1999; Brandes and Haldon 2000; Crow 2009; Ivison 2007; Niewöhner 2007, 2008.
3 For a good general survey, refer to Kennedy 2004 and also Kaegi (1992).
4 Lilie 1976. For a detailed survey of the warfare, annual raiding and related matters, Brooks 1898; Kaegi 2008; Ahrweiler 1962; Christides 2011. For a general history of the period up to 850, see Auzépy 2008.
5 Prigent 2020; Haldon 2016b; Haldon and Brubaker 2011, especially chapter 11.
6 Kennedy 1995; Haldon and Kennedy 1980: 106–11; Bonner 1992, 1996; Bosworth 1992; see also Haldon 2014: commentary to Leo, *Taktika*, Constitution XVIII, 654–60.
7 Described in detail by the tenth-century geographer Kudāma ibn Ja'far, in De Goeje 1870ff. VI: 199.
8 Haldon and Kennedy 1980: 113–15. For a detailed account of Byzantine–Arab warfare in the ninth to tenth centuries, with translated sources, see the still useful Brooks 1900; Vasiliev 1950/1968; and for a general survey of the military organisation of the early *caliphate*: Kennedy 2001.
9 See Haldon 1999: 13–33; 2014: commentary to Leo, *Taktika*, Constitution xviii, §105, 508–16; Stouraitis 2011; Laiou 1993, 2006; Dagron 2009; and the essays in Koder and Stouraitis 2012; Miller and Nesbitt 1995.
10 Khoury 1982: 5–7; Haldon 2014: 33–7; 2018 with further literature and sources. Yet the Church remained reluctant, and the patriarch Polyeuktos rejected a proposal of the Emperor Nikephoros II Phokas (r. 963–969) that soldiers who fell fighting for the empire should be counted among the martyrs. See detailed discussion in Kolia-Dermitzaki 1991: 132–41.
11 Kaegi 1964; Haldon 1999, 2002; for Islamic arms and armour in the Umayyad period (660–751), Nicolle 1997. For Leo VI's *Taktika*: Dennis 2010, with extensive commentary in Haldon 2014.
12 McGeer 1995; Haldon 2014: commentary to Leo, *Taktika*, constitution xviii.
13 Howard-Johnston 1983; Dagron and Mihaescu 1986: 28–135 (text); Dagron 1995, 1997. See also Dennis 1985: 37–239 (text: 144–238).
14 On the various types of warships in use by both sides, see in general Pryor and Jeffreys 2006: 123–304; Haldon 2014, commentary to Leo, *Taktika*, constitution xix, 8; 61–8; 247–52.
15 See Haldon 2014 for commentary to Leo, *Taktika*, constitution xix, devoted to naval warfare and related matters; Pryor and Jeffreys 2006 for a detailed account of naval warfare from the seventh century onwards as well as discussion of the various types of warships in use; Eickhoff 1966; Christides 2002; Koder 1999.
16 For the first text: Chatzelis and Harris 2017 (critical edition: Dain 1938); for the second: McGeer 1995: 3–59 (text), 61–78 (notes).
17 On Byzantine weapons and armour, see Dawson 2002 for the tenth century and later; and Haldon 2002 for the sixth to tenth century. Detailed discussion available also in Grotowski 2010.

18 McGeer 1995 for a detailed account; Haldon 1999 for broader context and survey.
19 See, for example, the survey of the forces of the Hamdanid *Emirs* of Aleppo, the main enemies of the Eastern Roman Empire in Northern Syria during the first half of the tenth century, in McGeer 1995: 229–48. While dealing with the second half of the tenth century, this account applies in essentials just as much to the preceding century. See also Schwarzer 1991. Leo VI's advice: Leo, *Taktika*, Const. xx, §206. 1060–2. For detailed commentary on Islamic warfare against the Byzantines in the tenth century, see Haldon 2014: commentary to Leo, *Taktika*, constitution xviii, 495–695.
20 *Liudprand of Cremona, Legatio* §39 (ed. Becker 1915: 195–6); Dagron 1997: 42–7.
21 See e.g., Leo, *Taktika*, constitution xiv, §101.
22 For discussion and sources of the Byzantine apocalyptic tradition: Brandes 1990, 1991, 2007.

Bibliography

Ahrweiler, H., 'L'Asie mineure et les invasions arabes,' *Revue Historique*, vol. 227 (1962), pp. 1–32.
Ahrweiler, H., *Byzance et la mer: la marine de guerre, la politique at les institutions maritimes de Byzance aux VII- XV siècles* (Paris: PUF, 1966).
Auzépy, M.-F., 'State of emergency (700–850),' in J. Shepard (ed.), *Cambridge History of the Byzantine Empire ca. 500–1492* (Cambridge: Cambridge University Press, 2008), pp. 251–91.
Bashear, S., 'Apocalyptic and Other Materials on Early Muslim-Byzantine Wars: A Review of Arabic Sources,' *Journal of the Royal Asiatic Society*, 3rd series, vol. 1 (1991), pp. 173–207.
Becker, J., *Liutprandi episcopi Cremonensis opera* (Leipzig: Teubner, 1915).
Bonner, M., 'Some Observations Concerning the Early Development of Jihad on the Arab–Byzantine Frontier,' *Studia Islamica*, 75 (1992), pp. 5–31.
Bonner, M., 'The Naming of the Frontier: 'awāsim, thughūr and the Arab Geographers,' *Bulletin of the School of Oriental and African Studies*, vol. 57, no. 1 (1994), pp. 17–24.
Bonner, M., *Aristocratic Violence and Holy War. Studies in the Jihad and the Arab-Byzantine Frontier* (New Haven: Yale University Press, 1996).
Bosworth, C.E., 'The City of Tarsus and the Arab–Byzantine Frontiers in Early and Middle Abbasid Times,' *Oriens*, vol. 33 (1992), pp. 268–86.
Brandes, W., 'Die apokalyptische Literatur,' in F. Winkelmann, W. Brandes (eds.), *Quellen zur Geschichte des frühen Byzanz (4.-9. Jahrhundert). Bestand und Probleme* (Berlin: Akad. d. Wiss. der DDR, 1990), pp. 305–22.
Brandes, W., 'Endzeitvorstellungen und Lebenstrost in mittelbyzantinischer Zeit (7.–9. Jahrhundert),' in *Poikila Byzantina* 11, *Varia* 3 (Berlin: Habelt, 1991), pp. 9–62.
Brandes, W., 'Byzantine Towns in the Seventh and Eighth Century—Different Sources, Different Histories?,' in G.P. Brogiolo and B. Ward-Perkins (eds.), *The Idea and Ideal of the Town between Late Antiquity and the Early Middle Ages, The Transformation of the Roman World 4* (Leiden: Brill, 1999), pp. 25–57.
Brandes, W., 'Die Belagerung Konstantinopels 717/718 als apokalyptisches Ereignis. Zu einer Interpolation im griechischen Text der Pseudo-Methodios-Apokalypse,' in K. Belke, E. Kislinger, A. Külzer and M. Stassinopoulou (eds.), *Byzantina Mediterranea: Festschrift für Johannes Koder zum 65. Geburtstag* (Vienna: Böhlau, 2007), pp. 65–91.
Brandes, W. and Haldon, J.F., 'Towns, Tax and Transformation: State, Cities and Their Hinterlands in the East Roman World, ca. 500–800,' in G.P. Brogiolo, N. Gauthier and N. Christie (eds.), *Towns and their Territories between Late Antiquity and the Early Middle Ages, The Transformation of the Roman World 9* (Leiden: Brill, 2000), pp. 141–72.
Brooks, E.W., 'The Arabs in Asia Minor (641–750), from Arabic Sources,' *Journal of Hellenic Studies*, vol. 18 (1898), pp. 182–208.
Brooks, E.W., 'Byzantines and Arabs in the time of the Early Abbasids, 1,' *English Historical Review*, vol. 15 (1900), pp. 737–39.
Brubaker, L. and Haldon, J.F., *Byzantium in the Iconoclast Era, ca. 680–850: A History* (Cambridge: Cambridge University Press, 2011).
Chatzelis, G. and Harris, J., *A Tenth-century Byzantine Military Manual: The Sylloge Tacticorum* (Abingdon/New York: Routledge, 2017).

Christides, V., 'The Naval Engagement at Dhat as-Sawari AH 34/AD 655–656: A Classical Example of Naval Warfare Incompetence,' *Byzantina*, vol. 13 (1985), pp. 1331–45.

Christides, V., 'Arab-Byzantine Struggle in the Sea: Naval Tactics (7th–11th c. AD), Theory and Practice,' in Y. Yousef al-Hijjri and V. Christides (eds.), *Aspects of Arab Seafaring: An Attempt to Fill in the Gaps in Maritime History* (Athens: Institute for Graeco-Oriental and African Studies/Kuwait Foundation for the Advancement of Science, 2002), pp. 87–106.

Christides, V., 'The Second Arab Siege of Constantinople (717–718?): Logistics and Naval Power,' in D. Bumazhnov, E. Grypheou, T.B. Sailors and A. Toepel (eds.), *Bibel, Byzanz und christlicher Orient: Festschrift für Stephen Gerö zum 65. Geburtstag* (Leuven/Paris/Waldpole, MA: Peeters, 2011), pp. 511–33.

Crow, J., 'Byzantine Castles or Fortified Places in Pontus and Paphlagonia,' in T. Vorderstrasse and J. Roodenberg (eds.), *Archaeology of the Countryside in Medieval Anatolia* (Leiden: Brill, 2009), pp. 25–43.

Dagron, G., 'Le combattant byzantin à la frontière du Taurus: guérilla et société frontalière,' in *Le combattant au Moyen Âge, XVIIIe Congrès de la société des historiens médiévistes de l'enseignement supérieur public* (Montpellier/Paris: Publications de la Sorbonne, 1995), pp. 37–43.

Dagron, G., 'Apprivoiser la guerre. Byzantins et arabes ennemis intimes,' in N. Oikonomidès (ed.), Τό εμπόλεμο Βυζάντιο *(Byzantium at War)* (Athens: Idryma Goulandrē-Chorn, 1997), pp. 37–49.

Dagron, G., 'Légitimer la guerre à Byzance,' *Mélanges de l'université Saint-Joseph*, vol. 62 (2009), pp. 1–19.

Dagron, G. and Mihaescu, H., *Le traité sur la Guérilla (De velitatione) de l'empereur Nicéphore Phocas (963–969). Texte établi par Gilbert Dagron et Haralambie Mihaescu, trad. et comm. par G. Dagron* (Paris: Editions du CNRS, 1986).

Dain, A., *Sylloge Tacticorum, quae olim 'inedita Leonis Tactica' dicebatur* (Paris: Société d'édition, 'Les Belles lettres,' 1938).

Dawson, T., 'Suntagma hoplôn: the equipment of regular Byzantine troops, c. 950–c. 1204,' in D. Nicolle (ed.), *A Companion to Medieval Arms and Armour* (Woodbridge: Boydell, 2002), pp. 81–90.

De Goeje, M.-J. 1870, *Bibliotheca Geographorum Arabicorum* (Leiden: Brill, 1870).

Nunc continuata consultantibus, R. Blachère et al. (Leiden: Brill, 1938).

Dennis, G.T., *Three Byzantine Military Treatises: Text, trans. and notes. Corpus Fontium Historiae Byzantinae 25* (Washington DC: Dumbarton Oaks, 1985).

Dennis, G.T., *The Taktika of Leo VI. Text, Translation and Commentary. Corpus Fontium Historiae Byzantinae 49* (Washington DC: Dumbarton Oaks, 2010).

Eastwood, W.J., Haldon, J.F., Gümüşçü, O., Yiğitbaşioğlu, H. and England, A., 'Integrating palaeoecological and archaeo-historical records: Land use and landscape change in Cappadocia (central Turkey) since late Antiquity,' in T. Vorderstrasse and J. Roodenberg, eds., *Archaeology of the Countryside in Medieval Anatolia* (Leiden : Brill, 2009), pp. 45–69.

Eickhoff, E., *Seekrieg und Seepolitik zwischen Islam und Abendland* (Berlin: De Gruyter, 1966).

el Cheikh, N.M., *Byzantium Viewed by the Arabs* (Cambridge, MA; London: Harvard University Press, 2004).

Garrood, W., 'The Byzantine Conquest of Cilicia and the Hamdanids of Aleppo, 959–965,' *Anatolian Studies*, vol. 58 (2008), pp. 127–40.

Grotowski, P.Ł., *Arms and Armour of the Warrior Saints: Tradition and Innovation in Byzantine Iconography (843–1261)* (Leiden: Brill, 2010).

Haldon, J.F., *Byzantium in the Seventh Century: The Transformation of a Culture* (Cambridge: Cambridge University Press, 1997).

Haldon, J.F., *Warfare, State and Society in the Byzantine World 565–1204* (London: UCL Press, 1999).

Haldon, J.F., 'Some aspects of Early Byzantine Arms and Armour,' in D. Nicolle (ed.), *Companion to Medieval Arms and Armour* (Woodridge: Boydell, 2002), pp. 65–87.

Haldon, J.F., 'Approaches to an Alternative Military History of the Period ca. 1025–1071,' in E. Chrysos (ed.), *Byzantium in the Eleventh Century* (Athens: Institouto Vyzantinōn Ereunōn, Ethniko Hidryma Ereunōn, 2003), pp. 45–74.

Haldon, J.F., 'Social transformation in the 6th–9th century East,' in W. Bowden, A. Gutteridge and C. Machado (eds.), *Social and Political Life in Late Antiquity, Late Antique Archaeology 3.1* (Leiden: Brill, 2006), pp. 603–47.

Haldon, J.F., '"Cappadocia Will Be Given Over to Ruin and Become a Desert": Environmental Evidence for Historically-Attested Events in the 7th–10th Centuries,' in K. Belke, E. Kislinger, A. Külzer and M. Stassinopoulou (eds.), *Byzantina Mediterranea. Festschrift für Johannes Koder zum 65, Geburtstag* (Vienna: Böhlau, 2007), pp. 215–30.

Haldon, J.F., *A Critical Commentary on the Taktika of Leo VI* (Washington DC: Dumbarton Oaks, 2014).

Haldon, J.F., *The Empire that Would Not Die: The Paradox of East Roman Survival, 640–740* (Cambridge MA: Harvard University Press, 2016a).

Haldon, J.F., 'A Context for Two "Evil Deeds": Nikephoros I and the Origins of the Themata,' in O. Delouis, S. Métivier and P. Pagès (eds.), *Le saint, le moine et le paysan. Mélanges d'histoire Byzantine offerts à Michel Kaplan, Byzantina Sorbonensia 29* (Paris: Publications de la Sorbonne, 2016b), pp. 245–65.

Haldon, J.F., 'Eastern Roman (Byzantine) Views on Islam and on Jihād, ca. 900 CE: A Papal Connection?,' in R. Balzaretti, J. Barrow and P. Skinner (eds.), *Italy and Medieval Europe: Papers for Chris Wickham on the Occasion of his 65th Birthday* (Oxford: Oxford University Press, 2018), pp. 476–85.

Haldon, J.F., Hewes, C. and Lacey, A., 'Greek Fire Revisited: Recent and Current Research,' in E. Jeffreys (ed.), *Byzantine Style, Religion and Civilization: In Honour of Sir Steven Runciman* (Cambridge: Cambridge University Press, 2006), pp. 290–325.

Haldon, J.F. and Kennedy, H., 'The Arab-Byzantine Frontier in the Eighth and Ninth Centuries: Military Organisation and Society in the Borderlands,' *Zbornik Radova Vizantološkog Instituta*, vol. 19 (1980), pp. 79–116.

Howard-Johnston, J., 'Byzantine Anzitene,' in S. Mitchell (ed.), *Armies and Frontiers in Roman and Byzantine Anatolia, British Archaeological Reports, Int. ser. 156* (Oxford: BAR, 1983), pp. 239–90.

Howard-Johnston, J., *Witnesses to a World Crisis: Historians and Histories of the Middle East in the Seventh Century* (Oxford: Oxford University Press, 2010).

Ivison, E., 'Amorium in the Byzantine Dark Ages (Seventh to Ninth Centuries),' in J. Henning (ed.), *Post-Roman Towns, Trade and Settlement in Europe and Byzantium 2: Byzantium, Pliska, and the Balkans, Millenium-Studien 5.2* (Berlin/New York: De Gruyter, 2007), pp. 26–59.

Kaegi, W.E., 'The Contribution of Archery to the Turkish Conquest of Anatolia,' *Speculum*, vol. 39 (1964), pp. 96–108.

Kaegi, W.E., *Byzantium and the Early Islamic Conquests* (Cambridge: Cambridge University Press, 1992).

Kaegi, W.E., 'Confronting Islam: Emperors versus Caliphs (641–c. 850),' in J. Shepard (ed.), *Cambridge History of the Byzantine Empire ca. 500–1492* (Cambridge: Cambridge University Press, 2008), pp. 365–94.

Kennedy, H., 'The Financing of the Military in the Early Islamic State,' in Av. Cameron (ed.), *States, Resources and Armies: Papers of the Third Workshop on Late Antiquity and Early Islam* (Princeton: Darwin Press, 1995), pp. 361–78.

Kennedy, H., *The Armies of the Caliphs: Military and Society in the Early Islamic State* (London/New York: UCL Press, 2001).

Kennedy, H., *The Prophet and the Age of the Caliphates. The Islamic Near East from the Sixth to the Eleventh Century* (London/Harlow/New York: Longman, 2004).

Khoury, A.T., *Apologétique byzantine contre l'Islam (VIIIe-XIIIe siècles)* (Leiden: Brill, 1982).

Koder, J., 'Aspekte der thalassokratia der Byzantiner in der Ägäis,' in E. Chrysos, D. Letsios, H.A. Richter and R. Stupperich (eds.), *Griechenland und das Meer* (Mannheim-Möhnsee: Bibliopolis, 1999), pp. 101–9.

Koder, J. and Stouraitis, I. (eds.), *Byzantine War Ideology between Roman Imperial Concept and Christian Religion* (Vienna: Verlag der Österreichischen Akademie der Wissenschaften, 2012).

Kolia-Dermitzaki, A., Ὁ βυζαντινός "ἱερός πόλεμος". Ἡ ἔννοια καί ἡ προβολή τοῦ θρησκευτικοῦ πολέμου στό Βυζάντιο [Byzantine 'Holy War.' Its Meaning and Promotion in Byzantium]. (Athens: Basilopoulos, 1991).

Krausmüller, D., 'Killing at God's Command: Niketas Byzantios' Polemic against Islam and the Christian Tradition of Divinely-Sanctioned Murder,' *Al-Masāq*, vol. 16 (2004), pp. 163–76.

Kühn, H.J., *Die Byzantinische Armee im 10. Jahrhundert* (Vienna: Verlag Fassbaender, 1991).

Laiou, A.E., 'On Just War in Byzantium,' in J.S. Langdon, St.W. Reinert, J.S. Allen and C.P. Ioannides (eds.), *To Hellenikon. Studies in Honor of Speros Vryonis Jr* (New Rochelle, NY: Artistide D. Caratzas, 1993), pp. 153–74.

Laiou, A.E., 'The Just War of Eastern Christians and the Holy War of the Crusaders,' in R. Sorabji and D. Rodin (eds.), *The Ethics of War. Shared Problems in Different Traditions* (Aldershot/Burlington: Ashgate, 2006), pp. 30–43.

Lee, B.D., 'Theological Diplomacy in the Middle Byzantine period -Propaganda War between Constantinople and Caliphate or interfaith Dialogue?,' in A. Kralides and A. Gkoutzioukostas (eds.), *Byzantium and the Arab World: Encounter of Civilizations* (Thessaloniki: Aristotle University of Thessaloniki, 2013), pp. 269–80.

Lilie, R.J., *Die Byzantinische Reaktion auf die Ausbreitung der Araber, Miscellanea Byzantina Monacensia 22* (Munich: Institute for Byzantine Studies, 1976).

Mango, C. 'The Development of Constantinople as an Urban Centre,' in *17th International Byzantine Congress, Major Papers* (New Rochelle, NY: Aristide Caratzas, 1986), pp. 118–36.

McGeer, E., 'Infantry versus Cavalry: The Byzantine Response,' *Revue des Etudes Byzantines*, vol. 46 (1988), pp. 135–45.

McGeer, E., 'The Syntaxis Armatorum Quadrata: A Tenth-century Tactical Blueprint,' *Revue des Etudes Byzantines*, vol. 50 (1992), pp. 219–29.

McGeer, E., *Sowing the Dragon's Teeth. Byzantine Warfare in the Tenth Century* (Washington DC: Dumbarton Oaks, 1995).

McGeer, E., 'Military Texts,' in E. Jeffreys, J.F. Haldon and R. Cormack (eds.), *The Oxford Handbook of Byzantine Studies* (Oxford: Oxford University Press, 2008), pp. 907–14.

Métivier, S., 'L'organisation de la frontière arabo-byzantine en Cappadoce (VIIIe-IXe siècle),' in E. Cuozzo, V. Déroche, A. Peters-Custot and V. Prigent (eds.), *Puer Apuliae: Mélanges offerts à Jean-Marie Martin* (Paris: Association des amis du centre d'histoire et civilisation de Byzance, 2008), pp. 433–54.

Miller, T.S. and Nesbitt, J.S. (eds.), *Peace and War in Byzantium* (Washington DC: Dumbarton Oaks, 1995).

Mottahedeh, R.P. and al-Sayyid, R., 'The Idea of the Jihād in Islam before the Crusaders,' in A.E. Laiou and R.P. Mottahedeh (eds.), *The Crusades from the Perspective of Byzantium and the Muslim World* (Washington DC: Dumbarton Oaks, 2001), pp. 23–9.

Nicolle, D., *Early Medieval Islamic Arms and Armour* (Madrid: Instituto de Estudios sobre Armas Antiguas, Consejo Superior de Investigaciones Cientificas, Patronato Menendez y Pelayo, 1976).

Nicolle, D., 'Byzantine and Islamic Arms and Armour: Evidence for Mutual Influence,' *Graeco-Arabica*, vol. 4 (1991), pp. 299–325.

Nicolle, D., 'Arms of the Umayyad Era: Military Technology in a Time of Change,' in Y. Lev (ed.), *War and Society in the Eastern Mediterranean: 7th-15th Centuries* (Leiden/New York/Köln: Brill, 1997), pp. 9–100.

Niewöhner, P., 'Archäologie und die, Dunklen Jahrhunderte im Byzantinischen Anatolien,' in J. Henning (ed.), *Post-Roman Towns, Trade and Settlement in Europe and Byzantium 2: Byzantium, Pliska, and the Balkans, Millenium-Studien 5.2* (Berlin/New York: De Gruyter, 2007), pp. 119–57.

Niewöhner, P., 'Sind die Mauern die Stadt?,' *Archäologischer Anzeiger*, Issue 1 (2008), pp. 181–201.

O'Sullivan, S., 'Sebeos' Account of an Arab Attack on Constantinople in 654,' *Byzantine and Modern Greek Studies*, vol. 28 (2004), pp. 67–88.

Prigent, V. 'Retour sur l'origine des thèmes byzantins,' *Travaux et Mémoires*, vol. 24/2 (2020), pp. 105–36.

Pryor, J. and Jeffreys, E., *The Age of the Δρόμων. The Byzantine Navy ca 500–1204.* (Leiden/Boston: Brill, 2006).

Schwarzer, J.K., 'Arms from an Eleventh-century Shipwreck,' *Graeco-Arabica*, vol. 4 (1991), pp. 327–50.

Stouraitis, I., 'Jihad and Crusade: Byzantine Positions towards the Notions of "Holy War",' *Byzantina Symmeikta*, vol. 21 (2011), pp. 11–63.

Sullivan, D., 'Byzantine Military Manuals: Prescriptions, Practice and Pedagogy,' in P. Stephenson (ed.), *The Byzantine World* (London/New York: Routledge, 2010), pp. 149–61.

Sypiański, J., 'Arabo-Byzantine Relations in the 9th and 10th Centuries as an Area of Cultural Rivalry,' in A. Kralides and A. Gkoutzioukostas (eds.), *Byzantium and the Arab World* (Thessaloniki: Aristotle University of Thessaloniki, 2013), pp. 465–78.

Vasiliev, A.A., *Byzance et les Arabes* I: *La dynastie d'Amorium (820–867)*; II: *Les relations politiques de Byzance et des Arabes à l'époque de la dynastie macédonienne (Les empereurs Basile I, Léon le Sage et Constantin VII Porphyrogénète) (867–959)*, Fr. edn. H. Grégoire and M. Canard. (Brussels: Institut de philologie et d'histoire orientales, 1950/1968).

Whittow, M. *The Making of Orthodox Byzantium: 600–1025* (London: Macmillan, 1996).

Zaki, A.R., 'Medieval Arab Arms,' in R. Elgood (ed.), *Islamic Arms and Armour* (London: Scolar Press, 1979), pp. 202–12.

11

CRUSADING ARMIES AND THE TURKS IN THE MIDDLE EAST

John D. Hosler

The Crusades were the framework for two centuries of conflict between Western and Eastern armies in the Middle East. Arabic and Greco-Latinate sources broadly separate the adversaries into the two categories of 'Franks' and 'Turks,' respectively, which has served to obscure the diversity of the military forces in the region. 'Franks' could be French but they were also Norman, Italian, English, Danish, and so on; some might be crusaders, and others might be Christian settlers in the region (MacEvitt 2018). Likewise, a 'Turk' might well be a Kurd, an Arab or a *mamluk*; although in this case the term usually, at least, denoted a soldier at war, whereas the term 'Saracen' usually indicated Muslims settled in the region. The nature of Frankish and Turkish armies in the period increases in complexity when one scratches below the surface of generalised chronicle accounts. Complexity also arises in the magnitude of martial affairs in the Levant, broadly construed. The campaigns of the major Crusades produced the most well-known battles and supply well-worn anecdotes. But there were many smaller Crusades and other important campaigns that add to the illustration of war in the period. Sound studies have recognised the diversity of the forces arrayed in time and space as well as the commonalities of martial affairs in a time of broad continuity in terms of strategy, operations and tactics.

Context

Serious Turkish threats manifested against the Byzantine Empire in the eleventh century. The most famous is the victory of Alp Arslan, Sultan of the Seljuks, over the army of Emperor Romanus IV at the Battle of Manzikert in 1071 (Kaldellis 2017: 246–8). Further progress of the Turks continued thereafter in two principal directions. The first largely consisted of attacks upon rival Muslim regional powers, the Shia Fatimid *Caliphate* in Egypt and the Sunni Abbasid *Caliphate* in Baghdad. Sweeps along the Levantine Coast resulted in the displacement of the Fatimids from its environs. In 1073, the Kwarezmian mercenary general, Atsiz ibn Abaq, captured Jerusalem, although he did not sack it. The Fatimid *vizier* Badr al-Afdal retook the city in 1098, right on the eve of the arrival of the Western crusaders. Other Turkish incursions steadily added lands in the Jazira and Syria and, in the context of already declining Abbasid territorial control and influence, the Seljuk power base grew (Peacock 2015).

DOI: 10.4324/9780429437915-15

The second major thrust was westward, through Asia Minor and towards Constantinople. Turkish progress was steady but by no means inexorable and not entirely military in form. Nicaea fell in 1081 when its governor, Sulayman, usurped control of the city; Antioch fell four years later when its governor, Philaretos Braakhamios, converted to Islam. The early 1090s saw the loss of Cappadocia and the city of Nikomedia to the Turkish warlords Abu'l Kasim, Danishmen and Çaka, and by 1094 the western coast of Asia Minor was in Turkish hands (Frankopan 2012: chs. 3–6). These losses persuaded the Byzantine Emperor Alexios Comnenus to turn to the West for military assistance, and this was a driver of the campaigns that would become known as the First Crusade.

The primary and contributing causes of the First Crusade are many, varied and still hotly debated. Pope Urban II's 1095 sermon at Clermont evidentially emphasised not only the plight of the East but also purported torture and murder of Christians there and the possibility of recovering Jerusalem; the latter had been a carrot shrewdly dangled by Alexios in his messaging campaign (Frankopan 2012: 92–5).[1] Jerusalem was a Christian right, it was believed, because the Jews had been disinherited for their role in the crucifixion of Jesus Christ, a charge validated by Titus' sacking of the Second Temple in 70 (Yuval 2006: 38–49). Roman control (and thus, Christian control, post-313) of the city endured until Heraclius lost it to *Caliph* 'Umar ibn al-Khattab in 638. Formerly Christian lands could be recovered, with mid-eleventh-century efforts in Sicily and the burgeoning *Reconquista* in Spain demonstrating proof of concept. Urban offered an indulgence to Christian warriors who, by striving in a holy task for God, could receive a remittance of penance (Bysted 2015).[2] Such papal backing was exceptional but not exclusive: despite local episcopal prohibitions on inter-Christian violence (the Peace and Truce of God), Pope Alexander II, at the urging of Cardinal Hildebrand, sponsored Duke William of Normandy's conquest of England in 1066. Later, Hildebrand, now Gregory VII, argued for the possibility of obtaining salvation through Christian warfare, and sacred violence itself turned on theological notions of love and charity, even for one's enemies as they fell (*Correspondence* [1932]1990: 60–1; Robinson [1973]2011: 191–2; Riley-Smith 1980). A conglomeration of all the above factors contributed to the origin of the First Crusade, and Alexios' call for assistance was the spark that set the tinder alight.

Thrust and Parry

For their part, the Turks had some knowledge of the incursions into Iberia in the eleventh century but were probably broadly unaware of specific Christian motivations until after the First Crusade (Christie 2016: 311–23). Its campaigns are well known in the broad strokes: the capture of Nicaea (with Byzantine support); a stirring relief action at Dorylaeum; the dismal siege of Antioch, followed by the defeat of Kerbogha's Turkish army outside its walls; and the infamous siege and sack of Jerusalem in 1099 (France 1994).[3] Muslim intellectual reactions to the stunningly successful war trickled in. In the mind of an early poet, Islam itself was under attack (Latiff 2017: 70–3). By 1105, at least one well-informed Muslim jurist discerned a grand Christian strategy of retaking Sicily, Al-Andalus, and the Levant; this theme was eventually adopted by the Mosul historian Ibn al-Athir and expanded conceptually into a perceived assault on the entire Dar al-Islam. Beyond that, al-Athir also claimed the Crusade to have originated with an invitation from the Shia Fatimids in Egypt, who employed it as a foil against the (Sunni) Seljuks, laying bare inter-Islamic tensions in the region (Kedar 2014: 74; Frenkel 2011: 29).

The capture of Jerusalem in 1099 led to the establishment of a Western presence in the Levant and naturally provoked a string of Muslim military responses. The four 'Crusader States' (Kingdom of Jerusalem, Counties of Tripoli and Edessa, and the Principality of Antioch) each tended to patrol within their own sectors, and it was more common for the army of the Kingdom of Jerusalem to arrive in relief than the other way around (Buck 2017: 119–21). Such assistance could be of critical importance, and when delayed, the result could be disastrous, such as Roger of Salerno's crushing defeat to Ilghazi, Atabeg of Aleppo, at the Battle of the Field of Blood in 1119 (Morton 2018: 96–7). Turkish prowess was impressively displayed throughout the early to mid-twelfth century, with the capture of Edessa by Imad ad-Din Zengi's army in 1144 serving as both a capstone of active resistance and a primary motivation for the call of the Second Crusade.[4]

This is not to say that Muslims were unified in their efforts. Rough dichotomies of Christian-versus-Muslim are broadly true for the period but can also be misleading because settlers had to deal with their counterparts in everyday affairs. Interactions between antagonistic rivals could also be placid or even mutually beneficial, as demonstrated by a tradition of commercial treaties between the later Ayyubid Dynasty and Italian merchants (Wolff 1969a: 697). Moreover, inter-Turkish rivalries were apparent from very early on. These predominated into the reign of Zengi's second son, Nur al-Din, but he managed to defeat the Franks at the Battle of Inab in 1149 and consolidate power in Syria against both Sunni rivals and the Shia Ismaili assassins. This he did while utilising the call to *jihad* to gain military support, which also enabled him to initiate diplomatic and military incursions against the Fatimids (Hillenbrand 1999: 116–18).

Twelfth-century Arabic poetry extolled the virtues of this *jihad* and incorporated it into a call for the destruction of the remaining Crusader States. This would be in two steps: first, to overthrow the Fatimids and unify the Islamic faith; second, to recover Jerusalem. The latter naturally became primary, following Saladin's deposition of the Fatimid *Caliph* in 1171, and then became reality when he retook the city in 1187 (Dajani-Shakeel 1976: 108). His role in both is not fully straightforward. Saladin's interest in Jerusalem, for example, does not seem to have been piqued until as late as 1186 (Partner 1996: 339). His well-known unification of Muslims under the Ayyubid banner was also less certain than commonly thought and depended a great deal on manifold diplomatic overtures to *emirs* in Syria and due deference to the Muslim merchant class in the Levantine cities and towns. At times his alliances cracked but they never shattered, and his triumph over the Jerusalem army at the Horns of Hattin in July 1187 was followed by his capture of the Holy City the following October and an impressive subsequent campaign towards Antioch in the north.

However, on the heels of his strategic stalemate against the armies of the Third Crusade and his premature death in early 1193, Saladin's Ayyubid successors were less adept. Jerusalem had been retained, but Western control of the Levantine ports was problematic. Moreover, disunity returned in conflicts between his sons and his brother, al-'Adil Sayf al-Din, the latter of whom was finally crowned sultan on 4 August 1200 after spending years fracturing the sibling alliances. Al-Adil's rule held to the Fifth Crusade, but Muslim coalitions were exhausted by the constant military service (Wolff 1969b: 695–711). In the wake of that war (in which soldiers under King John of Brienne took Damietta in 1219 in a spectacular siege but foolishly declined a settlement that included their reception of Jerusalem), Saladin's great-nephew, al-Salih Najm al-Din Ayyub, augmented the strength of local forces with *mamluk* slave-soldiers (Powell 1986; Mylod et al. 2017). The market had been flooded with Turkish slaves in the wake of the 1230s Mongol campaigns in the Caucasus, and al-Salih gobbled them up. When he became master of both Syria and Egypt in 1240, he imprisoned former Ayyubid officers and replaced them with *mamluks*. These were crucial in the defeat of the army of Louis IX of France at the Battle of Mansourah (1249) during the Seventh Crusade. Later, in a convoluted process full of intrigue and violence, one of those *mamluks*, Aybak al-Turkmani, was named the *atabeg* and married al-Salih's widow, Shajar al-Durr. In an instant, the Ayyubid line was ended and *mamluk* control of Egypt began (Amitai 2016: 325–6).

Under the Mamluk Sultanate, both the Mongol and Western presence in the Levant were soundly turned aside. The former was dealt with in short order. Following the sack of Baghdad in 1258 and massacre of hundreds of thousands of its inhabitants (followed by the sack of Aleppo and surrender of Damascus), the victor, Hulagu Khan, departed for Azerbaijan and left his general Ketbuqa to defend Syria. Ketbuqa's Mongol Army was in turn intercepted and crushed at the Battle of 'Ayn Jalut by the Mamluk Sultan, Qutuz, and his general Baybars on 3 September 1260 (Amitai-Preiss 1995: 26–48; Jackson 2017: 128–9). Regarding the crusading ventures, the Eighth Crusade (the second of Louis IX) was hampered from the start. Hulagu had actually written to Louis to request he attack the Mamluk Sultanate, but whether the king ever received the letter is unclear and, at any rate, the campaign did not begin until 1270 (DeVries 2014: 218–19). Louis' army landed at Tunis with expectations of marching eastward to Egypt, but he immediately fell ill and died. Edward Plantagenet (the future King Edward I of England) carried on the campaign by sailing to Acre with eight ships and 300 knights. His limited success resulted in a treaty with now-Sultan Baybars in May 1272, but he could neither succor Jerusalem nor turn away the greater *mamluk* presence (Lower 2018: 174, 180–1; Jordan 2016). In 1291, Sultan al-Ashraf Khalil captured Acre, the last surviving Latin stronghold in the Levant—the crusader States, one of the great laboratories of Eastern–Western military history, had met their end.

Military Organisation and Sustainment

European armies obviously faced different challenges campaigning to, and in, the Middle East from their erstwhile foes, but it would be a mistake to categorise one as harder than the other. Summoning, mustering and maintaining armies was very difficult in the Middle Ages, a period in which political and military leaders were just starting to experiment with centralised recruitment, standing forces and regularised pay. Logistics were then, as now, rather complicated and required due consideration of sustainment methods in both permissible and non-permissible environments. As the Crusades—and the responses to them—were essentially coalition wars, commanders faced the additional task of managing factions over long campaign durations that often exceeded customary soldierly terms of service.

Crusading armies were by necessity amalgamations of soldiers hailing from disparate regions and serving their local elites with whom they mustered. No single Western power could win back Jerusalem on its own; indeed, as R.C. Smail astutely observed, 'No Christian ruler of the twelfth century had an army at his disposal which met his needs in full' (Smail 1995: 97). In 1094, England's population could provide a maximum of 20,000 infantry and 4,000–6,000 cavalry, according to dubious numbers in the sources. The reality was likely much lower, and in any case, a ruler could not strip his land bare of all combat power. Neither would all subjects answer a call to arms, nor could all men of status, such as knights, fulfill their obligations to serve (Morillo 1992: 48–50). Arguments for larger army numbers have generally not been well accepted. Jonathan Riley-Smith has argued that there were probably between 1,200 and 5,000 knights total, from all regions, at the end of the First Crusade in 1099 (Riley-Smith 2002: 13–28). Consequently, coalitions had to be built, and Christian sermons were a practical conduit for stirring the hearts and sensibilities of warriors in disparate European lands.

These coalitions seem to have increased in diversity throughout the twelfth century, then decreased in the thirteenth as crusading trended towards being ventures of single monarchs. Principal components of the armies of the First Crusade were French, Normans from Southern Italy and soldiers from the Low Countries. The Second Crusade featured two main groups mustered in France and the Empire, augmented by a fleet sailing from Dartmouth with perhaps 10,000 English, Flemish, French and Germans who successfully besieged Lisbon in 1147. Third Crusade participants came from French, English and Norman levies, stragglers from Frederick Barbarossa's imperial army, Danes, Welsh, Hungarians, Flemings, Catalonians and Italians from Pisa, Genoa and Venice (Tibble 2018). In the thirteenth century, imperial struggles in the Empire and the Crusade against the Albigensians in Languedoc diverted attention to home fronts and crusading numbers declined, never again to reach the 60,000 range of the first three major efforts (Wolff 1969b: 377–8; Marvin 2008). The participation base narrowed as well. Frederick II's 1227 army for the Sixth Crusade was dominated by Germans and mercenaries, and Louis IX's 1248 host for the Seventh Crusade numbered only 15,000 primarily French soldiers (Abulafia 1988: 165–6; Tyerman 2006: 786–7). Always augmenting deployed armies, of course, were settled Christian warriors in the crusader states and contingents from the Hospitaller, Templar, Teutonic and other military-religious Orders (Marshall 1992: 56–7).

That such diverse forces were able to fight together in a coordinated fashion is striking and a testament to a common warrior culture—indeed, the same could be said for coalition Muslim armies—but perhaps more impressive is that crusading armies were effectively sustained over massive distances in adverse conditions. In the early going, the stretch from the West to Constantinople was self-funded in the sense that individual knights raised personal funds and carried supplies along; these were augmented by pre-arranged fairs and markets along the pilgrim routes (Murray 2018: 185–201). Once in Asia Minor, Byzantine logistical assistance was critical and helped determine or condition strategy and rates of march (Bachrach 2006: 80–1; France 2006: 53–4). Supplies often failed and, without recourse to major Muslim markets, Western armies experienced severe privation, particularly during winter months, as at Antioch in 1097–1098 and Acre in 1189–1190. In the thirteenth century, formal, pre-arranged supply agreements appeared, and for the Fourth and Fifth Crusades the evidence of their details is quite good (Madden 2006: 209–28; Wolff 1969a: 387–8).

Once in the Levant, crusaders could tap Christian markets, but as the crusader states dwindled in size over the Ayyubid Period, such opportunities necessarily lessened. Correspondingly, seaborne resupply, always important, became more critical. A broad naval strategy to counter this eventually emerged, most prominently in the reign of Saladin and then the Mamluk Sultanate. Saladin worked to build up the Alexandrian fleet. This, combined with his successful campaign against every Levantine port south of Tripoli (excepting Tyre) in 1187 ensured safe harbours for its ships. Given more time to operate, his fleet might have actively patrolled the Eastern Mediterranean and provided a buttress against Western invasions, but the bulk of Alexandrian fleet's ships were captured in Acre's port in 1191

John D. Hosler

(Ehrenkreutz 1955: 111, 115). In the late thirteenth century, the *mamluk* forces razed the Levantine harbours in order to deny the crusaders safe berths. Beirut remained intact, but most of its fortifications were destroyed (Fuess 2001: 47–8).

Muslims in the High Middle Ages were not unified in any reasonable sense of the word, and their armies reflected this. Abbasid dominance reached its zenith in the ninth century but then steadily declined. A revival came to an end in 908, and the *caliphate* plunged into chaos anew in the reign of *Caliph* Muqtadir (d. 932): 909 saw the rise of the Fatimids in Tunisia, and they took Egypt in 969. In Spain, Umayyad descendants established the *Caliphate of Cordoba* in 931 (Kennedy 2004: 296–8). The Turkish incursions of the eleventh century then essentially finished off the Abbasid's political power, although the religious significance of the *caliphate* remained until 1258.

Apparent unity under the Seljuk Turks can also be misleading. Their sweeping away of local Arab dynasties in Syria created localised discontentment, and a civil war among the Seljuks in 1092 fractured their unity. The armies of the First Crusade capitalised by defeating the fragmented Turkish forces arrayed separately against them, particularly at Antioch against Ridwan, Duqaq and Kerbogha (Morton 2018: 62–3). Thereafter, certain Turkish warlords opposed Seljuk diktats and established themselves as powerful *atabegs* working within the Seljuk enterprise, with Zengi and Nur al-Din as chief examples. Other tensions and inter-Muslim strife emerged as well. Thomas Asbridge has detailed the enterprises, for example, of Dubais ibn Sadaqa, an Arab Bedouin who opposed the Seljuks and even raided the Abbasid capital of Baghdad (Asbridge 2013: 78). The militant activities of other non-state actors also played a role, most famously the subterfuges of the Ismaili assassins (Mallett 2016: 133–47; Hillenbrand 1995: 205–20).

There was a greater degree of organisation and unity in Egypt. But the size of the Fatimid army there is beset with source problems. Extant figures are impossibly large for the period: 100,000 purported regulars, bolstered by twice that number of auxiliaries. The reality was much lower, probably no more than 50,000 in total strength (Beshir 1978: 44–5; Lev 1997: 128–33). Still, the Fatimids ran an impressive military system that was organised and well-financed, with regular monthly payments from centralised funding mechanisms that would astound historians of modern warfare accustomed only to Western models. They drew on soldiers from diverse regions of the Maghreb and the Middle East for soldiers, and the enslaved components could be particularly sizable, so much so that Saladin killed as many as 30,000 of them in the 1171 'Battle of the Blacks' (Beshir 1978: 38–45; Bacharach 1981: 487–8; Eddé 2011: 40–3; Lev 1999: 81–4).

Numbers are more reliable and specific for the reign of Saladin. Following his toppling of the Fatimids, Saladin could rely upon a large base army from Egypt. A formal review in 1171 counted 194 available *tulbs* (units numbering between 70 and 200 horsemen and led by an officer and trumpeter), amounting to approximately 14,000 cavalry, most of them *tawashi*, the heavy variant (Gibb 1962: 76; Bosworth 1975: 70–1). Added to these were allies from Syria and the Jazira, as well as auxiliary Turks, Kurds, Arabs, Africans and assorted *mujahideen* (Gibb 1962: 78–84; France 2016: 145–56). Total numbers may have swelled to 40,000 after Hattin, although that would have been unusually large (France 2015: 82). Unlike the crusaders, who had serious long-range sustainment issues, Saladin and other Muslim leaders could better manage by actively courting the merchant class and thereby supply these larger numbers. And unlike the Fatimids, his control of Syria alleviated the logistical problem of supplying expeditionary forces across the Sinai and in the forward base of Ascalon (Nicolle 2007: 117).

Strategy, Operations and Tactics

Although Muslim writers perceived early crusading activities as part of a Mediterranean-wide Christian reconquest plan, such grand strategising was only a dream, at best, in the minds of certain prelates. Adherents of the so-called 'pluralist' school of thought see crusading targets as having expanded only gradually beyond the Levant and Spain and into the Maghreb and North-Eastern Europe (Housley 2006: 3–4). In a strictly military sense, all the Crusades sought to defeat or destroy non-Christian opposition and occupy or establish places of strategic advantage. Doing so meant careful preservation of limited manpower and territorial control via fortified cities and castles, two notions that were part-and-parcel with strategy in Western Europe.

Battle is always risky, but in the Middle Ages armies were not easily replaced, if at all. Since levies were drawn from the West, expeditionary armies had to survive until reinforcements arrived. But these were never assured and, in some cases, were not even arranged. The armies of the First Crusade were whittled from perhaps 60,000 strong (including non-combatants) to one-sixth that size in the course of three years. They were very fortunate to capture Jerusalem with that reduced force (France 1994: 142). The Germans suffered major losses at Dorylaeum during the Second Crusade, but at least Conrad III could expect help coming from Louis VII's French army; when that was heavily reduced at Mount Cadmus, the prospects for the entire affair dimmed severely (Marvin 2018: 30–3). The destruction of the Jerusalem Army at Hattin was instantly recognised as catastrophic. When Pope Urban III heard the news, he reputedly died from a heart attack on the spot, and his successor, Gregory VIII, issued the papal bull for a new Crusade before news of Jerusalem's subsequent fall even reached Rome—it was a presumed *fait accompli* (Bird et al. 2013: 4). The crusaders at the Siege of Acre, the Third Crusade's opening engagement, took heavy losses in the neighbourhood of nearly 30,000 but were able to make up for all them with seaborne reinforcements. Still, the loss of so many men doomed prospects when additional losses prohibited Richard the Lionheart's march on Jerusalem.

Muslim armies could reinforce more quickly because fresh recruits were closer, but they too faced similar issues. Soldiers on winter rest were unenthusiastic about joining a campaign. Even in good weather, delays were possible, as exemplified by the case of Tughtakin of Damascus: both the Fatimids in Egypt (in 1103) and the Seljuks in Syria (in 1109) promised him reinforcements that failed to arrive. An 1105 alliance between Tughtakin and Egypt fared better, but their combined forces were nonetheless defeated by the Franks near Ascalon (Cobb 2014: 108, 117). Saladin took serious losses during the siege of Acre and, at one point, sent his *qadi* Baha' al-Din Ibn Shaddad on a long recruiting mission into Syria and Mesopotamia (Hosler 2018: 39). One source claims Saladin may have taken as many as 7,000 casualties after Arsuf in 1191, and while that may be an overestimate it is clear that the losses hurt. After withdrawing, Saladin reportedly 'spent the rest of the day in that camp writing letters to his far-flung territories to mobilise the rest of his forces' (*Itinerarium* 1997: 257; *Rare and Excellent* 2002: 177). Muslim commanders without Saladin's impressive territorial control obviously had fewer options for restoring reduced armies.

Control of territory via fortified positions was also important. Of prime interest were the Levantine ports, which were well defended by stone walls and harbour chains. Unless their garrisons could be compelled to surrender, these towns were quite difficult to take. Acre held out against the Christians for two years during the Third Crusade, and, prior to that war, Tyre, which had been reinforced by Conrad of Montferrat, frustrated Saladin's every move against it (Lyons and Jackson 1982: 279–83). Freestanding castles could also hold their own if well garrisoned. These took many forms and were conditioned by the crusaders' financial resources, which were small at the beginning but gradually grew and were eventually buttressed by the wealth of the militant orders. Simple two or three-storey towers gave way to concentric fortifications like Belvoir, then spur and hilltop castles like the Teutonic castle of Montfort (Galilee), the sort that might better withstand Ayyubid and *mamluk* counterweight trebuchets (*Montfort* 2017: 3–5). Still, with a large army, proper artillery and appropriate sustainment for both, Muslim armies were able to breach even the stoutest walls, with the most famous example being Baybar's taking of the Hospitaller fortress Krak des Chevaliers (Syria) in 1271 (Fulton 2018: 264–72).

Major open-field engagements and castle sieges were impressive, but they took time. Moving from target to target was important for morale and the preservation of combat power in a non-permissive environment. The Westerners' long spell at Antioch nearly cost them the entire First Crusade, as their supplies dwindled and Turkish forces gathered in the surrounding region. Bohemond of Taranto had to exhort his men not to flee in the face of this material deprivation; infamously, Count Stephen of Blois did anyway, abandoning the crusade and returning to France (*Robert the Monk* 2005: 126). Likewise, while Saladin is famous for his grand victory over the Jerusalem Army at Hattin, in general he preferred quick victories and abbreviated sieges over long, drawn-out affairs. Recurring success bolstered the confidence and loyalty of his *emirs* and, equally important, their willingness to remain in the field (Hamblin 1992). Lengthy affairs tested their fortitude. At Acre, several *emirs* attempted to leave before its conclusion, citing inactivity, the need to tend to their *iqtaat* (rights to extract land revenues), or their lack of proper equipment for the winter months; one defected outright (Hosler 2018: 85–8).

Nevertheless, battles did commence. Due to the rough similarities between Western and Eastern weapons systems, crusading warfare is often characterised by contrastive styles of combat tactics (France 1997: 169–70). In general, these featured tight ranks of Western infantry, both in formation and in their famous 'fighting marches,' the latter of which was a form of both protection and manoeuvre warfare. Infantry was to hold firm against the harassing missile tactics of Muslim cavalry: *caracoles* ('Parthian shot'), feigned retreats and the *undique* ('crescent moon,' double envelopment) were designed to draw out soldiers into smaller, looser groups that could be surrounded and destroyed. If the lines held, however, the Muslim horses would tire, thus opening the way for a well-timed heavy cavalry charge into them. Such tactics were difficult to execute and required tremendous training and discipline. John France has noted in comparison that even eighteenth- and nineteenth-century cavalry charges often went awry: as with the construction and sustainment of coalition armies, then, in the period of the Crusades both 'Frankish' and 'Turkish' medieval armies found innovative solutions to ubiquitous military problems (France 2009: 159).

Notes

1 There is debate on Western knowledge of conditions in the East; see Morton 2014: 47–68.
2 For eschatological aspects, see Rubenstein 2011.
3 Muslim characterisations of the massacre were mixed; see Hirschler 2014: 37–76.
4 Emphasised in Eugene III's crusading bull but not by Bernard of Clairvaux, who stressed Zengi's threat to Jerusalem; see Phillips 2007: 52, 71.

Bibliography

Abulafia, David, *Frederick II: A Medieval Emperor* (Oxford: Oxford University Press, 1988).

Amitai, Reuven, 'The Early Mamlūks and the End of the Crusader Presence in Syria (1250–1291),' in A.J. Boas (ed.), *The Crusader World* (Abingdon: Routledge, 2016), pp. 324–45.

Amitai-Preiss, Reuven, *Mongols and Mamluks: The Mamluk-Ilkhanid War, 1260–1281* (Cambridge: Cambridge University Press, 1995).

Asbridge, Thomas S., 'How the Crusades Could Have Been Won: King Baldwin II of Jerusalem's Campaigns against Aleppo (1124–5) and Damascus (1129),' *Journal of Medieval Military History*, vol. 11 (2013), pp. 73–93.

Bacharach, Jere L., 'African Military Slaves in the Medieval Middle East: The Cases of Iraq (869–955) and Egypt (868–1171),' *International Journal of Middle East Studies*, vol. 13, no. 4 (1981), pp. 471–95.

Bachrach, Bernard S., 'Crusader Logistics: From Victory at Nicaea to resupply at Dorylaion,' in J.H. Pryor (ed.), *Logistics of Warfare in the Age of the Crusades* (Aldershot: Ashgate, 2006), pp. 43–62.

Beshir, B.J., 'Fatimid Military Organization,' *Zeitschrift Geschichte und Kultur des Islamischen Orients*, 55 (1978), pp. 37–56.

Bird, J., Peters, E., and Powell, J.M. (eds.), *Crusade and Christendom: Annotated Documents in Translation from Innocent III to the Fall of Acre, 1187–1291* (Philadelphia: University of Pennsylvania Press, 2013).

Bosworth, C.E., 'Recruitment, Muster, and Review in Medieval Islamic Armies,' in V.J. Parry and M.E. Yapp (eds.), *War, Technology and Society in the Middle East* (London: Oxford University Press, 1975), pp. 59–77.

Buck, Andrew D., *The Principality of Antioch and its Frontiers in the Twelfth Century* (Woodbridge: Boydell, 2017).

Bysted, Ane. *The Crusade Indulgence: Spiritual Rewards and the Theology of the Crusades, c. 1095–1216* (Leiden and Boston: Brill, 2015).

Christie, Niall, 'An Illusion of Ignorance? Muslims of the Middle East and the Franks before the Crusades,' in A.J. Boas (ed.), *The Crusader World* (Abingdon: Routledge, 2016), pp. 311–23.

The Chronicle of the Third Crusade: The Itinerarium peregrinorum et gesta regis Ricardi, tr. H.J. Nicholson (Farnham: Ashgate, 1997).

Cobb, Paul M., *The Race for Paradise: An Islamic History of the Crusades* (Oxford: Oxford University Press, 2014).

The Correspondence of Pope Gregory VII: Selected Letters from the Registrum, tr. With an Introduction and Notes by E. Emerton (1932, reprint, New York: Columbia University Press, 1990).

Dajani-Shakeel, Hadia, 'Jihād in Twelfth-century Arabic Poetry: A Moral and Religious Force to Counter the Crusades,' *Muslim World*, vol. 66 (1976), pp. 96–113.

DeVries, Kelly, 'Meet the Mongols: Dealing with Mamluk Victory and Mongol Defeat in the Middle East in 1260,' in S. John and N. Morton (eds.), *Crusading and Warfare in the Middle Ages, Realities and Representations: Essays in Honour of John France* (Farnham: Ashgate, 2014), pp. 218–19.

Eddé, Anne-Marie, *Saladin*, tr. J.M. Todd (Cambridge: Belknap/Harvard, 2011).

Ehrenkreutz, A.S., 'The Place of Saladin in the Naval History of the Mediterranean in the Middle Ages,' *Journal of the American Oriental Society*, vol. 75, no. 2 (1955), pp. 100–16.

France, John, *Victory in the East: A Military History of the First Crusade* (Cambridge: Cambridge University Press, 1994).

France, John, 'Technology and the Success of the First Crusade,' in Y. Lev (ed.), *War and Society in the Eastern Mediterranean, 7th-15th Centuries* (Leiden/Boston: Brill, 1997), pp. 163–76.

France, John, 'Logistics and the Second Crusade,' in J.H. Pryor (ed.), *Logistics of Warfare in the Age of the Crusades* (Aldershot: Ashgate, 2006), pp. 77–93.

France, John, 'A Changing Balance: Cavalry and Infantry, 1000–1300,' *Revista de história das ideas*, vol. 30 (2009), pp. 153–77.

France, John, *Great Battles: Hattin* (Oxford: Oxford University Press, 2015).

France, John, 'Egypt, the Jazira and Jerusalem: Middle-Eastern Tensions and the Latin States in the Twelfth Century,' in M. Sinibaldi, et al (eds.), *Crusader Landscapes in the Medieval Levant: The Archaeology and History of the Latin East* (Cardiff: University of Wales Press, 2016), pp. 145–56.

Frankopan, Peter, *The First Crusade: The Call from the East* (Cambridge: Belknap/Harvard, 2012).

Frenkel, Yehoshua, 'Muslim Responses to the Frankish Dominion in the Near East, 1098–1291,' in C. Kostick (ed.), *The Crusades and the Near East: Cultural Histories* (London: Routledge, 2011), pp. 27–54.

Fuess, Albrecht, 'Rotting Ships and Razed Harbours: The Naval Policy of the Mamluks,' *Mamlūk Studies Review*, vol. 5 (2001), pp. 45–71.

Fulton, Michael S., *Artillery in the Era of the Crusades: Siege Warfare and the Development of Trebuchet Technology* (Leiden/Boston: Brill, 2018).

Gibb, Hamilton A.R., *Studies on the Civilization of Islam* (Boston: Beacon Press, 1962).

Hamblin, William J., 'Saladin and Muslim Military Theory,' in B.J. Kedar (ed.), *The Horns of Hattin: Proceedings of the Second Conference of the Society for the Study of the Crusades and the Latin East* (Jerusalem: Yad Itzak Ben-Zvi, 1992), pp. 228–38.

Hillenbrand, Carole, 'The Power Struggle between the Saljuqs and the Isma'ilis of Alamut, 497–518/1094–1124: The Saljuq Perspective,' in F. Daftary (ed.). *Studies in Isma'ili History* (Cambridge: Cambridge University Press, 1995), pp. 205–20.

Hillenbrand, Carole, *The Crusades: Islamic Perspectives* (New York/London: Routledge, 1999).

Hirschler, Konrad, 'The Jerusalem Conquest of 492/1099 in the Medieval Arabic Historiography of the Crusades: From Regional Plurality to Islamic Narrative,' *Crusades*, vol. 13 (2014), pp. 37–76.

Hosler, John D., *The Siege of Acre: Saladin, Richard the Lionheart, and the Battle that Decided the Third Crusade* (London/New Haven: Yale University Press, 2018).

Housley, Norman, *Contesting the Crusades* (Oxford: Blackwell, 2006).

Jackson, P. (ed.), *The Seventh Crusade, 1244–1254: Sources and Documents* (Farnham: Ashgate, 2009).

Jackson, Peter, *The Mongols and the Islamic World: From Conquest to Conversion* (London/New Haven: Yale University Press, 2017).

Jordan, William Chester, *Louis IX and the Challenge of the Crusade: A Study in Rulership* (Princeton: Princeton University Press, 2016).

Kaldellis, Anthony, *Streams of Gold, Rivers of Blood: The Rise and Fall of Byzantium, 955 A.D. to the First Crusade* (Oxford: Oxford University Press, 2017).

Kedar, Benjamin Z., 'An Early Muslim Reaction to the First Crusade?' in S. John, and N. Morton (eds.), *Crusading and Warfare in the Middle Ages, Realities and Representations: Essays in Honour of John France* (Farnham: Ashgate, 2014), pp. 69–74.

Kennedy, Hugh, *The Court of the Caliphs: The Rise and Fall of Islam's Greatest Dynasty* (London: Weidenfeld & Nicolson, 2004).

Latiff, Osman, *The Cutting Edge of the Poet's Sword: Muslim Poetic Responses to the Crusades* (Leiden/Boston: Brill, 2017).

Lev, Yaacov, 'Regime, Army and Society in Medieval Egypt, 9th–12th centuries,' in Y. Lev (ed.), *War and Society in the Eastern Mediterranean, 7th-15th Centuries* (Leiden/Boston: Brill, 1997), pp. 115–52.

Lev, Yaacov, *Saladin in Egypt* (Leiden/Boston: Brill, 1999).

Lower, Michael, *The Tunis Crusade of 1270: A Mediterranean History* (Oxford: Oxford University Press, 2018).

Lyons, Malcolm Cameron and Jackson, D.E.P., *Saladin: The Politics of the Holy War* (Cambridge: Cambridge University Press, 1982).

MacEvitt, Christopher, 'What Was Crusader about the Crusader States?' *Al-Masāq*, vol. 30 (2018), pp. 317–30.

Madden, Thomas F., 'Food and the Fourth Crusade: A New Approach to the "Diversion Question",' in J.H. Pryor (ed.), *Logistics of Warfare in the Age of the Crusades* (Aldershot: Ashgate, 2006), pp. 209–28.

Mallett, Alex, 'Muslim Responses to Western Intervention: A Comparative Study of the Crusades and Post-2003 Iraq,' *Journal of Medieval Military History*, vol. 14 (2016), pp. 133–47.

Marshall, Christopher, *Warfare in the Latin East, 1192–1291* (Cambridge: Cambridge University Press, 1992).

Marvin, Laurence W., *The Occitan War: A Military and Political History of the Albigensian Crusade, 1209–1218* (Cambridge: Cambridge University Press, 2008).

Marvin, Laurence W., 'Louis VII as General on the Second Crusade: A Failure of Command, Control and Communication,' in M.L. Bardot and L.W. Marvin (eds.), *Louis VII and his World* (Leiden/Boston: Brill, 2018), pp. 29–49.

Montfort: History, Early Research and Recent Studies of the Principal Fortress of the Teutonic Order in the Latin East (eds.), A.J. Boas and R.G. Khamisy (Leiden/Boston: Brill, 2017).

Morillo, Stephen, *Warfare under the Anglo-Norman Kings, 1066–1135* (Woodbridge: Boydell, 1992).

Morton, Nicholas, 'Encountering the Turks, the First Crusaders' Foreknowledge of their Enemy: Some Preliminary Findings,' in S. John and N. Morton (eds.), *Crusading and Warfare in the Middle Ages, Realities and Representations: Essays in Honour of John France* (Farnham: Ashgate, 2014), pp. 47–68.

Morton, Nicholas, *The Field of Blood: The Battle for Aleppo and the Remaking of the Medieval Middle East* (New York: Basic Books, 2018).

Murray, Alan V., 'The Middle Ground: The Passage of Crusade Armies to the Holy Land by Land and Sea (1096–1204),' in G. Theotokis and A. Yildiz (eds.), *A Military History of the Mediterranean Sea: Aspects of War, Diplomacy, and Military Elites* (Leiden/Boston: Brill, 2018), pp. 185–201.

Mylod, E.J., Perry, Guy, Smith, Thomas and Vandeburie, Jan (eds.), *Fifth Crusade in Context: The Crusading Movement in the Early Thirteenth Century* (New York/London: Routledge, 2017).

Nicolle, David, *Crusader Warfare*, vol. II, *Muslims, Mongols and the Struggle against the Crusades* (London: Hambledon Continuum, 2007).

Partner, Peter, 'Holy War, Crusade and Jihād: An Attempt to define some Problems,' in M. Balard (ed.), *Autour de la premièr croisade: Actes du Colloque de la Society for the Study of the Crusades and the Latin East (Clermont-Ferrand, 22–25 juin 1995)* (Paris: Publications de la Sorbonne, 1996), pp. 333–43.

Peacock, A.C.S. *The Great Seljuk Empire* (Edinburgh: Edinburgh University Press, 2015).

Phillips, Jonathan, *The Second Crusade: Extending the Frontiers of Christendom* (London/New Haven: Yale University Press, 2007).

Powell, James M., *Anatomy of a Crusade: 1213–1221* (Philadelphia: University of Pennsylvania Press, 1986).

The Rare and Excellent History of Saladin, tr. D.S. Richards (Farnham: Ashgate, 2002).

Riley-Smith, Jonathan, 'Crusading as an Act of Love,' *History*, vol. 65 (1980), pp. 177–92.

Riley-Smith, Jonathan, 'Casualties and the Number of Knights on the First Crusade,' *Crusades*, vol. 1 (2002), pp. 13–28.

Robert the Monk's History of the First Crusade: Historia Iherosolimitana, tr. C. Sweetenham (Farnham: Ashgate, 2005).

Robinson, I.S., 'Gregory VII and the Soldiers of Christ,' *History*, vol. 58 (1973), pp. 169–92.

Rubenstein, Jay, *Armies of Heaven: The First Crusade and the Quest for Apocalypse* (New York: Basic, 2011).

Smail, R.C., *Crusading Warfare: 1097–1193*, 2nd ed. (Cambridge: Cambridge University Press, 1995).

Tibble, Steve, *The Crusader Armies: 1099–1187* (London/New Haven: Yale University Press, 2018).

Tyerman, Christopher, *God's War: A New History of the Crusades* (Cambridge: Belknap/Harvard, 2006).

Wolff, R.L., 'The Fifth Crusade,' in H.W. Hazard and K.M. Setton (eds.), *A History of the Crusades,* vol. 2, *The Later Crusades, 1189–1311* (Madison: University of Wisconsin Press, 1969a), pp. 377–428.

Wolff, R.L., 'The Aiyūbids,' in H.W. Hazard and K.M. Setton (eds.), *A History of the Crusades,* vol. 2, *The Later Crusades, 1189–1311* (Madison: University of Wisconsin Press, 1969b), pp. 693–714.

Yuval, Israel Jacob, *Two Nations in Your Womb: Perceptions of Jews and Christians in Late Antiquity and the Middle Ages*, tr. B. Harshav and J. Chipman (Berkeley: University of California Press, 2006).

12

THE LIMITS OF MONGOL MILITARY POWER

Timothy May

Introduction

In the thirteenth century, the Mongols irrupted out of the steppes as an unstoppable force. Under the leadership of Chinggis Khan (Chingiz), the various tribes of Mongolia were transformed into an indomitable military force that rarely knew defeat. The basis for the success of the Mongol Empire rested in their armies of disciplined horse archers. The lifestyle of the nomads of the Central Eurasian Steppes provided each nomad with essential skills from years of hunting, herding and archery. With excellent equestrian and archery skills taught almost from birth, each nomad had a modicum of military skills that most sedentary people lacked outside of the martial classes. Yet, the Mongols' military transcended individual skills. Highly disciplined not only as units but also in manoeuvres as a result of military reforms introduced by Chingiz Khan as well as training derived from cultural practices such as group hunts, the Mongol military practised a form of steppe warfare that was substantially different from that of their predecessors (May 2016: 42–9; May 2020: 72–98). Both the Jin, the Khwarazmian Empire, as well as the Rus' were intimately familiar with steppe warfare in all of its facets but found few answers against the Mongols.

Expansion of the Mongol Empire

With a highly disciplined army of horse archers, the Mongols possessed mobility and firepower that out-manoeuvred and overwhelmed the kingdom of Xi Xia (1209) and steadily reduced the Jin Empire (1211–1234). Indeed, the Jin Empire of Northern China and Manchuria was granted a reprieve when Chinggis Khan abruptly turned his army west to to invade the Central Asian Khwarazmian Empire in 1219 after the governor of Otrar, a Khwarazmian border town, massacred a Mongol-sponsored caravan. Both the Jin and the Khwarazmian Empire possessed armies substantially larger than those of the Mongols and were quite familiar with steppe warfare. Yet against the Mongols, they found few answers. Even when defeated in battle, the Mongols quickly recovered and avenged any losses.

The Mongols' string of victories encouraged disgruntled elements to join the Mongols, but they also acquired new skilled individuals like siege engineers as well as new weapons,

DOI: 10.4324/9780429437915-16

such as trebuchets. The Mongols also incorporated the defeated into their armies. For the populace this usually meant they became arrow fodder, marching before the Mongols to fill ditches with debris (and their bodies) as well as manning siege equipment (May 2016: 77–9). Yet, the Mongols also recognised that their defeated opponents still had much to offer militarily. As the Mongols primary acted as horse archers, they gladly accepted the services of those who could serve as heavy cavalry. Although the Mongols did not use infantry themselves, they saw the purpose of it, particularly in regions with terrain ill-suited for cavalry warfare. Thus, while the Mongols did not serve as infantry, they had regiments of Chinese infantry. In a similar fashion, they did not attempt to transform the Chinese into horse archers but allowed them to fight in a manner familiar to them (May 2016: 38–40). The Mongols did not attempt to put square pegs into round holes.

Despite Chingiz Khan's death in 1227, the Mongol Empire continued to expand. Under the leadership of Chingiz Khan's third son, Ögödei Qa'an (r. 1229–1241), the Mongols ended the Jin Empire in 1234. Not long afterwards, they began skirmishing with the Song Empire (960–1279) after the latter sought to seize territory that the Jin had previously taken from the Song. Meanwhile, Mongol armies entered the Middle East and Afghanistan to bring these regions under their control. Beginning in 1236, a large Mongol army under the command of Sübe'edei (Subedei), perhaps the best-known Mongol general, marched west bringing the Eurasian Steppes and Rus' principalities under Mongol control. Although the Mongols invaded Hungary, Poland and Bulgaria, they did not occupy the region for reasons that are still disputed (Jackson 2005: 71–4; Pow 2019: 301–21; Rogers 1996: 3–26). While Hungary and Poland remained in shambles, Bulgaria formally submitted and became a European extension of the Mongol Empire (Jackson 2005: 103).

One of the reasons why the Mongols may have halted their invasion of Europe was the death of Ögödei in 1241. His death and the subsequent regency of his wife, Töregene (r. 1241–1246), largely brought the conquests to a halt. The only major territorial acquisition was the conquest of the Seljuk Sultanate of Rum (located in modern Turkey) and the Byzantine state of Trebizond, thus extending the Mongol Empire to the Mediterranean Sea and the southern shores of the Black Sea. The reign of Güyük Qa'an (r. 1246–1248) did not produce any conquests, although armies were dispatched. He planned to resume the conquest of Europe, but died in the steppes of Kazakhstan, bringing the campaign to a halt. Three years passed before a new Qa'an sat on the throne.

With the election of Möngke (r. 1251–1259), the eldest son of Chinggis Khan's fourth son Tolui, the Mongol military reached its zenith. The Mongol Empire enrolled approximately a million nomads in the military (Allsen 1987: 189–216). This number easily doubled when considering sedentary troops from the conquered territories. With Möngke firmly in power, Mongol armies marched forth once again. Möngke's brother Hülegü led an army of approximately 150,000 men into Iran to bring any power that failed to recognise Mongol dominion. These include the Ismailis (the so-called Assassins), the Abbasid *Caliphate* and the Ayyubid principalities in Syria. Other armies marched south to eliminate the Song Empire. While Möngke led one army, his brother Khubilai led another against the Kingdom of Dali, located in the modern People's Republic of China's province of Yunnan.

While the Mongols conquered Dali to open another front and other armies pressed the Song Empire from multiple directions, the conquests slowed due to the mountainous terrain and stout defences. Meanwhile, in the Middle East, Hülegü methodically eliminated the Ismailis in 1256 and then the Abbasid *Caliphate* in 1258. On both occasions, instead of seeing the Islamic world unite to thwart the Mongols, Muslim potentates brought their armies to join the Mongols. When the Mongols invaded Syria in 1259 and 1260, the Ayyubids found themselves isolated. Events in 1259, however, dramatically changed the course of history.

Möngke died while laying siege to a city in Sichuan. His armies withdrew to Mongolia in order to select a new ruler. Khubilai continued to press the attack in an ironic move as he had only reluctantly joined the invasion. Meanwhile in the west, while a Mongol force occupied Syria, Hülegü withdrew most of his army to the Mughan steppe in Azerbaijan. He remained there upon hearing of Möngke's death. Normally after the death of a Qa'an, a date would be set for the new *quriltai*, which would select the successor.

Indeed, Möngke's death signalled the end of a unified Mongol Empire. Civil war erupted between Ariq-Böke (Möngke's youngest brother and his regent) and Khubilai (another brother). The *quriltai* selected Ariq-Böke as the ruler but Khubilai, who did not attend the *quriltai*, then held his own *quriltai* where he (unsurprisingly) was named the new Qa'an. Civil war ensued. The civil war in the east allowed the other Mongol princes further west to choose sides or pursue their own agendas. Soon, Hülegü found himself at war with the Jochid Ulus, the Mongol appanage in the Pontic and Caspian steppes, which also claimed the Mughan steppe. Rather than conquering the world, the Mongols sought to conquer themselves (Jackson 1978; May 2018a: 335–51; May 2018b: 343–58).

Despite the Mongols' seemingly inexorable march across Eurasia, there was indeed a limit to Mongol power. Mongol military supremacy began to fray in 1260. A few months after occupying Syria, and just two years after the destruction of the Abbasid *Caliphate*, a relatively small battle took place at 'Ayn Jalut in modern Israel. The Mongols' opponent was the Mamluk Sultanate of former slave soldiers who had established control over Egypt in 1250 in the wake of King Louis IX's (r. 1226–1270) failed Seventh Crusade. Unlike many states, the *mamluk* regime did not wait for the Mongols to invade; instead, they took advantage of a lull in Mongol activity caused by the death of Möngke Qa'an and invaded Mongol Syria. Other setbacks occurred farther east. In 1281, the Mongols invaded Japan. Although they had invaded Japan in 1274, it was but part of a larger strategy to conquer the Song Empire in south China. The Mongols did not intend to conquer Japan at that time. Khubilai Qa'an (r. 1260–1291), however, made every effort to add Japan to his empire during the second invasion, but a combination of factors thwarted his intent. He also met defeat in Southeast Asia. Much like the Americans, French and Japanese, the Mongols found Southeast Asia a difficult place for war. Their experience mirrored those that came after them, and for good reason. While they did experience some success, the Mongols also learned that the efforts to pacify and directly rule the region were beyond their ability and thus they settled for a compromise, which in many ways actually proved more beneficial. While the Mongols did not conquer Europe, there is little evidence to suggest that they could not. Syria, Japan and Southeast Asia, however, offer more inviting lessons on the limits of Mongol military power.

Syria

The conflict between the Mongols and the *Mamluks* had a considerable impact on the region and drastically changed Syria. Indeed, the war that began in 1260 only ended in 1322. It must be kept in mind that the war was not one between the Mamluk Sultanate (1250–1517) and the Mongol Empire, but rather the Mongol Ilkhanate (1260–1335), which dominated much of the Middle East from Central Anatolia to Afghanistan, with the Upper Euphrates River serving as a border between the two states. Between 1260 and 1323, the war consisted of both hot and cold conflicts, espionage and occasional attempts at peace. While the *mamluks* must be considered the victors in the conflict, even though peace eventually occurred, it is clear that the Mongols never viewed their military not up to the task (Amitai 2013: 31–2; Amitai 2016: 46–8). When one examines the course of war, the *mamluk* victories were rarely overwhelming and the *mamluks* often demonstrated a reluctance to directly engage the Mongols except when defending their own territory. Nonetheless, and indeed because of the *mamluk* reluctance, the Mongols consistently failed in their efforts to reestablish control in Syria.

The Mongol strategy of conquest never faltered. Even after 'Ayn Jalut, and until the very end of the war between the Ilkhanate and the Mamluk Sultanate, the Mongols never wavered in their belief that Köke Möngke Tengri (The Blue Eternal Heaven) had bequeathed the earth to Chingiz Khan and his descendants (Bira 2004; Amitai 2013: 39–61; Jackson 2005: 45–7). It was their right to rule and refusal to accept Mongol rule placed one in a state of rebellion against Heaven. Thus, the Mongols invaded with their armies of horse archers. Occasionally they also used allies. Indeed, Armenians led by King Hethum of Cilicia (r. 1226–1270) as well as knights from the Principality of Antioch led by Prince Bohemund VI (r. 1252–1268) accompanied the Mongols at Aleppo and were present when Damascus opened its gates in 1260. With the dissolution of the Mongol Empire after 1260, while the Ilkhanid Mongols believed it was their divine right to rule Syria, they also found it increasingly difficult to do so, often because their Mongol neighbours often attacked the Ilkhanate's borders. The Jochids frequently attacked across the Caucasus Mountains while the Chaghatayids occasionally invaded Khurasan. From the perspective of the Ilkhans, both of these were a much greater threat and higher priority than the *mamluks*, a reasonable premise as defence of one's territory should take precedence over conquest.

To counter the Ilkhanid forces, the Mamluk Sultanate focused on a defensive strategy. This began with Sultan Baybars (r. 1260–1277), who seized the throne shortly after 'Ayn Jalut. In order to control the newly acquired lands in Syria, Baybars appreciated the need for more manpower. In addition to securing the loyalty of local princes, he also increased the numbers of *mamluks* as well as mobilising troops from those who fled from Ilkhanid territories, including Mongols (Amitai 1995: 71–3).[1] He did not seek to replace all rulers with *mamluks*' governors, only those who did not demonstrate fealty. This was made easier by fear of the Mongols (Amitai 1995: 50). Additionally, he recruited Turkmen nomads as well as Bedouin to guard the frontiers and collect intelligence (Amitai 1995: 64–71; Amitai 2013: 19–20). Baybars also strengthened fortifications along the Euphrates River, particularly at fords. Part of this response included burning the pastures on the *mamluks*' side of the Euphrates, thus denying the Mongols the necessary grass for their horses. As the *mamluks*' horses were stable-fed, the loss of pasture affected them less (Amitai 1995: 76–7; Raphael 2011: 73–80; Amitai 2013: 21–4). Baybars also committed to an effective system of communications. However, this had been in practice prior to the Mamluk Sultanate, but he unified and extended it. Through a system of signal fires which could be spotted from other fortresses, messenger pigeons, as well as riders, the *mamluks* could respond to Mongol incursions rapidly (Amitai 1995: 74–6; Silverstein 2007: 176–8).

Due to their comparably fewer numbers, the *mamluks* did not take the offensive against the Ilkhanate, at least not directly. Rather, they sought to eliminate clients of the Ilkhanate. Their focus was on the Principality of Antioch, Cilicia and the Seljuks of Rum, all of which directly bordered the Mamluk Sultanate (Amitai 1995: 157–78; Amitai 2013: 18–19). Additionally, during lulls in the *Mamluk*–Ilkhanid conflict, the *mamluks* targeted the strongholds of the Kingdom of Jerusalem. Largely, restricted to the coastline, the Franks were not an offensive threat to the *mamluks*. Indeed, they sought to ally with the *mamluks* against the Mongols in 1260, as the Mongol invasion of Europe in the 1240s still in their mind as well as their own contacts with the Mongols in the Middle East was sufficient to convince most Franks of the Mongol threat (Jackson 2005: 117–22; Amitai 1995: 24). Furthermore, they wrongly assumed that a continued existence of accommodation might be reached with the *mamluks* as had existed during the Ayyubid period (1174–1250). Although the *mamluks* made treaties with the Franks, they also reduced their strongholds one by one (Amitai-Preiss 2005: 68; Lower 2018: 26–8, 71–3). Baybars focused on eliminating the castles of the Military Orders, rightly realising that any truce with them was temporary and that the death of a particular leader would not substantially alter the relationship—the Military Orders simply selected another leader who followed the mandates of the Order. These castles also tended to exist in the interior, and threatened *mamluk* lines of communication.

As the *mamluks* came to power during the Seventh Crusade (1248–1250), led by King Louis IX of France (r. 1226–1270) which came close to conquering Egypt, the *mamluks* were well aware of a Crusade's potential to bring a large army against them. Indeed, Louis' ill-fated Eighth Crusade greatly worried Baybars, who must have given a sigh of relief when it turned to Tunis instead of Egypt. For this reason, the *mamluks* sought to eliminate the Franks in Syria. Taking the coastal cities was imperative, to deny the Crusaders a landing point and base of operations. When the *mamluks* captured a coastal city, they razed the walls (Irwin 1986; Lower 2018: 71–3, 116, 177–82). While this made it easier for a Crusader army to capture a city, it also made it difficult for them to defend such urban centres after occupation. In many cases, the Mamluk Sultanate also destroyed the city, determined to make the Syrian coast indefensible by the Franks. With these regions then reduced, they allowed nomads (Turkmen, Bedouins, Mongols who deserted the Ilkhanate) to nomadise in the region, thus serving as a defensive force not dependent on fortifications. The greatest fear of the *mamluks* was the arrival of a Crusading army operating in conjunction with the Ilkhanate. While this was discussed, and indeed the Eighth Crusade had this potential, it never manifested. On the rare occasions where it was attempted, such as during Prince Edward's (the Black Prince) arrival in the Middle East in the wake of the Eighth Crusade, the Europeans lacked sufficient numbers to achieve anything or the Mongols were otherwise occupied to launch a massive invasion.

Additionally, the *mamluks* took a keen interest in training. The *mamluks* existed solely for war, unlike the average Mongol warrior, who in times of peace returned to his flocks and herds. While the Mongols may have participated in training, the *mamluks'* training regimen exceeded that. Furthermore, it was not just a handful of men, such as European knights compared to men at arms. Thus, the individual *Mamluk* was an elite warrior. While they did not comprise the entire Mamluk Sultanate Army, they remained the core of it. Man for man, the *mamluks* were superior, although by how much remains debatable. Furthermore, they understood that if they failed, the Mongols were unlikely to show clemency. They literally fought for their lives, thus providing ample motivation.

Although Ghazan conquered Syria in 1299, the Mongols failed to hold it. Besides the initial conquest, this was the only occasion that Mongol forces had successfully conquered Syria despite several other attempts. Ghazan invaded three times with only the first time being successful. Despite their failure, the Ilkhanid Mongols do not appear to have doubted that they could conquer Syria. As demonstrated with their success in 1299, the Mongol militarily could defeat the *Mamluk* Army. Even Baybars, one of the best commanders of the late thirteenth century, demonstrated a reluctance to engage the Mongols. So why did the Mongols fail to conquer the region?

The key factor was the outbreak of civil war throughout the Mongol Empire. With the death of Möngke in 1259, his brothers Ariq-Böke and Khubilai vied for the throne, with Khubilai Khan emerging the victor in 1265. Hülegü and Berke, the ruler of the Jochid Ulus (Golden Horde), largely sat on the sidelines. Although Hülegü sided with Khubilai beginning in 1262, the civil war undermined the unity of the empire. In 1262, Berke invaded Ilkhanid territories in an attempt to reclaim Azerbaijan, which had previously been under Jochid authority. The loss of the region also denied the Jochid access to the trade city of Tabriz as well as other trade routes in the Middle East. Repeated invasions served as a continual distraction for the Ilkhanids. This war became institutionalised. Like the *Mamluk* War, there were periodic détentes, but the Jochids never abandoned their claims to Azerbaijan and, indeed, Janibeg Khan (1342–1357) conquered it in 1357 after the Ilkhanate collapsed. The Jochids, however, promptly lost it after Janibeg died returning north. His son, Berdibeg, abandoned Tabriz in order to secure the Jochid throne in Sarai.

This war had another facet to it. The Jochids also formed an alliance with the Mamluk Sultanate. As the *mamluk* regime needed slaves, which came from Jochid territories, a convivial relationship was necessary. Both needed Constantinople, which controlled the trade routes between Egypt and Jochid territory, thus when the Byzantines reclaimed Constantinople in 1261, it benefitted both (Saunders 1977; Amitai-Preiss 1995). It is likely that the Jochid threat also prevented the Ilkhans from asserting hegemony over Constantinople.

To the north east, the Ilkhans bordered the Chaghatayid Ulus. While the Chaghatayids were unlikely to overrun the Ilkhanate, they did vie for control of Khurasan. Despite plans for coordinating an invasion of the Mamluk Sultanate in 1270 with King Louis IX (probably planned for after the captured of Tunis), the Ilkhan Abaqa marched east to deal with an invasion from the Chaghatayid khan, Baraq. Abaqa decisively defeated him at Herat and then crossed the Amu Darya to sack Bukhara as a warning against other invasions (Biran 2002: 175–220). This campaign, however, prevented him from aiding the Franks. As a result, the paltry forces of Prince Edward accomplished little. Baybars easily defeated the small force that Abaqa sent to aid him (Amitai-Preiss 1995: 125–6). The fact that the *mamluks* timed their own attacks against Ilkhanid clients when the Ilkhans were engaged with their cousins further illustrates their difficulties against the Mamluk Sultanate. While the *mamluks* were not willing to risk offensive actions against the Mongols, they steadily reduced Ilkhanid offensive capabilities by eliminating vassals adjacent to *mamluk* territory, thus eliminating potential troops and logistical aid, but also reducing the intelligence-gathering capabilities of the Ilkhanate. While the civil wars did not cause the Ilkhanate to lose territory, they did, however, reduce their ability to expand into new territory as they never considered the Mamluk Sultanate to be a threat on a par with either the Jochids or the Chaghatayids.

Although the Mongols invaded Syria several times, they only conquered it twice. While the efforts of the *mamluk* regime must not be dismissed, a key factor in the Mongols' failure was connected to climatic conditions. As their method of warfare was dependent on horse archers and mobility, the Mongols brought spare horses with them. While numbers vary, scholars have determined that, on average, each Mongol warrior brought five spare mounts with him. As they were pasture fed, the availability of grass was always a concern, as 10,000 men equalled 50,000 horses. So, 20,000 men could have 100,000 horses. One might wonder why the Mongols did not shift their strategy and use other types of troops such as infantry, stable-fed horses (which still required fodder but not as much pasture) or other types of troops. While it has been argued that pasture was an issue, based on Hülegü's own comment in a letter to King Louis IX, Reuven Amitai argues that sufficient pasture existed, but finding enough water to support the massive horse herds of the Mongols posed greater challenges (Morgan 1985: 231–5; Amitai-Preiss 1995: 225–9; Amitai 2013: 34; Meyvaert 1980).

While it is not implicitly stated, it is clear that the Mongols had some doubts about the reliability of other troops against the *mamluk* forces. At 'Ayn Jalut, an Ayyubid prince who had submitted to the Mongols, fled the battlefield, exposing a Mongol flank (Amitai-Preiss 1995; Thorau 1985). They may have been concerned with the reliability of other Muslim troops as well. Although Muslim potentates joined the Mongols against the Abbasid *Caliphate*, that occurred in an era that Mongol invincibility was unquestioned and the feckless nature of the *Caliph* Must'asim was well known. With the Mamluk Sultanate establishing its credentials as 'Defenders of the Faith' and an effective propaganda campaign, the Mongols did not trust their Muslim subjects. Indeed, at Abulstayn, when the Mongols caught the *mamluk* force departing Rum, the Mongol generals did not use local Seljuk troops because they did not trust them. When the Mongols conquered Syria, Georgians, Armenians and troops from Antioch marched alongside them. With the Jochids an omnipresent threat, the Mongols could ill afford to shift Armenian and Georgian troops for an invasion of Syria. Armenians from Cilicia were still available, but *mamluk* depredations against them clearly took a toll there. While Antioch could provide only small numbers of troops, it was lost to the Mongols in 1268.

Yet climate affected the Mongols in ways other than simply dictating troop numbers. It also determined the length of a campaign. When the Mongols conquered Syria in 1260, approximately only 10,000 troops stayed in Syria due to pasture and water. The majority of these were stationed in the Biqa' Valley in Lebanon, which had sufficient pasture for them. Unfortunately, it was difficult to support more than this number. After Ghazan conquered Syria in 1299, he ultimately withdrew. He realised that despite winning a decisive victory, he could not continue to advance to Egypt to eliminate the *mamluk* regime as he lacked the ability to station sufficient numbers of troops in Syria. He even sent emissaries to Europe offering to hand over Jerusalem to them if they would provide troops. One suspects that he was aiming for Syria to remain an Ilkhanid province and not an independent kingdom. Nevertheless, no aid came from Europe and thus Syria remained in *mamluk* hands.

Japan

While Syria stymied the Mongols in the west, Japan thwarted Khubilai Khan's ambitions in the east. The Mongol Khan launched two invasions in 1274 and 1281. The first had limited goals while in the second Khubilai clearly sought to conquer Japan. The conflict started as an extension of the Mongols' war with the Song Empire (960–1279). As the robust Song economy allowed the Song to continue to prosecute the war, despite losing militarily, Khubilai sought to hamper the Song's economy by destroying their trade with Japan. Since Hakata served as the primary trading point, Khubilai focused his efforts there. Initially, Khubilai attempted to secure Japan's submission working under the Mongol ideology that Tengri (Heaven) had bequeathed the earth to the Mongols. This failed, and so the Mongols invaded. They successfully destroyed Hakata, but a storm damaged much of their fleet, forcing them to withdraw. It is unclear if they had planned to attack other locations, but Hakata was certainly their primary target (Conlon 2001; Delgado 2008; Rossabi 2009; Sasaki 2015).

Not long after the defeat of the Song, the Khubilai Khan decreed that the Mongols would invade Japan again. Again, emissaries departed for Japan demanding the Emperor's submission. Their deaths sealed the invasion. Aware of the Mongols' intent, the Japanese then prepared their defences, making Hakata Bay the centrepiece of their resistance. The Mongols meanwhile prepared a fleet that remained the largest in history until the D-Day invasion in 1944. Two fleets were launched, one from Korea and the other from China. They were to rendezvous at Tshushima and Iki, two islands between Korea and Japan. The fleet from China missed the rendezvous date, thus causing the fleet from Korea to proceed with the invasion. The Mongols quickly overran the two islands and proceeded to Hakata, which was now heavily fortified (Rossabi 2009; Conlon 2001; Delgado 2008; Sasaki 2015).

While the Mongols attacked Hakata, the fleet from China eventually arrived. As the Mongols were still engaged at Hakata, the second fleet moved to find another landing, hoping to take the Japanese unaware. This fleet headed to Imari Bay. Here an army comprising Mongol cavalry and Chinese infantry landed and made its way inland, where a two-month stalemate began (Rossabi 2009; Conlon 2001; Delgado 2008; Sasaki 2015).

In 1274, the Japanese realised that the Mongols would invade after they rejected their demands for submission. The fact that they would attack Hakata was also not a surprise due to its proximity to Korea and China, as well as the volume of trade that it conducted with the Song Empire. Here the Japanese engaged the Mongols with mixed success. While the Mongols withdrew, it was the only claim that the Japanese could make to substantiate a victory. On the battlefield, the Mongols dominated, relying on unit tactics as opposed to the *samurai*'s emphasis on individual prowess and valour.

The 1281 invasion demonstrated that the Japanese learned much from their encounter with the Mongols. In the interim, the Japanese focused on training their warriors to function as a unit. Understanding that while an individual *samurai* might be superior to the Chinese, Koreans and Mongol warriors, he was not superior to an *arban* (unit of ten). Additionally, the Japanese commanders realised that the Mongols were most formidable in open space where their cavalry had room to operate. Thus, the best option was to deny the Mongols any opportunity to deploy their forces. To this end, they built a wall at Hakata along the beach, placed roughly at the edge of the high tide area. Thus, the Mongols could land, but without breaching the wall they could not establish a beachhead. Also, the placement of the wall allowed the Japanese to shoot with their own siege engines at the Mongol ships, and their landing crafts. Finally, as the Mongols were denied any spot to land, it meant the Japanese also had access to the beach and thus could launch naval counterattacks. These were small boat hit and run attacks, in which *samurai* attempted to swarm a Mongol ship at night; killing, causing havoc, and then escaping under the cover of darkness. To counter this, the Mongols formed their ships into floating fortresses by chaining them together. Thus, men from other ships could come to the aid of another, preventing a single ship from being isolated. Furthermore, should a ship become too damaged, it could be cut from the fortress and pushed away. But, this made the entire fleet vulnerable to fire attacks, as the Mongol ships were virtually sitting in the bay. Further, the hit and run night attacks by the *samurai* permitted them to seek valour while also demoralising the Mongol troops.

The Mongols failed to conquer Japan. During both the invasions, climatic factors mitigated the Mongol invasions. Unlike in Syria where the dry climate limited pasture and water, storms struck the Mongol fleets engaged in the invasion of Japan. On the first occasion the storms did sufficient damage to prevent the Mongols from pressing their advantage against the Japanese. Satisfied with the destruction to Hakata, they withdrew. During the second invasion, the storm, which produced a tsunami, had a devastating effect. The fleet in Hakata Bay survived largely intact while the fleet from China in Imari Bay fared far worse. The difference in their fates as largely due to the difference in the mode of construction of the ships. The Korean ships were of high quality, whereas the Chinese ships were not. Many of the Chinese ships were actually intended for river warfare and thus had shallow drafts. Others were hastily repaired Song ships. Few were in a condition to weather a storm of great intensity, much less a tsunami. The storm's damage increased due to the Mongol Navy's tactics of forming the floating fortress. While it provided a better defence against the Japanese raiders, the storm made the ships crash into each other, and if one began to founder, those chained to it were at risk. Whereas the Korean Fleet limped back to Koryo, the Chinese fleet drifted to the bottom of the sea. Those troops stranded on the shore could not escape. The Japanese forces reduced them, killing or enslaving any survivors (Delgado 2008; Sasaki 2015).

Although the storm determined the final outcome of the battle, a Mongol victory was not inevitable. In 1281, the Japanese were much better prepared and they demonstrated that they could hold their own against the Mongols. As individual warriors, a *samurai* would have been a match for any individual within the Mongol Army. Whereas the unit tactics that proved so successful for the Mongols in 1274, the Japanese could now also operate with unit tactics. While the Mongols had more experience with this, many of their troops, particularly among the Chinese infantry, lacked élan. Many had been soldiers in the Song dynasty and now fighting for their conquerors with the enthusiasm one might expect. Although they may have also realised they were fighting for their lives, one could not expect a high morale or a complete change in attitude from the recently defeated. Thus, even without the storm, the Japanese demonstrated that the conquest of Japan would have been a slow and difficult process.

Southeast Asia

The final demonstration of the limits of Mongol military power occurred in Southeast Asia. Most of their campaigns in this region ended unsatisfactorily. Four kingdoms faced the Mongol threat and each responded differently. With the conquest of the Kingdom of Dali (modern Yunnan Province), the Mongol Empire (Yuan Dynasty) now bordered Burma or the Kingdom of Myan, also referred to as the Kingdom of Pagan, so named after its capital. Border hostilities occurred and the Mongols experienced considerable success against Myan, even facing much larger armies that included numerous war elephants. The Mongols also experienced considerable success in Champa (South Vietnam). Here they invaded by sea. In Java and Annam or Dai Viet, the Mongols faced numerous military challenges and failed to conquer these kingdoms.

The Mongols of the Yuan Empire still adhered to the ideology that Köke Möngke Tengri had bequeathed the earth to Chingiz Khan and his descendants, but with the conquest of the Song Empire in 1279, Khubilai Khan modified the Mongols' ideology of conquest. Khubilai sought to restore the tribute system that the Song formerly imposed on its neighbours. Over the years, the neighbouring states ceased to send tribute thus failing to recognise the supremacy of the Song Empire. Having conquered the Song, the Mongols sought to impose this system and this desire factored into their strategy.

Mongol ideology alone did not bring the Mongols into conflict with Pagan in modern Myanmar (1044–1287). The ruler, Narathihapate (r. 1256–1287) rejected Mongol overtures of submission and tribute, which if accepted most likely would have left the kingdom unmolested. The Mongols demanded tribute twice, once in 1271 and then in 1273. Mongol military action did not take place until 1277, after the Mongols had made significant progress against the Song Empire. An expeditionary army led by Nasir al-Din, the Governor of Dali, defeated the Pagan forces at Ngasunggyan. Here King Narathihapate attempted to awe the much smaller Mongol forces with numerous war elephants. Initially, it had some success as elephants' scent made the Mongols' horses unruly. Nasir al-Din dismounted his men and posted them within the forest, rather than directly deploying them in the open field that they had originally occupied. The trees would slow the advance of the elephants while the Mongols launched volley after volley at these beasts. The continuous fire from the Mongols' powerful bows maddened the elephants so that they rampaged through the Pagan Army to escape (Polo 2015). Afterwards, the battle became a simple mop-up campaign. After the conquest of the Song Empire in 1279, Mongol forces from Yunnan continued to press into Pagan territory. In 1283–1285, they pushed south. Narathihapate abandoned any resistance and fled. One of his sons assassinated him in 1287. Mongol offensives resumed in 1287. They may have reached Pagan, although the evidence is unclear. What is certain is that the Mongols adhered to their tsunami strategy in which they invaded but only permanently occupied a much more limited swath of territory, leaving a buffer zone between them and any opponents (May 2015, 2017). Their invasion did succeed in breaking the power of Pagan. Although sporadic invasions took place until 1301, the Mongols never sought to conquer the entirety, withdrawing only after tribute payments were made.

The desire to extract tribute from Champa for the Mongols was a continuity of the old Chinese tribute system. Against Champa, a Mongol Fleet sailed to the kingdom while Mongol officials also asked the ruler of Dai Viet to permit Mongol forces to march through his kingdom to join the Mongol armada in Champa. The Prince Tran Thanh Tong wisely understood that the Mongols expected his complete submission and rejected the Mongol overtures. Meanwhile in Champa, the Mongol forces landed and successfully conquered the kingdom (Buell 2009; Delgado 2008).

Using Prince Tran's refusal to aid them against Champa, the Mongols then invaded Dai Viet from the north and Sögetü's troops from the south, before moving north into Dai Viet. The Mongols quickly conquered the cities. Despite their success, the Mongols found out, much as later invaders learned, that controlling the countryside was much more difficult. The thick jungles and mountains rendered the cavalry impotent, and the insalubrious climate was not to the liking of the Mongol warriors or their horses. Despite the extensive use of infantry, the Mongols could not fully establish their control in Dai Viet. Eventually they withdrew, but only after Prince Tran agreed to send tribute to the Yuan emperor (Buell 2009; Delgado 2008).

The final front in Southeast Asia was in Java. As the Mongols restored the tribute system and also promoted trade, they soon had contact with the kingdoms found on the island. One prince formally submitted to Khubilai Khan and agreed to serve as the Mongols' *daruqachi* if they would aid him in defeating his enemies, including King Kertanagara who rejected Mongol demands for submission. Although the Mongols successfully defeated all who opposed them on Java, their ally, Raden Vijaya, now devoid of enemies, betrayed the Mongols and ambushed them. The Mongols fought their way back to their ships, with great loss (Bade 2013). They left Java and reported back to Khubilai. Although the Great Khan was livid and wanted to order another fleet to sail to Java, his advisors calmed him and convinced him that the conquests would not be worth the expense.

Assessment

The Mongols had mixed results in Southeast Asia. There is evidence that the Mongols performed tolerably well in the region, but also experienced difficulty against emerging new forms of warfare. In Burma, King Narathihapate engaged the Mongols in pitched battles trusting that his numerous war elephants gave him a decisive edge against the Mongol cavalry, not only due to their thick hides (over which they wore armour), but also because horses become nervous, and therefore unruly, around elephants. Even horses trained to be docile, like those of the Mongols, could not overcome this behaviour. Yet, the Mongols proved to be adept at finding solutions to this. Time and time again, the Mongols demonstrated that they mastered the battlefield against the Pagan Kingdom's army.

The Mongols experienced similar success in Champa. Indeed, it is difficult to determine if Champa possessed a coherent strategy against the Mongols. They did not oppose the Mongol landing and, whatever resistance existed, the Mongols swiftly crushed it. Their two-pronged invasion of Dai Viet, however, met a much more determined opponent than King Narathihapate. Rather than depending on shock and awe accompanied by braggadocio, Prince Tran understood the Mongols' intention and did not conceive them as a distant threat. Indeed, as Dai Viet bordered the former Song Empire, he was quite familiar with the Mongols' reputation and their abilities. Furthermore, Prince Tran was not a new ruler attempting to establish his legitimacy. He was cautious but strategic and his tactics became the inspiration for resistance by modern Vietnamese against French, Japanese, American and Chinese invaders.

Rather than engage the Mongols directly, Prince Tran abandoned the cities. This is not to say that there was no resistance, but rather, Tran did not put all of his eggs in one basket, believing that he could stave off the Mongol conquest. The strategic retreat bought time for him to mobilise the countryside and establish bases in the jungle and mountains, from where he could negate the Mongols' mobility advantage as well as their archery. The jungles not only effectively neutralised cavalry manoeuvres but also eliminated Mongol superiority in archery. The canopy of the jungles prevented the Mongols from using kill zones and effective volleys from a distance. Instead, the fighting would be much closer. While the Mongols could still use their bows, many of the advantages that the Mongol composite bow had were neutered. Using ambushes, Prince Tran waged a war of attrition. The Mongols could control the cities and the coast. But, without control of the countryside, Prince Tran jeopardised their communications as well as logistics, especially food supplies. Furthermore, he also realised that so long as the Mongols controlled the sea-lanes, they could bring not only reinforcements but also supplies. The superior resources of the Yuan Empire could bring an end to Dai Viet's resistance. Thus, after defeating the Mongols on land through guerrilla war, Tran also sought victory at sea, which he achieved.

While some at the Yuan Court wanted to avenge the defeats, Prince Tran neutralised them by then offering tribute to the Great Khan. In doing so, he gave the Mongols what they wanted: recognition of their authority through tribute, thus legitimising Mongol overlordship. Even though Prince Tran won militarily, he realised that victory needed to be quick—the longer the war continued, the more Dai Viet would suffer. Thus, after achieving military victory but by also giving the Mongols a political victory, he could spare his kingdom from further devastation. He surely remained on guard against further incursions, but in sending tribute payments, he convinced the Mongols that trade and tribute had more to offer than war.

The invasion of Java is a masterful piece of duplicity on the part of a section of the Javanese power elite. Bring in an outside force to destroy your enemies and then destroy them, hoping that the long distance and bitter defeat would be bitter enough for the Mongols not to return. It proved effective, but also ran the risk of failure. Indeed, the Song Dynasty, driven from North China in the tenth century by the Khitans, who then formed the Liao Empire, encouraged the Jurchen (a Manchurian people) to invade the Liao Empire. The Song imagined that they would use these barbarians and then be able to reclaim their lost territory in China. Unfortunately for the Song, the Jurchen proved too adept at warfare and politics. Not only did they destroy the Liao, but they also pummelled the Song, taking even more territory. The Song then attempted the same strategy with the Mongols. And this strategy backfired on the Song, resulting in Mongol conquest of their empire. While the Javanese ruler promoted trade with the Mongols, he had no desire to be ruled by them.

In the case of Dai Viet and Java, the Mongols encountered two leaders who developed cunning plans. While the defeat in Java was rapid, this was also due to the distance, which limited the Mongols' ability to reinforce their original force or launch a counterattack with fresh forces. The fact that Dai Viet successfully resisted despite being on the Mongol border is testimony to the effectiveness of Prince Tran's strategy. Yet, other factors contributed to the Mongol defeat.

As in the previous examples, climate proved influential. The Mongols found the climate of Southeast Asia insalubrious. Heat and humidity took a toll on both the Mongols and their horses. While evidence is limited, it is certain that tropical diseases also played a role in undercutting the Mongols in Dai Viet. We have less information on the Mongol condition in Burma. The humidity also had other ill effects. It weakened the Mongol bow. A composite bow's stiffness determines its draw weight, and thus its power. As humidity seeps into the bow materials (wood, horn, sinew), it makes them less stiff. With a lesser draw weight, the arrows shot from the bow thus have less penetrating power and range. With less power, the arrow not only had less power to penetrate armour, but also to cut through branches and leaves (in forested terrain), making it more likely miss its target. Additionally, constant rainfall made it even more difficult for any cavalry or large infantry formations to practise mobile war at a rapid pace, as the men and horses would quickly turn a rain-soaked terrain into a morass.

Unlike the *samurai* and *mamluks*, there is no indication that the Javanese or Dai Viet had possessed superior martial prowess. This is not to say that they lacked military skills, but rather they lacked a large cavalry corps such as the Mamluk Sultanate or even a social class such as the *samurai*, where martial skills were part and parcel of their lives. Still, they learned how to fight effectively in their territory, giving them a significant edge over the Mongols and their Chinese infantry.

Finally, the trade factor must be considered. Khubilai Khan died in 1291. His grandson Temur Öljeitü came to the throne after him. After some brief invasions of Burma, Mongol expansion ended. Whereas Khubilai envisioned his fleets allowing him to conquer the world, his grandson took a more prudent look at the coffers and realised that seaborne expansion was detrimental. Furthermore, he was more concerned with ongoing wars with other Mongol rulers. So, instead, he pursued trade. The South China Sea was, during the Yuan Empire, a Mongol Sea. The Mongol fleets did not sail in new conquests, but to trade as well as curb piracy, thus allowing merchants to come to ports such as Guangzhou in China. A more militant ruler than Temür Öljeitü might have sought to renew expansion. Temür Öljeitü, even after making peace with his relatives, had no interest in it, preferring to rule his own state.

Note

1 These Mongol refugees, known as *wafidiyya*, usually entered the Mamluk Sultanate as a result of power shifts that arose with new rulers. If their faction lost, they departed to other Mongol states or the Mamluk Sultanate, which welcomed them not only for the intelligence they brought, but also for the manpower. Conversely, some *mamluks* fled to the Mongols due to the turbulent nature of politics in Egypt.

Bibliography

Allsen, Thomas T., *Mongol Imperialism: The Policies of the Grand Qan Möngke in China, Russia, and the Islamic lands, 1251–1259* (Berkeley: University of California Press, 1987).

Amitai, Reuven, *Holy War and Rapprochement: Studies in the Relations between the Mamluk Sultanate and the Mongol Ilkhanate (1260–1335)* (Turnhout: Brepols, 2013).

Amitai, Reuven, 'Continuity and Change in the Mongol Army of the Ilkhanate,' in Bruno De Nicola and Charles Melville (eds.), *The Mongols' Middle East: Continuity and Transformation in Ilkhanid Iran* (Leiden: Brill, 2016), pp. 38–52.

Amitai-Preiss, Reuven, *Mongols and Mamluks: The Mamluk-Ilkhanid War, 1260–1281* (Cambridge: Cambridge University Press, 1995).

Amitai-Preiss, Reuven, 'The Conquest of Arsuf by Baybars: Political and Military Aspects,' *Mamluk Studies*, vol. 9 (2005), pp. 61–83.

Bade, David, *Of Palm Wine, Women and War: The Mongolian Naval Expedition to Java in the 13th Century* (Singapore: Institute of Southeast Asian Studies, 2013).

Bira, Shagdarin, 'Mongolian Tenggerism and Modern Globalism. A Retrospective Outlook on Globalization,' *Journal of the Royal Asiatic Society*, 3rd series, vol. 14, no. 1 (2004), pp. 3–12.

Biran, Michal, 'The Battle of Herat (1270): A Case of Inter-Mongol Warfare,' in Nicola Di Cosmo (ed.), *Warfare in Inner Asian History (500–1800)* (Leiden, Brill, 2002), pp. 175–220.

Buell, Paul D., 'Indochina, Vietnamese Nationalism, and the Mongols,' in Volker Rybatzki et al. (eds.), *The Early Mongols, Language, Culture, and History* (Bloomington: Indiana University Press, 2009).

Conlon, Thomas D., *In Little Need of Divine Intervention: Takezaki Suenaga's Scrolls of the Mongol Invasions of Japan* (Ithaca: Cornell East Asia Program, 2001).

Delgado, James P., *Khubilai Khan's Lost Fleet* (Berkeley: University of California Press, 2008).

Irwin, Robert, *The Middle East in the Middle Ages: The Early Mamluk Sultanate 1250–1382* (Carbondale: Southern Illinois University Press, 1986).

Jackson, Peter, 'The Dissolution of the Mongol Empire,' *Central Asiatic Journal*, vol. 32 (1978), pp. 178–244.

Jackson, Peter, *The Mongols and the West: 1221–1410* (Harlow: Longman, 2005).

Lower, Michael, *The Tunis Crusade of 1270: A Mediterranean History* (Cambridge: Cambridge University Press, 2018).

May, Timothy, 'The Mongol Art of War and the Tsunami Strategy,' Золотоордынская цивилизация. Научный ежегодник [Golden Horde Civilization. Research Annual], vol. 2 (2015), pp. 31–7.

May, Timothy, *The Mongol Art of War* (Barnsley: Pen & Sword, 2016).

May, Timothy, 'Grand Strategy in the Mongol Empire,' *Acta Historica Mongolici*, vol. 16 (2017), pp. 78–105.

May, Timothy, *The Mongol Empire* (Edinburgh: Edinburg University Press, 2018a).

May, Timothy, 'Race to the Throne: Thoughts on Ariq-Böke's and Khubilai's Claims to the Mongol Throne,' in George Bilavschi and Dan Aparaschivei (eds.), *Studiea Mediaevalia Europea et Orientalia: Miscellanea in Honorem Professoris Emeriti Victor Spinei Oblata* (Bucuresti, Editura Academiei Romane, 2018b), pp. 343–58.

May, Timothy, 'Herding the Enemy: Culture in Nomadic Warfare,' in Wayne E. Lee (ed.), *Warfare and Culture in World History*, 2nd ed. (New York: New York University Press, 2020), pp. 72–100.

Meyvaert, P., 'An Unknown Letter of Hulagu, Il-Khan of Persia, to King Louis IX of France,' *Viator*, vol. 11 (1980), pp. 245–59.

Morgan, David O., 'The Mongols in Syria, 1260–1300,' in P.W. Edbury (ed.), *Crusade and Settlement* (Cardiff: University College Cardiff Press, 1985), pp. 231–5.

Polo, Marco, *The Travels*, tr. Nigel Cliff (New York: Penguin, 2015).

Pow, Stephen, 'Climatic and Environmental Limiting Factors in the Mongol Empire's Westward Expansion: Exploring Causes for the Mongol Withdrawal from Hungary in 1242,' in L.E. Yang et al. (eds.), *Socio-Environmental Dynamics along the Historical Silk Road* (Cham: Springer, 2019), pp. 301–21.

Raphael, Kate, *Muslim Fortresses in the Levant: Between Crusaders and Mongols* (New York: Routledge, 2011).

Rogers, Greg S., 'An Examination of Historian's Explanations for the Mongol Withdrawal from East Central Europe,' *East European Quarterly*, vol. 30, no. 1 (1996), pp. 3–26.

Rossabi, Morris, *Khubilai Khan, his Life and Times*, 20th Anniversary edition (Berkeley: University of California Press, 2009).

Sasaki, Randall J., *The Origins of the Lost Fleet of the Mongol Empire* (College Station, TX: Texas A&M University Press, 2015).

Saunders, J.J. 'The Mongol Defeat at Ain Jalut and the Restoration of the Greek Empire,' in Geoffrey Rice (ed), *Muslims and Mongols: Essays on Medieval Asia* (Christchurch: Whitcoulls for the University of Canterbury, 1977), pp. 67–76.

Silverstein, Adam J., *Postal Systems in the Pre-Modern Islamic World* (Cambridge: Cambridge University Press, 2007).

Thorau, Peter, 'The Battle of ʿAyn Jalut: A Re-examination,' in P.W. Edbury (ed.), *Crusade and Settlement* (Cardiff: University College Cardiff Press, 1985), pp. 236–41.

13

THE CYCLE OF EMPIRE IN CENTRAL ASIA

The Frontier and Beyond

Andrew de la Garza

In the late tenth and early eleventh centuries, the scholar Abul Qasim Ferdowsi used a variety of existing sources, from written chronicles to poems and folk tales, to compile the *Shahnameh*, a comprehensive history of the Persian people from the Stone Age to the dawn of the medieval era that told the story of their rise from savagery to civilisation. One of the most famous figures in this book is Rostam, a legendary hero from Persia's distant past. He was the prince of Zabolistan (a small state probably located in modern Afghanistan) and a vassal of the larger neighbouring Kingdom of Iran. He was both a formidable combatant with a wide array of weapons, while mounted or on foot, and a gifted tactician and commander. His two most widely used epithets, 'Dastan' ('trickster') and 'Piltan' ('body like an elephant') summed up his unique combination of brains and brawn. Rostam and his troops were a valuable asset for the kings of Iran, assisting in many military operations and protecting their state from threats both mundane and supernatural—from enemy kings and barbarian hordes to dragons and demons. Yet despite his stalwart service and his demonstrable honour and integrity, his relations with his superiors were often difficult. While Iranian kings needed Rostam, they also feared him, dreading the day when he would turn his turn his considerable talents against them. The growing paranoia of Iran's ruling dynasty eventually led to the invasion of Zabolistan, the exile of Rostam and, finally, his assassination (*Shahnameh*; Ferdowsi 2004).

The story of Rostam, like many other events in the early chapters of the *Shahnameh*, is more fiction than history, based on folktales and legends dating back to the dawn of the Iron Age and incorporating obviously unrealistic elements of swords and sorcery fantasy. Rostam and the other major characters are, at most, based on greatly distorted or composite versions of actual historical figures. These accounts, however, illustrate a very real dynamic in world history, one that was especially prominent in the history of Central Asia—the evolution of frontier empires. Such empires have their origins in encounters between nomadic, tribal peoples and sedentary, civilised states. The nomads either migrated to the frontier of an existing state, or that state expanded to meet them. While living on the frontier the nomads entered into a symbiotic—if often confrontational—relationship with their civilised neighbours. Nomadic, pastoral peoples were mostly generalists, with all able-bodied men and women performing a variety of roles. Sedentary agricultural communities, however, supported specialised workers who could manufacture items—tools, weapons, luxury goods—that nomads

DOI: 10.4324/9780429437915-17

could not easily make for themselves. Such items were traded for commodities that the nomads had in abundance. The most obvious of these was livestock. Pastoralists sustained themselves by raising horses, cattle, camels and other animals, but they could also produce a surplus to be traded for other goods. Nomads, occupying land that was suboptimal for agriculture but adequate for grazing, essentially maintained reserve herds of animals for their civilised neighbours. The optimal arrangement for sedentary communities was to keep only as many animals on hand as were immediately needed for sustenance or for civilian and military labour—keeping reserves nearby was an inefficient use of land that could otherwise be used for agriculture. Instead they traded with nomads for additional animals as needed. Another valuable resource that nomads could provide was military service. Pastoralists lived, worked and fought on horseback. Effective cavalry and scouts were always in demand and, especially in and around Central Asia, nomadic horsemen became even more formidable by mastering the composite bow and mounted archery. Horse archers, with their ability to move and attack simultaneously, were the most dangerous combatants on premodern battlefields.

Through both cooperation and conflict, sometimes trading and sometimes raiding, nomads constantly learned from their civilised neighbours. Their generalist approach to labour, lack of specialised roles and the constant need to react to changing conditions on their travels made them very adaptable and flexible. They readily mastered new skills and concepts, often learning with greater ease than sedentary peoples bound by roles, rules and tradition. Over time they learned agriculture, how to use—and then make—new technology, military science and statecraft. They built larger communities, cities and, finally, states of their own. In this process of transition they became especially formidable, combining civilised knowledge and expertise with nomadic adaptability and warrior culture. Such groups were essentially latent empires. Latent frontier empires became emergent when some adversity—war, civil conflict, natural disasters—overtook their more civilised neighbours. Sometimes they exploited such an opportunity by invading from the outside, but often they took power from the inside after becoming an essential part of a state's or empire's military establishment. While nomadic troops were valued for their general military skills—horsemanship, archery, scouting—they were also recruited to meet two specific needs. One of these related to external threats. Nomadic migrations and invasions were a serious problem for many premodern states. A common solution was for governments to ally with certain nomadic tribes and use them to keep the others in check. The standard bargain was land, citizenship and protection in exchange for military service, which might take the form of tribal warbands serving as auxiliaries or individual members of a tribe joining the regular army. States as diverse as the Roman Empire, the Arab Muslim Empire and a succession of Chinese dynasties used this plan. While effective in the short term, it often had unintended consequences, creating enclaves on the frontier inhabited by communities and leaders pursuing their own agendas. Another application for nomads as military recruits involved a state's internal politics. The exotic nature of foreigners, specifically their lack of connection to any existing faction inside the state, made them a unique asset. Elite foreign troops, often mercenaries or military slaves, were expected to be loyal only to the leader and state, capable of intervening impartially in the factional disputes that were all too often central to premodern politics. Such units, typically composed of Turks and other Central Asian peoples, played the same role in many Middle Eastern and Asian states that the Praetorian Guard did in Rome. Like the Praetorians, however, such groups, instead of being above faction often became factions of their own and potential threats to the state. Nomadic troops honed their skills in elite units like these, learning new weapons and tactics and learning to fight—and lead—not in warbands of dozens or hundreds but in armies of thousands. Many frontier empires had their origins in mutinies or military coups led by such men.

The rise, rule and decline of frontier empires shaped events in Central Asia and adjacent regions from antiquity through the early modern era. Indo-Europeans launched some of the earliest frontier empires from the borders of Mesopotamia and India—Persians, Indo-Aryans and Parthians. They would be followed by groups like the Macedonians, Arabs, Mongols, Manchus and especially the Turks, who were especially prolific builders of frontier empires all along the Silk Road, from China to Eastern Europe. The Seljuqs, Ghaznavids and the assorted dynasties of the Delhi Sultanate were prominent examples of Turkish frontier empires. The frontier dynamic was also important in Western Europe, before that continent was fully settled and civilised. The Romans rose to prominence after centuries of obscurity as a minor tribal people on the margin of Hellenistic Civilisation. Centuries later, at the end of the Roman era, Germanic tribes like the Franks built frontier empires that would form much of the foundation of medieval Europe. The frontier dynamic was less pronounced in the Americas and sub-Saharan Africa. The absence of or limited access to large livestock denied nomadic peoples an essential method of gaining economic and military strength, tilting the balance of power more in favour of sedentary, civilised polities. Some premodern frontier empires, such as the Aztecs, did emerge, but the pace of empire building quickened after European contact and its disruption of existing states and hierarchies (and, in the Americas, the introduction of the horse). Groups like the Comanche, Lakota and Zulus carved out their own steppe empires, until they were checked by the development of industry and modern military technology (Gump 1996).

The frontier dynamic is also apparent in the fictionalised history of the *Shahnameh*. Many of the major factions depicted in the story represented different stages in the life cycle of frontier empires. The Kingdom of Iran was a mature frontier empire, stable, wealthy and powerful. The nomadic tribal groups on its borders, depicted as the major antagonists, like the devil-worshiping Mazandaranis and, especially, the clever and ruthless Turanis of Central Asia (anachronistically described as 'Turks' but actually uncivilised Indo-Europeans like the Scythians) were latent empires, learning from both trade and conflict with their civilised neighbours. Most importantly to the story, 'half civilised' but more advanced border states such as Rostam's Zabolistan could, in the right circumstances, become emergent frontier empires very quickly. Iranian kings relied on Rostam as a military leader, but they also feared him. They understood the risk because their kingdom had also originated as a frontier empire. The earliest chapters of the *Shahnameh* describe how early Persians overthrew 'Arab' (actually Mesopotamian) tyrants and built a state of their own. In fiction, Rostam's decency and sense of honour forestalled him from turning on his masters and finally led to him becoming the victim instead. In reality, few leaders on the frontier passed up the chance to prey on a vulnerable state. The later chapters of the book, more grounded in documented historical events, describe the final stage in the life of a frontier empire—decline and fall. They explain how two Persian empires—the Achaemenids and the Sassanids—were overtaken by the Macedonians and Arabs, builders of new frontier empires. The *Shahnameh* would become a classic work not just for Persians but for other peoples across the Middle East and Asia—because it told a story very similar to their own histories.

The life of a frontier empire was typically brief. Geography was the most obvious problem. Government became difficult as states grew to span out of natural boundaries and include multiple geographic spheres of influence. Centralised leadership was a struggle when information took weeks or months to travel across an empire. Yet most frontier empires had other flaws, rooted in the tribal culture and politics of their founders. Government, for most nomadic and pastoral peoples, was traditionally decentralised. Their nominal rulers led as first among equals, governing by cooperation and consensus with the leaders and people under them. Empire builders like Alexander or Chingiz Khan ruled as true tyrants, but they were aberrations. Their deaths were usually followed by a regression to the mean, especially if there was no potential successor with their obvious ability and charisma. The problems of succession were compounded by a lack of certainty. Most tribal peoples did not practise primogeniture but instead relied on appanage, a process in which everyone within the inner circle of the former ruler—close relatives, most trusted lieutenants—had a plausible claim to being the new leader.

Central Asian peoples like the Turks and Mongols especially emphasised the concept of lineage, so that anyone in the male line of descent from an illustrious ruler like Chingiz Khan or Timur was assumed to have inherited the special traits that qualified him for leadership. If multiple candidates existed, negotiation was often the only option—between the contestants themselves or adjudicated by a council. If the deliberations failed to produce a result, the consequences could include a destructive civil war or simply a peaceful agreement to divide the state. Many frontier empires, including Alexander's and Chingiz Khan's, fragmented in a series of succession crises. These states were also prone to division because they typically had no unifying principle or coherent political ideology beyond the cult of a leader. The appeal of tribal or ethnic identity, or anything approximating modern nationalism, also diminished as empires grew larger and more diverse. Military elites and their leaders were often self-interested and opportunistic, prioritising their ambitions over those of the state as a whole.

Frontier empires in their infancy were usually inclusive because one goal of tribal warfare was the assimilation of the defeated, as individuals or, in the case of decisive victory, entire tribes. Nomadic warbands were often motley assemblages of born tribesmen, recruits and conscripts who gradually developed a shared identity and culture. On a larger scale, the success of frontier empires and their acquisition of military, economic and political power encouraged the assimilation of outsiders. Through means both violent—captivity, slavery, government coercion—and non-violent—trade and the proliferation of religion, philosophy, art and popular culture—they encouraged foreigners to become more like themselves. The concept of race as a law of physics, as constant as gravity or the speed of light, is a very recent development, barely two centuries old. In premodern settings ethnicity was usually inherited, but in the right circumstances it became information that could be learned—that could go viral. The rise to power of groups like the ancient Indo-Europeans, the Hellenised Macedonians, Han Chinese, Arabs and Turks allowed relatively small populations of empire builders to expand across continents, becoming multitudes not simply by biological reproduction but by shaping others into versions of themselves. This process, however, was the work of generations and did not always proceed according to the more urgent timelines of state politics. Frontier empires and their leaders, as they became more sedentary and civilised, sometimes became trapped in an awkward transitional phase where the founding group ossified into a closed elite. This was problematic because, as they grew, these states became dependent on the contributions of an increasingly diverse population. Armies evolved from nomadic cavalry to complex organisations that included infantry, siegecraft, engineers and logisticians. Effective government also required an extensive civilian bureaucracy and the contributions of private contractors and businesses.

Most of these skilled workers came not from the founding group but from the sedentary civilisations that they allied with or conquered. This often led to social tension. Many people from nomadic cultures, even if they had been sedentary for generations, saw themselves as superior to the supposedly soft and decadent descendants of older civilisations. In and around Central Asia, Turks and Mongols identified as *qazzaqs* (travellers) and saw themselves as dynamic, virile and free people, in contrast to the Persians and other sedentary neighbours, the so-called Tajiks. If the vital contributions of people outside the founding group were not adequately rewarded, if their social mobility was constrained and they were not allowed to share in the rewards of the state's success, they might decide instead to undermine that state. Perhaps the most famous example of this problem of the 'glass ceiling,' another chronic failure mode of frontier empires, was the rise of Timur. A general in the army of the Chaghatai Khanate (a successor state of Chingiz Khan's empire), he was a gifted commander but also a Turk in a state with a closed elite of Mongols. When further rewards, promotions and recognition were denied to him, he responded by leading a coup against the Mongol king and becoming a military dictator (Soucek 2000).

Some frontier empires, however, escaped all of these traps and survived for centuries, not years or decades. The original Persian Empire was one of the first such states to do so and thus would serve as an inspiration and model for later generations of empire builders. The founders of frontier empires, lacking experience in government and statesmanship on a large scale, often consciously imitated earlier, successful empires and adopted both their practical organisation and their idiom and iconography. Persia was an obvious standard in Central Asia and South Asia as well as much of the Middle East. Elsewhere the Greeks, Romans and Chinese were used to draft a blueprint for empire. What all of those earlier states and civilisations had in common was the development of a coherent political ideology and unifying principle. The first Persian Empire developed an ideology based on several closely related principles—Absolutism, the Enlightened Despot and the Divine Right of kings (those terms were widely used during the Enlightenment, but the concepts that they referred to were much older). It was based on the proposition that the best form of government—and the best way to do the most good for the most people—was an authoritarian state led by a single exceptional individual. This person was both exceptionally gifted and exceptionally virtuous—a genius and a person of character and integrity who would put the interests of the state and the people above his own. In this conception, obligation went both ways. The people had a duty to be loyal; the leader had a duty to serve them. The ruler also had to answer to an even higher authority. He was granted divine favour, literally on a mission from God to make his country and the world a better place. For the Persians, the Zoroastrian faith played a central role in a political system based on the Enlightened Despot. Zoroastrianism was unusual for that era because it was monotheistic and especially because it was based on morality and ethics and promoted worship of a deity who was revered as the embodiment of goodness. It was the duty of a devout Zoroastrian to pursue good and reject evil. A Zoroastrian ruler would have the same obligation.

In many premodern civilisations, the organisation and politics of the supernatural world reflected those of the real world, with gods and mortal rulers behaving similarly. This was true in the Persian Empire, and it was true for many of the Mesopotamian states that they overtook during their expansion. Most Mesopotamian religions envisioned a pantheon of gods that had supernatural powers but human failings. They were often selfish, cruel, jealous and foolish. Their human followers worshipped them not because they were exemplary or morally superior but because they were powerful, with the ability to offer rewards or inflict punishment.

Mesopotamians obeyed their mortal rulers for similar reasons. Politics was deeply cynical and transactional. Mesopotamian states and empires were led by tyrants and oligarchs, and they often produced results that were dystopian even by premodern standards. The tales of decadence and atrocity recounted in the first chapters of the *Shahnameh* were not entirely fabrications. The Persians expanded into Mesopotamia after an extended period of instability and conflict in that region. Their empire grew to fill a military and political power vacuum, but it also filled a moral vacuum. Ideas were as essential to its rise as weapons or wealth. The idea of the state as a living thing, an entity that existed to serve the greater good and that could inspire respect and affection as well as fear and greed, was a compelling one to people whose leaders had failed them time and again. While Persian emperors often did not live up to the ideal of the Enlightened Despot in reality as well as they did in theory, this ideology still kept people invested in their state and its government. The social contract was a simple one—peace, prosperity and dignity in exchange for loyalty. Peace and prosperity were ensured by the empire's physical might—armies of both soldiers and bureaucrats, wealth and infrastructure—that defended it from enemies, enforced the rule of law and facilitated trade. Dignity was assured by the empire's official policy of tolerance. Citizens were allowed to maintain their original language, religion and culture. Non-Persians were integrated into the imperial government at all levels. While no. one was forced to become Persian, many people did assimilate, motivated by admiration or by simple ambition. Assimilation was an obvious way to improve economic, social and political status (Waters 2014).

The three great early modern Muslim empires—Ottomans, Safavids and Mughals—all had Central Asian roots. They also emulated many aspects of earlier Persian empires. By no coincidence, these empires greatly outlasted the typical lifespan of the Central Asian frontier empires that came before them, surviving for centuries instead of years or decades. While the founders of these states could trace their ancestry back to nomads from the steppes, they themselves were descended from people who had over generations become more and more sedentary and civilised. Civilised habits and attitudes were no longer seen as a weakness but as a positive good.

Perhaps the most striking example of this dynamic, one that would be influential in the development of all three empires, was the so-called 'Timurid Renaissance' in Central Asia during the fifteenth and early sixteenth centuries. Timur himself, who at various times in his life was a bandit, a mercenary and a warlord, was no scholar. Yet during his rise to power and construction of an empire he craved legitimacy. He wanted to be seen as an enlightened leader and statesman, not simply a strongman. To build that image he surrounded himself with the trappings of civilisation, gathered wholesale during his conquests and campaigns. Works of art were bought or simply stolen. Monuments were disassembled and shipped back to his capital city of Samarkand. If that was not possible, they were recreated—or improved upon—from scratch. There was no shortage of skilled workers with which to do so. Timur also assembled, by bribery or by coercion, an army of artists, scholars, scientists, engineers and other experts to make and maintain his wonders. It is debatable whether or not Timur was ever completely civilised, or if he truly internalised any of this display. His descendants however, who grew up surrounded by all of these marvels and who received their childhood lessons from world-renowned scholars in their fields, were different. Many of them were patrons of the arts and sciences or became scholars themselves. This was a serious business in the tense political environment of Central Asia in that era. Timur's empire, like many frontier empires before it, did not long survive the death of its indispensable founding figure. A series of succession crises fragmented it into a collection of smaller city states, most of them based on important trading or manufacturing depots on the Silk Road. Their Timurid rulers competed fiercely for territory and control of trade. They also competed for status, which in this setting could be earned by hiring the brightest minds and producing the finest works of art, architecture and scholarship. The ensuing intellectual arms race, waged by leaders drawing on abundant resources gained from their position as a nerve centre of world trade, was very similar to the better known Renaissance taking place at the same time thousands of miles to the west. It included not just advances in the arts but inquiries into philosophy, including an extended debate on the purpose of the state and the ideal form of government. In Central Asia this discussion would be based not just on a return to the Greco-Roman classics but on the history and literature of Persia. The so-called 'Mirror of Kings,' a genre of instructional literature for aspiring statesmen dating back to the first Persian Empire and before, all the way to the prehistory of the *Shahnameh*, became popular for both Timurid potentates and the rulers of the larger empires that emerged on the margins of Central Asia (Dale 2004).

These concepts, as well as more tangible innovations, would be essential to the development of the Ottomans, Safavids and Mughals, but those empires in their early days were not so enlightened. They shared many of the common and potentially fatal flaws of earlier frontier empires—decentralised government, dependence on an unruly elite of traditional military leaders and the founding group's disinclination to fully share the rewards of success with an increasingly diverse population and workforce, both civilian and military. As a result, each empire would experience an existential crisis. For the Ottomans it would be a crushing defeat at the hands of Timur, the captivity and death of an emperor and civil war. A century later the Safavids would in turn be devastated by Ottoman military victories, a resource shortage and deadly succession conflicts. The Mughals were nearly doomed by a rebellion led by Afghans, a group who believed, not without reason, that they had been unfairly denied the rewards they had earned for their contributions to the empire.

The Ottomans originated as one of the many successor states of an earlier frontier empire, the Seljuqs. The dynasty founded by Osman would initially be successful based on traditional Turkish political and military practice. The crisis of the early fifteenth century, caused both by foreign enemies and internal division, brutally exposed the weaknesses of that system. The army and the political elites that it supported were still heavily reliant on local strongmen, tribal chieftains and their poorly organised levies and warbands. Imposing order, logic and professional standards on both its military and civilian elites was an essential process in building a stronger empire in the aftermath of crisis. This started with the continued development of the Janissary corps and its civilian auxiliaries. *Janissaries* were conscripted at a young age from non-Muslim groups on the empire's western frontier and then subjected to years of rigorous training and indoctrination—preparation to be not just skilled soldiers but leaders. This practice was very similar to the organisation of earlier military slave formations and elite foreign units, designed to create a group unattached to any existing faction and capable of impartially enforcing the orders of the ruler and the protecting the interests of the state. For the rulers of the Ottoman Empire, they were an essential check on the founding elite of Turkish chieftains and their *sipahi* cavalry. Beyond maintaining order, the *Janissaries* and a corps of professional bureaucrats (who were often conscripted from the same pool as *Janissaries* and who integrated those soldiers into their ranks when they were physically disabled by age or injury) changed the mindset of the empire. They became a technocratic elite that competed with and overmatched an older aristocracy based on family and tribe affiliation. Leadership by trained professionals in both military and civilian roles enabled mastery of complex systems. The Ottomans built a diverse army that combined the components of traditional Central Asian military practice—cavalry, mounted archery, mobility—with elements sourced from all over Europe and the Middle East—gunpowder weapons including small arms and artillery, siegecraft, field fortifications, effective infantry tactics—into a highly effective combined arms system. Its integration with the civilian bureaucracy made it resilient as well as powerful, supported by a rationalised scheme of logistics, procurement and infrastructure.

The empire's civil service deployed a comprehensive code of law and regulations, rational taxation and central economic planning. The material structure of government was supported by a coherent and humane political ideology based on absolutism and the Enlightened Despot. The Ottoman Sultan evolved into a true emperor, as with the Persians and earlier Muslim empires. Further, religion became an essential part of his legitimacy. The sultan was expected to live and rule as a good and devout Muslim, doing God's work to protect the empire and its people. The Ottoman defeat of the Egyptian *Mamluks* in the early sixteenth century and the resulting acquisition of the Muslim holy sites at Mecca and Medina allowed the sultan to assert Divine Right by assuming the role of *caliph*, the secular guardian of the Muslim faith and its believers. The empire, however, also provided assurances for non-Muslims. Like the Persian Empire and earlier Islamic states, the Ottomans enforced an official policy of tolerance and a social contract based on loyalty in exchange for peace and prosperity. While non-Turks and non-Muslims (except those conscripted into government service) were not forced to abandon their religion, language or culture, assimilation was still a tempting proposition for the ambitious who sought advancement in government or private business (Murphey 1999; Imber 2002).

The Safavid Empire in Iran was built from the start around a positive and optimistic unifying ideology, in this case the teachings of the Safavis, a militant order of Sufi Muslims. Ismail, the empire's founder and first emperor, presented himself as a messianic figure who could restore peace and justice to a region in crisis. Iran had experienced more than a century of political division, war and lawlessness after the collapse of the Ilkhanid Empire (another successor state of the Mongol Empire), conquest and then abandonment by Timur and his successors and unending conflict between a collection of Persian and Turkish city states and tribal confederations. The Safavids exploited this political and moral vacuum, promising to overthrow the tyrants and oligarchs who were preying on the country and create a just and humane government based on Islamic principles and led by an enlightened and divinely inspired ruler. This movement attracted and mobilised many different groups. The Qizilbash ('red heads,' from their distinctive red caps) shock troops of this revolutionary state mostly fought as Central Asian style cavalry, but their membership was ethnically diverse. Their leader Ismail, who became the first Safavid Emperor, was the literal embodiment of this dynamic, with Persian, Turkish, Georgian and Greek ancestry.

While the Safavids succeeded in reuniting Iran, their success would prove difficult to sustain. They were threatened on three sides—to the north by the Uzbeks and other Turkish tribal confederations, to the east by the Mughals and to the west by the Ottomans. This last enemy would prove to be the most dangerous by far. The Ottomans' victory at the battle of Chaldiran in 1514 gained them control of Mesopotamia and precipitated an existential crisis for the Safavids. Mesopotamia was both an abundant agricultural region and an intersection of trade routes, and its resources were essential for the Safavids, as they had been for previous

Persian empires. For this reason Ctesiphon, the capital of the Sassanid Empire, was actually located in modern Iraq, near the future site of Baghdad. The loss of Mesopotamia caused an extended economic recession in the empire, which was then confined to borders roughly equivalent to those of the modern state of Iran. That region was still mostly wilderness and, before petroleum became an important commodity, had few valuable natural resources. An even more dangerous problem was a series of succession crises, some of which escalated to civil war. One weakness that the Safavid Empire inherited from earlier frontier empires was a lack of certainty in succession. As descendants of the preceding ruler, male members of Ismail's illustrious bloodline all presumably had the special qualities required in a ruler and thus at least a theoretical claim to the throne. As the sixteenth century progressed, few of the disputes between them would be resolved amicably. This problem was aggravated by the disorganisation of two institutions essential to the state—the Safavi Sufi order and the Qizilbash militias. Individual units and their leaders often worked independently and at cross purposes, competing with both each other and the central government and taking sides in succession disputes.

The reign of Shah Abbas at the start of the seventeenth century would finally end this extended crisis and create a more sustainable empire. Abbas, who grew up in a court devastated by repeated conflicts over succession and who had lost several close relatives to political violence, was obsessed with order and control—and with restoring those conditions to his state. He reformed both the military and civil service, building both on the examples of earlier Persian empires and hard lessons learned from the Ottomans. He continued the development of the *ghulams*, units of conscripts and military slaves selected from non-Muslim groups on the frontier in conscious imitation of the Ottoman *Janissaries*. As with the Ottomans, these troops were maintained as a check on the founding elites—in this case the Qizilbash and other tribal militias. They were also a vehicle for adopting new technology and tactics. The Safavids supplemented their traditional cavalry forces with infantry and gunpowder weapons, mastering the tools that the Ottomans had used so effectively against them. Abbas also implemented a massive infrastructure programme, building roads, aqueducts, way stations for travellers and, most notably, an entirely new capital city at Isfahan. Much of the new construction was meant to facilitate trade. Abbas revitalised the economy of the resource starved empire through an ambitious central planning scheme. He identified new strategic resources, most notably silk, that the Safavids could produce and trade on a global market and built a network of state-licensed merchants and trading posts that operated across Central Asia, the Middle East and Europe. For Safavid citizens with the relevant skills and connections, participation was mandatory. Traders and their houses, especially Georgians and Armenians, were conscripted *en masse* as if they were soldiers (Babaie 2004; Dale 2010).

The Mughals were also innovators. This was apparent from the start with the sophisticated military organisation developed by the empire's founder Babur and his successors. It was perhaps the most refined variant of the combined arms system first created by the Ottomans, using both traditional nomadic methods—mounted archers, heavy cavalry, tactics based on manoeuvre and mobility—and components from sedentary civilisations in Europe, the Middle East and South Asia—light infantry armed with both bows and muskets, heavy infantry, field fortifications and artillery. Yet a hybrid army was not enough to secure this state—especially after other states and factions in India began to imitate and then master these technologies and tactics. Military innovation aside, under its first two emperors Babur and Humayun the Mughal Empire appeared to be simply the latest dynasty of the Delhi Sultanate, founded by a new generation of adventurers and upstarts. A nearly fatal flaw that it shared with earlier dynasties and frontier empires was a closed elite, in this case composed of the Turkish and Mongol associates of Babur and their descendants. Despite an increasingly diverse military, civil service and economy, they held a monopoly on the highest ranks in government. That glass ceiling antagonised many other groups, most notably Afghans, who had been the ruling elite under the previous Lodhi Dynasty. Many of them remained in the military and in government, doing essential work for the empire, but the best and brightest of them were not fully rewarded for their contributions with positions of real power and prestige. Their grievances were mobilised by the warlord Sher Shah Suri who led a rebellion that nearly destroyed the Mughals and forced Humayun to temporarily abandon India.

Humayun's successor Akbar developed more enlightened politics that would finally stabilise and secure the empire. Like Abbas, he had survived a difficult childhood in the aftermath of defeat and civil war and was obsessed with restoring order and stability. He understood that reliance on the founding elite as the basis of the state was unsustainable, but he was also sceptical of simply creating another closed elite like the *Janissaries* to contain them. Living in India, he was well aware of the dark history of the earlier Delhi Sultanate—a series of mutinies and coups by military slaves and other elite foreign troops leading to a grim succession of caudillos and military juntas—and the lessons it held about the perils of such a plan. Both the Ottomans and Safavids would eventually learn those lessons the hard way. His solution was to build a system based on open access and meritocracy in which diverse groups from all over South Asia and beyond joined both the military and civil service and progressed through their ranks. Under Mughal law no hereditary aristocracy other than the royal family was recognised. Noble titles were instead earned by achieving sufficient rank in government service. Military officers and civil servants ascended a ladder of numerical ranks, with clearly defined rules, expectations and targets for promotion (or demotion). Higher ranking officers held both civil service and military titles, and were expected to perform duties in both capacities. While the Mughals never developed a true general staff in the modern sense, the close integration of military and civilian leadership produced an army that could impose its will on the battlefield and still sustain itself in any terrain or climate, with logical and effective systems for logistics, procurement, recruitment and training.

The workforce that managed this organisation became more diverse over time. Like the Ottomans, the Mughals made tolerance and equality under the law official policy. Not only South Asians of all sorts but strivers from Central Asia, the Middle East, Africa and Europe—Muslims, Hindus, Christians and Jews—joined the ranks of the Mughal military and civil service. Others were drawn to the Empire in pursuit of opportunity—merchants, artists, scholars and missionaries. With natural resources and human capital far exceeding those available to the Ottomans or Safavids—and with effective programmes of regulation and infrastructure to manage them—the Mughal Empire became a global centre of trade, culture and innovation. Yet opportunities for material success were not the only inspiration. Under Akbar the empire developed the so-called 'Akbari Constitution,' a positive and optimistic political ideology based on the 'Enlightened Despot.' While it was never embodied in a single document, it presented a social contract that would have been familiar to Cyrus the Great or Louis XIV—loyalty from the people in exchange for peace, prosperity and rights guaranteed by the rule of law. Citizens of the empire became more invested in and attached to their state, and the most ambitious of them aspired to membership in the Mughal elite. Instead of assimilating into a specific ethnicity or religion, they adopted the language, manners, dress and martial training of the *sharifi* (exalted) leaders who had earned nobility and wealth in imperial service (de la Garza 2016: Gommans 2002: Streusand 1996).

While the three great early modern Muslim Empires effectively combined innovation and the lessons of the past to greatly outlast the normal lifespan of earlier frontier empires, their success was not sustainable indefinitely. For the Ottomans and Safavids especially, new elites eventually became as problematic as those they had replaced. Instead of becoming a solution to faction, groups like the *Janissaries* and Safavid *ghulams*—isolated from the mainstream of society and subjected to intensive, even cult-like, indoctrination—became factions of their own, more loyal to their leaders and each other than to the ruler. Like the Praetorians in Rome, they engaged in subversion, sabotage and open violence against the state. The Safavids also created a civilian version of the Praetorian Guard, a Muslim clergy regulated and empowered by the state. Religious conflict—between factions within the Safavi order, between various Muslim sects and between Muslims and non-Muslims—had contributed to the disorder of the empire's first decades. After Abbas, Safavid rulers decided that religious conformity based on Shia Muslim theology and doctrine was essential to maintaining social order. That policy, however, led to other problems. Like their military counterparts, the Muslim *ulema* became a powerful and disruptive faction of their own. The growing intolerance of the Safavids and their Shia clergy alienated religious and ethnic minorities, dividing the empire instead of creating a unifying principle (Babaie 2004; Dale 2010).

The rulers of the empires themselves became less reliable over time. All of these empires made the same critical mistake as many other states before and after them—they combined a political ideology and system based on the Enlightened Despot with hereditary succession. When the pool of potential successors is restricted, at best, to a few dozen people, simple probability dictates that over time few rulers will have the exceptional ability required to be an enlightened despot in reality as well as in theory. The traditional Central Asian model of open succession, which was supposed to ensure that the best of these candidates became new leaders, brought its own problems. Both the Ottoman and Safavid Empires experienced a series of destructive succession crises. They finally resorted to keeping potential heirs confined to the court, but that practice led to rulers with no real training or practical experience. The first Mughal Emperors attempted to enforce primogeniture, but that policy was later abandoned. The Mughals never sequestered their heirs, continuing with the traditional method of training princes by giving them real-world experience in high-ranking military and government positions. Such preparation also allowed them to build powerbases for the conflicts to come. Succession crises became more frequent and damaging later in the empire's history. Other problems for the Mughals were not structural but contingent—a series of bad choices, especially those made by the Emperor Aurangzeb. His decisions to expand the empire beyond its natural geographic boundaries (resulting in a decades long war of attrition in the Deccan) and to expand the power of the state in service of a new ideology informed by Sunni Muslim orthodoxy expended both material and political resources. Individuals and communities grew more distant from a state that no longer honoured the social contract created by Akbar, one that increasingly restricted their freedoms, taxed their wealth and judged them for their presumed disloyalty based on their ethnicity or religion. Aurangzeb's long reign ended with an empire that was increasingly divided by faction and tribe and that would be, for the first time in two centuries, vulnerable to external threats (Dale 2010).

Even while these empires were still strong, they had to deal with the perils of success. While adversity drove innovation, strength and stability encouraged conservatism. The tools and techniques that built the empires were seen as the ideal, difficult or impossible to improve upon. Over time the pace of innovation in technology and organisation slowed. This problem was most pronounced for the Ottoman Empire, which survived into the nineteenth and twentieth centuries, long after the Safavids and Mughals had fallen. Military, political and mercantile elites that had been innovative in a previous era struggled to understand new ideas, inventions and systems. Most tellingly, they were hesitant to integrate or share power with the aspiring elites that sought wealth and status by mastering new technology. This fundamental conflict left the Ottomans and other Asian states and empires unable to effectively contain the rise of the Western powers (Huff 2011).

Imperialism and the creation of European colonial empires doomed the last descendants of Central Asian empire builders and prevented others from following that path. The Industrial Revolution and the tools that it produced—railroads, steamships, telegraphs, repeating rifles—would be impossible to fully replicate or control by aspiring warlords and emperors on remote frontiers continents and oceans away from the centres of European power. Groups like the Comanche and Zulus briefly commanded empires on the steppes, but they would never reach the heights attained by the Persians, Turks, Mongols and their descendants. With their fall and with the consolidation of Western power, the age of frontier empires was over.

Bibliography

Babaie, Sussan, *Slaves of the Shah: New Elites of Safavid Iran* (London: I.B. Tauris, 2004).

Dale, Stephen F., *The Garden of the Eight Paradises: Babur and the Culture of Empire in Central Asia, Afghanistan and India (1483–1530)* (Leiden: Brill, 2004).

Dale, Stephen F., *The Muslim Empires of the Ottomans, Safavids and Mughals* (New York: Cambridge University Press, 2010).

de la Garza, Andrew, *The Mughal Empire at War: Babur, Akbar and the Indian Military Revolution, 1500–1605* (New York: Routledge, 2016).

Ferdowsi, Abolqasem, trans. Dick Davis, *Shahnameh* (New York: Penguin, 2004).

Gommans, Jos, *Mughal Warfare: Indian Frontiers and High Roads to Empire, 1500–1700* (New York: Routledge, 2002).

Gump, James O., *The Dust Rose Like Smoke: The Subjugation of the Zulu and the Sioux* (Lincoln: University of Nebraska Press, 1996).

Huff, Toby E., *Intellectual Curiosity and the Scientific Revolution: A Global Perspective* (New York: Cambridge University Press, 2011).

Imber, Colin, *The Ottoman Empire, 1300–1650: The Structure of Power* (New York: Palgrave Macmillan, 2002).

Murphey, Rhoads, *Ottoman Warfare, 1500–1700* (New Brunswick: Rutgers University Press, 1999).

Soucek, Svat, *A History of Inner Asia* (New York: Cambridge University Press Press, 2000).

Streusand, Douglas E., *The Formation of the Mughal Empire* (New Delhi: Oxford University Press, 1996).

Waters, Matt, *Ancient Persia: A Concise History of the Achaemenid Empire, 550–330 BCE* (New York: Cambridge University Press, 2014).

PART IV

Naval Warfare and Piracy in the Pre-Industrial World

14

THE EGYPTIAN ORIGINS OF AMPHIBIOUS WARFARE

Out of Africa

Mark Charles Fissel

Ancient Egyptians as Pioneers of Amphibious Ways of War

Ancient Egypt's topography fostered the development of a full panoply of amphibious warfare practices. 'Orientalist' Nilocentric environmental determinism, however, has obscured how ancient Egypt's theatre of amphibious warfare encompassed a diverse aquatic system, not a solitary river. The Nile, one of the world's largest and longest rivers, was richly augmented with tributaries. These included countless vanished rivers, streams and havens, many recently rediscovered by satellite imagery (Parcak, Mumford, and Childs 2017: 1–2, 5–6, 18–22; Bunbury and Jeffreys 2011: 65–75; Blouin 2017: 1–11; Hassan 1997: 52, 59–62, 66–7). Artificial canals further enmeshed a navigable and expansive network. The Nile estuary was unusual in a region possessing few river mouth ports or deep water harbours to facilitate the interface between sea and hinterland (Trim 2007: 366). Nilotic geography nourished Egypt's fluvial-centric civilisation and partially remedied the limitations of early sailing techniques (and sails themselves), e.g., an incapacity to tack in contrary winds. The Nile flows north; Nile winds sometimes blow south (to be captured by square sails). Shipbuilding advances then enabled watercraft's evolution from riverine to 'seagoing,' capable of navigating the Red Sea and the littoral of the Eastern Mediterranean (Ward 2010: 42–9; Ward 2009: 9–16; Ward 2004: 22–3; Landström 1970: 35–55, 60–89, 98–115, 122–39; Fabre 2004/5: xiii–42).

Amphibious ways of war demonstrate how and why the Egyptians mastered shallow-draught ship building, sailing techniques, aquatic tactics, etc., which in turn nurtured the expansionist designs of the pharaohs. For example, amphibious warfare became Egypt's tool for controlling rivers and ports, and securing trade routes in Nubia and the Levant. Sneferu (ca. 2613–ca. 2589 BCE), founder of the Fourth Dynasty of the Old Kingdom, established influence over the Levantine littoral whilst waging combined operations in Nubia and Libya (Campagno 2004: 694–7). Juan Carlos Moreno García opines that

> the foreign policy followed by Sneferu seems inspired by the traditional aim of eliminating any rival in areas strategic to Egyptian interests. . . . His campaigns against the

DOI: 10.4324/9780429437915-19

Libyans and his maritime expeditions to the Levant were also concomitant with a deep reorganisation of the Delta.

(García 2010: 7–8)

Sneferu blended riverine and littoral warfare through a strategy based upon mobilisation of the Delta's divergent waterscapes and landscapes. Pharaohs intermingled trade and war, their boats serving dual purposes were designed and built accordingly. Sehure (or Sahure, ca. 2553–ca. 2541) likewise meshed amphibious force projection with commercial enterprises (such as his flotilla sent to 'Punt' in East Africa [ca. 2540]). 'Travelling boats' were facilely converted for combat (Landström 1970: 110, 112).

Because Old Kingdom Egypt had neither a standing navy nor a standing army, modern, conceptual compartmentalisations based upon function cannot be superimposed. Hieroglyphs approximating 'army' and 'expedition' possess elastic meanings (Vencl 1984: 130; Marcus 2007: 140 item 22, 143–4; Wachsmann 1998: 10). Between ca. 3700 and ca. 1069, the pharaoh's merchant marines engaged in at least 149 recorded 'maritime operations,' according to tables assembled by Gregory Phillip Gilbert. These 'paramilitary' ventures intermingled military, commercial and political objectives. The overwhelming majority were 'combined operations'; 72 actions, Gilbert calculates, did or likely could have included 'marines,' whom he distinguishes from purely naval forces, or raiders, or the 'army.' Quantification, like narrative sequencing, of Egyptian sources is conjectural. Even when defined narrowly, 'marines' are ubiquitous over the millennia (Gilbert 2008: 111–34). Riverine warriors of the 3000s BCE had been prototypal amphibious fighters, exploiting readily fluvial propulsion. Symbiotic linkages amalgamated advances in commercial expansion and state formation with sail-powered technology and technique.

Egyptian ship design derived from centuries of conveyance along the Nile and its estuaries, and produced vessels of varying size and function that became increasingly sturdy and 'seaworthy' (e.g., due to technological improvements such as 'thick planks shaped to interlock with one another along their edges and deep, unpegged mortise-and-tenon joints') (Ward 2010a: 48). Improved sail transport generated wealth, increasing available resources for community and state use, and ultimately enabled logistical feats such as moving armed forces up and down the Nile, and supplying those troops systematically (Mark 2013: 33). Conveyance of freightage and soldiers utilised identical types of vessels. Strategic and tactical use of shallow draught bulk carriers added new dimensions to Middle Kingdom (ca. 2055–ca. 1650, Eleventh to Fourteenth Dynasties) riverine amphibious warfare. Cargo ships, workhorses of the maritime economy, were adapted for military use, initially for Nile expeditions (e.g., against Nubia); they later hugged coastlines as ferries for combatants *en route* to the Levant. Eighteenth Dynasty freighters measured 102 feet in length, and 34 feet wide, akin in size, though not fabrication and function, to Roman Republican warcraft (which boasted superior hull construction and greater armament). 'The Egyptians used their fleet to transport troops. They rapidly took over the enemy's territory owing to this method of transportation' (Spalinger 2005: 3). The capacious conveyors of the New Kingdom disgorged more numerous and better supplied marine contingents.

Ship design evolved in tandem with riverine fighting. 'Marines' upon disembarkation discovered that oars and rowing benches presented disconcerting hurdles in combat. Usually a narrow causeway ran lengthwise, stem to stern. Broad decks, requisite for unencumbered ship to ship combat, were rarities. Lacking decks and keels, Old Kingdom boats could not easily sustain serious combat against one another, or routinely challenge deep saltwater. Little enough space existed for provisions and water, let alone room for fighting. Slightly wider decks might exist at bow and stern, but decking amid ship accommodated oarsmen. The absence of keels, minimal deck space, and the rudimentary nature of square sails, inhibited open water navigation and true naval warfare. Archaic sail design persisted into the New Kingdom (after ca. 1550), when Egyptian nautical designers improved rigging (e.g., the brailed sail), conspicuously during the reigns of Amenophis III (ca. 1390–ca. 1352) and Akhenaten (ca. 1352–ca. 1336) (Vinson 1994: 28, 41; Landström 1970: 43–8, 68, 111–12; Fabre 2004/5: 110–23; Wachsmann 1998: 241–6, 251–4; Emanuel 2015: 199–202). Nevertheless, the barges sent by Hatshepsut to Punt (that navigated the African coastline ca. 1493) remained fairly narrow due to continued reliance upon rowers. Ultimately, the Egyptians built watercraft with 'castles' fore and aft (from the 1100s BCE) for archers and spearmen. Hence while the offensive capabilities of these ships as well as their range and manoeuvrability were rudimentary, their variegated crews still developed amphibious ways of war. Hybrid personnel and functionalist competencies made possible hybrid missions.

According to Juan Carlos Moreno García,

> a non-permanent army did not imply the absence of specialists or even of elaborate strategies: besieging operations, amphibious attacks and encircling maneuvers suggest an experienced military mind, an impression reinforced by the use of specialized corps (archers, infantry armed with battle-axes, spearmen, etc.).
>
> *(García 2010: 6, 32)*

Amphibious warfare was empowered, too, by the combat-readiness of construction personnel and related support services:

> a large part of the working force . . . were equipped to fight. And the use of more specialized troops on expeditions is not excluded. . . . [T]he expedition leader ('chancellor of the god') and overseer of the army simultaneously commanded the troops called *mnf3.t* and *ḥy,-* and that these soldiers accompanied him on the missions with which he was entrusted.
>
> *(Fischer 1959: 268–9)*

Diverse forces from far-flung regions (e.g., the Nubian units) mobilised; they were funded and supplied by municipalities, rural areas and religious centres. Since the Fifth Dynasty, recruitment had rested upon logistical underpinnings in the eastern Delta, a staging area for expeditions. The Old Kingdom, then, had created adjustable amphibious prototypes in techniques, technology and institutionalisation that Middle Kingdom rulers improved upon and bestowed to early New Kingdom pharaohs.

Riverine Warfare's Legacy

Riverine warfare, stretching back to Old Kingdom Egypt, subsequently combined with the essentially amphibious nature of Middle Kingdom warfare. It flowered in Thutmose III's spectacular littoral and riverine operations (ca. 1471–ca. 1462), and in vigorous estuarine campaigns of Ramesses III (during his fifth, eighth and eleventh regnal years, ca. 1183–ca. 1177?).

Rapid deployments and swift incursions during the disorders of the Middle Kingdom multiplied operational techniques: (a) boarding parties, (b) tactical manoeuvring propelled by sails as well as rowers, (c) the crippling of enemy oars by ramming or hacking, and (d) conflagrations ignited either whilst in combat or when the enemy had vacated his vessels (Hamblin 2006: 455). A doctrine of harassment and domination was refined in numerous Nubian campaigns. Logistically, Mentuhotep I (ca. 2180–ca. 2140) and his successors used transports to insert 'marine' forces that secured riverbanks (and thus water passage) and surprised adversaries. Middle Kingdom 'marines' were thus more than raiders. Tactically, the speed of Middle Kingdom amphibious warfare enabled 'marines' to infiltrate walled areas and occupy settlements. Ubiquitous watchmen, upon the ramparts and as scouts perched on the cabins of vessels, personify the agile tactics of riverine engagements. The Theban tomb paintings of Intef, an 'overseer of troops' of the Eleventh Dynasty, portray 'marines' combining shock (shipboard infantry bearing shields whilst hefting axes/maces, in one instance poised upon the prow) with missile (archers interspersed liberally among heavy infantrymen and oarsmen) (Figure 14.1; Bestock 2018: 239–41; Vinson 1994: 35).

When ramming and boarding an enemy boat, heavy infantry cut and bludgeoned their way aboard; or when descending upon a defended river bank, gained a foothold. Archers and heavy foot soldiers shifted positions, dictated by mutable tactical situations. If opposing vessels needed to be peppered with arrows before a collision, bowmen stepped forward. If the disembarkation on a riverbank faced resistance, the bowmen slipped forward and unloosed their arrows to thin the enemy ranks. This classic assault technique of combining shock and missile was complemented by the third boat painted in Intef's sepulchre. That watercraft's warriors are exclusively bowmen, holding aloft sheaths of arrows and gripping their bows. These figures may simply be embellishing motifs. However, the proven accuracy in the rendering of the vessel suggests equal reliability in depictions of those aboard it (Roberts 2016: 60–8; Gaballa 1976: 39, 148, note 42). The archers' vessel could be deployed wholly for the launching of barrages of arrows (hence the emphasis on the abundance of arrows available, raised high). Here is yet another tactical technique, a boat dedicated to applying a proportionally greater shower of missiles without hazarding engagement by boarding or disembarkation. Intef's visual representations echo inscriptions recorded in the tomb of Tefibi, likely fighting in the service of the claimant Wahkare Khety (during the First Intermediate period of the Middle Kingdom, ca. 2080, near Thebes) (Gilbert 2008: 44–5).

Tefibi 'reached the [Nile's] east side, sailing upstream . . . (and encountered vessels of the Theban) confederacy (battling them as) far as the fortress of the port of the South. . . . I reached the west side, sailing upstream; there came another (contingent of watercraft) . . . with another army from (the Theban) confederacy.' Tefibi fought these, too. If one qualifies Tefibi's invocation of the traditional hyperbole, he responded to the tactical surprise posed by the appearance of another contingent of enemy vessels that had remained undetected. The translations impart a sense of surprise, urgency and sharp combat, though some of this is of course the bold use of formulaic language that conveys a sense of heroic struggle: '(The enemy commander) hastened to battle . . . I ceased not to fight to the end, making use of the south wind as well as the north wind, of the east wind as well as the west wind.' Allusion to the four quadrants suggests the physical breadth of the engagement, and to tacking and harnessing wind power in oared vessels that manoeuvred briskly for advantageous position amidst the river currents. Tefibi targeted command and control, intending to disable as many of the enemy's ships as possible: '(The enemy commander was struck and) fell into the water;

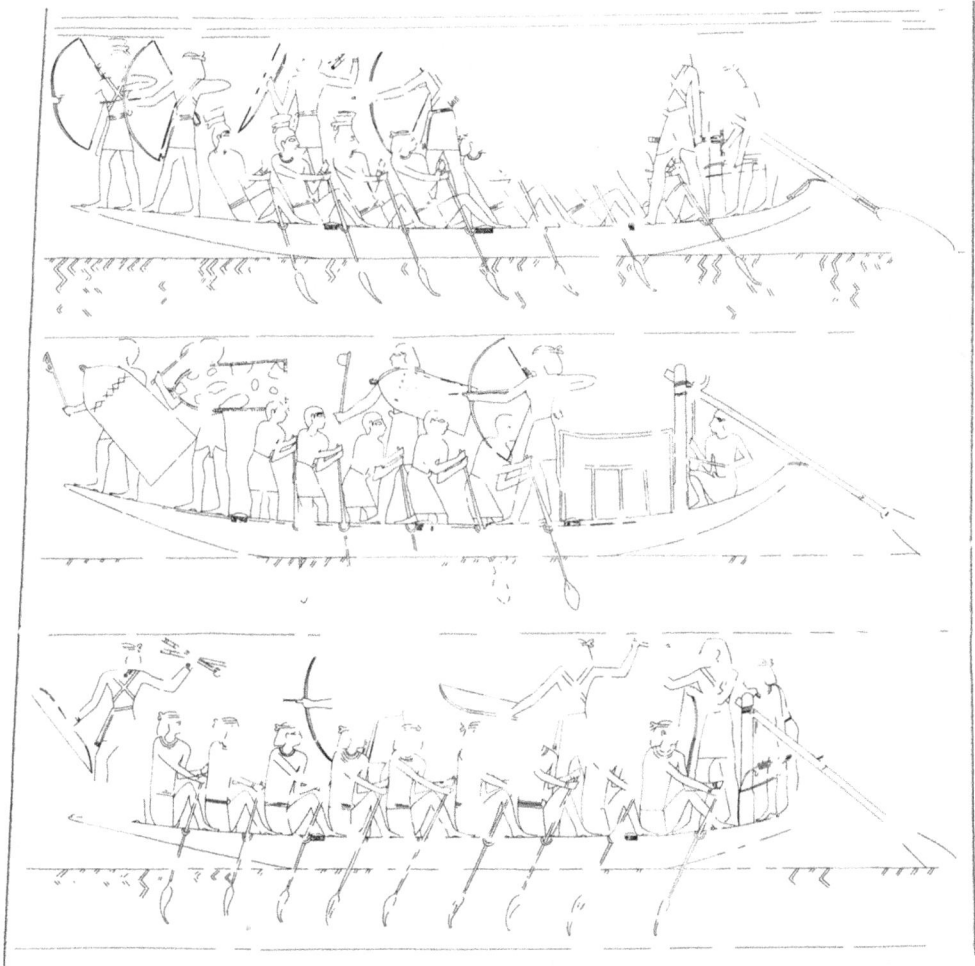

Figure 14.1 Riverine warfare ca. the Eleventh Dynasty (ca. 2134–ca. 1991) depicted in the tomb of Intef at Thebes (reproduced courtesy of the German Archaeological Institute, Cairo). Tafel 14: Kriegsschiff mitt Infanterie an Bord. Pfeiler I b, 27 ff., in Jaroš-Deckert, Brigitte: *Grabung im Asaif (1963–1970). Band V: Das Grab des Jnj-i?tj.f. Die Wandmalereien de XI. Dynastie* (Archäologische Veröffentlichungen 12), Mainz 1984. Dr. Daniela Rosenow and Dr. Mustafa Tupev of the DAI Cairo graciously assisted in identifying and making available this image. According to Gaballa (1976: 39), this 'scene of river warfare is the first representation of this kind so far known since the Gebel el-Arak knife handle.'

Source: Line drawing by Reginald Coleman, © The German Archaeological Institute, Cairo (DAI Cairo).

his ships ran aground.' Once enemy vessels were crippled at the interface of land and water, land-based (presumably) forces set upon the crews and wrought final destruction: 'Fire was set' aboard the beached boats (Gilbert 2008: 45). Tefibi's son, too, exploited speed, mobility and unseen dispositions. 'You (the god) did convey him (the king) up-river. . . . The land trembled, Middle Egypt feared, all the people were in terror, the villages in panic. . . . (because of the marines who) descended by water and landed' (Gilbert 2008: 45).

In Gilbert's analysis, Khety's strategic practice as carried out by Tefibi senior and junior was 'a maritime security operation, which sounds more like a constabulary task where the fleet was used as a peacekeeping force and to support counterinsurgency operations' (Gilbert 2008: 46, 48). During the Intermediate Period of the Middle Kingdom (ca. 2040), in the vicinity of Abydos, Thebes and Koptos, internecine conflicts between Intef I (Intef the Elder, Eleventh Dynasty, nomarch of Thebes) and Ankhtifi (nomarch of Herakliopolis, ca. 2100) produced firsthand accounts of assaults upon riverbank positions. Riverbanks were used tactically to coordinate land-based forces with waterborne attack. Enemy vessels were run aground, savaged and burnt, as demonstrated by Tefibi's forces. Fortifications were approached by water, swiftly and decisively, as in the case of Ankhtifi's marines laying siege. Ankhtifi 'sailed downstream' to reconnoiter the positions of the 'Theban and Coptite nomes . . . (and) seized the fortresses of Armant. . . . I sailed upstream to demolish their fortress with the strong troops of Hefat.' Then Ankhtifi

> sailed downstream with my trusty and strong troops. I landed on the west of the Theban nome, the van of the fleet on the hill of Semekhsen, the rear of the fleet in the domain of Tjemy. . . . I landed on the east of the Theban nome, the rear of the fleet at the tomb of Imbi, the van of the fleet at Shay-sega. I lay siege to its walls.
> *(Gilbert 2008: 44; see also Manassa 2013: 14; Edwards 2016; and Lichtheim 1988)*

Surmounting topographical interfaces, the 'marines' pursued from rivers into lakes, across sandbars, plunging into marshes and wetlands. Tjehemau, a 'Nubian mercenary' who served Montuhotep I around 2050, commanded riverine warcraft and encountered enemy forces 'standing on the riverbank' near Thebes. 'They planned fighting. The opposition fell, fleeing because of me.' Tjehemau then ventured onto Lake Faiyum and engaged troops occupying a sandbank proximate to where a river fed the lake (Hamblin 2006: 383, 452). Egyptian martial practitioners saw their topography literally and figuratively as a fluid geography where their tactics were mutable, like the physical features they encountered. The Middle Kingdom's military 'was amphibious' in all facets of the term (Spalinger 2005: 6).

Riverine and lacustrine combined operations decided Egypt's civil wars. Internal rivalries of the Middle Kingdom waned. Pharaohs then reoccupied resource-rich Nubian territories via fluvial highways. Senusret I (ca. 1965–1920, second pharaoh of the Twelfth Dynasty) and Senusret III (ca. 1878–ca. 1839, fifth pharaoh of the Twelfth Dynasty) waged 'almost yearly campaigns' employing riverine amphibious warfare 'not only to crush dissent, but also to ensure the "milking" of produce and resources' (Redford 2003: 195). Nomes mobilised according to their military capabilities, e.g., when the Oryx Nome's Amenemhat led 400 troops, and then 'six hundred of all the bravest,' on successive riverine expeditions into Nubia under command of a vizier appointed by Senusret I (Fields 2007: 11). Senusret III's eight amphibious expeditions consolidated Egypt's southern borders, bolstering the garrison and fortifications at the Semna gorge to sustain incursions into Nubia (Gilbert 2008: 67; Fields 2007: 41).

At the close of the late Seventeenth Dynasty, Egypt's 'brown-water navy' patrolled the Nile, fighting Nubians to the South and the Hyksos in the North. The latter had long inhabited the Nile Delta (including Memphis and the strategic entrepôt of Avaris). A resurgent Egypt, however, coveted Avaris as 'a gateway to the ancient Near East.' As a 'prime location for a naval base with associated ship construction and harbour facilities,' it was targeted by pharaohs reconnecting Egypt with Levantine trade routes (Gilbert 2008: 48; see also Bietak 2017: 56–60; and Bietak 2009a: 15–17; and Bietak 2009b: 16–17; and Marcus 2006: 187–9). Amphibious power projection pressured Hyksos-occupied Avaris via domination of riverbanks and waterways. 'Kamose arranged his fleet to lay siege to the Hyksos capital . . . allowing his elite troops to secure both sides of the river at Avaris' (Spalinger 2005: 3). Kamose recorded his assailment of Avaris:

> I moored at Perdjedken. . . . I commanded the fleet assembled one behind the other. I put the prow of one at the rudder of another, with my bodyguard, flying upon the water like a falcon. My own ship of gold at the head of it. I was like a divine falcon in front of them. I set a valiant *mek*-ship probing towards the river bank, a *djat*-ship following it.
>
> *(Gilbert 2008: 49)*

Kamose's consolidation exploited the interface between river and land, both tactically and strategically. In Spalinger's words, 'Indeed if a town or even a city resisted, all that Kamose would have to do is bypass it and to attack one to the immediate north, thereby isolating the enemy in a pocket' (Spalinger 2005: 3). In another account, Kamose describes his amphibious actions.

> I sailed down (the Nile) as a champion to overthrow the Asiatics (Hyksos). . . . troops of Mazoi (probably Nubian mercenaries) being on top of our cabins to spy out the Asiatics in order to destroy their positions. . . . I turned back the Asiatics who had encroached. . . . I spent the night in my boat. . . . When day dawned I was down on him (Teti, the Hyksos leader) like a hawk. . . . I overthrew him, I destroyed his wall, I slew his folk, I caused his wife to go down to the river-bank.
>
> *(Säve-Söderbergh 1946: 1–2)*

Kamose died in the course of driving the Hyksos from Egypt, bequeathing the fulfilment of his campaign against the Hyksos to his son Ahmose.

Ahmose, son of Ebana, so-named after the pharaoh he served (Ahmose I ca. 1539–ca. 1514 BCE), reveals the 'marine' nature of the capture of Avaris. Interchangeability between land service and shipboard experience was routine because the weapons and tactics of land warfare closely matched those of aquatic warfare (Emanuel 2015: 202).

> I served as an officer . . . on the ship. The Fighting Bull in the time of the lord of the two lands Nebpehtyra (Ahmose I). . . . I was transferred to the northern fleet because I was brave. I followed the king on foot when he travelled in his chariot. When one besieged the town of Avaris, I was courageous on foot in front of his majesty. Then I was promoted to the ship 'Appearing-in-Memphis' . . . fighting upon the water in the canal of Avaris.
>
> *(Gilbert 2008: 43–62, particularly 51, 148, note 132).*

The scene shifts south, to the Nubian theatre, where amphibious operations such as the transfer by land of watercraft around cataracts, is documented: '(A)fter his majesty had slain the Mentyu of Asia, he sailed south to Khent hen nefer, to destroy the bowmen of Nubia. Then his majesty made a great slaughter of them' (Gilbert 2008: 68, 150, note 169). Ahmose, son of Ebana, continued his amphibious service under Ahmose I's successors.

> I conveyed by ship the King of Upper and Lower Egypt, Djeserkara (Amenhotep I). . . . sailing south to Kush to extend the boundaries of Egypt. Then his majesty slew that Nubian bowman in the middle of his army. . . . Then I rowed the King of Upper and Lower Egypt, Aakheperkaura (Thutmose I) . . . when he was sailing south to *Khent hen nefer*, in order to put down the strife throughout the foreign lands, and to expel the intruders from the desert region. Then I was valiant in front of him in the bad water in the hauling of the ships over the cataract. Then one appointed me to be 'Captain of Sailors'.
>
> *(Gilbert 2008: 68, 150, note 170)*

Vessels were hauled overland or shunted through canals around cataracts (such as that excavated near Aswan during the Sixth Dynasty by Weni) in combined operations where foot soldiers trod the riverbanks whilst boats navigated upriver (Gilbert 2008: 68, note 171; see also Hamblin 2006: 395).

Subjugation of the Hyksos, transformation of the aquatic complex at Avaris and appropriation of the resources of the Nubians and the Hyksos (e.g., the latter's prized lumber) laid the groundwork for a state-sponsored system for waging amphibious warfare (e.g., Ramesses III's mobilisation in defence of the Delta). Royal dockyards (Perunefer, Memphis, Avaris, etc.) stockpiled timber (whilst trade routes in Lebanon, Syria and Palestine remained open). '(T) his wood supply from Lebanon was a kind of monopoly held by the Egyptians and yearly guaranteed. The main agent in charge of this multifarious task was the Egyptian army' (Mizrachy 2012: 27). Construction of infrastructure to sustain marine combat comprised a facet of Egyptian state formation, achieved spectacularly by Ahmose I, then Amenhotep I. Avaris was reconstructed by Ahmose, 'including a new waterway and citadel' and ultimately eclipsed by Memphis as part of a system which strengthened the pharaoh's 'hold on the eastern Delta trade routes' and beyond (Gilbert 2008: 59, 148, note 128; see also Krol 2015: 297). Avaris experienced a resurgence during 'the Ramesside period (as) the harbour of Piramesse, the capital of the Nineteenth and Twentieth Dynasties' (Herbich and Forstner-Muller 2013: 262). Delta ports and harbours figured more importantly than their counterparts in modern history: 'Ports were important in the preindustrial world because there were no urban industrial centers for the generation of wealth' (Trim and Fissel 2007: 431; see also Stager 2001; and Fabre 2004/5: 45–75). Amphibious warfare was the primary means by which ports were taken, thus their strategic indispensability in Egyptian designs on the Levant, Byblos being a paradigmatic case (Marcus 2006: 187–9; Trim and Fissel 2007: 431; Kilani 2017).

Maritime gateways cradled riches and power, nuclei around which commerce, military force and political power orbited. Often 'crucial liminal points between zones' of influence, they were key to control of coasts and hinterland. Preferably, craft proceeded within sight of the shore until a safe destination, a 'harbour,' was reached. Warcraft, especially of medium size, followed coastlines in daylight, frequently availing themselves of depots and ports that could feed their armed hosts and replenish supplies such as fresh water (Mizrachy 2012: 27–8; see also Hamblin 2006: 367, 399–400). Egyptians manoeuvred prudently through the seaboard interface. Even when detouring through open water, routes between the Nile Delta and the Levantine Coast were essentially 'inshore,' hence amphibious. Adapting Middle Kingdom transports, along with their passages, reallocated resources ingeniously.

Commercial vessels became military transports, and hubs of trade (Byblos, Arvad, etc.), strategic depots that augmented the progress of waterborne military force. Vessels designed for the carrying trade were retrofitted for inshore warfare. Centuries of traversing aquatic environments generated an array of watercraft ('Byblos ships,' *aHaw*-boats, galleys, rafts, etc.) used for commerce. 'A "Byblos ship" meant quite simply a sea-going ship' (Landström 1970: 63). The Egyptian 'navy' braved the deep *en route* to Syria–Palestine when prevailing winds so dictated, and thus sped the fleet. Ezra Marcus reckons that:

> A direct sail from the shores of the eastern Delta to the modern border of Lebanon and Syria covers a distance of approximately 270 nautical miles. A vessel sailing at 3 to 6 knots (nautical miles per hour) would make that voyage in 45 to 90 hours, i.e., 2–4 days. In contrast, a ship's course that brought the vessels as close to the shore as possible would cover approximately 377 nautical miles in 63 to 126 hours, or 2.5–5 days. Naturally, ships would not have traveled in such straight lines, and if they called at ports along the way or were waylaid by inclement weather, the distance covered and the time would have increased commensurately.
>
> *(Marcus 2007: 146)*

Sailing seasons, too, circumscribed voyages. Egyptian 'overseas' expeditions kept close to shore for safety and logistical necessity, especially at the beginning and end of journeys. Levantine voyages lasted three weeks or more, possibly prolonged for months, inclusive of nightly dockings and anchorages. Vessels inched along the littoral for the most part, crews oftentimes disembarking to sleep and eat, re-embarking in the morning. The larger the ship the more difficult the 'dockings.' 'Coasting' was not without its inherent perils. Diverging from the coastline might become necessary due to uncertainty regarding the location of reefs and such in shallow water, or to escape winds that might blow vessels against a rocky shore. Once in blue water, crews depended upon dead reckoning in clear skies for direction (Fabre 2004/5: 5, 7–9). Stormy weather, which materialised rapidly, obscured celestial bodies used for navigation, generating rough seas capable of capsizing vessels. In sum, Egyptian force projection in the Mediterranean was most often inshore or littoral.

Despite the daunting challenges of the Mediterranean, Egyptians maintained dominion through the strategic mobility that amphibious warfare and mercantile wealth made possible. The frequency and sustainability of campaigns resulted from the harnessing of amphibious capabilities to deploy and withdraw troops from theatres of operations without squandering time and resources on overland journeys from Egypt. Thutmosid campaigns, discussed below, with their sudden appearance of rested and well-equipped armies, startled Near Eastern powers. The sea, via mastery of littoral navigation, provided an element of surprise that kept client states obedient and enemies apprehensive. Grafting riverine expertise onto seagoing amphibious capabilities extended the pharaoh's reach, climaxing in Thutmose III's Euphrates campaign.

Logistical, Strategical, Tactical and Institutional Legacies of Riverine Amphibious Warfare: Thutmose III's Euphrates Campaign

Around 1481–1469, Thutmose III blunted Mitanni's threat to Egyptian commerce in Lebanon, Syria and Palestine, which imperiled Egypt itself. Successive Thutmosid preemptive strikes perfected the competencies by which warriors moved by land, and then by amphibious means, to the Levant. Historians commend Thutmose III's victory at Megiddo (ca. 1481), and the Euphrates campaign in his thirty-third regnal year (ca. 1471). Around Thutmose's twenty-ninth and thirtieth regnal year, 'the Syrian coast was reached for the first time during the fifth expedition (which secured) an operative base easily reached by sea' (Säve-Söderbergh 1946: 36).

Thutmose's first amphibious landing (campaign six, thirtieth regnal year) seized Arvad and perhaps disembarked 3,000 'marines.' An army of approximately 10,000 in campaign seven (thirty-first regnal year) took Lebanese-Syrian ports (e.g., Simyra). A network of harbours sustained Thutmose's grand strategy, safeguarding trade routes/supply lines, and preventing seizure of harbours by an enemy. For example, a hostile Prince of Tunip coveted Ullaza and that strategic hub changed hands, finally re-secured by Thutmose's forces through amphibious descent upon the Tunipian garrison (Wachsmann 1998: 10; Gabriel 2009a: 155–7). Ullaza (the likely disembarkation point for expedition eight) was roughly a sister haven to Simyra, flanked by Ardata, just inland. To the north, Arvad anchored Egyptian domination. South of the Ullaza-Simyra-Ardata triad stood storied Byblos, an Egyptian base of operations (or 'partner port') since at least the expeditions of Amenemhet II (ca. 1929–ca. 1895), where pharaohs stockpiled Lebanese cedar, their preferred timber for shipbuilding (Mizrachy 2012: 27; Kilani 2017; Marcus 2007: 137; Wachsmann 1998: 9). 'From the Old Kingdom onward, there are references to "Byblos ships," apparently a specific type of seaworthy vessel used to make the run to the harbour of that city in North Syria' (Kelder and Cline 2018: 25–6; see also Landström 1970: 63, 89; and Fabre 2004/5: 93). Byblos, Arvad, Simyra and Ullaza plugged in to a commercial infrastructure interlaced with the Nile's estuarine dockyards.

The coast subjugated, Thutmose's eighth (or, Euphrates) campaign (thirty-third regnal year) wafted its substantial expeditionary force to Ullaza most likely, where construction of portable flatboats and river craft, and the amassing victuals from royal storehouses, ensued. Workshops affiliated with the fortress of Ullaza worked in tandem with Byblos shipwrights (who appear to have produced 'cut boat parts') (Marcus 2007: 153; see also Morris 2005: 153–7).

> [T]he Gebel Barkal stela seems to specify a location which might have served as a center for the wood production activity, that is the seat of the Egyptian garrison at Ullaza . . . [whose] fortress was used as a 'shipyard' for construction of ceremonial boats intended to be delivered to Egypt in their entirety (?). In such a stronghold the Egyptians could have implemented their experience in techniques of boats building in the Nile Valley for hundreds of years.
>
> *(Mizrachy 2012: 27; see also Landström 1970: 63–4)*

The pharaoh's unilateral control of armed force and supervision of overseas commerce, coupled with the longevity of Egyptian amphibious practices, exploited a 'near state monopoly on long-distance trade' that facilitated institutional and logistical expertise (Kelder and Cline 2018: 25–6). Henenu and Intefiqer, for example, constructed vessels upon the shoreline for Punt-destined expeditions (Hamblin 2006: 389, 397). Manufactured 'boat parts' and 'kits,' components hauled to the Red Sea, exemplified riverine and littoral amphibious warfare technology adapted for the Euphrates. '[S]hips were built on the Nile, dismantled and transported in parts down to the sea for re-assembly. In this way they avoided hauling unnecessary timber along the difficult route, and could quickly build up their ships on the inhospitable coast' (Landström 1970: 63–4). Cheryl Ward has detailed

> the assembly of seagoing ship 'kits' and staging of months-long voyages by thousands of men on an intermittent basis. These seagoing ships, first constructed in Nile dock-yards of imported cedar of Lebanon, were disassembled and then carried in pieces by men and donkeys across 145 km (90 miles) of the Eastern Desert to the shore of the Red Sea.
>
> *(Ward 2012: 221–2)*

Logistical expertise and technical skill enabled force projection across diverse aquatic theatres of operations, testifying to Egypt's trailblazing of amphibious warfare.

Bureaucratic maturity and its systematic gathering, storing, and dispatching of 'naval stores,' also made possible deployment of boat-bridges or flat-boat rafts. Craftsmen manu-factured components of shallow-draught vessels, subsequently transported dismantled in ox-drawn wagons.

> I had many vessels . . . built . . . in the neighbourhood of the Lady of Byblos [*m hAw tA nbt Kpnj*], which were then placed on carts and oxen dragged them. They travelled before my Majesty to cross that great river which flows between this foreign land and Naharin (Mittani).
>
> *(Cumming 1982: 2, number 365, line 1232)*

These 'assault' craft were possibly '*aHaw*-boats,' a 'generic appellation for one of the most prevalent types of Egyptian boats used in Egypt, also used for riverine and maritime bat-tles' (Mizrachy 2012: 31–2). The 'fording rafts' and pontoons of antiquity (e.g., Assyr-ian pontoon-trains) were nonexistent in this era. '*aHaw*-boats' were multi-purpose craft of simple enough construction that they could be disassembled into components light enough and small enough to fit on a dray, yet sufficiently durable to be lashed together to establish a bridge across a river's strong current. Toting these 'rafts' and watercraft, Thutmose com-menced a 35-day northeast march. Venturing past Qatna, Tunip and Aleppo towards the territory of Mitanni, Thutmose ultimately 'crossed the Euphrates and trampled (down) the towns on both its banks' (Cumming 1982: 7, number 366, line 1246). Thutmose's engi-neers ferried materials and personnel across the river, fashioning bridges from watercraft. Richard Gabriel conjectures the watercraft were 'rafts,' a conclusion contradicted by Yosef Mizrachy. Conveyance of a 15,000-man army via ten flatboats might have consumed three days (Gabriel 2009a: 172). Relying on a 'boat-bridge' would modify that timetable, and small islands likely protruded into the midst of the Euphrates, which would have assisted in anchoring vessels. Either way, the eighth campaign witnessed 'the first instance known to history of the use of boats to transport an invading army across a river' (Faulkner 1946: 40). Thutmose's traversement by army, beasts of burden and baggage is heralded as the 'first amphibious river crossing' in military history (Gabriel 2009b: 47; see elaboration in Redford 2003: 225).

Once Thutmose occupied the eastern bank of the river, the Mitanni contingent and their allies fought, then fled east from the Euphrates. 'The numerous army of Mitanni was over-thrown in the space of an hour,' according to the Gebel Barkal stela (Cumming 1982: 2, number 365, line 1230; 'mSa aSA n mTn sxr(.w) m km n wnwt'). However, had the full army of the Mitanni been in the vicinity, they would have attacked in the midst of the Egyptian river crossing, when the columns were vulnerable. The force Thutmose defeated may have consisted of local troops loyal to the Mitanni King, or a scouting party. With their adversaries dispersed, the Egyptians commanded both banks of the river and were free to navigate the Euphrates, which they did in a fashion resembling the Nubian Wars. Evidence of the application of Nilotic riverine warfare can be gleaned from accounts of (a) the scattering of the Mitanni confederation in the wake of Carcemish, and (b) tactical movements following the Euphrates crossing. The Annals Inscription chronicles that Thutmose III '(went) by *skdwt* for *itrw*-measure in pursuit of them' ([. . .]*n.f itrw n skdwt m s3.sn* . . .)' which imports a transit across water (Mizrachy 2012: 31, 47, note 146).

A mobile aquatic force went

> upon the water as a tactically constructed maneuver, conducted within the confines of time and distance. A relatively short-term sailing recognized by the *skdwt*-episode could have been used in order to speed up the movement of the Egyptian army in its pursuit after the fleeing Mitannians.
>
> *(Mizrachy 2012: 31)*

In short, proven amphibious techniques on the Nile were acclimated to the Euphrates. The Egyptians proceeded in the manner of the Middle Kingdom civil wars and countless forays into the Upper Nile against Nubian settlements, their riverine force reducing towns, raiding settlements, making their presence known in a riverine context that denoted boundaries and the extension of sovereign power. Stelae were raised, essentially boundary markers and proclamations of power projection, to intimidate Qatna, Senzar (on the Orontes), Tunip and other recalcitrant cities. Acts of violence were simultaneously symbolic and viscerally real (Bestock 2018: 1–12, 264–8; Matić 2017: 7–24; Spalinger 1982: 47; Heinz 2001; Spalinger 2011; Moscati 1963: 16).

While amphibious warfare empowers strategic mobility, it remains subject to the limitations of scale imposed upon forces moving rapidly. Thutmose III's amphibious campaigns could not sack the distant Mitanni capital of Washshuganni, nor indefinitely occupy vast territories. Given that the exact size of Thutmose's forces can only be conjectural,

> surmising a vast conveyance of *aHaw*-boats across the Syrian desert, which aims at meeting the needs of transporting a considerable army across the Euphrates to the Mitannian territory, is less convincing than perceiving a careful use of them, focusing on aggressive action upon a certain segment of the Euphrates flow. In fact, nothing in the sources hints at an objective of reaching the Euphrates other than acting within the confines of the river valley.
>
> *(Mizrachy 2012: 32)*

Combined operations were not intended as a large-scale invasion or a means to occupy territory.

The expeditions climaxing in the Euphrates campaign and after reveal (a) proven doctrines, (b) refined aquatic tactics, (c) institutional competencies and (d) martial techniques 'out of Africa,' in other words as developed on the Nile and across its Delta. Riverine warfare as practised in the civil wars and against the Nubians (e.g., combined operations seizing riverbanks, facilitating intimidation via spectacle and enabling harassment). The Nubian par-adigm rendered 'a permanent presence in the form of depots and garrison posts to transfer the old-fashioned "sphere of influence" into something resembling an empire' (Redford 2003: 195; also Mumford 2018: n.p.).

Thutmose's eighth campaign was, therefore, a show of force utilising riverine amphibious warfare practices, tools and tactics developed in Egypt, not a decisive battle but a demon-stration of the logistical, strategic and tactical proficiencies of the Egyptian state. Its precur-sors were riverine *expedition chevauchées*. The strategic limitations inherent in the Euphrates campaign should not obscure the importance of an enhanced amphibious capability (Frayne 2015: 74–88). Traditional land-based Levantine '*expedition chevauchées*' required vast ar-mies. Counting 'smiths,' 'carpenters' and transport personnel, a traditional land expedition may have numbered 20,000 humans and 7,000 animals, inclusive of chariot forces requisite to defend the long columns. Thutmose III's incremental mobilisations deployed in a distant theatre of operations significant numbers of troops, i.e., perhaps fielding 15,000 effectives in the Euphrates campaign, carrying three infantry 'divisions' of 5,000 apiece, 1,000 horse and 500 chariots to the shores of the Eastern Mediterranean in just over a week (though quantification of Egyptian forces is highly speculative). Amphibious operations reinvented traditional land campaigns by increasing both the speed and scale of power projection, giving offensive warfare a greater edge, making decisive outcomes (in other words, victory) more attainable (Gilbert 2006: n.p.; Gabriel 2009b: 42–7). Marching 600 miles from Egypt might consume a month and a half to five months. Furthermore, exhausting overland marches had limited capacity for transporting materials, e.g., a 'siege train.' Due to inherent freight-age limitations, offensive operations were at a disadvantage when facing fortifications. The inferiority of siegecraft, logistically and even tactically (because tactical failure had permit-ted pharaohs' enemies to ensconce themselves and force a siege, e.g., Megiddo) inhibited 'decisive' victory as well as territorial occupation. The logistical challenge is exemplified by the necessity of successive campaigns to establish a sufficient foothold to launch, ultimately, Thutmose III's amphibious effort (riverine as well as littoral). Once in place, however, utili-sation and domination of the Byblos-Ullaza-Arvad-Simyra network grafted a 'chariot-based military' upon the land–sea interface (Spalinger 2013: 405). Expansive and sometimes arid terrain was overcome more easily. Egypt now could strike potently, within geographical con-straints, across expanses of water. His strategy of amphibious descent exerted influence upon the Levant at least through Thutmose's forty-second regnal year, ca. 1462. Thutmose III waged at least 17 major campaigns, land based and amphibious, the latter being inshore, littoral and riverine.

Estuarine Warfare and Defending the Delta

Violent raids and territorial incursions had displaced coastal populations throughout the Eastern Mediterranean, coincident with late Bronze Age socio-political instability, spurring Egypt's improvements in defensive estuarine warfare. Commencing ca. 1208, 'Sea Peoples' (from north, east and west) challenged Egypt's estuarine bastion. Seti I (ca. 1290–1279), Ramesses II (ca. 1279–1213) and Merenptah (ca. 1213–1203) had defended the Nile Delta against North African invaders (Spalinger 2005: 249–60; Cline and O'Connor 2012b). Seti I repulsed Libyans in Western Delta estuaries (Gilbert 2008: 53). Ramesses II applied littoral tactics against Libyans, establishing a garrison on North Africa's coastline, at Zawiyet Umm el Rakham. Merenptah also conducted combined operations within the Delta (D'Amato and

Salimbeti 2015: 7–8; Lesko 1980: 83–6). Conflicts with marauders climaxed with Ramesses III's victories ca. 1183–1177.

Estuarine amphibious warfare not only traversed interfaces but surmounted variegated aquatic networks. The ancient Nile Delta was more branched than at present. Its network was in continuous development because of the accumulation of fluvial deposits with every flood of the Nile and the changes of sea level. Egypt's volatile waterscapes hosted 40,000 years of amphibious conflict, whilst humankind lacked the technological capabilities to run rough-shod over topography (Torab 1996: abstract; Van Peer et al. 1998). Mutable Nilotic landscapes shaped consciousness and language. A unique terminology and ideology developed in Egypt concerning maritime environments and activities, observes Veronica Morriss (2012: 73, 78–79). Old and Middle Kingdom language reveals that cognitive awareness of 'river' and 'sea' was not binary. Maritime space was conflated with the environmental fluidity of the Delta.

> During antiquity the inundated regions of the Nile Delta were connected to the Mediterranean Sea via the coastal cordon of river mouths, lakes, and lagoons. The connectivity of the river and the sea is reflected in the amalgamation of meaning behind the terms for these environments. The Nile was in many ways a sea with its seasonal conditions, fluctuating winds and currents, and the imminent threat of shipwreck.
>
> *(Fabre 2004/5: 27–30)*

Thus, Egyptian amphibious practices were of necessity multifaceted and adaptable. Estuarine warfare dovetailed littoral and riverine Asiatic amphibious warfare with Egypt's Nubian riverine art of war. Nilotic fluvial systems, natural and artificial, were contoured to nurture Egypt's synthesis of war and commerce. Pharaohs constructed canals, manmade harbours and waterfronts located well inland. David Trim notes,

> the dimensions and suitability for traffic of the entire fluvial system [are] as important as the size or navigability of its core river alone. . . . The riverine components of the fluvial system are both route networks and lines of force. The system's nodal points . . . are thus critically important in strategy, logistics and operational art.
>
> *(Trim 2007: 363–4)*

Wetlands, marshes, swamps and bogs can be militarised landscapes, especially tactically. The Delta, the apex of the operational pyramid, became the school for amphibious warfare, an estuarine recruiting ground where Thutmose III enlisted a 'sizeable contingent' of troops (Redford 2003: 196).

Amphibious warfare became increasingly decisive, partly due to a confluence of technological innovations, specifically iron swords, oared war-galleys, brailed sailing rigs, 'top-mounted' crow's nests, 'loose-footed' sails and incorporation of decking that facilitated vessel-to-vessel combat whilst shielding rowers (Spalinger 2005: 249; Emanuel 2018: 242; Wachsmann 1998: 25–32, 163–76; D'Amato et al. 2015: 8–11, 14–16, 20–5, 33–44, 46–60; Fabre 2004/5: 104–28; Landström 1970: 110–15; Wood 2012: 5–15). Chester Starr classifies Ramesses III's defence of the Delta ca. 1200 BC as climaxing in 'the first known sea battle in ancient history' (and claims that another would not occur until 'half a millennium' had passed) (Starr 1989: 14; see also Emanuel 2014: 23–4; and Emanuel 2015: 193). 'In earlier history, vessels were used for transportation of troops, horses and arms or as mobile bases during warfare, but not for combat' (Gnirs 2013: 716). Was it a 'sea battle'? Ramesses III's triumphs over the invaders are depicted at the mortuary temple of Medinet

Habu (Figure 14.2; Redford 2018; Sales 2012: 92; Roberts 2016: 60–8; Wachsmann 1998: 29–32, 166–74; Shaw 2010: 81–2; Edgerton and Wilson 1936). However, Donald Redford cautions that 'Medinet Habu is not a grab-bag out of which one can pull isolated facts to support preconceived notions' (Redford 2018: xi).

Figure 14.2   The Medinet Habu relief (north wall) illustrating Ramesses III's defence of the Delta. Epigraphic Survey Negative 995, OIP 8 (MHI Historical Records), pl. 36.

Source: Courtesy of the Oriental Institute of the University of Chicago. Invaluable assistance provided by Susan Allison, Associate Registrar of the OI.

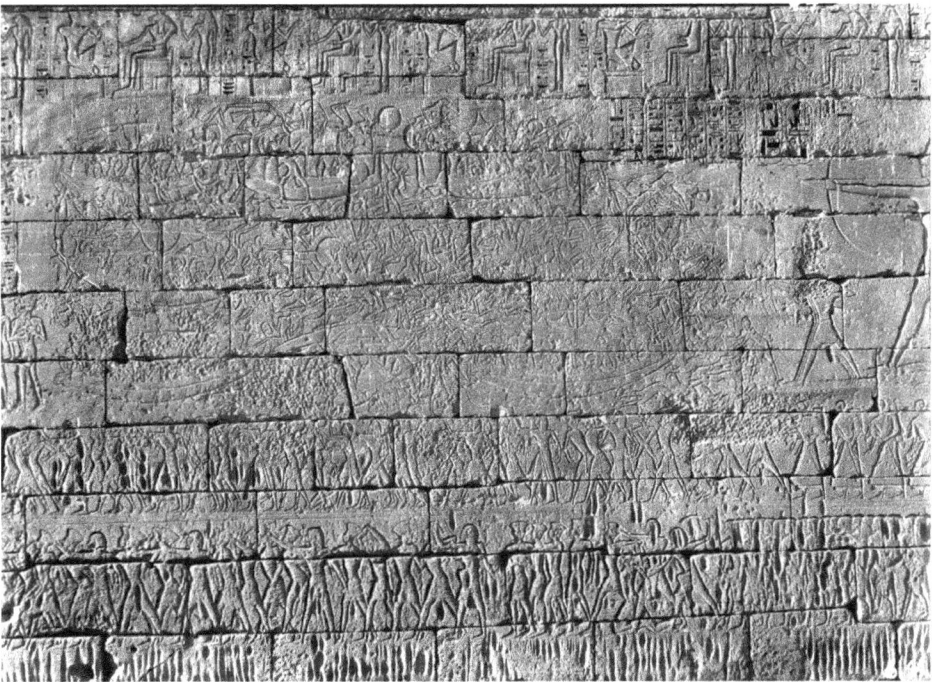

Figure 14.3   Medinet Habu images depicting combat among the vessels of the Egyptians and their adversaries, the 'Sea Peoples.' Detail from Epigraphic Survey Negative 64.

Source: Courtesy of the Oriental Institute of the University of Chicago.

Figure 14.4 Close up of salient and partially protruding figure of Egyptian (upper left corner in Figure 14.3) smiting a 'Sea Peoples' warrior. Iconographical grappling between crews of watercraft is symbolic but does contain potentially realistic technical elements relevant to the study of amphibious warfare. Detail from Epigraphic Survey Negative 0956, OIP 8 (MH I Historical Records), pl. 50B.

Source: Courtesy of the Oriental Institute of the University of Chicago.

Figure 14.5 Line drawing of shallow-draught vessel (its context in Figure 14.3 and its commander in Figure 14.4) waging estuarine amphibious warfare (note Egyptian combatants accoutered in a way in which one might classify them as 'marines'). Epigraphic Survey Negative 881, copy from P. 16337/N. 8166.

Source: Courtesy of the Oriental Institute of the University of Chicago.

In confronting incursion into the Delta, the pharaoh coordinated defensive amphibious warfare with riverine combined operations. The Sea Peoples penetrated the channels of the Nile mouths. But a net had been laid stealthily for them to ensnare them as they entered the Nile mouths (Redford 2018: 27–8; Edgerton and Wilson 1936: 41–2). That 'net' included land-based forces as well as boats (Figures 14.3 and 14.5), the ancient equivalent of combined operations in defence of an estuary. Egyptian coastal craft pressed the invaders towards the shoreline, where Egyptian chariots and infantry unloosed arrows (Figures 14.6 and 14.7); Sales 2012: 92; Cifolla 1991: 9–57, 1988: 275–306).

They came on, with fire prepared in their van, straight against Egypt. . . . I organized my frontier . . . made ready ahead of them [with] chiefs, garrison-commanders, and *maryannu*. I prepared the Nile-mouths like a strong wall with a fleet of warships, transports, and skiffs, prepared and manned completely, stem to stern, with shock-troops with their weapons. The infantry was made up of every choice pick of the land of Egypt. . . . the flame flared against them at the Nile mouths. A dado [Redford translates as a stockade or corral] of lances engulfed them on the coast, [they were] dragged up, cast, and thrown down on the shore. They were slain and made into heaps from head to tail, their ships and property like flotsam and jetsam!

(Redford 2018: 36–7)

Although deck-to-deck fighting *may* have commenced in saltwater, these engagements were clearly inshore, estuarine and littoral, not 'naval.' Gilbert classifies this as warfare in the littorals:

Ramesses III's sea battle was a trap set to catch a raiding fleet of Sea People in narrow waters between the king and his archers ashore and an Egyptian fleet that attacked the Sea People from behind. . . . [T]his naval battle was what would today be termed action by maritime forces in the littorals.

(Gilbert 2008: 54)

At least one clash occurred with combatants 'within sight of each other just beyond the mouth of the Pelusiac branch' (Redford 2000: 13; see also Evian 2016: 163; as well as Cline and O'Connor 2016: 136–7). Mustering land forces for aquatic defence via estuarine and riverine warfare epitomises defensive amphibious operations. Indeed, the pharaoh is depicted in an act of defensive amphibious warfare, at the water's edge possibly, launching arrows, flanked by archers (Figures 14.6 and 14.7; Edgerton and Wilson 1936: 41).

Though likely geographically and chronologically separate, the land and sea battles pictured constituted combined operations, most explicitly use of land-based archers (Spalinger 2011: 194 note 20; O'Connor 2012: 209–70; Cifolla 1988: 300–4). Ramesses exploited missile weapons both on *terra firma* and afloat. Early on he had commissioned at least three types of seaworthy vessels, replete with bowmen (Wachsmann 1998: 10–11, 26–7; Hamblin 2006: 454). The Harris Papyrus (and carved reliefs) communicate amphibious warfare doctrine. '(T)he tactic of parallel land and sea assaults seems to have been the *modus operandi*' (Emanuel 2018: 246). The invaders encountered Ramesses' defensive web 'while afloat but still in sight of land (as in the Medinet Habu relief), or even in the open water. . . . Ramesses III's records at Medinet Habu and in the Great

Figure 14.6   Medinet Habu relief depicting Ramesses III (figuratively at the water's edge?) engaged in defensive amphibious warfare, the pharaoh as a 'a great wall for sheltering Egypt so that there may come no land to injure' (Edgerton and Wilson 1936: 41; also addressed in Redford 2018: 27). Epigraphic Survey Negative 630.

Source: Courtesy of the Oriental Institute of the University of Chicago.

Harris Papyrus (infer) systematic, coordinated land and sea campaigns by a confederation of tribes, for the purpose of a strategic objective' (Emanuel 2018: 247, 249). The clashes become even more 'estuarine' if the Sea Peoples descended from 'forward bases' within the Delta (though Redford's transliteration of item three rejects this interpretation) (D'Amato and Salimbeti 2015: 52; Redford 2000: 12). Ramesses deployed, prototypically, the classic triad of anti-landing defence: combatants afloat, opposition at the shoreline and mobile ground forces (foot soldiers, and likely chariots). However, do reliefs and texts convey operational history? Deciphering iconographic representation is as treacherous as squeezing ancient Egyptian military records into the linearity, concepts and terminological constructs of later eras. Egyptian primary sources can for modern purposes complicate rather than clarify historical reality (Shaw 2005: 77, 79, 81–4; Spalinger 1982: 9–15, 20–1, 26, 31, 37–9, 46–7, 76, 97, 237–41; Lundh 2002; Moscati 1963: 71–88). The reliefs at Medinet Habu, lauded as splendid representations of 'sea power' are a case in point (Gaballa 1976: 122–3; Spalinger 2011; Heinz 2001; Redford 2018).

Figure 14.7 The northern invaders have been 'capsized and overwhelmed. . . . Their weapons are scattered in the sea. His arrow pierces him whom he has wished among them' (Edgerton and Wilson 1936: 41; also addressed in Redford 2018: 27). Reproduction of a line drawing of the relief partially depicted in Figure 14.6, by Alfred Bollacher, catalogued as Epigraphic Survey Negative 2101, OIP 8 (MH I Historical Records), pl. 37.

Source: Courtesy of the Oriental Institute of the University of Chicago.

Conceptualising and Representing Ancient Amphibious Warfare

The extent and complexity of Egyptian amphibious ways of war are underappreciated partly because archaeological remains (especially organic materials evidential of human activities upon the water) survive in haphazard fashion (Vencl 1984: 116–18, 123–31). Additionally, written and iconographic records derive from a culture profoundly different from our own. The paucity of amphibious iconography, for example for riverine operations between the era of the Gebel el-Arak knife handle (3300–3200? BCE) and the tomb of Intef (ca. 2000? BCE) leaves a void (Assmann 2002: ix–283). Representations of Middle Kingdom-era 'sea battles' are absent. Interpreting and articulating Egypt's military achievements are further complicated, as postmodernism demonstrates, because language describing the past distorts as it explains. One term that grates against postmodern sensibilities is 'sea power.' Amphibious warfare's geographically specialised modes (littoral, inshore, riverine, lacustrine and estuarine) advanced mightily in ancient Egypt before 'naval' warfare was practicable, or that 'sea power' existed. The latter terms apply to blue-water warfare, fought in saltwater, oftentimes vessel-to-vessel, upon decks, in deep and open waters, systematically. 'Sea power' projects force via 'naval' warfare, which requires vessels' logistical autonomy, enables independent action, and provides substantial weapons platforms. Certainly ca. 1500, Minoans and Assyrians had the ability to wage 'naval' warfare, so the flotillas of Thutmose III and Ramesses III (and the 'Sea Peoples') could undertake seaborne engagements. By the fifth century BCE, standing navies plied the Mediterranean (Wood 2012: 26–9).

Even then, in the ancient Mediterranean, 'naval' battles were most frequently fought in sight of land, as the ships were utterly dependent upon nearby land-based supply systems. Gregory Gilbert, an authority on sea power and Egyptology, credits the Egyptians as 'first to conduct amphibious operations,' for example as early as vizier Weni 'leading an amphibious assault behind the rear of an enemy force in Southern Palestine around 2300 BCE'

(Gilbert 2008: 109). In riverine campaigns to take Avaris, Kamose and Ahmose combined strategic and tactical amphibious operations to 'blockade an enemy stronghold' as well as exploit occupation of riverbanks to achieve regional control (Gilbert 2008: 109). Above, we invoked Spalinger's characterisation of the Middle Kingdom's military as 'amphibious' (Spalinger 2005: 6). By the time of the New Kingdom, complex amphibious grand strategies became feasible. Landings via inshore and littoral operations attained greater scale and efficiency. Building upon centuries of technical knowledge in prefabrication of shallow-draught vessels, boats were designed, fashioned and transported overland, launched on the Euphrates, and riverine operations conducted. Tactical application of riverine amphibious warfare especially had been refined during the Middle Kingdom period, later broadened by Ramesses III's estuarine defence.

Utilising a formula for 'sea power' different from that presented here, Gilbert proposes a comparatively holistic definition:

> For Ancient Egypt, I believe that the sea power mindset involved using physical, demographic, geographic, economic as well as military resources related to the river Nile, its delta, its tributaries and the adjacent waters. It is a counterpoint to the western continental mindset. To me sea power is much more than 'naval,' 'blue-water' or 'the sea.'
>
> *(Personal communication)*

The ancient Egyptian 'mindset' and its conception of aquatic environments lends weight to Gilbert's broader definition. The 'Great Green' was inextricably one interconnected world of water, in which land–sea interfaces were quite naturally more relevant than, say, ocean depth. The earliest human conflicts occurred along rivers and shorelines, in wetlands and estuaries, more frequently than upon the sea, especially out of sight of land. Amphibious operations preceded deep-water fleets. Littoral, inshore, riverine, lacustrine and estuarine warfare gestated 'sea power,' however defined. In summation, diverse forms of amphibious warfare were mastered by the Egyptians before the eras of blue-water warfare. Coming to terms with geography (in a Braudelian sense), the ancient world's blend of commerce and armed aggression served as a catalyst for Egyptian state formation. The result was the world's prototypical amphibious warfare.

Bibliography

Andelkovic, Branislav, 'Models of State Formation in Predynastic Egypt,' *Archaeology of Early Northeastern Africa: Studies in African Archaeology* 9 (Poznań: Poznań Archaeological Museum, 2006), pp. 593–609.

Andelkovic, Branislav, 'Political Organization of Egypt in the Predynastic Period,' in Emily Teeter (ed.), *Before the Pyramids: The Origins of Egyptian Civilization* (Chicago: Oriental Institute Museum Publications, University of Chicago, 2011), pp. 25–32.

Assmann, Jan, *The Mind of Egypt: History and Meaning in the Time of the Pharaohs* (New York: Metropolitan, 2002).

Barako, Tristan, *The Seaborne Migration of the Philistines*. A thesis presented by Tristan Joseph Barako. The Department of Near Eastern Languages and Civilizations degree of Doctor of Philosophy, Near Eastern Languages and Civilizations, Cambridge, Mass: Harvard University, (2001).

Bestock, Laurel, *Violence and Power in Ancient Egypt: Image and Ideology Before the New Kingdom* (Abingdon/New York: Routledge, 2018).

Bietak, Manfred, *Avaris: The Capital of the Hyksos—Recent Excavations at Tell el-Dab'a* (London: British Museum, 1996).

Bietak, Manfred, 'Perunefer: The Principal New Kingdom Naval Base,' *Egyptian Archaeology*, vol. 34 (2009a), pp. 15–17.

Bietak, Manfred, 'Perunefer: An Update,' *Egyptian Archaeology*, vol. 35 (2009b), pp. 16–17.

Bietak, Manfred, 'Harbours and Coastal Military Bases in Egypt in the Second Millenium B.C.,' in Harco Willems and Jan-Michael Dahms (eds.), *The Nile: Natural and Cultural Landscape in Egypt* (Mainz: Mainz Historical Cultural Sciences, 2017), pp. 53–70.

Blouin, Katherine, 'Beyond the Nile: Orientalism, Environmental History, and Ancient Egypt's Mareotide (Northwestern Nile Delta),' *History Compass* (2017), pp. 1–11.

Breasted, James, *A History of the Ancient Egyptians* (New York: Charles Scribner's Sons, 1908).

Breasted, James, *A History of Egypt from the Earliest Times to the Persian Conquest* (New York: Charles Scribner's Sons, 1909).

Bunbury, Judith, and Jeffreys, David, 'Real and Literary Landscapes in Ancient Egypt,' *Cambridge Archaeological Journal*, vol. 21, no 1 (2011), pp. 65–75.

Campagno, Marcelo, 'In the Beginning Was the War: Conflict and the Emergence of the Egyptian State,' in M. Chlodnicki, Krzysztof Cialowicz, Terry Friedman, and S. Hendrick (eds.), *Egypt at Its Origins: Proceedings of the International Conference 'Origin of the State. Predynastic and Early Dynastic Egypt' (Orientalia Lovaniensia Analecta) Papers given at Krakow 2002* (Leuven, Paris, and Dudley Mass: Uitgeverij Peeters en Departement Oosterse Studies, 2004), pp. 689–703.

Casson, Lionel, *The Ancient Mariners: Seafarers and Sea Fighters of the Mediterranean in Ancient Times*, second edition (Princeton New Jersey: Princeton University Press, 1991).

Casson, Lionel, *Ships and Seamanship in the Ancient World* (Baltimore/London: Johns Hopkins, 1995).

Cavillier, Giacomo, *Il faraone guerriero. Il Sovrani del nuovo regno alla conquista dell'Asia tra mito strategia bellica e realta archeologica* (Turin: Tirrenia Stampatori, 2001).

Cifolla, Barbara, 'Ramses III and the Sea Peoples: A Structural Analysis of the Medinet Habu Inscriptions,' *Orientalia*, vol. 57 (1988), pp. 275–306.

Cifolla, Barbara, 'The Terminology of Ramses III's Historical Records: With a Formal Analysis of the War Scenes,' *Orientalia*, vol. 60 (1991), pp. 9–57.

Cline, Eric H., 'The Sea Peoples,' in Jeffrey Spier, Timothy Potts, and Sara E. Cole (eds.), *Beyond the Nile: Egypt and the Classical World* (Los Angeles: J. Paul Getty Museum, 2018), pp. 29–34.

Cline, Eric H. and O'Connor, David, *Thutmose III: A New Biography* (Ann Arbor: University of Michigan Press, 2006).

Cline, Eric H. and O'Connor, David, *Ramesses III: The Life and Times of Egypt's Last Hero* (Ann Arbor: University of Michigan Press, 2012a).

Cline, Eric H. and O'Connor, David, 'The Sea Peoples,' in Eric H. Cline, and David O'Connor, *Ramesses III: The Life and Times of Egypt's Last Hero* (Ann Arbor: University of Michigan Press, 2012b), pp. 180–208.

Cline, Eric and O'Connor, David, 'The Mystery of the "Sea Peoples,"' in David O'Connor and Stephen Quirke (eds.), *Mysterious Lands* (Abingdon/New York: Routledge, 2016), pp. 107–38.

Cumming, Barbara (ed.), *Egyptian Historical Records of the Later Eighteenth Dynasty, Fascicle 1* (Warminster: Aris and Phillips, 1982).

D'Amato, Raffaele and Salimbeti, Andrea, *Sea Peoples of the Bronze Age Mediterranean c. 1400 BC–1000 BC* (Oxford: Osprey, 2015).

Edgerton, William F. and Wilson, John A. (eds.), *Historical Records of Ramses III: The Texts in Medinet Habu Volumes I and II* (Chicago: University of Chicago Press, 1936).

Edwards, Katrina Jane, '"I Was a Man of Whom there Was no Other!" A Linguistic Review of the "Autobiographical" Inscriptions of Ankhtyfy,' Master of Arts thesis, University of Auckland (2016).

Emanuel, Jeffrey, 'The Sea Peoples, Egypt, and the Aegean: Transference of Maritime Technology in the Late Bronze–Early Iron Transition (LH IIIB–C),' *Aegean Studies*, vol. 1 (2014), pp. 21–56.

Emanuel, Jeffrey, 'The Late Bronze–Early Iron Age Transition: Changes in Warriors and Warfare and the Earliest Recorded Naval Battles,' in Geoff Lee, Helene Whittaker and Graham Wrightson (eds.), *Ancient Warfare: Introducing Current Research*, vol. 1 (Newcastle-upon-Tyne: Cambridge Scholars, 2015), pp. 191–209.

Emanuel, Jeffrey, 'Differentiating Naval Warfare and Piracy in the Late Bronze–Early Iron Age Mediterranean: Possibility or Pipe Dream?,' in Łukasz Niesiołowski-Spanò and Marek Węcowski (eds.), *Change, Continuity, and Connectivity North-Eastern Mediterranean at the turn of the Bronze Age and in the Early Iron Age* (Wiesbaden: Harrassowitz Verlag, 2018). [An electronic version differing from the pagination of the print version was used for this chapter.]

Evian, Shirly Ben-Dor, *The Battles between Ramesses III and the 'Sea-Peoples': When, Where and Who? An Iconic Analysis of the Egyptian Reliefs* (Tel Aviv: University Central Library, E. Sourasky Library/ The Neiman Library, 2016).

Fabre, David, *Seafaring in Ancient Egypt* (London: Periplus, 2004/5).

Faulkner, Raymond Oliver, 'Egyptian Seagoing Ships,' *The Journal of Egyptian Archaeology*, vol. 26 (February 1941), pp. 3–9.

Faulkner, Raymond Oliver, 'The Euphrates Campaign of Tuthmosis III,' *The Journal of Egyptian Archaeology*, vol. 32 (December 1946), pp. 39–42.

Faulkner, Raymond Oliver, 'Egyptian Military Organization,' *The Journal of Egyptian Archaeology*, vol. 39 (December 1953), pp. 32–47.

Fields, Nic, *Soldier of the Pharaoh, Middle Kingdom Egypt 2055–1650 BC* (Oxford: Osprey, 2007).

Fischer, Henry, 'A Scribe of the Army in a Saqqara Mastaba of the Early Fifth Dynasty,' *Journal of Near Eastern Studies*, vol. 18, no. 4 (October 1959), pp. 233–72.

Frayne, Douglas, 'Thutmose III's Great Syrian Campaign: Tracing the Steps of the Egyptian Pharaoh in Western Syria, Part I: From Idlib to Aleppo,' in T. Harrison, E. Banning, and S. Klassen (eds.), *Walls of the Prince: Egyptian Interactions with Southwest Asia in Antiquity, Essays in Honour of John S. Holladay, Jr.* (Leiden: Brill 2015), pp. 74–88.

Gaballa, G.A., *Narrative in Egyptian Art* (Mainz: Verlag Philipp Von Zabern, 1976).

Gabriel, Richard, *Thutmose III* (Dulles, VA: Potomac, 2009a).

Gabriel, Richard, 'Amphibious Pharaoh,' *Military History*, vol. 26, no. 4 (October–November 2009b), pp. 42–7.

García, Juan Carlos Moreno, 'War in Old Kingdom Egypt (2686–2125 BCE),' in J. Vidal (ed.), *Studies on War in the Ancient Near East* (Münster: Ugarit Verlag, 2010), pp. 1–43.

García, Juan Carlos Moreno (ed.), *Ancient Egyptian Administration (Handbook of Oriental Studies: Section 1; The Near and Middle East)* (Leiden: Brill 2013).

Gilbert, Gregory Phillip, *Weapons, Warriors, and Warfare in Early Egypt* (Oxford: BAR International, 2004).

Gilbert, Gregory Phillip, 'Ancient Egyptian Joint Operations in Lebanon under Thutmose III (1451–1438 BCE),' *Semaphore*, Issue 16 (August 2006), unpaginated.

Gilbert, Gregory Phillip, *Ancient Egyptian Sea Power and the Origin of Maritime Forces* (Canberra: Sea Power Centre Australia, 2008).

Gnirs, Andrea M. 'Coping with the Army: The Military and the State in the New Kingdom,' in Juan Carlos Moreno García (ed.), *Ancient Egyptian Administration* (Leiden: Brill, 2013), pp. 639–718.

Groenewegen-Frankfort, H.A., *Arrest and Movement: An Essay on Space and Time in the Representational Art of the Ancient Near East* (New York: Hacker Art Books 1972).

Hamblin, William J., *Warfare in the Ancient Near East to 1600 BC: Holy Warriors at the Dawn of History* (Abingdon/New York: Routledge, 2006).

Hassan, Fekri, 'The Dynamics of a Riverine Civilization: A Geoarchaeological Perspective on the Nile Valley, Egypt,' *World Archaeology*, vol. 29, no. 1 (1997), pp. 51–74.

Hayes, Jack, 'From Great Green Walls to Deadly Mires: Wetlands as Military Environments and Ecosystems in Chinese History,' *Water History* (Dordrecht and Berlin: Springer Science and Business Media, 2013). https://www.academia.edu/30427009/From_Great_Green_Walls_to_deadly_mires_wetlands_as_military_environments_and_ecosystems_in_Chinese_history

Heagren, Brett, '"Siege Warfare" in Ancient Egypt, as Derived from Select Royal and Private Battle Scenes,' in Jeremy Armstrong (ed.), *Circum Mare: Themes in Ancient Warfare* (Leiden: Brill, 2016), pp. 237–61.

Heinz, Susanna Constanze, *Die Feldzugsdarstellungen des Neuen Reiches: Eine Bildanalyse* (Wien: Österreichische Akademie der Wissenschaften, 2001).

Herbich, Tomasz, Forstner-Muller, Irene, 'Small Harbours in the Nile Delta. The Case of Tell el-Dabca,' *Institut des Cultures Mediterraneennes et Orientalesde l'Academie Polonaise des Sciences Études et Travaux*, vol. 26 (2013), pp. 257–72.

James, Peter, 'The Levantine War-Records of Ramesses III: Changing Attitudes, Past, Present and Future,' *Antiguo Oriente*, vol. 15 (2017), pp. 57–148.

Kelder, Jorrit, and Cline, Eric, 'In the Midst of the "Great Green": Egypto-Aegean Trade and Exchange,' in Jeffrey Spier, Timothy Potts and Sara Cole (eds.), *Beyond the Nile: Egypt and the Classical World* (Los Angeles: J. Paul Getty Museum 2018), pp. 24–8.

Kelder, Jorrit, Cole, Sarah and Cline, Eric, 'Memphis, Minos, and Mycenae: Bronze Age Contact between Egypt and the Aegean,' in Jeffrey Spier, Timothy Potts and Sara E. Cole (eds.), *Beyond the Nile: Egypt and the Classical World* (Los Angeles: J. Paul Getty Museum, 2018), pp. 9–17.

Kilani, Marwan, 'Byblos in the Late Bronze Age: Interactions between the Levantine and Egyptian Worlds,' Oxford D.Phil. thesis (2017).

Köhler, Christiana, 'The Rise of the Egyptian State,' in Teeter, Emily (ed.), *Before the Pyramids: The Origins of Egyptian Civilization* (Chicago: Oriental Institute Museum Publications, 2011), pp. 123–5.

Krol, Alexey '"White Walls" of Memphis at Kom Tuman,' in P. Kousoulis and N. Lazaridis (eds.), *Proceedings of the Tenth International Congress of Egyptologists, University of the Aegean, Rhodes, 22–29 May 2008*, vol. 1 (Leuven/Paris: Uitgeverij Peeters, 2015), pp. 295–303.

Landström, Björn, *Ships of the Pharaohs: 4000 Years of Egyptian Shipbuilding* (Stockholm: Doubleday 1970).

Leclant, Jean, 'Fouilles et travaux en Égypte et au Soudan, 1963–1964,' *Orientalia*, New Series, vol. 34, no. 2 (1965), pp. 175–232.

Lesko, Leonard H., 'The Wars of Ramses III,' *Serapis*, vol. 6 (1980), pp. 83–6.

Lichtheim, Miriam, *Ancient Egyptian Autobiographies Chiefly of the Middle Kingdom* (Zurich: Vandenhoeck and Ruprecht, 1988).

Lundh, Patrik, *Actor and Event: Military Activity in Ancient Egyptian Narrative Texts from Tuthmosis to Merenptah*, PhD thesis (Archaeology and Ancient History) (Uppsala: Akademitryck, 2002).

Manassa, Colleen, *Imagining the Past: Historical Fiction in the New Kingdom* (Oxford: Oxford University Press, 2013).

Marcus, Ezra S., 'Venice on the Nile? On the Maritime Character of Tell el Dab'a/Avaris,' in Ernst Czerny, Irmgard Hein, Hermann Hunger, Dagmar Melman and Angela Schwab (eds.), *Timelines: Studies in Honour of Manfred Bietak*, vol. 2 (Leuven: Uitgeverij Peeters, 2006), pp. 187–90.

Marcus, Ezra S., 'Amenemhet II and the Sea: Maritime Aspects of the Mit Rahina (Memphis) Inscription,' *Ägypten und Levante*, vol. 17 (2007), pp. 137–90.

Mark, Samuel, 'The Earliest Sailboats in Egypt and their Influence on the Development of Trade, Seafaring in the Red Sea, and State Development,' *Journal of Ancient Egyptian Interconnections*, vol. 5, no. 1 (2013), pp. 28–37.

Matić, Uros, 'Scorched Earth: Violence and Landscape in New Kingdom Egyptian Representations of War,' *Journal of Historical Researches*, vol. 28 (2017), pp. 7–28.

Mizrachy, Yosef, 'The Eighth Campaign of Thutmose III Revisited,' *Journal of Ancient Egyptian Interconnections*, vol. 4, no. 2 (2012), pp. 24–52.

Morris, Elene Fowles, *The Architecture of Imperialism: Military Bases and the Evolution of Foreign Policy in Egypt's New Kingdom* (Leiden: Brill, 2005).

Morriss, Veronica, 'Islands in the Nile Sea,' Texas A&M MA thesis (Anthropology) College Station (2012).

Moscati, Sabatino, *Historical Art in the Ancient Near East no. 8 Studi Semitici* (Università di Roma: Rome, 1963).

Mumford, Gregory, 'The Late Bronze Age Collapse and the Sea Peoples' Migrations,' in J.S. Greer, J.W. Hilber and J.H. Walton (eds.), *Behind the Scenes of the Old Testament: Historical, Cultural, and Social Contexts of Ancient Israel* (Grand Rapids: Baker Academic Press, 2018), pp. 260–71.

Mumford, Gregory, 'Egypt and Nubia as Trade Partners and Military Adversaries,' in R. Dubie and J. Hutchinson (eds.), *World History: Ancient and Medieval Eras* (ABC-CLIO: 2018/2019) unpaginated submission, https://www.academia.edu/34472289/IN-PRESS_Egypt_and_Nubia_as_Trade_Partners_and_Military_Adversaries_G._Mumford_pp._xx-xx_in_R._Dubie_and_J._Hutchinson_eds._World_History_Ancient_and_Medieval_Eras._Santa_Barbara_ABC-CLIO_Inc._3_500_words_Accepted_Dec._2018_pending_online_publication_2019_

Nederhof, Mark-Jan, 'Gebel Barkal Stela of Tuthmosis III,' https://mjn.host.cs.st-andrews.ac.uk/egyptian/texts/corpus/pdf/GebelBarkalTuthmosisIII.pdf https://www.academia.edu/5301278/Gebel_Barkal_stela_of_Tuthmosis_III_by_Mark-Jan_Nedehof

Nelson, H.H., 'The Naval Battle Pictured at Medinet Habu,' *Journal of Near Eastern Studies*, 2 (1943), pp. 40–55.

O'Connor, David, 'The Mortuary Temple of Ramesses III at Medinet Habu,' in Eric Cline and David O'Connor (eds.), *Ramesses III: The Life and Times of Egypt's Last Hero* (Ann Arbor: University of Michigan Press, 2012), pp. 209–70.

O'Connor, David and Cline, Eric H (eds.), *Amenhotep III: Perspectives on His Reign* (Ann Arbor: University of Michigan Press, 2001).

Parcak, Sarah, *Archaeology from Space: How the Future Shapes Our Past* (Henry Holt and Company: New York, 2019).

Parcak, Sarah, Mumford, Gregory and Childs, Chase (eds.), 'Using Open Access Satellite Data Alongside Ground Based Remote Sensing: An Assessment, with Case Studies from Egypt's Delta,' *Geosciences*, vol. 7, no. 4 (2017), doi:10.3390/geosciences7040094

Peters, Scott M., 'Decoding the Medinet Habu Inscriptions: The Ideological Subtext of Ramesses III's War Accounts,' Columbia University senior thesis (History) (2011).

Petrie, W. Flinders, *A History of Egypt during the XVIIth and XVIIIth Dynasties* (Charles Scribner's Sons: New York, 1897).

Petrie, W. Flinders, *A History of Egypt from the XIXth to the XXXth Dynasties* (Charles Scribner's Sons: New York, 1905).

Redford, Donald B., *History and Chronology of the Eighteenth Dynasty of Egypt: Seven Studies* (Toronto: University of Toronto Press, 1967).

Redford, Donald B., 'Egypt and Western Asia in the Late New Kingdom: An Overview,' in Eliezer Oren (ed.), *The Sea Peoples and Their World: A Reassessment* (University of Pennsylvania Press: Philadelphia, 2000), pp. 1–20.

Redford, Donald B., *The Wars in Syria and Palestine of Thutmose III* (Leiden: Brill, 2003).

Redford, Donald B., 'The Northern Wars of Thutmose III,' in Eric H. Cline and David O'Connor (eds.), *Thutmose III: A New Biography* (Ann Arbor: University of Michigan Press, 2005), pp. 325–43.

Redford, Donald B., *The Medinet Habu Records of the Foreign Wars of Ramesses III* (Leiden: Brill, 2018).

Roberts, R. Gareth, 'Identity, Choice, and the Year 8 Reliefs of Ramesses III at Medinet Habu,' in Christoph Bachhuber and R. Gareth Roberts (eds.), *Forces of Transformation: The End of the Bronze Age in the Mediterranean* (Oxford: Oxbow, 2016), pp. 60–8.

Robertson, Joshua, *Ramesside Inscriptions Historical and Biographical Volume IX* (Wallasey: Abercromby Press, 2018).

Sales, José das Candeias, 'The Smiting of the Enemies Scenes in the Mortuary Temple of Ramses III at Medinet Habu,' *The Journal of Oriental and Ancient History*, vol. 1 (2012), pp. 85–116.

Sanders, Nancy K., *The Sea Peoples: Warriors of the Ancient Mediterranean, 1250–1150 B.C.* (London: Thames and Hudson, 1985).

Säve-Söderbergh, Torgny, *The Navy of the Eighteenth Egyptian Dynasty* (Uppsala: Uppsla Universitets Arsskrift, 1946).

Shaw, Ian, 'Socio-economic and Iconographic Contexts for Egyptian Military Technology: The Knowledge Economy and "Technology Transfer" in Late Bronze Age Warfare,' in M. Wissa (ed.), *The Knowledge Economy and Technological Capabilities: Egypt, the Near East and the Mediterranean 2nd millennium B.C.—1st millennium A.D. Proceedings of a Conference Held at the Maison de la Chimie Paris, France 9–10 December 2005, Supplementa* (Barcelona: Aula Orientalis, 2010), pp. 77–85.

Spalinger, Anthony John, *Aspects of the Military Documents of the Ancient Egyptians* (New Haven: Yale University Press, 1982).

Spalinger, Anthony John, *War in Ancient Egypt* (Blackwell: Oxford, 2005).

Spalinger, Anthony John, 'Covetous Eyes South: The Background to Egypt's Domination over Nubia by the Reign of Thutmose III,' in Eric Cline and David O'Connor (eds.), *Thutmose III: A New Biography* (Ann Arbor: University of Michigan Press 2006), pp. 344–69.

Spalinger, Anthony John, *Icons of Power: A Strategy of Reinterpretation* (Prague: Charles University Press, 2011).

Spalinger, Anthony John, 'The Organisation of the Pharaonic Army (Old to New Kingdom),' in Juan Carlos Moreno García (ed.), *Ancient Egyptian Administration* (Leiden: Brill, 2013), pp. 393–478.

Spalinger, Anthony John, 'Simple Words, Simple Pictures: The Link Between the Snapshots of Battle and the War Diary Entries in Ancient Egypt,' in Jeremy Armstrong (ed.), *Circum Mare: Themes in Ancient Warfare* (Leiden: Brill, 2016), pp. 13–33.

Stager, Lawrence E., 'Port Power in the Early and Middle Bronze Age: The Organization of Maritime Trade and Hinterland Production,' in S.R. Wolff (ed.), *Studies in the Archaeology of Israel and Neighboring Lands in Memory of Douglas L. Esse* (Chicago: Oriental Institute of the University of Chicago, 2001), pp. 625–38.

Starr, Chester, *The Influence of Sea Power on Ancient History* (Oxford: Oxford University Press, 1989).

Torab, Magdy, 'A Geomorphological Map of the Ancient Branches of the Nile Delta,' *The Bulletin of The Society of Cartographers*, vol. 30, Part 2 (1996), abstract.

Trim, David J.B., 'Medieval and Early-Modern Inshore, Estuarine, Riverine and Lacustrine Warfare,' in D.J.B. Trim and Mark Charles Fissel (eds.), *Amphibious Warfare 1000–1700: Commerce, State Formation and European Expansion* (Leiden: Brill, 2007), pp. 357–419.

Trim, David and Fissel, Mark Charles (eds.), *Amphibious Warfare 1000–1700: Commerce, State Formation and European Expansion* (Leiden: Brill, 2007).

Van Peer, Philip, Yury Demidenko, Elena Garcea, Marcel Otte, Nicolas Rolland, Avraham Ronen, and Romuald Schild, 'The Nile Corridor and the Out-of-Africa Model: An Examination of the Archaeological Record,' *Current Anthropology*, vol. 39, no. 2, supplement (June 1998), S115–S140.

Vencl, Slavomil, 'War and Warfare in Archaeology,' *Journal of Anthropological Archaeology*, vol. 3, no. 2 (May 1984), pp. 116–32.

Vinson, Steve, *Egyptian Boats and Ships* (Princes Risborough, Bucks: Shire Publications, 1994).

Wachsmann, Shelley, 'The Ships of the Sea Peoples,' *The International Journal of Nautical Archaeology and Underwater Exploration*, vol. 11, no. 4 (1982), pp. 291–304.

Wachsmann, Shelley, *Seagoing Ships and Seamanship in the Bronze Age Levant* (College Station: Texas A&M Press, 1998).

Ward, Cheryl, 'Boatbuilding in Ancient Egypt,' in Frederick Hocker and Cheryl Ward (eds.), *The Philosophy of Shipbuilding* (College Station: Texas A&M University Press, 2004), pp. 12–24.

Ward, Cheryl, 'Evidence for Egyptian Seafaring,' in R. Bockius (ed.), *Between the Seas: Transfer and Exchange in Nautical Technology: Proceedings of the Eleventh International Symposium on Boat and Ship Archaeology, Mainz 2006* (RGZM–Tagungen: Mainz, 2009), pp. 9–16.

Ward, Cheryl, 'From River to Sea: Evidence for Egyptian Seafaring Ships,' *Journal of Ancient Egyptian Interconnections*, vol. 2, no. 3 (2010a), pp. 42–9.

Ward, Cheryl, 'Supersized Egyptian Ships,' *International Journal of Nautical Archaeology*, vol. 39, no. 2 (2010b), 387–435.

Ward, Cheryl, 'Building Pharaoh's Ships: Cedar, Incense and Sailing the Great Green,' *British Museum Studies in Ancient Egypt and Sudan*, vol. 18 (2012), pp. 217–32. http://www.britishmuseum.org/research/online_journals/bmsaes/issue_18/ward.aspx

Ward, Cheryl, and Zazzaro, Chiara, 'Evidence for Pharaonic Seagoing Ships at Mersa/Wadi Gawasis, Egypt,' *International Journal of Nautical Archaeology*, vol. 39, no. 1 (March 2010), pp. 27–43.

Wood, Adrian, *Warships of the Ancient World 3000–500 BC* (Oxford: Osprey, 2012).

15

CHINA'S PIRATE WARS

1522–1810

Robert J. Antony

Although pirates have been active in the South China Sea throughout history, the golden age of Chinese piracy began in the sixteenth century and lasted until the early nineteenth century. Over those three centuries there was an unprecedented growth in Chinese piracy unsurpassed in size and scope anywhere else in the world. Whereas during the golden age of piracy in the West between 1650 and 1720 the number of pirates had never exceeded 5,500, in China pirates numbered in the tens of thousands. China's golden age of piracy swelled in three great waves: first there were the merchant-pirates of the mid-Ming period from 1522 to 1574; this period corresponded to what is generally referred to as the period of *wokou* piracy. They were followed by bands of rebels, merchants and pirates during the Ming-Qing transition between 1600 and 1684. During this cycle piracy was symptomatic of the political anarchy, economic instability and social dislocations of the era. Many of the 'sea rebels' (*haikou*) combined commerce with piracy and insurgency. During the third wave, in the mid-Qing period from 1780 to 1810, once again there were several large pirate leagues. Chinese piracy had reached a crescendo during the third great wave, when over 70,000 poor fishermen and sailors engaged in piracy as a means of survival. All three great pirate cycles were characterised by the rise of huge leagues whose power surpassed that of the imperial state in China's maritime world. Never before in history had piracy been so powerful and menacing.

This chapter examines China's pirate wars between 1522 and 1810. According to Kenneth Swope, warfare and violence were central features of this period, and wars against pirates, bandits and rebels were always at the forefront of the government's concerns (Swope 2020: 119–37). During these three centuries, imperial China's naval strategy was markedly conservative; its aim was coastal defence and protecting coastlines and hinterlands, rather than offensive campaigns at sea. In fighting pirates, however, Ming and Qing navies faced intractable systemic problems. Even at the best of times, the military establishment was hard pressed to combat piracy. Imperial forces were neither structurally nor technically equipped to handle the large-scale piracy of this period. Military strategy was decisively land centred, defensive and highly localised. It consisted mainly of constructing and manning guard posts, batteries, watch towers and signal posts at intervals along the coastline, as well as maintaining small flotillas of warships for coastal patrols. In effect, the military installations and coastal patrols marked the limits of China's sovereignty on the seas. The defensive land-centred

DOI: 10.4324/9780429437915-20

strategy precluded building a blue water navy capable of operating effectively on the high seas. This, of course, was very different from European states, which roughly at the same time were earnestly building ocean-going navies to protect their merchant fleets from pirates and rival countries in waters far away from home. In Late Imperial China, because military campaigns alone were never sufficient to eliminate piracy, they had to be coupled with various other measures—embargoes, imperial amnesties, appeasements, forced coastal evacuations, privateer navies and various harsh judicial processes.

Pirate Wars of the Mid-Ming Dynasty

The golden age of Chinese piracy started in the middle of the Ming Dynasty during the reign of the Jiajing Emperor (r. 1522–1566). Beginning in the early 1520s, China witnessed a steady growth in the number of piracies along the coast. Within two decades numerous small gangs of pirates had expanded into larger, better organised fleets. Cresting in the 1550s, at a time when the Ming Navy was in decline and the dynasty was facing a serious threat on the northern frontier from resurgent Mongol armies under Altan Khan, pirates presented a grave challenge to imperial authority in the southern coastal provinces. At the height of their power they even threatened the major cities of Nanjing, Suzhou, Hangzhou and Guangzhou. This was also a time of civil war in Japan, when sailors and masterless *samurai* took to the seas as raiders. Characterised by Ming officials as 'dwarf outlaws' (*wokou*), a term used pejoratively for Japanese pirates, they in fact included virtually all pirates operating in China's waters. *Wokou* bands were composed of motley crews of Chinese, Japanese, Southeast Asian, European and even African rovers and renegades.[1] Over 80 per cent of the pirates, in fact, were Chinese. Using swift vessels, they plundered shipping and pillaged ports from China to Malaya (So 1975).

A major underlying reason for the sudden upsurge in piracy in the 1520s was the Jiajing Emperor's determination to enforce rigidly the existing sea bans and to enact tough new ones. The imperial court outlawed all private overseas trade and instead hopelessly attempted to restrict maritime commerce within the narrow confines of the tributary system. Anyone caught building large ocean-going junks, trading with foreigners, travelling abroad without authorisation or colluding with smugglers was to be treated as a pirate and when caught executed. Instead of curbing illegal activities, however, the bans actually furthered them. Because tribute missions were infrequent and the amounts of imports and exports severely limited, they satisfied neither Chinese nor foreign merchants. As a result, illicit trade quickly expanded along the coast to meet the growing demands and increasingly more people on sea and shore came to depend on illicit trade for their livelihoods. Piracy became the most vivid expression of opposition to official maritime policies and the most important means of conducting seaborne trade in this era (Antony 2003: 22–3).

The Jiajing Emperor's enforcement of the sea bans criminalised large segments of the maritime population. Gradually all strata of coastal society, from fishermen and sailors to merchants and gentry, became involved in illegal enterprises. Pirate gangs received protection from influential families and local officials, while many merchants and gentry financed the activities of pirate bands. One of the striking features of Chinese piracy in this era was that the most powerful leaders, including Xu Dong, Wang Zhi, Hong Dizhen, Wu Ping and Lin Daoqian, all had merchant backgrounds. These merchant-pirates mixed trade with smuggling and pillaging, organised large fleets and established bases on offshore islands, such as Shuangyu near the walled city of Ningbo. On these island strongholds they traded silks, brocades, pearls, copper, porcelains, spices, Japanese swords, firearms and various foreign items from such places as Patani, Malabar, Europe and Japan. Table 15.1 outlines the distribution of major pirate bases on the south China coast between 1530 and 1661. By the 1550s the wokou disturbances had greatly intensified along the shores of Zhejiang and Fujian, and in the following decade had spread further southward into Guangdong waters. Numbering in the tens of thousands, pirates plundered coastal shipping, pillaged villages and market towns, and attacked government garrisons with impunity (Antony 2003: 24–7; Chin 2010: 46–8).

Jiajing's harsh anti-maritime policies, however, were not sufficiently backed up with a strong military presence along the coast. In fact, just as *wokou* raids were intensifying, Ming naval capacity hit an all-time low. There was indeed something of an anti-martial attitude among court officials, with a stress of civilian over military talent in the selection of officers. According to Tan Lun (1520–1577), a high-ranking naval officer in Zhejiang, most officers were men lacking military ability who had purchased their positions in expectations of making huge profits by trading with pirates and smugglers. Because soldiers were poorly trained and underpaid, their morale was low and they were incompetent in combat. What is more, guard units were deployed in scattered fragments, each one numbering anywhere from 70 to 1,200 soldiers. Another official named Bu Tatong (1509–1555) reported that over half of the coastal defence forces in Fujian in the early 1550s had deserted and abandoned their

Table 15.1 Distribution of Major Pirate Bases, 1530–1661

Years	Bases	Location	Leaders
1530–1548	Shuangyu	Zhoushan	Xu Dong (1530–1548)
1548–1552	Liegang	Zhoushan	Wang Zhi (1548–1551), Xu Hai (1552)
1552–1555	Zhelin	Shanghai	Xu Hai (1552), Wang Zhi (1554), Lin Bichuan (1553–1555)
1557	Cengang	Zhoushan	Wang Zhi (1557)
1555–1558	Wuyu	Amoy	Hong Dizhen (1555–1558)
1558–1560	Yuegang	Zhangzhou	Hong Dizhen (1558), Xu Xichi (1558–1560)
1559	Haitanshan	Putian	Hong Dizhen (1559)
1559–1569	Meiling	Zhangzhou	Wu Ping (1559–1564), Xu Xichi and Zeng Yiben (1567–1569)
1564–1571, 1605–1661	Nan'ao	Shantou	Wu Ping (1564), Lin Daoqian (1566), Yang Zhi (1571), Zheng Zhilong (1605–1661)
1568–1575, 1620–1629	Raoping	Chaozhou	Lin Daoqian (1568–1574), Lin Feng (1573–1575), Zhu Cailao (1620–1629)
1570–1574	Chenghai	Chenghai	Zhu Liangbao (1570–1574)
1574	Penghu	Taiwan	Lin Feng (1574)
1605–1625	Beigang	Taiwan	Yan Siqi (1605–1625)
1615–1635	Jieshi	Shanwei	Liu Xiang (1615–1635)
1550s–1620s	Matsuura	Satsuma	Wang Zhi (1551–1552), Li Dan (1605–1625), Yan Siqi (1605–1625)
1555–1556	Goto	Satsuma	Wang Zhi (1555–1556), Xie Lao (1556)

Source: Ho 2011: 95–6.

Table 15.2 Bu Tatong's Account of Coastal Defence Units in Fujian, ca. 1550

Defence unit	Original number of troops	Number of missing troops	Current number of troops
Fenghuo	4,068	3,000	1,068
Xiaocheng	4,402	2,383	2,019
Nanri	4,700	2,557	2,143
Wuyu	3,427	1,468	1,959
Tongshan	1,822	1,192	630
Xuanzhong	1,133	476	657

Source: So 1975: 139.

posts (see Table 15.2). To meet shortfalls, regular soldiers had to be supplemented with mercenaries and local commanders frequently had to enlist fishing and trading junks as privateer fleets to reinforce the regular navy. In 1564, for example, the regional commander in Guangdong, Yu Dayou, enlisted some 300 Portuguese soldiers in Macao and leased a number of private junks to help put down a local disturbance and to fight the pirates. Several years later when the pirate Zeng Yiben attacked Guangzhou he was repulsed by the combined force of Portuguese mercenaries from Macao and local Chinese privateers. Manpower deficiencies aside, there also was the persistent problem of insufficient armaments and warships (So 1975: 135–8; Hucker 1974: 285–7; Cheng 2013: 15–16).

In 1547, the Jiajing Emperor appointed Zhu Wan (1494–1550) as Grand Coordinator in charge of all coastal military affairs in Zhejiang and Fujian, the areas most ravaged by pirates. After much wavering and indecision, finally coastal defence had become a high governmental priority. After taking up his new post, Zhu Wan discovered that pirates, Portuguese and Japanese traders, and powerful local gentry families were actively collaborating with one another in smuggling and pirate enterprises. He immediately set into motion a series of new measures to enforce the sea bans to eliminate the pirate trade, and carried out several vigorous military campaigns against pirates. By effectively deploying available troops against illegal trading centres in both provinces, within a year Zhu Wan's policies began to show great success in disrupting the infrastructure of the clandestine trade. Most significantly, Zhu Wan completely destroyed the *wokou* entrepôt on Shuangyu Island in 1548. In the next year at Zhoumaxi his army attacked and captured a major pirate leader, Li Goutou (Dog Head Li) and 96 followers, who were summarily executed. Unsurprisingly, Zhu Wan's policies were unpopular among the prominent coastal families and many local officials who had vested interests in maintaining the status quo. Ultimately, they succeeded in having the emperor impeach Zhu Wan, who, in disgrace, committed suicide in 1550 (Higgins 1981: 10–13, 139–41; Ho 2011: 86–7).

After Zhu Wan's death, anti-piracy campaigns wavered and piracy became more rampant than ever. Wang Zhi, who hailed from a prominent family of salt merchants in Anhui, became the most powerful pirate. During the 1550s he commanded a huge, well-armed merchant fleet that operated from bases in Kyushu. In particular, he became active in the illicit gunpowder and firearms trade between Japan, China and Southeast Asia and, as several historians have noted, he was instrumental in the diffusion of modern weaponry into Japan (Petrucci 2010: 63–4; So 1975: 148; Chonlaworn 2017: 190–1). By that time, too, his syndicate had been so successful that he was able to once again establish fortified bases on the coast and offshore islands, and even to offer protection to Chinese, Portuguese and Japanese traders operating illegally in China. Following repeated defeats between 1553 and 1554, imperial troops refused to engage pirates and remained inside their protected forts, thereby leaving the countryside open to pillage. By 1555, Wang Zhi commanded hundreds of ships and several thousand men armed with modern guns and cannons. He appeared to be invincible (Zheng [1562] 2007: 3: 1–8; Antony 2007: 108–9).

Although the government preferred military campaigns of 'annihilation' (*jiaomie*) against pirates and rebels, such solutions were not always successful or feasible. So officials frequently and simultaneously adopted another approach called 'pacification' (*zhaofu*), whereby dissidents who surrendered were offered imperial pardons, monetary rewards, official titles and government offices. Pacification was actually an age old policy that dated back to the Han dynasty. Under the Ming Dynasty, pacification policies were occasionally used in dealing with *wokou* pirates. According to Charles Hucker, Chinese governments had traditionally responded to serious hostilities in two ways: a straightforward military solution called extermination or an indirect politico-economic solution called pacification, suggesting the 'summoning and appeasing' of dissidents. Pacification, nonetheless, was always supported by real, though muted, threats of military action. Although a controversial policy, pacification was 'a normal means of coping with the disaffected' (Hucker 1974: 274).

Hu Zongxian (1512–1565), a key official in Zhejiang and Fujian in the late 1550s, effectively employed annihilation and pacification policies in dealing with several powerful pirate leaders, such as Xu Hai and Wang Zhi. While Hu favoured a policy of reinstating maritime trade to help restore the coastal economies and at the same time alienate the poor from joining the pirates, the emperor favoured continuation of rigid sea bans and the prosecution of an aggressive war against pirates. Attempting to sidestep court policies, Hu conducted military campaigns, while at the same time negotiating with several pirate leaders in an effort to cause dissention within their ranks. He also began to cooperate with well-connected local gentry and merchant families, including several of their family members into his inner circle of advisers. In dealing with the pirate chief Xu Hai, Hu Zongxian manipulated the pirate boss through the skilful combination of tricks, bribes, negotiations and military campaigns that eventually ended in Xu's death in battle in September 1556. Similarly, Hu sent personal envoys to Japan to coax Wang Zhi into surrendering in return for a pardon and promises of a military commission. After prolonged negotiations, finally in 1557 Wang Zhi surrendered to Ming officials in Hangzhou expecting to receive an imperial pardon, but after allowing him to languish in prison for over a year the throne decided to execute him in 1560. After Xu Hai and Wang Zhi had been defeated, the government pacified a number of their subordinates with pardons and naval commissions; later they proved instrumental in defeating pirates based in the Zhoushan Archipelago. Gradually, by 1559, the *wokou* disturbances in Zhejiang had subsided and the remaining pirate bands scattered further south along the Fujian and Guangdong Coasts (Hucker 1974: 273, 287–9; So 1975: 107–10; Fitzpatrick 1979: 12–13, 20–3).

Little by little, a number of military officers and troops became more accustomed to warfare and developed successful techniques and strategies for defeating the pirates. Among the most successful of these new leaders was Qi Jiguang (1528–1588), who had been born into a prominent military family in Shandong. After soundly defeating pirates on the Shandong Coast in 1553, he was quickly transferred to fight pirates in Zhejiang. Together with two other capable commanders, Yu Dayou (1503–1579) and Tan Lun, he led their troops to a decisive victory at Cengang in 1558. With the situation now under control, Qi concentrated on training and disciplining his troops into an effective army. Rather than relying on the regular army or mercenary troops, he recruited stalwarts from local communities, thereby forging peasants and miners into well-trained soldiers who willingly defended their homeland. It is important to note that Qi also was one of the first commanders to organise an elite corps of musketeers trained in the volley technique, which proved effective in fighting pirates in the1560s. At the same time, he also oversaw the construction of a new fleet of over 40 warships, which he equipped with cannon, fire-arms, rockets and flamethrowers, as well as the building of new coastal fortifications and city walls. Once the *wokou* were defeated in Zhejiang, Qi turned his attention to Fujian. In a series of brilliant campaigns on sea and land between 1562 and 1565, his elite forces eliminated the pirate menace in the province (So 1975: 146–50; Swope 2017: 170–1; Andrade 2017: 79–80, 87).

Wokou raids, however, continued along the Guangdong Coast until the end of the century, but after 1570 the number of incidents dramatically decreased. It had taken Ming authorities over 20 years to bring the *wokou* disturbances under control. After the Jiajing Emperor died in 1567, his successors slowly and cautiously began to open up overseas trade with Southeast Asia but not with Japan.[2] Several major pirate leaders also were killed in battle, surrendered to the state or fled to Southeast Asia. In addition, the political reunification of Japan under the Tokugawa *Shogunate* in the latter part of the sixteenth century did much to curtail Japanese piracy in the whole region. In 1574, a Ming Fleet chased the last of the *wokou* pirates all the way to the Philippines. As China's economy finally stabilised in the last decades of the sixteenth century, large-scale piracy diminished for the time being (Antony 2003: 27–8).

Pirate Wars of the Ming-Qing Transition

After a hiatus of roughly 50 years, a new wave of large-scale piracy surged forth during the Ming-Qing transition between 1600 and 1684, with piracy reaching a peak in the 1640s to 1660s. Often characterising pirates in official accounts as 'sea rebels,' the pirate upsurge was symptomatic of the general crisis in China which accompanied the change of dynasties. Given the economic and political anarchy of this period, a clear distinction between piracy, rebellion and trade was impossible. As Chen Shunxi noted in his diary in 1679, in his home area in Western Guangdong, soldiers, local militias, pirates and bandits were all fighting one another and indiscriminately pillaging villages and towns (Chen 2010: 45). The Zheng family, taking advantage of the instability, built a piratical empire in the South China Sea based on combinations of trade, piracy and political manipulation. Other pirates joined forces with bandits and rebels to attack markets and walled cities. European traders, with the continued support of their governments, also took advantage of the turmoil in China to pillage ports and merchant ships (Cheng 2013; Hang 2015).

The piracy of this period resulted from a combination of adverse political, military, economic and ecological factors. By the start of the seventeenth century, the relative stability and prosperity of the previous several decades had been abruptly shattered. Externally, Manchu incursions on the north-eastern frontier challenged Ming sovereignty and forced the government to commit large amounts of money and troops to defend the border, thereby weakening its military presence along the coast. Internally, corruption, factionalism, incompetence and fiscal bankruptcy crippled the Ming state. Except for a brief period between 1631 and

1632, after 1628 the Ming government once again prohibited overseas trade. At the same time, the Chinese economy stagnated and then declined, as foreign trade came to a standstill and prices skyrocketed. As the military and economic crisis deepened, social unrest increased; there were numerous tenant and worker revolts, food riots and bandit disturbances throughout the southern provinces. For several decades, Fujian and Guangdong found themselves at the centre of fighting, pillaging and devastation. Dispossessed by one side or another, people did not have time to recuperate before soldiers or pirates came back to raid again. Local society had become highly militarised, as communities built stone blockhouses and formed militias armed with firearms and cannons (Antony 2003: 28–30).

At this time too, Europeans were aggressively waging wars with one another across the globe, and as their conflicts spread into Asian waters, they also plundered Asian ships and ports. By the early seventeenth century, the Dutch and British had replaced the Portuguese and Spanish as the dominant Western powers in Asia, and like their predecessors they also combined trade with piracy. Their main objective was to disrupt the Chinese junk trade with Spanish-controlled Manila. Encouraged by their sovereigns, Dutch and British adventurers took advantage of the political vacuum in China to plunder Chinese junks and occasionally even coastal settlements. In 1622 the Dutch attacked Macao and then seized the Pescadores, looting and burning villages and kidnapping hundreds of people to sell into slavery. In 1624 they occupied Taiwan, and from this base conducted trade and piratical expeditions until Zheng Chenggong (Koxinga) expelled them in 1661. The Dutch actually obtained much of the silks and chinaware so vital to their trade with Japan through plunder. European freebooters continued to seize Chinese and other vessels in the South China Sea into the early eighteenth century. Their presence contributed further to destabilising the coast, not only because they intercepted ships trading with China, Japan and Southeast Asia, but also because they supplied weapons to pirates and fenced their booty. Pirates, therefore, never lacked the money, supplies and weapons necessary for carrying on their activities (Antony 2007: 38–9).

Events of the first half of the seventeenth century, in many respects, represent continuations of unresolved issues of the previous century. As in the earlier period of *wokou* disturbances, the most skilful pirates generally associated with smuggling syndicates that conducted trade throughout the South China Sea. For example, in the first 30 years of the seventeenth century, over 10 pirate-merchant fleets operated, competed and coexisted in the waters of South-Eastern China. They alternated plundering one another and merchant ships with business ventures to Nan'ao, Penghu, Taiwan, Macao, Japan and the Philippines. However, unlike the earlier *wokou* period, the large-scale piracy of the early seventeenth century focused more on plundering ships at sea than coastal settlements.

The new Qing state inherited much of the previous Ming political, legal and military institutions with all their strengths and weaknesses. During their conquest of China, the Manchus focused mainly on land warfare and paid far less attention to naval affairs. Although they had inherited the Ming Navy, it remained undermanned and neglected. While the navy was relatively effective at keeping the peace in inland waterways, it was ineffective at coastal security. When faced with large pirate bands, officials often preferred to incorporate them into the navy rather than to fight them. Large numbers of warships remained unseaworthy and obsolete—as late as the nineteenth century, warships still were modelled after those of the Ming Dynasty. As in the past, great importance was placed on defending the maritime zone with garrisons and fortifications on land (Lee 2017: 239–42; Lococo 2012: 125).

Beginning in the 1620s, the Zheng family, first under Zheng Zhilong and then under his son Zheng Chenggong, began building a sizable piratical empire that within 20 years came to dominate trade in the South China Sea. Zheng Zhilong, who began his career as a merchant, smuggler and pirate in Southern Fujian, displayed remarkable organisational skills and an uncanny knack for manipulating officials. For a time he lived in Macao, where he converted to Christianity; later he went to Manila, then Nagasaki, and Taiwan. Over many years he slowly built up a huge network that included not only his relatives and fellow countrymen, but also many Japanese and European traders and officials. After repeatedly defeating imperial fleets sent out to destroy him, in 1628 he surrendered to the Ming Court in exchange for a pardon and commission as a 'patrolling admiral' (*youji jiangjun*) in the imperial navy with orders to clear the coast of pirates. At that time, he had a fleet of about 1,000 ships and commanded over 20,000 fighting men. He also had an elite corps of 300 to 600 musketeers, composed largely of Africans who wore European uniforms and trained in Dutch drilling techniques. As a Ming official, Zheng now was able to legitimately and systematically exterminate his rivals, such as Liu Xiang, whom he defeated off the Leizhou Peninsula in 1635 (*Haikang xianzhi* [1938]1974, 44: 21b). By then, Zheng already dominated the lucrative Fujian–Taiwan trade network from strongholds on Xiamen (Amoy) and neighbouring islands. He levied protection fees on merchant junks and plundered those vessels that had refused to pay. Reportedly, too, many high-ranking officials in Fujian and even in the imperial court were on his payroll (Shao 1961: 131; Cheng 2013: 44–7, 73–5; Hang 2015: 51–7, 87–8). By the late 1630s, Zheng had eliminated all his rivals and had become so powerful that he was like 'a whale swallowing up the sea' (*Ming Qing shiliao wubian* 1972: 7b).

With the fall of the Ming Dynasty in 1644, Zheng Zhilong vacillated for another two years before finally surrendering to the Manchu conquerors. Unfortunately, Zheng had misjudged the Qing ruler who placed him under house arrest in Beijing until 1661, when he was finally executed. Many of his clansmen, including his son, Zheng Chenggong, however, continued to resist the Qing in the name of Ming loyalism. Taking advantage of the political turmoil, Zheng Chenggong expanded his power base in South China, and by 1651 he commanded the Zheng family organisation. From dozens of bases along the Southern Coast of China (refer to Table 15.3), for the next ten years his fleets monopolised shipping in Fujian, Guangdong, Taiwan, the Philippines and much of South-East Asia. Zheng financed his huge maritime empire through trade, robbery and extortion. At his peak he commanded over 150,000 soldiers, had a navy of some 5,000 ships, and a trading conglomerate valued at roughly 4 million taels of silver. Under the capable leadership of Zheng Zhilong and Zheng Chenggong, what began as a disparate band of pirates and smugglers was gradually transformed into a huge multinational commercial network and an informal state. They were among the richest and most powerful men in the world (Hang 2015: 41, 74, 78, 84, 88–9).

The Qing Navy had great difficulties overcoming the Zheng forces, which had overwhelming naval strength. By 1655, after several failures to get Zheng Chenggong to surrender, his forces began to expand northward from bases in Fujian into Zhejiang. The next year he destroyed a Qing naval fleet off Quemoy (Jinmen Island), prompting the government to adopt a less aggressive strategy. On the one hand, the Court reintroduced a series of harsh sea bans, yet on the other, also offered amnesties and rewards to pirates who surrendered. This was the age-old strategy of divide and conquer, which slowly began to show results, as a number of Zheng's lieutenants 'returned allegiance' to the Qing emperor. Failing to take

Robert J. Antony

Table 15.3 Zheng Chenggong's Key Maritime Bases, ca. 1658

Type	Base Area	County/District	Province
Island	Xiamen (Amoy)	Tong'an	Fujian
Island	Jinmen (Quemoy)	Tong'an	Fujian
Coastal	Haicheng	Haicheng	Fujian
Coastal	Zhangpu	Zhangpu	Fujian
Coastal	Tong'an	Tong'an	Fujian
Coastal	Zhao'an	Zhao'an	Fujian
Interior	Pinghe	Pinghe	Fujian
Interior	Changtai	Changtai	Fujian
Island	Nan'ao	Chenghai	Guangdong
Coastal	Chaoyang	Chenghai	Guangdong
Coastal	Jieyang	Jieyang	Guangdong
Coastal	Tiantai	Tiantai	Zhejiang
Coastal	Taiping	Wenling	Zhejiang
Coastal	Haimenwei	Haimenwei	Zhejiang
Coastal	Panshiwei	Panshiwei	Zhejiang

Source: Ho 2011: 145.

Nanjing after a prolonged siege, Zheng retreated back to his base in Xiamen. Then in June 1660 over 1,000 Manchu and Zheng warships clashed at Amoy Bay in one of the most brutal and bloody naval battles ever fought in East Asia. An estimated 10,000 soldiers on both sides died. Not long afterwards, in 1661, Zheng withdrew to Taiwan, where he drove the Dutch away. After his unexpected death six months later, his heirs continued to resist the Manchus until Shi Lang, a subordinate of Zheng Chenggong who had defected in 1646, seized Taiwan for the Qing. In July 1683, after extensive preparation, Shi Lang led a force of 300 warships and 20,000 soldiers from Fujian, soundly defeating the Zheng forces in a battle off the Pescadores. Two months later, Zheng Keshuang, the grandson of Zheng Chenggong, surrendered and Taiwan was incorporated into the Qing Empire (Lorge 2012: 94; Cheng 2013: 202; Hang 2015: 115–24, 132–43).

In the meantime, after the Ming Dynasty collapsed, anarchy quickly spread everywhere and piracy escalated out of control. The new Manchu rulers responded with even harsher sea bans than their Ming predecessors. Imperial decrees commanded officials in coastal areas to burn all boats, to prohibit the construction of large sea junks, and to bar the purchase of foreign made vessels. Chinese merchants were also prohibited from setting out to sea under pain of death. Still unable to curb piracy, beginning in 1661, the government adopted a scorched earth policy, forcing all residents along the coast from Shandong to Guangdong to relocate inland at a distance of 10 to 15 miles. All houses and property within that no-man's zone were destroyed and anyone caught trying to return to the coast was beheaded. A proclamation posted in Chenghai County, Guangdong, in March 1662 explained this gruesome measure:

All villages outside the boundary [i.e., fifteen miles from the coast], all places on the seacoast, must obey and move inland. . . . All villages and residents outside the boundary, hearing this, must spread the order and move immediately inside the border; no

250

hesitation or resistance will be tolerated. Once you have removed' behind the border-line, you shall not cross it to farm the land. You shall not cross it to build houses in which to live. Violators will be executed for treason. . . . If you commoners linger, delay, or wait and see, you will be exterminated as rebels.

(Ho 2011: 204–5)

By the time the coastal depopulation programme ended in 1683 hundreds of thousands of people had been displaced, at least 200,000 people had died, and about a million acres of land had been abandoned and laid to waste (Ho 2011: 16, 129, 220–3, 255, 282).

Such draconian measures did not destroy the pirates, but rather devastated the maritime communities which depended on the sea for their livelihood. Those policies, in fact, drove many people into piracy and open rebellion. In Guangdong, two brothers, Su Cheng and Su Li, who had been smugglers, organised a fleet of pirates whose stated purpose was the establishment of a new dynasty. Although some displaced farmers had joined their cause, most of their followers were fishermen and smugglers. Before its collapse in 1664, the Su organisation controlled a 300-mile stretch of coastline between Haifeng and Chaozhou. To the south around Guangzhou, in 1663 and 1664, Zhou Yu and Li Rong led bands of displaced Dan (Tanka) fishermen in a pirate uprising that shook the economic core of the province (Antony 2003: 33–4). In the area along the Sino-Vietnamese border, other Ming loyalists and pirates, such as Yang Yandi and Chen Shangchuan, continued to resist the Manchus and engage in acts of piracy for nearly 20 years from 1663 to 1681 (*Haikang xianzhi* [1938]1974, 45: 4a–6b; also see Antony 2014a; and Hang 2017). Perhaps the actions of these rebel pirates were motivated less by Ming loyalism than by survival instincts in the face of the harsh Manchu policies. Above all else, the stringent Qing policies and protracted wars totally disrupted the flow of trade, which during these years moved away from China's coastal ports to Taiwan, Macao and to other trading centres in Asia.

Other significant changes were at hand with regards to piracy and pirate suppression in China and elsewhere around the globe. Between 1680 and 1730, attitudes and public policies toward piracy and trade had changed dramatically in both China and the West. In 1684, just one year after the Qing military had crushed the remnants of Zheng's heirs on Taiwan and had finally secured control over all of China, the Kangxi Emperor (r. 1662–1722) rescinded most of the sea bans. Convinced that national security depended on the prosperity and stability of the southern coastal provinces, the imperial court legalised the overseas junk trade and opened up several ports to foreign commerce.[3] An interconnected network of customs offices, sub-county yamens and military posts was established, or re-established, all along the southern littoral, and the civil officials and military officers who manned these posts had the specific tasks of regulating trade and curbing piracy and smuggling. In the 1720s the Qing government also began enacting a series of laws to protect private property, including harsh new laws against piracy. The new measures brought stability and prosperity to coastal South China, and piracy declined. Thereafter, most Chinese gentry and sea merchants, who now had a vested interest in maintaining the existing system, became bulwarks of support in pirate suppression. It was also about the same time that Western merchants began putting pressure on their home governments to suppress piracy and officials responded by passing tough anti-piracy laws and by building navies to protect their merchant ships on the high seas. Western piracy soon waned in Asia as it did elsewhere around the globe (Antony 2003: 165).

Pirate Wars of the Mid-Qing Period

The next great pirate wave, in the late eighteenth and early nineteenth centuries, belonged to an age of indigenous piracy throughout Asia. From Malabar to the Malay Archipelago to the South China Coast, piracy was once again on the rise. Between 1780 and 1810, several competing pirate leagues, composed of self-contained fleets that functioned independently of one another, infested the South China Sea. Throughout those years, bands of pirates repeatedly plundered native and foreign shipping, pillaged coastal villages, markets and ports, and exercised hegemonic power over maritime society through highly institutionalised systems of protection rackets. Once again the South China littoral had slipped away from the control of the imperial state into the hands of huge pirate leagues (Murray 1987; Antony 2003).

The intensification of piracy along the South China Coast at that time resulted from several factors. A combination of changing socio-economic conditions, especially China's population explosion and the increased trade in Southeast Asia and the West, were the most important underlying causes. The expansion and prosperity of commerce in the entire region acted as a catalyst for maritime predation among the poverty stricken fishermen and sailors living in Southern China. Paradoxically this was an 'age of prosperity,' but one in which wealth was unevenly distributed. Despite the flourishing economy, population pressure intensified competition and kept wages low for most seafarers. It is hardly surprising that the majority of pirates in this era were common sailors and fishermen who engaged in occasional piracy in order to survive in an increasingly harsh and competitive world. Piracy arose not because of a general immiseration of society, but rather because of the strains that prosperity had placed on the more marginal elements of Chinese society. The pirates of this era fought not to destroy commerce but rather to gain a more equitable share in it (Antony 2003: 54–81).

Piracy was one of the many social disorders to plague China during this period. After a century of relative calm, in the 1780s large-scale piracy reappeared in the South China Sea when Tay Son rebels in Vietnam began to sanction raids into Chinese waters to bolster their revenues and increase their manpower with the addition of Chinese pirates. Although not the cause of the upsurge in piracy, the Vietnamese rebellion significantly contributed to its expansion and professionalisation. Rebel leaders actively recruited bands of Chinese outlaws, supplied them with warships and provisions, provided them with safe harbours, and rewarded them with official ranks and titles so that they would engage in piracy as a means of obtaining revenues for their cause. Pirate fleets, numbering as many as 200 vessels, set out every spring and summer from bases along the Sino-Vietnamese border for Chinese waters, and returned each autumn laden with booty that they shared with their Tay Son patrons. Their most valued targets were large Chinese trading junks and Western merchantmen; in 1793, for example, pirates captured the Portuguese ship *Flore do Mar* off Macao Roads, killing all but four of its crew. Table 15.4 sketches pirate attacks against foreigners between 1793 and 1810. Besides providing Chinese pirates with safe havens, ships and weapons, Tay Son leaders also offered them a semblance of legitimacy and respectability. Sponsorship transformed pirates from being mere robbers and outlaws into legitimate naval forces (Antony 2014b).

With the collapse of the Tay Son Rebellion in 1802, Chinese pirates scattered back into Chinese waters and over the next several years recuperated and became stronger and better organised than ever. By 1805 several large pirate associations, totalling over 80,000 individuals, had emerged to dominate the maritime region from Zhejiang to Vietnam. Cai Qian and Zhu Fen were the most powerful pirate chiefs in Zhejiang, Fujian and Taiwan, while Wushi

Table 15.4 Pirate Attacks on Foreigners and Foreign Ships, 1793–1810

Date (mm/yyyy)	Place	Particulars	Consequences
–/1793	Near Macao	Pirates attacked the Portuguese ship *Flore do Mar*	All but 4 crew members killed
–/1796	Near Macao	A Portuguese ship plundered by pirates	Pirates killed all non-Chinese onboard
–/1796	Near Macao	Pirates attacked British ship *Kennett*	Most of the crew killed
–/1800	Whampoa	Pirates attempted to board British schooner *Providence*	Repulsed by the schooner's crew
–/1804	Taipa	Pirates occupied Taipa anchorage and threaten Macao	Reduced Macao's stock to two-day supply of rice
10/1804	Near Lintin	Pirates disrupted communications between Lintin and Macao	Difficulties in supplying foreign ships anchored at Lintin
–/1805	Near Macao	Pirates captured Portuguese brig returning from Manila	Several crew held captive for over a year
08/1805	Near Macao	Chop boat belonging to Dobell and Biddle plundered by pirates	Both men barely escaped but all their belongings taken
12/1806	Near Macao	John Turner and 5 Lascars captured by pirates and held for 5 months	Ransom valued at $6,000 paid to pirates
02/1807	Macao	Pirates attempted to land and attack Guia Fort	Repulsed by Macao soldiers
–/1808	Near Macao	Pirates attacked the British ship *Dover*	Ship escaped with little damage
–/1808	Near Macao	Pirates captured American schooner *Pilgrim* en route from Manila	8–9 crewmen held captive for several months
02/1809	Near Cheunpi	An officer and 2 sailors of the *Royal George* in the ship's yawl attacked by pirates	
02/1809	Near Macao	Pirates captured Portuguese-flagged brig	Pirates refitted brig for pirating; in September the brig was retaken
08/1809	Near Macao	Pirates attacked American ship *Atahualpa*	
09/1809	Near Macao	Pirates captured Richard Glasspoole and 6 sailors from the *Marquis of Ely*	Ransom of $4,200 and supplies paid to pirates
09/1809	Mouth of Pearl River	Pirates blockaded 3 Siamese tribute junks	
09/1809	Near Macao	Pirates captured brig belonging to Portuguese governor of Timor	
01/1810	Near Whampoa	Pirates attacked a small foreign boat	Three chests carried away; boat's crew thrown overboard; one *lascar* drowned

Sources: 'Chinese Pirates' 1834: 71, 79; Glasspoole 1831: 123; Morse 1926: 2: 422–5; 3: 7–8, 32, 63, 95, 116–17, 123; Andrade 1835: 34.

Er, Donghai Ba, Zheng Yi, and later his widow Zheng Yi Sao, and Zhang Bao dominated the waters around Guangdong and Vietnam. The reinvigorated pirates established new bases along the coast and outer islands, and soon extended their domination over most of the fishing and coastal trade, as well as over many villages and market towns through an extensive protection racket based on extortion, bribery and terror. Even Western merchants paid protection fees to pirates. For the next five years these formidable pirate leagues constituted a level of control over South China's maritime society that transcended that of the state and local elites (Morse 1926: 3: 8; Antony 2003: 118–20).

After 1805 the pirates, not the imperial navy, had military superiority at sea. For the most part, Qing warships were antiquated and cumbersome. They also were significantly fewer in number and greatly inferior to many of the swifter and heavily armed pirate warships. Although Fujian and Guangdong both had official quotas of 130 to 150 naval vessels, these numbers were rarely met. Ships destroyed by storms or pirates were infrequently replaced while others were in constant disrepair. After Nayancheng (1764–1833) arrived in Guangzhou as the new governor-general in 1805, he reported to the throne that there were fewer than 20,000 naval personnel and only 83 warships in service. Four years later, damage and destruction left the Guangdong Navy with fewer than 70 seaworthy warships. At the same time the pirates, who numbered in the tens of thousands, had hundreds of ships, some with crews of as many as 200 sailors. While naval cannons had shot weighing only one catty (*jin*), pirates had heavier shot of 14 catties (Wei 2006: 97–102; Chang 1983: 37, 131; Antony 1994: 21). In 1809, Captain Francis Austin of the H.M.S. *St. Albans* assessed Chinese naval armaments: 'At present their breastwork is too low, their guns not capable of being pointed but by moving the vessel, and so ill-cast and mounted as to be evidently unsafe to fire' (Murray 1979: 372). In battle after battle, the pirates outmanoeuvred and outgunned the imperial navy.

Because of intrinsic weaknesses, Qing officials routinely hired mercenaries and leased civilian Chinese and foreign ships to augment regular coastal and naval forces. Mercenaries were enlisted to man guard posts and forts wherever regular troops were deficient. In 1805 Nayancheng levied a fleet of 120 fishing and salt junks, promising their owners rewards of silver and official titles. Also that year, Portuguese officials in Macao outfitted two cruisers which accompanied 20 Chinese warships for patrol duty in the Pearl River Estuary; however, nothing came of this first Sino-Portuguese venture. Later in 1809, the new governor-general, Bailing (1748–1816), imposed a stringent embargo along the coast, prohibited merchant junks from leaving port, and hired over 60 fishing boats and hundreds of mercenaries to supplement the navy. Still unable to eliminate pirates in the Pearl River Estuary, Bailing turned to the British, who supplied the Chinese Navy with two vessels for patrolling the vicinity of Guangzhou. One of these ships, the *Mercury*, had 20 cannon and 50 American volunteers, and was successful in destroying eight pirate junks in one engagement. This, however, was a minor victory that did little to curb piracy. At the end of the year it still was apparent that there was little that the Qing Navy, even with the aid of mercenaries and a few armed foreign vessels, could do to eradicate piracy (Andrade 1835: 23; Chang 1983: 37; Antony 2006: 10).

Officials in Fujian and Guangdong therefore also encouraged the local communities to adopt self-defence measures to combat pirates. In Fujian, Governor Ruan Yuan (1764–1849) not only ordered the construction of new warships but also sanctioned the gentry in coastal communities to organise militia and to drill militiamen in the use of muskets (Wei 2006: 90–4). In Guangdong, both Nayancheng and Bailing exhorted village leaders to organise militia, recruit local stalwarts, build palisades, and procure firearms, cannons and other

Table 15.5 Pirate Invasion of the Pearl River Estuary, 1809

Date (mm/dd–)	Place	Particulars	Consequences
07/20–21	Weijiame	Zhang Baozai battles against Brigade-General Xu Tinggui	Pirate Zongbing Bao (White Flag) killed; Xu also killed; navy defeated, losing 25 war junks
08/06–07	Xiaolan and Hengdang	Guo Podai battles against Xiao Shitai	Villagers defeated
08/09–12?	Zini and Sanshan	Guo Podai attacks government installation and villages in Xiangshan	Pirates burn down customs house, blockade river, demand over $10,000 tribute from villagers, women and children held for ransom, several hundred villagers killed
08/16–26	Pingzhou, Mazhou, Shawan, Dieshidun and Huanglian	Guo Podai attacks villages in Panyu, Nanhai and Shunde	Pirates defeated by local militia and braves
08/16?	Xiangshan	Zhang Baozai attacks government post	Pirates destroy fort near Macao and drive Qing war junks further upriver to Huangpu
08/18–21	Baotangxia	Militia defeat pirates; over 100 pirates killed	
	Zhang Baozai attacks village in Dongguan		
08/23–24?	Laocun	Zhang Baozai continues attacks in Dongguan	Pirates defeated; over 100 pirates killed and 7 leaders captured
08/27?	Xintang	Zhang Baozai attacks village in Zengcheng	
09/08–10?	Jiaoxiang and Dafen	Zhang Baozai attacks delta islands and villages in Dongguan	Pirates kill several thousand villagers and kidnap women and children for ransom; Canton declares martial law
09/14	Sisha and Xinzao	Zhang Baozai attacks villages and markets in Panyu	Pirates defeated by local militia and braves; over 100 pirates killed
09/15	Near Huangpu	Zhang Baozai battles with two Portuguese war ships	Portuguese defeated
09/16–20	Pearl River delta	Naval battle with pirates	Combined English and Chinese forces defeat pirates
09/25–26	Shating	Zhang Baozai continues attacks in Panyu	Pirates burn down customs house, kill several hundred villagers, and kidnap 400 people; villagers pay $600 tribute to pirates

(Continued)

Table 15.5 (Continued)

Date (mm/dd–)	Place	Particulars	Consequences
10/01–03	Chencun	Zhang Baozai continues attacks in Panyu	Villagers resist but are defeated; over 250 women and children kidnapped for ransom; village burned down; pirates withdraw after 1,000 braves appear
10/05–08	Lanshi, Ganjiao, Beihai and Fojiao	Zhang Baozai continues attacks in Panyu and Nanhai	Villagers defeated; nearby military post burns down; women and children ransomed for over 15,000 taels silver
10/20–22	Near Shawan	Pirates battle against Commander-in-Chief Sun Chuanmou	Pirates defeat naval force
11/01–02	Dahuangpu	Zhang Baozai, Zheng Yi Sao and Guo Podai step up attack with over 300 junks	Naval forces and villagers defeated
11/04–05	Dayushan (Lantao)	Zhang Baozai and Zheng Yi Sao battle with combined Qing and Portuguese force	Pirates defeat combined forces
11/18–28	Near Dongyong, near Dayushan	Combined Chinese and Portuguese naval forces battle with pirates	Pirates defeat combined forces

Sources: GZD (14559) JQ14.06.19, (15184) JQ14.08.23; *Dongguan xianzhi* [1921]1968: 33: 25b–26a; *Xiangshan xianzhi* [1873]1961: 6: 79b, 8: 58b; *Shunde xianzhi* [1853]1974: 27: 4a–b, 20b, 31: 18b–19a; *Panyu xianzhi* 1871: 22: 16b–17a; *Nanhai xianzhi* [1872]1967: 14: 21b–22a; Yuan 1830: 1: 13a–24a; Andrade 1835: 34–45.

weapons. As Peng Zhaolin, the magistrate of Xiangshan County from 1805 to 1810, explained: 'Since villages along the coast are crowded like the scales on a fish, soldiers have difficulty guarding everywhere. If the people do not defend themselves, how can they not avoid calamity?' (*Xiangshan xianzhi* [1873]1961: 2: 39a–b). In and of themselves, these community-based self-defence efforts did not militarily put a stop to large-scale piracy. They were important, nonetheless, because they helped to isolate and weaken pirates, who, cut off from supplies and aid on land, could not sustain themselves indefinitely.

In the meantime, the Jiaqing Emperor (r. 1796–1820) stubbornly demanded that the navy pursue a vigorous 'sea war' (*haizhan*) against pirates. Although the great Fujian pirate chieftains Cai Qian and Zhu Fen were killed in sea battles in 1809 and their fleets soon afterwards scattered, further to the south in Guangdong the pirates maintained the upper hand. For the imperial navy the low point came in late 1808 and 1809. First, pirates under Zhang Bao annihilated Lieutenant-Colonel Lin Fa's fleet in a battle near the mouth of the Pearl River. This was followed in the following summer by the defeat of Commander Xu Tinggui's fleet near Macao. Over 1,000 sailors and officers, including Xu, were killed in this one fiasco. It was later reported that defeated troops fled in all directions, looting villages along the way. At the end of the year, a combined Sino-Portuguese taskforce also engaged in several undecided battles with Zhang Bao and other pirates near Lantao Island. The situation had become quite bleak. In repeated campaigns in 1808 and 1809, the Qing Navy had been nearly decimated, leaving only 14 warships and a few requisitioned merchant junks left to protect Guangzhou and the rich Pearl River Delta (Andrade 1835: 45–6, 52; Morse 1926, 3: 117–18, 122; Chang 1983: 153–7; Antony 1994: 24–7).

Having repeatedly defeated the imperial navy, the pirates had virtual control over the Guangdong Coast and even many inland villages and towns. Throughout the summer of 1809, gangs of pirates swarmed deep into the Pearl River Delta, even penetrating within a few miles of Guangzhou itself (refer to Table 15.5). Unable to stop the pirates militarily, the emperor finally initiated a new pacification policy, coaxing pirates to surrender in exchange for pardons and rewards. In April 1810, the most powerful leaders, Zhang Bao and Zheng Yi Sao, surrendered with over 17,000 followers, including 5,000 women and children, together with roughly 280 junks and some 12,000 firearms. The government quickly rewarded Zhang Bao with a naval commission and sent him to fight the remaining pirates in Western Guangdong, particularly the formidable fleets commanded by Wushi Er and Donghai Ba. Once they were defeated, further resistance quickly crumbled (*Xiangshan xianzhi* [1873]1961, 8: 61a–b; *Haikang xianzhi* [1938]1974, 45: 17a–18a; Yuan Yonglun 1830, 18–19; Morse 1926, 3: 123, 144). For all practical purposes, the golden age of piracy in China had come to an end.

Conclusion

Peter Lorge has argued that although over the past 2,000 years naval warfare was crucial to the creation and unification of the Chinese empire, nonetheless it was largely limited to operations on inland waters and, beginning in the Ming Dynasty, markedly defensive. 'The sea was of limited military importance' (Lorge 2012: 81–2, 91). Thus, during the golden age of Chinese piracy, both the Ming and Qing military establishments were ill prepared to effectively handle large-scale piracy. For much of the time between 1522 and 1810, huge pirate leagues, not the government, actually controlled the waters off China's coast. Government volatility and weaknesses at sea allowed pirate bands to grow in size and strength, and become militarily well organised and economically independent. Such leaders as Wang Zhi, Zheng Zhilong and Zhang Bao, at different times, commanded tens of thousands of pirates and fleets of hundreds of warships. From fortified strongholds on offshore islands, and in coastal harbours, pirates set up trading hubs and controlled shipping lanes through the systematic use of violence, extortion and bribery. They effectively challenged the imperial state's sovereignty and became an integral component of maritime society.

Although the imperial state preferred to destroy pirates through aggressive sea wars, the navy was rarely up for the task. In battle after battle, on both sea and shore, pirates usually outnumbered, outmanoeuvred and outgunned imperial forces. In the war against piracy the state had to rely on non-military measures, which included embargoes, imperial pardons, appeasements, forced removal of coastal populations, mercenaries and privateer navies, as well as various harsh judicial processes and the use of extensive summary executions. Over the 300 years' period covered in this chapter, the government's most successful policy was the ancient stratagem of annihilation and pacification, a carrot-and-stick approach that coupled military campaigns with liberal offers of amnesties. The aim was to divide and conquer one's enemies. Once pirates surrendered, they were rewarded with military commissions and sent out to fight their former comrades. In the end, officials preferred to incorporate pirates into the imperial navy rather than fight them. After Zhang Bao surrendered and joined the imperial navy in 1810, soon afterwards large-scale piracy came to an end along the China coast, but petty piracy has continued to remain a problem well into modern times.

Notes

1 Although mention of *wokou* raids on the China coast date back to the early thirteenth century in the Yuan Dynasty, they only became a major problem in the mid-sixteenth century. By and large, important differences between the *wokou* activities of the Mongol and Ming periods were: in the former case, Japanese constituted the majority of pirates and they mostly raided the Korean Coast; in the latter case, the pirates were multinational in composition and they mostly raided the Chinese Coast.
2 Actually, although the Ming Court relaxed its bans on maritime trade after 1567, nevertheless, in 1572, 1592, 1626 and 1628, it re-enacted prohibitions that continued to restrict legal trade.
3 In 1717 there was a short-lived ban on trade with Southeast Asia.

Bibliography

Andrade, José Ignacio, *Memoria dos feitos macaenses contra os piratas da China: e da entrada violenta dos inglezes na cadade de Macáo* (Segunda ediça Lisboa: Typografia lisbonense, 1835).
Andrade, Tonio, 'The Arquebus Volley Technique in China, c. 1560s: Evidence from the Writings of Qi Jiguang,' in Y.H. Teddy Sim (ed.), *The Maritime Defence of China: Ming General Qi Jiguang and Beyond* (Singapore: Springer, 2017), pp. 73–92.
Antony, Robert J., 'Pacification of the Seas: Qing Anti-Piracy Policies in Guangdong, 1794–1810,' *Journal of Oriental Studies*, vol. 32, no. 1 (1994), pp. 16–35.
Antony, Robert J., *Like Froth Floating on the Sea: The World of Pirates and Seafarers in Late Imperial South China* (Berkeley: University of California, Institute for East Asian Studies, China Monograph Series, 2003).
Antony, Robert J., 'State, Community, and Pirate Suppression in Guangdong Province, 1809–1810,' *Late Imperial China*, vol. 27, no. 1 (2006), pp. 1–30.
Antony, Robert J., *Pirates in the Age of Sail* (New York: W.W. Norton, 2007).
Antony, Robert J., '"Righteous Yang": Pirate, Rebel, and Hero on the Sino-Vietnamese Water Frontier, 1644–1684,' *Cross-Currents: East Asian History and Culture Review* (E-Journal), no. 11 (2014a).

Antony, Robert J., 'Maritime Violence and State Formation in Vietnam: Piracy and the Tay Son Rebellion, 1771–1802,' in Stefan Amirell and Leos Muller (eds.), *Persistent Piracy: Maritime Violence and State-Formation in Global Historical Perspective* (London: Palgrave, 2014b), pp. 113–30.

Chang, Thomas C.S., 'Ts'ai Ch'ien, the Pirate Who Dominates the Seas: A Study of Coastal Piracy in China, 1795–1810,' PhD dissertation, University of Arizona, 1983.

Chen Shunxi, *Luanli jianwen lu* [A record of the chaos and abandonment seen and heard], in *Ming Qing Guangdong xijian biji qizhong* (Guangzhou: Guangdong renmin chubanshe, 2010).

Cheng Wei-chung, *War, Trade and Piracy in the China Seas, 1622–1683* (Leiden: Brill, 2013).

Chin, James K., 'Merchants, Smugglers, and Pirates: Multinational Clandestine Trade on the South China Coast, 1520–50,' in Robert J. Antony (ed.), *Elusive Pirates, Pervasive Smugglers: Violence and Clandestine Trade in the Greater China Seas* (Hong Kong: Hong Kong University Press, 2010), pp. 43–58.

'Chinese Pirates,' *Chinese Repository*, vol. 3 (June 1834), pp. 432–7.

Chonlaworn, Piyada, 'Rebel with a Cause: Chinese Merchant-Pirates in Southeast Asia in the 16th Century,' in Y.H. Teddy Sim (ed.), *The Maritime Defence of China: Ming General Qi Jiguang and Beyond* (Singapore: Springer, 2017), pp. 187–200.

Dongguan xianzhi [Gazetteer of Dongguan county] (1921, reprint, Taibei: Taiwan xuesheng shuju, 1968).

Fitzpatrick, Merrilyn, 'Local Interests and the Anti-Pirate Administration in China's Southeast, 1555–1565,' *Ch'ing-shih wen-t'I*, vol. 4 (1979), pp. 1–50.

Glasspoole, Richard, 'A Brief Narrative of My Captivity and Treatment amongst the Ladrones,' in Charles Neumann (tr.), *History of the Pirates* (London: Oriental Translation Fund, 1831), pp. 97–128.

GZD. *Gongzhongdang zhupi zouzhe* [Palace memorials] (Palace Museum: Taibei, n.d.).

Haikang xianzhi [Gazetteer of Haikang county] (1938, reprint, Taibei: Chengwen chubanshe, 1974).

Hang, Xing, *Conflict and Commerce in Maritime East Asia: The Zheng Family and the Shaping of the Modern World, c. 1620–1720* (Cambridge: Cambridge University Press, 2015).

Hang, Xing, 'Leizhou Pirates and the Making of the Mekong Delta,' in Robert Antony and Angela Schottenhammer (eds.), *Beyond the Silk Roads: New Discourses on China's Role in East Asian Maritime History* (Wiesbaden: Harrassowitz Verlag, 2017), pp. 115–32.

Higgins, Roland L., 'Piracy and Coastal Defense in the Ming Period, Government Response to Coastal Disturbances, 1523–1549,' PhD dissertation, University of Minnesota, 1981.

Ho, Dahpon David, 'Sealords Live in Vain: Fujian and the Making of a Maritime Frontier in Seventeenth-Century China,' PhD dissertation, University of California, San Diego, 2011.

Hucker, Charles O., 'Hu Tsung-hsien's Campaign against Hsü Hai, 1556,' in Frank A. Kierman, Jr. and John King Fairbank (eds.), *Chinese Ways in Warfare* (Cambridge, MA: Harvard University Press, 1974), pp. 273–307.

Lee, Chi-Lin, 'An Analysis on the Development of Ming-Qing Maritime Defense and Navy,' in Y.H. Teddy Sim (ed.), *The Maritime Defence of China: Ming General Qi Jiguang and Beyond* (Singapore: Springer, 2017), pp. 233–43.

Lococo, Paul, Jr., 'The Qing Empire,' in David A. Graff and Robin Higham (eds.), *A Military History of China*, Updated Edition (Lexington: University Press of Kentucky, 2012), pp. 115–33.

Lorge, Peter, 'Water Forces and Naval Operations,' in David A. Graff and Robin Higham (eds.), *A Military History of China*, Updated Edition (Lexington: University Press of Kentucky, 2012), pp. 81–96.

Lorge, Peter, 'The Martial Arts in Qi Jiguang's Military Training,' in Y.H. Teddy Sim (ed.), *The Maritime Defence of China: Ming General Qi Jiguang and Beyond* (Singapore: Springer, 2017), pp. 59–72.

Ming Qing shiliao wubian [Historical materials on the Ming and Qing periods, fifth series] (Taipei: Academia Sinica, 1972).

Montalto de Jesus, C.A., *Historic Macao* (Hong Kong: Kelly and Walsh, 1902).

Morse, Hosea Ballou, *The Chronicles of the East India Company Trading to China, 1635–1834*, 5 vols (Cambridge: Harvard University Press, 1926).

Murray, Dian H., 'Sea Bandits: A Study of Piracy in Early Nineteenth-Century China,' PhD dissertation, Cornell University, 1979.

Murray, Dian H., *Pirates of the South China Coast, 1790–1810* (Stanford: Stanford University Press, 1987).

Nanhai xianzhi [Gazetteer of Nanhai county] (1872, reprint, Taibei: Chengwen chubanshe, 1967).

Panyu xianzhi [Gazetteer of Panyu county] (Guang ji tang [Harvard Yenching Library] 1871).

Petrucci, Maria Grazia. 'Pirates, Gunpowder, and Christianity in Late Sixteenth-Century Japan,' in Robert J. Antony (ed.), *Elusive Pirates, Pervasive Smugglers: Violence and Clandestine Trade in the Greater China Seas* (Hong Kong: Hong Kong University Press, 2010), pp. 59–72.

Shao Tingcai, *Dongnan jishi* [A Record of Events in the Southeast], in Taiwan wenxian congkan, no. 96 (Taibei: Taiwan yinhang, 1961).

Shunde xianzhi [Gazetteer of Shunde county] (1853, reprint, Taibei: Chengwen chubanshe, 1974).

So, Kwan-wai, *Japanese Piracy in Ming China during the Sixteenth Century* (East Lansing: Michigan State University Press, 1975).

Swope, Kenneth, 'Cutting Dwarf Pirates Down to Size: Amphibious Warfare in 16th-Century East Asia,' in Y.H. Teddy Sim (ed.), *The Maritime Defence of China: Ming General Qi Jiguang and Beyond* (Singapore: Springer, 2017), pp. 163–86.

Swope, Kenneth, 'Chinese Ways of Warfare, 1500–1800,' in Robert Antony, Stuart Carroll and Caroline Dodds Pennock (eds.), *The Cambridge World History of Violence*, vol. 3, *1500–1800* (Cambridge: Cambridge University Press, 2020), pp. 119–37.

Wei, Betty Peh-T'i., *Ruan Yuan, 1764–1849: The Life and Work of a Major Scholar-Official in Nineteenth-Century China before the Opium War* (Hong Kong: Hong Kong University Press, 2006).

Xiangshan xianzhi [Gazetteer of Xiangshan county] (1873, reprint, Taibei: Zhongshan tongxianghui, 1961).

Yuan Yonglun, 1830. *Jinghai fenji* [A record of the pacification of pirates], in *Fieldwork and Documents: South China Research Resource Station Newsletter*, vol. 46 (2007).

Zheng Ruozeng, *Chouhai tubian* [Illustrated compendium on maritime defence] (1562, reprint, Beijing: Zhonghua shuju, 2007).

16

RISE OF BRITISH NAVAL POWER IN THE INDIAN OCEAN

Amarendra Kumar

Introduction

The lucrative prospects of oceanic trade with the East encouraged maritime nations like England and the Netherlands to authorise commercial companies to explore the possibilities of trade and commerce. The British East India Company (EIC) in 1600, and the Dutch East India Company (VOC) in 1602, thus, set out for the Indian Ocean through the route via the Cape of Good Hope—a monopoly of the Portuguese Crown since 1498. The Portuguese *Estado-da-India* was bound to treat this as a violation of its alleged exclusive rights to the Orient, guaranteed by a Papal Bull. Hence, the *Estado* opposed the 'intruders' by sea, and placed every obstacle in their way upon land (Rawlinson 1920: 1). However, as the century progressed, *Estado*'s naval challenge was gradually eliminated in the Indian Ocean by these protestant nations. The British Royal Navy went even further to purge the influence of the VOC and the French by actively supporting the EIC's Bombay Marine to maintain its control in the East in the eighteenth century. By the mid-nineteenth century, British control over the sea lanes and strategic points across the Indian Ocean was complete.

It is necessary to have an idea of the Indian Ocean as an entity in order to appreciate the rise of the British Navy. As depicted on maps and globes, physically the Indian Ocean appears bifurcated on its northern frontier by the Indian subcontinent which effectively divides it into two related halves: one dominated on its northwestern side by the Arabian Sea, the other on its northeastern side by the Bay of Bengal. Whether viewed from Eastern Africa or South-East Asia, India is what some scholars have called the fulcrum around which the Indian Ocean gravitates (Alpers 2014: 5–6). In his seminal work *The Arabian Seas*, R.J. Barendse (2002: 3) has divided the Indian Ocean into three parts: the seas of and around the Indonesian Archipelago; the Bay of Bengal Area; and the Western Indian Ocean, which included the Arabian Sea and the interior of the Persian Gulf and Red Sea. Pearson (2015: 2), however, points out that in terms of commonalities arising out of expanse, geographical and related features, the general term Indian Ocean ignores the fact that there is a monsoon Indian Ocean, down to about 10° S, and a very different one below this, where cold fronts rather than predictable monsoons prevail.

DOI: 10.4324/9780429437915-21

The rise of British naval power in the Indian Ocean had been a gradual process. The seventeenth century may be regarded as its formative phase—characterised by EIC's strategy of diplomacy combined with limited arm twisting of the native polities, largely to achieve commercial ends. The actual consolidation took place during the eighteenth century with the direct involvement of the British Royal Navy as it confronted and subdued various European and local rivals. The direct involvement of the Royal Navy proved to be the game changer as it gradually enforced total control over the sea routes and key points in the Indian Ocean. As noted in the preface of the first volume of *The Oxford History of the British Empire*, 'a shape was imposed on what had been accomplished by chance only after state authorities came to appreciate the commercial importance of the various colonies, fortified posts, and trading routes throughout the world that had been established by private adventurers' (Canny, 1998: xi). Jan Glete (2001: 82) also noted that 'without the resources of the state, the battle-fleet was not possible.' The 'battlefleet', thus, was the crucial factor that manipulated the interdependence of the actions of the European seas and the Indian Ocean. In fact, the outcome of the engagements in the European seas had a profound bearing upon the course of action in the Indian Ocean conflicts.

I

The role of sea power cannot be denied in the way England's position in the world of long-distance trade and colonisation underwent transformation. The English government had been charting monopolistic trading companies to explore the world markets in the 1550s, but the earlier attempts of the English to break into the Spanish trade with Central and South America had resulted in wars and failure. Their attempts at colonisation also proved futile. England worked upon the plan of creating a permanent navy, as distinct from the assortment of merchant vessels, a bit late. It was not until the reign of Henry VIII (1509–1547) that the first Navy Board was established in 1546. The sailing vessels were accordingly adopted as the major units of the King's fleet (Stevens and Westcott 1944: 105). Due emphasis was placed upon seaworthiness, manoeuvring qualities, strength and size of the ships with a plan to reduce the number of non-combatant personnel on board. As a result, ships of the characteristic proportions of the galleon type (of 400–500 tons displacement) began to be produced in England and were also equipped with effective firearms onboard (Stevens and Westcott 1944: 106–7). The heroic exploits of Charles Drake against the Spanish colonies in the Pacific, together with the inking of an alliance with the Netherlands in 1585, had raised Queen Elizabeth's ambitions and strengthened her resolve to challenge the supremacy of the Spanish Armada. With the destruction of the Spanish Armada in 1588, the Englishmen could move freely beyond their own seas. The defeat of the Spanish Armada, according to G.V. Scammell (1981: 458), corresponded to an armed and aggressive beginning of the British Navy.

Sea power was first exercised in the Indian Ocean by the Portuguese *Estado-da-India*. The *Estado*, in fact, was 'a radically new idea of the role of the state at sea to Asia.' As noted by K.M. Panikkar (1945: 39), the significance of Vasco Da-Gama's entry into the Indian Ocean laid not in the navigational achievements of the Europeans, but in their policy of looking upon the seas as their possession. The littoral polities of the Indian Ocean, used to free navigation of the seas till then, resisted the Portuguese notion of the possession of the sea. The Zamorin of Calicut, thus, not only confronted the Portuguese armada, but also brought about a global (naval) alliance to repel the Portuguese aggression on sea on an urgent basis

(Krishna Aiyyar 1999: 167–81). Elsewhere, the Ottomans fought bitterly to prevent the Portuguese from gaining a secure position in the Red Sea. But, in the long run, the Ottoman failure to develop a deepwater navy capable of operating in the wider Indian Ocean, or a navy to operate in the Persian Gulf, left the Portuguese in a general position of strength. The Portuguese Fleet consolidated its hold over several of the islands at the mouth of the Persian Gulf, including Kish, Kishm (Qeshm) and Hormuz. The Ottoman flotilla under Piri Reis, despite its best efforts, failed to seize Hormuz from the Portuguese in 1551 (MacDougall 2017: 190). Pearson argues that in addition to the ability to make use of organised violence at sea, a favourably located chain of fortified points at Hormuz, Goa and Malacca etc., across the Indian Ocean, further enabled the Portuguese to effectively control the sea lanes. In strategic language, the expanse of the *Estado* during the sixteenth century was

> a position facing north, resting on the water throughout, although with footholds on shore, and extending over a front of roughly 5,000 miles along the northern coasts of the Indian Ocean, from Aden on the left or western hand to Malacca on the right or east.
>
> *(Ballard [1927]1984: 124)*

It was only upon the arrival of Dutch and English trading Companies in the Asian waters during the early seventeenth century that the Portuguese sea power in the Indian Ocean was seriously challenged. The VOC, in fact, carried the Luso–Dutch rivalry to Asian waters and created frequent disruptions in Portuguese commercial activities. The EIC, on the other hand, championing the cause of freedom of navigation, forcefully challenged the Portuguese monopoly of the Indian Oceanic route. The EIC's Marine force, as we shall discuss later, assisted the Shah Abbas of Persia to capture Hormuz from the Portuguese in 1622. The joint Anglo-Persian fleet, after the assault and capture of the Fort Kishm on the mouth of the Persian Gulf, enforced a siege of Hormuz till the time its garrison surrendered (MacDougall 2017: 191). The EIC's acrimony towards the *Estado* was also because of its repeated failure initially to open a factory at Surat. The *Estado*'s representatives in the Mughal Court had been constantly planning against the EIC's efforts to obtain such permission from the Emperor. The awe of the Portuguese armada—proficient in making use of violence at sea for so long—also perhaps deterred the Mughals from making a decision in favour of the EIC. The EIC, thus, was left with the only option of exhibiting its naval prowess through a direct contest with the *Estado*'s armada in the Indian surroundings. In the ensuing contest in 1612, the myth of the Portuguese invincibility on the sea was thoroughly exposed by Captain Best's squadron right in front of Surat (at Swally-hole) as the Portuguese armada was comprehensively beaten in full public view (Anderson 1856: 16–17). The EIC was subsequently granted permission to open a factory at Surat. Admiral Ballard ([1927]1984: 161) described the sound of the British guns as 'prophetic'—the full portent of which none of the hearers, including Best, had the faintest notion. The fall of Hormuz further slackened the Portuguese grip, but the EIC had no plans to stake its claims over the Indian Ocean. The mercantile concerns of a joint stock company prevented it from investing in a Portuguese-type 'armada-flotilla' system. The Company's merchantmen were sufficiently armed for self-defence on the high seas. However, the Company's administrators felt the need to protect its trade, commerce and settlements in the East. Hence, a small Marine force known as the 'Honourable East India Company's Marine' was formed at Surat in 1612. According to Charles Low (1877: 16–17), in those early days, when the Company contended for its very existence with rival associations and hostile nationalities, it found in the Marine the only champion to fight its battles. This was a small local force of grabs and galivats, mounting from two to five or six guns, and officered by volunteers from the Company's ships, who traded as well as fought. By 1615, it consisted of 'ten grabs and galivats' (Low 1877: 16–17, 24). Considering the magnitude of the threat attracted by the EIC in the East, the English Crown authorised it to 'fight against all enemies of the state.' Through several new charters in the 1650s, the Company was also allowed many of the same prerogatives—monopoly, minting, treatymaking and troop rising—as the VOC had enjoyed for a long time (Ballard [1927]1984: 175; Winius and Vink 1994: 56).

Throughout the seventeenth century the EIC and the VOC remained fiercely hostile to each other in the Indian Ocean arena. However, their hostility was interspersed with periods of short-term reconciliations in tune with the political understandings reached out between their respective mother countries in Europe. It was during such intervals that both the Companies acted in concert and directed their energies to undermine the Portuguese position in the East. During the second decade of the seventeenth century, for example, joint Anglo–Dutch actions were directed against the Portuguese and Spanish possessions in the East. Goa, in particular, was continuously harassed and blockaded by the Anglo–Dutch fleet (Ballard [1927]1984: 162–7). It was at that juncture, when the possibility of any relief from Goa reaching Hormuz was negligible, that the attack on Hormuz was planned and successfully executed by the EIC's Marine forces—consisting of a squadron of five ships and four Pinnaces. The fall of Hormuz brought commercial benefits to the EIC also, as it was decided that, in future, there would be an equal division of customs dues of the port, and the English would be free of all duties in perpetuity (Sykes 1915: 277–9). The Portuguese, afterwards, clung firmly to Muscat in the Persian Gulf, till it was captured by the Omani Sultanate in 1651. Hormuz, on the other hand, gradually withered away, as the Persians transferred the commercial activities of Hormuz to Gombroon, which was renamed 'Bandar Abbas.' Soon afterwards, the EIC and the VOC got engaged in a stiff competition for the control of Persia (Jackson and Lockhart 1986: 394, 449–50). The Anglo–Dutch hostility reached a new climax in the Indian Ocean, when, in February 1623, the chief factor of the EIC in Amboyna was killed in cold blood along with nine other Englishmen, for 'conspiring to take the Dutch Fort.' This incident played an important role in English politics under the early Stuarts, and influenced Anglo–Dutch relations during the century (Bassett 1960; Chancey 1998). The VOC, in the meantime, continued to be extremely ruthless on the decaying Portuguese power in the Indian Ocean. Not only was Goa subjected to frequent blockades, but the Portuguese strongholds of Batticaloa and Trincomalee in Sri Lanka were occupied with the help of the local ruler. In Sri Lanka, Galle became the main port and headquarters of the VOC in 1640. The next year, in alliance with the Johor King, the VOC seized Malacca from the Portuguese (Ballard [1927]1984: 188–9). Very soon, the VOC linked its capital at Batavia with Colombo and Cape Town. Colombo, on the southwest coast of Sri Lanka, represented the midway node between the two extremes of Indonesia and the Cape, while Cape Town provisioned VOC ships (Alpers 2014: 93). The decline of the Portuguese in the Indian Ocean in the seventeenth century, argues Jean Glete (2001: 88), was because of the fact that the Portuguese had 'never fought an Asian power who attacked their bases and sea lines of communication in a systematic way. They never met an Asian maritime strategy to defeat them throughout Asia.' Until then, the Portuguese were always able to use the sea lines of communication to concentrate their very limited resources on the area for the time being under threat. The Portuguese, unable to fight on two fronts simultaneously, prevented further losses by signing a peace deal with the English in 1634, which deescalated tensions between them in the Indian Ocean (Mukherjee 1973: 93–100; Roy 2016: 70–1).

Meanwhile England, France and the Netherlands had been embroiled in an intense rivalry in Europe. England, therefore, augmented its naval strength to match her adversaries on the sea. According to estimates, by 1633 the Royal Navy had 50 ships of 23,995 tons, carrying in all 1,430 guns. Five of these were first-rate ships of approximately 1,060 tons each (Hannay 1898: 172). Charles I's policy of 'ship-money fleets' in the 1630s further helped in the creation of a standing, peacetime navy for England. The state, thereafter, became progressively less dependent on the hiring or commandeering of merchant vessels on the outbreak of war; its ships were increasingly its own in times of war as well as peace. From the 1660s, the English Parliament grew accustomed to voting large sums in direct taxation specifically for the navy (Aylmer 2001). By the 1650s the English were building their vessels rather larger and with a higher ratio of guns per ton than their Dutch rivals, but in all these countries there was a general consolidation of the warship at under 1,000 tons and capable of carrying between 30 and 80 guns of different sizes. By the time the line of battle was firmly established as the standard tactical formation during the 1660s, the purpose-built line-of-battle ship ranging eventually up to 2,000 tons, and carrying 120 large cannons, came to occupy the centre stage (Harding 2001e: 40). Further, during the reign of Charles II (r. 1649–1685), a special corps of soldiers was established for service in the fleet during the Second Anglo–Dutch War in 1664. This was the Admiral's, or, as it was called from the colour of its uniform, the Yellow Regiment. It was the first corps of marines proper of which we are aware. Soldiers had been largely employed in the fleet before, but it does not appear that any attempt had been made to distinguish between the soldiers who served in the King's ships and the soldiers who were available for all military services. The Admiral's Regiment was especially devoted for the fleet (Hannay 1898: 317–18).

In Asia, the English and the Portuguese further reinforced their relations after the marriage between Catherine of Portugal and Charles II of England in 1661. As dowry, Charles II received the island of Bombay on the western coast of India, which was subsequently given to the EIC in 1668. Realising the strategic value of Bombay, the EIC developed it as a fortified settlement, which replaced Surat as its headquarters in India in 1686. The Company's Marine force was also accordingly rechristened 'The Bombay Marine.' It should be noted that in India territorial beginnings of the EIC had been made in 1639 when it acquired Madras from the ruler of Chandragiri. Madras, subsequently developed as a fortified settlement spanning about 6 square miles, ceased to be subordinate to Bantam, as it had been until then, and was elevated to the position of an independent trading station by 1657 (Roy 2016: 74). Later on, in 1690, the EIC obtained zamindari rights of three Bengal villages, namely Gobindpur, Sutanati and Kalikata. These coastal villages were also developed by the Company into fortified trading settlements overlooking the Bay of Bengal. By the close of the seventeenth century, with India as the centre of its activities in the Indian Ocean, the EIC possessed three important fortified settlements respectively on the western, southern and eastern coasts of India. The Bombay Marine kept the Indian waters secure for the Company's trading vessels as far as possible.

Notwithstanding the growth and expansion of the British Navy in Europe, it cannot be denied that the seventeenth century belonged to the Dutch in the East. It was because the VOC had the largest presence of armed vessels and men of war beyond the Cape of Good Hope to defend their vast commercial enterprise. The Dutch political compulsions in Europe, however, checked the VOC from creating impediments in the English consolidation in India even when EIC's strength in comparison was insufficient and inferior. Hence, we find that during the Anglo–Dutch Wars in Europe, the stronger and numerically superior VOC fleet did not attack the EIC commerce or settlements in the Indian Ocean. The VOC, perhaps aware of the unsettled conditions in Europe, was confident that if victory attended their efforts on the coast of England, it could be extended right up to the Indian coasts by capturing the English settlements easily. Conversely, any gain at the cost of the rival EIC in the periphery would have meant little in the absence of a favourable outcome of the main theatre of conflict in Europe. For example, Australia—discovered, though not occupied, by the Dutch—was ceded to Britain at the end of the Second Anglo–Dutch War. Admiral Ballard ([1927]1984: 203–8) has beautifully summed up this situation thus:

> Through the issue of the stubbornly contested battles which then took place within the sight of the cliffs of Sussex and Kent, all possibility of a Dutch monopoly of eastern trade based on an assured command of the Indian Ocean vanished; whereby the whole future history of the intercourse between east and west was materially affected.

In two decades after the conclusion of the Second Anglo–Dutch War in 1667, the English had doubled their merchant shipping (Silburn 1912: 217–18). With the conclusion of the Third Anglo–Dutch War in 1674, the two nations rather chose to cooperate (Treaty of Navigation and Commerce) in naval matters against a common enemy—France. Shortly after William III's marriage to Mary in 1677, discussions had begun that touched on the specific issues of Anglo-Dutch military and naval cooperation. In addition to establishing a specific ratio of force sizes, the agreement established the clear precedent for a proportionately higher number of English ships. A treaty was finally signed on 21 May 1689, the very day that England declared war on France, though it was backdated three weeks to 29 April. This agreement was the detailed document that, with slight modification, governed Anglo–Dutch naval cooperation for the entire 25-year period, till 1714 (Hattendorf 2012: 65–81). The cordial relations with the Portuguese and the Dutch, hence, enabled the British to concentrate on the French navigation in the Indian Ocean. A brief overview of the local politics of India at that point of time would not be out of place here.

II

The seventeenth century saw the rise of local powers in Asia (from Oman to Malacca) with functional navies—capable of challenging the European merchantmen in close proximity to their shores. However, since these were essentially land-based powers, as part of a strategy to protect their political and economic interests, they also used to put economic and military pressure on the European settlements within or near their territories. In retaliation, the Europeans harassed the native shipping—though infrequently, owing to their larger commercial interests. The Europeans were mindful of the fact that, without the willing cooperation of the local elites, they could not penetrate deep into the hinterland to exploit the source of spices or other goods. Moreover, their own limitations on land also obliged them to play safe by way of accommodating the native powers—many a time through the assured supply of arms and other logistics. Scammell (1988, and 1991: 68–72) explains this in terms of the success of the Portuguese to rule by exploiting internal divisions among the Asians, with indigenous support.

During the sixteenth century, the indigenous navies were rarely assigned a real military role. Hence, as noted above, they never challenged or threatened Portuguese command of the sea. For example, the Mughals never showed any pretensions to dominate sea lanes or navigation, and were content with plying their trading vessels as well as annual pilgrim ships under the European protection by obtaining their so called 'navigational or protection passes' (Rothermund 2014: 63–75; Mathew 1995). In contrast, the Maratha Navy— founded by Shivaji during the second half of the seventeenth century—aimed to achieve 'mastery over the sea' (*Rajniti* 1929: 48). The contemporary French traveller, Abbe Carre, noted that the Maratha Fleet was well complemented by marine forts and naval bases on strategic locations along the Konkan coast (Sen 1977: 154–5). It was divided into two fighting squadrons under a naval commander. The Maratha Navy refused to accept the exclusivity of European claims over the Konkan waters. Not only did the Maratha ships refuse to buy the European passes, but Shivaji also insisted that Maratha navigational passes must be purchased by everyone, irrespective of their nationalities, for safe navigation (Apte 1973: 201–3, and 1967: 255–61). The navigational passes, thus, became a perpetual source of conflict in this part of the Indian Ocean. An ironic consequence of this competition in the Indian Ocean, according to Alpers (2014: 86), is that every player regarded itself as the rightful claimant to domination and its rivals as pirates.

Apparently, for the Portuguese Viceroy, a *cartaz* signalled a legitimate merchant ship, while its British equivalent amounted to a legal dodge to enable piracy. By contrast, to indigenous merchants and sailors it was the Portuguese who were the pirates. Peninsular India also saw the rise of an organised navy under the rulers of Mysore during the last quarter of the eighteenth century (MacDougall 2014). The situation on the seafront, hence, stood in sharp contrast to the previous centuries as the navies of the littoral powers were now capable of jeopardising the commerce of the mercantile companies within the local surroundings. The European Companies, on the other hand, were also aware that none was capable of threatening their sea lines of communication—the mainstay of their power in Asia—in a systematic way. What these Companies, therefore, needed was only a minor reworking of their strategy vis-à-vis the local potentates. Hence, at times, we find them participating in the local wars indirectly, while, at others, they preferred to play arbiters in their power struggle. The EIC, for example, always ensured that balance of power was maintained between the Mughals and the Marathas in their naval struggles. It served the interests of the European companies better when the littoral polities were in the ferment, and especially when there was no final outcome of their mutual rivalries. It gave them more scope for manipulating the situation to further their own interests (Winius and Vink 1994: 70). It must not be assumed, however, that such manipulations were simply a one-way traffic. There are instances of the littoral polities trying to take advantage of the mutual rivalries among the Companies vying for better sites on the coast for settlement and trading privileges. Shivaji, for example, played upon the realistic fears of the EIC about a possible Dutch invasion of Bombay. During the Mughal–Maratha struggle, thus, Shivaji would put pressure on the EIC to desist from helping the Mughal Navy, lest he would offer (naval) support to the VOC to capture Bombay (Kail 1981: 77; Paranjape 1931, vol. 1: Letter no. 345). Almost a decade later, when Captain Richard Keigwin of the EIC rebelled against his own Company's administration in 1683, and gained control over Bombay, the tactful Marathas took advantage of the situation and signed a pact of friendship with the rebel Bombay government. Bombay, accordingly, withdrew the facility of annual wintering to the Mughal Fleet in its harbour, thereby considerably eliminating the chances of Mughal naval incursions on Maratha posts for the time being (Kumar 2014: 320–4; Desai 1974: 33).

In addition to incidents of indirect involvement in the local conflicts, many a time the EIC directly clashed with the some of the local powers. For example, in 1679, when Shivaji had occupied and subsequently fortified a small rocky projection (by the name of Khanderi) located at the entrance of the channel leading to Bombay harbour, the Bombay Marine, through a direct naval action, attempted to dislodge the Marathas, but failed miserably. To overcome the Maratha challenge, the EIC's Marine forged an alliance with the Siddi of Janjira—the Mughal naval Commander on the west-coast, and an avowed enemy of the Marathas—but to no avail (Paranjape 1931, vol. 2, letter nos. 364, 365, 367, 369, 373, 374, 377, 378, 379; Sarkar 1973: 270–5). The EIC's audacity against the native powers reached a new high when its administrators didn't think even twice while giving offense to the Mughal Emperor Aurangzeb. As a result, in 1690, the Mughal Admiral carried out an invasion and blockade of Bombay. It ought to be remembered that it was only in 1687 that the EIC had shifted its headquarters from Surat to Bombay. The Mughals compelled the besieged governor of Bombay to sign a humiliating treaty (Ovington 1929: 93–4). These incidents exposed the vulnerability of the EIC in India as the Bombay garrison and Marine forces were unable to hold ground against a strong, land-based force complemented with a workable flotilla. The Company urgently needed to strengthen its Marine forces and also explore the possibilities of expanding its land base in India.

Early in the eighteenth century, the EIC invoked the attention of the home government by requisitioning the service of the Royal Marine to 'suppress pirate nations.' The heavy losses suffered by the Company in fighting the native polities, particularly the Marathas, had compelled them to brand the latter as pirates early in the eighteenth century. The Maratha Navy was then under the command of *Sarkhel* (Admiral) Kanhoji Angre. His fleet, based at Kolaba, 'kept a vigilant eye on all sailing vessels' from Surat to Goa (Sen 1973: 258; Elliot 2010). The Maratha Navy had extended absolute control all along the western seaboard, as it had successfully neutralized the challenge of the Mughal Navy by 1715. It was at this juncture that Charles Boone took charge as the new Governor of Bombay. Boone resolved to eliminate the Maratha opposition on the sea for good, and accordingly placed orders for the construction of new, purpose-built war vessels capable of manoeuvring in the coastal waters (Low 1877, vol. 1: 94). In 1716, the Bombay Marine, with one ship of 32 guns, four grab ships mounting between 20 and 28 guns, and 20 grabs and galivats, carrying between 5 and 12 guns, launched an offensive on Maratha strongholds along the coast, which continued till 1721 (Low 1877, vol. 1: 96). Boone's diplomacy had also gained him allies like the Portuguese and other enemies of the Marathas. However, the Bombay Marine failed hopelessly to achieve its objectives. Not only did one of its locally constructed 'floating castles' experience design faults which rendered it out of action, but the stratagem of Kanhoji Angre also denied the Bombay Marine any advantage against the Marathas. A contemporary account of the struggle between Kanhoji Angre and the EIC has been penned by Clement Downing (1923: 1–61), who was part of the EIC's squadron employed to attack the Marathas. The Bombay Marine's failure in India apparently convinced the British government of the necessity to increase the deployed strength of the Royal Navy in the Indian Ocean.[1] After Kanhoji's death in 1729, his sons and successors continued his legacy and kept the banner of the Maratha Navy high in the Arabian Sea. The Maratha Confederacy had extended its control over the Northern and Central Indian plains during the same time. In this process, under the leadership of the *Peshwa*, they had ejected the Portuguese from Bassein in 1739. The accomplishment of the Marathas on the Konkan Coast, argues Lakshmi Subramanian, could not be neglected by the EIC, which had been forced to remain a silent spectator to these events (Subramanian 1981). An unfortunate consequence of the inland expansion of the Marathas, however, was a gradual neglect of their navy. Since the navy was no longer a priority in the larger plans of the Maratha state, it failed to capitalise upon the gain offered by the exit of the Portuguese from Bassein. The Maratha Navy was further weakened due to a bitter feud over succession within the Angre family. Keeping a close watch over the situation, the EIC continued to augment its naval strength, in the meantime. By 1742, the Bombay Marine had one ship of 24 guns, four of 18 guns, six bomb-ketches and 28 galivats, employing nearly 100 officers and from 1,700 to 2,000 men (Low 1877 vol. 1: 118).

It was only when the relations between the *Peshwa* and the Maratha Admiral deteriorated to the point of no return that the EIC got the chance to establish its absolute naval control on the west coast. Proposing naval assistance to the *Peshwa*, the EIC planned a joint campaign to capture Maratha naval bases of Suvarnadurg and Vijaydurg (also known as the Gheria fortress). The *Peshwa* spearheaded the campaign on the land, while the Company's Marine simultaneously launched offensives from the sea. Suvarnadurg was thus reduced in 1755, but, Gheria was larger and comparatively stronger. It had successfully survived Boone's attempts three decades before. While the allies were pondering upon a future course of action, a squadron of the Royal Navy arrived at Bombay. The Company grabbed the opportunity with both hands and roped in the services of the Royal Navy for the reduction of Gheria. A joint fleet under the command of Admiral Charles Watson of the Royal Navy—the then Commander-in-Chief of the East Indies—comprising the vessels of the Bombay Marine and the Royal Navy, proceeded towards Gheria early in 1756 (Ballard [1927]1984: 242–3). In the course of action, Gheria Fort was pounded from the sea side and the Maratha vessels in the harbor were set on fire. Tulaji Angre, the Maratha Admiral, was captured by the *Peshwa*'s army as it entered the fort from the land side. The Maratha Navy was doomed forever, leaving the field open to the Bombay Marine. By the 1760s, Low (1877 vol. 1: 123) writes, 'with increased numbers, improved discipline, and a regular uniform, the Bombay Marine became a little Navy, although it did not assume that name.'

III

The victory against the *Nawab* of Bengal—a French ally—in the Battle of Plassey in 1757, saw the Company's status bcoming elevated to that of a political power in India. The EIC's success in Bengal was facilitated by the same squadron of the Royal Navy which had destroyed the Maratha Navy a year before. As the Royal Navy captured Chandannagar—a French settlement on the banks of the River Hooghly—it also set into motion the Anglo–French struggle in the Indian Ocean. As an extension of the Europe's Seven Years' War, it introduced to the region a disruptive new level of direct European state interest and involvement. In the first place, the ensuing wars were now to be waged by Crown forces instead of privately owned ships; second, they were inspired as much by political as commercial ends (Ballard [1927]1984: 242–3; Phillips and Sharman 2015: 166–201).

The outcome of the Anglo–French struggle in India, also known as the Third Carnatic War, had been far from encouraging for the French. They were decisively defeated by the English in the Battle of Wandiwash in 1760, and had to surrender even Pondicherry—their

Indian headquarters. Most of their settlements, however, were restored as per the terms of the Treaty of Paris (1763). But, the English ensured that their fortifications were demolished and garrison strength drastically reduced so that they no longer remained a challenge (Majumdar and Dighe 1977: 334–5). In the meantime, the Dutch were also comprehensively defeated in the Battle of Bidera by the Company's forces in 1759; thereby severely restricting the latter's naval operations in India (Malleson 1883: 108–24). The defeat of the French and the Dutch allowed the EIC to concentrate on expansionist wars against the Indian polities like the Marathas, Mysore and Hyderabad.

But the outbreak of the War of American Independence (1775–1783) put the British plans in jeopardy again. Their involvement in America gave the French and the Dutch an opportunity to exploit the situation in India by attempting to regain their lost ground (Bryant 2013: 258). The loss of half its American colonies and the rising importance of Asian trade to Britain in the late eighteenth century, argues John McAleer (2017: 39), brought about the apparent reorientation of the British towards the East, as there was now a need to find equivalents for the Asian empire. The Royal Navy, therefore, stepped up its vigilance of the Indian Ocean sea lanes. In 1778, the EIC's Supreme Council in Calcutta and Madras accordingly resolved that the 'French presence in India must be eradicated.' Even though the EIC was not going to gain much commercially in the case of a military action on the French, the latter could not be simply allowed to undermine British political power in India by allying with its enemies. Pondicherry, thus, capitulated to a joint operation by the Royal Navy under Admiral Vernon and the Madras Army commanded by General Hector Munro in October 1778. Further, as legitimate targets in a war between the European nations, the French factory at Mahe was captured the following year (Bryant 2013: 285–6). The declaration of the Anglo–Dutch War (1780–1784) in Europe further intensified competition in the Indian Ocean. The VOC was further harassed by the EIC forces alongside its bases in India (Negapatam) and Ceylon (Trincomalee). Negapatam was captured (12 November 1781) in a well-executed attack, largely by marines and sailors from the fleet, commanded by Munro, whereas Trimcomalee fell to an amphibious force of Company troops escorted across to Ceylon by Admiral Edward Hughes (Commander the East Indies Station of the Royal Navy) two months later (Phillips and Sharman 2015: 299–300).

In India, Haider Ali—the ruler of Mysore—emboldened by the promise of French help, planned to besiege and take English strongholds like Madras. But, in the absence of French help (from the sea side), he could not execute the plan, and subsequently withdrew. In 1782, a French fleet commanded by Admiral Bailli de Suffren captured Trincomalee from the English and handed it over to the Dutch—now French allies (Barua 2011: 22–40; Sivasundaram 2013: 72). Though the Royal Navy put up a tough fight to Suffren's fleet throughout, the odds always seemed to be in favour of the French till the time Suffren operated in the Indian Ocean. During the war, the French ships were provisioned, repaired and refitted at the Dutch ports of Batavia and Trincomalee. The French, however, failed to gain the upper hand against the British, largely due to the consequences of the actions in Europe. Two French convoys bound for India (as reinforcement to Suffren's fleet) under Soulanges and de Guichen were intercepted and captured by the Royal Navy in the Bay of Biscay. Soon after, the French Navy suffered another defeat at British hands (Richmond 1931: 315). Rodger (2006: 169–83) attributes the British success to their strategy of dominating home waters, especially the Western approaches. This was the best method of guarding against the risk of invasion, protecting British trade, and interfering with the trade and naval operations of the enemy since the 1740s. Notwithstanding the odds, and the accompanying frustration because of his incompetent and disloyal subordinates, Admiral Suffren persisted in his mission and extended his siege of the EIC's coastal pockets in India. But, at the time when the French appeared stronger on land as well as on water, and when it seemed as if they would gain a decisive upper hand against the English in India, the news of Anglo–French peace in Europe reached India in 1783 (Frazer 1896: 148). The British, thus, were literally bailed out of the most critical of the situations they had faced against the French in India. The Anglo–French understanding ensured that the status quo was maintained vis-à-vis overseas possessions. Pondicherry and Mahe were accordingly restored to the French Company (Ballard [1927]1984: 310–11; Bryant 2013: 310–11).

During the second half of the eighteenth century, the EIC was also looking for settlements in Southeast Asia, especially in the China Sea or the Indonesian Archipelago. A settlement in the region would not only supply the Southeast Asian produce the Company required for Chinese trade, but would also eventually supersede Dutch Melaka and Batavia as provisioning points and refitting stations for China-bound East Indiamen (Das Gupta 1939: 1402–31). Moreover, the exigencies arising out of the War of American Independence—particularly the French and Dutch hostilities in Southeast Asia—required the British government to look for a suitable outpost for the Royal Navy along the Malay Peninsula on a priority basis. Interestingly, some English adventurers, along with English 'private country traders' like Francis Light, accordingly, played prominent roles in the acquisition of Penang from the Sultan of Kedah in 1786, on lease from his country in return for military assistance to the Sultan against his enemies (Andaya and Andaya 1982: 99–102). During the peace negotiations at the end of the American War, control/retention of strategic points like the Cape of Good Hope, Egypt, Ceylon and Mauritius was discussed. Of these, Mauritius, which had served as a French base for war in India during the War of the Austrian Succession and the Seven Years' War, was of particular interest to the British. It is estimated that between 1780 and 1783 more than 100 French warships left the island supplied with arms, men and food as Port Louis (as French Mauritius was known then) became the French capital in the Indian Ocean after the fall of Pondicherry (Jackson 2001: 20).

The Pitts India Act of 1784 brought some crucial changes in the jurisdiction of the EIC's Indian administration with a view to safeguard British political and commercial interests in the East. The British government took over direct responsibility of political and military matters, leaving commercial affairs exclusively to the Company. The Napoleonic wars in the last decade of the eighteenth century further strengthened the resolve of the British policy makers to arrange for the safety and security of British overseas colonies. Even though the Royal Navy's Western and East Indies Squadrons were positioned overseas to protect colonies and trade, argues N.A.M. Rodger (2006: 169–183), 'they were too small, and not primarily designed or disposed for naval warfare against European enemies.' The outcome of their efforts depended entirely upon British domination of European waters. The French invasion of the Netherlands in 1794, and the organisation of the (French-controlled) Batavian Republic simultaneously, had raised concerns about the safety of the British possessions in the East. The very thought of the French replacing the Dutch at the latter's bases in the East made the British policy makers uneasy. Hence, before the French succeeded the Dutch in the East, the Royal Navy sprung into action (Malleson 1884). In August 1796, the Royal Navy invaded and occupied the Cape of Good Hope. Trincomalee, which the English had lost earlier to the French, was also retaken. As per records, the British had set up four hospitals for military services in Trincomalee by 1800 (Sivasundaram 2013: 258–9). William V of the Netherlands, exiled himself to London following the French invasion, had instructed the Dutch governors overseas to transfer their territories to British safekeeping. As a result, Britain ended up occupying the Cape, Ceylon, Padang (Sumatra), Malacca, Ambon and Banda (Stockwell 2009: 371–94). In the meantime, the British drove the French out of Pondicherry for good, and occupied Perim Island off the Yemen coast in 1799, in the hope of blocking a future move of Napoleon from Egypt to India (Kearney 2004: 131). Though Anglo–French peace returned with the Treaty of Amiens in 1802, the English, with the Cape and Ceylon in their possession, still eyed up control of Mauritius—one of the strategic French posts in the Indian Ocean. The Napoleonic wars provided the English an opportunity to launch an attack on Mauritius. In 1810, the Royal Navy, after enforcing a blockade of Mauritius, managed to land 10,000 soldiers of the EIC, and captured it (Jackson 2001: 20–1). However, it was only after the downfall of Napoleon that Britain's control over Mauritius was recognised by the Treaty of Paris, 1814. In the meantime, Dutch Java was also attacked and captured in 1811. In all these undertakings, the use of the EIC's army well complemented the prowess of the Royal Navy. Thus, by 1815, the EIC was South Asia's paramount territorial power. St Helena, the Cape and Mauritius formed a chain of British way stations on the route to India, acting as a 'sub-network' within—as well as a gateway to—the Indian Ocean (Phillips and Sharman 2015: 179, 185; McAleer 2017: 239).

After the fall of Napoleon, Anglo–Dutch relations were also realigned as England was looking for a a junior partner in the region. Accordingly, Java and their other East Indies interests were returned to the Dutch by 1816. Later on, in 1824, through a treaty, Anglo–Dutch differences were further settled. The Dutch recognised Singapore as a British possession and exchanged Malacca for Bencoolen. This treaty divided the Malay Archipelago into two spheres of influence. Two years later, Penang, Malacca and Singapore were amalgamated as the Straits Settlements under the Bengal Presidency (Stockwell 2009: 374–5; Phillips and Sharman 2015: 185, 195; Bethune Cook 1918). With this, apparently the European challenge to the EIC in the Eastern Indian Ocean was over, and Britain moved a step closer towards establishing its hegemony over the Indian Ocean. What remained now was to ensure complete security of communications between Europe and the Indian Ocean possessions, for which some local coastal polities further needed to be pacified. For example, a major challenge to British control of the Arabian Sea came from the Qawasim (an Arab tribe) based at Ras al-Khayma, a headland of the Arabian Peninsula near the Strait of Hormuz. As the Qawasimis were also giving offence to the Sultanate of Oman—an ally of the Britain through a Friendship treaty of 1798—the British forces, acting on behalf of Oman, seized Ras al-Khayma in December 1819. In 1820, a General Treaty of Peace was imposed, which, as noted by Lord Curzon at the end of the century, was 'to secure the maritime peace of the Gulf.' The treaty granted the British the right to build a fortified factory at the Omani capital of Muscat and forbade similar footholds by their French and Dutch rivals (Alpers 2014: 105).

Steam navigation was fast becoming popular in the nineteenth century as it considerably reduced the time of communication between Europe and India. The shortest routes were through the Persian Gulf and the Red Sea. While the Persian Gulf route was protected by the Bombay Marine, the Red Sea route had not evoked much interest till the time Sir John Malcom, the Governor of Bombay (November 1827–December 1830), ordered the Bombay Marine to survey the Red Sea and establish coal depots between Bombay and Suez, in preparation for a possible steam service. In any case, both the routes needed to have coaling stations in between, for the steam ships. To provide more readily the facilities needed for steam communication, the Marine Force was partially reorganised and transformed as it came to be known as the Indian Navy from 1 May 1830 (Low 1877 vol. 1: 532). The Navy's work, after this date, consisted not merely in the charting of maritime channels and coast lines, but in the pacification of the maritime native states. In 1836, when an English ship from Madras was harassed and its belongings plundered in the vicinity of Aden, two warships and 800 soldiers of the Company forcibly seized the city in January 1839. Taking Aden was but the first step in Britain's growing involvement in the Middle East (Headrick 2011). Subsequently, Sir Charles Napier sent a fleet of warships in 1840–1841 to take over Egypt, Palestine and Mesopotamia. The Red Sea route was thus secured too (Phillips and Sharman 2015: 198).

Throughout the Indian Ocean littoral, the (British) Raj braided British suzerainty into the fabric of local polities. For example, without alienating the residual sovereignty of the so-called Trucial Sheikhdoms (now the United Arab Emirates) the British signed a Perpetual Maritime Truce in 1853, making the Sheiks permanent allies of Britain. Initially concluded to help the Bombay Marine to suppress piracy in the Western Indian Ocean, the agreement was later extended in 1892 to an 'exclusivity agreement,' whereby the Sheikhdoms were incorporated into the Raj as 'protected' states (Alpers 2014: 105). In the Eastern Indian Ocean, the vast Chinese resources tempted the British to use superior technology to force open the China trade. The Royal Navy launched the *Nemesis*—the largest iron steamer yet built. It was a typical steam vessel having a flat bottom, fitted with masts and sails, which enabled it to navigate across oceans and steam up rivers and in shallow waters. On board, it had two pivot-mounted 32- pounders, five six-pounders, ten smaller cannons and a rocket-tube. The *Nemesis* successfully subdued the last challenge on the sea in the Eastern Indian Ocean in 1839. China was overpowered and the way to Guangzhou was opened (Headrick 2011). The story of the rise of British naval power in the Indian Ocean would be incomplete without an appreciation of the role played by the EIC's Marine in the charting of maritime channels and coastlines to meet strategic requirements. The post of a Marine Surveyor-General was created by Lord Wellesley who sent *Antelope* and *Panther* for the survey of the Red Sea in 1803–1804. These surveys, however, ended by 1838 (Hoskins 1928: 192–3; Low 1877 vol. 2: 70–2, 75).

Conclusion

We may conclude by stating that British naval power evolved in a gradual manner in the Indian Ocean. The beginnings could be traced from the formation of EIC's Indian Marine for the safety and security of its commercial settlements, merchant ships and navigation against the perils of pirate attacks, commercial competitors or other local challenges. As the EIC was tightening its political grip over India during the eighteenth century, the active support of the Royal Navy facilitated its consolidation. The loss of America shifted British attention towards the East, late in the eighteenth century. By securing European waters, the Royal Navy gave a mortal blow to the French competition in the East and dislodged them from all the strategic points facilitating control of the Indian Ocean. The VOC had gone bankrupt by then, and so would not have posed any serious challenge to British dominance. The residual littoral powers were subsequently pacified through a cautious mix of diplomacy and force during the nineteenth century, thus paving the way for the absolute control of the British Navy in the Indian Ocean.

Note

1 I am indebted to Dr David Wilson, Lecturer in Maritime and Scottish History, University of Strathclyde, Glasgow (UK), for showing me the graph table depicting the deployment of the Royal Navy men-of-war during this period. The table was a part of his presentation on 'A War Against Piracy? Local Agency and the Limits of Imperial Authority in Oceanic Spaces throughout the British Atlantic, 1716–1726,' at The Hakluyt Society Symposium, 2019, on 'Rethinking Power in Maritime Encounters 1400–1900,' at Leiden University, the Netherlands on 5–6 September 2019.

Bibliography

Alpers, Edward A., *The Indian Ocean in World History* (New York: Oxford University Press, 2014).

Andaya, Barbara Watson and Andaya, Leonard Y., *A History of Malayasia* (reprint, London: Macmillan Education, 1982).

Anderson, Philip, A.M., *The English in Western India*, 2nd revised ed. (London: Smith, Elder & Co., 1856).

Apte, B.K., 'Sovereignty of the Sea as Practiced in the Maratha Period,' *Proceedings of the Indian History Congress*, vol. 29, no. 1 (1967), pp. 255–61.

Apte, B.K., *A History of the Maratha Navy and Merchant Ships* (Bombay: State Board for Literature and Culture, 1973).

Aylmer, G.E., 'Navy State, Trade, and Empire,' in Nicholas Canny (ed.), *The Oxford History of the British Empire*, vol. 1 (Oxford: Oxford University Press, 2001), pp. 467–81.

Ballard, G.A., *Rulers of the Indian Ocean* (1927, reprint, Delhi: Neeraj Publishing House, 1984).

Barendse, R.J., *The Arabian Seas: The Indian Ocean World in the Seventeenth Century* (New Delhi: Vision Books, 2002).

Barua, Pradeep P., 'Maritime Trade, Seapower and the Anglo-Mysore Wars, 1767–1799,' *The Historian*, vol. 73, no. 1 (2011), pp. 22–40.

Bassett, D.K., 'The "Amboyna Massacre" of 1623,' *Journal of Southeast Asian History*, vol. 1 (1960), pp. 1–19.

Bethune Cook, J.A., *Sir Thomas Stamford Raffles: Founder of Singapore, 1819* (London: Arthur H. Stockwell, 1918).

Bryant, G.J., *The Emergence of British Power in India 1600–1784: A Grand Strategic Interpretation* (Woodbridge: The Boydell Press, 2013).

Canny, Nicholas (ed.)., *The Oxford History of the British Empire*, vol. 1, *The Origins of Empire* (Oxford: Oxford University Press, 1998).

Chancey, Karen, 'The Amboyna Massacre in English Politics, 1624–1632,' *Albion: A Quarterly Journal Concerned with British Studies*, vol. 30, no. 4 (1998), pp. 583–98.

Das Gupta, S.N., 'The English East India Company's Quest for Settlements in the East Indies in the Mid-Eighteenth Century,' *Proceedings of the Indian History Congress*, vol. 3 (1939), pp. 1402–31.

Desai, W.S., *Bombay and the Marathas upto 1774* (New Delhi: Munshiram Manoharlal, 1974).

Downing, Clement, *A Compendious History of the Indian Wars with an Account of the Rise, Progress, Strength, and Forces of Angria the Pyrate* (1737, reprint, Oxford: Oxford University Press, 1923).

Elliot, Derek L., 'Pirates, Polities and Companies: Global Politics on the Konkan Littoral c. 1690–1756,' Working Papers no. 136/10, 2010, Department of Economic History: London School of Economics.

Frazer, R.W., *British India*, 3rd ed. (London: T. Fisher Unwin, 1896).

Glete, Jan, *Warfare at Sea, 1500–1650: Maritime Conflicts and the Transformation of Europe* (New York: Routledge, 2001).

Hannay, David, *A Short History of the Royal Navy, 1217–1688* (London: Methuen & Co., 1898).

Harding, Richard, *Seapower and Naval Warfare, 1650–1830*, Taylor & Francis e-library, 2001 [e-book].

Hattendorf, John B., *Talking about Naval History: A Collection of Essays* (Washington D.C.: US Government Printing Office, 2012).

Headrick, Daniel R., *Power Over Peoples: Technology, Environments, and Western Imperialism, 1400 to the Present* (Princeton: Princeton University Press, 2011).

Hoskins, Halford Lancaster, *British Routes to India* (New York: Longmans, Greens & Co., 1928).

Jackson, Ashley, *War and Empire in Mauritius and the Indian Ocean* (New York: Palgrave, 2001).

Jackson, Peter and Lockhart, Laurence (eds.), *The Cambridge History of Iran*, vol. 6, *The Timurid and Safavid Period* (Cambridge: Cambridge University Press, 1986).

Kail, Owen C., *The Dutch in India* (Delhi: Macmillan India, 1981).

Kearney, Milo, *The Indian Ocean in World History* (New York: Routledge, 2004).

Krishna Aiyyar, K.V., *The Zamorins of Calicut* (1938, reprint, Calicut: Publication Division, University of Calicut, 1999).

Kumar, Amarendra, 'Keigwin's Bombay (1683–84) and the Maratha–Siddi Naval Conflict,' *Proceedings of the Indian History Congress*, vol. 75 (2014), pp. 320–4.

Low, Charles Rathbone, *History of the Indian Navy (1613–1863)*, 2 vols (London: Richard Bentley & Son, 1877).

MacDougall, Philip, *Naval Resistance to Britain's Growing Power in India, 1660–1800: The Saffron Banner and the Tiger of Mysore* (Woodbridge: The Boydell Press, 2014).

MacDougall, Philip, *Islamic Seapower During the age of Fighting Sail* (Woodbridge: Boydell and Brewer, 2017).

Majumdar, R.C. and Dighe, V.G. (eds.), *The History and Culture of the Indian People*, vol. VIII, *The Maratha Supremacy* (Bombay: Bharatiya Vidya Bhavan, 1977).

Malleson, G.B., *The Decisive Battles of India: From 1746 to 1849 Inclusive* (London: W.H. Allen & Co., 1883).

Malleson, G.B., *Final French Struggles in India and on the Indian Seas*, new ed. (London: W.H. Allen & Co., 1884).

Mathew, K.S., 'Taxation in the Coastal Towns,' in K.S. Mathew (ed.), *Mariners, Merchants and Oceans* (New Delhi: Manohar, 1995), pp. 143–7.

McAleer, John, *Britain's Maritime Empire: Southern Africa, the South Atlantic and the Indian Ocean, 1763–1820* (Cambridge: Cambridge University Press, 2017).

Mukherjee, Ramkrishna, *The Rise and Fall of the East India Company: A Sociological Appraisal* (n.d., reprint, Bombay: Popular Prakashan, 1973).

Ovington, J. (tr. by H.G. Rawlinson), *A Voyage to Surat in the Year 1689* (Oxford: Oxford University Press, 1929).

Panikkar, K.M., *India and the Indian Ocean: An Essay on the Influence of Sea Power on Indian History* (London: George Allen & Unwin, 1945).

Paranjape, B.G. (ed.), *English Records on Shivaji (1659–1682)*, 2 vols (Poona: Shiv Charitra Karyalaya, 1931).

Pearson, M.N. (ed.), *Trade, Circulation and Flow in the Indian Ocean World* (Hampshire: Palgrave MacMillan, 2015).

Phillips, Andrew and Sharman, J.C., *International Order in Diversity: War Trade and Rule in the Indian Ocean* (Cambridge: Cambridge University Press, 2015).

Rajniti of Ramchandra Pant Amatya, *A Royal Edict on The Principles of State Policy and Organisation*, tr. by S.V. Puntambekar (Madras: Diocesan Press, 1929).

Rawlinson, H.G., *British Beginnings in Western India: 1579–1657* (London: Oxford University Press, 1920).

Richmond, Herbert, *The Navy in India (1763–1783)* (London: Ernest Benn, 1931).

Rodger, N.A.M., 'Sea-Power and Empire, 1688–1793,' in P.J. Marshall (ed.), *The Oxford History of the British Empire,* vol. 2, *The Eighteenth Century* (2001, reprint, Oxford: Oxford University Press, 2006), pp. 169–83.

Rothermund, Dietmar, *Violent Traders: Europeans in the Age of Mercantilism* (New Delhi: Manohar, 2014).

Roy, Tirthankar, *The East India Company: The World's Most Powerful Corporation* (2012, reprint, New Delhi: Penguin Books India, 2016).

Sarkar, Jadunath, *Shivaji and His Times* (1919, reprint, New Delhi: Orient Longman, 1973).

Scammell, G.V., *The World Encompassed: The First European Maritime Empires c. 800–1650* (Berkeley and Los Angles: University of California Press, 1981).

Scammell, G.V., 'The Pillars of Empire: Indigenous Assistance and the Survival of the "Estado da India" c. 1600–1700,' *Modern Asian Studies*, vol. 22, no. 3 (1988), pp. 473–89.

Scammell, G.V., *The First Imperial Age: European Overseas Expansion c. 1400–1715* (1989, reprint, London/New York: Routledge, 1991).

Sen, S.N., 'Half a Century of the Maratha Navy,' *Journal of Indian History*, vol. 10 (1973), pp. 252–65.

Sen, S.N., *Foreign Biographies on Shivaji* (Calcutta: K.P. Bagchi, 1977).

Silburn, P.A., *The Evolution of Sea-Power* (London: Longmans, Green and Co., 1912).

Sivasundaram, Sujit, *Islanded: Britain, Sri Lanka & the Bounds of an Indian Ocean Colony* (Chicago: The University of Chicago Press, 2013).

Stevens, William Oliver and Westcott, Allan, *A History of Sea Power* (New York: Doubleday, Doran & Company, 1944)

Stockwell, A.J., 'British Expansion and Rule in South-East Asia,' in Andrew Porter (ed.), *Oxford History of the British Empire*, vol. 3, *The Nineteenth Century* (1999, reprint, Oxford: Oxford University Press, 2009), pp. 371–94.

Subramanian, Lakshmi, 'Bombay and the West Coast in the 1740s,' *Indian Economic and Social History Review*, vol. 18, no. 2 (1981), pp. 189–216.

Sykes, P.M. 1915. *A History of Persia*, vol. 2 (London: MacMillan & Co., 1915).

Winius, George D. and Vink, Marcus P.M., *The Merchant-Warrior Pacified: The VOC and Its Changing Political Economy in India* (Delhi: Oxford University Press, 1994).

PART V

The Impact of Gunpowder

17

ECONOMY AND GUNPOWDER WEAPONS IN WESTERN AND CENTRAL EUROPE

Philip T. Hoffman

Western and Central Europe were politically fragmented and had been so, almost without interruption, from the collapse of the Roman Empire in the West. Like other splintered parts of the globe, the polities in this part of Europe fought incessantly. In the early Middle Ages, most of the leaders of these polities were what we would today call warlords, controlling realms that possessed little permanent tax revenue. To reward the followers who battled on their behalf, the leaders gave them not money from taxes but grants of land with powers over people. Over time, these leaders became the lords and princes of medieval history, and to fund their fighting they eventually managed (typically by negotiating with elites) to impose permanent taxes. This achievement allowed them to turn their administratively weak dominions into strong states able to impose heavy taxes by the standards of the preindustrial world. Most of this tax revenue paid for continued conflict. In the nineteenth century, economic growth and representative institutions gave leaders more to spend, even during lulls in the fighting. The result was that by 1913—admittedly at the end of an arms race building up to the First World War—European powers and the United States (itself a relatively small military spender) accounted for nearly 85 per cent of the world's military expenditure (Eloranta 2007: 260).

From the end of the Middle Ages, this military expenditure went increasingly (and by the seventeenth century almost exclusively) on fighting with the technology of gunpowder weapons; that is, the technology of firearms, artillery, ships armed with guns, and the older weapons (swords, pikes, cavalry lances) that proved essential for gunpowder warfare through at least the sixteenth century. The gunpowder technology, it turns out, was ideally suited to conquest abroad. It helped a small number of Europeans to topple the Aztec and Inca Empires, and allowed them to gain toeholds in Asia by building fortresses that rebuffed assaults by enormous armies. In the late nineteenth century, more advanced gunpowder weapons and the invention of remedies against malaria made it possible for the Europeans to colonise Africa. By 1914, 84 per cent of the world was, or had been, a European colony.

Gunpowder technology was expensive—particularly the fortifications and the navies—and the costs rose as armies grew in size. By the First World War, millions of soldiers were being mobilised and supplied via the railroads. Surely all warfare and mass mobilisation and

DOI: 10.4324/9780429437915-23

the enormous spending on gunpowder technology had an economic effect. Further, the economy had an effect on the technology. What, then, were those effects?

They turn out to have been both direct and indirect. Some of the direct positive effects (on the metal-working industry, on ship design and navigation) were swamped by much larger direct negative effects, particularly in agriculture. The indirect effects, though, were more important. In particular, warfare with gunpowder weapons not only helped build stronger states; it also fostered positive interaction among cities that contributed to industrialisation and that seems to have been unique to Europe. But the most important indirect effect, at least within Europe, was the financial innovations devised to meet the costs of war and conquest that even heavy taxes could not fund: the first equity market and the first liquid markets for government debt.

Those innovations and the other indirect effects turn out to have been far more important for the European economies than all the colonies the Europeans created. Outside of Europe, colonisation by Europeans often harmed native economies. Within Europe, though, the colonies added little to economic growth and were likely to have been an economic drag.

Gunpowder Weapons in Western and Central Europe

Gunpowder and guns were invented in China. Gunpowder reached Europe in the thirteenth century (it was first mentioned in Europe in 1267), and the first guns (tiny handheld cannons, as in China) appeared early in the 1300s, and certainly by 1326.[1] These early guns were used, along with other defences, to protect cities from attackers who would try to scale city walls or undermine them. They began to appear on ships, once again just as in China. But then the development of gunpowder weapons in Western and Central Europe took off in a different direction from that in East Asia. Larger guns—cannons or artillery—were developed in Europe because they could punch holes in the walls protecting medieval cities. In China such a technique would have failed, because walls there were too thick. In Europe, by contrast, walls around cities and fortresses tended to be high and thin. Assaults with artillery paid off and were employed successfully in the Hundred Years War between England and France (1337–1453) and in the French invasion of Italy in 1494.

The cannons in Europe were improved constantly. They sprouted trunnions—stubby axles that emerge from the sides—so that they could be mounted on wheeled carriages. Ships began to be armed with heavier artillery too. Specialised carriages and water-tight gun ports were developed to speed artillery fire on ships and to make it possible to mount a growing number of cannons on multiple decks. As crews learned to shoot more rapidly, bombardment began to replace the boarding that had long been the goal of naval battles. In England, that transformation of naval warfare was already underway by the late sixteenth century.[2]

Those were far from the only advances with gunpowder technology in Western and Central Europe. Architects and military engineers devised fortifications that could resist artillery bombardment, starting with bastions that allowed defenders to fire upon attackers no matter how close they got to the walls of a fort or city. In response, attackers subjected cities and forts to lengthy sieges and built their own fortifications to defend themselves from counterattack by a relief army. They also invented clever siege tactics that could eventually overcome even the best fortifications: for instance, zig-zag trenches that would defeat the defenders' cross fire and rolling shields that would let sappers approach walls and mine them (Pepper and Adams 1986; Parker 1996; Ostwald 2007).

Handheld firearms also progressed. Lighter and more reliable matchlock muskets began to replace the small handheld cannons in the second half of the fifteenth century. In the seventeenth century, flintlock muskets equipped with bayonets pushed aside matchlocks. The flintlocks fired more rapidly than matchlocks and were less likely to misfire. The bayonets made it possible to do away with the pike men who had protected the infantry from cavalry charges and give virtually all infantry men guns. Firing rates jumped, and at sea, so did naval firepower (Hoffman 2015).

What drove all this sustained innovation—something unknown in most other sectors of preindustrial economies—was not just the incessant warfare in Western Europe but the focus on gunpowder technology and the heavy spending on it. It also helped that states were small and military experts circulated freely across Europe. Rulers hired experts or bought arms even from enemy countries, with the Holy Roman Emperor Charles V, for instance, being unable to stop gunsmiths in Nürnberg from selling to his arch enemy, the French King Francis I. Armies and navies quickly learned, even from their enemies, and advances therefore spread rapidly. Western and Central Europe were the only parts of Eurasia where these conditions so favourable to improving gunpowder technology happened to prevail all of the time. Although Asian and Middle Eastern States certainly advanced the technology, they would then falter or fall back. In the end they simply could not keep up with the Europeans' relentless pace.[3]

As a result, the Europeans took the lead in advancing gunpowder technology. The gap was apparent when Europeans arrived in China in the sixteenth century with matchlocks and armed ships. Their guns were copied, and attempts were made to imitate their armed ships too. As far as the guns were concerned, the Chinese caught up, but their ships lacked the European vessels' fire power and their ability to sail against the wind. Chinese siege tactics lagged behind too, and in 1800 their guns were still matchlocks, not the better flintlocks that had been standard in European armies for a century or more.

The gap in military technology widened even more in the late eighteenth and nineteenth century, as useful knowledge from the Enlightenment was applied to the gunpowder technology and as Europe industrialised. In the eighteenth century, for example, the British Navy experimented with metal sheathing for hulls in order to reduce the damage done by shipworms. Experiments—typical of the Enlightenment—showed that copper worked best but only if the ships' nails and fittings were replaced with a copper alloy. As a result, ships could stay at sea longer, and their speed increased because the copper kept the hull free of weeds and barnacles. Similarly, in the nineteenth century, advances in science, engineering and metal working during the Industrial Revolution yielded cheap accurate rifles, machine guns that made soldiers 80 times as lethal as the best flintlock musketeer, field artillery that was about 1,000 times deadlier than Napoleonic-era field cannons, and totally new weapons such as the torpedo, a masterpiece of physics and engineering.[4]

How did these improvements to gunpowder technology facilitate conquest and colonisation? There are numerous examples of how the technology helped small European forces to defeat numerically superior enemies or establish forts abroad. Consider, for instance, the conquistador Hernan Cortes. Thanks to muskets and artillery mounted in brigantines, and also the lancers on horseback, he managed to take the Aztec capital Mexico City. More important, gunpowder technology won him native allies who realised that his weapons complemented their numbers. Together, they could overthrow the Aztecs; separately, that would have been impossible. Similarly, in Southeast Asia, the European fortifications at the Portuguese fort in Malacca held off attacks by forces that outnumbered the Portuguese and their allies 10 to 1.

Similar forts spread throughout Southeast Asia, where they stored food, merchants' goods and ships' provisions, and allowed the Portuguese to prey on local shipping (Hoffman 2015: 8–13). And in the late nineteenth century, better firearms—including machine guns—helped Europeans colonise Africa, once they had effective remedies against malaria. After the conquests, gunpowder technology made it possible for relatively small numbers of Europeans to control territory (Pakenham 1991; Headrick 2010).

To be sure, gunpowder technology was not the only advantage the Europeans had here. Disease played a role in the Americas, and funds to pay allied native troops and their European officers were critical in the initial conquest of India. Co-opting native leaders was important too. But without gunpowder technology, the conquests and imperialism would have been impossible.

Direct Economic Effect of War with Gunpowder Weapons

What impact did all the spending on war have on the economy, particularly before the Industrial Revolution? It stimulated sustained technological advances in gunpowder technology. Did it do the same outside the military sector?

The answer is no; at least if we look for a direct connection between military inventions and new civilian technology. Before the Industrial Revolution, sustained technological change was unknown outside the military sector. There were certainly periodic technological advances throughout human history, but follow up was lacking, either because of a lack of fundamental knowledge or because information about the new technology did not circulate widely. The new technology might not spread, and improvements to build upon it would peter out. A farmer might adopt a new crop, for example, but only his neighbours would imitate him, and no one would then add a second novel crop.

So, before the Industrial Revolution there was at best sporadic technical change outside the military sector, and little contribution by the military to civilian technology either. War against Spain and the Muslims may have motivated the Portuguese to develop a ship that could sail in any ocean and to discover routes and navigational techniques that let them sail around Africa and on to South Asia. But demand for gold and spices continuously spurred the Portuguese on too (Headrick 2010: 20–32).

These limited additions to civilian technology were probably offset by all the damage warfare did to the economy in Western and Central Europe and all the costs it imposed on the civilian economy. Warfare interrupted trade on land and at sea, and arming merchant ships raised the expense of transportation. Worse yet, civilians had to flee the violence at the hands of undisciplined soldiers, who stole crops and farm animals and threatened or killed the unarmed, whether they were friend or foe. Typically, the fleeing civilians sought refuge in walled cities, but there many of them became sick and died. If their farms were abandoned for a year or more, the fields would have overgrown with weeds and brush. The land would have to be cleared again before planting could resume, and agricultural productivity could fall by 25 per cent for as long as a generation (Hoffman 1996: 185–6; 2015: 208; Gutmann 1980; Lynn 1997: 415–34).

What about during the British Industrial Revolution? It broke free from these constraints for the first time in human history, with rapid and continuing technology change in a number of industries, but particularly textiles, metals and transportation. Did all the warfare and spending on it directly contribute inventions to the Industrial Revolution? Here, the answer is yes; at least when it came to metals and the iron industry. In that sector, major inventors such as Henry Cort and John Wilkinson also supplied the British military, with Wilkinson devising boring machines that could manufacture cast iron cannons or steam engine cylinders for James Watt. The far larger and equally dynamic textile sector, however, had no connection to the military (Hoffman 2015: 209–10; Haber and Lamoreaux 2021, chapter 3). So during the Industrial Revolution, the military's direct contribution to civilian technology was limited too.

One might think that the twentieth century would be different, with defence spending spurring new technologies that could then be applied to civilian life. Examples immediately spring to mind, from jet engines to radar. But if those military technologies had an economic impact, then they would have stimulated economic growth, and a link would exist between defence spending and economic growth rates. There is little robust evidence for such a link, however, at least with data from after the Second World War (Ram 1995: 251–74). Furthermore, the technologies that military spending contributed to, such as the jet engines and radar, might well have developed without war.

Indirect Effects

The indirect effects that warfare had on the economy in Western and Central Europe were more important. They included states that were stronger, in the sense that they could raise more tax revenue from their economies and so pay for the huge expenses of fighting with gunpowder technology; financial innovations that allowed states to spread the funding of a war over time so that it did not all have to be paid for as the fighting went on; and last, but not least, growing urbanisation that linked nearby cities in Europe in a way that was unique in Eurasia. In addition, one could even claim (although there are problems with the argument) that these indirect consequences help explain why Western Europe industrialised before the rest of the world and why, within Western Europe, Britain was the first economy to do so.

The stronger states were gunpowder war's biggest impact in Western and Central Europe. In the Middle Ages, European polities by and large lacked substantial permanent tax revenue. Temporary taxes could be imposed when wars were being fought, in order to hire mercenaries and supplement the forces supplied by a prince's vassals and troops furnished by cities. But a truce would put a stop to the taxation, leaving soldiers furloughed and unpaid. They would then forage in the countryside and hold innocent civilians for ransom—a major problem, for example, during the Hundred Years War. Substantial permanent taxation was the solution, and in Western Europe the solution was usually reached via negotiation between the prince and elites, often during a war. When, during the Hundred Years War, the French King John II was defeated by the English and taken prisoner in 1356, there was a temporary peacetime tax imposed to pay for his ransom. By negotiating with elites, the King's son managed to get that levy turned into a permanent tax to stop the large numbers of marauding troops who were wreaking havoc throughout the country. Elsewhere, permanent taxation took longer. In Brandenburg-Prussia, for example, it came after the country had been ravaged during the Thirty Years War (1618–1648). By offering nobles greater power over their serfs, the ruler of Brandenburg Prussia got a representative assembly of towns and nobles to grant him a temporary tax in order to fund an army. He then used the army to quash resistance to making the taxes permanent, and bought the nobles' assent by offering them positions as officials or army officers (Hoffman 2015: 54, 120–1, 134–6).

Taxes rose even higher in the nineteenth century. In part, the greater tax revenue resulted from increasing incomes and economic growth. But much of the soaring revenues stemmed from efforts to make taxes uniform across an entire country and from negotiations with elites in national assemblies—efforts and negotiations often undertaken during wars. It may seem surprising, but giving a national assembly powers of the purse actually boosted tax revenue (Dincecco 2009, 2011).

The result was that the small states in Western Europe imposed far heavier taxes than in the Ottoman Empire or Imperial China. Comparisons have to be taken with a grain of salt because surviving tax figures concern revenue under control of the central government and not tax money that was collected and spent locally. They also do not reflect the large sums that could be raised during emergencies. Still, by this yardstick, per capita taxes were far higher in Western and Central Europe, and by the eighteenth century the difference was an order of magnitude. In 1750–1799, per capita taxes in France, Spain, Britain and the Netherlands were more than 15 times what they were in China, when measured in silver, and much higher than in the Ottoman Empire too. The gap in per capita taxes is also large when measured in days of unskilled labour. The central governments of France and Britain were each collecting more tax revenue than China despite its much larger population, which was some ten times that of France and over 20 times that of Britain. Imperial China and the Ottoman Empire had never developed representative institutions, and that was a major reason why they collected so much less per capita tax revenue than the European states. Their size turns out to have been an obstacle both to high taxes and to the development of representative institutions. In addition, China had an early bureaucracy which could collect taxes and provide information about what taxpayers could pay without rising up in revolt. European countries had no such bureaucracy, at least initially. That forced leaders to negotiate with elites, and the negotiations gave rise to representative institutions.[5]

Beyond their enormous tax revenues, Western European states also had the advantage in the sphere of financial innovations that helped them finance long-term military ventures and related commercial undertakings. During wars, innovations such as government bonds spread the costs of fighting over multiple years when tax increases during the during the hostilities would not suffice. If the bonds were secure and could be traded on exchanges, as in eighteenth-century Britain, the government's cost of borrowing fell because lenders knew they could easily be sold. The commercial projects benefitting from financial innovations included expeditions of conquest mounted by private mercantile enterprises that traded and waged war abroad on behalf of countries—firms like the Dutch East India Company and the English East India Company. The Dutch company was in fact the world's first joint stock company with an indefinite life span and tradable shares. That financial step in a new direction raised enormous sums for Dutch military and trading ventures in Southeast Asia. In Western Europe, other countries tried to imitate its success. But outside of Western Europe, similar financial innovations were lacking (Neal 1990; Gelderblom, de Jong et al. 2013; Hoffman 2015: 169–70).

There was one other contribution that war made to the economy in Western Europe: it concentrated manufacturing (and especially the artisanal production common before industrialisation) inside cities. In China, and perhaps in other large empires, much more manufacturing took place out in the countryside, where entrepreneurs could hire inexpensive labour during the agricultural off season. Such rural manufacturing or cottage industry certainly existed in Europe, with females in peasant households spinning in the winter and helping to take in the grain harvest in the summer and fall. But it seemed rarer in Western and Central Europe, and the reason, so it has been argued, was that more frequent warfare in Europe drove entrepreneurs to move their capital-intensive manufacturing operations into cities, where walls provided protection against rampaging land armies (Rosenthal and Wong 2011). Fields could not be moved into the city, but machines, tools and skilled artisans could, and workshops could be constructed behind urban walls too.

With less access to cheap off-season labour than in the countryside, wages in cities would be higher; the cost of urban housing would augment wages too. If Robert Allen is correct, the elevated urban wages would then encourage efforts to substitute capital for labour or, in other words, to invent machines. For Allen, that is one of the major reasons why Britain was the first economy in the world to industrialise. He would add that wages in Britain were also boosted by all the intercontinental trade Britain engaged in; trade that Britain dominated (at least in the eighteenth century) because of its victories in war.[6]

Here Joel Mokyr would disagree. In his view, Britain industrialised first because it had access to useful information from the Enlightenment and skilled mechanics and engineers who could put that information into practice by building, inventing and maintaining machines. Much of the rest of Western Europe had access to useful Enlightenment knowledge too, but only Britain had an abundant supply of skilled mechanics and engineers—essential human capital, in the language of economics. They had this human capital because they trained more apprentices and their apprentices were not limited by a restrictive guild system. There is evidence to support Mokyr's claim. If Allen were correct, unskilled workers would have tried to migrate from France or Germany to Britain, but no indication exists that they did. Skilled mechanics and engineers, by contrast, did try to leave Britain for the continent, despite legislation that barred them from departing. This is just what one would expect if continental Europe had had a limited supply of mechanics and engineers (Mokyr 2009; Kelly, Mokyr et al. 2012; Mokyr 2017).

The movement of manufacturing into European cities likely contributed to industrialisation in another, different way. The cities that grew in Europe tended to expand in clusters; population growth of nearby cities tended to be correlated from the Middle Ages onward. Within Eurasia, that correlation was unique to Western Europe. It was particularly strong where representative institutions existed and could help merchants and entrepreneurs bargain for freedom from tolls and from other regulations that restricted their ability to manufacture goods and trade in what they made. Such representative institutions existed in the Low Countries and in Britain, and they helped clusters of manufacturing cities thrive. The clusters in turn made it easy for entrepreneurs like Richard Arkwright, the famed inventor of the water frame spinning machine and the textile mill, to find mechanics and skilled artisans who could construct and maintain the machines he wanted to make.[7]

Conquests, the Slave Trade and War's Role in the Great Divergence

One definitive path connected the economy to gunpowder war in Western and Central Europe, or so historians claim. It traces economic growth in Europe back to the continent's colonies and its involvement in the slave trade. Because war with gunpowder weapons gave Europeans the technology needed for conquests abroad, historians maintain, it brought riches to Europe, from the silver mined in the Americas to the profits earned from the traffic in slaves. Beyond the flow of income, it also brought resources, from timber and food to the cheap cotton produced by slave labour in the Americas. Resources were an essential prerequisite for Europe's industrialisation. Food and timber freed Europe from the Malthusian constraints of a limited supply of land that would have choked off economic growth. Without imported food and timber, most Europeans would have had to toil as farmers in the countryside, and Western and Central Europe would never have grown rich. In short, there would not have been a 'great divergence' between, on the one hand, a prosperous Western and Central Europe (and European clones like the United States) and, on the other, a much poorer rest of the world.[8]

This line of argument is influential, but on closer inspection, it falls apart. The riches from the American silver mines funded Spanish wars in Europe, not economic growth, and if the profits from the slave trade had been the fuel that empowered the Industrial Revolution, then Portugal would have been the first Western European country to industrialise, for over 12 times as many slaves were landed in its colony of Brazil than in North America.[9] As for food imported from colonies, this was a minor part of the calorie intake in Western and Central Europe until the second half of the nineteenth century, when railroads and cheap trans Atlantic steamships brought American grain across the Atlantic. But by then industrialisation had already begun in Belgium and France, and it had been underway since the eighteenth century in Britain, all without food imports, which could not have been a prerequisite for industrialisation. To be sure, there was some food shipped to Europe before 1850. But it was by and large sugar and tea for middle- or high-income consumers in Western Europe, not necessities that warded off famines.[10] And, in any case, famines had disappeared from England, France and Italy before 1800.

The argument about cotton and slavery also collapses if we look at the evidence. The British Industrial Revolution was much more than cotton textiles. It was iron, transportation, arms, wool and linen textiles, and a wide variety of hardware (Mokyr 2009). And cotton did not have to be produced by slaves on American plantations. It was cultivated by peasants and small farmers in many parts of the world, including many free white farmers in the United States after the Civil War. It did not have to be produced by slaves or in colonies either.

Conclusion

In Western and Central Europe, war with gunpowder weapons stimulated innovation in armaments, the metals sector, and ocean shipping and navigation. It also created strong states in Europe and spurred financial innovations such as the stock market and government bond markets. Its overall effect on the economy, however, was likely negative. It killed people, interrupted trade and destroyed capital, from livestock to the enormous amount of labour invested in clearing fields. True, it also gave Europe colonies and the resources the colonies produced. But the economies of the colonies often suffered a great deal of harm, and if Britain is an example, imperialism ended up being a drag on the mother country's economy too (Davis and Huttenback 1986; Acemoglu, Johnson et al. 2002; Engerman and Sokoloff 2012).

The clearest connection between gunpowder warfare and the economy went in the reverse direction, from new technology in the industrialising parts of Central and Western Europe to the way wars were fought. The industrial revolution made possible rifled artillery, needle guns, railroads, and those inventions changed warfare, with rail transport making it possible to mobilise and supply larger armies than ever before. But before those inventions could be used effectively, tactics and military organisations had to adapt, and even political institutions had to change (Showalter 1976; Onorato, Scheve et al. 2014). It is no surprise that extensions of the franchise in Europe tended to parallel the mass mobilisation for the armies that railroads supplied.

Notes

1 For the origin of gunpowder weapons in China and their arrival in Europe, see Andrade 2016 and the sources he cites. His book also gives an up-to-date account of the different path that the early development of firearms took in Europe and China. ·

2 For a description of these changes, the particular example of the English Navy, and an argument about why there was so much innovation, see Parker 1996; Martin and Parker 1999; Hoffman 2015.

3 The argument for this and the two following paragraphs is taken from Hoffman 2015, with the example from Nürnberg coming from Willers 1973: 237–301. For a different perspective and comparisons with the rest of Eurasia, see Chase 2003; Andrade 2010; Andrade 2011.

4 These last examples are taken from Dupuy 1984; Dupuy 1985; Headrick 2010; Epstein 2014.

5 For the tax comparisons, see Brandt, Ma et al. 2014, table 3; Hoffman 2015: 50–1. The arguments about the state size, taxes and the development of representative institutions come from Stasavage 2009 and Stasavage 2016 and Sng 2014.

6 Here I am weaving together the arguments in Allen 2009 and Rosenthal and Wong 2011.

7 The argument here is from Cox 2017, and the example of Arkwright and the textile industry comes from Fitton 2012; Cookson 2018.

8 For a perceptive and critical review of this literature, see Vries 2013. The seminal work on the Great Divergence is Pomeranz 2000. Beckert 2015 is a recent version of the argument about cotton.

9 Klein 2010, Appendix A.2, pp. 216–17.

10 Findlay and O'Rourke 2007: 328–9; Broadberry, Campbell et al. 2014: 286–9. One new crop from the new world—the potato—did help to feed growing urban populations in Europe, particularly in Ireland and the Low Countries. But British consumers spurned potatoes for more costly bread and exotic luxury goods such as tea, coffee and sugar. And elsewhere potatoes did not ward off famine; rather, they led people to marry earlier and have more children, leaving them vulnerable if the potato crop failed, as in Ireland. Nunn and Qian 2011; Berger 2019; Gráda 2019; Mokyr 2013; and Mokyr 2009: 178–82.

Bibliography

Acemoglu, D., Johnson, S., et al., 'Reversal of Fortune: Geography and Institutions in the Making of the Modern World Income Distribution,' *The Quarterly Journal of Economics*, vol. 117, no. 4 (2002), pp. 1231–94.

Allen, R.C., *The British Industrial Revolution in Global Perspective* (Cambridge: Cambridge University Press, 2009).

Andrade, T., 'Beyond Guns, Germs, Steel: European Expansion and Maritime Asia, 1400–1750,' *Journal of Early Modern History*, vol. 14 (2010), pp. 165–86.

Andrade, T., *Lost Colony: The Untold Story of China's First Great Victory over the West* (Princeton: Princeton University Press, 2011).

Andrade, T., *The Gunpowder Age: China, Military Innovation, and the Rise of the West in World History, 900–1900* (Princeton: Princeton University Press, 2016).

Beckert, S., *Empire of Cotton: A Global History* (New York: Vintage, 2015).

Berger, T., 'Adopting a New Technology: Potatoes and Population Growth in the Periphery,' *The Economic History Review*, vol. 72, no. 3 (2019), pp. 869–96.

Brandt, Loren, Ma, Debin, and Rawski, Thomas G., 'From Divergence to Convergence: Re-evaluating the History Behind China's Economic Boom,' *Journal of Economic Literature*, vol. 52, no. 1 (2014), pp. 45–123.

Broadberry, S., Campbell, B. et al. (2014). 'British Economic Growth, 1270–1870: An Output-based Approach,' 2014. Retrieved 5 March 2022 from https://gpih.ucdavis.edu.

Chase, Kenneth, *Firearms: A Global History to 1700* (Cambridge/New York: Cambridge University Press, 2003).

Cookson, G., *The Age of Machinery: Engineering the Industrial Revolution, 1770–1850* (Woodbridge: Boydell and Brewer, 2018).

Cox, G.W., 'Political Institutions, Economic Liberty, and the Great Divergence,' *Journal of Economic History*, vol. 77, no. 3 (2017), pp. 724–55.

Davis, L.E. and Huttenback, R.A., *Mammon and the Pursuit of Empire: The Political Economy of British Imperialism, 1860–1912* (Cambridge: Cambridge University Press, 1986).

Dincecco, M., 'Fiscal Centralization, Limited Government, and Public Revenues in Europe, 1650–1913,' *Journal of Economic History*, vol. 69 (2009), pp. 48–103.

Dincecco, M., *Political Transformations and Public Finances: Europe, 1650–1913* (Cambridge: Cambridge University Press, 2011).

Dupuy, T.N., *The Evolution of Weapons and Warfare* (New York: Da Capo, 1984).

Dupuy, T.N., *Numbers, Predictions and War: Using History to Evaluate Combat Factors and Predict the Outcome of Battles* (Fairfax, VA: Hero, 1985).

Eloranta, J., 'From the Great Illusion to the Great War: Military Spending Behaviour of the Great Powers, 1870–1913,' *European Review of Economic History*, vol. 11, no. 2 (2007), pp. 255–83.

Engerman, S.L. and Sokoloff, K.L., *Economic Development in the Americas since 1500: Endowments and Institutions* (Cambridge: Cambridge University Press, 2012).

Epstein, K., *Torpedo: Inventing the Military-Industrial Complex in the United States and Great Britain* (Cambridge, MA: Harvard University Press, 2014).

Findlay, R. and O'Rourke, K.H., *Power and Plenty: Trade, War, and the World Economy in the Second Millennium* (Princeton: Princeton University Press, 2007).

Fitton, R.S., *The Arkwrights: Spinners of Fortune* (Matlock: Derwent Valley Mills Educational Trust, 2012).

Gelderblom, O., de Jong, A. et al., 'The Formative Years of the Modern Corporation: The Dutch East India Company VOC, 1602–1623,' *The Journal of Economic History*, vol. 73, no. 4 (2013), pp. 1050–76.

Gráda, C.Ó., 'The Next World and the New World: Relief, Migration, and the Great Irish Famine,' *The Journal of Economic History*, vol. 79, no. 2 (2019), pp. 319–55.

Gutmann, M.P., *War and Rural Life in the Early Modern Low Countries* (Princeton: Princeton University Press, 1980).

Haber, S. and Lamoreaux, N. (eds.), *The Battle over Patents: History and Politics of Innovation* (Oxford: Oxford University Press, 2021).

Headrick, D.R., *Power over Peoples: Technology, Environments, and Western Imperialism, 1400 to the Present* (Princeton: Princeton University Press, 2010).

Hoffman, P.T., *Growth in a Traditional Society* (Princeton: Princeton University Press, 1996).

Hoffman, P.T., *Why Did Europe Conquer the World?* (Princeton: Princeton University Press, 2015).

Kelly, M., Mokyr, J. et al., *Precocious Albion: Human Capability and the British Industrial Revolution* (Brussels: Centre for Economic Policy Research, 2012).

Klein, H.S., *The Atlantic Slave Trade* (Cambridge: Cambridge University Press, 2010).

Lynn, J.A., *Giant of the grand siècle: the French Army, 1610–1715* (Cambridge/New York: Cambridge University Press, 1997).

Martin, C. and Parker, Goeffrey, *The Spanish Armada* (Manchester: Manchester University Press, 1999).

Mokyr, J., *The Enlightened Economy: An Economic History of Britain 1700–1850* (New Haven, Yale: Yale University Press, 2009).

Mokyr, J., *Why Ireland Starved: A Quantitative and Analytical History of the Irish Economy, 1800–1850* (London: Routledge, 2013).

Mokyr, J., *A Culture of Growth: The Origins of the Modern Economy* (Princeton: Princeton University Press, 2017).

Neal, L., *The Rise of Financial Capitalism: International Capital Markets in the Age of Reason* (Cambridge: Cambridge University Press, 1990).

Nunn, N. and Qian, N., 'The Potato's Contribution to Population and Urbanization: Evidence from a Historical Experiment,' *The Quarterly Journal of Economics*, vol. 126, no. 2 (2011), pp. 593–650.

Onorato, M., Scheve, K. et al., 'Technology and the Era of the Mass Army,' *Journal of Economic History*, vol. 74, no. 2 (2014), pp. 449–81.

Ostwald, J.M., *Vauban under Siege: Engineering Efficiency and Martial Vigor in the War of the Spanish Succession* (Leiden: Brill, 2007).

Pakenham, T., *The Scramble for Africa* (New York: Random House, 1991).

Parker, Geoffrey, *The Military Revolution: Military Innovation and the Rise of the West, 1500–1800* (Cambridge/New York: Cambridge University Press, 1996).

Pepper, S. and Adams, N., *Firearms and Fortifications: Military Architecture and Siege Warfare in Sixteenth-Century Siena* (Chicago: University of Chicago Press, 1986).

Pomeranz, Kenneth, *The Great Divergence: China, Europe, and the Making of the Modern World Economy* (Princeton, NJ: Princeton University Press, 2000).

Ram, R., 'Defense Expenditure and Economic Growth,' in K. Hartley and T. Sandler (eds.), *Handbook of Defense Economics* (Amsterdam: Elsevier, 1995), pp. 251–74.

Rosenthal, J.-L. and Wong, R.B., *Before and Beyond Divergence: The Politics of Economic Change in China and Europe* (Cambridge, MA: Harvard University Press, 2011).

Showalter, Dennis, *Soldiers, Technology, and the Unification of Germany* (Hamden, CT: Archon, 1976).

Sng, T.-H., 'Size and Dynastic Decline: The Principal Agent Problem in Late Imperial China 1700–1850,' *Explorations in Economic History*, vol. 54 (October 2014), pp. 107–27.

Stasavage, D., *Geographic Scale and Constitutional Control: The Example of European State Formation* (New York: New York University Press, 2009).

Stasavage, D., 'Representation and Consent: Why They Arose in Europe and Not Elsewhere,' *Annual Review of Political Science*, vol. 19 (2016), pp. 145–62.

Vries, P., *Escaping Poverty: The Origins of Modern Economic Growth* (Vienna/Goettingen: Vienna University Press, 2013).

Willers, J.K.W., *Die Nürnberger Handfeurerwaffe bis zur Mitte des 16. Jahrhunderts: Entwicklung, Herstellung, Absatz nach archivalischen Quellen* (Nürnberg: Stadtarchiv Nürnberg, 1973).

18

GUNPOWDER AND THE RISE OF MUSCOVY

ca.1200–ca. 1800

Kaushik Roy

Introduction

The Grand Duchy of Muscovy was a minor power in North-East Russia during the late thirteenth century. To the west, Muscovy had to tackle Poland, Lithuania and Sweden. And to the south lay the steppe nomads and the Ottoman Empire (established in 1299). Only its eastern frontier was free from the threat of any big power. However, by the eighteenth century, Muscovy or Rus, or imperial Russia, emerged as one of the strongest powers in Eurasia. The role of gunpowder weapons in this transformation cannot be ignored. Many scholars have looked at this transformation through the paradigm of the Military Revolution. For instance, Marshall Poe, in an article, asserts that a Military Revolution occurred in Muscovy in the first half of the seventeenth century, which transformed this principality from a medieval entity to a modern state (Poe 1996: 603–18). In a similar way, Michael C. Paul asserts that between the late sixteenth and early seventeenth centuries, the Russian Army underwent radical tactical and organisational changes which had a tremendous impact on the society. This paved the way for modernisation/Westernisation under Peter the Great (1672–1725; Paul 2004). Several others have pointed to the role of individual rulers (like Ivan IV, Peter the Great and Catherine II) as an explanation for the rise of Muscovy (Waliszewski 2006).

In my opinion, the military rise of Muscovy was a slow and gradual process which lasted for over five centuries. And this process was dependent both on contingent and structural factors. So, this chapter takes a different approach. Instead of the overused concept of Military Revolution or emphasising the role of great individuals overtly, I will opt for an analysis of the various interrelated themes which played a crucial role in the rise of Muscovy. The first four themes deal with different military branches and the last theme focuses on the state–society relationship in generating military power. Furthermore, this piece attempts to study the growth of Muscovite military power from a comparative perspective. Within each theme, a chronological order is maintained as far as possible.

Cavalry

One may argue that the history of Russia started with the migration of the Eastern Slavs into Ukraine between the sixth and eighth centuries. The heart of the Kievan state was the

DOI: 10.4324/9780429437915-24

Dnepr Valley, along the river route that ran from Novgorod in the north (the gateway to the Baltic) down south, past Kiev to the Black Sea. Kiev became the capital in the ninth century. Russia's conversion to Christianity came at the end of the tenth century. Following the baptism of the Grand Prince Vladimir in 988, Kiev became the seat of the metropolitan Russian Orthodox Church. However, separatist tendencies were present in outlying regions like Novgorod, Vladimir-Suzdalia and Galicia-Volhynia. The early Rus (inhabitants of these regions including Muscovy) were descendants of the Scandinavians (Halperin 1987: 10; Shaw 2006: 19–20). The principal threat for Muscovy lay in the south.

The Golden Horde/Kipchak Khanate ruled over South Russia from 1240 to 1480. Since the Mongols were pastoral nomads, they did not annex or settle in the Russian forest zone. At the turn of the fourteenth century, Islam became the state religion of the Golden Horde (Halperin 1987: 7). By the 1390s, the Golden Horde was seriously weakened by two factors: Black Death and Timur's invasions (Halperin 1987: 29).

The steppe nomadic military forces consisted primarily of cavalry equipped with composite reflex bows and lances. Against them, sword- and lance-equipped Lithuanian and Polish cavalry failed. So, Muscovy mimicked the nomadic force and built its ground force around a core of mounted horse archers. Still, the Muscovite mounted archers were not as good as the steppe nomadic mounted archers in terms of speed, manoeuvrability and archery. Nevertheless, the Muscovite mounted archers, equipped with reflex composite bows and deployed like a steppe nomadic army in the battlefield divided into five groups (centre, front, rear, left and right), were able to defeat their East European opponents, who were equipped like Western knights. For instance, in 1456, 5,000 Novgord cavalry equipped with lances and covered in heavy armour were defeated by the Muscovite horse archers. The Muscovite mounted archery took a heavy toll on the Novgord's mounts. When the Muscovite mounted archers started shooting arrows, the Novgord's horses took fright and threw off their riders from their saddles. Once the Novgord cavaliers clad in heavy armour fell to the ground, they were helpless as they could neither move nor use their heavy lances (Ostrowski 2002: 23).

By the 1420s, Crimea had broken away from the Kipchak Khanate, and by the 1440s, Kazan on the Upper Volga had also become independent. The Kipchak Khans' power was limited to Lower Volga centred around the old Golden Horde capital of Sarai (Davies 2007: 2). The disintegration of the Kipchak Khanate, in the long run, paved the way for the annexation of its successor polities by Muscovy. However, the steppe nomads still remained a power to reckon with till the reign of Catherine II (r. 1762–1796). The symbiotic alliance between Muscovy and the Crimean Khanate between 1470 and 1509 enabled Grand Prince Ivan III (r. 1462–1505) to secure Muscovy against attack by the Kipchak Horde and gain territory at the expense of Lithuania. The alliance with Muscovy also enabled the Crimean Khans to exact tribute from Poland–Lithuania (Davies 2007: 13).

From 1510 onwards, relations between the Crimean Khanate and Muscovy deteriorated. In 1571, Khan of Crimea Devlet Girei invaded with an army of 40,000 Crimeans, Nogais and Circassians and burned much of Moscow. This raid resulted in the death of 80,000 Russians, and 150,000 captives were carried away. The Khan's army comprised primarily cavalry and a few artillery pieces supplied by the Ottomans (Davies 2007: 16–17). The Crimean Khanate lasted till 1783 and carried out raids almost annually. The potential military strength of the khanate came to about 250,000 cavalry (Ostapchuk 2012: 158–9). It must be noted that slow-moving infantry armed with pikes and slow-firing matchlocks with limited range and cumbersome field artillery were nearly useless against fast-paced dispersed nomadic cavalry warfare. To check wide-ranging steppe nomadic cavalry raids, Muscovy had to use cavalry as well as build fortresses and defensive lines. During the sixteenth century, Muscovy's enemies in the south and in the west also deployed primarily cavalry forces. For instance, in 1567, the Grand Duchy of Lithuania had 24,400 cavalry and only 3,600 infantry (Kupisz 2012: 64).

In the mid-sixteenth century, the core of Muscovy's force comprised cavalry provided by the provincial landlords. Most of the large landholders of Muscovy were *pomeshchiki*, i.e., owners of landed estates (*pomestia*) granted to them by the government. Muscovy's *pomeste* system was initially organised by Ivan III. When he annexed Novgorod in 1478, the landed properties owned by the city's boyars, monasteries and bishops were confiscated and then re-distributed to the men who relocated to the Novgorod lands and served the grand prince as military servicemen. In the sixteenth century, especially under Vasily III (r. 1505–1533), this system spread further within the territories of Muscovy. Thus, the *pomeste* system was a form of conditional land tenure. In exchange for enjoyment of the landed estates, the owners had to utilise the income derived from such lands to serve the grand prince fully equipped and mounted on horses during campaigns. Thus, the *pomeste* system was somewhat equivalent to the Abbasid *Caliphate*'s *iqta*, Mughal *jagir* and the Ottoman *timar* system. Probably, the *pomeste* system was influenced by the Islamised steppe nomads' *iqta* system. However, by the end of the sixteenth century, the *pomeste* system was failing. Rising taxes resulted in depeas-antisation (flight of the peasants) and the *pomeste* owners failed to get the surplus required for buying mounts, weapons and provisions (Martin 2012: 20–1). The *timar* and the *jagir* system at least worked in the fertile lands of Anatolia, Mesopotamia and North India; but such land tenures for mounted military service were unsuitable in the infertile land of Russia with a sparse population. Whatever may have been the limitations of the *pomeste* cavalry, it must be noted that this mechanism provided the grand princes of Muscovy with a mobile military force under their direct control which could be mobilised at a moment's notice. Before the introduction of *pomeste* cavalry, the grand princes were dependent on their relatives' and vassals' military contingents for survival (Ostrowski 2006: 215). So, *pomeste* cavalry not only increased the power of the grand prince, but also aided in the expansion of Muscovy.

In the second half of the seventeenth century, the Muscovite cavalry replaced their composite reflex bows with firearms (pistols and carbines) and swords. This was done, despite the fact that it was extremely difficult to use a carbine from horseback, particularly when the mount was trotting or galloping. And pistols had a very short range. Further, till the mid-nineteenth century, in terms of rate of fire, accuracy and penetrative power, the composite reflex bow was superior to the smoothbore musket. Still, by the 1680s, the Muscovite cavalry had rejected the composite reflex bow in favour of firearms. This was due to the decline of the military power of the steppe nomads in South Russia. Muscovy no longer required a large number of mounted archers to confront the deadly steppe nomadic cavalry. Furthermore, for countering nomadic cavalry forays, instead of relying solely on the symmetrical response of deploying cavalry forces, Muscovy also opted for the construction of defensive lines. Again, as the Golden Horde's successor polities were disappearing, the source of good composite bows was drying up. After all, the nomads made the best bows. And it was dangerous for Muscovy to depend on their enemies for the supply of essential military equipment. Furthermore, it took a lot of time and energy to produce a good mounted archer out of a sedentary peasant. Since the czars/tsars were intent on increasing the size of the army quickly, in the course of the seventeenth century they opted for gunpowder infantry supplemented by sword-equipped cavalry. A peasant could be trained easily and quickly in firing musket, and traditionally the Rus noblemen had fought with swords mounted on horseback. Finally, as Donald Ostrowski argues, the Muscovite cavalry experienced a shift in the tactical doctrine.

In battle, they were no more the premier arm but played a supplementary role of guarding the gunpowder infantry and then launching mopping up operations (Ostrowski 2010: 513–34).

Infantry

Ivan IV (1530–1584) did not neglect the importance of disciplined gunpowder-equipped infantry. The first Westernised infantry formations, i.e., 1,000 men equipped with matchlocks, were established in 1512. In the 1540s and 1550s, he established *streltsy* (musketeers) as an elite corps stationed in Moscow and provincial cities. The musketeers were maintained and outfitted at government expense. By the middle of the seventeenth century, their service had become lifelong and hereditary. They became an elite military caste enjoying certain privileges and resided with their families in their own settlements (*slabody*) within Moscow. Between 1632 and 1681, the strength of the *streltsy* rose from 34,000 to more than 50,000. From a comparative perspective, the number of handgun-equipped *Janissaries* between 1567 and 1680 rose from 12,798 to 54,222. The *streltsy* behaved like the *Janissaries*. In 1698 the *streltsy* mutinied due to arrears of pay and the bitterness felt due to military service in winter. Peter the Great (r. 1682–1725), however, was able to suppress the rebellion (Ralston 1990: 17; Agoston 1999: 135; Herd 2001: 110–30). For setting up Westernised infantry, the tsars, like the rulers of premodern India, depended on foreign mercenaries. In 1632 there were more than 1,500 German mercenaries in the Russian Army (Ralston 1990: 20).

In the late sixteenth century, the Russian commanders were reluctant to deploy infantry musketeers in the open field, preferring to place them in static defensive positions strengthened with natural obstacles like riverbanks, forests, etc. At times, the Rus commanders also deployed their gunpowder infantry behind manmade obstacles like earthworks, palisades, gabions, wagons and *guliai gorod*. The *guliai gorod* were portable wooden walls carried on wagons. In the battlefield, they were bolted with iron nails and the musket-equipped infantry were stationed behind them (Davies 2012b: 93–4). The *guliai gorod* was a typical Russian innovation. Instead of using pikemen for protecting the gunpowder infantry, while the latter were reloading their handheld firearms against hostile cavalry charge, the Muscovite infantry used *guliai gorod* (Davies 1999: 154). In this respect, the Muscovite infantry tactics differed from the Western model.

By the time of Poltava (28 June 1709), the Russian infantry had undergone substantial managerial and technical improvements. Bayonet, which was in use by the European infantry at the end of the seventeenth century, was adopted by the Russian infantry just before Poltava. In 1708, Petrine infantry was equipped with a .78 calibre, 6.3 kg smoothbore muzzle loading musket with a range of 750 to 1,000 feet (Hellie 2005: 237–8). The infantry equipped with flintlock muskets displayed good regimental and battalion level manoeuvrability in moving from column to files and vice versa, both on the line and during march. In addition, Peter's infantry was capable of forming infantry squares. Furthermore, the infantry was capable of providing interlocking and enfilading fire (Brown 2009/10: 111). The 1755 infantry drill regulation stressed the development of firepower over shock action. Like the Prussian infantry, Russian infantry in accordance with this drill regulation was deployed in the battlefield in three lines. Further, the regulation emphasised the delivery of frontal, cross and oblique fire (Menning 2005: 278).

Nevertheless, it must be noted that in the Petrine army, the infantry, compared with the Western European armies, played a somewhat reduced role. In the Western European armies, infantry comprised 75 per cent of the force, but in Peter's army foot soldiers comprised less than 60 per cent of the field force (Stevens 2002: 153). There is a reason behind such a development. Historian Robert I. Frost argues that in North-East Europe between 1550 and 1720, Western-style gunpowder infantry organised in linear fashion was not always successful. In fact, the most feared force of the Thirty Years' War (1618–1648), the Swedish Army depended primarily on 'cold steel,' i.e., a cavalry charge with swords for delivering the killing blow in the battlefield. And the excellent Polish hussars were a woe to many Western-style, infantry-centric armies (Frost 2014). Neither the artillery nor gunpowder infantry, but rather the dragoons (units able to tackle both hostile infantry and cavalry) raised by Peter were responsible for the decisive victory at Poltava over the Swedes. The dragoons were equipped with carbines, pistols and broad swords. Further, the dragoon regiments had organic 3-pounder field artillery support (Ostrowski 2009/10: 81–106).

Artillery, Forts and Fortresses

Gunpowder weapons changed both the face of battle as well as the contours of military architecture. Gunpowder first emerged in China and the Mongols came to know about it during Chingiz Khan's war with the Song Empire in Central and South China. Probably, the Russians acquired their knowledge of gunpowder weapons from the Mongols who introduced this weapon technology in India, Persia and Western Eurasia. In 1376, while besieging the city of Great Bolgar on the Volga River, the Russian forces encountered firearms including clay grenades. In 1380, the Russians used small calibre guns against the Tatars (steppe nomads). These may have been some sort of flame thrower. Muscovy adopted *tufaks* (small cannon and mortar) from the steppe nomads. The German knights in Livonia started using cannons in 1382 and the Lithuanians in 1384. Probably, the Russians picked up the knowledge of casting cannons from the Lithuanians (Esper 1969: 187–8). Small calibre pieces made of rolled sheet iron reinforced with iron bands, made by the Germans and used by the Livonians, were also introduced in Russia. By 1494, Muscovy had established a cannon casting foundry and a powder yard near Kremlin to manufacture bronze guns and munitions. Granulated powder and iron cannonballs were manufactured in the 1520s. Horse-drawn wheeled gun carriages also became available at this time (Davies 2002b: 17). In response to the threat posed by Muscovy and Sweden, during the 1580s, Poland updated its artillery with the aid of Hungarian and German mercenaries (Duffy 1997: 167–8).

Until the mid-seventeenth century, like the Ottomans and the Mughals, the Russians also preferred large numbers of heavy bombards for besieging forts. Such guns however were not very successful in breaching the walls of forts. The Muscovite siege techniques involved lobbing incendiary shots over the fortress walls to start fires among the buildings within the fort and then capturing the walls of the fortress by launching infantry assaults (Davies 2012a: 3). The conquest of Kazan in 1552 by Muscovy was the first step in the gradual destruction of the steppe nomadic power. The ramparts of the town were constructed with wood and earth and it was garrisoned by 30,000 troops. Between 23 August and 2 October 1552, 150 pieces of ordnances (serpentines, falconets, mortars, etc.) of Ivan IV were able to demolish the walls. He followed the artillery barrage with a successful infantry assault (Waliszewski 2006: 153–4).

The eighteenth century witnessed more emphasis on firepower. In the 1740s, regimental artillery was increased to four guns and two mortars per infantry regiment (Menning 2002a: 52). What was important was that the Russian artillery branch was not in a state of stasis but experienced continuous innovation. Some of the innovations in the sphere of artillery were successful, but others were not. The light cannons of the Russians during the 1750s were below Western European standards in terms of quality. The barrels were rather heavy and the carriages were clumsy. There were no centralised manufacturing units for constructing standardised carriages. The carriages of the light field guns were often made by the regiments themselves (Duffy 1981: 68). Again, tiny field mortars were designed by the artillery officer Mikhail Vasilevich Danilov, who was influenced by Saint-Remy's *Mémoires d'Ártillerie*. These pieces were sent to the Russian Army in 1757 but soon proved useless in combat conditions. However, Peter Ivanovich Shuvalov's secret howitzer was a successful project. He designed a howitzer with a bore that was not round but oval. And this piece scattered its small shot widely. This howitzer had spindly wheels and screw elevation. However, the loading of these howitzers was a slow process and especially dangerous against a cavalry charge. These secret howitzers were introduced in 1758 and proved useful against the Prussian Army of Frederick the Great. They remained in service till the 1780s (Duffy 1981: 69–71).

During the medieval period, in treeless South Russian steppe, important cities like Kiev and Belgorod were fortified with stones and ramparts composed of unfired bricks (Shepard 2006: 47–68). But, in North-East Russia, due to the prevalence of forest, fortifications were made of wood which was highly vulnerable to fire. Gradually, Muscovy shifted to stone architecture. In 1368, a three-day siege of Moscow by Lithuania and Tver failed, thanks to the stone walls protecting the city (Martin 2006: 166). During the sixteenth century, the new European military architecture was based on mutually supporting bastions which were low and relatively invulnerable to artillery attack, yet studded with well-hidden cannons to sweep all assault routes with a withering enfilading fire. The rampart-bastion-artillery defended fort system was part of the so-called 'European Military Revolution' (Arnold 1999: 31, 33).

Trace italienne came slowly to North-Eastern Europe. In 1580, Pskov possessed strong ramparts with a surrounding palisade (Waliszewski 2006: 294). Smolensk commanded the main road from East Europe to Moscow. The Smolensk fortress, rebuilt by Muscovy in the 1590s, was the largest in Europe. More than 4 miles in length, the citadel's crenellated walls were between 10 and 23 feet thick and between 26 and 40 feet tall, and studded with 38 towers. The foundations were dug more than 12 feet deep and a system of revetted listening galleries projected 30 feet from the foot of the walls (Duffy 1997: 169; Brown 2012: 278). When 12,000 Poles started the Siege of Smolensk on 19 September 1609, the Russian garrison, under their experienced and skilful engineers, excavated secret galleries, dug under the foundations and burst the mines constructed by the Polish besiegers. In other words, the Muscovite garrison was adept at countermining practices (Duffy 1997: 169).

Besides forts, Muscovy also constructed long lines of fortification like the Chinese Wall, Antonine and Hadrian Walls, and the Roman lines along Danube. Muscovy's fortified lines were designed to check the forays of the steppe nomadic cavalry raids. The construction of successive fortified lines backed by armed peasant militia enabled Muscovy to colonise the steppe in the long run. Before the founding of Kozlov in 1635, the 120-km stretch of forest and steppe between the Voronezh River in the west and the Tsna River in the east was vulnerable to attacks by the steppe nomads. Voronezh was a garrison town but it was 180 km from Kozlov. A small garrison town was Lebedian, which was 70 km from Kozlov. The wall was constructed of earth and garrisoned with peasant militia equipped with firearms (Davies 1992: 482–6).

We are told that the Habsburg Army underwent a Scientific Revolution during the late seventeenth and early eighteenth centuries. In the field of military engineering and siege warfare, significant advances occurred both in the spheres of scientific-technical and managerial branches (Lund 2012: 199–248). How far was Russia successful in this regard?

Printed manuals on gunnery, tactics, drill, fortification and siegecraft spread military techniques far more rapidly, writes Jeremy Black (1999), than word of mouth or manuscript. The first Russian printed book on military matters was published in 1647. It was an infantry manual which emphasised training and included a large number of sketches and plans to teach drill. The manual also noted the importance of military education. Black continues that literacy and printing had other values. They fostered discussion of military organisation and methods and encouraged a sense of system. In 1680, the official report of Russian ordnance listed all arms and ammunition as well as the level of current production (Black 1999: 20). The Russian Army's huge engineering corps, with equipment for building pontoon bridges and constructing roads, impressed the Western observers (Keep 2002: 201).

Navy

Peter the Great can be termed the 'father of the Russian Navy.' Peter, like Kaiser Wilhelm II (1859–1941), was a naval enthusiast. What is more, Peter himself was an expert on naval and maritime matters and a licensed boat builder (Kaganov and Monas 1991: 759). However, Peter did not establish a Blue Water Fleet, but a Brown Water Fleet geared for littoral warfare. The two maritime theatres that interested him were the Black Sea and the Baltic. In the summer of 1695, Peter launched an attack against the Turkish fortress of Azov, which withstood the siege, thanks to the regular supply of food and weapons to the garrison by the Turkish Fleet. This failure emboldened Peter to establish a Russian Fleet in the Black Sea. The tsar immediately set up a naval shipyard at Voronezh on the river Danube for constructing galleys and ships. The galleys were built on the model of a Dutch galley which Peter had ordered from Holland. The new fleet enabled Peter to take Azov on 18 July 1696 (Grey 1961: 627; Salvo 2015: 43).

Lack of naval power prevented Muscovy from capturing Reval and Riga and controlling the Eastern Baltic against Sweden. Peter depended on foreign mercenaries from Britain, France and Holland to build his ships (Moulton 2005: 54–5). Ships were constructed at Archangel and St. Petersburg. Timber for manufacturing the ships was brought from Kazan. In addition, canals and roads were constructed in order to facilitate transportation of the shipbuilding materials from the interior of the country to St. Petersburg (Moulton 2005: 56). In 1703, Peter established St. Petersburg because he wanted a seaport on the Baltic. He not only made this city his naval headquarters, but also the capital of his empire. Furthermore, a system of canals was constructed to feed the city and to bring goods from the hinterland, which were then exported to Europe (Jones 1984: 413–33).

The School of Mathematics and Navigation was established by Peter at Moscow in 1701. This school was modelled on the Royal Mathematical School of Christ's Hospital at London founded by Charles II in 1673. Most of the mathematical masters of the Royal Navy were alumni of Christ's Hospital. Following the practice of this school, the Dartmouth School introduced mathematics and navigation in 1679. When Peter came to England in 1698, he was influenced by these British schools. Peter expressed his wish of establishing such a school in Moscow. Vice-Admiral David Mitchell, who attended to Peter, recommended Henry

Farquharson, a mathematician cum astronomer of Aberdeen University as a candidate for establishing Peter's school. On 14 January 1701, Peter appointed Farquharson as the Director of the Moscow School of Navigation and Mathematics. Besides naval personnel, this school also trained artillery officers, civil servants, clerks, craftsmen, etc. (Hans 1951: 532–6).

For the battleships, the standard tactic, as evolved in the Atlantic during the seventeenth century, was to move in the line ahead formation. By sailing in this formation, the vulnerable stern area was protected and the line presented the enemy with a long line of gun batteries (Harding 1999: 105). During the latter stages of the Great Northern War, especially during the Russian campaigns in Finland (1713–1715), Peter had learnt that old-fashioned ships like the galleys, flat-bottomed transports and launches gave the Russian Navy a decided advantage in the Northern Baltic Sea, especially along its rocky coast and the shoals. Russian galleys and transport ships could easily operate in such maritime terrain while the larger gunpowder ships of the British and Swedish navies designed for fighting in accordance with line tactics in the high sea were at a disadvantage. This became clear during the 1720 campaign (Moulton 2005: 64). Russian naval supremacy in the Baltic opened the coast of Sweden to frequent and devastating raids by Russian galley-borne amphibious forces (Frost 2014: 296). In 1725, Russia had 34 ships of the line, 9 frigates, 34 smaller sailing ships and several hundred galleys in the Baltic. The Baltic Fleet resulted in an annual expenditure of 1.5 million rubles (Kipp 2002: 155). This sort of expenditure could be sustained due to the transformation of Muscovy into a centralised military polity.

State, Society and Military Power

Russia's rise to power was related to the transformation which Muscovy experienced during the time period under review. The military capacity of a state is reflected in the size of the armed forces it could maintain and the ability to feed and equip the armed forces both during peace and war. Maintenance of a permanent standing army armed with state of the art weapon systems and supplying it during a long drawn campaign away from the core territory, required disposable demographic and financial resources and a bureaucratic infrastructure on the part of the state. Gathering resources for feeding Mars represents the litmus test for judging the ability of a state to project power. Furthermore, early modern Europe was in the throes of a Scientific Revolution, which transformed the nature of armed forces and the dynamics of warfare. The initiation and implementation of the Scientific Revolution required the establishment of institutions of higher learning with state patronage. In all these regards, the Western European polities including Muscovy succeeded. To explain this success, scholars have introduced various types of concept to explain the nature of Muscovite polity.

Marshall Poe asserts that Muscovy was a despotism. The authority of the tsar was not limited by human law. The tsar was the author of worldly law and not the subject of it. The Muscovites believed that they were the tsar's slaves and orphans. Tsardom was understood as the natural order of things. Partly, this was because the Muscovites (who were mostly illiterate peasants tilling infertile land) had no understanding of Classical European political thought. And the cities were fortresses, rather than being places of culture and commerce. The Muscovite elite constructed a state which mobilised scarce resources mainly for military purpose (Poe 2002: 473–86).

Poe is right in several aspects. However, the rule of the tsars faced serious rebellions from the peasants and nobles. The boyars, the middling and lesser nobility, remained a strongly entrenched class. And in China also, the peasants believed that the rule of the emperor represented mandate of the heaven. Yet, the Chinese peasants repeatedly rebelled. Prussia was also a polity geared for mobilising demographic and economic resources for war. Poland, which was a premier opponent of Muscovy, failed militarily in the seventeenth century because the nobles defeated all the centralising attempts of the Polish monarchy to tax them and raise forces under the direct control of the royalty (Frost 2014: 249). However, the tsars succeeded in bringing the boyars under autocratic control.

Marshall Poe and Eric Lohr, in a jointly written piece, categorise Muscovy as a Garrison State. One of the characteristics of the Muscovite Garrison State was the prevalence of state-driven military enterprises rather than presence of substantial commercial enterprises. And most of the people were serfs. Further, travel in and out of Russia was strictly controlled (Poe and Lohr 2002: 1–5). Many of these characteristics were indeed present in Russia. However, the Garrison State model is generally introduced by political scientists to categorise modern military dictatorship and militarism in civil society.

Was Muscovy then a fiscal-military state? Early modern Europe, according to a group of scholars, witnessed the transformation of the domain state into a fiscal-military state. In a domain state, the ruler maintained himself from his own resources supplemented by occasional grants from corporate social groups like clergy and burghers during dire emergencies. However, the rising expenditure of warfare (both demographic and economic) forced the state to tap into individual rather than collective wealth. The result, writes Peter Wilson, was expansion of the taxation apparatus and greater intervention in social life. Thus emerged the fiscal-military polity (Wilson 2009: 95–124).

Gabor Agoston persuasively argues that all attempts by the Ottoman central government to initiate even limited reforms for expanding the power of the 'limited monarchy' were defeated by a coalition of *Janissaries*, religious elites, guilds and local power brokers who had much to gain from a devolution of power, away from the Ottoman centre. In contrast, Muscovy evolved as a military-fiscal state rather than a fiscal-military state because acquisition of military manpower, livestock and grain was done more by conscription than by taxation (Agoston 2011: 281–319). In contrast, Janet Hartley claims that Russia, from the late seventeenth century, evolved as a fiscal-military state. Some of its characteristics were the huge size of the army and that almost half of the expenditure of the state went to the armed forces (Hartley 2009: 125–45). Initially, John Brewer applied the concept of the fiscal-military state to explain the rise of Britain in the early modern era. Two of the characteristics of Britain were the presence of a strong parliament and liquid cash obtained from mercantile capitalism (Brewer 1989). These two features were missing in Petrine Russia. Moreover, there is no point in nit-picking between fiscal-militarism and military-fiscalism.

The early modern era in Western Eurasia witnessed the expansion of the population and wealth in general. Some of the polities were able to tap into these sources and create an expanding military establishment. Among these successful polities, some were democratic (Britain), others were quasi-autocratic (Habsburg Empire) and a few were fully autocratic (Russia and Prussia). So, Russia, to my mind, could be classified as an autocratic military state. The Muscovite autocracy under the leadership of the successive tsars was geared for conducting warfare by exploiting the available resources of its domain to the maximum. Let us have a look at the empirical indices.

In the first half of the sixteenth century, the standing army of the Habsburg monarchy represented 1.25 per cent of the population and in the case of France, the ratio was 2 per cent (Palffy 2012: 53). According to one historian, during the first half of the seventeenth century, Muscovy had a population of 8 million and about one-eighth of the productive resources went on paying the army (Hellie 2002: 66). In the same period, the tsar could mobilise about 200,000 soldiers for campaiging in the vast theatre stretching from the Baltic in the north to Ukraine in the south (Brown 2002: 120–1). In 1630, the paper strength of the Habsburg Imperial Army came to about 150,000 men (Adams 1990: 34). In 1700, the Baltic powers like Denmark and Sweden had 70,000 and 100,000 troops respectively. The East European powers like Prussia and Poland had 41,000 and 35,000 soldiers in their ranks (Jespersen 1999: 194). Historian Brian Davies calculates that, between 1630 and 1700, the size of the Russian Army rose from 90,000 to 245,000 personnel. Of the latter, due to logistical and administrative constraints, in a particular theatre, the tsar could bring 100,000 men, and of them, 50,000 could be present in a field army (Davies 2013: 62). Between 1705 and 1713, Russia drafted 337,196 men out of a total population of 15 million and the size of the armed forces came to about 265,000 men (Hellie 1990: 94–5). As a point of comparison, between the 1670s and the 1780s, French peacetime troop levels stood at 160,000 (Lynn 1990: 3). In the 1760s, 3.3 per cent of all the eligible males of Muscovy were in the army. The average for Europe during the same period was 1.5 per cent (Aksan 1993: 226). By 1789, the size of the Habsburg Army jumped to 314,800 personnel (Wilson 1999: 80). Under Tsar Paul I (r. 1796–1801), the size of the Russian Army (including regulars and irregulars) numbered about 450,000 men (Menning 2002b: 84). One can conclude that Russia had the largest army in Europe.

Muscovite logistics were an amalgam of state and private enterprise. In 1556, the Muscovite Court ordered that the *pomeste* landholders had to provide three to four months' supplies, which should be brought to the mobilisation point in their own carts and wagons (Stevens 2012: 126). Forced requisition was common. Private merchants who sold foodstuffs to the soldiers for cash also accompanied the army. Such a supply system gave the army only a limited operating radius. It was also a risky endeavour. In the case of bad weather and the enemy's scorched earth policy, the army might be deprived of supplies and would ultimately disintegrate.

Gradually, Muscovy created a bureaucracy to feed, clothe and equip its armed forces. Muscovy established state granaries where provisions for the campaign forces were collected. For instance, in the mid-sixteenth century, there were 18 tsarist granaries at Voronezh. Supplies were also maintained at the forward-based state depots which were established just before launching a campaign. For example, during the 1552 Kazan Campaign, goods and military supplies were stockpiled at Sviiazhsk (a fortress which was built in 1551 for functioning as a supply depot), upriver from the besieged fortress. In most cases, river boats were used to send supplies to the Russian units going on campaigns (Smith 1993: 35–65). In 1678, the relief army sent by Muscovy under Romodanovskii, against the Ottoman force laying siege to Chigirin, was provisioned by grain from the government granaries at Kiev, Chernigov and Briansk (Davies 2002a: 101). For supplying the army, the central administration grew during the seventeenth century. From 45 chancelleries (*prikazy*), the number grew to 55. In the 1690s, clerks and officials at Moscow numbered 2,700. And the number of local officials at the provincial level rose from 500 to 1,900 (Bushkovitch 2002: 32). Peter set up the GKK, which was responsible for the troops' pay, uniform, equipment, weapons and munition. And the GPK was charged with obtaining food and forage for the army (Keep 2005: 254).

With regard to extraction of resources from the society to sustain the military apparatus, one of the chief competitors of Muscovy, the Ottoman Empire, failed miserably. During the second half of the eighteenth century, in Europe, the federative contractual armies were being replaced with state-managed centralised bureaucratic forces. However, the Ottoman Army became highly decentralised and was mostly organised and financed at the local level by the governors, tax collectors and village collectives. The Ottoman centre devolved the responsibilities for taxation to provincial notables and their entourages, who acquired semi-permanent, annually renewed rights to taxation. As the Ottoman government lacked the financial resources to build a force under its direct control, its supply system in Balkan also failed. The *Janissaries* became localised, indisciplined, prone to mutiny and spent time in activities other than preparing for war (Aksan 2012: 324–45). The position of the *Janissaries* was similar to the legions of late Roman Empire. These legions stationed permanently in the frontier provinces engaged in local commercial activities and became loyal to the provincial landholders rather than to Rome.

Peter the Great, like Stalin, established the infrastructure of a command economy with the sole aim of transforming backward Russia into a strong military power. At the beginning of the seventeenth century, Russia had practically no domestic metal industry. All domestic iron ore came from swamp ore and it was neither voluminous nor high grade enough to produce reliable weaponry (Kotilaine 2002: 67). Peter established colleges or ministries of commerce, mines and manufacturing. The Manufacturing College was established in 1719 to manage the state factories, to arrange their transfer in certain cases to private ownership, to build new factories, organise labour supplies and to give loans to manufacturers on favourable terms. The Mining College was established in the same year to manage mines, to supervise the processing of ores and to arrange loans for the construction of mining enterprises. All production was strictly controlled by these colleges (Matley 1981: 411). Peter's strenuous attempts to make Muscovy self-sufficient in arms production were successful to a certain extent. Between 1701 and 1711, domestic production of handguns rose from 6,000 to 40,000 per year (Ralston 1990: 26).

In addition to setting up industries, Peter also played a crucial role in establishing institutions of higher learning in Russia. The role of the naval college was discussed earlier. Peter's Russia also experienced scientific development in the field of cartography. Till 1711, the *Prikaz* was in charge of ordering and producing maps. From 1711 till 1765, the Senate was in charge of generating cartographic materials. Following Peter's decree, the Russian Academy of Sciences was founded in 1727. In 1739, a separate Geographical Department was created at the Academy to produce a general map of Russia (Tolmacheva 2000: 52–3). Qing China lacked any department for producing modern maps. As a result, the Manchu elite had no conception of the intrusion by the 'foreign devils' along different points of the 'Heavenly Kingdom.' Hence, Qing China's response to Western colonialism remained fragmentary (Mosca 2013).

Conclusion

The transformation of Muscovy from a marginal power in North-Eastern Europe into a major player in Eurasia was an uncertain process and occurred in halting stages. Geographical luck played a crucial role in the survival of Muscovy in the initial stage. Kiev could have posed a serious challenge to the rise of Muscovy. However, Kiev lay in the direct path of the steppe nomads and within the operating range of the latter's army. So, Kiev was destroyed by the Mongols in 1240. But, Muscovy lay far in the north in the forest zone. The steppe nomads were not interested in settling down there nor was the region lucrative enough for permanent annexation. Further, the forest zone of Muscovy was unsuitable for permanent stationing of large numbers of steppe nomadic cavalry.

Second, unlike the Ottoman Sultans or the Mughal *Padshahs*, the tsars were able to bring the landowners under direct control. This was possible because Russia was an infertile land with a scarce population. So, the landowners of Russia lacked the demographic and economic assets to successfully challenge the tsars' programme of constructing a military autocratic state in bits and pieces. The very weaknesses of the Russian landowners enabled the tsars, unlike their Ottoman and Mughal counterparts, to establish a military autocracy geared for meeting military requirements.

Third, Russia, unlike China and India, lacked a long coastline studded with large numbers of warm water ports. Nor could Muscovy be approached easily by a high sea fleet. The Baltic gave access to only a limited coastal area of North-East Russia. The Black Sea is enclosed by land on every side and is connected to the inland Mediterranean Sea through the narrow Dardanelles Strait. So, Western maritime powers' Blue Water fleets could not approach Muscovy in strength for establishing maritime colonialism. Nor was Muscovy productive enough, as India was, in commercial and agricultural terms, to entice the colonisers.

The amalgam of the above-mentioned indigenous and extraneous factors enabled Muscovy to survive and gather strength till it emerged as a behemoth in the nineteenth century. In Muscovy's steady rise to power, the navy played a marginal role. Though the navy aided the tsars to gain the coastline of the Baltic and the Black Sea, it still played second fiddle to the growth of Muscovy's power. Till the end of the Romanovs during the Bolshevik Revolution, Russia primarily remained a land power. And among the military branches, cavalry, though no more the premier arm in the eighteenth century, remained important. We can conclude by saying that gunpowder was one of the many important factors like geography and politics that resulted in the acceleration of Muscovy to the strongest power in East Europe at the beginning of the nineteenth century.

Bibliography

Adams, Simon, 'Tactics or Politics? "The Military Revolution" and the Hapsburg Hegemony, 1525–1648,' in John A. Lynn (ed.), *Tools of War: Instruments, Ideas, and Institutions of Warfare, 1445–1871* (Urbana/Chicago: University of Illinois Press, 1990), pp. 28–52.

Agoston, Gabor, 'Ottoman Warfare in Europe: 1453–1826,' in Jeremy Black (ed.), *European Warfare: 1453–1815* (London: Macmillan, 1999), pp. 118–44.

Agoston, Gabor, 'Military Transformation in the Ottoman Empire and Russia, 1500–1800,' *Kritika: Explorations in Russian and Eurasian History*, vol. 12, no. 2 (2011), pp. 281–319.

Aksan, Virginia, 'The One-Eyed Fighting the Blind: Mobilization, Supply, and Command in the Russo-Turkish War of 1768–1774,' *International History Review*, vol. 15, no. 2 (1993), pp. 221–38.

Aksan, Virginia, 'Ottoman Military Power in the Eighteenth Century,' in Brian J. Davies, (ed.), *Warfare in Eastern Europe: 1500–1800* (Leiden/Boston: Brill, 2012), pp. 315–47.

Arnold, Thomas F., 'War in the Sixteenth-Century Europe: Revolution and Renaissance,' in Jeremy Black (ed.), *European Warfare: 1453–1815* (London: Macmillan, 1999), pp. 23–44.

Black, Jeremy, 'Introduction,' in Jeremy Black (ed.), *European Warfare: 1453–1815* (London: Macmillan, 1999), pp. 1–22.

Brewer, John, *The Sinews of Power: War, Money, and the English State, 1688–1783* (London and Boston: Unwin Hyman, 1989).

Brown, Peter B., 'Tsar Aleksei Mikhailovich: Muscovite Military Command Style and Legacy to Russian Military History,' in Eric Lohr and Marshall Poe (eds.), *The Military and Society in Russia: 1450–1917* (Leiden/Boston: Brill, 2002), pp. 119–45.

Brown, Peter B., 'Gazing Anew at Poltava: Perspectives from the Military Revolution Controversy, Comparative History, and Decision-Making Doctrines,' *Harvard Ukrainian Studies*, vol. 31, no. 1/4 (2009/10), pp. 107–33.

Brown, Peter B., 'Command and Control in the Seventeenth-Century Russian Army,' in Brian J. Davies, (ed.), *Warfare in Eastern Europe: 1500–1800* (Leiden/Boston: Brill, 2012), pp. 249–313.

Bushkovitch, Paul, 'The Romanov Transformation, 1613–1725,' in Frederick W. Kagan and Robin Higham (eds.), *The Military History of Tsarist Russia* (New York: Palgrave, 2002), pp. 31–45.

Davies, Brian, 'Village into Garrison: The Militarized Peasant Communities of Southern Muscovy,' *Russian Review*, vol. 51, no. 4 (1992), pp. 481–501.

Davies, Brian, 'The Development of Russian Military Power: 1453–1815,' in Jeremy Black (ed.), *European Warfare: 1453–1815* (London: Macmillan, 1999), pp. 145–79.

Davies, Brian, 'The Second Chigirin Campaign (1678): Late Muscovite Military Power in Transition,' in Eric Lohr and Marshall Poe (eds.), *The Military and Society in Russia: 1450–1917* (Leiden/Boston: Brill, 2002a), pp. 97–118.

Davies, Brian, 'The Foundations of Muscovite Military Power, 1453–1613,' in Frederick W. Kagan and Robin Higham (eds.), *The Military History of Tsarist Russia* (New York: Palgrave, 2002b), pp. 11–30.

Davies, Brian, *Warfare, State and Society on the Black Sea Steppe: 1500–1700* (London/New York: Routledge, 2007).

Davies, Brian, 'Introduction,' in Brian J. Davies, (ed.), *Warfare in Eastern Europe: 1500–1800* (Leiden/Boston: Brill, 2012a), pp. 1–17.

Davies, Brian, 'Guliai-gorod, Wagenburg, and Tabor Tactics in 16th-17th Century Muscovy and Eastern Europe,' in Brian J. Davies, (ed.), *Warfare in Eastern Europe: 1500–1800* (Leiden/Boston: Brill, 2012b), pp. 93–108.

Davies, Brian, *Empire and Military Revolution in Eastern Europe: Russia's Turkish Wars in the Eighteenth Century* (2011, reprint, London: Bloomsbury, 2013).

Duffy, Christopher, *Russia's Military Way to the West: Origins and Nature of Russian Military Power, 1700–1800* (London/Boston: Routledge & Keegan Paul, 1981).

Duffy, Christopher, *Siege Warfare: The Fortress in the Early Modern World, 1494–1600* (1979, reprint, London/New York: Routledge, 1997).

Esper, Thomas, 'Military Self-Sufficiency and Weapons Technology in Muscovite Russia,' *Slavic Review*, vol. 28, no. 2 (1969), pp. 185–208.

Frost, Robert I., *The Northern Wars: War, State and Society in Northeastern Europe, 1558–1721* (2000, reprint, London/New York: Routledge, 2014).

Grey, Ian, 'Peter the Great and the Creation of the Russian Navy,' *History Today*, vol. 11, no. 9 (1961), pp. 625–31.

Halperin, Charles J., *Russia and the Golden Horde: The Mongol Impact on Medieval Russian History* (1985, reprint, Bloomington, IN: Indiana University Press, 1987).

Hans, Nicholas, 'The Moscow School of Mathematics and Navigation (1701),' *Slavonic and East European Review*, vol. 29, no. 73 (1951), pp. 532–6.

Harding, Richard, 'Naval Warfare: 1453–1815,' in Jeremy Black (ed.), *European Warfare: 1453–1815* (London: Macmillan, 1999), pp. 96–117.

Hartley, Janet, 'Russia as a Fiscal-Military State: 1689–1825,' in Christopher Storrs (ed.), *The Fiscal-Military State in Eighteenth-Century Europe: Essays in Honour of P.G.M. Dickson* (Aldershot: Ashgate, 2009), pp. 125–45.

Hellie, Richard, 'Warfare, Changing Military Technology, and the Evolution of Muscovite Society,' in John A. Lynn (ed.), *Tools of War: Instruments, Ideas, and Institutions of Warfare, 1445–1871* (Urbana/Chicago: University of Illinois Press, 1990), pp. 74–99.

Hellie, Richard, 'The Costs of Muscovite Military Defense and Expansion,' in Eric Lohr and Marshall Poe (eds.), *The Military and Society in Russia: 1450–1917* (Leiden/Boston: Brill, 2002), pp. 41–66.

Hellie, Richard, 'The Petrine Army: Continuity, Change, and Impact,' in Jeremy Black (ed.), *Warfare in Europe: 1650–1792* (Aldershot: Ashgate, 2005), pp. 235–51.

Herd, Graeme P., 'Modernizing the Muscovite Military: The Systemic Shock of 1698,' *Journal of Slavic Military Studies*, vol. 14, no. 4 (2001), pp. 110–30.

Jespersen, Knud J.V., 'Warfare and Society in the Baltic: 1500–1800,' in Jeremy Black (ed.), *European Warfare: 1453–1815* (London: Macmillan, 1999), pp. 180–200.

Jones, Robert E., 'Getting the Goods to St. Petersburg: Water Transport from the Interior 1703–1811,' *Slavic Review*, vol. 43, no. 3 (1984), pp. 413–33.

Kaganov, Grigorii and Monas, Sidney, '"As in the Ship of Peter",' *Slavic Review*, vol. 50, no. 4 (1991), pp. 755–67.

Keep, John L.H., 'The Russian Army in the Seven Years War,' in Eric Lohr and Marshall Poe (eds.), *The Military and Society in Russia: 1450–1917* (Leiden/Boston: Brill, 2002), pp. 197–220.

Keep, John, 'Feeding the Troops: Russian Army Supply Policies during the Seven Years War,' in Jeremy Black (ed.), *Warfare in Europe: 1650–1792* (Aldershot: Ashgate, 2005), pp. 253–73.

Kipp, Jacob W., 'The Imperial Russian Navy, 1696–1900: The Ambiguous Legacy of Peter's "Second Arm",' in Frederick W. Kagan and Robin Higham (eds.), *The Military History of Tsarist Russia* (New York: Palgrave, 2002), pp. 151–81.

Kotilaine, J.T., 'In Defense of the Realm: Russian Arms Trade and Production in the Seventeenth and Early Eighteenth Century,' in Eric Lohr and Marshall Poe (eds.), *The Military and Society in Russia: 1450–1917* (Leiden/Boston: Brill, 2002), pp. 67–95.

Kupisz, Dariusz, 'The Polish-Lithuanian Military in the Reign of King Stefan Bathory (1576–86),' in Brian J. Davies, (ed.), *Warfare in Eastern Europe: 1500–1800* (Leiden/Boston: Brill, 2012), pp. 63–92.

Lund, Erik A., 'The Generation of 1683: The Scientific Revolution and Generalship in the Habsburg Army, 1686–1723,' in Brian J. Davies (ed.), *Warfare in Eastern Europe: 1500–1800* (Leiden/Boston: Brill, 2012), pp. 199–248.

Lynn, John A., 'The Pattern of Army Growth: 1445–1945,' in John A. Lynn (ed.), *Tools of War: Instruments, Ideas, and Institutions of Warfare, 1445–1871* (Urbana/Chicago: University of Illinois Press, 1990), pp. 1–27.

Martin, Janet, 'The Emergence of Moscow (1359–1462),' in Maureen Perrie (ed.), *The Cambridge History of Russia*, vol. 1, *From Early Rus' to 1689* (Cambridge: Cambridge University Press, 2006), pp. 158–87.

Martin, Janet, 'Economic Effectiveness of the Muscovite *Pomeste* System: An Examination of Estate Incomes and Military Expenses in the mid-16th Century,' in Brian J. Davies, (ed.), *Warfare in Eastern Europe: 1500–1800* (Leiden/Boston: Brill, 2012), pp. 19–34.

Matley, Ian M., 'Defense Manufactures of St. Petersburg, 1703–1730,' *Geographical Review*, vol. 71, no. 4 (1981), pp. 411–26.

Menning, Bruce W., 'The Imperial Russian Army, 1725–1796', in Frederick W. Kagan and Robin Higham (eds.), *The Military History of Tsarist Russia* (New York: Palgrave, 2002a), pp. 47–75.

Menning, Bruce W., 'Paul I and Catherine II's Military Legacy, 1762–1801', in Frederick W. Kagan and Robin Higham (eds.), *The Military History of Tsarist Russia* (New York: Palgrave, 2002b), pp. 77–105.

Menning, Bruce W., 'Russian Military Innovation in the Second Half of the Eighteenth Century,' in Jeremy Black (ed.), *Warfare in Europe: 1650–1792* (Aldershot: Ashgate, 2005), pp. 275–93.

Mosca, Matthew W., *From Frontier Policy to Foreign Policy: The Question of India and the Transformation of Geopolitics in Qing China* (Stanford, California: Stanford University Press, 2013).

Moulton, James R., *Peter the Great and the Russian Military Campaigns during the Final Years of the Great Northern War, 1719–1721* (Lanham, Maryland: University Press of America, 2005).

Ostapchuk, Victor, 'Crimean Tatar Long-Range Campaigns: The View from Remmal Khoja's History of Sahib Gerey Khan,' in Brian J. Davies, (ed.), *Warfare in Eastern Europe: 1500–1800* (Leiden/Boston: Brill, 2012), pp. 147–71.

Ostrowski, Donald, 'Troop Mobilization by the Muscovite Grand Princes (1313–1533),' in Eric Lohr and Marshall Poe (eds.), *The Military and Society in Russia: 1450–1917* (Leiden/Boston: Brill, 2002), pp. 19–40.

Ostrowski, Donald, 'The Growth of Muscovy (1462–1533),' in Maureen Perrie (ed.), *The Cambridge History of Russia*, vol. 1, *From Early Rus' to 1689* (Cambridge: Cambridge University Press, 2006), pp. 213–39.

Ostrowski, Donald, 'Peter's Dragoons: How the Russians Won at Poltava,' *Harvard Ukrainian Studies*, vol. 31, no. 1/4 (2009/10), pp. 81–106.

Ostrowski, Donald, 'The Replacement of the Composite Reflex Bow by Firearms in the Muscovite Cavalry,' *Kritika: Explorations in Russian and Eurasian History*, New Series, vol. 11, no. 3 (2010), pp. 513–34.

Palffy, Geza, 'The Habsburg Defense System in Hungary against the Ottomans in the Sixteenth Century: A Catalyst of Military Development in Central Europe,' in Brian J. Davies, (ed.), *Warfare in Eastern Europe: 1500–1800* (Leiden/Boston: Brill, 2012), pp. 35–61.

Paul, Michael, 'The Military Revolution in Russia, 1550–1682', *Journal of Military History*, vol. 68, no. 1 (2004), pp. 9–46.

Poe, Marshall, 'The Consequences of the Military Revolution in Muscovy: A Comparative Perspective', *Comparative Studies in Society and History*, vol. 38, no. 4 (1996), pp. 603–18.

Poe, Marshall, 'The Truth about Muscovy', *Kritika: Explorations in Russian and Eurasian History*, New Series, vol. 3, no. 3 (2002), pp. 473–86.

Poe, Marshall and Lohr, Eric, 'Introduction: The Role of War in Russian History,' in Eric Lohr and Marshall Poe (eds.), *The Military and Society in Russia: 1450–1917* (Leiden/Boston: Brill, 2002), pp. 1–15.

Ralston, David B., *Importing the European Army: The Introduction of European Military Techniques into the Extra-European World, 1600–1914* (Chicago/London: University of Chicago Press, 1990).

Salvo, Maria Di, 'About Peter the Great's Ship Predestinatsiia,' in Maria Di Salvo, Daniel H. Kaiser and Valerie A. Kivelson (eds.), *Word and Image in Russian History: Essays in Honour of Gary Marker* (Brighton: Academic Studies Press, 2015), pp. 43–54.

Shaw, Denis J.B., 'Russia's Geographical Environment,' in Maureen Perrie (ed.), *The Cambridge History of Russia*, vol. 1, *From Early Rus' to 1689* (Cambridge: Cambridge University Press, 2006), pp. 19–43.

Shepard, Jonathan, 'The Origins of Rus (c. 900–1015),' in Maureen Perrie (ed.), *The Cambridge History of Russia*, vol. 1, *From Early Rus' to 1689* (Cambridge: Cambridge University Press, 2006), pp. 47–72.

Smith, Dianne L., 'Muscovite Logistics, 1462–1598,' *Slavonic and East European Review*, vol. 71, no. 1 (1993), pp. 35–65.

Stevens, Carol B., 'Evaluating Peter's Army: The Impact of Internal Organization,' in Eric Lohr and Marshall Poe (eds.), *The Military and Society in Russia: 1450–1917* (Leiden/Boston: Brill, 2002), pp. 147–71.

Stevens, Carol B., 'Food and Supply: Logistics and the Early Modern Russian Army,' in Brian J. Davies, (ed.), *Warfare in Eastern Europe: 1500–1800* (Leiden/Boston: Brill, 2012), pp. 119–46.

Tolmacheva, Marina, 'The Early Russian Exploration and Mapping of the Chinese Frontier,' *Cahiers du Monde russe*, vol. 41, no. 1 (2000), pp. 41–56.

Waliszewski, Kazimierz, *Ivan the Terrible*, tr. from the French by Lady Mary Loyd (1904, reprint, Stroud: Nonsuch 2006).

Wilson, Peter, 'Warfare in the Old Regime: 1648–1789,' in Jeremy Black (ed.), *European Warfare: 1453–1815* (London: Macmillan, 1999), pp. 69–95.

Wilson, Peter H., 'Prussia as a Fiscal-Military State, 1640–1806,' in Christopher Storrs (ed.), *The Fiscal-Military State in Eighteenth-Century Europe: Essays in Honour of P.G.M. Dickson* (Aldershot: Ashgate, 2009), pp. 95–124.

19

SMALLPOX AND WAR
IN AMERICA

Erica Charters

Introduction

Could it not be Contrived to Send the <u>Small Pox</u> among those disaffected Tribes of Indians? We must on this occasion Use Every Stratagem in our power to Reduce them.

(Add MS 21634, f. 243, n.d.)

This undated note from British Commander-in-Chief in North America, Jeffrey Amherst, forms the basis of an iconic episode in military history and the history of medicine. Alongside further correspondence between Amherst and one of his officers (Henry Bouquet), and an ominous receipt for blankets and a handkerchief 'taken from people in the Hospital to Convey the Small pox to the Indians,' it provides evidence for the infamous incident in which the British Army distributed blankets infected with smallpox among hostile Native Americans (Bouquet Papers Add MS 21634, f. 241, Add MS 21654, f. 168; War Office Papers WO 34/40, ff. 305–9). The incident is widely regarded as the first documented case of biological warfare. It is often also taken to encapsulate the nature of British–indigenous war in the Americas, if not the ruthless nature of European military colonisation in America more broadly (Wheelis 1999; Fenn 2000; Ward 2001; Ranlet 2000; Dixon 2005). Pulitzer-Prize winner poet Paul Muldoon, for example, employs Bouquet, Amherst and smallpox blankets to portray British colonialism in his poem 'Meeting the British.'

The role of smallpox during war in the Americas was not simply one of British biological warfare. Smallpox, like most diseases, could affect the outcome of military engagements by significantly diminishing manpower among European as well as Native American forces. Because of its distinctively horrific symptoms and contagiousness, it was highly feared among indigenous and European populations alike. It was therefore common for rivals to manipulate such fears with wartime accusations of deliberate smallpox spreading in the hope of shifting military alliances and weakening enemy resolve and recruitment. Yet there is little doubt that British officers did attempt to spread smallpox among indigenous populations during the 1763–1764 Pontiac's War: British correspondence makes it clear that Amherst's 'Stratagem' had been implemented.

Although the spread of smallpox was difficult to control, European armies used inoculation from the early eighteenth century to render soldiers immune to smallpox, incorporating differential immunity into the conduct of war. As with other diseases, smallpox was not simply a biological organism that followed a natural course. Rather, it was a feature of military operations because war—namely the movement and dispersion of soldiers and the realities of campaigning—facilitated its spread. By the same token, smallpox could also be constrained through human agency, such as medical intervention or military discipline. Smallpox thus makes clear the social and cultural contexts of disease during war, showing how disease was a central part of warfare; feared, rationalised and manipulated by military personnel.

Smallpox and the Nature of Early Contact

Although it was known that disease was a factor in early contact between indigenous populations and Europeans in America, it was Alfred Crosby's 1972 *The Columbian Exchange* that outlined the central role of disease—and smallpox in particular—in facilitating and shaping European colonisation of the Americas. Writing before environmental history developed, Crosby made novel use of a range of sources—drawing on evidence from archaeology, demography, epidemiology, botany and physical anthropology—to make his case that European arrival and conquest of the Americas was a biological event, above and before all else. It was disease, and most decisively smallpox, which determined military conquest. As he explained to readers, 'We have so long been hypnotized by the daring of the conquistador that we have overlooked the importance of his biological allies' (Crosby 2003: 69).

Crosby's model succeeded in highlighting the central role of disease in imperial conquest. As the demographic anthropologist Henry Dobyns summarised, 'Viruses and germs constituted the true shock troops with which the Old World battered the New' (Dobyns 1983: 24). European diseases, beginning with the initial smallpox epidemic in the 1520s, spread across the Americas: for example, the initial smallpox epidemic ranged from present-day Chile across most of Central and North America. Given that indigenous populations had no previous exposure to European diseases such as smallpox, measles, influenza and typhus, they were uniquely and entirely susceptible to these pathogens. Outbreaks not only caused massive mortality, but also upset existing political, social and cultural orders, which further exacerbated mortality rates, resulting in indigenous population decline estimated at more than 50 per cent (Dobyns 1983; Baker and Kealhofer 1996; Henige 1998). In Crosby's account, the role of European diseases, often epitomised by the most lethal disease of smallpox, displaces the role of war—and indeed of humans—in the conquest of America.

While Crosby's *Columbian Exchange* was seminal in sowing the fields of global environmental and disease history, its broad overview—tackling some hundreds of years of history across the Americas—has led specialists to offer corrections and refinements to his model of conquest via disease. In particular, regional studies have pointed out that aboriginal populations were not identical in their reactions to disease outbreaks, identifying a range of epidemiological, political, social and military responses to European contact. For example, because disease requires carriers, and because Europeans tended to travel along water systems, aboriginal river societies suffered higher rates of disease and depopulation than those who lived inland. By the late sixteenth century, populations that had previously been the military powerhouses along the Lower Mississippi River were displaced by upland societies—such as the Chickasaws and Choctaws—whose inland locations insulated them from disease outbreaks.

Such inland and forest societies took advantage of the decline of their previously powerful riverine enemies, moving into their settlements and creating new military alliances and practices (Dobyns 1983 pt. 3; Johnson and Lehmann 1996). In this model, European disease resulted in depopulation, but it thereby also resulted in re-settlement, migration and new types of warfare. Archaeological, osteological and ethnohistorical evidence has thus shown how many Native American groups adapted to post-Columbian conditions, highlighting variety, flexibility and resilience among indigenous societies (Kunitz 1994; Baker and Kealhofer 1996; Smith 1987b).

Scholars have also challenged the Columbian Exchange model by pointing to the significant role of war and colonial violence in indigenous depopulation. For example, burial sites at a sixteenth-century Coosa chiefdom village (in modern day southeast United States) reveal wounds and fatalities from European steel weapons. Archaeological analysis suggests mass destruction and depopulation due to war—likely Hernando de Soto's 1540 raids and expedition—and not epidemic disease (Blakely and Detweiler-Blakely 1989; Blakely and Mathews 1990). More generally, historian Paul Kelton argues that the Columbian Exchange model, which sees disease as an indiscriminate biological phenomenon—a kind of ravenous wildfire—obscures the crucial role of Europeans in facilitating, accelerating and shaping its spread. Aggressive slave raiding and conflict disrupted indigenous communities' social, political and economic structures, producing famine, malnutrition and dislocation, which made them especially vulnerable to European epidemic diseases when they finally arrived (Kelton 2007). In other words, disease acted in league with warfare rather than replacing it as the agent of conquest.

It was the nature of European contact and conflict that therefore also shaped the course of disease. Whereas short-term expeditions did not necessarily result in the spread of epidemic diseases such as smallpox, it was long-term invasive contact that led to outbreaks of lethal and contagious disease. For example, Spanish and English activity—such as slave raids and resource destruction—in the American southeast in the mid-sixteenth and seventeenth centuries, inflamed cycles of Native American conflict and created pressure points through subsistence crises. Invading Europeans thus reshaped indigenous diets and geopolitics (for example, by pushing into close proximity tribes that had previously maintained buffer zones between them, which had also acted as epidemiological barriers), thereby creating epidemiological vulnerabilities and encouraging the spread of epidemic disease (Kelton 2007; Kelton 2015; Jones 2003). Such models of interdependence and interaction between humans and disease restore human agency to narratives of conquest that otherwise focus on microbes. Not only do these accounts frame Europeans as active agents—not passive biological carriers—of demographic catastrophe in early modern America, but indigenous populations also emerge as adaptive societies that had agency in their interactions with disease pathogens.

Smallpox and Immunity

As Crosby recognised, diagnosing disease through historical records is fraught with difficulties. However, given smallpox's distinctive appearance and symptoms, it is fairly certain that Spanish chroniclers' use of *viruelas* refers to this particular disease. Smallpox was well known to Europeans; indeed, it was endemic across much of Europe. In early modern London, for example, smallpox broke out every three to four years. As a result, most adults from densely populated European areas were survivors of the disease, and therefore immune to it. Smallpox was thus known in Europe as a childhood disease, albeit a highly lethal one.

Smallpox is caused by infection with the *variola* virus, one of the orthopoxviruses. It usually takes around ten days between infection and the first symptoms of fever, intense headache and malaise, to manifest themselves. At this point, the characteristic rash and its pustules spread over the face and entire body, growing in size and severity across seven to ten days, until either death or recovery ensues—the latter involving gradual scabbing of the lesions, often leaving deep scarring (pockmarks) and sometimes secondary complications such as blindness and male infertility. In the 'ordinary' type of smallpox, survival meant life-long immunity to the disease, with fatality rates ranging between 10 and 30 per cent. But in more lethal strains of the disease, or where secondary infections of pustules or the respiratory tract occurred, fatality rates could be much higher (Riley 2010; Hopkins 2002; Fenner et al. 1988, ch. 1; Shaffa 1972). Moreover, the symptoms and course of the disease are particularly unpleasant. As a description from an outbreak in 1958 recorded:

> The majority of patients had fully developed smallpox in the supportive stage, with confluent pustules covering the entire body. The head was usually covered by what appeared to be a single pustule; the nose and the lips were glued together. When the tightly filled vesicles burst, the pus soaked through the bedsheet, became smeared on the blanket and formed thick, yellowish scabs and crusts on the skin. . . . Wails and groans filled the rooms.
>
> *(Fenner et al. 1988: 27)*.

There was, and still is, no cure for smallpox. The best treatment, both in the sixteenth century and the twenty-first century, is careful nursing, hydration and rest. However, such care was also the most likely way to catch the disease: transmission of the infection most often takes place through the respiratory tract: by inhaling the virus via the exhalations of a sick person—through droplets in their speech, coughing or sneezing. In other words, close and sustained proximity through bedside nursing is the most effective way to catch the disease (Riley 2010: 453–6). Smallpox, unlike many diseases, has no animal reservoirs or carriers: it is transmitted only by humans. This in part explains why it was able to be eradicated in the twentieth century; it also explains why the spread of the disease is inextricably linked to the movement of humans.

Fear of catching a loathsome disease was countered by the knowledge that surviving smallpox meant one was immune for life. This differential immunity—the fact that most Europeans arriving in the Americas had already survived the disease in their childhood and were thus immune to it—played a critical role in shaping the impact of smallpox on local populations in America. Although there are slight differences in immunity to smallpox through inherited or genetic differences, the central way in which individuals become immune to this fearsome disease is through acquired immunity—by recovering from the virus and thereby enjoying immunity to it for the remainder of one's life. As Crosby and other historians have pointed out, this differential immunity between arriving Europeans and indigenous populations added to the already substantial physical toll of the disease. Native Americans suffered from a disfiguring and debilitating disease while Europeans appeared invulnerable (Jones 2003; Riley 2010; Curtin 1968 on differential immunity).

In the eighteenth century, Europeans also began practising inoculation—the purposeful infection of an individual with what was usually a mild form of smallpox—thereby rendering the individual immune, and avoiding the risk of future infection by a more lethal strain. Inoculation should not be confused with vaccination, which was developed in the 1790s.

Whereas inoculation infects individuals with the smallpox virus itself, producing the actual disease in the inoculated person, vaccination delivers either a different strain of the virus (initially cowpox—part of the genus orthopoxvirus) or an attenuated version of the disease. Vaccination produces antibodies and thereby resistance to the disease in the vaccinated individual, but not the actual disease. Inoculation, by contrast, is a highly risky procedure: not only are inoculated individuals infected with actual smallpox, and thus suffer the full course of the disease, they also become sources of infection themselves, able thereby to spread the disease. Due to these serious risks, inoculation was usually practised only when an epidemic of smallpox had broken out nearby, making the risks of inoculation (with a milder strain of smallpox) tolerable by comparison (Miller 1957; Smith 1987a; Rusnock 2002, chs. 2–3).

After its introduction in Europe, inoculation was also discussed and occasionally practised in America. Just as in Europe, the practice was highly contentious. Most famous were the public and published arguments in 1720s Boston, in which Cotton Mather played a leading role in arguing for its medical effectiveness while simultaneously denouncing critics of inoculation as Satan's minions (Breen 1991; Barrett 1942). Regardless of Mather's fiery rhetoric, inoculation was often questioned not only on religious grounds, but also on the basis of statistical and medical efficacy. Although it was likely practised among slave populations from the 1720s (Stewart 1985), it was not widely implemented—either in Europe or America—until at least the 1760s and 1770s (Razzell 1977; Miller 1957; Smith 1987a; Fenn 2001, ch. 5; Few 2017; Gronim 2006; Becker 2004).

Smallpox, Fear and War

Because immunity to smallpox is acquired by infection, and not passed on to offspring, American colonists were just as susceptible to smallpox as were indigenous populations. That is, since smallpox was endemic (recurring every few years) in early modern European cities and towns, most Europeans acquired their smallpox immunity by surviving the infection when children. In America, by contrast, smallpox was not endemic because American settlements were not as densely populated. With outbreaks occurring only every 20 years or so, those who had grown up in the colonies likely had no immunity to it. When smallpox did break out, colonists and indigenous populations were similarly vulnerable.

Historians agree that smallpox was one of the most feared diseases of the early modern period, especially once the incidence of bubonic plague declined across the western world from the seventeenth century onwards. With its loathsome symptoms, fatality rates that could surpass 30 per cent among those infected, and permanent disfigurement for survivors, the disease caused particular horror and dread. Europeans often commented on the intense fear with which colonial and indigenous populations held the disease. In 1757, British Commander-in-Chief Lord Loudoun noted that smallpox was spreading among some of the colonial regiments, and even though 'the kind is good and very few die,' he noted 'The Terror People have for it in this country is inexpressible' (quoted in Charters 2014: 45).

As with disease in general, smallpox epidemics in America tended to break out during wars. War brought troops from places such as Europe and the Caribbean Islands, where smallpox was endemic. Warfare also increased the movement of people throughout the continent, allowing infected people to carry the disease to distant and hitherto insulated populations. Seventeenth-century smallpox epidemics thus followed migration and cycles of conflict, flaring up every few years from the 1640s onwards. By the eighteenth century, smallpox epidemics coincided with European wars, tangibly demonstrating the intertwined nature of warfare and disease. In Charlestown, South Carolina, for example, major outbreaks of smallpox followed the pattern of European colonial wars (McCandless 2011: 2–6). Outbreaks in 1697–1698, 1711–1712, 1738–1739, 1759–1760 and 1763 coincided with the Nine Years War (1688–1697) (known as King William's War in America), the War of the Spanish Succession (1701–1714) (known as Queen Anne's War [1702–1713] in America), the War of the Austrian Succession (1740–1748) (encompassing the War of Jenkins' Ear (1739–1748)) and the Seven Years War (1756–1763) (known as the French and Indian War [1754–1763] in America). Likewise, a smallpox epidemic in 1775–1781 coincided with the War of American Independence (1775–1783).

Geographically, as well, smallpox followed the spread of conflict. In the 1680s and 1690s, it appeared among North-East tribes in conflict zones, along the French and British frontiers and along the Gulf Coast. Likewise, during the Seven Years War, smallpox reflected the progress of the war, spreading throughout Canada and New England, to Virginia, the Carolinas and the Southeast, also spreading among the Iroquois, Potawatomi, Wea, Kickapoo, Miami, Shawnee and Arikara, and reaching as far as the North-West coast by the middle of the 1760s (Dobyns 1983: 15). In her study of the smallpox epidemic of 1775–1782, Elizabeth Fenn documents how the movement of soldiers, veterans and Native allies during the War of American Independence spread smallpox widely throughout America (Fenn 2001).

As a result, it was not only civilians who feared the outbreaks of epidemic disease that war inevitably brought; soldiers feared infection even more. Outbreaks of smallpox thus encouraged desertion and discouraged recruitment (Charters 2014, ch. 1; Becker 2004). Indeed, D. Peter MacLeod (1992) has traced how the incidence of smallpox—rather than military performance—shaped whether Native Americans allied themselves to Europeans during the conflicts of the 1750s. He explains that 'smallpox imposed a rhythm of its own upon the intensity of Amerindian participation in the war' (MacLeod 1992: 53). Given that smallpox epidemics spread by the conflict likely killed more than did combat among Iroquois communities, such fears are understandable (MacLeod 2012: 7–8).

These fears were also taken into account by European military administrators. During American campaigns, French officers—reliant on indigenous warriors—were well aware of the danger of alienating the support of their Native American allies through outbreaks of smallpox. The Marquis de Montcalm, commander of French forces in North America, reported anxiously from Montreal in spring 1758 that smallpox had broken out among Native Americans in the Pays d'en Haut (Upper Country of New France). While he noted that 'it is a terrible loss for us,' Montcalm further explained that it was France's good fortune that the tribes believed it was spread by the English (A1 3498, 18 April 1758). In fact, this belief was propagated by the French themselves in order to maintain their military alliances, even though it was more likely that the French—not the British—were responsible for the transmission of smallpox among their indigenous allies (MacLeod 1992: 50–1; Balvay 2006 on French–Native American military alliances).

Inoculation and the Armed Forces

Military officials not only incorporated fears of smallpox into their military strategy by spreading rumours regarding enemy efforts at deliberately introducing smallpox; they also incorporated differential immunity through the use of inoculation. As mentioned, inoculation—the deliberate infection with what is hoped is a mild strain of smallpox—was practised across the eighteenth century until the safer method of vaccination was developed in the 1790s.

Although inoculation was relatively preferable to contracting smallpox 'naturally,' it was still a highly risky operation. Statistically, those who caught smallpox naturally had a 10 to 30 per cent chance of dying from the disease; those who were inoculated experienced fatality rates of between 1 and 3 per cent (Rusnock 2002, pt. 1; Miller, 1957; see above on fatality rates).

The statistical difference between inoculation and natural smallpox, however, hides a gulf of uncertainty. The eighteenth-century controversy over inoculation was also a debate about quantification, itself a new practice. Alongside religious and moral scruples over the purposeful introduction of infection into what was otherwise a healthy body, contemporaries also argued over different mathematical representations of lives lost and, indeed, how to measure lives and their value. As scholars of risk studies recognise, people assess risk not through a general framework of statistical probability, but by weighing up individual circumstances and narrative reports (Slovic 2000, 2010; Sunstein 2008). Someone choosing inoculation did so in the context of having heard about those who had died from it, and knowing that there was a certain outcome of sickness and a real risk of death. Consequently, individuals who chose deliberate infection with smallpox via inoculation—usually a parent choosing to infect his or her child—would accept its risks only when there was news of a much more lethal smallpox epidemic nearby. As historians have pointed out, inoculation was accepted most widely among elites, who considered it the cutting edge of scientific progress, and implemented most broadly among institutionalised populations—such as orphans or soldiers—whose welfare was rationalised and decided by elite authorities on their behalf (Rusnock 2002, ch. 4; Smith 1987a, ch. 3; Miller 1957, chs. 4–6).

Given that soldiers and sailors regularly subsumed their discipline and health to general orders, and that their bodies were in many ways already the property of the state, experimentation and intervention were a part of European military medicine (Vess 1974; Keel 2001; Harrison 2010; Charters 2014). It was military cadets, for example, on whom inoculation was first trialled before being conducted on Russia's Catherine II in 1768 (Dimsdale 1781: 20–1). Likewise, even though inoculation was not widespread in eighteenth-century France, in 1769 the Duke of Choiseul instituted inoculation at the La Flèche Military School, the military institution that trained cadets for admission to the French *École Militaire* (Lebrun 1971: 286–7).

Because inoculation was more widespread among eighteenth-century armed forces than among civilian populations, many historians have assumed that eighteenth-century European forces serving in the American colonies were entirely immune to smallpox, either having survived the disease in Europe or having been inoculated upon enlistment. Such assumed differential immunity explains the strategic role of smallpox during wartime: as discussed, indigenous and colonial-born troops were uniquely susceptible to the disease; a difference that could be integrated into military strategy (Becker 2004; Fenn 2001, ch. 3). Yet, such immunological differences were not so stark. For example, although the majority of eighteenth-century British soldiers were immune to smallpox, the British military physician Richard Brocklesby estimated that, during the Seven Years War, a substantial proportion of British troops—around 20 per cent—had not had smallpox, and so were vulnerable to it (Charters 2014: 44).

During the War of American Independence, George Washington's 1777 decision to inoculate American troops is seen as an attempt to match immunity levels among British soldiers. Smallpox was a problem for both sides from the start of the war, with smallpox in 1775 Boston affecting both the Continental and British armies. Yet it was during the Quebec campaign of 1775–1776 when smallpox most severely affected Continental military strategy. It discouraged American recruitment and even spurred on some soldiers to self-inoculate, although this was expressly forbidden for fear that it would further spread the disease among the troops. As the American General Horatio Gates declared at the end of the campaign, 'As fine an Army as has ever marched into Canada has this year been entirely ruined with smallpox' (Becker 2004: 420). It was not only sick rates exceeding 30 per cent that plagued American forces; Continental officers and soldiers believed the differential immunity between British and American soldiers was part of British military strategy, with inoculated British soldiers and inoculation itself wielded against the American invaders. As a result, beginning in early 1777, Washington implemented mass inoculation in the Continental Army. Historians credit this policy not only with containing the problem of smallpox among Continental forces for the remainder of the war, but also with growing acceptance of inoculation among the American civilian population (Becker 2004; Fenn 2001, chs. 2–3; Blake 1953).

Yet the role of smallpox during the War of American Independence is not only one of successful containment through inoculation. Many Continental officers and troops were convinced that the British took advantage of their relative immunity by purposefully spreading the disease in conflict zones. According to American accounts, the British inoculated Canadian civilians for the purpose of spreading smallpox during the Quebec campaign—they released inoculated prisoners and freed enslaved people among American troops in order to spread the loathsome disease to a population known to be vulnerable to it, thus using disease as a weapon during a bitter civil war (Becker 2004; Fenn 2001 chs. 3–4). Thomas Jefferson was convinced by such claims, stating that smallpox 'was sent into our army designedly by the commanding officer in Quebec' (Becker 2004: 409).

While there is little doubt that British officers and troops alike were pleased to see smallpox aiding their campaigns, there is no evidence that the British manipulated smallpox as an infective tool of war. Indeed, it is clear that not all British troops were immune to smallpox: some suffered from the disease, and others were inoculated during the war. For example, in 1776, after the disastrous American invasion of Canada, British soldiers in Canada who had not already had the disease were ordered to be inoculated precisely because the Americans had 'left some sick in the small-pox on their quitting Montreal' (Reide 1793: 5–6; Marshall 2006: 105–6).[1] British military medical practitioners also offered inoculation to local civilian populations (free of charge for the poor) at the request of civilians and their local governors (Reide 1793: 27–8, 33–6).

Smallpox and immunity to smallpox, whether differential or self-cultivated via inoculation, clearly played a role in warfare in America. Yet it is also clear that fear of smallpox and rumours of its use as a biological weapon were central to its role in conflicts. Smallpox thus must be understood not only as a biological entity, but also as a disease with intense psychological and political impact. This was linked to its longer history as a decisive factor in the European conquest of America, and how differences in immunity could shape its ravages.

Rumours, War and Biological Weapons

Wars are environments ripe for unverified accounts: people are hungry for information, and yet confirmed news is difficult to establish as circumstances and events rapidly shift. As scholars of rumour also point out, such stories are also central to establishing and reaffirming identifies, most notably the difference between 'us' and 'them.' Rumours of atrocities committed by the other side are thus common during wartime, circulated by civilians and military personnel alike. This does not dismiss their significance; such accounts capture deeply held beliefs and concerns, even if their content is questionable (Dowd 2015).

Rumours and accusations of the intentional spread of smallpox are sprinkled throughout indigenous accounts of their relationship with Europeans. Such accounts were based in the reality that Europeans had, initially, been responsible for introducing the disease to the Americas, as well as the fact that the disease spread via commercial and military networks in which Europeans played a crucial role. Such rumours also drew on European desires and ambitions to take over Native American territories, and indigenous concerns over their physical susceptibility to what they considered a foreign, European disease, even in the eighteenth century. For example, an Ottawa account of the smallpox of 1757 recorded that it was sold to them by the British in Montreal during wartime, concealed in a tin box:

> [A]fter they reached home they opened the box; but behold there was another tin box inside, smaller. They took it out and opened the second box, and behold, still there was another box inside of the second box, smaller yet. So they kept on this way till they came to a very small box, which was not more than an inch long; and then they opened the last one they found nothing but mouldy particles in this last little box.
>
> *(MacLeod 1992: 50)*

It is from these mouldy particles that the dreaded smallpox reportedly arose. Reminiscent of the story of Pandora's Box and tales of poisoned gifts more generally, the Ottawa smallpox narrative captures how the disease was linked to war and rooted in European perfidy, as well as in the complex culpability of Indigenous–European contact itself (Mayor 1995; Fenn 2000; Dowd 2015, ch. 2; Myrillas-Brazeau 2019).

Such rumours among Native populations troubled Europeans not only because of their questionable accuracy, but also because of the damage they could do to military alliances. As mentioned, smallpox influenced Native American communities' decisions over going to war and choosing sides in European conflicts. As a British military officer despaired from Savannah in the midst of the War of American Independence, 'The Indians being in great dread of the Small pox—which at present rages in all the lower parts of this Province—are desirous of going away which it will be difficult for me to prevent' (PRO 30/55/22 2 March 1780 f. 243). Likewise, when French officials questioned the Tunica as to why they attacked an English convoy in the spring of 1764, they explained that it was not only to defend their land—whereas the French had never given them any diseases, the English earned their enmity because they 'killed almost all of our children by the pox which they brought' (SP 78/263 f. 65 21 Marh 1764). As in American rumours during the War of American Independence, the incidence of smallpox was understood as the product of intentional design by the British, reinforcing views of British imperial treachery and corruption.

Yet there is one documented incident of the purposeful spread of smallpox by British officials during wartime. As quoted at the outset, in 1763, in the midst of Pontiac's War (1763–1766), British Commander-in-Chief Jeffrey Amherst exchanged correspondence in July of 1763 (in the form of unofficial postscripts) with Colonel Henry Bouquet, discussing their desire to spread smallpox among the Native Americans fighting them in the vicinity of Fort Pitt (Western Pennsylvania) (WO 34/40 ff. 305–9; Add MS 21634, f. 241, f. 243). Yet an actual attempt to spread smallpox by means of handkerchiefs and a blanket taken from the British military smallpox hospital had already been implemented by those based at the fort some weeks earlier. Journal entries by American colonist and military captain William Trent, as well as accounts submitted by Swiss-born British military officer Simeon Ecuyer, establish beyond doubt that the items were given to two Delawares in June 1763 (Volwiler 1924: 400; Dixon 2005: 150–5; Fenn 2000; Add MS 21654, f. 168).

While payment for the handkerchiefs and blanket was approved by Bouquet almost a year later, it is notable that it was described as part of 'extraordinary' expenditures by then-acting Commander-in-Chief, Thomas Gage (Add MS 21654, f. 168 r.v.). There is also no documentary evidence linking the desires of Bouquet and Amherst, expressed in July of 1763 to 'extirpate this execrable race,' to the actions of Trent and Ecuyer in June 1763 (Add MS 21634 f. 241), or indeed to any implemented official policy. Historians have also pointed out that smallpox does not transmit well via blankets, and that outbreaks of smallpox among Native American societies of the Ohio Valley did not correspond to the gift of infected blankets (Dixon 2005: 150–5). The British attempt to deliberately infect their Native American enemies in the summer of 1763 almost certainly failed.

What remains, then, is the potency of the attempt, alongside the repugnant language of two British military officers. This one incident also carries the weight of the many other accusations and rumours of deliberate infection during wartime. Biological warfare is almost universally considered immoral and unjust, as well as abhorrent and cowardly. The documented incident at Fort Pitt in 1763 thus carries condemnation of all involved, even if unsuccessful. But the abhorrence and immorality of biological warfare is also what gives accusations and stories of its practice their power. Whether circulated by Americans during their war against the British, by the British among their Canadian allies against the Americans, by the French among their Native American allies against the British, or by indigenous societies against Europeans, these accounts of smallpox as a tool of war capture the intense fear of the disease, its symbiotic relationship with war, its history as part of the process of European conquest of America, and the way in which it could discriminate between Europeans and indigenous and colonial populations.

Conclusion

Smallpox and immunity to smallpox continued to play a role in nineteenth- and twentieth-century American wars. Not only did smallpox reach epidemic proportions during the Spanish–American War (1898) and later campaigns in Puerto Rico and the Philippines, but smallpox vaccinations—carried out by military personnel on civilian populations—were part of American operations, often used to demonstrate the humanitarian nature of campaigns, and thus to justify the wars themselves (Willrich 2011, ch. 4; Cirillo 2004). And, although smallpox was declared eradicated in 1980, thanks to a global vaccination campaign headed by the World Health Organization, fears continue of its use as a biological weapon. Given that the virus lives on in two laboratories (one in the United States and one in Russia), such fears are not entirely unfounded. In 2002–2003, the American administration implemented a preventative campaign against potential smallpox bioterrorism: as a result, nearly 500,000 military personnel were vaccinated against smallpox. What is interesting to note is that such vaccinations were compulsory; by contrast, frontline American medical and emergency workers, who were urged—but not compelled—to be vaccinated, mostly refused (Willrich 2011: 341–2). Once again, armed forces were at the vanguard of the smallpox front, and became the one segment of society that underwent artificially induced immunity.

As historians of disease have long recognised, the way societies respond to disease is shaped by cultural and social factors alongside biological ones (Rosenberg 1992). With regard to smallpox, the method of contagion—requiring person-to-person transmission—as well as its odious symptoms shaped how it was conceptualised and regarded. Because of differences in immunity, and because of the sporadic appearance of smallpox in America throughout the sixteenth, seventeenth and eighteenth centuries, it was associated with invading Europeans and their wars: associations that had basis in fact. European colonisers sometimes used the destructive nature of smallpox among indigenous populations to rationalise their conquest of Native American lands; likewise, Native American accounts of intentional spreading by Europeans articulate anxieties about European colonial ambitions and Native America vulnerability. At the same time, the role of smallpox in indigenous military alliances is a useful reminder that Native Americans were military agents, able to decide for themselves when—and whether—to join with Europeans in their conflicts. Indeed, Amherst's and Bouquet's ominous and abhorrent postscripts on deliberate smallpox spreading attest to their frustrations at the success of Native American military campaigns, which stymied the British officers throughout 1763 (Dowd 2002).

Since its arrival in America, smallpox has been inextricably bound up with the nature of military and imperial conquest. Appearing during wars, and spread by their movements, it makes more sense to see smallpox as being part of conflict—rather than replacing it, as crude models of the Columbian Exchange can posit. Smallpox was in symbiosis with war: it made wars more deadly, shaping military alliances and hostilities, and warfare made the disease more deadly, increasing its reach and inflaming its spread. Yet smallpox's role in wartime also lies in its rhetorical and emotional power: fear of the disease and accusations of its deliberate spread by enemy forces have also played a role in conflicts for hundreds of years. Fear of the disease interacted with conflict and enmity to shape rumours and allegations, both to accuse and to justify. This reveals that the disease was not simply a biological agent in military conflict: it was manipulated by humans to produce immunity; to affect recruitment positively and negatively; to generate fears, enmity and alliances; and to shape the nature of war itself.

Note

1 As Marshall and British records demonstrate, inoculation was not compulsory though strongly advised: it was billed as an 'extra' service in medical accounts, and it was also ordered that those inoculated 'have as little intercourse as possible with those men who refuse to be inoculated' (Boston orders 1 December 1775, WO 36/1), showing that some soldiers rejected inoculation.

Bibliography

Archival Sources

Bouquet Correspondence, British Library (London)
Add MS 21634
Add MS 21654
British Military and Colonial Correspondence, The British National Archives (London)
PRO 30/55/22
SP 78/263
WO 34/40
WO 36/1

French Military Correspondence, Service historique de la Défense (Paris)
A1 3498

Printed Works

Baker, Brenda and Kealhofer, Lisa, 'Assessing the Impact of European Contact on Aboriginal Populations,' in Brenda Baker and Lisa Kealhofer (eds.), *Bioarchaeology of Native American Adaptation in the Spanish Borderlands* (Gainsville: University Press of Florida, 1996), pp. 1–13.

Balvay, Arnaud, *L'épée et la plume: Amérindiens et soldats des troupes de la marine en Louisiane et au Pays d'en haut, 1683–1763* (Quebec: Les Presses de l'Université Laval, 2006).

Barrett, John T., 'The Inoculation Controversy in Puritan New England,' *Bulletin of the History of Medicine*, vol. 12, no. 1 (1942), pp. 169–90.

Becker, Ann, 'Smallpox in Washington's Army: Strategic Implications of the Disease during the American Revolutionary War,' *Journal of Military History*, vol. 68 (2004), pp. 381–430.

Blake, John, 'Smallpox Inoculation in Colonial Boston,' *Journal of the History of Medicine and Allied Sciences*, vol. 8, no. 3 (1953), pp. 284–300.

Blakely, Robert L. and Detweiler-Blakely, Bettina, 'The Impact of European Diseases in the Sixteenth-century Southeast: A Case Study,' *Midcontinental Journal of Archaeology*, vol. 14, no. 1 (1989), pp. 62–89.

Blakely, Robert L. and Mathews, David S., 'Bioarchaeological Evidence for a Spanish-Native American Conflict in the Sixteenth-century Southeast,' *American Antiquity*, vol. 55, no. 4 (1990), pp. 718–44.

Breen, Louise A., 'Cotton Mather, the "Angelical Ministry," and Inoculation,' *Journal of the History of Medicine and Allied Sciences*, vol. 46 (1991), pp. 333–57.

Charters, Erica, *Disease, War, and the Imperial State: The Welfare of the British Armed Forces during the Seven Years' War* (Chicago: University of Chicago Press, 2014).

Cirillo, Vincent J., *Bullets and Bacilli: The Spanish-American War and Military Medicine* (New Brunswick: Rutgers University Press, 2004).

Crosby, Alfred, *The Columbian Exchange: Biological and Cultural Consequences of 1492* (1972, reprint, Westport: Praeger, 2003).

Curtin, Philip, 'Epidemiology and the Slave Trade,' *Political Science Quarterly*, vol. 83, no. 2 (1968), pp. 190–216.

Dimsdale, Thomas, *Tracts, on Inoculation, Written and Published at St. Petersburg in the Year 1768. . . .* (London: W. Owen, Carnan and Newbury, 1781).

Dixon, David, *Never Come to Peace Again: Pontiac's Uprising and the Fate of the British Empire in North America* (Norman: University of Oklahoma Press, 2005).

Dobyns, Henry, *Their Number Become Thinned: Native American Population Dynamics in Eastern North America* (Knoxville: University of Tennessee Press, 1983).

Dowd, Gregory Evans, *War Under Heaven: Pontiac, the Indian Nations, and the British Empire* (Baltimore: The Johns Hopkins University Press, 2002).

Dowd, Gregory Evans, *Groundless: Rumors, Legends, and Hoaxes on the Early American Frontier* (Baltimore: Johns Hopkins University Press, 2015).

Fenn, Elizabeth, 'Biological Warfare in Eighteenth-Century North America: Beyond Jeffrey Amherst,' *Journal of American History*, vol. 68, no. 4 (2000), pp. 1552–80.

Fenn, Elizabeth, *Pox Americana: The Great Smallpox Epidemic of 1775–82* (New York: Hill and Wang, 2001).

Fenner, D. et al., *Smallpox and Its Eradication* (Geneva: World Health Organization, 1988).

Few, Martha, 'Medical Humanitarianism and Smallpox Inoculation in Eighteenth-century Guatemala,' *Historical Social Research*, vol. 37, no. 3 (2017), pp. 303–17.

Gronim, Sara, 'Imagining Inoculation: Smallpox, the Body, and Social Relations of Healing in the Eighteenth Century,' *Bulletin of the History of Medicine*, vol. 80, no. 2 (2006), pp. 247–68.

Harrison, Mark, *Medicine in an Age of Commerce and Empire: Britain and Its Tropical Colonies, 1660–1830* (Oxford: Oxford University Press, 2010).

Henige, David, *Numbers from Nowhere: The American Indian Contact Population Debate* (Norman: University of Oklahoma Press, 1998).

Hopkins, Donald R., *The Greatest Killer: Smallpox in History* (London: University of Chicago Press, 2002).

Johnson, Jay and Lehmann, Geoffrey, 'Sociopolitical Devolution in Northeast Mississippi and the Timing of the de Soto Entrada,' in Brenda Baker and Lisa Kealhofer (eds.), *Bioarchaeology of Native American Adaptation in the Spanish Borderlands* (Gainsville: University Press of Florida, 1996), pp. 38–55.

Jones, David S., 'Virgin Soils Revisited,' *William and Mary Quarterly*, vol. 60, no. 4 (2003), pp. 703–42.

Keel, Othmar, *L'Avènement de la médecine clinique modern en Europe: 1750–1815* (Montreal: Les Presses de l'Université de Montréal, 2001).

Kelton, Paul, *Epidemics and Enslavement: Biological Catastrophe in the Native Southeast 1492–1715* (Lincoln: University of Nebraska Press, 2007).

Kelton, Paul, *Cherokee Medicine, Colonial Germs: An Indigenous Nation's Fight against Smallpox, 1518–1824* (Norman: University of Oklahoma Press, 2015).

Kunitz, Stephen, *Disease and Social Diversity: The European Impact on the Health of Non-Europeans* (Oxford: Oxford University Press, 1994).

Lebrun, François, *Les hommes et la mort en Anjou aux 17e et 18e siècles: essai de démographie et de psychologie historiques* (Paris-La-Haye: Mouton: 1971).

MacLeod, Peter D., 'Microbes and Muskets: Smallpox and the Participation of the Amerindian Allies of New France in the Seven Years' War,' *Ethnohistory*, vol. 39, no. 1 (1992), pp. 42–64.

MacLeod, Peter D., *The Canadian Iroquois and the Seven Years' War* (Toronto: Dundurn, 2012).

Marshall, Tabitha, 'The Health of the British Soldier in America, 1775–1781' Unpublished PhD Dissertation (McMaster University: Canada, 2006).

Mayor, Adrienne, 'The Nessus Shirt in the New World: Smallpox Blankets in History and Legend,' *The Journal of American Folklore*, vol. 108, no. 42 (Winter 1995), pp. 54–77.

McCandless, Peter, *Slavery, Disease, and Suffering in the Southern Lowcountry* (Cambridge: Cambridge University Press, 2011).

Miller, Genevieve, *The Adoption of Inoculation for Smallpox in England and France* (Philadelphia: University of Pennsylvania Press, 1957).

Myrillas-Brazeau, Angeliki, 'Fear of Smallpox in Colonial New France,' unpublished conference paper presented at the University of Oxford, 20 June 2019.

Ranlet, P., 'The British, the Indians, and Smallpox: What Actually Happened at Fort Pitt in 1763?' *Pennsylvania History*, vol. 67, no. 3 (2000), pp. 427–41.

Razzell, Peter, *The Conquest of Smallpox: The Impact of Inoculation on Smallpox Mortality in Eighteenth Century Britain* (Lewes, Sussex: Caliban Books, 1977).

Reide, Thomas, *A View of the Diseases of the Army in Great Britain, America, the West Indies. . . .* (London: J. Johnson, 1793).

Riley, James C., 'Smallpox and American Indians Revisited,' *Journal of the History of Medicine and Allied Sciences*, vol. 65, no. 4 (2010), pp. 445–77.

Rosenberg, Charles, *Explaining Epidemics and Other Studies in the History of Medicine* (Cambridge: Cambridge University Press, 1992).

Rusnock, Andrea, *Vital Accounts: Quantifying Health and Population in Eighteenth-century England and France* (Cambridge: Cambridge University Press, 2002).

Shaffa, Ehsan, 'Case Fatality Ratios in Smallpox' (Geneva: World Health Organization, 1972): https://apps.who.int/iris/handle/10665/67493

Slovic, Paul, *The Perception of Risk* (London: Routledge, 2000).

Slovic, Paul, *The Feeling of Risk: New Perspectives on Risk Perception* (London: Routledge, 2010).

Smith, J.R., *The Speckled Monster: Smallpox in England, 1670–1970* (Chelmsford: Essex Record Office, 1987a).

Smith, Marvin, *Archaeology of Aboriginal Culture Change in the Interior Southeast: Depopulation during the Early Historic Period* (Gainsville: University of Florida Press, 1987b).

Stewart, Larry, 'The Edge of Utility: Slaves and Smallpox in the Early Eighteenth Century,' *Medical History*, vol. 29 (1985), pp. 54–70.

Sunstein, Cass, *Laws of Fear: Beyond the Precautionary Principle* (Cambridge: Cambridge University Press, 2008).

Vess, David, *Medical Revolutionaries in France: 1789–1796* (Gainsville: University Presses of Florida, 1974).

Volwiler, A.T., 'William Trent's Journal at Fort Pitt, 1763,' *The Mississippi Valley Historical Review*, vol. 11, no. 3 (1924), pp. 390–413.

Ward, Matthew, 'The Microbes of War: The British Army and Epidemic Disease among the Ohio Indians, 1758–1765,' in D.C. Skaggs et al. (eds.), *The Sixty Years' War for the Great Lakes, 1754–1814* (East Lansing: Michigan State University Press, 2001), pp. 63–78.

Wheelis, M., 'Biological Warfare before 1914,' in E. Geissler (ed.), *Biological and Toxin Weapons: Research, Development and Use from the Middle Ages to 1945* (Oxford: Oxford University Press, 1999), pp. 20–32.

Willrich, Michael, *Pox: An American History* (New York: Penguin, 2011).

20

GUNPOWDER AND THE NORTH AMERICAN INDIAN WAY OF WAR

Kaushik Roy

Introduction

The traditional view advocates that before the coming of the Europeans, warfare among the native tribes of North America was in a state of stasis and not casualty prone. The argument runs that the introduction of gunpowder technology by the Europeans in North America initiated a Military Revolution among the Indians. This in turn made inter-tribal warfare and conflict between the European settlers and the North American Indians bloody. Scholars also debate on whether Indian versus European warfare could be categorised as Total War or not. I argue that warfare among the natives of North America before 1492 experienced certain evolutionary changes. Instead of using the much cliched concept of Military Revolution, I propose that the forces of both the Europeans and American Indians experienced a Military Synthesis. Military Revolution refers to the adoption of a new piece of technology which results in a quantum jump in the military effectiveness of the concerned military organisation (Parker 1988). In contrast, the concept of Military Synthesis means a process by which a particular society selects some elements from the opposing military system and fuses them with some of their own traditional military techniques. It is to be noted that the concerned society neither fully adopts new or foreign military elements nor does it reject totally its own traditional war-making apparatus. Thus, Military Synthesis gives rise to a hybrid military system which represents both continuity and change (Roy 2005: 651–90). I argue that the natives of North America and the Europeans both adopted elements from the opposing military tradition and also adapted their own military culture in response to different require-ments in the new strategic scenario. The result was that both the American Indians and the Europeans came up with hybrid military forces. Also, the impact of war which the Europeans waged against the Indians in certain aspects could be termed as total.

Total War is a heuristic device derived from Carl von Clausewitz's concept of Absolute War. Total War stands for radical expansion in the scope and intensity of war. In such con-flicts, the aim of the belligerents becomes absolute: total destruction of the hostile state sys-tem, including its social structure and cultural fabric. Total War involves total mobilisation of the economy and society for military purpose. In the case of Total War, even non-combatants of the hostile party become legitimate targets. Therefore, in such conflicts, the line between

DOI: 10.4324/9780429437915-26

combatants and non-combatants vanishes. Hence, genocide/ethnic cleansing is an integral feature of Total War. However, Total War never occurs in reality. Frictions in the real world always limit the totality of war. Even the world's most horrendous conflict, the Second World War, in certain aspects was not total. Total War is an idee. Scholars can measure conflicts in the real world against the notion of Total War (Chickering 1999: 13–28). This chapter will compare the colonial war in North America against the theoretical model of Total War.

In this chapter, terms such as indigenous tribes of North America, American Indians and Amerindians have been used interchangeably. The first section of this chapter portrays the nature of warfare as it evolved among the American Indians before the coming of the Europeans. The second section studies how the impact of gunpowder caused changes in the way the American Indians fought amongst themselves. And the third section narrates how and why the Hybrid Amerindian Way of War failed against the Europeans. Actually, war between the indigenous tribes and the Europeans and war among the Amerindians were intertwined. The European colonists often made alliances with different Amerindian communities while attacking specific American Indian tribes. However, for the sake of brevity, we will discuss these two themes separately.

Warfare in the Pre-Columbian Era

Human beings entered North America about 15,000 years ago (Barr 2006: 4). They came from North-East Asia over the land bridge that used to connect Asia with North America. This land bridge covered the area now known as the Bering and Chukchi Seas (Stannard 1993: 8–9).

Doyne Dawson asserts that primitive societies were always in a state of war. Development of agriculture resulted in an increasing population density and the rise of permanent human settlements. Hunting-gathering could sustain one person per 100 square km. Agriculture could sustain a population density 50 times more than foraging economy. Domesticated crops became an important source of food by 500 BCE. By 200 CE, maize and squash cultivation had become widespread in North America. In 8000 BCE, there were about 100,000 Paleo-Indians in North America (Cordell and Smith 1996: 234–5; Smith 1989: 1569; Snow 1996: 142; Dawson 1996: 1–28; Dawson 2001: 71). Thanks to the gradual expansion of agriculture, the population of North America before 1492 was about 12 million (Starkey 2004: 238). The gradual development of agriculture enabled the deployment of a larger number of warriors, which in turn, increased the frequency and lethality of warfare.

Noted anthropologist Lawrence H. Keeley claims that prehistoric war was brutal. He continues that relative casualty rates in combat within primitive societies reached a level almost equal to that of modern wars (Keeley 1997). Keeley asserts that prehistoric societies practised Total War even before the arrival of the Europeans and the introduction of gunpowder technology. He notes: 'Total-war strategies were manifested in the plunder of wealth and food; destruction of houses, fields, and other means of production; and killing or capture of women and children. All these were common features of primitive warfare' (Keeley 1997: 48). Similarly, Azar Gat, in his global survey of evolution of warfare, writes that pre-state conflicts were endemic and lethal (Gat 2006: 11–25).

Warfare was rampant among the Aleut, Alutiiq and North-West Coast Indians. It occurred between two clans of the same tribe and also between various groups of tribes. The objectives of warfare were revenge, injury, insult and slave taking. Combat comprised both the raiding and sieges of forts. Along the North Pacific Coast of North America, forts were constructed about 1,500 years ago. Each clan owned a village and each village had a fort. A fort was made of rocks, surrounded with a trench and palisades made of wood. Almost 1,000 years ago, there was an intensification of warfare in this region due to an increase in population, expansion of the whale economy, pressure exerted by the Eskimos and the introduction of new technologies (bows and arrows which supplemented long knives, and bone and stone clubs) (Moss et al. 1992: 73–90).

Conflicts in the Eastern Woodlands occurred due to a desire for revenge and enhancement of status as well as for acquiring meat and deer hides. Competition within and between societies played an important role in the development of Mississippian chiefdoms around a millennium ago. They constituted the most complex societies in the prehistoric eastern part of North America. Chiefdoms are complex tribes with a ranked social structure. They usually have two ranks, commoners and nobles, and a formal political leadership in the form of a hereditary chief. The chief has religious and redistributive functions. However, chiefdoms lacked coercive bureaucracy (Dawson 2001: 63). After 1300 CE, thanks to the spread of corn agriculture, there was a rapid expansion of population among the Iroquoian-speaking peoples who lived around the Lower Great Lakes. Some of their settlements had up to 2,000 people. By the fifteenth century, the communities started uniting to form leagues or confederacies (Trigger and Swagerty 1996: 327). By this time, intergroup fighting had shifted from minor ambushes to attacks by a large number of well-organised warriors with conquest as their principal objective. This large-scale combat was initiated by the highly ranked members of these societies as they had the most to gain from a successful conclusion of their warlike endeavours (Milner 1999: 108).

The Aboriginal society of Australia engaged in four types of combat: battles, ritual trials, raids for women and revenge attacks (Connor 2013: 12). In precolonial South-East Asia, the principal objective of war was acquiring labour and not land. Due to an abundance of fertile land and a relatively sparse population, acquisition of land was not considered to be significant. Rather, capturing slaves through launching surprise raids was the norm. In fact, acquisition of slaves also raised the social status of successful warriors (Rodriguez 2003: 151).

The five Iroquois tribes (Cayuga, Mohawk, Oneida, Onondaga and Seneca) who inhabited the north-eastern part of North America were engaged in fighting with each other and against other non-Iroquois tribes. Raids were common (Barr 2006: 7). Capturing enemy manpower was an important objective of war among the Amerindian tribes. Besides raiding, the prehistoric American Indians also engaged in siege warfare. Assaults on fortified villages were common. Between 1100 and 1300 CE, most of the villages of the north-eastern part of North America were protected with circular or oval palisades, and some had exterior ditches. Few villages possessed double-walled palisades. A double-walled palisade structure comprised two palisade lines positioned adjacent to each other (Keener 1999: 781).

Palisades, fortified hill tops, walls on hill slopes, towers, tunnels, movable ladders, etc. were common features of siege war in the south-western part of North America from ca. 700 CE. Palisades were made with upright logs or stones. Towers were constructed with stones. They had both ceremonial as well as defensive functions. Sometimes, a series of towers was constructed for defence. Forts at the hill tops surrounded with stone walls were occasionally used only during an attack by the enemy force. At times, a particular village was designated as a guard village, where the people of the surrounding habitation took refuge and the warriors made a last stand during an enemy invasion or raid (Farmer 1957: 249–50).

Scalping was an ancient custom which existed in North America long before the arrival of the Europeans. This tradition continued among the indigenous tribes of North America even in the colonial era. Scalping involved the removal of a patch of hair-covered skin from the crown of the head of a fallen enemy. Among the Pawnee, scalps symbolised life power. The Pawnee considered scalps as trophies of war. Their religious beliefs sanctioned taking and sacrificing scalps. After the Pawnee warriors returned from the warpath with scalps, the latter became an essential item for fertility rituals. The demand for scalps occasionally resulted in war and also influenced battle tactics (Logt 2008: 71–85). The Pueblos, did not indulge in scalping on a large scale. Usually one or two scalps were taken in each fight and these were the scalps of the bravest enemy warrior. The scalps were brought back to the village and given to the men who wished to be initiated into warrior society (Haas and Creamer 1997: 247). Just as scalping was an important objective of war among the Amerindians, headhunting was also one of the principal objectives of conflict among the indigenous tribes of Ambonia (Knaap 2003: 165–92). Between the fifteenth and eighteenth centuries, the lowland Burmese and the highland Shans followed the practice of taking the heads of the masses of both the dead and wounded enemy soldiers (Charney 2004: 2).

Wayne E. Lee rightly notes that the Amerindian tribes followed a highly ritualised but at the same time lethal warfare. The Amerindian polities were small-scale units. Due to their low demographic resources, their preferred tactics were to surprise and ambush the enemy parties or the hostile villages. Sustained small-scale ambushes could produce a large number of casualties in the long run (Lee 2011: 52–5).

Before the coming of the Europeans, the natives of North America fought with clubs, bows, arrows and stone hatchets. William H. Holmes says that the tomahawk emerged as a weapon much later. The European observers applied the name tomahawk to the celt-hatchet and the grooved axe, which were similar to the English hatchet. The celt-hatchet was not generally used as a weapon of war by the tribes of Virginia. In fact, it was only in the seventeenth century that the Indians began to use stone hatchets and small axes instead of clubs in war. The tomahawk was actually a war club and the Europeans later used this term mainly to denote a small battle axe. The Europeans applied the name tomahawk to a variety of tools: stone hatchet, grooved axe, spiked club, etc. (Holmes 1908: 264–76).

Like the ancient Spartans, the Apache had their traditional form of military training. An Apache boy became a man and a warrior with the completion of military training known as *dikohe*, especially for conducting raids. The training started at the age of 14. It involved tremendous physical hardships like running early in the morning, jumping in the cold water of streams, eating cold food, etc. The objective of such training was to inure the young men to physical hardship and to make them agile. The village *shaman* also guided the young men completing *dikohe*. After completing *dikohe*, the young males were allowed to marry (Opler and Hoijer 1940: 617–21). One can say that the tribes of North America were quite militaristic in nature.

Gunpowder and Warfare among the Indians

David J. Silverman asserts that an arms race occurred among the indigenous tribes of the north-west part of North America not due to the pressure of colonialism but due to the threat posed by the Iroquois (Silverman 2016: 23). This assertion requires partial qualification. Both intertribal and European-American Indian contest played a crucial role in initiating the arms race amongst the Amerindians.

Keith F. Otterbein writes that tactical innovations, along with the possession of a larger number of firearms, played an important role in continuous Iroquois victories against their

Indian opponents in the seventeenth century. In 1609, the Algonquins acquired match-locks. For the next 25 years, they had an advantage over the Mohawks. The Mohawks, faced with superior firepower, waged guerrilla warfare: conducting tactical retreat and launching sudden ambushes. They also practised sudden rushes and 'bear hug' tactics which involved close quarter combat, before their enemies could reload and fire their matchlocks. The Mohawks also took advantage of the terrain. The Allegheny Plateau and the Adirondack Mountains of Upper New York State were ideally suited for ambushes and raids. However, Iroquois military methods changed when they started acquiring muskets (Otterbein 1964: 56–8).

Their arrival in the St Lawrence region during the seventeenth century brought the French into contact with the Huron Confederacy on the north-eastern extremity of Lake Huron. The Hurons became trading partners of the French. They exchanged beaver furs for firearms. However, the French were able to provide firearms to the Hurons in limited numbers. In 1614, the Dutch established a trading post at Fort Orange (Albany) on the Hudson River, as well as a trading relationship with the Iroquois Confederacy. The Huron and the Iroquois were traditional enemies. The Dutch were more than willing to provide firearms and gunpowder to the Iroquois in return for furs. The Huron–Iroquois War of 1648–1649 resulted in the victory of the Iroquois (Lee 2007: 707–9). By the 1660s, thanks to the supply of specially manufactured Dutch flintlocks for the American Indian tribes, the flintlock musket became the principal weapon for the Iroquois ambush parties. The Iroquois tactics were to fire one or two volleys at their targets from an ambush position, then rush to the distracted enemy and finish them off in hand-to-hand combat with tomahawks and clubs (Silverman 2016: 27–8). Here, we find a synthesis of new weapons (muskets) and new techniques (volley fire) with traditional weapons (clubs, tomahawks) and old tactics (rushing at the distracted enemy for close-quarter combat).

James D. Rice asserts that due to low demographic resources, the Powhatans relied on ambushes and sudden raids in order to minimise casualties. But, in accordance with their strategy of conquest and colonisation, the Powhatans also conducted open battles and ex-tended sieges during the sixteenth and early seventeenth centuries (Rice 2020: 23–7, 30, 32). In fact, fighting decisive battles in accordance with the strategy of conquest was also practised by the Iroquois. In March 1649, the Iroquois defeated and almost destroyed the Hurons. The main engagements took place near Georgian Bay. The Iroquois invaded Hu-ronland. On the dawn of 16 March, the Iroquois suddenly attacked the Huron village named St Ignace. The Iroquois force numbered 1,000 and was well equipped with firearms. They had started their march from their homeland in autumn, sustained themselves in the winter by hunting and approached the Huron village during the night to launch a deadly surprise assault. Many Huron men, women and children were killed and the rest were taken captive. The Iroquois made St Ignace their base camp for regrouping. Meanwhile, a second Iroquois force attacked and burnt the Huron village named Saint Louis. Besides being surprised, the Huron villages protected with wooden palisades fell to the steel axe- and firearms-equipped Iroquois attacking force. Furthermore, tension between Christian and traditionalist Hurons also weakened their defensive capacity (Otterbein 1979: 141–52).

Indigenous people facing handheld firearms-equipped enemies avoided open battles in the field and abandoned wooden armour. This was because wooden armour was useless against bullets and open field battle resulted in a huge number of casualties. However, the spread of firearms resulted in an increase of sieges of fortified villages. Indigenous invaders equipped with firearms and steel cutting tools could breach the traditional defensive structures. In response, indigenous fortifications experienced innovations. Circular palisades were replaced with straight wall fortifications which gave the defenders equipped with firearms a clear shot at the attackers (Silverman 2016: 23–4).

Thus, we see that in cases of both field battles and siege war, the American Indians integrated firearms with their traditional set of weaponry and evolved composite tactical formulas in accordance with the terrain and the changing strategic requirements. The American Indian military system adopted certain elements from the foreigners and adapted their own method to meet the new strategic and tactical requirements. This process of military hybridisation was also evident, as we will discuss in the next section, during their confrontation with the Europeans.

American Indians against the Europeans: Military Revolution or Total War or Hybrid Warfare?

A group of scholars argues that the European colonisation resulted in a sort of genocide in Australasia and in the Americas (Richards 2013: 29, 36–7). David E. Stannard writes that the conquest of the New World by the Europeans was akin to a holocaust. The European invaders, through military and economic means, deliberately eliminated a large chunk of the native population (Stannard 1993). From the opposite end, Robert M. Utley challenges the argument that the Americans initiated a genocide of the Indians. He writes:

> No more than a tiny proportion of the white population of the United States, mainly in the West, ever advocated such a measure. No government official ever seriously proposed it. The occasional rhetoric of generals such as William T. Sherman and Philip H. Sheridan was just that: angry reactions to particular events involving particular Indian groups.
>
> *(Utley 1999: 399)*

However, certain aspects of the war as practised by both the Amerindians and the Europeans anticipated certain features of Total War. For instance, both the Indians and the Europeans not merely aimed to kill the combatants but also deliberately destroyed the food, clothing, shelter and transportation networks. Thus, civilians and the civilian infrastructure became targets of attack (Chickering and Forster 1999: 7–8).

Armstrong Starkey claims:

> European soldiers brought the new weapons and techniques . . . with them to North America and by 1675 had provoked a military revolution of a sort among Native Americans, a revolution that for 140 years gave them a tactical advantage over their more numerous and wealthier opponents.
>
> *(Starkey 1998: viii)*

Patrick Malone goes further and argues that the European colonists introduced the Indians to a new technology which enabled the latter to wage a 'form of total warfare' which was previously outside their aboriginal experience (Malone 2000: 24). Malone continues: 'The widespread use of fire arrows and torches against English houses was one demonstration of the Indians' willingness to practice total warfare' (Malone 2000: 80).

Adam J. Hirsch, in an article, claims that the clash of culture was responsible for the increasing lethality of European–Amerindian conflicts. Initially, both the Europeans and the Amerindians aimed to limit conflicts. The Europeans aimed to wage a battle-centric war in the style of the eighteenth-century European paradigm of war. The objective was to destroy the hostile force in a decisive battle. However, the American Indians refused to meet them in the battlefield by massing their forces for a decisive confrontation. Rather, the Amerindians cut off the European stragglers and attacked the detached posts. Faced with a guerrilla war, the Europeans, in accordance with the principles of counterinsurgency, attacked the food, shelter and non-combatants among the Indians. The Amerindians, especially the Pequots, initially followed a policy of Limited War. They did not aim to destroy all the European settlements even when they had the opportunity. Their objective was to attack and then bargain from a position of strength. But, when the American Indians found that the Europeans responded with brute force, attacked their villages and took more of their lands, then the Indian response to war also became total. For Adam Hirsch, this transition from Limited War to Total War occurred between the Pequot War (1636–1638) and King Philip's War (1675–1778) (Hirsch 1988: 1187–1212). Even if Hirsch's argument is correct, then one can say that the cultural clash was the product of material forces: hunger for more land, resources and inflow of human bodies from the Old World. Let us see the evolution of the Indians' Hybrid War against the Europeans.

Despite royal proclamations and colonial laws, the French, Dutch and English traders provided the tribes of New England with matchlocks and flintlocks in exchange for furs and wampum (Malone 1997: 232). By the late seventeenth century, the Amerindians refused to mass their forces while encountering the European forces. Instead, the North American tribes fought in a loose coordinated fashion and targeted individuals and small groups. They conducted ambushes, took captives and also engaged in ritualised demonstrative display of violence (Plank 2011: 221, 223). This method of fighting was known as the Skulking Way of War.

It would be too simplistic to argue for the evolution of a single homogeneous American Indian Way of War in response to European aggression. In the eighteenth century, the American Indians occasionally fought in a half moon formation, a technique they acquired from their communal hunts (Eid 1988: 147–58). The point to be noted is that even the steppe nomads used half moon tactics, a technique which they learnt from their own communal hunts. Wayne Lee says that besides the Skulking and Sneaking Way of War, the Indians also practised the Cutting Way of War, i.e., cutting off the stragglers or enemy columns, especially the supply wagons, etc. What is important is that the Indian tactics, both in the case of field and siege war, continued to evolve in response to the threat posed by European colonialism. For instance, in 1712, the Second Tuscarora War started. The English–Indian forces attacked the Tuscarora villages along the Neuse River and the Contentnea Creek in eastern North Carolina. Threatened by cannons and mines, the Tuscaroras constructed underground bunkers, counter-trenches, etc. (Lee 2004: 713–20).

Amerindians needed access to European weapons to defend themselves not only against the Europeans but also against more powerful indigenous groups. In addition, the natives also required metal for the manufacture of hunting tools and cooking implements. By the late seventeenth century, much of the natives' traditional hunting practice had died. As a result, the Amerindians relied on European firearms for hunting (Pulsipher 2011: 25, 33).

The failure of the North American tribes to establish a self-sustaining metal manufacturing infrastructure ensured their doom in the long run. The Indians were able to repair firearms and occasionally manufacture shots. But they could not produce flintlock muskets in large quantities to arm all their young males. In the swampy and forest terrain of North America, use of cannons was limited. Nevertheless, light cannons and mortars played a significant role in successful sieges of Indian villages. And the American Indians were not capable of manufacturing such instruments of war. The very fact that they were dependent on the Europeans, whom they were fighting, for the supply of war materials was a prime cause of their failure. Once the Europeans cut off the supply of firearms and gunpowder, the Amerindian tribes were left in a terrible situation. The Indians could excel over the Europeans in the use of flintlocks in the forest environment. In terms of tactics, they had an edge over their European adversaries. But their real failure was in the field of ordnance production—i.e., logistics of warfare.

Timothy J. Shannon asserts that the Indians' Skulking Way of War was a product of the Military Revolution which occurred among the Indians during the seventeenth century. And this method of warfare proved resistant to the challenges posed by the new style of warfare which the Europeans brought to North America during the second half of the eighteenth century (Shannon 2010: 139). Actually, the Plains Indians created a method of Hybrid War in the eighteenth century by adopting horses and breechloaders from the Europeans (Vandervort 2006: 6–7). But in the late nineteenth century, the so called Horse-Gun Pattern of Plains Indian Warfare also fell prey to the numbers game: the United States had more men, horses and rifles. And the Cheyenne, Sioux and Arapaho tribes lacked the scientific technological base for manufacturing breechloaders on a mass scale.

The new type of war that the Europeans practised in the second half of the eighteenth century, as noted by Shannon, was actually the product of an amalgamation of West European techniques of war with the American Indian Way of War. And I called this amalgamated war making Hybrid War. In the eighteenth century, in South Asia also, the British outwitted the indigenous powers by constructing a hybrid military organisation by amalgamating Western military elements with certain Indian military techniques (Roy 2011: 195–218).

The Europeans also accepted certain elements of Native American military culture, which they integrated within their own reformed military system. By the eighteenth century, enemy scalps were prized trophies among the whites as well as among the Eastern Algonguian tribes and the Iroquois (Ward 1999: 200). The European colonists borrowed several items of the American Indians' military paraphernalia. Some examples are as follows: the use of paddle birchbark canoes for navigating the creeks; snowshoes and moccasins for travelling in the deep snow during winter. These became part and parcel of the colonial militias. The hatchet or tomahawk became an essential sidearm, replacing the sword in many units. Also, Indian corn parched and beaten to nocake became the standard iron ration during a long march across difficult terrain (Malone 2000: 98). During the 1812 War, the British, unlike the Americans, employed Indian auxiliaries. And this factor, to a great extent, allowed the British and their Indian allies to maintain control over the area that stretched from Mackinac Island to the Mississippi River (Jung 2012: 27).

Infighting and tensions between the natives facilitated Western expansion in Asia, Africa and the New World. In 1700, the governor of New England and New York was afraid that a general uprising of the Amerindians might destroy the settlements of the European colonisers who were at that time demographically weak. In north-eastern North America, till 1725, European influence, and presence, was quite limited (Baker and Reid 2004: 77–106). Thanks to British diplomacy, the Delaware and the Iroquois did not aid the Shawnee tribe in 1774 during the Dunmore's War (Downes 1934: 327–9). Low demographic resources, coupled with the small size of indigenous polities, limited the number of warriors that the Indian chiefs could bring to the field. Some Indian chiefs were aware of this limitation. They tried to overcome it by attempting to forge diplomatic alliances with other tribes. One prime example of this sort of effort was the six-month tour of the Indian country by Tecumseh of the Shawnee tribe, between August 1811 and January 1812. Tecumseh's objective (though he ultimately failed) was to form a multitribal confederacy to resist the Indians' cessions of land to European colonists and to forge a nationalist Amerindian identity. Tecumseh was for the resistance of the American cultural effort to sedenterise the tribes. He advised the Creek Indians to kill those old chiefs who advocated peace with the Americans (Sugden 1986: 273–88).

Demographic superiority of the Europeans played an important role in their victory over the Amerindians. One author rightly notes that the American War of Independence was a real disaster for the Indian tribes. For instance, the Shawnees occupied the region between Virginia and Kentucky. And between 1775 and 1790, some 80,000 non-Indians poured into the Shawnee country (Calloway 1992: 41, 47). In the second half of the nineteenth century, in the American 'wild west,' there were only 3,500 Cheyennes and 6,000 Apaches (Vandervort 2006: 13–14, 33).

Before European colonisation, New England's Indian population at its maximum came to about 144,000. By 1670, on the eve of King Philip's War, this number had fallen to 8,700. At that time, the Europeans numbered over 50,000 (Starkey 1998: 7). Soon after the English started colonising Carolina in 1670, major epidemics hit this region at least once in every generation: 1698, 1718, 1738 and 1759. A variety of less virulent illnesses almost never left the native settlements. The indigenous population of North America, long isolated from the rest of the world, lacked immunity to the pathogens introduced by the European intruders. By 1759, some 60 per cent of the original native population was wiped out by diseases (Merrell 1984: 542–3). One of the most dangerous diseases which devastated Amerindian society was smallpox. The average mortality was 30 per cent of those infected. The American Indians contracted these diseases from close contact with the Europeans and also from clothing acquired from the latter (Boyd 1994: 5–40).

Demographic superiority, coupled with technological advances and tactical evolution on the part of the Americans, sounded the death knell of the Indians. While the American Indian population registered drastic decline, the inflow of a huge number of Caucasoid people from overseas raised the population of North America in the nineteenth century. Between 1840 and 1860, the population of North America soared from 17 million to 31 million (Gates 2001: 115). During the second half of the nineteenth century, use of steamers for carrying supplies along the rivers increased the operative range of the American punitive expeditionary force. Further, the use of skirmishers to protect the supply wagons by the American field commanders thwarted the Indians' hit and run techniques. In 1868, the US Army realised that, in addition to the regulars, they needed frontiersmen for fighting the Indians: 'the children of the west' who could fight the Indians on their own terms. The frontiersmen were volunteers, armed with rifles and mounted on horses. Like the Indian tribes, they were supposed to travel light with minimum rations and fall upon farmsteads of the hostiles (Athearn 1948: 272–84; Vandervort 2006: 3–4). The United States continued to acquire lands of the Indians by military conquest till 1871 when Congress ordered its termination (Ward 1999: 209).

In terms of casualties, war with the Indians amounted to Limited War for the Europeans and later the Americans. According to one calculation, between the 1860s and 1880s, the

US Army suffered only 2,000 casualties in the Indian Wars and killed about 6,000 Indians (Tucker 2003: 121). But for the Indians, war with the European colonists had certain elements of Total War. Not only had they lost their land, but their lives, movable property and their culture (way of life) faced extinction at the hands of the European colonists. Low demographic resources and the small size of their polities allowed the Indian tribes to mobilise only a limited number of males for combat. Thus, even if, in terms of absolute number, the number of Amerindians killed in war was limited compared with the great human carnage that characterised the modern Western world, in relative terms the loss of a few hundred killed had a devastating effect on the Amerindian polities. At the same time, both the American Indians and the Europeans fought Hybrid War as both military systems borrowed selected elements from its opponents and integrated them with their traditional military techniques to counter the hostiles' combat superiority.

Conclusion

Warfare in the pre-European phase in North America was evolving over the millennia. It was endemic and bloody but laced with rituals. The Amerindians enthusiastically adopted firearms. The arms race in colonial North America among the indigenous tribes was both due to the threat posed by European colonisers as well as the dangers posed by other firearms-equipped indigenous tribes. The introduction of gunpowder increased the lethality of war. For the American Indians, the European warfare in North America was a case of Total War. This is because the Europeans carried out a quasi-genocide among the American Indians and conquered their land. The livestock, food crops and non-combatants among the Indians were legitimate targets of attack for the European settlers. The net result was that only a small section among the Indians survived in the Caucasoid-dominated United States of America as a marginal group. But, for the European intruders, it was a case of Limited War due to relatively small number of males mobilised and the small number of casualties suffered in wars against the Indians. Further, the indigenous tribes of North America were never able to pose an existential threat to the overseas settlers.

During the late seventeenth century, both the Indians and the Europeans had perfected a hybrid military culture by adopting certain elements from the opposing military tradition and then amalgamating the imported elements with their own military techniques. However, the Amerindians were not able to sustain their hybrid military culture because, unlike their European opponents, they lacked a productive base for manufacturing firearms and cannons on a large scale. One can say that in spite of exhibiting a high learning curve, the Amerindians failed because of their low technological base. Despite the recent overemphasis on culture in academic scholarship, we cannot totally neglect the role of hardware while discussing conflicts among societies. To sum up, gunpowder was a crucial factor in the victory of the Europeans. However, gunpowder alone was not adequate. Demographic and political factors also acted in conjunction with gunpowder technology. The natives of North America failed to construct large polities. Initially they played off various European powers like the French, British and Spanish for acquiring firearms and gunpowder. Once Britain was out after the American War of Independence, the defeat of the Indians was merely a question of time. Along with gunpowder weapons manufacturing technology, political sophistication and superior demographic resources of the Europeans, coupled with the import of diseases, resulted in the collapse of American Indian resistance.

Bibliography

Athearn, Robert G., 'A Winter Campaign against the Sioux,' *The Mississippi Valley Historical Review*, vol. 35, no. 2 (1948), pp. 272–84.

Baker, Emerson W. and Reid, John G., 'Amerindian Power in the Early Modern Northeast,' *William and Mary Quarterly*, Third Series, vol. 61, no. 1 (2004), pp. 77–106.

Barr, Daniel P., *Unconquered: The Iroquois League at War in Colonial America* (Westport, Connecticut/London: Praeger, 2006).

Boyd, Robert, 'Smallpox in the Pacific Northwest: The First Epidemics,' *British Columbian Quarterly*, no. 101 (Spring 1994), pp. 5–40.

Calloway, Colin G., '"We Have Always Been the Frontier": The American Revolution in Shawnee Country,' *American Indian Quarterly*, vol. 16, no. 1 (1992), pp. 39–52.

Charney, Michael, *Southeast Asian Warfare: 1300–1900* (Leiden/Boston: Brill, 2004).

Chickering, Roger, 'Total War: The Use and Abuse of a Concept,' in Manfred F. Boemeke, Roger Chickering and Stig Forster (eds.), *Anticipating Total War: The German and American Experiences, 1871–1914* (Cambridge: German Historical Institute Washington DC in association with Cambridge University Press, 1999), pp. 13–28.

Chickering, Roger and Forster, Stig, 'Introduction,' in Manfred F. Boemeke, Roger Chickering and Stig Forster (eds.), *Anticipating Total War: The German and American Experiences, 1871–1914* (Cambridge: German Historical Institute Washington DC in association with Cambridge University Press, 1999), pp. 1–9.

Connor, John, 'Traditional Indigenous Warfare,' in Craig Stockings and John Connor (eds.), *Before the Anzac Dawn: A Military History of Australia to 1915* (Sydney: University of New South Wales Press, 2013), pp. 8–20.

Cordell, Linda S. and Smith, Bruce D., 'Indigenous Farmers,' in Bruce G. Trigger and Wilcomb E. Washburn (eds.), *The Cambridge History of the Native Peoples of the Americas*, vol. 1, *North America*, Part 1 (Cambridge/New York: Cambridge University Press, 1996), pp. 201–66.

Dawson, Doyne, 'The Origins of War: Biological and Anthropological Theories,' *History and Theory*, vol. 35, no. 1 (1996), pp. 1–28.

Dawson, Doyne, *The First Armies* (London: Cassell, 2001).

Downes, Randolph C., 'Dunmore's War: An Interpretation,' *The Mississippi Valley Historical Review*, vol. 21, no. 3 (1934), pp. 311–30.

Eid, Leroy V., '"A Kind of Running Fight": Indian Battlefield Tactics in the Late Eighteenth Century,' *Western Pennsylvanian Historical Magazine*, vol. 71, no. 2 (1988), pp. 147–71.

Farmer, Malcom F., 'A Suggested Typology of Defensive Systems of the Southwest,' *Southwestern Journal of Anthropology*, vol. 13, no. 3 (1957), pp. 249–66.

Gat, Azar, *War in Human Civilization* (Oxford/New York: Oxford University Press, 2006).

Gates, David, *Warfare in the Nineteenth Century* (Houndmills, Basingstoke: Palgrave, 2001).

Haas, Jonathan, and Creamer Winifred, 'Warfare among the Pueblos: Myth, History and Ethnography,' *Ethnohistory*, vol. 44, no. 2 (1997), pp. 235–61.

Hirsch, Adam J., 'The Collision of Military Cultures in Seventeenth-Century New England,' *Journal of American History*, vol. 74, no. 4 (1988), pp. 1187–1212.

Holmes, William H., 'The Tomahawk,' *American Anthropologist*, vol. 10, no. 2 (1908), pp. 264–76.

Jung, Patrick J., 'Toward the Black Hawk War: The Sauk and Fox Indians and the War of 1812,' *Michigan Historical Review*, vol. 38, no. 1, Special Issue (2012), pp. 27–52.

Keeley, Lawrence H., *War before Civilization: The Myth of the Peaceful Savage* (1996, reprint, New York/Oxford: Oxford University Press, 1997).

Keener, Craig S., 'An Ethnohistorical Analysis of Iroquois Assault Tactics used Against Fortified Settlements of the Northeast in the Seventeenth Century,' *Ethnohistory*, vol. 46, no. 4 (1999), pp. 777–807.

Knaap, Gerrit, 'Headhunting, Carnage and Armed Peace in Ambonia,' *Journal of the Economic and Social History of the Orient*, vol. 6, Part 2 (2003), pp. 165–92.

Lee, Wayne E., 'Fortify, Fight, or Flee: Tuscarora and Cherokee Defensive Warfare and Military Culture Adaptation,' *Journal of Military History*, vol. 68, no. 3 (2004), pp. 713–70.

Lee, Wayne E., 'Peace Chiefs and Blood Revenge: Patterns of Restraint in Native American Warfare, 1500–1800,' *Journal of Military History*, vol. 71, no. 3 (2007), pp. 707–41.

Lee, Wayne E., 'The Military Revolution of Native North America: Firearms, Forts and Polities,' in Wayne E. Lee (ed.), *Empires and Indigenes: Intercultural Alliance, Imperial Expansion, and Warfare in the Early Modern World* (New York/London: New York University Press, 2011), pp. 49–79.

Logt, Mark van de, '"The Power of the Heavens shall Eat of my Smoke": The Significance of Scalping in Pawnee Warfare,' *Journal of Military History*, vol. 72, no. 1 (2008), pp. 71–104.

Malone, Patrick M., 'Changing Military Technology among the Indians of Southern New England, 1600–1677,' in Douglas M. Peers (ed.), *Warfare and Empires: Contact and Conflict between European and non-European Military and Maritime Forces and Cultures* (Aldershot: Ashgate 1997), pp. 229–44.

Malone, Patrick M., *The Skulking Way of War: Technology and Tactics among the New England Indians* (1991, reprint, Maryland: Madison Books, 2000).

Merrell, James H., 'The Indians' New World: The Catawba Experience,' *William and Mary Quarterly*, Third Series, vol. 41, no. 4 (1984), pp. 537–65.

Milner, George R., 'Warfare in Prehistoric and Early Historic Eastern North America,' *Journal of Archeological Research*, vol. 7, no. 2 (1999), pp. 105–51.

Moss, Madonna L. and Erlandson, Jon M., 'Forts, Refuge Rocks, and Defensive Sites: The Antiquity of Warfare along the North Pacific Coast of North America,' *Arctic Anthropology*, vol. 29, no. 2 (1992), pp. 73–90.

Opler, Morris Edward and Hoijer, Harry, 'The Raid and War-Path Language of the Chiricahua Apache,' *American Anthropologist*, New Series, vol. 42, no. 4, Part 1 (1940), pp. 617–34.

Otterbein, Keith F., 'Why the Iroquois Won: An Analysis of Iroquois Military Tactics,' *Ethnohistory*, vol. 11, no. 1 (1964), pp. 56–63.

Otterbein, Keith F., 'Huron vs. Iroquois: A Case Study in Inter-Tribal Warfare,' *Ethnohistory*, vol. 26, no. 2 (1979), pp. 141–52.

Parker, Geoffrey, *The Military Revolution: Military Innovation and the Rise of the West, 1500–1800* (Cambridge: Cambridge University Press, 1988).

Plank, Geoffrey, 'Deploying Tribes and Clans: Mohawks in Nova Scotia and Scottish Highlanders in Georgia,' in Wayne E. Lee (ed.), *Empires and Indigenes: Intercultural Alliance, Imperial Expansion, and Warfare in the Early Modern World* (New York/London: New York University Press, 2011), pp. 221–49.

Pulsipher, Jenny Hale, 'Gaining the Diplomatic Edge: Kinship, Trade, Ritual, and Religion in Amerindian Alliances in Early North America,' in Wayne E. Lee (ed.), *Empires and Indigenes: Intercultural Alliance, Imperial Expansion, and Warfare in the Early Modern World* (New York/London: New York University Press, 2011), pp. 19–47.

Rice, James D., 'War and Politics: Powhatan Expansionism and the Problem of Native American Warfare,' *William and Mary Quarterly*, vol. 77, no. 1 (2020), pp. 3–32.

Richards, Jonathan, 'Frontier Warfare in Australia,' in Craig Stockings & John Connor (eds.), *Before the Anzac Dawn: A Military History of Australia to 1915* (Sydney: University of New South Wales Press, 2013), pp. 21–38.

Rodriguez, Felice Noelle, 'Juan De Salcedo Joins the Native Form of Warfare,' *Journal of the Economic and Social History of the Orient*, vol. 46, no. 2 (2003), pp. 143–63.

Roy, Kaushik, 'Military Synthesis in South Asia: Armies, Warfare, and Indian Society, c. 1740–1849,' *Journal of Military History*, vol. 69, no. 3 (2005), pp. 651–90.

Roy, Kaushik, 'The Hybrid Military Establishment of the East India Company in South Asia: 1750–1849,' *Journal of Global History*, vol. 6 (2011), pp. 195–218.

Shannon, Timothy J., 'The Native American Way of War in the Age of Revolutions, 1754–1814,' in Roger Chickering and Stig Forster (eds.), *War in an Age of Revolution, 1775–1815* (Cambridge/New York: German Historical Institute Washington DC and Cambridge University Press, 2010), pp. 137–57.

Silverman, David J., *Thundersticks: Firearms and the Violent Transformation of Native America* (Cambridge, Massachusetts/London: Belknap Press, 2016).

Smith, Bruce D., 'Origins of Agriculture in Eastern North America,' *Science*, New Series, vol. 246, no. 4937 (1989), pp. 1566–71.

Snow, Dean R., 'The First Americans and the Differentiation of Hunter-Gatherer Cultures,' in Bruce G. Trigger and Wilcomb E. Washburn (eds.), *The Cambridge History of the Native Peoples of the*

Americas, vol. 1, *North America*, Part 1 (Cambridge/New York: Cambridge University Press, 1996), pp. 125–99.

Stannard, David E., *American Holocaust: The Conquest of the New World* (1992, reprint, New York/Oxford: Oxford University Press, 1993).

Starkey, Armstrong, *European and Native American Warfare, 1675–1815* (London: UCL Press, 1998).

Starkey, Armstrong, 'European-Native American Warfare in North America, 1513–1815,' in Jeremy Black (ed.), *War in the Early Modern World: 1450–1815* (1999, reprint, London/New York: Routledge, 2004), pp. 237–62.

Sugden, John, 'Early Pan-Indianism: Tecumseh's Tour of the Indian Country, 1811–1812,' *American Indian Quarterly*, vol. 10, no. 4 (1986), pp. 273–304.

Trigger, Bruce G. and Swagerty, William R., 'Entertaining Strangers: North America in the Sixteenth Century,' in Bruce G. Trigger and Wilcomb E. Washburn (eds.), *The Cambridge History of the Native Peoples of the Americas*, vol. 1, *North America*, Part 1 (Cambridge/New York: Cambridge University Press, 1996), pp. 325–98.

Tucker, Spencer, 'The United States Military, 1815–2000,' in Jeremy Black (ed.), *War in the Modern World since 1815* (London/New York: Routledge, 2003), pp. 116–41.

Utley, Robert M., 'Total War on the American Indian Frontier,' in Manfred F. Boemeke, Roger Chickering and Stig Forster (eds.), *Anticipating Total War: The German and American Experiences, 1871–1914* (Cambridge: German Historical Institute Washington DC in association with Cambridge University Press, 1999), pp. 399–414.

Vandervort, Bruce, *Indian Wars of Mexico, Canada and the United States, 1812–1900* (New York/London: Routledge, 2006).

Ward, Harry M., *The War for Independence and the Transformation of American Society* (London: UCL Press, 1999).

21

FIREARMS AND MARITIME GUNPOWDER STATES OF ASIA

Michael W. Charney

Scholars have long attributed the emergence in early modern maritime Asia of powerful and more centralised states than their predecessors to the impact of the development and spread of firearms after the use of gunpowder by the Mongols. Western and Central Europe may have pursued the rapid development of firearms to compensate for the loss of manpower due to the ravages of the Bubonic Plague. But once developed, Asian rulers also raced to obtain them. Firearms in early modern Asia were by no means a maritime monopoly, nor were effective firearms derived solely from European sources. Indeed, there were overland supplies of guns and gunpowder. Gunpowder technology also spread within the interior of the continent. However states on maritime trade routes gained new firearms more quickly and often in greater numbers than their landlocked peers. The growth of maritime revenues in the fifteenth and sixteenth centuries certainly made it easier for maritime states to afford the new weapons. The flow of trade between South and Southeast Asia and the increasingly significant international markets in Western Europe and China made controlling the ports along these trade routes attractive, particularly in the 'Age of Commerce' (Reid 1995). The early modern period also saw a boom in international demand for the region's own assortment of commodities, spices and certain breeds of pepper that grew only in the region, particularly in the Indonesian Archipelago. Asian states and European empires would fight for control of both the Straits of Melaka and these commodities for centuries.

This chapter looks broadly at the intersection of firearms with the emergence and collapse of the maritime empires in Asia between the fifteenth and nineteenth centuries. The historiography that has examined the role of firearms in building these states has often been fractured by the lenses of area studies, so that Mughal India has been viewed as both an Asian gunpowder state and an Islamic gunpowder empire, alongside the Ottomans and the Safavids (Hodgson 1974; McNeill 1993). Nevertheless, scholars have generally agreed that firearms were an important element in securing stronger central control within these states and empires. Certainly, these states were stronger and more centralised than the short-lived Mongol Empire they followed and thus lasted longer. Many of them represented the last indigenous states in their regions prior to the introduction of colonial rule.

We can identify three main phases in the transition to gunpowder armies. The first was marked by the spread of gunpowder across Eurasia by the Mongols, their destruction of

DOI: 10.4324/9780429437915-27

existing empires of very significant size and development, and a period of political revival which, in the wake of Mongol collapse, saw the region dominated by small, more aggressive, and often more sea-dependent political centres that in many areas encountered Chinese- and Indian-made firearms, which were crude but novel and impressive. In the second phase, roughly covering the sixteenth century, the smaller warlords of the period, whether local kings in Burma, sultans in Northern India, or *daimyo* in Japan, fought each other for supremacy, increasingly seeking to acquire better European-made firearms. This led to the third phase, which saw political reconsolidation in the form of large empires, such as Burma, the Mughals, the Tokugawa *Shogunate* and the Qing that would carve up much of maritime Asia between them from the seventeenth century, until their demise at the hands of the Europeans.

The introduction of gunpowder in the fourteenth century would aid many of the successor states to the fallen classical kingdoms to pursue warfare on a very large scale. At the same time, the pull factor of spices and other opportunities saw Europeans intervening in Asian warfare (invited partly because of their firearms), in ways that helped some indigenous kingdoms to thrive and led others to their ruin. Gunpowder and related factors led to military change in Asia from roughly the fourteenth century to the nineteenth century. This period witnessed the modification of gunpowder weapons in significant ways. Mainland states developed large river and coastal fleets armed with cannon, and armies increasingly required a degree of organisation to apply firearms effectively in battle and to sustain logistics. Elephants gradually succumbed to the lead slug and later the bullet as faster and more agile horses became the safer, and preferred, mount. Firearms gradually replaced to some degree spears and lances and even bows and arrows. Shields and armour also disappeared once effective firearms were being wielded by their enemies. Changes of this kind were also taking place throughout the Indonesian Archipelago, but additionally, the island world was experiencing greater and increasing intrusion and involvement of Europeans. Asians were fighting in increasing numbers in alliance with or in the employ of Europeans, and indigenous fortifications were succumbing not to land-based artillery but to cannonade from men-of-war. The present chapter thus explores the ways in which firearms impacted the military history of Asia, emphasising changes in armies and their organisational fabric, over the course of five centuries.

Phase One: Early Firearms and the Rise of New States in the Fourteenth to Early Sixteenth Centuries

Maritime Asia covers an expansive geographic area and extremely varied physical topographical and climate regimes, which impacted the ways in which wars were fought, as did disparities in the geographical concentration of people. Areas with larger concentrations of people, such as the Near East, the Indian Subcontinent and East Asia developed very large standing armies that fought for control of land or foreign policy. Regardless of their maritime presence, many of the large gunpowder states of these areas had significant interiors dominated by mountain valleys, extensive deserts and wide grassy plains. Nomadic tribesmen were an ever-present threat over many centuries (and often the source of the ruling dynasties), making fast, agile and tall horses more important than they would be in Southeast Asia for example and arguably cavalry much more important than artillery or small arms.

In contrast to the lands to its north and northwest, Southeast Asia includes a vast territory made up of a mainland peninsula jutting south out of the rest of the Asian continent, linked to Australia and the Pacific by a range of large and small islands. The concentration and distribution of its demographic resources and its natural ecology helped determine what was important enough for competing parties to fight over. First, capturing or controlling other people was one motive and might make capturing rivals more important than killing them. This was particularly important where population resources were very low, even by already low regional standards. Some zones, such as the river valleys of the mainland or Java, would see high concentrations of people, making larger armies possible, and highlands and outer islands saw more scattered, less concentrated populations emerge, which limited the size of armies in those areas, although maritime revenues helped to offset the limits sometimes with paid mercenaries. As a result of the different contexts into which gunpowder and related weapons would be introduced, the impact of the latter was felt differently as well.

Under the Ming, post-Mongol China more vigorously embraced the warlike potential of gunpowder and firearms. China developed firing tubes vaguely comparable in principle to European muskets and cannon that operated like European bombards. China was also better interconnected with the maritime world of Asia than at any other time previously. A marked increase in maritime commerce attended the rise of the Ming in China after the Mongol interlude. During the heyday of China's involvement in maritime exploration in the early fifteenth century, expeditions that reached Africa also passed through Southeast Asia and the Indian Ocean, bringing these crude firearms with them. As the Ming refocused their armies on defending the frontiers from nomadic threats, securing the land frontiers led to the circulation of Chinese firearms overland in northern mainland Southeast Asia as well (Sun 2003). Whether via the Ottoman Turks, who used their guns on Constantinople in 1452, Central Asia by means of the Mughals, or overland from China, the Indian Subcontinent also produced its own crude guns and firearms. Soon, India sent not only firearms but mercenaries as well to other parts of Asia.

The lands between the Indian Subcontinent, China and Southeast Asia began the period being politically fragmented after the collapse of the classical states. The new states, which emerged in the latter's aftermath, were by comparison miniscule or at best a highly diminutive version of what had existed before. The Irrawaddy Valley was divided between Ava in the north and Pegu in the south and a surrounding ring of smaller polities often at war with each other. The Chao-Phraya and Mekong River Valleys fragmented into dozens of smaller polities at different times, although the big winner would eventually emerge as Ayudhya, later Siam, finally known today as Thailand. Vietnam in the eastern mainland broke into several parts; sometimes two, three or four political entities competed until the Tayson rebellion unified the Vietnam littoral in the late eighteenth century. The island world saw scores of coastal polities carve up what had been Sri Vijaya. For a region that had once had a handful of empires, this period saw perhaps hundreds of competing polities of various scales and temperaments, and they sought firearms from their neighbours (Lieberman 2004). Melaka had a vast store by the time the Portuguese first reached it in 1509.

During this period, a large and increasing volume of new resources, European and Muslim firearms and mercenaries became available to savvy local rulers as well. Maritime revenues helped to offset these limits, sometimes with paid mercenaries and firearms. The introduction of Chinese gunpowder in the fourteenth century may have helped many local rulers to impress their tributaries and emphasise their power through pyrotechnics. The explosive power and noise of gunpowder were something new, otherworldly, and shocking to rulers and their armies (and animals) alike. Over time, however, after the initial surprise was overstepped by familiarity and guns became much more commonplace, the shock value of gunpowder weapons lost its efficacy.

The weak strength of Chinese gunpowder, the crudeness of their firearms and their vulnerability to the moist conditions of Southeast Asian jungles meant that, however much they had potential value in East Asian warfare, demonstrated more easily against nomadic tribes on horseback, they simply did not represent in battle a serious challenge to the military effectiveness of a swordsman who honed his skill on a daily basis or the archer who could fire with greater accuracy from a longer range. The elephants that the Southeast Asian elite riders used for duels within battles could be trained to ignore the explosive sounds of gunpowder, unlike the horses of the nomadic tribes of China's frontiers. Southeast Asians in the period under review used their horses (ponies) more as mounted infantry than as cavalry. Chinese firearms necessitated little change to the various ways in which Southeast Asians fortified their defensive complexes, either through brickwork for permanent walls or felled trees for stockades. These new firearms also did little to encourage a change to the organisation of Southeast Asian armies, which remained mainly village levies with little in the way of a permanent standing army, beyond the royal bodyguard.

Phase Two: The Sixteenth Century

Over the longer term, it was more technically refined, powerful and increasingly accurate European firearms technology that would gradually make a difference between who won and who lost wars in the region. Change also came to how war was fought as well. Firearms changed what animals of war could be used in combat and how, and expedited the increasing importance of centrally trained regiments of musketeers and the rising use of warships.

Portuguese firearms and the Portuguese themselves would be sought by kingdoms throughout the coastal areas of the continent. The Portuguese brought a new generation of firearms developed in Europe. Although initially still relatively crude, difficult to manage and aim, and given to misfiring (with the wet environment dampening Portuguese powder as well), the firing drill, the solid hulls that provided better support for firing cannons than the flexible weaved hulls of many indigenous ships, combined with an indifference to killing as opposed to capturing an enemy (a major goal of indigenous warfare), made firearms in Portuguese hands much more deadly. Alongside servants of the Portuguese Crown came criminals, unemployed soldiers and the landless elements who began to settle on the fringes of the *Estado da India* (the Portuguese empire in Asia) and indigenous kingdoms in ports such as Dianga near Chittagong. These communities supplied hundreds of mercenaries to kings around the region and would play an important part in sixteenth-century campaigns, particularly in handling firearms. An entire Portuguese community would later be captured at Syriam in 1613 and deported into the Burmese heartland around Sagaing in Upper Burma, where they became the hereditary artillerymen for the Burmese Army for the next two centuries. Such opportunities were afforded in part by local Portuguese aggression that led to indigenous rulers turning on the Portuguese in the coasts and defeating them using firearms, such as the victory the King of Kandy, Vimaladharmasuriya I (r. 1590–1604), won over the Portuguese on the island of Sri Lanka.

As the sixteenth century wore on, other foreigners became an increasing part of the armies and navies of the maritime states of Asia. The Spanish introduced firearms to Japan, prompting a military revolution that for the better part of a century transformed warfare in the island empire. Muslim mercenaries from the Middle East were found throughout the armies of the Indian Ocean, and Ottoman mercenaries and guns were sent to the Sultanate of Aceh. The turn of the century brought the Dutch, who carried a war with the Hapsburgs into Asia, leading them to arm and support any state having problems with either the Spanish or the Portuguese. This resulted in the Portuguese and the Dutch arming opposite sides in the war between the Northern Trinh and the Southern Nguyen in Vietnam, and the Dutch aiding the Arakanese against Portuguese freebooters along the coasts of Burma. The seventeenth century also saw Asian communities responding to negative politics in their homelands scattered around the region, like the so-called Irish 'Wild Geese,' offering firearm skills they had by now picked up from the Europeans. These included not just Japanese Christians escaping Tokugawa persecution but also Makasarese from South Sulawesi. As explained below, though, the open market for foreigners who could use guns did not exist not simply because the foreigners (whether interregional or intraregional) were better skilled in the weapons, but for political reasons.

So many Europeans and Asians wandering the region, in search of an employer for their skills in firearms, makes it appear logical that knowledge of how to make gunpowder spread as well. Nevertheless, good recipes for gunpowder were rare and were kept a closely guarded secret. Gunpowder was often weak or ineffective and sometimes would not detonate at all. Ironically this was despite Burma being one of the few places in the world reputed for excellent stores of saltpetre. In one case, in 1610, the ruler of Toungoo had to dump boatloads of gunpowder because it was no longer any good (Charney 2004).

Much of the terrain covered by the Ottomans, the Mughals, the Qing and others indeed consisted of the plains of interior Eurasia, which favoured the integration of mobile artillery with cavalry. But several regions of their empires and certainly most of the areas covered by Southeast Asian gunpowder empires consisted of wetlands, jungles or difficult mountainous topography. In such areas, firearms were largely used at sea on ships and boats or when on land, against fortified positions in siege warfare. The sixteenth-century Japanese invasions led by Toyotomi Hideyoshi (r. 1586–1598) of the Korean Peninsula famously relied on ships to carry cannon and small arms for use against fortified Korean positions. The Koreans and the Chinese relied on a new kind of armed craft, the turtle boats, to help defend against Hideyoshi's shipboard cannon. The sixteenth century also saw many efforts to use ship-based guns against European vessels and against the Portuguese fortified position at Melaka. Large numbers of these floating fortified batteries succeeded in blockading the Portuguese port for significant spans of time. In such cases, galleys rather than conventional indigenous flexible hulled craft had to be used to withstand the force of the guns' recoil. But such vessels also required huge numbers of rowers and meant that each effort to fight the Europeans became incredibly costly, resulting in such campaigns being financially unsustainable.

On rivers, however, firearms found a more effective use in arming river boats, particularly across mainland coastal South and Southeast Asia. Here many of the river boats were solid hulled, each carved out of a tree trunk. Unlike the lashed or weaved ocean trading vessels, the solid hulls of these boats prevented them from shattering on the recoil of firing a cannon. Instead, a small hole was cut in the prow of the boat and a small cannon positioned to shoot forwards at the enemy. Aiming the gun was achieved by manoeuvring the boat. Although such vessels could not have fought against European warships on ocean waters, in the myriad river creeks of the coastal areas of the region, such vessels were more manoeuvrable and hence more effective against their frequently stranded European competitors. It was also much easier to restrict access to the river at the coasts than to resist foreign navies at sea. Early modern Burmese courts were able to demand that European vessels that wished to have access to the Irrawaddy had first to leave their cannon ashore, where they could be picked back up again on their way out of the river and the country (Charney 1997). The vast extent of the riverine islands of Meghna River Delta prevented the Bengal Sultanate and then the Mughals from exerting similar control in the North-Eastern Bay of Bengal. The deltaic terrain required the maintenance of fleets of locally made watercrafts bearing cannon.

Siege warfare also saw significant developments in Southeast Asia especially as regards skills in using gunpowder. Some gunpowder states, such as early modern Burma, saw the emergence of an effective indigenous mining and sapping skillset that made the Burmese much feared not only by the Asians but also by the Europeans in port areas. It was Burmese mining and explosives that helped to bring down Syriam in 1613. Nevertheless, not all Burmese sieges fared well. The Konbaung dynastic founder and feared warrior king, Alaunghpaya (r. 1752–1760), reportedly had guns cut from palm trees when he was besieging Ayudhya, the capital of Siam (modern Thailand) and, when the guns shattered on firing, was mortally wounded as a result. The Burmese would also develop skills in building wooden stockades which provided shelter but were more easily repaired or replaced than older brick fortifications when facing an enemy, such as the British in 1824–1826, who relied extensively on firearms (Charney 2004).

The Mughals introduced their own firearm-bearing forces at the First Battle of Panipat in 1526, aided by guidance from Ottoman advisors. These forces consisted of matchlockmen and small field artillery, against an enemy who had no firearms at all, and making use of the Ottoman practice of lining them along a line of carts connected by chains in front of the Mughal Army. The early Mughals would continue to apply firearms in new, effective ways, integrating them with mobility and adopting new technology such as swivel guns as well. The high point of Mughal reliance on firearms in war came under Akbar (r. 1556–1605), who took a great personal interest in firearms. During his reign, the Mughals increasingly relied upon siege artillery against fortified positions, aided as well by their ability in mining (Roy 2011: 30, 42–3; Black 2011: 15–16; Richards 1995: 27, 43). As we shall see below, the emphasis on the efficacy of gunpowder weapons in war would change as the Mughals increasingly used larger armies and more conventional military assets (war elephants, for example), relegating gunpowder weapons to secondary importance in their wars (Black 2011: 16; Richards 1995: 8, 28; Rosen 1996: 149; Edwardes 1995: 177; *Babur Nama* 1992: II, 564).

From the beginning, however, one limitation on the continuing development in the efficacy of artillery in war was that the prevailing cultural approaches to weapons and ritual and the continuity of pre-existing animist beliefs was that cannons were seen as objects of power per se and were imbued with charismatic power. It is the case that Charles Boxer overstated the idea that Asian rulers 'valued cannon rather for prestige and sacro-religious reasons than with any serious idea of using them offensively' (Boxer 1965: 171). Instead, efficacy regarding firearms has to be understood as culturally contingent; in the case of some Asian societies it was the ritual propitiation of the spirits in a moral system through which spiritual forces were controlled that determined the efficacy of weapons in battle, not their inherent technological qualities (Andaya 1994: I, 387).

It was certainly true that, for many rulers, possessing large cannon enhanced their legitimacy as ruler and demonstrated their power, and their efficacy was in this political context weighed by size rather than accuracy. In some societies, such cannon could be taken as linga and hence a representation of the king's sexual prowess; another sign of his superior store of charismatic power. The Ottoman Turks, the Mughals and the Siamese of the Ayudhya Kingdom all spent much energy and many resources on massive guns whose military utility was limited at best. In other kingdoms, such as that of eighteenth-century Cambodia, the rulers simply laid the barrels of their cannon upright against the walls of the palace, associating them with their rulership. Cannons were thus not discarded when technically outdated; offerings might be made to them before they were used, and they often received personal names. Incidentally, this approach would be applied in later centuries to all kinds of machinery, including locomotive engines. Such approaches complicated any notion of a military arms race comparable to that which propelled early modern Europe to the forefront of global military technology.

Phase Three: Political Consolidation and the New Empires from the Seventeenth Century

While firearms could be enormously useful in warfare per se, there is a pattern discernible in their relationship to dynastic cycles of rise and decline that shaped how they were used and what their impact would be. Kingdom builders, the warlike founders of dynasties, sought in firearms novel ways to outdo larger armies than their own. We see this in their application by mobile Mughal cavalry in Northern India and in the various campaigns of empire builders like Oda Nobunaga (r. 1573–1582) in Japan and Tabinshweihti (r. 1530–1550) in Burma. In such cases, firearms were acquired, then used in new ways, and, having had a great effect on the outcome of the campaigns, successor kings in the dynasties became worried about them as a potential and unnecessary threat to their positions. They were an unnecessary threat because these rulers, unlike the dynastic founders, had the strength of royal or imperial institutions to support them and did not need to rely on innovative new weapons in the same way. Indeed, it has been argued that the nomadic heritage of the Mughals required them to maintain a moving camp, and moving the entire court made a Mughal Army such a slow, plodding force that it became difficult to pull up and apply their massive siege guns when they could feasibly be used (Streusand 2001: 353; Gommans 2007: 13–15). Of course, someone else might make better use of these weapons to challenge them instead, and so established rulers saw in the spread of firearms a potential means of their undoing.

Over the course of dynasties then, 'gun policy' became more conservative. Various solutions to the firearms problem were attempted, some with more success than others. One solution was to rely on mercenaries to handle the firearms, since they relied on payment from the court and were less likely, it was believed, to be drawn into local politics. Another solution was to keep firearms in the hands of a particular royal service group, a hereditary endogamous group loyal to the sovereign who paid their wages and was the source of their exclusive status privileges. These groups were tightly controlled by the court. A third approach was undertaken in Japan, when the *Shogun* Ieyasu Tokugawa (r. 1603–1605) reduced the possession of guns to such an extent outside of the elite that he has widely been misunderstood as having banned guns altogether in early modern Japan.

This degree of control was difficult without monopolising the external sources of new firearms, the Europeans. Alliances and treaties with the Europeans helped for a time until mercenaries rebelled, as in Burma in 1603, or when other European states entered Asian waters to contest Portuguese and Spanish control: first the Dutch, then the French and then the British. The Dutch famously carried the ongoing Hapsburg wars into Asia, decimating Portuguese and Spanish forces alike, blockading Manila for example, on and off for 40 years, marking the period with several major naval battles. More seriously, the VOC was much more aggressive in destroying indigenous rivals in their effort to control the spice trade in the Indonesian Archipelago. One after another, the numerous independent coastal ports of the region were blockaded, bombarded and made untenable (Reid 1995). The wars between the Europeans in Asia also ensured that there was a market for Asian mercenaries, such as the Makasarese from South Sulawesi. There quickly emerged black markets on the fringes of European empires and indigenous kingdoms where a multinational assortment of mercenaries could be hired, firearms acquired and gunpowder bought. As firearms and gunpowder became more widely available and their circulation uncontrollable, rebellions in the eighteenth century nearly extinguished the remaining indigenous kingdoms of Asia.

The consequence of these Asian dynamics regarding gunpowder weapons was significant and put Asia on its own trajectory regarding the relationship between firearms and state centralisation and bureaucratisation than was experienced in Europe. Certainly, gunpowder and firearms were in different periods of Asian warfare very important to emergent states. But the dynamic mentioned above whereby guns would become so tightly controlled as to prevent their development or improvement by the state meant they became less important in interstate warfare as weapons (but ironically not as symbols of power) deprived maritime Asia of the full impact of the dynamics that in the West shaped, interactively, what could be characterised as true 'gunpowder states.' Asia, it has also been argued, seem to have seen less frequent, intense or disruptive intrastate wars, which made the martial competition between rulers and the ruled less significant in Asia than in Europe. In both cases, administrative refinements developed over time, making states stronger, more centralised and more sophisticated, but the intensity of military competition in Europe led to the more rapid development of state bureaucracies and military institutionalisation and a much more comprehensive revolution in the development and application of firearms.

In maritime Asia, it could even be argued that, instead of gunpowder states, there were merely several major proto-gunpowder states that emerged in which firearms never became quite so important for their technical efficiency as they did in the West. Instead, other factors remained important, and dominance over firearms and mercenaries could be partly offset by uneven command of other resources. In fact, even with the uneven availability of firearms, over the long term it was structural advantages, the combination of trade, territorial expansion and population growth, that remained key to the evolution of successful states in the region with the resources and organisational capacity to control particular river valleys or coastal areas. It was the combination of acumen in managing access, control and organisation of gunpowder resources and the management of other resources, such as manpower, revenues and agricultural produce, that supported the big winners; those states that, by the nineteenth century, had beaten all other indigenous competitors on the mainland, before a series of showdowns between European empires and the remaining indigenous states. One argument made in this chapter is that emphasising the impact of firearms per se is misleading, but instead that the real change to warfare came from a better management culture that evolved in response to the military defeats and failures of the early modern period than power per se of the cannon and musket. As the British at Islandlwana would learn in the Zulu Wars in 1879, a commander or an artilleryman was only as effective as his quartermaster. Similarly, in early modern Southeast Asia, the king may have had the authority to command, but he did not constitute the system that actually put men and weapons in the field in a way that could win battles.

Second, while the state learned how to control and use firearms, the emphasis was on control so that gunpowder development stagnated and sometimes disappeared. Sustained experimentation was naught. Those who specialised in the skills of marksmanship remained an isolated, self-contained group with little impact on the evolution of the state. There were no true bureaucracies of any size which evolved around the existence of guns and gunpowder aside from control mechanisms that were applied to political monitoring and control as well. These states may have thus become more authoritarian along the lines of the famous oriental despotisms of Western theory, but they did not evolve better financial machinery, better administrative practices beyond control; and certainly the mobilisation of firearms in warfare regressed rather than progressed, until a new war and a new dynastic challenger started the process all over again.

Stasis in firearms development had resumed in maritime Asia by the early nineteenth century. In those areas not under European rule, the final precolonial dynasties were in place. Although the final steps in European conquests would invite resistance from precolonial states and then from extra-state insurgents, this was generally ineffective. At the social level, firearms became ubiquitous from the late eighteenth century, 'imported in great quantities' by the English and the French (Sangermano 1893: 98). States tried to monopolise their control but so many firearms flooded in, and the availability due to a long period of intense interstate and intrastate war within the region meant firearms had spread much more widely by the end of the eighteenth century than at any time previously. This flood of firearms continued in the 1810s and 1820s as Western sources dumped huge numbers of outdated guns, presumably leftovers of the Napoleonic Wars, into Asian markets. An inability to control these firearms contributed to the undermining of the remaining states of the region, aiding the European conquests. Although earlier European firearms could and were copied by indigenous metalworkers, this became increasingly difficult as the early modern period wore on. Some processes such as rifling of the barrel were missed entirely in copies made in some Asian societies. Guns were becoming ever more complicated and powerful, and rural imitation copies could no longer compete in real warfare.

Conclusion

Gunpowder weapons would have a significant effect on Asian warfare. Changes in their application in warfare manifested themselves in different ways depending upon local climate, geography and the density of the population. However, the same kind of military revolution experienced in Western Europe, including the evolution of new and more efficient state institutions to support large, firearm-bearing standing armies, was absent or much more lightly felt throughout maritime Asia. One of the main reasons for this different outcome was perhaps that Europe 'got their first' and so was more advanced technologically by the time the European armed forces turned on the Asian states, particularly in the nineteenth century. However, another factor was the way in which Asian states attempted to hold the development of firearm technology back, out of the combined belief that they no longer needed better firearms to keep their military edge and the fear that the states or the societies they commanded were more vulnerable to disruption from threats posed by the uncontrolled circulation of firearms. Some Asian states, as a result, isolated possession of firearms to small hereditary groups tightly aligned with the throne; others, with little or slow technological improvements over time, as in Japan, imposed an extensive ban on firearms altogether.

Although many of the kingdoms of maritime Asia might constitute proto-gunpowder states, they by no means achieved the state development in response to gunpowder weapons experienced in Western Europe. They often possessed huge arsenals of rotting, centuries-old muskets and rusted cannon and piles of miscellaneous shot, and even when they tried to update their holdings in the early nineteenth century they were sourced old examples abandoned by the Western armies. More serious was the lack of development in strategy, tactics and marksmanship. Here the less intense interstate competition left a more deadly stamp. As a weapon that might be used as a political symbol, or to frighten domestic rebels, firearms had by the nineteenth century become mainly a weapon for show rather than a weapon to use against serious military threats.

Bibliography

Andaya, Leonard, 'Interactions with the Outside World and Adaptation in Southeast Asian Society, 1500–1800,' in Nicholas Tarling (ed.), *Cambridge History of Southeast Asia*, vol. 1 (Cambridge: Cambridge University Press, 1994), pp. 345–401.

The Babur-nama (Memoirs of Babur), By Zahiru'd-din Muhammad Babur Padshah, tr (London: Luzac & Co. 1922).

Black, Jeremy, *Beyond the Military Revolution: War in the Seventeenth Century World* (Houndmills: Palgrave MacMillan, 2011).

Boxer, Charles Ralph, 'Asian Potentates and European Artillery in the 16th–18th Centuries: A Footnote to Gibson-Hill,' *Journal of the Malaysian Branch of the Royal Asiatic Society*, vol. 38, no. 2 (1965), pp. 156–72.

Charney, Michael W., 'Shallow-draft Boats, Guns, and the Aye-ra-wa-ti: Continuity and Change in Ship Structure and River Warfare in Precolonial Myanma,' *Oriens Extremus*, vol. 40, no.1 (1997), pp. 16–63.

Charney, Michael W., *Southeast Asian Warfare: 1300–1900* (Leiden: Brill, 2004).

Edwardes, S.O., and Garrett, H.L.O., *Mughal Rule in India* (reprint, New Delhi: Atlantic Publishers, 1995).

Garza, Andrew de la, 'Mughals at War: Babur, Akbar and the Indian Military Revolution, 1500–1605,' unpublished PhD dissertation (Ohio University, 2010).

Gommans, J.J.L., *Mughal Warfare: Indian Frontiers and Highroads to Empire 1500–1700* (London: Routledge. 2003).

Gommans, J.J.L., 'Warhorse and Post-nomadic Empire in Asia, c. 1000–1800,' *Journal of Global History*, vol. 2 (2007), pp. 1–21.

Hodgson, Marshall G.S., (1974). *The Venture of Islam: Conscience and History in a World Civilization* (Chicago: University of Chicago Press, 1974).

Lieberman, Victor B., *Strange Parallels: Southeast Asia in Global Context, c. 800–1830*, vol. 1, *The Mainland* (Cambridge: Cambridge University Press, 2004).

Lorge, Peter A., *The Asian Military Revolution: From Gunpowder to the Bomb* (Cambridge: Cambridge University Press, 2008).

McNeill, William H., 'The Age of Gunpowder Empires, 1450–1800,' *Journal of Near Eastern Studies*, vol. 44 (1993), pp. 103–39.

Ralston, David B., *Importing the European Army: The Introduction of European Military Techniques and Institutions into the Extra-European World, 1600–1914* (Chicago: The University of Chicago Press, 1990).

Reid, Anthony, *Southeast Asia in the Age of Commerce: 1450–1680*, vol. 2, *Expansion and Crisis* (New Haven: Yale University Press, 1995).

Richards, John F., *The Mughal Empire, New Cambridge History of India* (Cambridge: Cambridge University Press, 1995).

Roberts, Michael, 'The Military Revolution, 1560–1660: An Inaugural Lecture Delivered Before the Queen's University of Belfast,' Queen's University of Belfast Lectures, New Series, no. 2 (1956), Belfast: Queen's University.

Rosen, Stephen Peter, *Societies and Military Power: India and Its Armies* (Ithaca: Cornell University Press, 1996).

Roy, Kaushik, *War, Culture and Society in Early Modern South Asia: 1740–1849* (Abingdon: Routledge, 2011).

Roy, Kaushik, *Warfare in Pre-British India: 1500 BCE to 1740 CE* (Abingdon: Routledge, 2015).

Sangermano, Vincenzo, *The Burmese Empire a Hundred Years Ago*, ed. John Jardine (Westminster: A. Constable and Company, 1893).

Streusand, Douglas, 'The Process of Expansion,' in Jos J.L. Gommans and Dirk H.A. Kolff (eds.), *Warfare and Weaponry in South Asia: 1000–1800* (Oxford: Oxford University Press, 2001), pp. 337–64.

Sun, Laichen, 'Military Technology Transfers from Ming China and the Emergence of Northern Mainland Southeast Asia (*c.* 1390–1527),' *Journal of Southeast Asian Studies*, vol. 34, no. 3 (October 2003), pp. 495–517.

Surakiat, Pamaree, 'Thai-Burmese Warfare during the Sixteenth Century and the Growth of the First Toungoo Empire,' *Journal of the Siam Society*, vol. 93 (2005), pp. 69–100.

PART VI

Transition from Industrial to Total War

22

NAPOLEON AND MASS WARFARE

Andrew Bamford

On 22 June 1812, the advance elements of the Grande Armée crossed the border into Russia. Led by the Emperor Napoleon I in person, this force totalled some 655,000 men, 200,000 animals and 1,422 guns. Including second-line formations moving up in support, it was organised into 13 army corps and a further four corps of reserve cavalry. Almost every nation of Europe was represented in its ranks; men from metropolitan France, from the new departments annexed over the previous two decades, from conquered nations impressed into French service, from allies and from vassals. Austria, Poland, Saxony, Westphalia and Bavaria each fielded a complete army corps; Italy and Prussia multiple divisions, and so on down to single battalions that represented the entire armed forces of tiny German duchies. Behind the cavalry vanguard and the marching infantry rumbled supporting forces; heavy guns to deal with the Russian fortresses, and whole battalions of supply troops to keep the others fed and equipped. Seemingly everything had been thought of: whilst some of the supply battalions had light carts to shuttle rations up to the advanced posts, others relied on oxen for motive power, with the logic that once the contents of the carts had been consumed, so too would be the animals that had hauled them (Chandler 1966: 753–9, 767, 852).

To all intents and purposes, the army that Napoleon led to war in 1812 represented the organisational pinnacle of the last two decades of military innovation. Methods that had grown out of expediency in the wars of the French Revolution had been refined in a further decade of conflict, and, imitation being the sincerest form of flattery, the armies that France had defeated had sought to remodel themselves in her image. Yet, in a matter of months, the Grande Armée would be destroyed, and the campaign would collapse into ignominy and tragedy. What is more, while the popular conception remains that it was the snows of 'General Winter' that destroyed Napoleon's hopes, the reality is that the bulk of the losses incurred by the Grande Armée were on its march into Russia, not on its horrific retreat. For all the preparations, the logistical arrangements soon collapsed and, with the Russians refusing battle and relying on scorched earth to do their work for them, men were soon dropping from hunger, thirst, sickness and exhaustion. By the time that the Russians stood at bay at Borodino, Napoleon could bring only 128,000 men against them, enough for only the most pyrrhic of victories (Mikaberidze 2007: 51–2). The snows of the eventual, inevitable, retreat served only to finish the job. The campaign of 1812, therefore, demonstrated both the

DOI: 10.4324/9780429437915-29

triumph of Napoleonic military organisation in the creation of the Grande Armée and, in its destruction, the crucial flaws in a system of mass warfare that had outstripped the logistical foundations on which it rested.

The idea of a mass mobilisation to place a whole nation on a war footing grew out of desperation during the early campaigns of the French Revolutionary Wars. Initial French attempts to export Revolutionary ideals fell apart largely on their own accord in a farce of mistrust and incompetence, and it soon became clear that the new liberties were threatened at home by advancing Austrian and Prussian armies. Furthermore, counter-Revolutionary activity around Lyon, in Provence, and, most significantly, in Western France, led to increasing instability. Politically, the answer was the Terror; militarily, however, different solutions were required as events demonstrated that Revolutionary zeal could not compensate for lack of men and materials.

The solution was the institution, ordered on 23 August 1793, of a *levée en masse*. Until then, France's armies had been composed of the remaining regulars of the old royal army, plus the politically inspired volunteer battalions of 1792. Now, however, all men were to serve: the young as soldiers; married men to forge weapons and manage transport; even the elderly were not spared, being tasked to keep up the morale of the troops and to 'preach the hatred of kings and the unity of the Republic.' Women and children were also to be mobilised, serving as nurses, making tents and turning lint into bandages. Later clauses went on to outline a series of measures, which, if implemented to the full, would have turned France into an armed camp, with provision for the supply of food, weapons and horses to the army (Elting 1989: 27, 37).

These developments, both the 1792 call for volunteers and the following year's *levée*, were not without theoretical precedent. Theorists Hypolite de Guibert (*Essai Général de Tactique*, 1770) and, in particular, Joseph Servan (*Le Soldat Citoyen*, 1781) recognised that standing armies would likely be insufficient to wage future wars and that motivated citizen soldiers would be required. As the events of the Revolutionary Wars would show, it would prove impossible to give the soldier all the rights of the citizen and still maintain discipline, but in these writings one can see the root of some of the Revolutionary edicts (Boycott-Brown 2001: 48–50). In reality, nothing like the envisaged nation in arms materialised, but from regulars, volunteers and draftees a new army was forged through a series of amalgamations. The first of these took place in 1793, creating demi-brigades in lieu of regiments on a planned basis of one battalion of regulars to two of volunteers. The former would provide steadiness and training, the latter Revolutionary zeal. A second round of amalgamations in 1796 combined existing units that had been worn down by combat and in essence put in place a regimental structure (though demi-brigade remained the terminology until 1803) that lasted until 1815 (Elting 1989: 33–4, 39–40, 207–9).

The first of these amalgamations was masterminded, as was much else, by the one military man on the ruling Committee of Public Safety, the Captain of Engineers, Lazare Carnot. Rightly credited as the 'Organiser of Victory,' Carnot did much to shape the French Army of this era whilst remaining largely detached from the murderous politics of the Terror. Indeed, so sufficiently divorced was he from the extremists that when the Terror was brought to an end by the coup of Thermidor (27 July 1794) he was soon part of the five-man Directory that was established to rule France. This continuity allowed Carnot to advance the careers of competent officers in place of some of the politically inspired appointments, replacements, and executions for the unsuccessful, that had seen the command of France's armies repeatedly change in the earlier years of the Revolutionary Wars (Rothenberg 1978: 35–6).

One such protégé was the young Napoleon Bonaparte, who obtained command of the *Armée d'Italie* in 1796. He had previously distinguished himself by directing the artillery at the Siege of Toulon in 1793, and in helping suppress a Royalist insurrection in Paris in October 1795 (Chandler 1966: 3–49).

The major land power remaining in the war was Austria, and it was her armies that France faced in Italy and, in larger numbers, on the Rhine. During the course of 1796 and 1797, by the lightning victories of Bonaparte in Italy and the bludgeon-work of Jourdan and Moreau on the Rhine, Austria was eventually obliged to sue for peace. Thereafter, Bonaparte was packed off to Egypt, largely to get him out of the way of the Directory who rightly feared his growing political ambitions. Plans for Oriental conquests turned sour, and in 1799 he returned to France and seized power in the coup of Brumaire (9 November 1799) by which he was established as First Consul and head of state. Bonaparte's justification for abandoning his men in Egypt was that France was in danger, and, indeed, this was largely true, for a Second Coalition consisting primarily of Austria, Russia and Britain was able to undo much of what had been achieved in France's earlier campaigns. Under the inspired leadership of the septuagenarian Aleksandr Suvarov, a Russo-Austrian joint force had cleared the French from Italy, but allied defeats in Holland and Switzerland caused the campaign to stall. This enabled Bonaparte to mount a new Italian campaign in 1800, forcing a crossing of the Alps and winning a decisive victory, after a near disaster in the early stages of the battle, at Marengo on 14 July. This was followed by Moreau's success at Hohenlinden that December. Peace was again on the cards, and a final settlement was reached at Amiens in 1802, which ended global hostilities. This enabled Bonaparte, as Consul-for-Life from 1802 and as the Emperor Napoleon from 1804, to consolidate his power and to begin the work of re-modelling France and her army (Chandler 1966: 53–130, 205–304; Boycott-Brown 2001: *passim*; Arnold 2005: *passim*).

For the most part, the French Army that Napoleon inherited, and the systems by which it was maintained, was the creation of the Revolutionary Wars. Other than the institution and expansion of the Imperial Guard and some tinkering with regimental establishments and internal organisation, the basic structure remained unaltered through to 1815. Key to this so far as manpower was concerned was the *Loi Jourdan* of 1798. Named after Jean-Baptiste Jourdan, a general under the Republic and later one of Napoleon's marshals, this legislation affirmed the old idea from the *Levée en Masse* that all citizens were soldiers and could be called upon in an emergency. However, absent of any such emergency the army would be kept up by voluntary enlistments and by calling up, as required, some or all of the annual class of conscripts who turned 20 in that year. If more men were available than the army required, a ballot was employed, but, even then, it was possible for those with the funds to pay someone else to serve in their stead in the event that they drew a 'bad' number. Conversely, however, as the wars went on, the demand for manpower meant that Napoleon was obliged to 'anticipate' a year's conscription class. This became a vicious circle, as, inevitably, it created shortfalls down the line and required further 'anticipations,' or, alternatively, calling up of additional older men from classes that had been previously balloted. A lack of hostilities outside the Iberian Peninsula in 1810 and 1811 allowed the system to at least partially recover, but the need to build up the army for Russia in 1812 meant more huge demands. Half of the 245,000 men levied in that year were intended for the *Garde Nationale*, but found themselves transferred en bloc to the line the following year when it became necessary to recreate the Grande Armée for the 1813 campaign. Before the wars were over, more 'anticipations' saw the 1814 and even the 1815 classes called up well before time (Elting 1989: 320–33).

The major continental armies all relied on conscription in one form or another, generally based on some variant of regional quotas; in Russia this had been in force since 1720, Prussia since 1733 and Austria since 1781 (Gooch 1980: 10–11). In those nations which fell under direct or indirect French control during the Napoleonic era—Italy, Holland, Naples and the various German states—similar models of organisation were adopted in order to facilitate the integration of their armies into the French military machine and, inevitably, with this came conscription along French lines (examples with respect to the German states can be found in Gill 2011: 64–70, 127–34, 178–87, 215–23, 248–75, 385–6, 411–23, 465–81). In the other major powers, however, change was less apparent, and for the most part the degree of change was directly proportionate to the level of defeat inflicted on that nation by France.

In Russia, with near limitless manpower to draw upon, little changed, although militia units were raised in 1812 and employed in the campaigns through to 1814 (two rare English translations of the Russian conscript experience can be found in Boland 2017: 13–15, 46–50, and a militia experience in Boland 2017: 58–125; see also Rothenberg 1978: 196–204). Austria, too, largely retained her old systems of recruitment, although 1809 saw the *Landwehr* embodied as a mass militia. Its better units fought with distinction, but such popular nationalist mobilisation did not sit well with a reactionary regime, and although the *Landwehr* was again called out in 1813–1814 it was fielded in far smaller numbers and more closely integrated into existing line regiments (Rothenberg 1978: 166–73). In Prussia, the case was different. A major defeat in 1806 led to a peace treaty that greatly restricted the size of the regular army. A limited amount of 'cheating' was possible by means of discharging trained men and bringing in new recruits, thus providing a nucleus of men with military training who could be recalled to the colours, but this was only possible on a limited scale. Thus, when Prussia re-entered the war against France in 1813 a huge expansion was necessary which was only possible through mass mobilisation. The regular army was doubled, but the bulk of the increase came from *Landwehr* and volunteer units whose formation, in the case of the latter in particular, was motivated by the same ideals of German nationalism that had frightened the Habsburg authorities after 1809. Expansion continued up to 1815, by which time many of the initial 1813-raised *Landwehr* units had acquired a veteran status that left little separating them from the line (Rothenberg 1978: 187–96). Ironically, this case of a nation in arms probably came closer to realising the ideals of Servan and Guibert than did anything in France.

Britain's case was different again. There, the focus until 1805 was on preventing a French invasion of the British Isles, which meant that priority went to the Royal Navy and to home defence forces. The era of the French Revolutionary Wars saw the raising of fencible infantry and cavalry—regular troops but ineligible for overseas service. After 1803, the main home defence role was undertaken by the militia, permanently embodied regiments raised by ballot. The difficulty when the invasion threat diminished and Britain shifted to a more active military role was that the home defence forces essentially became a manpower drain. Persuasion, backed up with bounties, was employed to get trained militiamen to transfer to the regulars, but this was never entirely successful. The resultant manpower shortfall in 1813–1814 was satisfactorily resolved not by new initiatives but through the timely return of peace. Had the war continued, it is hard to escape the conclusion that conscription would have become necessary, but the political consequences of this would have been problematic (for an overview see Cookson 1997: *passim*; the manpower situation is analysed in Bamford 2013: 127–76, 299–301).

Of course, population was not the only element involved in a national mobilisation. It was also necessary to utilise industrial and financial resources to wage war effectively, and, in this aspect, it was Britain that began the period with an advantage and retained—indeed, increased—it during the hostilities. As the birthplace of the Industrial Revolution, Britain was well paced to manufacture armaments on an increasingly organised scale. The dockyards, both royal and commercial, also represented major industrial enterprises. This is, of course, not to belittle the industrial development of the continental powers. France's arsenals were generally effective at arming her troops, as was Prussia's at Potsdam although the huge expansion of her army from 1813 swamped its capacity and much use was of necessity made of imported or captured weapons. Russia was probably the worst placed of the great powers, with little standardisation of small arms and lower quality all round, although, that said, her capacity to produce heavy cannon in large numbers, made easier by good access to indigenous iron deposits, was marked (Knight 2014: *passim*; on arsenals and weapons, see Elting 1989: 475–8; Rothenberg 1978: 195, 202).

Britain's maritime and imperial powerbase also gave her the economic upper-hand. While her campaigns in the West Indies were dismissed as 'filching sugar-islands,' her ability to isolate and capture enemy colonies (twice over, since they were nearly all given back after Amiens) and to monopolise maritime trade thereafter ensured a strong financial base. This in turn facilitated the financing of repeated continental coalitions and the supply of arms and equipment to her allies. French efforts to curtail this trade advantage by prohibiting European nations from trading with Britain as part of the so-called Continental System instituted through Napoleon's 1806 Berlin Decrees provoked widespread resistance, and led directly or indirectly to the French invasions of Spain and Portugal in 1807 and of Russia in 1812. Britain, too, created problems for herself by rigorous enforcement of her own blockade of France, and this helped provoke the War of 1812 with the United States, which dragged on until early 1815 and provided a major distraction for the British war effort.

The inevitable result of this sort of mass mobilisation, in whatever national variety it came, was that there were far more men in the field than had been the case in the conflicts of the previous century. Such large numbers required coordination, which brings us to the second of the three points to be addressed by this chapter, namely systems of command. There is a tendency when addressing the conduct of war between 1792 and 1815 to credit the generals of that era, and Napoleon in particular, with revolutionising the way in which armies were directed. Credit is therefore given for the systems of divisions and army corps without giving consideration to what had gone before, whereas much that was seemingly new in this era in fact built on existing doctrine and practice. In terms of foundations, the basic structure of military units was already established. Infantry was generally organised into regiments of two to four battalions; cavalry into regiments anything between two and ten squadrons strong. Artillery was organised into batteries of six to 12 cannon and/or howitzers. This much remained unchanged through to 1815. In terms of higher command, the grouping of regiments to form brigades under a junior general officer was also established practice. At a higher level, however, the brigades of the *ancien régime* were generally grouped into 'wings' or 'lines,' in accordance with a formal order of battle, or into 'columns' for an attack: only in the latter years of the Seven Years War did something like a divisional system begin to develop. More senior general officers headed such formations, but command arrangements above the brigade were often transient and staff functions, though well developed in many respects, were concentrated at army level (Bamford 2013: 179–80).

In the early years of the French Revolutionary Wars, the old organisational vagueness continued, made worse on the French side by the need to bring some sort of order to the vast number of newly raised units that were rushed to the front in 1792–1794. The vaunted French reliance on swarms of skirmishers operating in loose order, which spawned an increase in light infantry across all armies, was less a deliberate innovation and more a necessity for troops too badly drilled to fight in other ways. Nevertheless, out of this chaos, order eventually began to emerge and once the various amalgamations had created a more uniform army at a unit level so too did the higher organisation begin to stabilise. Thus, in the middle and late years of the Revolutionary Wars, a fairly typical model could be seen for a French combat division of two to three brigades totalling nine to 12 battalions of infantry, one to three regiments of cavalry, and supporting artillery. Where significant numbers of horsemen were available, regiments of heavy cavalry were concentrated into reserve formations for shock action on the battlefield, whilst light cavalry, sometimes combined with other light troops, was similarly separated out for outpost and advance guard work. Bonaparte, in Italy and in Egypt, tended to concentrate his cavalry to a far greater extent and this prefigured a more general move to separate divisions for cavalry in the decades to come. These divisions all had their own staffs to assist the divisional commander and thereby share the burden of logistical and operational planning (Griffith 1998: 156–65).

For the most part, divisional commanders reported directly to the army commander, but the beginnings of what would become a corps system began to develop in the larger armies towards the end of the period, as with the Armée du Rhin and Armée de Réserve in the 1800 campaign (Arnold 2005: 270–1, 274–5). When hostilities resumed in 1803, the initial French focus was on assembling an army to carry out an invasion of the British Isles, based in a series of camps along the coast of the English Channel, from Brittany to the Low Countries. These camps allowed for the reorganisation and training of what would become the Grande Armée. When it became clear in 1805 that the land forces of the Third Coalition—Austria and Russia, with very limited British help on the peripheries—were moving and that schemes for concentrating French naval power in the Channel were failing in the face of the wind, the weather and the Royal Navy, each of the camps of instruction furnished the forces to provide an army corps for the campaign to come (Chandler 1966: 321–32).

The Grande Armée as first constituted therefore comprised seven numbered corps, each of two to three infantry divisions plus either a brigade or a division of light cavalry. Each division had its own artillery, and each corps also had an artillery reserve of heavier guns. As the need arose, divisions could be shifted from one corps to another in response to operational requirements, or a new corps—temporary or permanent—created by taking divisions from two or more existing corps; in 1805, for example, a new VIII Corps was created under Marshal Mortier during the fighting on the Danube. Such cavalry as was left over, after each corps had received sufficient for scouting and screening, was grouped into a Cavalry Reserve under Marshal Murat, who had commanded the cavalry in Egypt and Italy and was now married to the Emperor's sister. Murat's command consisted of six divisions of heavy cavalry and dragoons. Lastly, the infantry and cavalry of the Imperial Guard formed the army's reserve, a role which would see it grow considerably in size as the wars went on (Rothenberg 1978: 126–8; Elting 1989: 57–9; on the role and expansion of the Imperial Guard during the era, see Uffindel 2007: *passim*).

The campaigns of 1805–1806 unquestionably demonstrated the strategic and operational superiority of the corps system when faced with opponents still using the methods of the previous war. The 1805 campaign began with Napoleon facing Austrian armies in Italy and

Germany. Leaving Marshal Massena with a small *Armée d'Italie* to hold down that theatre, Napoleon took as his objective the 72,000 men in south-central Germany nominally under the Archduke Ferdinand but in fact directed by his chief-of-staff, Karl Mack. Russian forces were marching westwards to join, but a window of opportunity remained for the French to act before they arrived; Prussia, from whom much was hoped by the allies, had remained neutral. Using Murat's cavalry as a screen, the Grande Armée advanced by rapid marches that brought it swiftly across the Rhine before the left wing continued its march to swing behind the Austrians at Ulm, cutting them off from their allies and lines of communication. There was hard fighting to get across the Danube to complete the encirclement, but in essence the campaign had been won by manoeuvring, and Mack surrendered while Ferdinand escaped with a small body of cavalry (Chandler 1966: 384–402).

There followed a campaign of a more conventional course as the French fought their way along the Danube to Vienna, before decisive battle took place at Austerlitz on 2 December 1805. By this time the Russians were in the field alongside a ragtag of surviving Austrian troops, and the Tsar's advisors pushed for a counteroffensive against French forces that were assumed to be overstretched. True enough, Napoleon had with him only I, IV and V Corps, plus the Guard and Murat's cavalry, but Marshal Davout was on his way to join with III Corps to give 75,000 men against 81,000 allies. Napoleon let the allies come to him, leaving his right weak in expectation that Davout would arrive in time to reinforce it, whilst massing his left for a counterattack. Things did not go completely to plan, but due to the flexibility inherent in the corps system it was possible to alter the balance of the battle so that the weight of the French attack was shifted to the centre after the left was checked. The result was a crushing victory that had Austria suing for peace and knocked Russia out of the war for the best part of a year (Goetz 2005: *passim*).

A similar level of operational superiority was demonstrated the following year when Prussia belatedly entered the fray. The opening stages of the 1806 campaign saw the two forces seeking each other out in the woods and hills of Southern Germany. As the French advanced northwards, each corps marched by parallel routes so as to spread the logistical burden but remain within supporting distance. Once contact was made, the corps that had located the enemy would fight a holding action whilst the remainder of the army concentrated. Again, things did not go entirely to plan, but, again, the flexibility of the system allowed the French to recover from any possible discomfiture. The issue this time was a failure of intelligence such that Napoleon believed himself to be facing the combined Prussian forces at Jena on 14 October when in fact he was facing only one of the two main Prussian armies. With numbers in his favour, the Emperor won a comfortable victory, but more impressive was the result of the twin battle fought simultaneously by Davout at Auerstadt in which III Corps defeated the second Prussian army (Chandler 1966: 452–506; Gallaher 2000: 119–35). The ability of a single corps to function as an army in miniature and fight against heavy odds was another advantage of the system, demonstrated most effectively here but also shown by Marshal Lannes in the opening stages of the Battle of Friedland, 14 June 1807, when the resurgent Russians were finally crushed after previously waging a bitter winter campaign in Poland, and by Davout again at Eckmühl in 1809 when Austria re-entered the war (Chandler 1966: 572–85, 680–94).

Austria and Russia in 1805 fought with little by the way of permanent military organisation above the brigade level, while the Prussians in 1806 and the Russians in 1806–1807 fielded large multi-arm divisions. Thereafter, however, the continental powers also began to adopt the corps system and thus much of the operational advantage that the French had enjoyed by means of superior organisation was lost. The Austrians in particular carried out a root-and-branch reform of their whole army between 1805 and 1809, under the aegis of the Archduke Charles. The experimentation with a popular *Landwehr* has already been noted, but the adoption of a corps system was also key. Going on the offensive in Germany and Italy in 1809 to capitalise on the diversion of French efforts to the new war in the Spanish Peninsula, the speed of the initial Austrian advance caused some discomfort before the situation could be restored and the Grande Armée, now including sizeable allied contingents, could concentrate for a counter-offensive. As the Austrians fell back, the fighting shifted to the Danube around Vienna. Here, Charles was able to hand Napoleon his first conclusive battlefield defeat at Aspern-Essling, on 21–22 May, when a rapid concentration against the French bridgehead nearly drove his outnumbered troops into the Danube. Revenge was had in July with the two-day battle of Wagram which effectively ended the campaign, but this was a stand-up fight and served to underline the switch to a more attritional style of warfare that had arguably begun as early as 1807 with the bloody battles of Eylau and Heilsburg against the Russians (the definitive account of the 1809 campaign in all aspects is Gill 2008–2010: *passim*; on Eylau and Heilsburg, see Chandler 1966: 535–51, 564–72).

Meanwhile, the French were also engaged in a war in the Iberian Peninsula, where an attempt to effect regime change in order to enforce the Continental System had led to mass popular uprisings. The Spanish field armies were mostly defeated by the end of 1811, but their guerrilla forces remained a major threat and ultimately in some cases grew in capability to the point of being able to match regular troops. Britain, meanwhile, sent increasing numbers of troops to the Peninsula, which ultimately grew into a major Anglo–Portuguese field army commanded from April 1809 by Sir Arthur Wellesley, ennobled later that year as Lord Wellington and ultimately raised to the dukedom of that name after a string of victories.

After the defeat of the semi-independent corps sent into the Peninsula in 1807, Napoleon's campaign there in 1808 saw the creation of an eight-corps Armée d'Espagne, taking three existing corps from the Grande Armée and assembling the remainder from the debris of the previous year's operations plus new conscripts and some allied troops. Thereafter, these corps were formed into a collection of regionally based armies, all of which eventually abandoned the corps echelon and operated with anything up to ten divisions reporting directly to the army commander. Only after Wellington's great victory at Vitoria, on 21 June 1813, which drove the French back over the Pyrenees, was a single Armée d'Espagne again formed under Marshal Soult, divided into three 'Lieutenancies' which were corps in all but name (the organisation of the French forces in the Peninsula can be tracked in the appendices to Oman 1902–1930).

Wellington's army likewise eschewed any formal corps organisation, with its basic operational component being a division of three brigades—two British and one Portuguese—with supporting artillery. The staff functions concentrated at corps level in other armies existed at the divisional echelon in this instance, allowing a limited degree of autonomous operation if required. It was with this organisation that Wellington won perhaps the most accomplished battlefield victory of his career, at Salamanca on 22 July 1812, in which a series of mutually supportive divisional blows rolled up the left flank of Marshal Marmont's Armée de Portugal in a classic oblique attack (Muir 2001: *passim*). Nevertheless, trusted subordinates such as Rowland Hill and Thomas Graham were increasingly given commands of multiple divisions, and when the time came to advance into Southern France in 1814 the army was divided into a Left and Right Wing and a Centre, the latter nominally under Sir William Beresford, the British general seconded to command the Portuguese Army, but in practice generally accompanied and directed by Wellington himself. In 1815, Wellington did adopt a formal corps organisation for the army that fought the Waterloo campaign, mixing Netherlands and Hanoverian troops with British, much as he had incorporated Portuguese troops in the Peninsula, but again keeping a Reserve Corps under his own direction (Bamford 2013: 187–204).

A major problem for the French in the Peninsula was the coordination of multiple armies and corps, something that overwhelmed the capabilities of Joseph Bonaparte who had been given the Spanish crown by his brother. Napoleon himself, however, faced a not dissimilar problem in the campaigns of 1812–1814 when the size of armies grew to the point that they required additional subordinate commanders of multi-corps formations. Results were mixed. Another Imperial sibling, Jerome Bonaparte, King of Westphalia, was a disaster commanding the southern wing of the Grande Armée in 1812; Marshal Ney was only partially successful in executing what should have been a crushing flank attack at Bautzen in May 1813; later that year Marshal Macdonald managed to get his Armée du Bober trapped with its back to the River Katzbach and suffered a heavy defeat at the hands of the Russo-Prussians under Blücher; Marshals Oudinot and Ney both failed as commanders of the Armée de Berlin in 1813, not only failing to capture the eponymous city but being defeated at Grossbeeren and Dennewitz respectively (Chandler 1966: 776, 889–98, 903–12; for Oudinot and Ney in 1813, see also Leggierre 2002: 160–76, 189–228).

In part, these 1813 failures were because the allied armies—Russians and Prussians in the spring 1813 campaign, joined by Austrians and Swedes for the decisive autumn campaign—had developed a means to take advantage of the French overstretch. In essence, when any one of the three allied armies was attacked by Napoleon in person, it would retreat; where Napoleon was absent, it would attack. Thus, even though Napoleon won at Dresden on 26–27 August, the defeats of Macdonald, Oudinot and Ney compromised his position and compelled him to fall back on Leipzig, where in the epic 'Battle of the Nations' on 16–19 October, the Grande Armée was crushed by the combined allied forces. That Napoleon was able to extricate himself and fight a renewed campaign in the spring of 1814 was testament to his own skills as a general, but the conclusion was never really in doubt and, after Paris fell and his marshals began to defect, the Emperor abdicated for the first time on 6 April 1814 (Chandler 1966: 903–1004). The vainglorious attempt at a comeback the following year, the Hundred Days that ended at Waterloo and led to a second and final abdication and exile to St Helena, offers little more that is instructive about the Napoleonic art of war. None of the major commanders were at their best, nor were the hastily assembled armies that they commanded on a par with those of earlier years. The campaign of Waterloo ended in French defeat for much the same reason as in 1813–1814; the allies were successful wherever Napoleon was not present, and were ultimately able to concentrate superior forces to crush him (Hussey 2017: *passim*).

If state mobilisation created ever larger armies, and organisation into divisions and corps enabled them to be controlled with increasing sophistication, the problem remained of keeping them fed and equipped. Notwithstanding the initial enthusiasm for exporting Liberty, Equality and Fraternity, the armies of Revolutionary France were soon obliged to support themselves with the resources of the territories that they conquered (Griffith 1998: 52–61). Such methods were all very well, but if allowed to become ingrained then they served to remove the responsibility for logistical support from the state and this in turn worked directly against the idea of mobilising a whole nation for war. French logistics remained functional during the campaigns of 1805–1806, and again in 1809, and thus allowed Napoleon to win his greatest victories. It would be fallacious, however, to assume that this says anything much about the quality of the Grande Armée's supply services, which were frequently corrupt and ineffective, being limited, even when staffed with honest personnel, by the inability of transport to keep up with the fast-marching troops. Rather, this success speaks of the fact that the French were operating in fertile territory and were thus able to subsist off the land; this extended even to appropriating enemy military resources, as with the use of captured Austrian cannon taken at Ulm and Austerlitz or the requisitioning en masse of Saxon cavalry mounts in the aftermath of Jena where the Saxons had been unwilling allies of the Prussians (Elting 1989: 315–16, 553–73).

Matters began to break down when the 1806 campaign moved into Poland, where the local resources were insufficient to support so many troops. More of the same was suffered by the French armies in the Peninsula, where forcible requisitions of supplies only exacerbated the popular hatred already felt for the occupying forces. The Peninsula Campaigns also saw the deliberate use of scorched earth tactics to exacerbate enemy logistical difficulties: obliged to fall back to defensive positions around Lisbon by Massena's numerically superior army attempting to invade Portugal, Wellington had all supplies carried off or destroyed so that the French were given the stark choice of starvation or attempting to storm the near-impregnable Lines of Torres Vedras. Unwilling to do either, Massena had to order a retreat (White 2019: *passim*). Much the same was done in Russia in 1812, although it is important to stress in this case that French logistical preparations were so inadequate that troops were already suffering from hunger whilst marching up through Poland, putting paid to the idea that Napoleon's logistics had successfully budgeted for a 20-day campaign (Chandler 1966: 767–75).

As already noted in the case of the French in Spain, subsistence by forced requisitions had consequences when it came to relations with the local population. Wellington, determined to keep the populace friendly when he invaded France in 1814, made a point of paying for all requisitions and left behind allied Spanish troops whose hatred of the French and inadequate logistics would have driven them to plunder. In this he was successful, whereas in Northern and Eastern France there was considerable popular resistance to rapacious invaders symbolised by the hated Russian Cossacks. In 1815, however, the Prussian Army—whose growth in combat power had outstripped by a large margin its logistical supports—sought to obtain supplies by force even in friendly Netherlands territory and caused major diplomatic problems by doing so, and continued to loot its way across France after Waterloo notwithstanding that the territory thus 'liberated' by the coalition armies was nominally that of their ally, the restored Louis XVIII (Hussey 2017: vol. 1, 123–38).

In conclusion, therefore, while the wars of 1792–1815 demonstrated substantial progress in the way that armies were raised, organised and commanded, that growth was uneven and produced a model of warfare which was based on inherent contradictions. Mass mobilisation of manpower created armies that required vast resources to support. To attempt to do so by conquest was ultimately a flawed endeavour, as the substitution of looting for logistics was only practical in Germany and Italy. In attempting to make war support war by shifting its burden to her neighbours, France under Napoleon ultimately created a situation where all the major powers were aligned against her, with British gold combining with continental manpower to create an unbeatable coalition. Thus, even the most innovative and flexible systems of command, which in any case were being replicated by her adversaries by the final years of war, were insufficient in themselves to stave off defeat. Such methods of warfare were only properly sustainable from a logistical perspective after the completion of the Industrial Revolution in the mid–late nineteenth century and, in particular, the development of railway networks. Thus, when the forces of Italian and German nationalism that had been tentatively unleashed in the campaigns of 1809–1815 led to renewed European warfare a half century later, the Napoleonic operational model could at last be perfected but in an era of breach-loading firearms where the tactical realities on the battlefield had shifted yet again.

Bibliography

Arnold, J.A., *Marengo and Hohenlinden* (Barnsley: Pen and Sword, 2005).

Bamford, A., *Sickness, Suffering, and the Sword* (Norman: University of Oklahoma Press, 2013).

Boland, D. (ed.), *Recollections from the Ranks: Three Russian Soldiers' Autobiographies from the Napoleonic Wars* (Solihull: Helion, 2017).

Boycott-Brown, M., *The Road to Rivoli* (London: Cassel, 2001).

Chandler, D. *The Campaigns of Napoleon* (London: Weidenfeld and Nicolson, 1966).

Cookson J.E., *The British Armed Nation 1793–1815* (Oxford: Oxford University Press,1997).

Elting, J., *Swords Around a Throne: Napoleon's Grande Armée* (London: Weidenfeld and Nicolson, 1989).

Gallaher, G., *The Iron Marshal* (London: Greenhill, 2000).

Gill, J., *1809 Thunder on the Danube*, 3 vols (Barnsley: Frontline 2008–2010).

Gill, J., *With Eagles to Glory*, 2nd ed (Barnsley: Frontline, 2011).

Goetz, R., *1805: Austerlitz* (Staplehurst: Spellmount, 2005).

Gooch, J., *Armies in Europe* (London: Routledge, 1980).

Griffith, P., *The Art of War of Revolutionary France 1789–1802* (London: Greenhill, 1998).

Hussey, J., *Waterloo: The Campaign of 1815*, 2 vols (London: Greenhill, 2017).

Knight, R., *Britain Against Napoleon: The Organization of Victory, 1793–1815* (London: Allen Lane, 2014).

Leggierre, M.V., *Napoleon and Berlin* (London: Tempus, 2002).

Mikaberidze, A., *The Battle of Borodino* (Barnsley: Pen and Sword, 2007).

Muir, R., *Salamanca 1812* (New Haven: Yale University Press, 2001).

Oman, C., *A History of the Peninsular War*, 7 vols (Oxford: Clarendon, 1902–1930).

Rothenberg, G.E., *The Art of Warfare in the Age of Napoleon* (Staplehurst: Spellmount, 1978).

Uffindel, A., *Napoleon's Immortals* (Staplehurst: Spellmount, 2007).

White, K., *The Key to Lisbon* (Warwick, Helion, 2019).

23

THE AMERICAN CIVIL WAR AND THE BEGINNING OF TOTAL WAR?

A. James Fuller

The Civil War was a modern, total war. This narrative dominated the literature for more than half a century, as historians pointed to the numbers and compiled evidence to support their view that the conflict was closer to the conflagrations of the twentieth century than it was to Napoleonic warfare: 620,000–750,000 dead and many tens of thousands of others maimed; staggering economic costs, especially in the defeated rebel states; the use of new military technologies; the shift of a whole society from peace to war, including the planned allocation of resources; and a conflict that included civilians as well as military personnel. Focused on the numbers of casualties, both in comparison with the death rates of other American wars as well as the percentages of pre-war society, historians continued to interpret the War of the Rebellion in terms that separated the conflict from previous wars, marking a dramatic shift in the nation and its warfare.

More recently, however, scholars have begun to challenge the now traditional interpretation of the Civil War as a modern, total war. In the past 25 years, historians have argued that the conflict was actually a limited war, far less destructive than the struggles of the twentieth century. Even as the total number of men killed in the rebellion was increased by more than 100,000 in ground-breaking scholarship, these revisionist writers insisted that the conflict should be seen as the last of the Napoleonic wars rather than foreshadowing the contests of the next century. At the same time, those who argue for the total war interpretation have pushed back against the challengers by more carefully defining their terms and by insisting that the Civil War became a total war for participants within those parameters. The result means a scholarly stalemate, with neither side surrendering to the other. But historians no longer accept the traditional narrative on its face, preferring a more nuanced version of that story or rejecting it outright. In popular history, however, the view of the Civil War as a modern, total war continues to dominate.

In considering the issue of the Civil War as total war, several questions must be raised. What was the traditional narrative and how did it come to be accepted so widely? When did the challenge to it arise and what caused that scepticism? What arguments did historians make in regard to their varying interpretations? What evidence did the two sides muster to support their arguments? How could respected scholars, all claiming to use the trusted

DOI: 10.4324/9780429437915-30

methods of the profession and to interpret the evidence in the context of the nineteenth century, come to such diametrically opposed interpretations of the Civil War?

Unsurprisingly, the view of the Civil War as a total war began in the post-Second World War era. John B. Walters, a Southerner and biographer of William Tecumseh Sherman, first applied the term in a 1948 article in the *Journal of Southern History* (Walters 1948). Taking a decidedly Lost Cause perspective, Walters argued that Sherman's March to the Sea in 1864 was the embodiment of a new kind of warfare and saw taking the war to the civilians as a hateful strategy that belied the North's insistence that the conflict was about restoring the Union and healing the wounds caused by sectionalism and rebellion. In one sense, then, the idea of the contest being a total war sprang from pro-Southern roots. But the concept was not simply the result of bias or a Lost Cause interpretation. Within a few years, T. Harry Williams, a Northerner and widely respected scholar, took up the argument in his celebrated 1952 work, *Lincoln and His Generals*, opening the book with the following assertion: 'The Civil War was the first of the modern total wars, and the American democracy was almost totally unready to fight it' (Williams 1952: 3).

In the last quarter of the twentieth century, historians took the modern, total war theory to its zenith, as James M. McPherson led a host of highly respected and talented scholars in recounting the narrative. McPherson most famously told the story of how the Civil War became a modern, total war (an apt argument given his broader interpretation involving modernisation theory) in his Pulitzer-Prize-winning synthesis titled *The Battle Cry of Freedom*, published in 1988 (McPherson 1988). Even as he argued for contingency and held that certain turning points caused the final outcome, McPherson outlined the move from limited war to total war in the course of the conflict.

Finding the roots of total war in the Civil War fit well with the Cold War era when Americans were coming to terms with the carnage of the global wars of their own century. For Southerners in the age of Jim Crow who were steeped in the Lost Cause, an anti-Northern argument about the roots of such warfare seemed logical. Meanwhile, in the nationalist view, connecting the ideological struggles of the present with the just war and righteous crusade of the past made perfect sense. For those who struggled to understand the mass death of the world wars, finding that even more Americans had died in the Civil War was somehow comforting. Then, in the Vietnam era and after, the total war interpretation became even more relevant as Americans tried to find precedents for guerrilla warfare and the so-called 'collateral damage' of civilians killed in war. The turn to 'history-from-the-bottom-up' brought the common soldier into focus rather than the generals, but the total war theory remained intact.

The scholars who took the modern, total war perspective found plenty of evidence to support their theories (Paludan 1988; Royster 1991; Faust 2008: 250–65). First, there were the sheer numbers of dead. Although it was impossible to calculate a precise total of Civil War dead, for about a century the accepted total was 620,000. Of these, roughly 360,000 were Union soldiers and 260,000 were Confederates. These numbers were compiled in the late-nineteenth and early-twentieth centuries, and historians on both sides of the total war debate agreed that they were problematic and inaccurate. The Confederate totals, especially, were questioned in light of inaccurate and lost records and the breakdown of Southern society and government in the latter months of the war. Still, 620,000 exceeded the losses the United States experienced in any other war, including the 407,000 Americans killed in the Second World War. The staggering number of dead men made the War of the Rebellion a slaughterhouse that foreshadowed the mass murder of the twentieth-century battlefield. When percentage of the population is considered, the losses were especially high on the Southern side. Indeed, as McPherson argued in an essay on the subject of how the Civil War became a total war, 'fully one quarter of the white men of military age in the Confederacy lost their lives' and 'nearly 4 percent of the Southern people, black and white, civilians and soldiers, died as a consequence of the war' (McPherson 1996: 66–7). Added to the dead were many tens of thousands of other casualties, including men maimed by war, as the amputations carried out in field hospitals in the name of stopping the spread of gangrene led to survivors without limbs. Especially murderous was the Overland Campaign of 1864. General Ulysses S. Grant wore down the Confederate Army of Northern Virginia commanded by General Robert E. Lee through attrition. In effectively bleeding the enemy to death, Grant remade American warfare, turning the conflict into a modern, total war (Wheelan 2014).

Second, the proponents of the total war interpretation calculated the economic costs of the war. Some numbers could be compiled, such as the amount of money the national government spent and the increase in the national debt (Goldin and Lewis 1975: 299–326). Estimates placed the amount spent at about $4.2 billion in 1860 dollars, with more than $2.2 billion in debt remaining at the end of the conflict. At its peak in 1865, the US government spent more than 11 per cent of the country's gross domestic production on the military (Stodola 2011).

But such numbers do not tell the whole story, as they only address government spending. There were many other costs involved, including the South's expenditures and the hidden costs of money diverted to the war from other pursuits. Beyond those costs, there were the economic consequences of the South being defeated. While the end of slavery and the resulting loss of slaveholder wealth could be celebrated as justly fulfilling the nation's commitment to its own libertarian principles, burned out towns and cities, destroyed farms and plantations, ripped up railroad lines and blown-up bridges attested to the staggering cost (Nelson 2012). As McPherson put it, the Union 'did all it could to devastate the enemy's economic resources as well as the morale of its home-front population' (McPherson 1996: 67). Photos of ruined buildings in Richmond, Virginia, eerily foreshadowed the devastation of cities in the wars of the next century.

The historians who argued for a modern, total war thirdly pointed to the use of technology. Among its many innovations, the Civil War introduced the rifle-musket, the minié ball, barbed wire, new kinds of landmines, better artillery, ironclad warships, repeating rifles, the submarine and the Gatling gun precursor to the modern machine gun. Some of these technologies were developed too late for full implementation or were used only on a very limited basis (Hattaway and Jones 1983; Weigley 2000). Still, the argument for technology seemed strongest when it came to the use of the rifle-musket and the minié ball, as these gave the ordinary infantryman superior firepower and a more destructive capability over what had been available in previous conflicts. Add to such technology the building of new kinds of fortifications and one understood why the conflict favoured the defensive position. With the use of trenches in places like Vicksburg, Mississippi and Petersburg, Virginia, the Civil War began to take on the look of the First World War rather than a Napoleonic contest.

The total war narrative also included the shift of the society from peace to war, especially when it came to the allocation of resources. As McPherson put it, 'The Civil War mobilized resources on a scale unmatched by any other event in American history except, perhaps, World War II' (McPherson 1996: 67). While the North did not shift entirely to a war economy, historians argued that the South did truly experience total war, as the Confederate government centralised power and attempted to plan and distribute resources in order to fight for independence (Sellers 1927; Thomas 1971; Hill 1936; Thomas 1979). Even the Union, however, experienced dramatic shifts toward total war that would have made the era seem like something different to the participants despite the economy not fully reaching the technical definitions used by scholars. The collaboration between the government and business, for example, meant that railroads and factories gave priority to the war effort. The shortage of labour created by the millions of men under arms meant that ordinary farm families experienced hardships caused by the war economy. When matters like the care of soldiers' families and shortages of resources like firewood and draft animals were added to the equation, a Northerner might well have considered the sacrifices made and thought of the war in terms not unlike those used in the twentieth century.

But the main factor in defining total war was the involvement of citizens politically and militarily. In such a conflict, non-combatants were targets of an enemy trying to destroy the capacity of the other side to make war. Of course, Sherman's March offered an obvious example for historians seeking to show how the Civil War involved civilians. Although the Lost Cause exaggerated the March to the Sea in regard to brutality inflicted on human beings, the destruction of property was real and became even more devastating once Sherman's columns crossed into South Carolina where they sought to punish the citizenry for having led the rebellion. While foraging soldiers 'paid' for what they took from citizens, the burning of buildings and the loss of food and destruction of crops brought the war to the civilians (Walters 1973). But it was not just Sherman who fought a total war. Rebel guerrilla activity spurred Union troops operating in Missouri to practise scorched earth policies designed to deprive the enemy of supplies, discourage the populace from helping the enemy, and break the other side's ability to make war. Eventually, Northern commander Thomas Ewing issued General Order No. 11, which forced thousands of residents in four Missouri counties to move and Union soldiers burned their homes to deprive the bushwhackers the support they needed to continue their guerrilla warfare (Stith 2008). Elsewhere, Union commanders Ulysses S. Grant and Philip H. Sheridan were also seen as generals who redefined the nature of warfare, targeting civilians in a war on the Southern economy as well as Confederate armies (Catton 1966; Heatwole 1998; Sutherland 1998; Wheelan 2012).

With a powerful body of evidence, then, the traditional total war narrative took hold of the imagination of not only academic historians but the American public (Burns 1990). But there were problems with the interpretation. One issue involved definitions. What was actually meant by 'total war' and 'modern war' when such terms were used? Typically, most historians who employed them did not really define them, but those who did usually painted with such broad strokes that the labels were vague and lacked specificity. This opened the door for the challengers who argued that the Civil War was *not* a modern, total war. These revisionists wrote in the era after the civil rights movement and most of their work came after the end of the Cold War. Some of their criticisms were shaped by the War on Terror in the early twenty-first century.

Influenced by different concerns and coming at the issue from different perspectives, they began to question the traditional interpretation. Without doubt, the leading voice expressing scepticism about the total war narrative was Mark E. Neely, Jr. A well-known and respected Lincoln scholar, Neely fired the first salvo against the total war interpretation in an influential 1991 article, 'Was the Civil War a Total War?' (Neely 1991). He offered a more comprehensive challenge in his 2007 book, *The Civil War and the Limits of Destruction* (Neely 2007). Neely pounced on the problem of definitions, arguing that historians had played fast and

loose with their terminology. The result was that 'total war' meant different things to different scholars at different times. He noted that American historians of the Civil War had taken
the term from those who wrote about twentieth-century warfare. While the origins of the
term were in the application of air power in the First World War, the label evolved over time,
especially following the Second World War. Since then, 'total war' had usually meant the
complete mobilisation of a country's economy for war. But it also meant attacking an enemy
country's economy to deprive them of the ability to make war. Although this was supposedly
part of modern warfare and something new in the Civil War, Neely argued that it was not
actually innovative, pointing to naval blockades as an example that predated the rebellion by
a long time. 'Total war' also meant the involvement of citizens, especially the targeting of
non-combatants. This was a 'modern' policy and an 'innovation' of the Civil War. But Neely
again disagreed, arguing that the way that such a policy was defined mattered. Some said the
involvement of citizens was only total war if the targeting of them was conducted 'without
limitation' or in an immoral fashion. Others said it was total war if all possible means of attack
were used against both military and civilian targets. Taking each definition in turn, Neely
argued that the Civil War was not a total war because citizens were only involved within
limitations—as long as they were white (Neely 1991: 9–13). In his 2007 book, he indicted
Americans for their racism, arguing that the brutal policies of total war were applied only
to non-whites, as murder and rape were carried out against African Americans and Native
Americans. But whiteness prevented the majority of Civil War soldiers from carrying out
such terrifying attacks on their white enemies. Others scholars seemed to confirm his view, as
they also argued that the brutality of war was racial in nature (Neely 2007; 140–97; Fellman
1989; Bartek 2010; Earle and Burke 2013).

Next, Neely addressed the arguments about the total number of deaths. Agreeing that
the accepted total of 620,000 was problematic and probably inaccurate, he was willing to use
it for the sake of argument. But whatever the actual number was, he discounted the impact
of casualty numbers for making the conflict a total war. First, he disagreed with the total
war theorists who said that the Civil War killed more Americans than any other war. Neely
pointed out the obvious problem of counting both sides as Americans. This meant that the
rebellion could not be compared to any other war since all other conflicts had the United
States fighting against another country. To make such comparisons valid, a scholar needed
to separate the two sides. This meant that the United States lost 360,000 men killed in the
Civil War and the Confederacy lost 260,000. Immediately the destruction of the war was
diminished, as the Second World War killed 407,000 Americans, a larger number than that
lost by either side in the rebellion. Second, Neely disagreed with the way that historians applied the totals, arguing that they should account for the fact that the vast majority of soldiers
died from disease rather than from battlefield wounds. Pointing out that the ratio of disease
to battlefield deaths was two to one, he recalculated the number of deaths. Disease killed
225,000 Union soldiers and 194,000 Confederates. This meant that the North lost 135,000
men and the South lost 66,000 men, for a total of 201,000 dead, less than half of the number
killed in the Second World War—a war in which the ratio of those killed by disease was much
smaller. Neely insisted that he was not trying to 'belittle the losses incurred in the Civil War'
but rather was attempting to show that 'they were not magnitudes greater than the losses in
any other conflict in which Americans fought' (Neely 2007: 207–16).

Revisionist scholars following Neely took on every aspect of the total war interpretation
with devastating criticisms. One major contention of the challengers was that Union policies
were actually limited and, thus, did not constitute total war. Sherman's March, long seen as

the epitome of the Civil War involvement of civilians, was reinterpreted in the late twentieth and early twenty-first century. While Walters had cast Sherman as carrying out terrorist tactics on the Southern home front, scholars like Joseph T. Glatthaar argued that the infamous general's campaign was carried out with remarkable restraint and that cruelty and brutality were not part of a systematic plan. Even the property targeted by Sherman's troops was chosen deliberately to deprive the South of war-making capacity rather than being indiscriminate or designed to hurt civilians (Glatthaar 1985). Of course, families did suffer, but it was not the immoral destruction of total war. General Sherman, long seen as the symbolic harbinger of the future of war, now became a paragon of restraint. In fact, the new scholarship found that Sherman was misunderstood and misconstrued (Trudeau 2008). Neely argued that Sherman had become the Civil War model of total war because of his rhetoric, saying that the general was a garrulous man whose non-stop speaking included violent threats and ideas that were never actually implemented. Because his vocabulary sounded so like the brutality of the twentieth century, historians latched on to Sherman's words and found him to be a treasure trove of quotations supporting their idea of a total war. But Neely found numerous examples of restraint and contextualised Sherman's threats, showing that they stemmed from the commander's frustrations at the moment. For every quotation from Sherman that supported total war, Neely found another that upheld limited warfare (Neely 1991: 14–19). General Sheridan similarly enjoyed a revision in the scholarship, becoming a paragon of restraint rather than an instrument of total war (Gallagher 2006).

The Neely-led assaults on the total war interpretation underscored the problems of such terminology and those who held to the traditional narrative recognised the validity of such criticisms. In his influential, prize-winning 1991 book, Charles Royster adopted the label 'destructive war' rather than 'total war,' which credited/blamed Confederate General Thomas J. 'Stonewall' Jackson as well as General Sherman for the innovations that led to a more devastating kind of modern warfare (Royster 1991). In his 1995 book, Mark Grimsley also changed terms, using 'hard war,' as he agreed that the Union did not carry out indiscriminate destruction but that the Northern policy moved from one of restraint to a much harsher way of war that was designed to take not only an economic toll but also a psychological one. Union 'hard war' measures aimed at demoralising Southern civilians to erode support for the Confederate cause while also destroying the economy that made the continuation of the rebellion possible. Although his nuanced interpretation nodded in the direction of a middle course between total war and the Neely challenge, Grimsley's conclusions reinforced the traditional narrative with a more precise terminology (Grimsley 1995).

Writers interested in the morality of the Civil War have also taken up the total war interpretation. In *Upon the Altar of the Nation*, Harry Stout criticised both sides for their 'moral misconduct' and blamed them for allowing the war to become 'a killing horror' justified as 'a moral crusade for "freedom" that would involve nothing less than a national "rebirth."' Accepting the total war theory, Stout wrote that the rebellion was 'a war waged deliberately on civilian populations with the full knowledge and compliance of commanders running all the way to the top.' Stout continued his 2006 condemnation, 'In this sense, the spirit of total war emerged quite clearly by 1864 and prepared Americans for the even more devastating total wars they would pursue in the twentieth century' (Stout 2006: xv–xxi).

But the revisionists continued their onslaught. In a 2011 article, Wayne Wei-Siang Hsieh argued that the 'Master Narrative' of total war was outdated and did not fit with Civil War scholarship. He accused the total war theorists of being engaged in American exceptionalism and ignoring developments in transnational history. These historians, Hsieh argued, had constructed a story that privileged the United States by making the Civil War the first chapter in modern warfare. But this ignored comparative analysis that clearly showed that the US experience was not exceptional. The result was a narrative that 'both exaggerates American influence on later events and obscures some of the most important aspects of Civil War violence, especially the war's connection to the demise of chattel slavery and its particular combination of lethality and limitation.' Hsieh repeated many of Neely's contentions, insisting that the Civil War was actually marked by restraint, that the most brutal aspects of the conflict were racist in nature, and that the total war interpretation rests on vague definitions and a preoccupation with violence. Building on Neely's earlier work, Hsieh attempted to dismantle the total war theory with comparative analysis that discounted both the destructiveness of the Civil War and its innovation as a modern war (Hsieh 2011). He also turned to more recent scholarship by historians like Earl J. Hess to challenge the total war view of technology. Hess studied the rifle-musket and argued that its impact on the Civil War was minimised by the terrain on many battlefields and that, even when soldiers had clear fields of fire, they usually lacked the training to make use of the weapon's greater accuracy (Hess 2008). Hess also examined Civil War fortifications and his conclusions varied greatly from the total war advocates who had argued that entrenchments were the consequence of new weapons making them seek better defensive positions. Instead, he contended that most fortifications and entrenchments were the result of sustained contact between the armies. Although elaborate earthworks were used in both the eastern and western theatres of the war, once Atlanta and Vicksburg fell, in the west the trenches were abandoned for a more mobile war in the open. This questioned the total war narrative that cast Civil War fortifications as forerunners of the trench warfare of the First World War (Hess 2007). Hsieh's own scholarship combined with other studies by Hess concluded that the length of the war was not the result of technology that favoured the defensive. Rather the conflict was prolonged by the military culture of the time, as training and tactics made the Civil War more like the Napoleonic wars than the struggles of the twentieth century. Battlefield results were rooted in the past and the institutions of the era rather than innovation and a new kind of warfare (Hsieh 2011).

Then new scholarship emerged to question and revise the total number of deaths in the Civil War. In his dissertation, and then in a widely acclaimed 2011 article, J. David Hacker offered an analysis of the total numbers of men killed based on census data and concluded that the actual number was at least 750,000 and may have reached 850,000 (Hacker 1999, 2011). The vast difference mostly stemmed from the undercounting of Confederate deaths from disease, but also from undercounting of Union losses and the exclusion of African-American soldiers and foreign-born troops. The results of Hacker's census-based counting meant that the war was far more destructive than previously thought. In his commentary introducing Hacker's article, McPherson proclaimed that he had 'been waiting for more than twenty-five years for an article like this one' that confirmed his contentions that the war killed more Americans than all of the nation's conflicts combined. He noted that Hacker's 750,000 estimate would equal 7.5 million dead in 2010 and argued that the new total 'calls into question Mark Neely's assertion that the Civil War was "remarkable for its traditional restraint"' (McPherson 2011: 309–10).

But the sceptics fought back against these new numbers. Nicholas Marshall fired off sharp criticism of what Neely had called the 'cult of violence' in an article entitled 'The Great Exaggeration: Death and the Civil War' (Marshall 2014). Marshall did not critique Hacker's analysis of census data, but instead argued that such numbers were being misunderstood as the proponents of the traditional total war narrative engaged in presentism, applying the standards of their own time to the past. Analysing nineteenth-century data, Marshall insisted that the total number of men killed in the war was actually not tremendously out of line with the death rates in times of peace. The incidence of disease, especially epidemics, and the higher rates of infant mortality meant that American families in the 1800s endured death much more regularly than those in the twentieth or twenty-first centuries. Building on the literature on death in American society in the nineteenth century, Marshall argued that Americans were much better prepared to deal with the impact of losing a family member than their descendants were in later wars. Using not only statistics but evidence drawn from diarists, he insisted that Americans were better trained to accept and cope with death in the Civil War era (Marshall 2014). In addition to experience teaching them to deal with death, Americans also found comfort in the notion that a soldier's death had meaning, that they died for a cause and a purpose that ennobled their sacrifice (Faust 2008; Schantz 2008; Clarke 2011).

In a published response to Marshall, Hacker undercut his critic's data, saying that he had cherry picked examples from uncorrected and inaccurate census sources. Furthermore, he pushed back against the charge that comparisons to other wars were presentism and argued that, regardless of the precise accuracy of the data, the Civil War killed more soldiers than any other American conflict. He also insisted that the war's human cost was unprecedented and that Americans were not prepared to deal with the loss of so many adult men of military age. Before the war, adult males between the ages of 15 and 44 made up an average of one in ten deaths. During the war, they accounted for one of every three deaths. Losing a healthy adult son or husband was not the same as losing an infant or elderly person. Still, Hacker agreed that more work needed to be done and concluded by pointing out that historians had not yet begun to consider the number of civilian deaths caused by the war, especially the number of African Americans who died in the transition from slavery to freedom. And the death of refugees—both black and white—had not yet been studied. And so the debate continued (Hacker 2014).

Far from sweeping aside the long-standing interpretation, the revisionist challenge ironically strengthened the modern, total war argument by pressing its proponents to be more precise and nuanced in their terminology. Still the challengers brought in significant new ideas, including the racial dimensions of violence. The result was a more complicated story. While they yielded ground to their critics, the proponents of total war did not surrender and continued to make the case of the conflict becoming something different over time. So, was the Civil War a total war? Historians today answer that question in different ways, but those differences reflect nearly 30 years of vigorous scholarly debate.

Bibliography

Bartek, James M., 'The Rhetoric of Destruction: Racial Identity and Noncombatant Immunity in the Civil War Era,' PhD dissertation (University of Kentucky, 2010).

Burns, Ken, *The Civil War* (Burbank, CA: PBS Home Video, 1990). Distributed by Paramount Home Entertainment, 2004.

Catton, Bruce, 'The Generalship of Ulysses S. Grant,' in Grady McWhiney (ed.), *Grant, Lee, Lincoln and the Radicals: Essays on Civil War Leadership* (New York: Harper Colophon, 1966).

Clarke, Frances M., *War Stories: Suffering and Sacrifice in the Civil War North* (Chicago: University of Chicago Press, 2011).

Earle, Jonathan and Mutti Burke, Diana, *Bleeding Kansas, Bleeding Missouri: The Long Civil War on the Border* (Lawrence, KS: University Press of Kansas, 2013).

Faust, Drew Gilpin, *This Republic of Suffering: Death and the American Civil War* (New York: Alfred A. Knopf, 2008).

Fellman, Michael, *Inside War: The Guerrilla Conflict in Missouri during the American Civil War* (New York: Oxford University Press, 1989).

Gallagher, Gary W. (ed.), *The Shenandoah Valley Campaign of 1864* (Chapel Hill, NC: University of North Carolina Press, 2006).

Glatthaar, Joseph, *The March to the Sea and Beyond: Sherman's Troops in the Savannah and Carolinas Campaigns* (Baton Rouge, LA: Louisiana State University Press, 1985).

Goldin, Claudia D. and Lewis, Frank D., 'The Economic Cost of the American Civil War: Estimates and Implications,' *Journal of Economic History*, vol. 35, no. 2 (1975), pp. 299–326.

Grimsley, Mark, *The Hard Hand of War: Union Military Policy Toward Southern Civilians, 1861–1865* (New York: Cambridge University Press, 1995).

Hacker, J. David, 'The Human Cost of the War: White Population in the United States, 1850–1880,' PhD dissertation (University of Minnesota, 1999).

Hacker, J. David, 'A Census-Based Count of the Civil War Dead,' *Civil War History*, vol. 57, no. 4 (December 2011), pp. 307–48.

Hacker, J. David, 'Has the Demographic Impact of Civil War Deaths Been Exaggerated?,' *Civil War History*, vol. 60, no. 5 (December 2014), pp. 453–8.

Hattaway, Herman and Jones, Archer, *How the North Won: A Military History of the Civil War* (Urbana: University of Illinois Press, 1983).

Heatwole, John L., *The Burning: Sheridan's Devastation of the Shenandoah Valley* (Charlottesville, VA: Howell Press, 1998).

Hess, Earl J., *Trench Warfare under Grant and Lee: Field Fortifications in the Overland Campaign* (Chapel Hill, NC: University of North Carolina Press, 2007).

Hess, Earl J., *The Rifle Musket in Civil War Combat: Reality and Myth* (Lawrence, KS: University Press of Kansas, 2008).

Hill, Louis B., *State Socialism in the Confederate States of America* (Charlottesville, VA: Historical Publishing Co., 1936).

Hsieh, Wayne Wei-Siang, 'Total War and the American Civil War Reconsidered: The End of an Outdated "Master Narrative,"' *Journal of the Civil War Era*, vol. 1, no. 3 (September 2011), pp. 394–408.

Marshall, Nicholas, 'The Great Exaggeration: Death and the Civil War,' *The Journal of the Civil War Era*, vol. 4, no. 1 (March 2014), pp. 3–27.

McPherson, James M., *The Battle Cry of Freedom: The Civil War Era* (New York: Oxford University Press, 1988).

McPherson, James M., 'From Limited to Total War, 1861–1865,' in James M. McPherson, *Drawn With the Sword: Reflections on the American Civil War* (New York: Oxford University Press, 1996), pp. 66–86.

McPherson, James M., 'Commentary on "A Census-Based Count of the Civil War Dead,"' *Civil War History*, vol. 57, no. 4 (December 2011), pp. 309–10.

Neely, Jr., Mark E., 'Was the Civil War a Total War?,' *Civil War History*, vol. 37, no. 1 (March 1991), pp. 5–28.

Neely, Jr., Mark E., *The Civil War and the Limits of Destruction* (Cambridge, MA: Harvard University Press, 2007).

Nelson, Megan Kate, *Ruin Nation: Destruction and the American Civil War* (Athens, GA: University of Georgia Press, 2012).

Paludan, Phillip S., *'People's Contest:' The Union and the Civil War* (New York: Harper & Row, 1988).

Royster, Charles, *The Destructive War: William Tecumseh Sherman, Stonewall Jackson, and the Americans* (New York: Vintage Books, 1991).

Schantz, Mark S., *Awaiting the Heavenly Country: The Civil War and America's Culture of Death* (Ithaca, NY: Cornell University Press, 2008).

Sellers, James L., 'The Economic Incidence of the Civil War in the South,' *Mississippi Valley Historical Review*, vol. 14 (1927), pp. 179–91.

Stith, Matthew M., 'At the Heart of Total War: Guerrillas, Civilians, and the Union Response in Jasper County, Missouri, 1861–1865,' *Military History of the West*, vol. 38, no. 1 (2008), pp. 1–27.

Stodola, Sara, 'Civil War at 150: Debt Lessons from Lincoln,' *The Fiscal Times*, 12 April 2011, https://www.thefiscaltimes.com/Articles/2011/04/12/Civil-War-at-150-Debt-Lessons-from-Lincoln, accessed 28 June 2019.

Stout, Harry S., *Upon the Altar of the Nation: A Moral History of the Civil War* (New York: Viking, 2006).

Sutherland, Daniel E., *The Emergence of Total War* (Buffalo Gap: State House Press, 1998).

Thomas, Emory M., *The Confederacy as a Revolutionary Experience* (Englewood Cliffs, NJ: Prentice Hall, 1971).

Thomas, Emory M., *The Confederate Nation: 1861–1865* (New York: Harper and Row, 1979).

Trudeau, Noah Andre, *Southern Storm: Sherman's March to the Sea* (New York: Harper Collins, 2008).

Walters, John B., 'General William T. Sherman and Total War,' *Journal of Southern History*, vol. 14 (November 1948), pp. 447–80.

Walters, John B., *Merchant of Terror: General Sherman and Total War* (Indianapolis: Bobbs-Merrill, 1973).

Wheelan, Joseph, *Terrible Swift Sword: The Life of General Philip H. Sheridan* (Boston, MA: De Capo Press, 2012).

Wheelan, Joseph, *Bloody Spring: Forty Days that Sealed the Confederacy's Fate* (Boston, MA: De Capo Press, 2014).

Weigley, Russel F., *The Great Civil War: A Military and Political History* (Bloomington, IN: Indiana University Press, 2000).

Williams, T. Harry, *Lincoln and His Generals* (New York: Alfred A. Knopf, 1952).

24

THE BRITISH ARMY, 1870–1918

From Small War to Trench War

Ian F.W. Beckett

The British Army stood apart from the experiences of most other states' armed forces in the late nineteenth century. By continental European standards, it was a small imperial constabulary recruited by voluntary means. It was also widely dispersed across the globe in overseas garrisons from Halifax, Nova Scotia to Hong Kong. The separate Indian Army gave Britain additional manpower but there were legal, financial and practical difficulties in deploying Indian troops outside boundaries that might be perceived as strictly necessary for the defence of India (Jeffery 1982). The white colonies of Australia, Canada and New Zealand became responsible for their own defence in 1871, while most other British possessions had locally raised forces but these were not substantial. It was also the case that the army's dispersion militated against the dissemination of learning even without the institutional and social factors inhibiting the growth of military professionalism. Much depended upon the deterrent of the Royal Navy, for Britain was first and foremost a global maritime power.

The largest force deployed between the end of the Crimean War in 1856 and the start of the South African War in 1899 was 35,000 men sent to Egypt in 1882: the actual field force was approximately 16,000 strong (Beckett 2007b). By comparison, in March 1864, Ulysses Grant assumed the responsibility for the direction of 533,000 men of the Union armies in the American Civil War (1861–1865), while the elder Helmuth von Moltke directed 850,000 men in the closing stages of the Franco–Prussian War (1870–1871).

Nevertheless, in one significant aspect, Britain enjoyed an unrivalled advantage over continental rivals in the wealth of practical experience of colonial campaigning. Between 1872 and 1899 the army fought 35 major overseas campaigns in addition to innumerable minor ones. There were 27 separate expeditions on the North-West Frontier of India alone between 1868 and 1908 (Corvi and Beckett 2009: 220; Nevill 1912: 404). The empire could not have been expanded to the extent that it was in the second half of the nineteenth century without soldiers harnessing advances in medicine, weaponry, technology and communications, or without displaying a degree of adaptability (Beckett 2003). Yet, as the South African War (1899–1902) demonstrated, much still needed to be achieved in order to shape a thoroughly modern army. Lessons from South Africa were ambiguous and did not automatically resolve existing institutional problems or clarify tactical debate. Moreover, the nature of the warfare confronting all armies between 1914 and 1918 posed a particular difficulty for an

DOI: 10.4324/9780429437915-31

army required to embrace mass mobilisation on an unprecedented scale. Yet, by 1918 Britain possessed an effective military instrument.

The Late Victorian Army

The British Army in the late nineteenth century operated against a background that impeded its ability to fulfil those military tasks required of it. Soldiers had to accept the limits imposed by parliamentary supremacy, not least financial constraints. As Henry Campbell-Bannerman, the Secretary of State for War, once remarked, his task was 'to avoid heroics and keep the estimates down' (Bond 1960: 523). For all the supposed resulting rationalisation of military organisation involved, both the Cardwell reforms of 1868–1872 and the later Haldane reforms of 1906–1908 were equally driven primarily by the desire for financial retrenchment (Spiers 1980a: 180; Spiers 1980b: 48–73).

It was pointed out in October 1883 that government revenues in 1863–1864 had been £70.2 million and the army estimates had been just over £15 million (21.45 per cent of the total). In 1882–1883, government revenues stood at £89 million but the army estimates were still only £15.4 million (or 17.36 per cent of the total) (Verner 1905: II, 312–13). The Royal Navy enjoyed far greater leverage in securing funding. Through the Naval Defence Act of 1889 it received £21.5 million, spread over five years, but the army just £600,000 (French 1990: 149–54).

The army also confronted socio-economic conditions and cultural attitudes that hindered recruitment within a system of voluntary enlistment. The country's male population aged between 15 and 24 doubled between 1859 and 1901 but the army's share of this age group remained virtually static at around a pitiful 1 per cent (Spiers 1980a: 38–9). The army's pay remained uncompetitive. The competition of the labour market when trade was good, the decline of the rural population and increasing Irish emigration to the United States after the 1840s all impacted upon recruitment. Overseas service, unsanitary barracks, harsh discipline, lack of recreational opportunities, the discouragement of marriage, the lack of training in trades and a complete lack of provision for veterans and reservists in civilian life added to the unattractive features of military life, albeit that the army's public image improved. A more militarised society emerged as a result of the increasing transmission of nationalist and imperialist themes in popular culture through the medium of the press, art, music, adult and junior literature, entertainment, including increasing military spectacle, and even the infant cinema. Nonetheless, a report on the army's health in 1909 revealed that over 90 per cent of those presenting themselves for medical inspection as recruits were without employment. By far the largest single category of recruit was classed as unskilled labourers, never dropping below 58.9 per cent of total recruits in any year between 1870 and 1900. Even then, most unemployed still shunned the army (Spiers 1992: 129–30).

The introduction of short service by Cardwell in 1870—six years with the colours and six with the reserve—was supposed not only to stimulate enlistment and reduce wastage, but also to attract a better class of recruit and produce a viable reserve. Paradoxically, it required the army to recruit more men than previously. Nor did it address the problem of imperial commitments when it was likely that the ranks would be filled with immature youths. The reserve increased from 3,000 men in 1869 to 80,000 by 1899 but it was only available for 'imminent national danger' until 1898. Cardwell also introduced 'localisation' in 1872 by which regular battalions were linked, one based at the home depot in order to help supply drafts to the other abroad. By 1885 there were only 54 battalions at home and 87 abroad.

Hugh Childers carried localisation to its logical conclusion of 'territorialisation' in 1881 by permanently linking regular battalions in new county regiments. Many regimental depots, however, were in Southern England, remote from the majority of potential recruits located in London or the industrial North and Midlands.

Strategy and operations were subject to political decisions. From the point of view of the army, there was no definite statement of its role beyond the annual Mutiny Act, which alone had offered some direction since 1689, albeit vague, in stating in its preamble that the army existed to preserve the balance of power in Europe. That clause was deleted in 1868 but, finally, in 1888 the Stanhope Memorandum ordered the army's priorities as aid to the civil power in the United Kingdom; the provision of drafts for India; the provision of garrisons for colonies and coaling stations; the provision of two corps for home defence; and the possible employment of one of the corps in a European war, although this was deemed so improbable that nothing need be done to prepare for such an eventuality (Beckett 1984).

There was little strategic agreement within the army; those of the Indian service arguing for massive and unrealistic reinforcements in order to meet any Russian advance into Central Asia with Afghanistan seen as a buffer zone. Britain intervened in Afghanistan none too successfully between 1839 and 1842, and again between 1878 and 1880 in an attempt to reduce perceived Russian influence at Kabul. By contrast, soldiers in Britain argued that any war fought against Russia should be primarily amphibious and aimed at peripheries such as the Baltic, the Black Sea or the Turkestan/Caspian area: any posture adopted by the Indian Army should be primarily defensive. Even more significantly, soldiers in Britain were obsessed with the threat of French invasion irrespective of the Royal Navy's claim as to its impossibility. There had been major invasion scares in 1846–1847, 1851–1852 and 1858–1859 and more occurred in 1871, 1882 and 1888. By 1899, the army was coming to accept that the Royal Navy could guarantee home defence but the fear remained in the public eye through continued invasion literature and press interest.

There was considerable opposition on the part of Liberal politicians to emulating continental practice and creating a General Staff. Campbell-Bannerman remarked in 1890 that a chief of staff should not 'be shut up in a room by himself in order that he might think' (Hamer 1970: 151). The post of Chief of Staff was recommended by the Hartington Commission in 1890 but there was also opposition from the army's commander-in-chief, the Duke of Cambridge, who feared diminution of his own position, and from his cousin, Queen Victoria (Hamer 1970: 135–47).

In terms of the application of military power, the Victorian Army is often seen as possessing what the military radical and theorist, J.F.C. Fuller, who fought in South Africa as a subaltern, characterised as a 'Brown Bess' mind (Fuller 1937: 19–20). There is much apparently to substantiate the image, not least the struggle of the Staff College for recognition after its foundation in 1858 or for the employment of its graduates.

But there were also real difficulties militating against the development of an intellectual community with the army widely dispersed and customary divisions between arms and regiments. There were also practical limitations. Lack of proper practice facilities in Britain was acute. An innovative camp of exercise was held at Chobham in 1853, and Aldershot was purchased as a major training base between 1854 and 1861, but there were then no more manoeuvres in Britain between 1853 and 1871, or between 1875 and 1890. The Military Manoeuvres Act (1897) prohibited any disturbance of antiquarian remains, places of historic interest or exceptional beauty, which severely impacted on any entrenchments, although the Military Land Act (1892) allowed for the purchase of 40,000 acres on Salisbury Plain.

Nonetheless, by 1899, there were a number of publishers offering military series for officers and there had been military journals from a much earlier period. It is difficult to assess how far military publications were read or disseminated. Books consumed or written by British officers might merely represent a consensus view, or be a synthesis of foreign treatises, lacking all originality. Edward Hamley complained in 1875 that British authors were often ignored by reviewers while 'any trumpery German pamphlet that comes out affords them food for endless comment' (Preston 1971: 64). *Modern Warfare as Influenced By Modern Artillery* (1864), the work of Patrick MacDougall, was heavily influenced by the Swiss-born interpreter of Napoleonic warfare, Antoine-Henri Jomini. Hamley's own *Operations of War*, which succeeded MacDougall's *The Theory of War* (1866) as the standard Staff College text, was purely Jominian. Another popular text, Robert Home's *Précis of Modern Tactics* (1873), was a digest of current continental manuals. Francis Clery produced a similar digest *Minor Tactics* (1875) that was heavily influenced by Prussian methods. Official infantry manuals also mirrored the 'ebb and flow' of continental practice (Spiers 1992: 250). No artillery tactical manual was published until 1892 but close attention was paid to German experience. The most contentious debate was between cavalry traditionalists and advocates of mounted infantry, which appeared more suited to colonial campaigning. A mounted infantry school was established in 1888 but closed in 1913.

A distinct 'continentalist' school of thought existed in the army that had little relevance to its most frequent employment in colonial warfare (Bailes 1988). The influential Henry Brackenbury suggested in 1869 that there was no need to draw on the 'poisonous wells' of French thought, while George Henderson remarked, 'We have no need to ask another nation to teach us to fight' (Bailes 1981: 38).

The practical requirements of actual campaigning found a distinctive British voice in the theory of 'small wars' as in George Younghusband's *Indian Frontier Warfare* (1898), and William Heneker's *Bush Warfare* (1907). It might be added that the 1877 edition of *Field Exercise and Evolution of Infantry* at least acknowledged the impact of colonial warfare with a new emphasis upon flexible open order. Further lessons were learned from campaigning in South Africa between 1877 and 1879 (Laband 2005: 79–82). Above all, there was Charles Callwell, whose *Small Wars: Their Principles and Practice*, first published in 1896, held the field until the publication of Charles Gwynn's *Imperial Policing* in 1934 (Beckett 2007c).

As already suggested, advances in medicine, technology and communication were all embraced by the army. The strategic use of railways for the defence of India was appreciated at an early stage and its importance borne out by the experience of the Indian Mutiny in 1857–1858. According to Sir John Lawrence, it was another technological innovation that had actually saved India, namely the electric telegraph: the length of telegraph line increased from 4,250 miles in 1856 to 17,500 miles by 1865. The Royal Engineers formed its first telegraph company in 1870 and deployed it to Ashanti (Asante) in 1873 (Headrick 1988: 72, 106–7, 121–2; 1991: 18, 52, 63–4).

The reconquest of the Sudan between 1896 and 1898 likewise demonstrated the significance of modern artillery, magazine rifles and machine guns. Winston Churchill wrote that the battle of Omdurman on 2 September 1898, in which some 11,000 dervishes were killed for the loss of just 48 British, Egyptian and Sudanese dead, represented the 'most signal triumph ever gained by the arms of science over barbarians' (Churchill 1933: 300). Machine guns had been attached to each brigade in Britain in 1893 when they did not form a permanent part of any other army. Similarly, Britain had led the way in introducing breech-loading artillery in 1868, though subsequently reverting to rifled muzzleloaders between 1870 and

1885. The British introduced magazine rifles only a matter of months after the Germans in the 1880s (Travers 1987: 63–4, 69; Spiers 1992: 241, 245). Equally, however, obsolete methods such as rallying squares worked perfectly well against indigenous opponents although most soldiers assumed that European opponents would not be as susceptible to the application of technology in warfare.

The South African War

In 1899 the army faced a new challenge against the Boers in South Africa. Many colonial campaigns were more a struggle against terrain and climate than indigenous opponents. Now, the British were pitted against a well-armed and highly mobile enemy of European descent. Rather than costing the British government no more than £10 million for a three- to four-month campaign utilising a maximum of 75,000 men, as originally expected, the war lasted 32 months and cost £230 million. In the end, Britain and the Empire fielded 448,000 men, and also employed conceivably as many of the 120,000 Africans, of whom 30,000 were armed. Yet the Boers never fielded more than 42,000 men and, following the defeat of their main field army in February 1900, never more than about 9,000 guerrillas (Beckett 2000).

The war's initial conventional phase encompassed British defeats at the Modder River (28 November 1899), Stormberg (10 December), Magersfontein (11 December) and Colenso (15 December), and at Spion Kop (23–4 January 1900), but culminated in the surrender of the Boer field army at Paardeburg (18–27 February 1900) and the fall of the Boer capitals of Bloemfontein and Pretoria in March and June 1900 respectively. The casualties in this conventional phase were unprecedented within living memory. The second phase began with the decision by younger Boer leaders on 17 March 1900 just after the fall of Bloemfontein to prolong the struggle by attacking British lines of communication rather than directly confronting the army advancing on Pretoria. That guerrilla phase did not conclude until the peace agreement reached in May 1902.

It has been argued that what went wrong in South Africa was not a refusal to assimilate new ideas, but a failure to act in accordance with the new teaching of the 1890s (Bailes 1988). There was recognition of what Henderson characterised as the 'second tactical revolution' introduced by the appearance in the 1880s and 1890s of magazine rifles, quick firing guns and smokeless powder. Many of those sent to Natal had experienced the great tribal uprising on the North-West Frontier in 1897. For them, the Boer War was not the first occasion on which the army had faced opponents armed with magazine rifles. This ignores the problem of dissemination, inadequate training and aversion to formal doctrine. The 1898 manoeuvres involving some 50,000 men illustrated significant tactical failings with a series of unsuccessful frontal assaults (Leeson 2008).

Lessons were learned but they were often ambiguous, not least from the effect of the peculiar atmospheric conditions on the *veldt*. The principal infantry lesson appeared to be the decisiveness of firepower. More extended order and manoeuvre was required to avoid its destructive effect although there was also emphasis upon the need to condition troops to face incoming fire in terms of the cultivation of high morale in keeping with prevalent Social Darwinist theory. Musketry training was given heightened emphasis although the significance accorded firepower and methodical preparation for attack in the new manuals, *Combined Training* in 1902 and *Infantry Training* in 1902, was scaled back in *Field Service Regulations* in 1909, which stressed the moral effect of the offensive spirit (Spiers 1981).

Extended order facilitated taking cover, which was certainly learned by those engaged. Examination of the Boer entrenchments after Paardeburg was a revelation. Yet, there was little post-war consensus on the utility of entrenchment (Jones 2012: 104–11). At least, *Instruction in Military Engineering Part I: Field Defences* in 1902, the *Manual of Military Engineering* in 1905 and the *Manual of Field Engineering* in 1911 provided the basis for adjustment to the conditions met for the first time on the Aisne in 1914.

New artillery tactics were evolved in the operations in Natal in February 1900, the so-called 'curtain fire' equating to what would later be characterised as 'creeping barrages.' Howitzers were also used in an indirect fire role although their potential was not fully capitalised upon until the Great War. In any case, infantry and artillery cooperation was difficult without adequate communications. It was emphasised in *Field Artillery Training 1902*, but only lip service was paid to it in *Combined Training* in 1905 and *Field Service Regulations* in 1909. Again, there was little post-war consensus (Jones 2012: 52–3, 118–19, 145). British longer-range bombardments such as those used at Colenso and Magersfontein—the heaviest British bombardment since the Crimean War—proved ineffective because there was no knowledge of the actual Boer positions. The war principally reinforced the existing desire to rearm with quick-firing guns. The Director-General of the Ordnance persuaded the government to authorise rearmament in January 1900 when the Boers possessed no such quick-firing guns. The artillery debate remained unresolved as to whether direct or indirect fire represented the likely future (Bowman and Connelly 2012: 78–83).

Even for the cavalry, the lessons were not as clear cut as often supposed since, in many instances, sword and lance had been just as effective as carbine. *Cavalry Training 1904 (Provisional)* swept away the emphasis upon cold steel only for the position to be spectacularly reversed in *Cavalry Training 1907*. A compromise 'hybrid' cavalry emerged from *Cavalry Training 1912* (Badsey 2008: 81–142).

Those methods evolved in the guerrilla phase of the war were highly effective with 're-concentration' of 116,000 Boer and 160,000 African civilians into 'concentration camps'; the construction of blockhouses along railway lines and at other key points; lavish use of barbed wire entanglements linked to the blockhouses crossing the *veldt*; and mobile mounted columns driving the commando between the lines of blockhouses. Guerrilla opponents, however, were not to be encountered again until after 1918.

The South African War is best regarded as a transitional military conflict. Apparent lessons were contradictory, particularly when viewed alongside those of the Russo–Japanese War (1904–1905), enabling different commentators to pick and choose examples to fit their own prior perceptions. The army was more of a learning organisation as a result of the significant military reforms in Britain after 1902, not least the establishment of the General Staff in 1905 and what has been characterised as a 'managerial revolution' (Gooch 1994). There were manoeuvres annually from 1902 and General Staff conferences were also instituted but the army remained tiny compared with continental conscript armies capable of mobilising millions of men at the outset of war.

The Great War

In 1914, all armies were confronted with conditions beyond their previous experience: it was often a painful transition. The British regular army consisted of merely 247,432 officers and men. There was just one peacetime corps headquarters—that at Aldershot—whereas there were 20 in France and 35 in Germany. Only three serving regulars had commanded at Aldershot: Sir John French, Sir Douglas Haig and Sir Horace Smith-Dorrien. There were just 12,738 regular officers, of whom only 908 had received staff training. The scale of the war was such that conscription had to be introduced in 1916 and, by 1918, a total of 5.7 million men had passed through the army, with 229,316 wartime commissions issued. Inevitably, it would take time to train the mass citizen army (Beckett and Corvi 2006: 2–3).

The learning process was compounded by the army's lack of any doctrinal uniformity. Although officers like Haig understood the balance to be struck between control and guidance of subordinates, there were many occasions on which superiors either interfered in the conduct of their subordinates when they should not have done, or equally did not intervene when they should have done (Beckett 1999). The lack of consistency bedevilled the army throughout the war. To give one example, on 1 July 1916, the first day of the Battle of the Somme, 83 British battalions went 'over the top.' Far from the popular image of men walking very slowly forward, 53 battalions were already in No Man's Land when the barrage lifted, ten intended to rush the opposing frontline in short bursts, and only 12 were to advance at a steady pace in a long line (Prior and Wilson 2005: 158–61). Although more sophisticated manuals for operational and tactical concepts were produced by the end of the war, it was still up to divisional or brigade commanders how far they adopted them. There was no Inspector General of Training until July 1918.

The war actually began much as expected, with even traditional cavalry charges in August 1914, although aerial reconnaissance also played a role. But in two respects, the first campaigns surprised all. First, there was unprecedented expenditure of ammunition so that all major armies suffered shell shortages. In the British case, the Russo–Japanese War had suggested the need for fire economy rather than more shell production and all pre-war expectations were far off the mark. By the time of the First Battle of Ypres in October 1914 batteries were being rationed in how many shells they could fire daily (Beckett, Bowman and Connelly 2017: 231–3). Second, there were far higher casualties than anticipated. The British estimate had been based on taking a mean between the South African and Russo–Japanese experiences. Instead of an assumed wastage rate of 65 to 75 per cent in the first 12 months, the reality was 63 per cent in the first three months (Beckett et al. 2017: 229).

Initially, the trenches dug on the Aisne in September 1914 were seen as a temporary expedient before the resumption of mobile operations, the so-called 'race to the sea' seeing each side leapfrogging to the north in order to regain mobility. But by November, it was clear that deadlock had arisen. As Sir John French wrote later, 'All my thoughts, all my prospective plans, all my possible alternatives of action, were concentrated upon war of movement and manpower' (Beckett et al. 2017: 207).

Soldiers now confronted managerial and technical problems. They were related in that overcoming managerial problems was arguably the key to finding technical solutions since it was the often hierarchical nature of armies that would be the obstacle to the necessary innovation. The British were slower to adapt than the French or Germans but then they began with that lack of doctrinal uniformity and without adequate mechanisms for the learning and dissemination of lessons. Ultimately, while General Headquarters (GHQ) remained resistant to change—not least once Haig succeeded French as commander-in-chief of the British Expeditionary Force (BEF) in December 1915—it was generated at lower levels as officers became more experienced. The key level for innovation was divisional command and, by 1918, the average age of divisional commanders had fallen to 47: some individuals were in their thirties (Beckett et al. 2017: 383). Unfortunately, Haig remained obsessed with the illusion that a complete breakthrough was possible, resulting in the prolongation of the Somme and Third Ypres Offensives in 1916 and 1917 with concomitant heavy losses.

Technology was not sufficiently advanced to achieve any breakthrough. 'Chateau generalship' was an unavoidable byproduct of the increased scale of warfare coupled with the lack of sufficient means of communication. Wireless was in its infancy. The best means available were telephone cables but these could be easily cut by shellfire. On 1 July 1916, Sir Henry Rawlinson, commanding Fourth Army, received 160 messages from corps headquarters during the course of the day but after seven hours was still unaware of the precise situation (Beckett and Corvi 2006: 177). GHQ calculated that a message could be relayed from battalion to corps in six hours but the reality was more like 12.

The British and their allies had to find a solution to deadlock, for the Germans had chosen where they stood in the retreat of 1914 and still occupied much of Belgium and the most productive parts of Northern France. They could, if they wished, stand on the defensive and mostly did so. The German Verdun Offensive in 1916 was not intended as an attempt to break the deadlock and it was only in the spring offensives of 1918 that the Germans applied themselves to the problems of breakthrough.

Artillery, which accounted for 67 per cent of all wartime casualties (Beckett 2007a: 223), was the key to the problem for it facilitated breaking down the opposing defences and covering the infantry while they crossed 'No Man's Land.' But with a static front, advancing infantry could only be protected to the extent of the range of that artillery: as guns got larger, defence increased in depth. Even a 'break in,' however, posed problems, for it was not easy to achieve surprise or to reinforce success. Reserves could not be retained close to the frontline or they would be hit by counterbattery fire. Moving men forward through trench systems was exceptionally difficult.

British artillery improved immensely by 1917 in terms of techniques such as sound ranging and flash spotting that enabled guns to be registered on specific targets in advance of any barrage. But unfortunately, the British made a fundamental error in 1915, which they did not correct until 1918. At Neuve Chappell in March 1915, 340 guns were available. Because of the shell shortage, everything available was fired off in 35 minutes over a narrow front of 2,000 yards. The concentration had a physiological as well as a physical effect and the German frontline was taken. Then the other problems of exploitation occurred and the offensive was closed down after two days. The lesson was the concentration of fire in a short period of time in a small area but it was assumed that firing more shells from more guns over a longer period would have greater impact. At Loos in September 1915, the British used 533 guns in a 48-hour bombardment over a frontage of 7,500 yards, and at the Somme 1,431 guns fired 1.7 million shells over seven days on a 22,000-yard front. Because these subsequent barrages were not as concentrated, the actual impact was diluted (Beckett 2007a: 224–5). Moreover, the German response was generally to dig deeper and to defend in more depth: sufficient Germans would always survive the opening bombardment to win the 'race to the parapet.'

Increasingly, it was recognised that there would always be occasions during which the infantry could not be covered by artillery. The solution was to give the infantry portable firepower and, by 1917, battalions were being reorganised with fewer riflemen and more men armed with rifle grenades, light mortars and the Lewis light machine gun. By 1917, the Lewis Guns had been ever more widely distributed so that, instead of a few for each battalion, they were now available down to section level. Better manuals were also appearing, such as *SS143* 'Instructions for the Training of Platoons for Offensive Action' in February 1917 and *SS135* 'Instructions for the Training of Divisions for Offensive Action' in

December 1917, but, as suggested earlier, there was no uniformity in application. There were continuing failures, for example, in the spring of 1918 in trying to emulate the German defence in-depth, contributing to the success of the German 'Michael' Offensive in March 1918.

Technical innovations were attempted but generally these were very much at a transitional phase. Gas was seen as a breakthrough weapon but, released from cylinders, as it was by the British for the first time at Loos, it proved fickle when the wind blew it back into the British lines. Soldiers became better protected against gas than any other weapon and it became possible to operate in a gas environment. Nonetheless, when put into shells, gas, if liberally mixed with high explosive, could be what would be termed a 'force multiplier.'

The most obvious mechanical attempt to develop a breakthrough weapon was the tank but tanks were chronically unreliable mechanically and could be knocked out by artillery and by anti-tank rifles that the Germans soon developed. Just how unreliable they were can be illustrated by the three major tank attacks of the war at Flers-Courcelette on 15 September 1917, Cambrai on 20 November 1917 and Amiens on 8 August 1918. At Flers, only nine of the 32 tanks that made the start line actually spearheaded the attack. Of the 378 tanks used at Cambrai, only 92 were still serviceable within three days. At Amiens, only six of 414 tanks were still operable after four days (Beckett 2007a: 238–9). Tanks could be used relatively successfully in conjunction with infantry, artillery and ground attack aircraft as some formations developed the beginnings of a rudimentary all-arms concept by 1918. Again, application was far from uniform (Boff 2010).

The '100 Days' campaign of 8 August to 11 November 1918 was extremely successful with the BEF taking 188,700 prisoners and 2,840 guns. The Germans had exhausted their remaining manpower resources in the spring offensives and the Allies now had crushing allied material superiority. In the successful assault on the Hindenburg Line on 29 September 1918 the British fired 945,012 shells, delivering 126 shells for every 500 yards of German line every minute for eight hours (Beckett 2007a: 235). The campaign also proved the efficacy of the so-called 'bite and hold' tactics that a number of commanders such as Rawlinson had glimpsed as early as 1915, seeking a series of more gradual 'step-by step' advances to unlock the front given the impossibility of total breakthrough (Harris and Marble 2008). The German frontline was still intact on 11 November 1918, although it had been forced far back. The technical means of converting the 'break-in' to breakthrough was never available, and the only arm of exploitation remained cavalry. After 1914, the greatest advance on any single day on the Western Front was just six miles at Amiens on 8 August 1918.

For all its remaining structural weaknesses, the British Army was far more combat effective in 1918 than previously. Ground won in 1918 was held and that ground was increasingly of strategic significance. There had been a learning process of sorts and even a degree of learning transfer from the Western Front to other theatres such as Gallipoli and Palestine (Fox-Glodden 2016). In the final analysis, there was enough quality, consistency and cohesion at different levels to mask weaknesses and to maximise strengths. As always with the British Army, the issue remained how far lessons would be genuinely learned given the aversion to formal doctrine and the reversion to 'imperial policing' in 1919.

Bibliography

Badsey, Stephen, *Doctrine and Reform in the British Cavalry, 1880–1918* (Aldershot: Ashgate, 2008).

Bailes, Howard, 'Patterns of Thought in the Late Victorian Army,' *Journal of Strategic Studies*, vol. 4 (1981), pp. 29–45.

Bailes, Howard, 'Technology and Tactics in the British Army, 1866–1900,' in Ronald Haycock and Keith Neilson (eds.), *Men, Machines and War* (Waterloo: Wilfrid Laurier Press, 1988), pp. 23–47.

Beckett, Ian, 'The Stanhope Memorandum of 1888: A Reinterpretation,' *Bulletin of the Institute of Historical Research*, vol. 57 (1984), pp. 240–7.

Beckett, Ian, 'Hubert Gough, Neill Malcolm and Command on the Western Front,' in Brian Bond (ed.), *Look to Your Front: Studies in the First World War* (Staplehurst: Spellmount, 1999), pp. 1–12.

Beckett, Ian, 'Britain's Imperial War: A Question of Totality?' *Joernaal vir Eietydse Geskieddenis*, vol. 25 (2000), pp. 1–29.

Beckett, Ian, 'Victorians at War: War, Technology and Change,' *Journal of the Society for Army Historical Research*, vol. 81 (2003), pp. 330–8.

Beckett, Ian, *The Great War* 2nd ed. (Harlow: Longman-Pearson, 2007a).

Beckett, Ian, 'Going to War: Southampton and Military Embarkation,' in Miles Taylor (ed.), *Southampton: Gateway to the British Empire* (London: IB Tauris, 2007b), pp. 149–66.

Beckett, Ian, 'Another British Way in Warfare: Charles Callwell and Small Wars,' in Ian Beckett (ed.), *Victorians at War: New Perspectives* London: Society for Army Historical Research Special Publication, 2007c), pp. 89–102.

Beckett, Ian, Bowman, Timothy, and Connelly, Mark, *The British Army and the First World War* (Cambridge: Cambridge University Press, 2017).

Beckett, Ian, and Corvi, Steven (eds.), *Haig's Generals* (Barnsley: Pen & Sword, 2006).

Boff, Jonathan, 'Combined Arms during the Hundred Days Campaign, August–November 1918,' *War in History*, vol. 17 (2010), pp. 459–78.

Bond, Brian, 'The Effect of the Cardwell Reforms in Army Organisation, 1874–1904,' *Journal of the Royal United Services Institute*, vol. 105 (1960), pp. 515–24.

Bowman, Timothy, and Connelly, Mark, *The Edwardian Army: Recruiting, Training and Deploying the British Army, 1902–14* (Oxford: Oxford University Press, 2012).

Churchill Winston S., *The River War* 3rd ed. (London: Eyre & Spottiswoode, 1933).

Corvi, Steven Corvi Beckett Ian (eds.), *Victoria's Generals* (Barnsley: Pen & Sword, 2009).

Fox-Glodden, Aimée, 'Beyond the Western Front: The Practice of Inter-theatre Learning in the British Army in the First World War,' *War in History*, vol. 23 (2016), pp. 190–209.

French, David, *The British Way in Warfare, 1688–2000* (London: Unwin Hyman, 1990).

Fuller, J.F.C., *The Last of the Gentleman's Wars* (London: Faber & Faber, 1937).

Gooch, John, 'Britain and the Boer War,' in George Andreopoulos and Harold Selesky (eds.), *The Aftermath of Defeat: Societies, Armed Forces, and the Challenge of Recovery* (New Haven, CT: Yale University Press, 1994), pp. 40–58.

Hamer, W.S., *The British Army: Civil-Military Relations, 1885–1905* (Oxford: Clarendon Press, 1970).

Harris, J. Paul, and Marble, Sanders, 'The Step-by-Step Approach: British Military Thought and Operational Method on the Western Front, 1915–17,' *War in History*, vol. 15 (2008), pp. 17–42.

Headrick, Daniel, *The Tentacles of Progress: Technology Transfer in the Age of Imperialism* (New York: Oxford University Press, 1988).

Headrick, Daniel, *The Invisible Weapon: Telecommunications and International Politics, 1851–1945* (New York: Oxford University Press, 1991).

Jeffery, Keith, 'The Eastern Arc of Empire: A Strategic View, 1850–1950,' *Journal of Strategic Studies*, vol. 5 (1982), pp. 531–45.

Jones, Spencer, *From Boer War to World War: Tactical Reform in the British Army, 1902–14* (Norman, OK: Oklahoma University Press, 2012).

Laband, John, *The Transvaal Rebellion: The First Boer War, 1880–81* (Harlow: Pearson Education, 2005).

Leeson, D.M., 'Playing at War: The British Military Manoeuvres of 1898,' *War in History*, vol. 15 (2008), pp. 432–61.

Nevill, Captain H.L, *Campaigns on the North West Frontier of India* (London: John Murray, 1912).

Preston, Adrian (ed.), *Sir Garnet Wolseley's South African Diaries (Natal) 1875* (Cape Town: A.A. Balkema, 1971).

Prior, Robin, and Wilson, Trevor, *Somme* (New Haven, CT: Yale University Press, 2005).

Spiers, Edward, *The Army and Society: 1815–1914* (London: Longman, 1980a).

Spiers, Edward, *Haldane: An Army Reformer* (Edinburgh: Edinburgh University Press, 1980b).

Spiers, Edward, 'Reforming the Infantry of the Line, 1900–14,' *Journal of the Society for Army Historical Research*, vol. 59 (1981), pp. 82–94.

Spiers, Edward, *The Late Victorian Army: 1868–1902* (Manchester: Manchester University Press, 1992).

Travers, Timothy, The *Killing Ground: The British Army, the Western Front and the Emergence of Modern Warfare* (London: Allen & Unwin, 1987).

Verner, Willoughby, *The Military Life of HRH George, Duke of Cambridge*, 2 vols (London: John Murray, 1905).

25

THEORY AND PRACTICE OF ARMOURED WAR DURING THE TWO WORLD WARS

Michael W. Charney

Motorised armoured vehicles, whether sporting a cannon or merely protecting transported infantry, did not emerge until the First World War. Nevertheless, imaginative thinking about doing so developed individually amongst different military personnel and engineers in France, Austria-Hungary and other military powers at the beginning of the twentieth century. In all cases, early proposals were rejected by the military establishments for a variety of reasons; in one case because they would be frightening to horses (Dredger 2017: 208). In peacetime it is often difficult to get political authorities to respond to new military innovations, but there is often much more receptivity when war has broken out, particularly when the war has proven difficult to win by conventional means. This was certainly the case with the tank.

Introduction of Armour in the First World War

The development of tanks was seen to end the stalemate on the Western Front in the First World War (1914–1918). The road to the development of the tank came with the outbreak of the First World War. The fast movement of armies after the opening of war in August 1914 soon gave way to stalemate and difficult, shell-pockmarked terrain that was hard to cross. In Britain, Lieutenant Ernest Dunlop Swinton read about trials of the caterpillar tractor in the United States that could cross difficult terrain. Swinton had the inspiration to armour such a tractor and arm it with a machine-gun. On 20 October 1914, Swinton took the idea to Colonel Maurice Hankey, a secretary to the British War Cabinet, and other members of the government, describing the kind of stalemate that had emerged on the Western Front as like a 'species of siege warfare,' which was likely to become permanent without technological intervention (Swinton 1933: 60, 309).[1] Certainly, the war was quickly demonstrating the need for a new weapon. Trench warfare, barbed wire and machine-guns had helped to slow down the gains an attacking army could make even with heavy numerical superiority. Air power was not yet technologically able to have a decisive impact on land battles. It became clear to some that the only way forward was to develop the technology that could help to protect troops while they moved close enough to enemy positions to be effective, including helping them get through the barbed wire and avoid being mowed down in no man's land by machine-guns.

DOI: 10.4324/9780429437915-32

Finally, in January 1915, Hankey delivered an amended proposal to the Prime Minister, Herbert Asquith and the First Lord of the Admiralty, Winston Churchill. Churchill responded positively and now championed development through his creation of the Admiralty Landships Committee (the early prototypes for the British tank were called landships, the name being changed in December 1915 to tank to help keep their purpose a secret) (Swinton 1933: 139, 161). The proposed vehicle would be powered by an engine and moved by bulletproof heavy rollers armed with a machine-gun. The vehicle could roll down barbed wire, defend itself in the process and open the way for infantry to sweep through the breach. Nevertheless, Churchill wished to keep the invention with the Admiralty and refused to share the research with the War Office. The impasse was finally broken when Swinton wrote a letter to John French, Commander in Chief of the British Expeditionary Force in France, describing the vehicle; French then pursued details on the invention from the Admiralty (Swinton 1933: 80–1, 139). Research on behalf of the War Office now propelled forward and the British had the world's first tank, the Mark I. The French also began to develop their own versions during the same period.

The British first introduced the Mark I to battle on the Somme on 15 September 1916. The Mark I was like other tanks being developed in France: slow, an easy target for artillery, and prone to breaking down and having to be abandoned on the battlefield. The same vehicle had two different kinds of armament. As first envisaged, the tank was supposed to be a 'machine-gun destroyer,' that is, a tank that could roll over barbed wire and target machine-gun nests with two 6 pounder guns and three Hotchkiss machine-guns. However, it was soon realised that the tank would be vulnerable to an attack by great numbers of infantry and hence was to be accompanied by other Mark I tanks armed with four Vickers machine-guns that would enable the 'consort' to fire 1,200 bullets a minute from each side. This division of labour led to these tanks being referred to as male tanks (the machine-gun destroyer) and female tanks (the consort). Confusingly, the prototype for the male tank inherited the name 'Mother,' despite being a male tank. Eventually, tanks would incorporate both functions in a single vehicle. Further work followed until the more reliable Mark IV was introduced. Mark I tanks nevertheless were being shared on other fronts, including eight which participated in the Second and Third Battles of Gaza (April and November 1917, respectively) in the Middle East (Swinton 1933: 184, 260).

French tank development had also emerged with the main intention of breeching trench lines and, like the British, focused first on heavy tank development, with several independent projects arriving at different outcomes; one being the army-designed Saint Chamond failing as a tank but finding success instead as an assault gun. Louis Renault offered a different vision of the tank in 1916, proposing, instead of a few heavy tanks, a great number of light, very mobile tanks, culminating in the Renault FT. This was introduced to battle in April 1917 in the Nivelle Offensive, and France then became the leading tank power of the war, producing more tanks than all other countries combined. The Renault FT would find a home in most of the tank armies of the world in the following two decades and derivatives would include the American M 1918, the Italian Fiat 300 and even the Soviet T-18. The Renault FT and the British Whippet (see below) raised questions about existing tank doctrine concerning what the role of the tank was after the enemy trench lines were breached and whether the tank could replace cavalry (Gilbert and Cansiere 2017: 20–2, 27).

While the British had taken the first steps towards the application of the tank in warfare, their training and doctrine had not caught up to the potentiality of the tank. They saw the tank as a way merely to break through trench lines rather than as a way to exploit the breakthroughs when made. As a result, rather than massing tanks, they put them in groups of eight to 30 and assigned them to army groups (Gilbert and Cansiere 2017: 16). The consequences were clearly shown when the British used 350–400 of the 29-ton Mark IV to break through German lines at the Battle of Cambrai on 20 November 1917, but were unprepared to take advantage of the potential exploitation, having allotted the task to the cavalry. Moreover, the Germans had experience of fighting against tanks from the Nivelle Offensive over half a year earlier and received specific training in anti-tank tactics. The attack was thus costly to the tank force, as 65 tanks were destroyed by the Germans, 71 suffered mechanical failure and 43 had to be abandoned. Worse, the Germans were able to retake the lost terrain (Gilbert and Cansiere 2017: 27). The British would eventually expand their thinking on the tank's role in warfare when they developed a light tank with greater tactical mobility—the more lightly armed, but much faster 'cruiser tank,' the Whippet, which was able to take on infantry, but not other tanks.

Although plans and ideas were circulated in the United States from the beginning of the war, the first American-made tank, the Holt Gas-Electric tank, was not developed until 1917. This was followed in late 1917 by the 45-ton 'War-Tank America' which was equipped with a flamethrower that could propel fuel oil 90 yards at the German pillboxes but arrived in France too late in 1918 to take part in hostilities. Also to arrive in France too late to take part in the war was a scaled down Renault FT tank, the Skeleton Tank. Rather than American-designed tanks, the American Expeditionary Force to France was then supposed to be instead provided with American-built versions of the Renault FT tank, which, as with the others, did not see combat (Zaloga 2017: 6–8, 10). As a result, the American Expeditionary Force (AEF) had to rely on loaned British and French tanks during their involvement on the Western Front. By the end of the war, the US tank corps had over 1,000 light tanks and over 100 heavy tanks. In the aftermath of the conflict, US interest in tanks waned and by 1926 the United States had about 500 mainly small 6-ton tanks in service (Zaloga 2017: 12–13).

Nevertheless, the First World War experience had long-term consequences, particularly in tank doctrine. Captain Dwight D. Eisenhower was one of the officers assigned to the heavy tank training school at Bovington and then was assigned to organise the cadre of the 302nd Tank Battalion, giving him intimate, if early, familiarity with the emerging use of armour in battle. More significantly, cavalry officer George S. Patton would be assigned the establishment of an American light tank school on 10 November 1917 and was the first American to learn to drive a tank. Patton also wrote a comprehensive report on every aspect of the tank, its technical aspects, how tanks should be maintained and repaired, tank tactics, doctrine, the tank's potential role in support, necessary training, and optimal organisation. In the latter case, his model of a five-tank platoon with a company of three platoons would be adopted by the US military for another century. Moreover, Patton envisaged tanks not only breaking through enemy lines but then transitioning into the mode of riding the enemy down, a role once reserved for the cavalry. From 20 August 1918, he would command the First Provisional Tank Brigade and would lead his brigade into battle at Metz on 12–15 September 1918 (Gilbert and Cansiere 2017: 31–3). Patton would go on to champion tank warfare in American military circles and emerged as one of the Allied generals most respected by his German opponents in the Second World War.

While Italy was one of the first countries to deploy armoured cars in war in 1911–1912, it entered the First World War without any tanks and would try hard to catch up. Like the Americans, the Italians studied French tanks and designed a heavy tank of their own, the

FIAT 2000, a 44-ton beast equipped with a powerful 65/17 howitzer and crewed by ten men, but this did not go beyond the testing of prototypes before the war ended.

By 1918, the Entente were committed to the tank as the decisive weapon that would win the war and had plans for 1919, had the war continued, to deploy over 30,000 tanks. By contrast to the Entente, the German High Command was slow to realise the importance of the tank, even after encountering them in battle. Germany was thus slower to adopt armoured warfare, at first forming a tank corps out of captured French and British tanks. The first German-built tank, built by Daimler, was not delivered until October 1917 and would not see action until 21 March 1918. This was the tall, slow-moving, seven-metre long, 36-ton A7V, protected by 30 mm of steel armour, armed with a 57-mm cannon and three Maxim machine-guns, operated by a crew of 18 and an officer. By the end of the war, the Germans had produced fewer than 64 tanks, compared with the thousands produced by the Allies (Gilbert and Cansiere 2017: 24). The First World War also saw other developments in armour that would be crucial in the decades ahead. These included the cousin of the tank: self-propelled guns. The Germans and others also developed the first generation of anti-tank weapons.

Theory and Experimentation in the 1920s and 1930s

The end of the First World War in 1918 saw Britain and France as the leaders in tank design, a position they would hold for the following decade. In the aftermath of the conflict, the Entente's various allies, particularly the United States, Japan and Italy, acquired increasingly obsolete British Whippet tanks and French Renault FT light tanks, which formed the basis of their own evolving tank corps and eventually the development of their own tank models.

Britain and France also led the world in the 1920s in organising and developing tactics for mechanised forces. Several major proponents of the development of the tank in the war continued to put unrelenting pressure on their governments and military establishments to focus resources on the tank after the conflict ended. John Frederick Charles Fuller opened the 1920s with a history of the role of the British tank in the First World War, *Tanks in the Great War (1914–1918)*, to remind the British public of its importance (Fuller 1920). Fuller was joined by other proponents of the tank in France, Britain and the United States, like Charles De Gaulle, and Basil Liddell Hart, in publishing works that would lay out various potentials and futures for the tank and often would influence each other (as well as Heinz Guderian, who learned about tanks through books, not personal experience, in the 1920s), although De Gaulle remained much more unappreciated than the others. Certainly, many of these proponents would remain active and influential in the 1930s. Nevertheless, despite their early and sustained lead in tank theory and development, France and Britain lost some ground in the 1920s. The World Trade Depression of 1929, combined with government austerity and the pursuit of efficiency in military economy in the 1930s, led to the slowed pace of tank development. There was also a shift in tank design to more emphasis on cheaper to build, light, very manoeuvrable tanks as opposed to the heavier tanks that had won the previous war.

Well into the early post-war period, the mainstream of Italian military thinking would continue to see cavalry as the main force to be used in exploitation of gaps in the enemy frontlines and reconnaissance into the mid-1920s. Nevertheless, development of tanks continued. Italy's first tank, the FIAT 3000, based on the French Renault FT, was produced from 1921 until 1935. The early development of Italian tank warfare doctrine was based on this model and the FIAT 3000 B variant of 1930 (eventually classified as the L 5/30). During much of the 1920s, the Italian tank force remained small, the tank force limited to one company-sized unit. A five-battalion tank regiment was formed in 1926 with each battalion having in theory 180 tanks, but in practice only 20. In October 1927, the Italian tank force became a separate branch of the land army (Capellano and Battistelli 2012: 6–7, 11). As Italy's mechanised warfare doctrine emerged, it focused on their motorised and mechanised units exploiting breaches of enemy lines made by artillery and infantry. In the 1930s, Italian military thinkers advocated using light tanks for reconnaissance as well as for war of manoeuvre. In Italy, the FIAT 3000 and 3000B were joined by the light tanks, the CV-33 (later L 3–33), from 1933; the CV-35 variant introduced in 1935; and the CV-38. These were modelled on Mark VIs bought from British Vickers Carden Lloyd. In support of these changes in technology and in doctrine, the Italians at first used tanks to support both infantry divisions and cavalry divisions, regrouping tanks into individual support battalions. The Italians then organised the first motorised brigade on the verge of the invasion of Ethiopia and refitted some of the CV-33s with flamethrowers (Pignato 2004: 7–8).

The 1930s saw the Soviet Union and Nazi Germany take the lead in tank design. The Soviets had an early start in post-war tank development due to the massive Russian Civil War that continued from 1917 until the early 1920s, while the Germans cooperated with the Soviets in the KAMA (after Kazan and Malbrandt) tank school, a secret, joint tank commander training school in Kazan in defiance of the Treaty of Versailles (Mitcham 2000: 304). Although agreed in 1926, it only commenced courses in 1929 and these came to an end with the rise to power of Adolf Hitler and the Nazis in Germany in 1933. While the Soviet Union industrialised and Germany rebuilt their industries, both used the school to develop new tactics and experiment with new tank designs. Like Britain, the Soviets also had an experimental mechanised brigade by the end of the 1920s to develop combined-arm tactics (Corum 1992: 190–3). The Red Army would wind up purchasing the patents for tank models invented by J. Walter Christie that had been rejected by the peacetime American military establishment. On the back of these models, the Soviets would build the tanks they would send to Spain in 1936.

The first precursor to the Second World War, the Italian invasion of Ethiopia in 1936 was generally ignored by tank theorists because it was seen as a colonial adventure that had little resemblance to the fighting that would take place in any war in Europe. What was hoped to be a better testing ground of the new leaders in design and theory was the Spanish Civil War, where the Soviets and the Nazis backed the Republican and Fascist forces respectively. Here, tanks, the cannon-armed Soviet T-26 light tank versus the machine-gun-armed Italian CV.3/35 and German PzKpfw I light tanks, were used in significant number for the first time since the First World War, mainly in 1937 (Zaloga 1999: 4–5). Generally, Soviet tanks fared better than the German and Italian light tanks, which struggled with the difficult Spanish terrain, but the poorly trained Spanish crews on both sides, and the hesitant commanders who deployed tanks in small numbers at any given time, limited the lessons that could be learnt of how tanks would perform in other contexts (Candil 2021: 128). The experiences in the Spanish Civil War showed, amongst other things, that cannon-armed tanks were superior to machine-gun-armed tanks and that light tanks were very vulnerable to anti-tank guns such as the German 37-mm gun (Zaloga 1999: 4–5).

The Italians modified their tank doctrine in response to lessons learnt from the Spanish Civil War. After fighting at Guadalajara in March 1937, the Italian tank forces pursuing retreating Republicans became separated from the infantry and artillery, and then were counterattacked by the Republicans who inflicted severe losses. This led the Italians to relegate armour to infantry support (Candil 2021: 98–9). Nevertheless, in 1937, the single motorised brigade was converted into an armoured brigade, followed by the creation of a cadre for a second armoured brigade. Further, in 1938, a new doctrine, 'The War of Rapid Course,' was issued under which the Italian Army would be rapidly mechanised, and emphasised high-speed, mobile warfare reliant on mechanised warfare and air power with motorised forces exploiting breeches in the hostile lines, thus paralleling some of the chief tactics of German *Blitzkrieg*. Finally, in 1939, the first and second armoured brigades were each transformed into an armoured division, forming the 131st Centauro and 132nd Ariete Armoured Divisions. Centauro would be deployed in Albania, which Italy had invaded on 7 April 1939, early in 1940, and would be used later for the invasion of Greece in 1941 (Pignato 2004: 3–4, 8).

For both Germany and the Soviet Union, the last half of the 1930s saw a race for building tanks and their best application. Hitler's Germany inherited the Panzer I and, breaking the terms of Versailles, Hitler ordered their mass production and tested them out in the Spanish Civil War. How the *Wehrmacht* would apply them in their own military organisation was debated. Proponents of panzers and panzer divisions, such as Major-General Oswald Lutz and his chief of staff, Oberst Heinz Guderian, had a face off with General of Artillery, Ludwig Beck, and Erich von Manstein, who favoured units of mobile assault guns in support of infantry divisions. In the offensive-oriented mood of the new *Wehrmacht* under Hitler, Lutz and Guderian won out, leading to the creation of the first Panzer Divisions, leaving the development of assault guns lagging behind (Forczyk 2010: 9–10). As expounded by Guderian in *Achtung-Panzer!*, tank formations of the past had been able to make breaks in enemy lines but not fully exploit them, because they were tied to slow-moving infantry on foot and horse-drawn artillery. His vision of the panzer division was a fully mechanised division, built around the panzer, that could move at speed with all the support the panzers needed (Horne 2007: 92).

Soviet tank design also benefited from the early experiences of the Spanish Civil War, which proved Soviet tanks to be poorly defended and highly flammable. The wrecks were sent back to the Soviet Union where tank engineer Mikhael Koshkin was put to the task of designing a new tank. Koshkin developed a model he had begun to contemplate in 1934 (hence the name T-34). It had sloped armour to protect the crew and ran on much less explosive diesel, innovations made on the basis of what was learned from the wrecks, but would run just as fast on its Christie chassis as the tanks being replaced and sported a huge cannon. The result would be the mainstay of the Soviet Army in the Second World War and was admitted by many tank commanders of the opposing German panzer forces to be the best tank in the world (Tucker-Jones 2021: 464–5).

Although lessons were also initially learned by the officers of the Soviet armoured corps, many of its best commanders would be slaughtered during Stalin's purge of the Soviet military from 1937 to 1939, including Marshal Mikhail Tukhachevsky (1893–1937) and its other main innovative military theorists (Glantz 1991: 25). Stalin had distrusted Tukhachevsky since the 1920s, but when he realised the need for an industrialised military, he put Tukhachevsky in charge of reforming the military. Tukhachevsky rightly believed that earlier plans to develop the Red Army had underestimated the industrial potential made possible under Stalin's first Five-Year Plan (1928–1932) and, as a result, it was dependent mainly on infantry formations. With the development of industry, an industrialised military was possible, including significant numbers of tanks and planes, and with these the development of combined arms operations. Where Tukhachevsky erred was in his exaggerated vision of what Soviet industry was yet able to support. He unrealistically envisaged the Red Army with possibly hundreds of thousands of planes and tanks, far outstripping pre-war Soviet security needs. Regardless, he and others shifted strategic thinking from a war of successive operations to the 'deepening of battle,' driving 150 divisions on a 450 km front 100–200 km against the opponent's rear and its logistical support (Glantz 1991: 24–5; Stone 1996: 1367, 1377–80).

While Japan watched tank development in Europe closely, it would not develop comparable forces until the 1930s. The Japanese began with research on a light tank model in 1925 but did not begin mass producing tanks until the 1930s. Japan had over a thousand tanks in 1937 when it first commenced its invasion of China, starting the Second Sino-Japanese War (1937–1945), the Imperial Japanese Army (IJA) expected tanks to be important only in support of infantry. They had only developed their first mechanised, combined arms formation, the 1st Independent Mixed Brigade, in 1934, and the generals in the field spread their tanks far and wide in a support role only, being organisationally within the infantry branch of the IJA. Disbanding the brigade and forming an entirely tank unit, the 1st Tank Group, the tanks were again used only to support infantry, leading to disaster against the Soviet armoured forces at Halka River and Nomonhan in July and September 1939. Another, the 3rd Tank Group, would be organised in Manchuria and yet another, the 2nd Tank Group, for the invasion of Malaya and Singapore. Despite the secondary place given to armour, in terms of numbers Japan rose to be the world's fifth largest tank power by the end of the 1930s, ahead of the United States, but behind the Soviet Union, Germany, France and Britain. Japan had 15 tank regiments in 1940 and would create seven more over the next two years, nine more in 1944, and an additional 15 in 1945 (Rottman and Takizawa 2008: 3–5, 8, 10).

The Second World War

No country used armour in battle as successfully as did Germany in the early years of the Second World War. It avoided a conflict with a country which did have a significant tank force, Czechoslovakia, as a result of the Munich Agreement in 1938 (and these tanks would later be absorbed into the German Army, supplementing their existing panzer forces). The Germans launched *Blitzkrieg* against Poland in September 1939, which was not very well supplied with tanks. Poland, dependent on cavalry that had trained against wooden models, fell in only a few weeks. Having built up their panzer forces to ten divisions, the Germans then prepared for the invasion of the Low Countries and France in the Spring of 1940. Technically, the French had more, and perhaps even better, tanks than the Germans. But while the Germans concentrated theirs into powerful panzer divisions that could operate on their own in the field independently of other kinds of divisions, the French divided theirs amongst infantry divisions per se. The French approach diluted the armour forces that could be deployed in any one place at any one time. This arrangement also prevented the French tanks from operating independently en masse. The Germans also surprised the French by circumventing the Maginot Line in the south and entering France through the less well protected north via Belgium. France fell after just six weeks.

The outbreak of the war in Europe put Japan further behind in tank development. Their arsenal relied mainly on older home designs of light tanks in both armament and defensive armour, which could only fare poorly in the field against other tanks. The IJA was slow to modernise and produce tanks for several reasons. First, this part of the imperial military assets was given much lower priority than the navy or the air force until the last year of the war, when the home islands came under threat of a potential land invasion. Even then, American strategic bombing had drastically reduced the Japanese industrial base. Second, unfortunate encounters with the Soviet Army in 1939 led to a non-aggression pact that in theory removed the only potential threat from a major armoured opponent in the region. Third, there was at first not as much urgency, given the demands on limited Japanese resources—when the Japanese did move against more powerful Western European and American forces in late 1941, on land they faced only the garrison forces that were stationed in colonial Southeast Asia.

Japan was impressed by rapid German successes with tanks in Poland and France. In April 1941, the Japanese finally made the armour branch a separate force within the army. An order of battle reorganisation in the field would occur in June 1942 when the 1st–3rd Tank Divisions were organised in Manchuria and, a month later, the Armor Army was created there including two tank divisions and a tank group (the understrength 4th Tank Division would be created at Chiba in Japan in July 1944). And the IJA signalled doctrinal change in September 1942 when it laid down that, rather than tanks supporting infantry, now tanks were the main striking force. Now every branch should cooperate to support the tank units, with the expectation of a war in the future with the Soviet Union and its armoured forces on the plains of East Asia. Elsewhere, on the Pacific islands, for example, Japanese tanks were too few and too lightly armoured to withstand the deployment of American airpower, anti-tank guns and M4 Sherman medium tanks (Rottman and Takizawa 2008: 6, 12, 16–17). Although the Japanese did develop better tanks and self-propelled guns later in the war, they were produced in too few numbers and were too late to have much of an effect.

As with the Japanese military, some of the major American commanders, such as Major-General George Lynch, considered tanks only to be useful in supporting infantry against machine-gun nests. Swift German victories in Poland in 1939 and France in 1940 changed these perspectives, and the United States created the Armoured Force on 10 July 1940 with a shift in interest from light tanks to the medium tanks that were at the centre of Germany's successes, focusing on the unfortunately designed M3 'Stuart' medium tank. Although proving itself (like the Crusader tank) inadequate in the hands of the British in the Western Desert in 1941 and in US Army hands in Tunisia in 1942, the tank remained in use, relegated to secondary tasks in Europe but showing greater success in the Pacific Theatre against the under-armoured Japanese, simply because there had been so many produced by this time (Zaloga 2017: 40–1; Zaloga 1999: 3).

German armour was soon kept busy in several areas that had not been planned for. In the case of Yugoslavia, the German ally turned against it, leading to the German invasion of Yugoslavia and its breakup into several minor fascist states and occupied areas. Italian adventures were particularly distracting for Germany's panzer forces. Italy had demonstrated its interest in the Balkans in 1939 with its invasion and annexation of Albania, which included the world's first application of an armoured division in war. By contrast, the Italian invasion of Greece failed, and required German intervention. The entrance of Italy into the war in 1940 had meant an expansion of the fighting to North Africa. This armoured war over vast expanses of desert, as it became, saw a sweeping war of movement, led by tanks on both sides. By the beginning of the Second World War, the Italian Army was mainly armed with weakly protected light tanks and, initially, both Italy and Britain fought a desert war in which both were reliant on relatively light armoured vehicles. The British deployed Crusaders, Matildas and Valentines, all light tanks that broke down easily, against the Italians. Italy dispatched three armoured divisions in total to North Africa, including Ariete in January 1941, its third armoured division, Littorio, in the early Spring of 1942, and elements of Centauro in late 1942. The Italians would also try to improve their armoured forces by developing a medium tank (the M13/40, M14/41 and M15/42) and a heavy tank, Ansaldo's 26-ton P26/40. About 1,800 of the M13/40 and its successors were built and used in North Africa but the heavy tank project would come too late to have significant effect before Italy's surrender in 1943 (only one had been built by this time), the 100 or so which were built in Italian factories under German occupation in 1943 and 1944 (none were produced in 1945) would be used instead by the Germans (Pignato 2004: 3–4, 52).

When the Italian forces stalled and began to be pushed back in the desert, Hitler despatched General Erwin Rommel and several German divisions to help bolster them. The Germans introduced heavy tanks to the war in the desert, which gave them an advantage in battle. From 1942, however, American-produced tanks began to reach British forces under their new commander in the desert, Field-Marshal Bernard Montgomery. Montgomery benefited from the build-up of forces under his predecessor, Field-Marshal Harold Alexander. Further, the new American tanks, like the Grants and Shermans, were more reliable and, in number, more effective than the Allies' previous tanks. Rommel, who would almost succeed in taking Egypt, was defeated in the Battle of El Alamein. Montgomery would go on to take Libya as well, while the Americans landed in North Africa. German forces, cornered in Tunisia, surrendered in North Africa in 1943.

The poor performance of two Soviet tank corps under Zhukov against Japanese forces at Khalkin-Gol River in August 1939, despite the victorious outcome of the battle, led Stalin to disband the tank corps in November 1939. Although after the failures of the Soviet–Finnish War (1939–1940), the Soviets would quickly try to repair the damage that had been done to operational art in the Red Army, the unavoidable legacy of the purges of the 1930s was that the Soviets would remain unprepared for the German invasion of June 1941 (Glantz 1991: 90–1).

Operation Barbarossa was the largest implementation of *Blitzkrieg*. *Blitzkrieg* tactics were again applied to great effect, but it was mainly faulty leadership after the officer purges and demoralisation among the troops that contributed to the massive defeats of Soviet forces in the first few months of the invasion, with millions of Soviet prisoners being taken. Fortunately, for the Soviets, the Germans had not planned on a long campaign, having hoped for a quick victory, and the German forces were unprepared for winter, particularly the harsh winters of Russia. Having reached the outskirts of Moscow, the German tank force froze, and their crews perished in the cold. Many other factors weighed in, including Hitler's decision to divert some of the panzers south to aid in the battle of Kiev.

As the Germans became bogged down in the north and the centre, the spring 1942 offensive saw German panzer divisions make rapid headway in the south, towards the oil rich Caucasus. The decision to take and hold Stalingrad, however, led to a catastrophic German defeat. In November 1942, Stalin unleashed reserves including a tank army and most of the new tank and mechanised corps against the Stalingrad Front. In this massive encirclement operation, the German Sixth Army and much of the Fourth Panzer Army were trapped inside Stalingrad, where they would slowly be decimated until the surrender of the survivors in February 1943. Stalin pushed his forces too far, and too quickly, where they were nearly caught in a trap themselves when the Germans tried to recommence offensive operations with an attack on the Kursk salient in July 1943 (Glantz 1991: 121–2). The ensuing German attack featured larger numbers of Tiger tanks and the new Panther tanks. In fact, Hitler had delayed the operation until the Panther tanks were ready and could be transported to the Front in significant numbers (Newton 2002: 27, 56). They represented an increasing reliance on Hitler's belief that *wunder waffe* would turn the tide back in Germany's favour. Nevertheless, leaked German plans allowed the Soviets time to prepare significant defences in depth. The conclusion of the battle, which was the largest tank battle in history, saw the Germans permanently lose the initiative on the Eastern Front as well as any realistic chance of a German victory.

The Allied invasion of Sicily in July 1943 and then the Italian mainland two months later saw the reintroduction of Western armour on the European continent but unlike in North Africa, the Italians fielded no armoured divisions, dividing their tanks instead into smaller units to greet them. Instead, armoured warfare in Italy from 1943–1945 was mainly a contest between the Allied and German forces.

Behind the increasing Allied successes in armoured warfare was the superior industrial might of the West and the Soviets, once the latter had recovered from the movement of their industry to Central Asia in 1941. American and Soviet industry could mass manufacture thousands of tanks each year, far outstripping German production, even after Albert Speer's reorganisation of Germany's war industries in 1944. But technological success was also important. The hard lessons of the early years of the war had contributed to improvements in tank design that helped to produce much better, more reliable, and more versatile machines. One of the keys to Soviet success was the T-34, which it manufactured in huge numbers. While the Germans would develop larger tanks carrying heavier armaments, the Allies could produce effective tanks in huge numbers, with better tactics. The improvements helped quantity overcome the technical German advantages in quality.

The End

Soviet tanks would play a significant role in the Soviet advances on the Eastern Front over the course of 1944. After Rumania switched sides, Germany began to re-equip its main remaining Balkan ally, Hungary, with German tanks so that they had a reasonable chance of standing their ground against Soviet armour. This culminated in the siege of Budapest in which Hungarian and German troops, including significant panzer forces, were surrounded.

The last major German tank offensive, the Ardennes Offensive (the Battle of the Bulge), was unleashed on 16 December 1944. By this time, the Reich was quickly running out of essential resources. It took considerable effort to find remaining supplies of fuel, men and air support. German plans included taking major Allied fuel depots and pillaging minor fuel depots along the way. After a week, heavy Allied resistance and counterattacks (in particular, by tanks under Patton), the drying up of oil reserves and the end of cloud cover, which allowed Allied planes, particularly P-47 fighter-bombers, to wreak havoc on the stranded German tanks, ended German hopes for victory and forced them to retreat, leaving their burning tanks behind (Barron 2014).

The German surrender, on 8 May 1945, left Japan as the major remaining Axis power in the war. Armour had played much less of a role in fighting in the Asia-Pacific thus far. Land battles until late 1944 had mainly been air; artillery and infantry engagements in jungle terrain South-East Asia and China had been out of reach of a significant Allied supply of tanks. From October 1944, however, the Allied invasion of Burma, and drive to the south, was aided by tanks. Nevertheless, the lack of mechanisation of Japanese forces in the theatre meant that there were no tank battles in the region on the scale of those seen in Europe.

In North-East Asia, however, the Soviets joined the war in August 1945 at a time that had been agreed among the Allies at Potsdam. Stalin had diverted significant armoured reinforcements to Siberia along the frontiers with the Japanese colony of Korea and its puppet state of Manchukuo. Catching the Japanese by surprise, at a time when Japan was reeling from the detonation of the first atomic bomb, Soviet armour swept through Japan's territory and occupied territory in Manchuria. Japan would quickly surrender as a result of these attacks and the dropping of the second atomic bomb shortly after. The war of armour globally had ended.

Conclusion

From a technology that was practically non-existent prior to the First World War, the tank had emerged by the end of the Second World War as the main battle weapon of land armies, as it has remained until the present. Tanks reoriented all other branches of the military into support functions for armour and became a measure of the strength and prowess of conventional armed forces. Yet, along the way it faced continual obstacles and resistance due to limitations of vision, conservative expenditures and misunderstandings of lessons learned in battle. Ultimately, it was a weapon that only progressed due to the efforts of individual pioneers in multiple countries who each pushed for its advancement.

Note

1 Nevertheless, while Swinton did circulate this idea, his role in furthering tank development further is regarded by some authorities as exaggerated (Harris 1995: 33–5).

Bibliography

Barron, Leo, *Patton at the Battle of the Bulge: How the General's Tanks Turned the Tide at Bastogne* (New York: Caliber, 2014).

Candil, Anthony J., *Tank Combat in Spain: Armored Warfare During the Spanish Civil War 1936–1939* (Philadelphia: Casemate, 2021).

Capellano, F. and Battistelli, P.P., *Italian Light Tanks 1919–45* (Botley: Osprey. 2012).

Corum, James S., *The Roots of Blitzkrieg: Hans Von Seeckt and German Military Reform* (Lawrence, Kansas: University Press of Kansas, 1992).

Dredger, John R., *Tactics and Procurement in the Habsburg Military, 1866–1918: Offensive Spending* (New York: Palgrave MacMillan, 2017).

Forczyk, Robert, *Erich von Manstein: Leadership, Strategy, Conflict* (Botley: Osprey, 2010).

Fuller, J.F.C., *Tanks in the Great War (1914–1918)* (London: John Murray, 1920).

Gilbert, Oscar E. and Cansiere, Romain, *Tanks: A Century of Tank Warfare* (Oxford: Casemate, 2017).

Glantz, David M., *Soviet Military Operational Art: In Pursuit of Deep Battle*, with a foreword by General Carl E. Vuono (New York: Frank Cass, 1991).

Harris, J.P., *Men, Ideas and Tanks: British Military Thought and Armoured Forces, 1903–1939* (Manchester: Manchester University Press, 1995).

Horne, Alistair, *To Lose a Battle: France 1940* (New York: Penguin, 2007).

Mitcham, Samuel W. Jr., *The Panzer Legions: A Guide to the German Army Tank Divisions of World War II* (Mechanicsburg, PA: Stackpole Books, 2000).

Newton, Steven H., *Kursk: The German View* (Cambridge, MA: De Capo Press, 2002).

Pignato, Nicola, *Italian Armored Vehicles of World War Two* (Carrollton, Texas: Squadron/Signal Publications, 2004).

Rottman, Gordon L. & Takizawa, Akira, *World War II Japanese Tank Tactics* (Oxford: Osprey Publishing, 2008).

Stone, David R., 'Tukhachevsky in Leningrad: Military Politics and Exile, 1928–31,' *Europe-Asia Studies*, vol. 48, no. 8 (December 1996), pp. 1365–86.

Swinton, Ernest D., *Eyewitness: Being Personal Reminiscences of Certain Phases of the Great War, Including the Genesis of the Tank* (Garden City, NY: Doubleday, Doran & Co., 1933).

Tucker-Jones, Anthony, *Stalin's Armour 1941–1945: Soviet Tanks at War* (Philadelphia: Pen & Sword, 2021).

Zaloga, Steven J., *M3 & M5 Stuart Light Tank 1940–1945* (Oxford: Osprey Publishing, 1999).

Zaloga, Steven J., *Early US Armor: Tanks 1916–40* (Oxford: Osprey, 2017).

26

RACE AND CONFLICT IN THE ASIA–PACIFIC REGION

1941–1945

Mark Johnston

Introduction

Racism was one of the defining factors of war in the Asia–Pacific Region in 1941–1945. Its exact role is a matter of explicit and implicit dispute, but racism contributed to the causes, the course and the outcomes of the conflict. The pioneering work in this field was John Dower's 1986 book, *War Without Mercy: Race and Power in the Pacific War*. Race was a consideration from the highest levels of politics and strategy, all the way down to the level of hand-to-hand combat, although the role of racism in the war's innumerable battlefield atrocities is especially contentious.

Racism and the Causes of the Conflict: 1941–1945

Overt racism existed in the pre-war cultures of all protagonists in the Asia–Pacific War. Assumptions about racial superiority were significant factors in the foreign policy of those protagonists, notably Japan on one side and the United States and the British Commonwealth on the other. These imperialist powers ruled over territory in which millions of inhabitants aspired for independence, but the rulers justified their empires largely on assumptions about white supremacy or, in Japan's case, being the 'leading race.'

A racist Western perception of a 'Yellow Peril' had focused on Japan since the Russo–Japanese War, and expressed itself in American and Australian legislation limiting or prohibiting Asian immigration and in the refusal of the Allied powers—especially Australia and the United States—to accept Japan's request for a statement supporting racial equality at the 1919 Paris Peace Conference. These expressions of racism stung the Japanese, as did the limits on the Japanese Navy set by the Washington Naval Conference and negative Western responses to Japan's invasion of China. Those responses eventually included embargoes on oil and other vital war-making resources; and when subsequently the Japanese decided to go to war with the Anglo-Americans in 1941, they justified this largely in terms of having been strangled by hypocritical Western imperialists. Japan had a point about hypocrisy in that, rather than perceiving the Japanese as following in their own imperialist footsteps, Western powers contrasted Japanese militarism, repression and irrationality to their own supposed democratic, peaceful and logical world views (Dower 1986: 29).

DOI: 10.4324/9780429437915-33

Japan itself had been following a race-inspired foreign policy in the 1930s. It had used overtly racist conceptions to justify its own imperialism since its occupation of Korea in 1910. Its decision to invade China, first by fully occupying Manchuria in 1931 and then invading China itself in 1937, had been based on an ultimately disastrous underestimate of the task involved. This underestimate, which evoked considerable sympathy for China in the West, owed much to a racist underestimate of China's capacity to resist. When Japan decided to go to war with the United States in 1941, its intelligence work on probable American psychological and economic responses to any attack was ludicrously optimistic. Its complacency was based on racial and cultural stereotyping that assumed the British would lose to the Germans while the American response would founder because of isolationism, domestic issues of race, labour and political factions, as well as capitalistic greed (Dower 1986: 259–60, 285). In turn the Western allies underestimated the Japanese ability to wage war, and this also contributed to the outbreak of war in 1941. There was no great urgency about preventing that war, largely because of race-based contempt for Japan's military abilities. And when the war began, the Allies were similarly overconfident about their chances of victory.

Racism and the Course of the Conflict: 1941–1945

The circumstances of the war's opening, with surprise attacks at Pearl Harbor and other places across a vast arc in the Pacific and Southeast Asia, created not only great shock but also intense anger among those attacked. Westerners, leaders and led, turned to racial stereotypes to condemn their 'treacherous' opponents, who were believed to have made a massive miscalculation of American strength. In 1941, Prime Minister John Curtin justified Australia's entry into the war against Japan in terms of the nation's commitment to maintaining the 'principle of a White Australia' (Johnston 2000: 85). In turn, a Japanese hope that in the wake of Pearl Harbor the Americans would not have the stomach for a long fight proved the inadequacy of their own race-based understanding of their now fully awakened foe. Racial stereotyping led the Allies to underestimate the technology and fighting qualities of their enemies. They believed that Japanese technology was inferior, an illusion well illustrated by the case of the Mitsubishi Zero aircraft, which would dominate the air war in the war's early stages.

Intelligence about Japanese air forces was scandalously inadequate on the eve of war in December 1941, so that training in Malaya was affected by lack of knowledge of Japanese organisation and aircraft. Type 0 or Zero fighters had been in use in China since July 1940, where they had virtually destroyed Chinese air power. The leading American airman in China, Claire Chennault, had warned of their capacity, and in May 1941 a Zero shot down by anti-aircraft guns in China had become available for inspection. Details of the Zero's performance were forwarded to the British command in Malaya in September 1941, but were not distributed before war broke out (Johnston 2011: 48). In October, Air Vice-Marshal Pulford, commanding the RAF in Malaya, said the Zero was on a par with the Buffalo fighter that was equipping Commonwealth squadrons and which would in fact prove completely inadequate in coming battles (Shores, Cull and Izawa 1992: 40).

Not only was their equipment considered second-rate, but the Japanese were widely regarded as physiologically incapable of being able aviators or marksmen (Worrall 2016). So established was this prejudice that when Japanese aircraft attacked the Philippines and Hong Kong in December 1941, Westerners initially assumed that they must be piloted by Germans or mercenaries (Dower 1986: 105). During early fighting near Mersing Bridge in Malaya in

January 1942, an Australian private, having his wounded shoulder dressed, was heard to say: 'Let me get back there—those Japs couldn't hit a bag of shit at a hundred yards' (Johnston 2000: 110). The classic stereotype of a Japanese soldier was well expressed by an Australian officer, who told a doctor of his first clash with the Japanese, on the Kokoda Trail: 'I couldn't believe it when I saw those big bastards bearing down on us. I thought they must have been Germans in disguise; I always pictured Japs as little toothy buggers with glasses' (Steward 1983: 107). These race-based prejudices had disastrous consequences for the Allies in the early stages of the Asia–Pacific War. The Japanese conquest of a vast swathe of territory in Asia and the Pacific, a despised enemy overcoming a powerful but complacent empire, constituted a 'world turned upside down,' and was without precedent since perhaps Alexander the Great's conquest of the Persian empire (Bayly and Harper 2005: xxix).

The early Japanese victories added to the Australian servicemen's race-based antipathy because of fear of Japanese invasion of their homeland. In January 1942, an Australian soldier serving in the Middle East wrote to his fiancée of his concern about the 'yellow horde,' saying 'my thoughts are full of smashing them, before they reach what they desire' (Johnston 2000: 84). An Australian private wrote from Papua in December 1942 that the 'vindictive hatred' Australians felt for the Japanese came largely because of fear of their 'policy of conquest brought so near to Australia' (Johnston 2000: 84). Early in 1943, General Thomas Blamey, the commander of Australian forces in the Pacific, tried to stir up hatred of the Japanese in veterans of the recent campaign by emphasising that the Australians were fighting to prevent the deaths of their families and the end of civilisation (Johnston 1943: 228). Similarly, US Admiral 'Bull' Halsey was notorious for his virulent racism, including the comment that the Japanese race was the product of monkeys mating with Chinese criminals (Dower 1986: 85).

Their initial successes won the Japanese much admiration in Asia for the strength and initiative they showed against westerners and their offer of a 'Greater East Asian Co-Prosperity Sphere.' But, as soon became apparent, the latter concept was a racist one, based on a hierarchy of races, with the Japanese—the 'Yamato race'—at the top. Thus, although in 1943 a committee of prominent Japanese civilians defined the war as 'the counteroffensive of the Oriental races against Occidental aggression,' Japanese dominance of local politics and economies and heavy-handed 'Japanisation' soon alienated many of those Asians who had initially welcomed them. The early Japanese victories led some Japanese leaders into complacency, as expressed in the phenomenon of 'victory disease.' An-oft repeated designation of themselves as the '*shido minzoku*,' or 'leading race', had the same consequence (Dower 1986: 59, 203).

Under the influence of such attitudes, Japanese officials toyed in 1942 and 1943 with visions of a greater empire, stretching as far as Australia and New Zealand. It was a vision with no place for consultation with other races, each of which must respect its 'proper place,' permanently below Japan (Dower 1986: 261, 264). Ba Maw, the wartime premier of Burma, initially welcomed the Japanese but later complained of their 'racial impositions,' which included arrogant behaviour such as demanding that civilians bow to them, and slapping or abusing any who showed any perceived impoliteness, not to mention ruthless brutality towards any supposedly serious offence (Dower 1986: 7, 286). One incident that typified the Yamato race outlook occurred near Gona, in Papua, in August 1942, when an English priest trying to surrender to invading Japanese forces was told by a Japanese officer: 'I have no time for whites' (Mayo 1977: 36).

Unlike Western racism in this conflict, Japanese racism tended to concentrate less on denigrating the enemy, though this was far from absent, and more on elevating themselves and emphasising their spiritual purity and superiority. The Japanese consciously sought to avoid talking about the conflict as a race war, because this was incompatible with claims to be holding the higher moral ground and with their alliance with Nazi Germany. In practice, the Japanese dehumanised their enemy in propaganda, typically calling them 'demons' or 'devils' and, while preaching magnanimity, treating them with ferocious violence (Dower 1986: 205, 216, 236, 261).

While the Japanese were misinterpreting the racial import and durability of their early victories, the Allied powers' distorted and race-based underestimation of their opponents' abilities gave way in the light of those victories to another distorted stereotype: that of the Japanese 'super soldier.' Overcoming this view took a year of intense fighting and Allied victories at places like Milne Bay, Guadalcanal and Midway. The commander of the 7th Australian Infantry Brigade at Milne Bay reported after the historic Allied victory there that destroying the enemy was 'a most effective way of demonstrating the superiority of the white race' (Johnston 2000: 86).

Whether it was in the time of the 'super soldier' or afterwards, Allied soldiers tended to see Japanese as animals rather than humans. For example, General Blamey called Japanese 'sub-humans' and a cross between 'the human being and the ape' (Johnston 1943: 207). Australian officers who had campaigned in New Guinea told soldiers preparing for the 1943 campaign that the Japanese was 'merely an educated animal' (Johnston 2000: 87). American troops expressed similar ideas, with metaphors of monkeys, vermin and insects most common, and many otherwise compassionate Western soldiers maintained attitudes towards the Japanese which today appear archaic and not far distant from Nazi terminology. There were some contrary voices in the West. When in Australia the government launched an intense hate campaign in March 1942, a prominent newspaper argued that Australians needed no stimulus to fight the Japanese aggressor, and certainly not 'a torrent of cheap abuse and futile efforts in emulation of Goebbels' (Charlton 1988: 34). The campaign was opposed by 54 per cent of Australians surveyed in a Gallup Poll on the issue (Dennis and Grey 2008: 324). A minority of Americans spoke up about the morality of killing helpless Japanese civilians from the air, even if the general opinion was that the Japanese enemy was reaping what he had sowed (Dower 1986: 41). Although leaders spoke in racist ways, not all of their constituents necessarily agreed: for example, General Blamey was not regarded as an oracle of wisdom by most of his troops, and Admiral Halsey, who made similar comments about the Japanese, was more popular but regarded as extreme in his views.

Atrocities

The Japanese entered the Asia–Pacific War in 1941 with a well-earned reputation for atrocities in China since 1937, most notably through its bombing and massacres of civilians, but also in publicised tales like those of Japanese officers competing to behead the most Chinese (Dower 1986: 42). The Imperial Japanese Army (IJA) continued that record everywhere it fought. Max Hastings observes that while Japanese racism now eclipsed British racism in Southeast Asia, 'Tokyo's new regime was characterised by a brutality such as the evicted imperialists, whatever their shortcomings, had never displayed' (Hastings 2011: 216). Especially outrageous to all military norms was the Japanese maltreatment of prisoners, who after the initial campaigns ended in early 1942, virtually never survived capture. Those captured in the early campaigns were treated in ways that violated the Geneva Convention, although Japan was not a signatory and thus felt under no obligation to observe its provisions. Japanese wartime excesses were practised on civilians throughout their possessions from 1941, including Singapore in 1942, where many Chinese civilians were massacred, and in Manila in 1945, where as many as 100,000 Filipino civilians were brutally murdered (Lewis and Steele 2001: 207). Some 60,000 Korean forced labourers died in wartime Japan, and 90,000 to 150,000 Burmese, Indian, Chinese and Malay civilians perished working on the Thai–Burma Railway (Dower 1986: 47). An estimated 200,000 women in Southeast Asia were forced into the role of 'comfort women,' and typically were raped 30 to 40 times daily. Only 30 per cent are estimated to have survived the war (Lewis and Steele 2001: 131–2). Japan held the initiative in the early stages of the Pacific War, and consequently set the tone for the fighting that followed (Linderman 1997: 354).

The reasons for this brutality as it applied to Allied troops derived largely from the Japanese military code which forbade surrender, but probably also owed something to a sense of superiority over other races and grievances for perceived racial slights. A perceptive Australian journalist wrote after the Papuan Campaign in 1943 that the fighting had been brutal largely because 'racial antipathy goes deeper' than in the war with the Germans. But he added that, 'Once create, as the Japanese have done, the fixed attitude that it is dishonourable to fall a prisoner of war, and barbarity becomes normality.' Fear and mercy were incompatible, he continued, and where armies did not share a code, there would be no chivalry (Tebbutt 1943: 321).

While Japanese behaviour seemed barbarous to the Allies, the Japanese in turn believed that Allies behaved barbarously, with prime examples being Americans souveniring body parts from dead Japanese (though this was forbidden by US military law), sinking hospital ships, shooting defenceless Japanese sailors in the water and airmen in parachutes, and bombing Japanese civilians. They also attributed these barbarities to Western racism (Dower 1986: 61–2, 243).

A major question is the degree to which race contributed to the Allies' response to the Japanese ferocity that they encountered from the opening of the Pacific War. Allied soldiers hated the Japanese more than they did their German, Italian and Vichy French opponents. This was partly derived from an inability to understand them. Most Allied soldiers found the Japanese language incomprehensible, but were also bewildered by their opponent's apparent willingness to die. An Australian official historian identified an oft-observed trait when nominating a feature of the Gona-Buna and Sanananda fighting as a 'fixity of purpose of the Japanese for most of whom death could be the only ending' (McCarthy 1959: 508). The Allies' estimate of the Japanese as 'animals' and 'vermin' made it easier to justify killing them—or 'hunting' and 'exterminating' them—and to quell moral qualms about doing so. Similarly, the Japanese interpretation of Allied soldiers as 'demons' or 'devils' promoted exterminationist language (Dower 1986: 255). But did the Allies' 'brutal beast' trope owe more to prejudice or to observation? Those who have sorted through the testimonies of Allied combat soldiers have identified the primary cause of their ruthlessness as not a sense of racial superiority but hatred based on fear and anger. Brutal acts were committed more often by Japanese than by any other enemy the Allies faced. For this reason, soldiers at the front had greater antipathy to Japanese than did troops further back. Allied unwillingness to take prisoners, for example, was partly a practical decision based on hard experience—sick or wounded Japanese often opened fire on approaching troops rather than surrender (McCarthy 1959: 442, 523). After an action at Sanananda, New Guinea, an Australian battalion commander reported: 'As in many other cases enemy wounded engaged our troops and had to be shot. This may give rise in future to Jap propaganda but they are doing it so consistently that our troops cannot take any chances' (McCarthy 1959: 521).

Those putting the interpretation that Allied hatred and ruthlessness were based more on experience than prejudice have been labelled 'apologists' for Allied atrocities, but the term might be more properly employed for those applying moral equivalence to a situation where far more atrocities were committed by one side, and from the outset (Cameron 2005: 558). Moreover, it is a conclusion that emerges from study of the writings of the soldiers of different Allied powers (Johnston 2000; Linderman 1997).

The same applies not only to soldiers but also to airmen and seamen. Stories of the execution of fliers—particularly those of participants in the Doolittle Raid in 1942 and that of Bill Newton VC in New Guinea in 1942—horrified civilian readers in the West, but the Japanese tendency to execute all airmen that they captured made for an especially vehement hatred among Allied aircrew (Johnston 2011: 435). Within six hours of the Pearl Harbour attack, the US Navy Department sent the order, 'Execute unrestricted air and submarine warfare against Japan.' Barney Sieglaff, who commanded the US submarine *Tautog*, gave a response said to be typical: 'After the carnage at Pearl Harbor—a sneak attack—who could have moral qualms about killing Japanese?' (Blair 1975: 106).

There was some chivalry at sea. In a most unusual incident, the Royal Australian Navy gave a full military funeral to the Japanese crew of a midget submarine that raided Sydney Harbour in May 1942. Similarly, a Japanese submarine commander is credited with permitting a lifeboat full of Australian aircrew to go unharmed after surfacing near them off Java in early 1942 (Johnston 2011: 127). On the other hand, hospital ships on both sides were attacked. By May 1943, the Japanese had asserted that eight of their hospital ships had been attacked by Allied warships or aircraft (Earhart 2008: 364). An Imperial General Headquarters communique of 20 February 1942 stated that, by contrast, its aircraft had in the recent raid on Darwin refrained from attacking a hospital ship in the harbour. In fact, the hospital ship, the Australian *Manunda*, had suffered a direct hit in the bombing, losing 12 killed and many more wounded (Gill 1968: 260). By contrast, during the fighting at Milne Bay, Papua, in September 1942, Japanese warships roaming in the harbour had on successive nights illuminated *Manunda* under their searchlights, but had each time refrained from opening fire (Gill 1968: 173). The Allies were outraged by the Japanese torpedoing of the Australian hospital ship *Centaur* off the Queensland coast in May 1943, in which 268 Australians perished (Gill 1968: 259). The American and British Chiefs of Staff agreed, however, that this was probably 'the act of an irresponsible commander' (Gill 1968: 260). Japanese claims that this commander, Hajime Nakagawa, did not know *Centaur* was a hospital ship ring hollow, especially as he repeatedly sought to annihilate survivors of his subsequent sinkings (Sturma 2011: 54). More credible was the Japanese response to Australian protests about the *Centaur* with counterclaims about American bombing attacks on the clearly marked Japanese hospital ship *Ural Maru*, off Guadalcanal in April 1943 (Earhart 2008: 364). Sieglaff claimed that as the Japanese used every naval and merchant ship they had for the war effort, the logic of 'total war' made every ship a legitimate target (Blair 1975: 106). 'Mush' Morton, a famous US submariner, notoriously described killing hundreds or even thousands of Japanese servicemen floating helpless in the water after his submarine, *Wahoo*, had sunk the Japanese transport *Buyo Maru* in January 1943. His motivation has been explained in part by his anger at stories of the bombing of the *Manunda* and of Japanese aircraft strafing the survivors in the sea (Sturma 2011: 46–7). Allied aircraft strafed Japanese floating in the sea in the aftermath of the sinking of

Japanese transports and warships in the Battle of the Bismarck Sea in March 1943. The motivation was primarily military rather than racial—based on a determination to prevent the Japanese from getting ashore to fight another day. Many Allied crews in this battle were also itching for revenge, some for previous campaigns but many for the Japanese machine-gunning of the crew of a B-17 bomber in their parachutes early in the battle (Gamble 2013: 310). The killing of unarmed survivors in the Bismarck Sea sickened some of the perpetrators (Johnston 2011: 278). Similarly, although Morton's post-battle killings were never condemned or openly endorsed by his superiors, they appalled other submariners, and few American captains followed his example (Blair 1975: 386).

Race undoubtedly *was* a factor in Allied brutality against the Japanese. Extracting gold teeth or other body parts from Japanese was far more prevalent than doing so from Germans—though that was not unknown (Johnston 2000: 82). The tiny minority who displayed a willingness to mutilate Japanese corpses probably acted at least partly from racist contempt. Nevertheless, racist prejudices, an inability to understand the suicidal attitudes of Japanese, and even the threat to one's homeland, did not goad Allied soldiers in the same way as personal experience, or personal expectation based on reports from other frontline soldiers. An astute Australian regimental historian says that not propaganda stories, but the physical evidence of Japanese atrocities was crucial in making veteran Australian troops hate the Japanese in a way they had not hated Italians and Germans (Haywood 1959: 153). After the first few months of the Pacific War, few prisoners were taken on either side. This was not Allied policy, especially as it soon became apparent how willing Japanese prisoners were to give away useful intelligence, but on both sides there grew a sense on the frontline that one must kill or be killed (Dower 1986: 10).

As the war continued, Japanese underestimation of their opponents continued to cost them dearly. They did not change their military codes because they believed Westerners would be unable to break them. For similar reasons they were lax about leaving behind documents on abandoned battlefields. They thought their willingness to fight to the death—apparent on battlefields from Buna and Tarawa to Okinawa—would so dishearten the Allies as to make them willing to accept a negotiated peace. This represented a failure to understand the enemy's desire for revenge, and exemplified their misunderstanding of their opponents (Dower 1986: 281). No doubt, atrocities reinforced the racist bigotry among frontline soldiers and even heightened it. Even when war turned against Japan and their losses were huge, there was still evidence that enraged Allied soldiers: the sight of dead comrades found to have been tortured or mutilated, and, in New Guinea evidence of cannibalism, frightened and infuriated Allied soldiers, as did the emaciated condition of any POWs who were rescued.

The peculiar circumstances in which Anglo-American frontline soldiers served gave them some reasons not to intensify but to moderate their racism. In combat, an exaggerated sense of martial superiority could easily explode in the faces of complacent individuals. When operating on the battlefield, lives depended on realism about the enemy's strengths and weaknesses. Training staff wanted their soldiers to be level-headed about Japanese strengths and weaknesses (Isby 1991: vi).

Yet while the life-and-death realities that Allied soldiers faced when confronting the Japanese tempered racism towards them in action, these realities also largely determined the character and intensity of their hatred for the Japanese. The racist language of the pre-war era was a convenient means for soldiers to express feelings that owed most to the unique circumstances of the frontline. As the vicious fighting at Sanananda drew to a close in January 1943, Trooper Ben Love put common thoughts into written words:

> What a peculiar manner these fanatical Jap soldiers display in their utter disregard of lives—their own as well as others. They say all Jap positions are now smashed, it is just a matter of mopping up. This mopping up costs lives against these mad-men.
>
> *(Johnston 2000: 89)*

The special character of Allied soldiers' hatred of the Japanese derived from the reality of the fighting more than the prejudices of civilian life.

Bombing

Another debate about race and the course of the war relates to strategic bombing. At least 320,000 Japanese civilians were killed by American bombing of Japan, and especially by incendiary bombing. Racism may have been a factor, but incendiary bombing had first been used not against Japan but against Germany. It was also practised on Japanese-held cities in China, thus killing many Chinese civilians. This suggests that few qualms were held about killing residents of any territory, provided that doing so could shorten the war. Mark Selden argued that the 'slaughter' the bombing of Japanese cities represented was attributable to a variety of reasons: 'technological breakthroughs, American nationalism, and the erosion of moral and political scruples about killing of civilians, *perhaps intensified by the racism that crystallized in the Pacific theatre*' (Selden 2009: 87; my italics). In short, racism was a factor, but not a prime motivator.

The dropping of the atomic bombs in August 1945, one of the most contentious issues in all of military history, had a racial dimension. The question has been asked whether the bomb would have been used on a German city had it been available in time. Regardless of whether race was a factor, there was also much hard-headed military and political thinking involved. One of the main issues was how many Allied servicemen would die at Japanese hands—either in an invasion of the homeland or as prisoners of war—if the war was not completed as quickly as possible.

The Outcomes of the Conflict: 1941–1945

Japanese claims to be fighting a war to liberate Asia proved to be a façade for their own hegemonic plans. Some Japanese soldiers and civilians, especially at the middle rank, did genuinely work to befriend conquered Asian peoples. For example, Colonel Suzuki Kenji, sometimes dubbed a 'Japanese Lawrence of Arabia,' and the Japanese Southern Agency played a key role in the genesis of the Burmese Independence Army, later Burma Defence Army (Dower 1986: 285). He was recalled to Japan for being too pro-independence. Similarly, Major Fujiwara Iwaichi, head of the 'F Agency,' saw himself as a T.E. Lawrence figure in facilitating the creation of an Indian National Army in 1942 (Bayly and Harper 2005: 8, 122). But every such instance of support was outweighed by many violations. Millions died under Japanese rule, many because of starvation resulting from Japan's indifference to the suffering of other Asians consequent on the diversion of agricultural resources to its own people (Hastings 2011: 416). H.P. Willmott claims that one of the reasons for Japan's defeat was her 'inability to recognize the force of any form of Asian nationalism other than her own; an inability to offer the people of her newly acquired territory anything other than a position of subservience and dependence' (Willmott 1999: 64). Whereas in 1941–1942 the Japanese were welcomed by many of the nationalist movements in countries they conquered, by 1945 resistance groups—some of which were broad enough to deserve the name 'popular movements'—were fighting the Japanese in most of those territories. For example, the Kachins, a hill tribe from Burma, had initially welcomed the Japanese as 'a race of super-Europeans,' but the depredations of their patrols—stealing livestock, recruiting forced labour and turning Baptist meeting halls into military brothels—led to a resistance movement that by late 1943 had killed more than 1,000 Japanese (Bayly and Harper 2005: 674–5).

Nevertheless, the Japanese, through their daring early successes, did puncture white supremacist myths and damage irreparably the white colonial administrations (Willmott 1999: 64). Their rhetoric about liberation inspired Asians wanting independence from colonial powers. After the war, the genie of nationalism proved impossible to return to the bottle in former British, Dutch, French and American colonies that had been conquered by Japan. For example, within five years the Philippines, Burma and Indonesia were independent states, and within two decades all the former colonial territories except Hong Kong had been lost, despite Western fights to hold them (Lewis and Steele 2001: 137). Once it was turned upside down, the world could not be fully returned to its original state. The Indian National Army, though it made little contribution to the war, contributed to Indian nationalism and independence in the aftermath. As one authoritative work puts it: 'war, the Bengal famine and the Indian National Army had made independence inevitable' (Bayly and Harper 2005: 734). In this sense, one role of race in the Asia–Pacific in 1941–1945 was to contribute indirectly to decolonisation.

Even though Japanese invasion never materialised in Australia, the threat of it improved the lot of the continent's indigenous people. Until war with Japan became a real possibility in 1941, Aboriginals were prevented from enlisting in large numbers, for Australian law demanded that Australian recruits be 'substantially of European origin or descent.' Successive governments referred to the eligibility of 'the better type of half-caste,' but the threat of Japanese invasion led to the formation of units such as the Torres Strait Light Infantry Battalion and the indigenous Northern Territory Special Reconnaissance Unit, in both of which the other ranks comprised indigenous Australians. Although these soldiers were not treated equally with white soldiers, they contributed to a recognition of Aboriginal civil rights in subsequent decades (Johnston 2006: 25). Similarly, the work of black troops of the US Engineer Corps on the Ledo Road from Assam to Burma was said to have improved race relations within the US armed forces (Bayly and Harper 2005: 547).

British racism was a factor in relations with their own peoples throughout the war. Prejudices against 'mixed race' people in India and Burma persisted, despite the fact that Anglo-Indians and Anglo-Burmese were unusually brave fighters and loyal workers (Bayly and Harper 2005: 341–2, 576). During the process by which the Indian Army expanded to become the largest volunteer army in history, the British had reluctantly to abandon officially a reliance on the racist concept of so-called 'martial races' (Bayly and Harper 2005: 702).

Conclusion

While considerations of race were connected to most aspects of the war in the Asia–Pacific region from 1941 to 1945, it was far from the only or main factor in its causes, course or outcomes. Issues of national interest played a major role in its causes, while political theories

Mark Johnston

such as nationalism and communism were huge factors driving governments, armies and popular movements. Many nations that fought or grew up as a result of the war made room for more than one ethnic group and focused on patriotism rather than racism. The war was won not by prejudice and anger but primarily by hard-headed thinking about strategy and about improvements in technology and tactics. The irrationalities of race, destructive though they were, probably led to fewer deaths than the cold logic of 'total war,' which all the major combatant nations saw as justifying the slaughter of civilians.

Bibliography

Bayly, Christopher, and Harper, Tim, *Forgotten Armies: Britain's Asian Empire and the War with Japan* (Harmondsworth: Penguin, 2005) (ebook version).
Blair Jr, Clay, *Silent Victory: The U.S. Submarine War against Japan* (Toronto: Bantam Books, 1975).
Cameron, Craig M., 'Race and Identity: The Culture of Combat in the Pacific War,' *International History Review*, vol. 27, no. 3 (2005), pp. 550–66.
Charlton, Peter, *War Against Japan 1941–1942* (Sydney: Time-Life Books, 1988).
Dennis, Peter, Grey, Jeffrey, Morris, Ewen and Prior, Robin, *The Oxford Companion to Australian Military History*, 2nd ed. (South Melbourne: Oxford University Press, 2008).
Dower, John W., *War Without Mercy: Race & Power in the Pacific War* (New York: Pantheon Books, 1986).
Earhart, David C., *Certain Victory: Images of World War II in the Japanese Media* (New York: Routledge, 2008).
Gamble, Bruce, *Fortress Rabaul: The Battle for the Southwest Pacific, January 1942–1943* (Minneapolis: Zenith Press, 2013).
Gill, G. Hermon, *Royal Australian Navy 1942–1945* (Canberra: Australian War Memorial, 1968).
Hastings, Max, *All Hell Let Loose: The World at War 1939–1945* (London: Harper Press, 2011).
Haywood, E.V., *Six Years in Support: Official History of the 2/1st Australian Field Regiment* (Sydney: Angus and Robertson, 1959).
Isby, David, Introduction to U.S. War Department, *Handbook on Japanese Military Forces* (London: Greenhill Books, 1991).
Johnston, George, *The Toughest Fighting in the World* (New York: Duell, Sloan and Pearce, 1943).
Johnston, Mark, *Fighting the Enemy: Australian Soldiers and their Adversaries in World War II* (Melbourne: Cambridge University Press, 2000).
Johnston, Mark, *Australia's Home Defence 1939–1945* (Canberra, A.C.T.: Department of Veterans' Affairs, 2006).
Johnston, Mark, *Whispering Death: Australian Airmen in the Pacific War* (Sydney: Allen & Unwin, 2011).
Lewis, Jonathan, and Steele, Ben, *Hell in the Pacific* (London: Channel 4 Books, 2001).
Linderman, Gerald F., *The World Within War: America's Combat Experience in World War II* (New York: The Free Press, 1997) (ebook version).
Mayo, Lida, *Bloody Buna* (London: New English Library, 1977).
McCarthy, Dudley, *South-West Pacific Area—First Year: Kokoda to Wau* (Canberra: Australian War Memorial, 1959).
Selden, Mark, 'A Forgotten Holocaust: U.S. Bombing Strategy, the Destruction of Japanese Cities, and the American Way of War from the Pacific War to Iraq,' in Yuka Tanaka and Marilyn, B. Young (eds.), *Bombing Civilians: A Twentieth-Century History* (New York: The New Press, 2009), pp. 77–96.
Shores, Christopher, Cull, Brian and Izawa, Yasuho, *Bloody Shambles: The Drift to War to the Fall of Singapore*, vol. 1 (London: Grub Street, 1992).
Steward, H.D., *Recollections of a Regimental Medical Officer* (Melbourne: Melbourne University Press, 1983).
Sturma, Michael, *Surface and Destroy: The Submarine Gun War in the Pacific* (Lexington: University Press of Kentucky, 2011).

Tebbutt, Geoffrey, Untitled memoir, 1943, Australian War Memorial MSS 1391.

Willmott, H.P., *The Second World War in the Far East* (London: Cassell, 1999).

Worrall, Simon, 'How Racism, Arrogance, and Incompetence led to Pearl Harbor,' *National Geographic* (2016), https://news.nationalgeographic.com/2016/12/countdown-pearl-harbor-attack-twomey-anniversary/.

27

MODERN NAVAL WARFARE

1815–2000

Andrew Lambert

Introduction

While naval warfare has occupied a prominent place in the history of conflict and diplomacy, it has tended to be waged between states with strikingly different political, economic and strategic views of the sea in national strategy. Seapower states tended to share inclusive political structures, which enable trade, capital and commerce access to the levers of power. That access was used to promote naval power and identity. They fought for sea control and economic advantage, while continental powers sought security and additional profitable territory. Continental navies were primarily used to limit the ability of sea powers to exploit the ocean. Since, as maritime strategist Julian Corbett observed in 1911: 'men live upon the land and not upon the sea,' it has always been easier to secure a strategic decision ashore than at sea (Corbett 1911: 14). However skilfully handled, naval power has exerted enormous influence on conflicts as varied as the Napoleonic War and the Falklands War of 1982.

Technology and Transformation of Naval Power

At the heart of the wars of the French Revolution and Empire was an asymmetric contest between the terrestrial ambitions and total war military methods of a radicalised French nation in arms and the limited maritime strategy of a British seapower state anxious to preserve the balance of power in Europe, and sustain its oceanic empire of commerce. Contemporaries saw this elemental clash of cultures as a replay of the Punic Wars (Lambert 2018: xiii–xv). In April 1814, British artist J.W. Turner responded to Napoleon's abdication with *The Rise of the Carthaginian Empire*, using the ancient maritime empire to mock the Emperor's assumption that the English, the new Carthaginians, would be defeated by his neo-Roman imperium. Turner's picture was the highlight of the 1815 Royal Academy exhibition; it told the British why they had triumphed, and where they should focus their energies going forward.

While making peace with France and the United States in 1814, Britain insisted on the right to stop and search neutral merchant shipping on the high seas, the legal basis of economic blockade. Without that right, secured through command of the sea, naval warfare would be futile, and Britain effectively disarmed. Sea power strategy secured Britain and its

DOI: 10.4324/9780429437915-34

expanding global trade, while representing British power. Britain had more ships and men than any combination of rivals, and a chain of bases stretching from the Thames to Australia. In 1815, Britain reinforced those bases with Malta, the Ionian Islands, Heligoland, Cape Town and Sri Lanka, while dispatching the ex-Emperor to St. Helena, an island way-station on the route to India.

British dominion of the oceans reflected a dynamic commercial sector, the influence of the City of London in national politics, and the need to expand trade and rebuild the economy after 22 years of war. Naval warfare would open new markets, protect existing trading relations, and ensure other actors, state or otherwise, recognised British authority. The fleet and dockyards were rebuilt, primarily as a deterrent (Lambert 1991: 1–12). The fleets of France, Russia and the United States in order of magnitude, if not quality, could not threaten Britain's command of the sea, although the first two decided how many battleships were needed.

Between 1815 and 1850 there were no conflicts between major naval powers. Naval power supported economic and territorial expansion. In 1816, Britain forced Algiers to abandon the corsairing and enslaving of Europeans, the fleet bombarded Algiers, and burnt the corsair fleet, using techniques pioneered in the previous 15 years, including cannon, mortars and Congreve rockets. The French conducted a similar operation at Veracruz in 1837. British attacks at Acre in 1840, and Sweaborg outside Helsinki in 1855 highlighted the ability of navies to exert influence on land.

The last sailing ship fleet battle, Navarino on 20 October 1827 was one-sided: British, French and Russian ships annihilated the Ottoman and Egyptian squadrons. Navarino secured Greek Independence, and brought an end to Aegean piracy, largely conducted by unpaid Greek 'privateers' (Pitcairn-Jones 1934). Britain also used sea power to suppress the Atlantic Slave Trade, a humanitarian measure that prompted other powers to fear Britannia was about to assume absolute dominion of the seas (Lambert 2006).

By 1840, steam-powered warships were changing the balance of power between land and sea. Steam ships had transformed strategic communications by 1830; increasing the speed and regularity of communication enabled London to control use of naval power at a distance, a process that accelerated after 1850 with the British submarine cable telegraph network. Admirals steadily lost their diplomatic and strategic autonomy (Kennedy 1971: 197–214).

By the mid 1840s the introduction of the screw propeller extended naval power into coastal waters. The first propeller warship, the sloop HMS *Rattler*, proved the concept, and in 1852 Britain decided that all future warships would have screw propulsion. This decision coincided with the emergence of the Second French Empire, the completion of the first full-powered steam battleship, *le Napoleon*, launching a technology-based naval arms race. In 1854, Britain and France declared war on Russia to defend the Ottoman Empire. Naval dominance ensured the war was fought in Russia, using Britain's strategy of amphibious power projection against major naval bases. Sevastopol Dockyard was captured and destroyed. Maritime logistics outperformed Russian land-based responses. The economic blockade quickly bankrupted Russia, and when Britain threatened to bombard St. Petersburg, the Tsar made peace. In 1856 the British paraded over 300 specialist ships created to attack Cronstadt and St. Petersburg before a host of foreign dignitaries, as a pre-emptive deterrent (Lambert 2011).

By 1858 the Anglo-French arms race had resumed. If Britain lost command of the sea it would lose its Great Power status and be open to invasion. Britain did not have the military power to conquer France. Predictably, Britain, richer and better equipped with shipyards and steam engineering facilities, won the race. This show of strength emphasised the cohesion of the seapower state.

The lessons of the Crimean War were rapidly digested in America, where the Official *Report on the Art of War in Europe*, published in 1859, providing details of Anglo–French ironclads and Russian sea mines (Delafield 1859). This information drove the rapid development of coastal and riverine warfare, which shaped an American Civil War (1861–1865), dominated by the strategic exploitation of water communications. Northern strategist General Winfield Scott's 'Anaconda Plan' envisaged cutting the Confederacy in half, by seizing the Mississippi, and imposing a 'British' economic blockade. The Confederacy, with few ships, sailors or shipyards, relied on their armies, which enabled the North to assemble overwhelming naval forces, moved by steam, and increasingly armoured, to overwhelm or bypass Confederate shore defences from Port Royal and New Orleans to Mobile and Wilmington. In 1862, Union General George McClellan attempted to replay the Sevastopol campaign, using amphibious lift to outflank the defences of Richmond with a superior army (Reed 1978). His campaign also witnessed the first, inconclusive, battle between two ironclads, CSS *Virginia* and USS *Monitor*, utterly unseaworthy coast assault craft. Britain and France were already operating first-class, ironclad capital ships.

Union dominance of the seas and rivers crippled the Confederacy. The Mississippi was secured in 1863, while the blockade cut vital imports and exports, although it remained far from watertight. In response the Confederacy created coastal ironclads and minefields: mines sank more Union warships than any other weapon. Naval forces and shore batteries protected Charleston for almost the entire war, despite frequent Union naval attacks. Charleston also witnessed the first sinking of a warship by a submarine: the CSS *Hunley*, an old boiler cylinder powered by hand cranks, armed with a mine on the end of a pole. The *Hunley* sank before it returned to base. The Confederates also used steam-powered commerce destroyers, sinking captured American merchant ships in a grim harbinger of twentieth-century U-Boat warfare. Captain Raphael Semmes of the CSS *Alabama* exploited his oceanographic expertise to loiter at key trade intersections, and the sailing ships came to him. His operations challenged existing British approaches to trade defence by convoy, which the Union Government exploited with a group of *Wampanoag*-type, high-speed, wooden-hulled steam cruisers, trying to deter British intervention in the conflict. Britain responded with larger, faster iron-hulled cruisers. The hastily built American vessels soon fell apart. In 1865 the newly re-United States rapidly demobilised its vast extemporised fleet of converted merchant vessels and coastal warships; it was too busy in the South and the West to worry about the sea. It would not resume serious naval activity until the late 1880s.

Production of iron, armour and heavy artillery fundamentally altered the naval strategic paradigm. Down to 1860 capital ships were wooden-hulled, three-masted vessels, latterly with screw propellers, armed with 90 to 130 small cannon. They were only effective in battles between roughly equal fleets. Minor naval powers needing coast defence relied on oared gunboats and coast fortresses. In the Crimean War, both were overwhelmed by a new coast assault system based on steam gunboats, mortar craft, rockets and armoured batteries. Iron allowed designers to build small, shallow draft ironclads with a few heavy guns to defend coastal waters, or attack forts, depending on the situation. The Crimean War batteries were followed by the USS *Monitor*, with two heavy guns a rotating turret. Although universally labelled 'coast defence ships,' a superior fleet like the Royal Navy was always going to use them offensively, against French, Russian or American naval assets.

In 1860 Russia, America and many smaller navies abandoned battleships building, and command of the sea, for integrated coast defences and oceanic raiders.

The ironclad fleet battle, between Austria and Italy off the Adriatic Island of Lissa in 1866, reprised old lessons. Superior leadership, skilled ship handling and superior morale crushed a larger, better equipped fleet lacking confidence and cohesion. The battle had no impact on a war that had already been settled at Sadowa in Central Europe. Only Britain and France built ironclad battlefleets, in another high-stakes Arms Race. Not only did Britain comprehensively out-build France, but it highlighted the message, naming an ironclad HMS *Northumberland* in memory of the ship that carried the Emperor Napoleon to his captivity on St. Helena. By 1865, France gave up the naval arms race to focus on the terrestrial challenge of Prussia. Britain responded by building battleships to attack forts, rather than fight other ships (Lambert 2012: 30). Naval power had no impact on the Franco–Prussian War 1870–1871. A superior French fleet blockaded the German North Sea and Baltic coasts, but offensive operations were hampered by extensive minefields. French amphibious troops were used to defend Paris, not attack Kiel. Defeated, stripped of provinces and saddled with a huge war indemnity, France lost its appetite for naval glory.

Naval warfare had never been significant in brief continental conflicts between heavily militarised states sharing a land border. Naval power worked when combatants were separated by a sea that could be commanded, and the war lasted long enough for economic pressure to weaken the continental state. The Crimean War was a good example. After 1871, Britain had no serious naval competitor. In 1878, British fleets blockaded a Russian attempt to seize Istanbul and the Dardanelles. The fleet on the spot was backed by a 'Baltic Fleet' at Spithead, threatening to attack St. Petersburg. Although it had minefields, torpedoes and forts, memories of 1856 ensured Russia backed down. The lesson was repeated with most of the same ships in 1885, halting a Russian invasion of Afghanistan. Powerless at sea, Russia, like other continental powers, sought legal measures to deny Britain access to the Baltic, without success.

In 1882, Britain's Mediterranean Fleet bombarded Alexandria, driving an insurgent army out of the port city. The follow-on amphibious operation secured control of the country and the strategic Suez Canal that linked Britain with its Asian and Australasian Empire. Britain used naval dominance to sustain a global combination of formal and informal empire, the latter based on capital loans and trading links, guaranteed by naval presence (Cain and Hopkins 1993).

While most states were content to leave the oceans to the British, imperial rivals like France and Russia focused on attacking British merchant shipping as a form of sea-denial. As Anglo–French tensions rose over Egypt, France adopted a *Jeune Ecole* strategy using torpedo boats and steam cruisers (Roksund 2007). However, critical technologies were immature: torpedoes were militarily insignificant before 1900. Small torpedo boats were unable to operate in moderate seas, while the first practical submarines, John Philip Holland's American design, only entered service around 1900.

Command of the sea remained a question of large seaworthy vessels with big coal bunkers, contemporary steam machinery being relatively inefficient. Those ships needed bases for fuel, maintenance and repair; only Britain created a global network, linked by submarine telegraph cable, because it depended on sea control. The City of London was quick to remind British Governments of the need for adequate naval defence (Smith 1991).

In the late 1880s, a 40-year period of unprecedented technological innovation reached a relative plateau, as several major powers sought overseas colonies to secure markets, outlets for emigrants, strategic assets and above all status symbols. These empires necessitated larger navies, to connect metropole and periphery, while the ability to build battleships became a critical indicator of technological prowess and great power status. States that imported their warships revealed their weakness. While Britain understood that warships were built to be used, and might be lost, other powers saw them as expensive prestige symbols, diplomatic counters to be stacked up in negotiations, rather than hazarded in action.

In 1890, Captain Alfred Thayer Mahan of the United States Navy (USN) produced a textbook for the new United States' Naval War College at Newport, Rhode Island, explaining why America needed a large navy. *The Influence of Sea Power upon History, 1660–1783* demonstrated that battleship-based command of the sea was the key to global power, without noting that it was also a unique asset. Translated into many languages, including German, Mahan's book validated the shift to battlefleet navies. These fleets consumed an increasing share of national expenditure, and were intimately connected with national prestige. Their names proclaimed national identity, and their architecture was distinctive. French battleships embodied radical change and republican virtues; those of the British stability, order and success, re-using names made famous by Nelson.

Steel hulls, triple expansion steam machinery, increased boiler performance, alloy steel armour and breech loading artillery provided a technology plateau for 25 years. Standard battleships displaced 12,000–15,000 tons, with four 12-inch and up to fourteen 6-inch weapons, and a maximum speed of 17 or 18 knots. They were intended to fight in linear formation, maximising broadside fire, outside the range of torpedoes. Battles would be settled by cumulative damage. Cruisers, which scouted for the fleet, and either escorted or attacked merchant shipping, were faster, and used smaller guns. Torpedo boats were steadily neutralised by larger gun-armed torpedo-boat-destroyers, later destroyers, developed by the Royal Navy in the early 1890s, to deny the English Channel to French flotillas. Relative technological stability enabled navies to build homogenous fleets. In 1889, Britain's Naval Defence Act created an all-new navy and overhauled the dockyards. It was funded by negotiating a reduction in the interest rate on the National Debt. When other great powers carried on building Britain repeated the measure in 1893. By 1897, the Royal Navy had a fleet of over 30 modern battleships, and 60 cruisers, more than double the strength of France and Russia. By 1898, the rising global stock of large warships transformed Britain's absolute dominance in 1880 into a qualified superiority, but that year's Fashoda Crisis still ended with a French climb-down.

The Franco–Russian naval challenge was replaced by the dramatic rise of German world power ambition, and a suitable battlefleet was projected in the 1897 Navy Law shaped by Admiral Tirpitz, to neutralise Britain in the event of a major European war. This overt challenge to British sea-control threatened the existence of an empire that depended on global communications.

The naval warfare of this era was dominated by imperial conflicts outside Europe. In the Sino–Japanese War of 1894–1895, modernised, militarised Japan secured control of the Korean Peninsula by defeating the Chinese off the Yalu River, capturing two important Chinese naval bases and destroying the Chinese fleet. While both countries relied on imported warships, Japan had learnt how to use them, from a British naval mission. The war encouraged Russia, France and Germany to seize Chinese territory, opening the way for an imperial carve-up. In the 'Boxer Rebellion' of 1900 the great powers used their fleets to protect their nationals in Beijing, and restrain one another's territorial avarice. The Spanish–American War of 1898 extended America's drive across the Pacific from newly annexed Hawaii to the Philippines, while dismantling Spain's Caribbean Empire. The battles in Manila Bay and Santiago de Cuba were as one-sided as the Sino–Japanese conflict: at Santiago, American battleships sank Spanish cruisers; at Manila, American cruisers sank Spanish gunboats.

The key naval conflict of the era, the Russo–Japanese War of 1904–1905, was fought over Korea and Manchuria. Russia had secured Port Arthur and assembled a powerful Asian fleet. Japan, which had used the Chinese war indemnity to buy six battleships and six large armoured cruisers, opened the war with a surprise torpedo attack at Port Arthur, crippling the Russian battlefleet. Japanese armies quickly secured Korea, and advanced on Port Arthur and Central Manchuria. Despite the Trans-Siberian Railway, Russia lost the land campaigns, logistically outperformed by Japanese sea transport. Port Arthur was taken by the Japanese Army, and the fleet sunk or scuttled. In desperation, Russia sent an ill-matched assembly of new and obsolete ships from the Baltic to Asia, where it was intercepted by a refreshed Japanese fleet in the Straits of Tsushima and annihilated. Peace followed. Japanese success reflected professionalism, experience, excellent equipment and quality personnel. Mines proved devastatingly effective in coastal waters, but torpedoes were less reliable. Battleships fought at unprecedented ranges, while Japanese armoured cruisers joined the battlefleet. Their Russian opposite numbers attacked Japanese communications. The naval conflict was still being digested when the next major war broke out (Corbett 1994).

Dreadnoughts and Grand Strategy: 1906–1945

To counter Germany's expanded Navy Laws, Britain signed an Entente with France in 1904, and with Russia in 1907, which enabled the three powers to maintain the European balance by forcing Germany to focus on land frontiers. Furthermore, the Navy Laws had an Achilles heel. Tirpitz had assumed continuing technological stasis, producing standard battleships at stable prices, ships that could pass through the strategic Kiel Canal linking the Baltic and the North Sea. First Sea Lord Admiral Sir John Fisher (1904–1910) changed the rules, shifting Britain's modern battleships to face Germany in the Channel and North Sea. By 1907, those new ships included Fisher's revolutionary all-big-gun battleship, HMS *Dreadnought*. With ten 12-inch guns, and a speed of 21 knots, it provided a quantum leap in fighting power. Steam turbine machinery increased power and greatly reduced maintenance, while new gun mountings and sights facilitated higher rates of fire, leading to synchronised targeting using mechanical computers. More significantly, *Dreadnought* halted battleship construction in Germany, which took a year to design a smaller design without turbines. The new ship forced the reconstruction of the Kiel Canal, a vast project completed in August 1914, the date Fisher always expected the next war to begin.

Fisher also hybridised battleships and armoured cruisers into an all-big-gun design, later christened the battlecruiser, which sacrificed armour for tactical and strategic speed. It was designed to dominate ocean shipping lanes and reinforce the battlefleet. In response, Germany built fast battleships, sacrificing endurance and seaworthiness for armour, because it had no ocean routes to secure.

Having drawn Germany into an arms race, Fisher prompted a public campaign in 1908–1909 that produced nine Dreadnoughts in a single year. He also steadily increased the size, armament and cost of the ships. Without money, shipyards and political flexibility to respond Germany lost the race. After 1912, Anglo–German relations began to improve: naval armaments did not cause the First World War. Fisher also pushed the development of submarines, and by 1910 they were capable of oceanic voyages.

In August 1914, Britain had command of the sea, and by December all German warships and merchant vessels outside the Reich had been sunk or interned. Fisher's battlecruisers, directed by strategic wireless, annihilated the last German squadron at the Battle of the Falkland Islands in December. Command of the sea would prove decisive, if the Entente could stabilise the war on land. It enabled Britain and her allies to access global manpower, raw materials, food and industrial output, and deny them to the Central Powers. British economic warfare included food, once Germany had nationalised food stocks (Halpern 1994).

The economic war continued, despite complaints from the neutral United States, which wanted to sell to both sides. In response, Germany used submarines, U-boats, to attack all shipping in British waters. After a U-boat sank the passenger liner *Lusitania* with heavy loss of life, America threatened to break off diplomatic relations. Germany backed down. In a war of words and contested legalities, Germany consistently abandoned the moral high ground to the British, who understood that the loss of economic opportunity would never outweigh the callous sacrifice of neutral civilian lives.

In 1915, Britain attempted to defeat Turkey and secure links with Russia by passing a fleet through the Dardanelles. The campaign was badly organised. Naval and military attacks faltered before resolute Turkish defences: the army had to be evacuated. Elsewhere, naval power secured Egypt, supported the Arab revolt, helped conquer Iraq and carried aid to Russia via the Arctic. In 1916, a French fleet forced the pro-German King of Greece to abdicate.

However, the primary war effort remained continental. The British Government followed French policy, sacrificing maritime strategy to mass continental warfare. Recalled to office in 1914, Admiral Fisher resigned in May 1915 when the Cabinet rejected his sea-based alternative, unwilling to challenge French strategy.

This left economic warfare as the primary offensive element in British strategy, backed by the Grand Fleet of Dreadnoughts based in the Orkney Islands, which kept the smaller German High Seas Fleet in the North Sea. In the summer of 1916, Admiral Scheer took his fleet to sea to attack British merchant shipping. Forewarned by signals intelligence, Admiral Jellicoe took the larger Grand Fleet to intercept. They met off the Jutland Peninsula in the only Dreadnought fleet battle on 31 May. The German battlefleet suffered heavy damage, but escaped in the smoke and haze, leaving the British with command of the sea. That failure led Germany to resume unrestricted U-boat warfare in spring 1917, accepting it might bring America into the war. When the United States joined the war, in April 1917, Britain increased diplomatic pressure on other neutrals to stop trading with Germany.

Although U-boats sank many ships, countermeasures, including minefields, convoy, air patrols and intelligence-led shipping diversions, steadily reduced losses. U-boat casualties became unsustainable, approaching 50 per cent by mid-1918. Unable to win the war on land or sea, desperately short of food and facing military collapse, Germany sued for terms in November 1918. Sea power had restricted Germany to Europe, cut off vital resources and enabled the allies to operate globally, including bringing over an American Army.

By 1918 Britain had created aircraft carriers, fitted anti-aircraft guns to their warships and developed submarine-detecting sonar. However, the main wartime threat to British sea power came from the United States, which launched a massive naval construction programme in 1916 to restrain British economic warfare. President Woodrow Wilson's demand for absolute

'Freedom of the Seas' would abolish the right to stop and search neutral merchant shipping, thereby securing American profits, rendering sea power strategically irrelevant and reducing Britain to the rank of a medium power.

After joining the war, Wilson ordered more battleships, hoping to bully Britain, which owed large sums to American banks, into conceding to his demands. At the Versailles Peace Conference, British Prime Minister Lloyd George bluntly refused. Wilson surrendered 'Freedom of the Seas' and in return Britain supported his League of Nations project. Most Americans opposed Wilson's policy: they did not want a massive navy, a war with Britain or a League of Nations. In 1919, America was not a seapower and Americans saw no need for a massive fleet. The army has always been America's primary defence (Beiriger 2017).

Three years later, America, Britain and Japan forestalled a costly battleship arms race in the Pacific with the Washington Treaty, which halted battleship construction for a decade, reduced fleet sizes on a 5.5.3 ratio, with France and Italy awarded 1.75, and restricted specification of all warships. The Washington process, which lasted until 1936, artificially constrained ship design, with devastating asymmetric consequences. The first post-Washington battleships only completed in the summer of 1940! Contemporary land and air armaments remained unlimited, with aircraft and tanks making spectacular progress. In 1939, German land and air forces defeated Poland, Denmark, Norway, Holland, Belgium and France in quick succession.

Anglo–French strategy had envisaged a re-run of the First World War, static land fronts, using global resources, secured by command of the sea to conduct economic warfare and air bombardment. Instead, France was overrun, and the Royal Navy, hampered by inadequate numbers and obsolescent assets, was defeated by German air power off Norway. Despite those defeats, Britain retained command of the sea, securing the British Isles, and denying Germany, Italy and later Japan the ability to co-operate. Too weak to challenge British, and later Anglo-American, command of the sea, their naval strategies were essentially negative.

Command of the sea facilitated the rapid shift of British resources from the North Atlantic to the Indian Ocean, or the Eastern Mediterranean. Food, fuel, raw materials and military hardware continued to flow into Britain, along with increasing amounts of North American military power. The 1918 convoy system was resumed in 1939: once again British Intelligence broke into German naval signals, while the German conquest of Western Europe shifted millions of tons of merchant shipping into the British war effort. In the Battle of the Atlantic, which the Germans waged as a crude tonnage war, the Norwegian Merchant Marine alone provided Britain with as much tonnage as the Germans would sink in 18 months (Faulkner 2012; Mawdesley 2019).

The Atlantic War was an attritional struggle: U-boats, aircraft and occasional surface raiders tried to sink merchant shipping, while Allied ships and aircraft tried to prevent losses and sink U-boats. The vital kill ratio was merchant ship tons sunk for each U-boat lost. Across the war this balance steadily moved against the Germans. The standard Type VII U-boat, an improved 1918 design, had limited underwater speed and endurance. By the spring of 1941, U-boats had inflicted serious losses in the North Atlantic, but the convoy system survived. On the eve of Operation Barbarossa, Hitler gambled that a major surface operation would persuade Britain to make peace. The new battleship *Bismarck* would destroy entire Atlantic convoys, forcing the merchant ships to disperse, making easy targets for U-boats. This obvious gambit met an overwhelming response. For the only time in the entire war, Britain focused every available naval resource, intercepting *Bismarck* in the Denmark Strait, between Greenland and Iceland. In a brief action, the battlecruiser HMS *Hood* was destroyed.

Bismarck was damaged; later crippled by carrier air strikes in the mid-Atlantic, it was sunk by two British battleships. This proved to be the turning point of the war: no more German battleships entered the Atlantic, enabling Britain to focus on the submarine threat.

By 1942, the massive expansion of Allied shipbuilding, and critical inter-war developments, including radar to detect and target ships and aircraft, had begun to reverse the impact of the Washington Treaty process. By 1943, new sensors and improved weapons significantly improved kill rates against air and submerged targets. These greatly expanded fleets had restored the strategic balance between land and sea by the middle of the war, enabling a series of massive sea-based power projection operations that led directly to the defeat of Italy, Germany and Japan.

In the Mediterranean, British sea power restricted the flow of supplies to Italian and German forces in North Africa and sustained the key island base of Malta. In spring 1943, Anglo–American amphibious forces captured Morocco and Algeria from the Vichy regime, and when North Africa was overrun the Royal Navy forced 250,000 Axis troops in Tunisia to surrender, a haul to equal Stalingrad. Those losses gravely weakened Italy, enabling the invasions of Sicily and Southern Italy that led Italy to surrender. By this stage, the Battle of the Atlantic had ended; the U-boats were withdrawn in May 1943, after morale-shattering losses approaching 75 per cent. That success facilitated Operation Bolero, shifting North American resources to Britain for the invasion of Western Europe.

By June 1944, the Soviet Union had defeated the *Wehrmacht* in the East, with critical strategic and political support provided by British Arctic Convoys, which shipped tanks, planes trucks, munitions and much more to Archangel and Murmansk, despite the best efforts of German air, sea and submarine forces based in Northern Norway. The political value of this support at the crisis points of the German invasion of the Soviet Union was immense.

In the Pacific, Japan's surprise air attack on Pearl Harbour demonstrated tactical prowess and strategic folly, starting a war Japan could not win. The 'underhand' method mobilised hitherto ambivalent Americans. While Japan quickly seized an Asia–Pacific empire, its primary focus remained the conquest of China, where millions of Japanese troops were deployed. The naval war was a secondary concern. On 4 June 1942, signals intelligence enabled three American carriers to ambush and sink four of Japan's six fleet carriers as they attacked the Midway Island. To exploit this victory, the Americans landed on Japanese-occupied Guadalcanal in the Solomon Islands, unwittingly starting an attritional campaign that eviscerated the Japanese naval air arm and exposed the inadequacy of its industrial base. By mid-1943, America had created an overwhelming aero–naval juggernaut that advanced from island to island, until it secured bases for land-based strategic bombing of the Japanese home islands. Meanwhile a ruthless submarine campaign annihilated the inadequately protected Japanese merchant fleet. The American invasion of the Philippines in the summer of 1944 led to a major battle around the Leyte Gulf: the Japanese outwitted the more numerous Americans but were ultimately defeated with devastating losses. In the spring of 1945, a one-way sortie by the battleship *Yamato*, the ancient name of Japan, to Okinawa was ended by carrier air attacks. The 68,000-ton warship, the largest yet built, capsized and exploded. American battleships spent most of the war escorting carriers and bombarding invasion beaches.

Navies in the Cold War Era

In August 1945, the three largest navies in the world belonged to the United States, Great Britain and Canada. Soviet naval options had been sacrificed to the critical land war.

The French Fleet was in ruin, as were those of other occupied European nations. Japan and Germany had no warships, while the Italian Navy was reduced to a cadre. This prompted the United States, a unique global hegemon, possessing the world's most powerful forces in all three environments, and a nuclear weapons monopoly, to demobilise the USN, create an independent air force, and unify defence around continental concepts. With bankrupt Britain losing the empire that had justified centuries of naval dominance, it seemed there was no threat at sea. Until the Cold War began the primary issues were continental and nuclear: NATO focused on Western Europe, not the Atlantic. The Korean War (1950–1954) reminded a nuclear-obsessed American Administration that naval and amphibious forces were the best tools to limited conflicts. American naval aviation revived, with massive supercarriers to operate jet aircraft. Submarines acquired nuclear propulsion, and then nuclear-tipped ballistic missiles, providing a credible second-strike to back land-based systems. The USN had a large role in the Vietnam conflict, from near continuous carrier air strikes to mining Haiphong Harbour, which forced North Vietnam to negotiate. During the Cuban Missile Crisis, the USN prevented Soviet missiles reaching the communist island.

The Soviet Union began rebuilding the navy in the late 1940s, to defend the maritime frontiers of an enlarged empire against Western carrier and ambitious forces. The surface fleet would operate under land-based air cover, with submarines deployed further forward. Surface-to-surface missiles promised improved kill ratios, with warheads large enough to cripple a carrier. The Polaris SLBMs obliged the Soviet Fleet to go further out to sea, with onboard air defence, to hunt missile submarines on an arc of 1,500 miles from Moscow and Leningrad. Longer-ranged Poseidon and Trident missiles ended this mission, leaving the Soviets to re-imagine their fleet as a regular naval force, although it was deployed to protect Soviet Ballistic Missile Submarine patrols. There were no plans or assets for a new Battle of the Atlantic: the Red Fleet remained as defensive as that of the Tsars.

The primary Soviet threat, massed air and missile attacks, focused American attention on anti-air and anti-missile systems. By the mid-1960s the USN could shoot down jet fighters up to 50 miles away. In the early 1980s, the micro-processor revolution enabled a multi-target AEGIS system that could address numerous targets at long range. By 2000, it could shoot down ballistic missiles. Unable to match this capability, and the astronomical cost of cutting-edge military technology, the Soviet Union collapsed. The point was emphasised in 1986 when the USN promulgated a 'New Maritime Strategy' deploying submarine, carrier and surface forces to attack the Soviet Ballistic missile submarines, confident it could manage the air and missile threat. This threat, not the visionary 'Star Wars' space-based missile defence system, was the cornerstone of Reagan's Second Cold War.

Naval warfare remained an option for many states, old and new, across the second half of the twentieth century. Although the Anglo–French Suez-Operation of 1956 was a diplomatic catastrophe, it was also a successful demonstration of amphibious power projection. Small-scale naval operations featured in all the Arab–Israeli Wars, while India and Pakistan made effective use of carriers and submarines respectively in 1971. In 1982, the Argentinian seizure of the Falkland Islands prompted a rapid sea-based response from Britain across thousands of miles of Atlantic Ocean. When the British nuclear attack submarine HMS *Conqueror* sank an Argentine warship, the remaining naval assets were withdrawn, leaving the defence of the islands to mainland-based aircraft. Two small British carriers operating VSTOL/STOVOL Sea Harrier fighters, backed by surface forces, defeated Argentine missile and bomb attacks, at some cost, and landed the Royal Marines and follow-on British Army units to recover the islands, with naval air and gunfire support. The distinctly maritime nature of this war, and the use of sophisticated missiles on both sides, made it the most significant naval event of the era.

In the Tanker War of the 1980s, the Soviet Union and the West co-operated to maintain the flow of oil from the Persian Gulf during the Iran–Iraq conflict. Regional instability has kept large American and British naval forces in position ever since. Naval activity in the First Gulf War of 1990–1991 has been dismissed as merely facilitating success in the air and on the ground. In reality the pre-war embargo, imposed by naval forces, seriously degraded Iraq. All the military hardware arrived by sea, even if the personnel flew in. Anglo–American naval forces in the Persian Gulf pushed up to the Kuwaiti Coast, threatening a major amphibious landing to distract Iraq from the land offensive; they also dealt with minefields and missile batteries. American Navy Tomahawk cruise missiles destroyed much of the Iraqi air defence system, after it had been jammed by US Navy Electronic warfare aircraft, enabling seven super carriers to conduct a high proportion of the overall air war effort, exploiting their mobility to prosecute new targets and reduce the need for air-to-air refuelling (Lambert 1991). In 2003, the naval mission was easier, because Iraq had not rebuilt its defences.

Future Trends

Carrier aviation remains the pinnacle of naval power, combining flexibility and reach. The United States continues to build super carriers, while the Royal Navy has acquired two new carriers, the largest ships it has ever operated, with fifth-generation fighter aircraft. Russia's only carrier, badly damaged in a dockyard accident, may never got to sea again, while China's carrier programme may have more to do with status that strategy.

Although many navies are shrinking, the world economy remains globalised, and the right to use the high seas for innocent passage is still under threat from pirates, economic parasite, terrorists, rogue states and the creeping continentalism of China and the Russian Federation, which seek to assert ocean dominion far beyond current international norms. Japan and Australia are expanding their fleets to meet the age-old challenge of continental exclusion. Authoritarian regimes still fear the 'corrupting sea.'

Bibliography

Beiriger, E.E., 'Building a Navy "Second to None": The U.S. Naval Act of 1916, American Attitudes Toward Great Britain, and the First World War,' *British Journal of Military History*, vol. 3, no. 3 (2017), pp. 4–29.

Cain, P. and Hopkins, A., *British Imperialism: Innovation and Enterprise: 1688–1914* (London: Longman, 1993).

Corbett, J.S., *Some Principles of Maritime Strategy* (London: Longman, 1911).

Corbett, J.S., *Maritime Operations in the Russo-Japanese War: 1904–1905* (Annapolis: USNIP, 1994).

Delafield, R (ed.), *The Art of War in Europe* (Washington: Government Printing Office, 1859).

Faulkner, M., *War at Sea: A Naval Atlas, 1939–1945* (Barnsley: Seaforth, 2012).

Halpern, P. *A Naval History of World War I* (Annapolis: Naval Institute Press, 1994) remains the standard account of the naval war.

Kennedy, P.M., 'Imperial Cable Communications and Strategy: 1870–1914,' *English Historical Review*, vol. LXXXVI (1971), pp. 728–52.

Lambert A.D., *The Last Sailing Battlefleet: Maintaining Naval Mastery 1815–1850* (London: Conway 1991).

Lambert, A.D., 'The Naval War: 1990–91,' in P. Grindal (ed.), *Opposing the Slavers: The Royal Navy's Campaign against the Atlantic Slave Trade* (London: IB Tauris, 2006), pp. 125–46.

Lambert, A.D., *The Crimean War: British Grand Strategy against Russia 1853–1856*, 2nd ed. (Farnham: Ashgate, 2011).

Lambert, A.D., *HMS Warrior 1860: Victoria's Ironclad Deterrent* (London: Conway, 2012).

Lambert, A.D., *Seapower States: Maritime Culture, Continental Empires and the Conflict that made the Modern World* (London: Yale University Press, 2018).

Mahan, A.T., *The Influence of Sea Power upon History: 1660–1783* (London: Sampson Low, 1890).

Mawdesley, E., *The War for the Seas: A Maritime History of World War II* (London: Yale University Press, 2019).

Pitcairn-Jones, C.G (ed.), *Piracy in the Levant 1827–8: Selected from the Papers of Admiral Sir Edward Codrington* (London: Navy Records Society, 1934).

Reed, R., *Combined Operations in the Civil War* (Annapolis: Naval Institute Press, 1978).

Roksund, A., *The Jeune Ecole: The Strategy of the Weak* (Leiden: Brill, 2007).

Smith, S.R.B., 'Public Opinion, the Navy and the City of London: The Drive for British Naval Expansion in the Late Nineteenth Century,' *War & Society*, vol. 9 (1991), pp. 29–50.

28

STRATEGIC AIR POWER IN THE TWO WORLD WARS

Mary Kathryn Barbier

Introduction

The term 'strategic air power' evokes images of skies filled with large aircraft releasing bombs over populated cities or open landscapes and creating fear and widespread death and destruction. The cost on the ground is huge, but few consider the cost in the sky. While historians write in general terms about the impact of anti-aircraft fire on aircraft, few discuss the psychological toll that sorties had on pilots and air crews. In 1949, 20th Century Fox Studios released *Twelve O'Clock High*, an iconic film starring Gregory Peck as the lead protagonist— General Frank Savage. Tasked with whipping a 'hard luck' bomber group into an efficient, disciplined unit, Savage struggled to get the men to continue to fly even though attacks by German fighters exacted a heavy cost on their daylight bombing raids. Director Henry King did an excellent job of highlighting the impact continuous bombing missions had on the pilots and crews, who were expected to function like unemotional, well-oiled machines. He demonstrated the flaw in that expectation in the film. Strong, unflappable Savage, who drove himself as hard as, if not harder than, the rest of the unit, suffered a breakdown and had to be relieved of his command temporarily. Although Savage eventually recovered, King focused public attention on the non-physical cost of air power.

When historians discuss strategic air power, it is easy to focus on outcomes, particularly when bombing campaigns supported operations on the ground or facilitated victory for one side. It is easy to ignore the human cost to civilian populations, to soldiers on the ground and sailors on the sea, and to pilots and air crews in the skies. Some scholars, however, are attempting to complete the story by examining the psychological toll or discussing the impact of air power on civilian populations. A good example of this new focus is Stephen Alan Bourque's *Beyond the Beach: The Allied War Against France* (2018). A big part of the story, however, is the evolution of strategic air power in the first half of the twentieth century. Although the theory was not put into practice until the Second World War, the origin of strategic air power is rooted in the First World War, if not earlier, as a result of the flying experiments of French aviator Henri Farman and the Wright brothers.

In *Air Power in the Age of Total War*, John Buckley suggested that an air power revolution occurred in the period 1908–1909. According to Buckley, 1908 was a banner year for

DOI: 10.4324/9780429437915-35

Wilbur Wright, who was in France when he remained airborne for three hours and travelled a record 80 miles. As Farman and the Wright brothers continued to make progress, military leadership in multiple countries took notice and tried to purchase aircrafts. The US Army acquired its first airplane in 1909 (Buckley 1999: 29–31). Within a few short years, however, enough advancements had been made for aircraft to play a role during the First World War.

The First World War

During the First World War, aviation developed in two ways. First was the vision of what air power could be—blowing up the enemy on his home ground. While there are a few examples of this occurring during the war—the Royal Naval Air Services bombing of Zeppelin bases in Cologne and Düsseldorf (22 September 1914 and 8 October 1914, respectively), the 4 December 1914 bombing of Freiburg im Breisgau by French air forces, and the 21 December 1914 German bombing of Kent, England—the vision lacked the technology to make it a reality. Consequently, capabilities resulted in frontline operational aircraft. In other words, air forces took an operational approach to their mission.

While many assume that the first hand-held bombs were dropped from an airplane during the First World War, it happened earlier. On 1 November 1911, Lieutenant Giulio Gavotti, who had been sent to Libya on a strategic reconnaissance mission without precise orders, 'decided to be the first person in the world to throw bombs from his airplane.' Luckily, he had a box of bombs in his plane. Using one hand to steer, Gavotti grabbed a bomb with the other. While flying, he inserted the detonator into the bomb. Arriving over his target, he held the bomb in one hand, 'pulled the trigger with his teeth,' and tossed the bomb over the side (Hippler 2013: 1). Although Gavotti tossed three more bombs out of his aircraft, he could not determine the effectiveness of his actions, nor could he anticipate the development of the strategic air power that would be unleashed during the Second World War. His actions did not immediately change the vision for air power in the summer of 1914, when militaries considered aircraft primarily in terms of a tool of reconnaissance. According to Walter Boyne, air power's initial impact was shaped and reflected in the prewar allocation of funds to air services of the original belligerents. In 1914, Germany and France earmarked roughly $26 million each, Russia $12 million, Britain $9 million and Austria–Hungary $318,000. Expenditures correlated to the availability of aircraft, pilots and maintenance crews, but despite the discrepancies in funding, Boyne argued that in August 1914, 'the air armies of the contending sides were approximately equal in size and capability' (Boyne 2003: 48–52).

By 1918, the use of air power had begun to evolve. Boyne noted that while aircraft were not decisive during the war, they still exerted influence, particularly in carrying out what became and remained their most important mission of intelligence gathering. He argued, for example, that the actionable intelligence, provided by British and French pilots conducting reconnaissance flights, prevented a German success during the Battle of the Marne and thereby contributed to the failure of the German Schlieffen plan (Boyne 2003: 55–63). In the Spring of 1915, British commanders used maps created from aerial photographs to plan and execute their first trench-warfare attack in the Battle of Neuve Chapelle.

The First World War's air war frequently evokes images of biplanes engaged in dogfights. Stories about Baron Manfred Albrecht Freiherr von Richthofen—the Red Baron—romanticised pilots in biplanes engaged in duels in the skies above the battlefield. Conjuring images of nineteenth-century cavalry charges, Prime Minister David Lloyd George called these daring pilots in their biplanes 'cavalry of the clouds' (Buckley 1999: 42). By 1917, however, focus was shifting from the individual ace, who captured public imagination, to large formations in which cooperation, rather than 'singular heroism' was the norm. This was the result of the increased size of air forces and exponential improvement in aircraft technology. Air forces were, consequently, able to adopt and execute a spectrum of new missions. Campaigns for air superiority, 'ground-attack operations' and 'strategic bombing offensive' did not just materialise during the latter conflict; they were tried during the former one (Buckley 1999: 42–4). And in tossing bombs out of their aircraft, pilots introduced new threats to enemy forces and resources on the ground.

The major players of the First World War did not all devise air strategies in the same ways or with the same end results. The air strategy dictated, or was influenced by, the aircraft—fighter planes or bombers. Both France and Germany devoted resources to the acquisition of reconnaissance aircraft. Accurate, timely intelligence allowed increased support for ground operations. Britain, on the other hand, focused on the strength of its empire—its navy and defence; therefore, the country grew its 'maritime air power' (Buckley 1999: 44; Muller 1998: 144–90). Although air forces carried out reconnaissance and tested strategic bombing, their greatest successes came from air support. There were tangible results, unlike strategic bombing that was in its infancy. In 'Close Air Support: The German, British, and American Experiences, 1918–1941,' Richard Muller provided an excellent case study, in which he argued that, by the end of the war, close air support had a significant impact on ground campaigns as well as battles (Muller 1998: 146). Credit for perfecting close air support falls to the German air service that focused on mastering an operational level of warfare, coupled with realistic training and a productive tradition of inter-arm cooperation.

Early in the war, the newly created German air service added a goal-oriented aircraft to its arsenal—observation and 'infantry contact' aircraft. The primary purpose was to provide actionable intelligence to ground units, but 'infantry aircraft' could occasionally engage the enemy on the ground. The Germans introduced the all-metal Junkers J 1 'furniture van.' Intended solely for 'close support missions,' the Junkers could withstand the 'small-arms' ground fire to which fabric-covered air frames were so vulnerable. As their air service fought against opposing air forces for air superiority, the Germans were persuaded to invest as well in a 'specialised escort aircraft' to provide protection for the 'infantry contact' units. These 'rugged two-seaters' were fast and carried 'powerful armament.' By the fall of 1917, Ernst von Höppner, the air service commander, ordered the formation of the escort craft into 'specialised ground attack units' that consisted of Hablerstadt CL II and IV attack planes. Muller argued that the way these planes were utilised was 'revolutionary.' The Germans' technological advances in aircraft led to significant 'doctrinal changes' (Muller 1998: 146–9; Boyne 2003: 47–74).

As was the case with the Germans, British integration of air and ground included 'artillery and infantry observation and contact aircraft.' Although they had some aircraft tasked with 'ground strafing,' the British would use observation and pursuit craft for 'ground support' when necessary. It was not until the summer of 1917, however, that leadership authorised aircraft fitted with bombs, 'carried in makeshift racks,' for 'direct attack' against German forces. Sir Hugh Trenchard initially wrote: 'these low-flying attacks that we had to make . . . were a wretched and dangerous business, and also pretty useless' (Muller 1998: 150). Despite Trenchard's pointed criticism, the RFC continued to launch these attacks and by mid-1918 had significantly improved its ability to 'deliver effective close air support to the army.' Although close support did not usually fall into the 'strategic' category, commanders on both sides had figured out how to use the existing airplanes in a strategic way (Gray 2015: 60–1).

Not content with aircraft that had limited task flexibility, the British developed other air-craft and expanded the ways in which they could take the war to the Germans. In addition to the Sopwith TF 2, 'a specialized trench strafer with downward-firing guns' came too late to see combat. Other aircraft widely used for close air support, the Sopwith Camel and the 'Bristol F2B two-seat fighter,' were improvised afterthoughts. While these aircraft increased the RFC's success rate, most of their battlefield achievements were the result of new meth-ods of utilising the existing planes. A few months before the war ended, their 'close sup-port work' climaxed. Experts evaluating 'close air support' concluded that its influence was 'beyond dispute.' The success of 'close air support' fostered the conclusion that air power in general would be a factor in conflicts in the future. What was unclear, however, was what that role would be (Muller 1998: 150–2). In order to figure this out, Trenchard and his col-leagues would debate intensively after 1918 what British air forces would look like and how they would be used in the wars of the (as it turned out, not-too-distant) future.

By the time the United States entered the fray, both its allies and Germany had developed air power operational doctrines. The Americans turned to the French for a 'training and practice' model, but they did not reach the performance level that the British attained by the end of the war. The experience that American air crews had by war's end was obviously significantly less than that of British pilots. That did not mean that the few aerial attacks on ground troops had insignificant effect. As might be expected, however, the Americans lagged behind both their British and German counterparts in some areas, such as communication between air and ground forces, but they made efforts to get up to speed, by conducting air–ground training exercises and developing methods for 'supporting infiltrating troops' and for dividing enemy infantry from their supporting tanks (Muller 1998: 153–5).

'Billy' Mitchell identified four special missions assigned to pursuit aviation over and above its main goal, which was to achieve 'air superiority.' Included in these missions were: 'close escort for observation planes, cooperation with day bombardment squadrons, attacks on en-emy observation balloons, and "attacks on ground troops"' (Muller 1998: 154). In addition, Mitchell advocated for the creation of a 'specialized branch of aviation' that could be tasked with solving 'particular operational problems.' Forward thinkers like Mitchell gave the US air forces an advantage moving forward, even if his superiors did not embrace all of his ideas. Furthermore, the country's 'growing aviation industry' coupled with the opportunity to learn lessons from 'the wartime experiences of all major belligerents' positioned the United States to take the lead in shaping its air forces to include 'a true close support component' (Muller 1998: 154–5).

Interwar Period

However, the specific incentives for technological research and development that enhanced close air support did not continue in the interwar period; neither did lessons learned dur-ing the war (Muller 1998: 146). Militaries having access to aircraft was one thing. Figuring out how to use them effectively was another thing altogether. During the interwar period, military officials had to figure out how to best utilise the new technologies that emerged during the previous conflict. For example, machine guns mounted in front of the pilot and synchronised with the movement of the propeller had been a game changer because they facilitated aerial battles between planes. But would dogfights in the next war look the same or would planes fly in large groups and take the war to the enemy? Could planes best be used to support ground offensives or defend homelands? Pilots could now drop increasingly more powerful bombs from aircraft—but to what ends? Air power theorists had to determine answers to three basic questions. What do we do? How do we do it? And why?

Historian John Buckley argued that the 'roots of strategic bombing theory' can be found in the pre-First World War period, when debates about aircraft being 'used in an entirely separate, or strategic, manner' started. In the postwar years, three men addressed the subject and devised air power theories, which they advocated to their superiors and colleagues with varying degrees of success. Billy Mitchell and Sir Hugh Trenchard were joined by Giulio Douhet because he 'developed a coherent doctrine of strategic air power,' as it became understood. Douhet was 'the first true air power philosopher' (Boyne 2003: 136; Hippler 2013: 27).

Born in Italy in May 1869, Douhet had acquired an interest in aviation by the early twentieth century. He thought that Italy should take the lead in adapting aircraft for military objectives. By 1910, Douhet, through his extensive publications on aeronautics, advocated for the construction of a fleet of bombers and of a 'powerful air fleet.' Doing so would allow Italy to defend itself against enemy air attacks. His superiors were less than pleased with his publication of 'Rules for the Use of Aircraft in War,' but Douhet continued to promote his ideas (Hippler 2013: 29–31; Buckley 1999: 37–8; Boyne 2003: 136; Millett 1998: 354).

During the First World War, Douhet, although assigned to an army staff position, continued to publish on air power. Over the next several years, his views on air power evolved. In some respects, he adapted Alfred Thayer Mahan's views on sea power to air power (Hippler 2013: 32–3, 40–1). He argued that having command of the air would not merely guarantee flexibility; it would allow an air force 'to be able to cut an enemy's army and navy off from their bases of operation and nullify their chances of winning the war' (Boyne 2003: 136–7; Hippler 2013: 40–1).

In 1921, Douhet published his seminal work, *Command of the Air*, in which he articulated his theory of strategic air power. Reiterating the importance of 'command of the air,' he argued that this was to be achieved by eliminating the enemy's ability to resist, not in air battles, and that air power would be the deciding factor in determining the outcome of the next war. By targeting 'communications, government, industry, transportation, and the will of the people,' air power would be victorious over an enemy. In addition to advocating the use of chemical weapons against an enemy population, Douhet supported the manufacture of a 'battle-plane' that could fly over long distances, deliver substantial payloads and thwart attacks by enemy fighter planes. His efforts initially led to Italy's creation of a small bomber force of long-range Caproni bombers (Boyne 2003: 94, 139; Hippler 2013: 145–8; Buckley 1999: 75–6).

Billy Mitchell read Douhet's works, was influenced by them, and like Douhet, Mitchell found himself at odds with his superiors and the President for neglecting US aerial forces (Boyne 2003: 139). In a 6 December 1919 statement to the *New York Times*, Mitchell claimed that, because of that neglect, the United States would be unable to meet any 'first class power in the air.' Firm in his convictions, he convinced Congress and the Navy to conduct tests to determine the validity of his theories, including the assertion that aircraft could drop bombs on ships at sea. Thwarting such an invasion was a central aspect of US strategic planning and funding. In 1921, the US Army Air Corps conducted a series of air strikes on the *Ostfriesland*, an ex-German battleship. During the final strike, a squadron commanded by Captain Lawson dropped 2,000-pound bombs, four of which hit in rapid succession, close to the *Ostfriesland*. The fourth bomb dealt the final blow, and the battleship sank.

Although successful, the exercise drew criticism because the German vessel was stationary during the airstrike (Jones 2010; Mitchell 1925: 71–4; Buckley 1999: 91). Mitchell, however, considered it definitive proof that aircraft had revolutionised warfare and added a new dimension to the old model of war conducted only on land and at sea. Air power was a game changer and demanded a revision of the existing rules of engagement and principles of strategy (Mitchell 1925: 6).

By November 1925, Mitchell's clashes with the Navy and War Departments reached a climax. Frustrated by the crash of a USN airship two months earlier, Mitchell released a statement in which he suggested that 'armchair admirals' did not 'care about air safety.' He was court-martialed. Convicted of insubordination, Mitchell resigned his commission, which ironically gave him the freedom to say what he wanted about the need for a strong independent US air force (Boyne 2003: 143).

While he was familiar with Douhet's theory, Mitchell agreed more with Hugh Trenchard's theories on air power and air warfare. Born in February 1873, Trenchard enrolled in a flying course in 1912, when flying was considered a pastime for the young and reckless. But when the Royal Aero Club presented him with his pilot's licence, Trenchard immediately received a new posting—to the newly established RFC—and a staff position at the Central Flying School in Wiltshire. Unlike many of his peers, he was quick to ascertain the 'unlimited military potential' of airplanes (Orange 2011; Buckley 1999: 35). With the outbreak of the First World War, Trenchard was tasked with organising the increasing squadron numbers in France. Within a few months, he was an operational commander on the front. In August 1915, he assumed command of the RFC in France.

Trenchard had definite ideas about how the RFC should be utilised. From the beginning, he pushed for improvements in aircraft designs and increased production of aircraft, which he believed should be equipped with certain technology—wireless sets, cameras and a devise to improve accuracy in hitting targets. He advocated a 'relentless and incessant' offensive strategy even if it meant a high casualty risk. Over the next few years, he also developed a theory called 'panacea bombing.' Rather than targeting frontline soldiers, bombers should attack transport networks and supply facilities. He believed that attacks on industrial centres would have a negative effect on civilian morale. In April 1918, British officials merged competing flying groups into the Royal Air Force (RAF). In the last months of the war, Trenchard oversaw bombing raids—both the nature of and the targets chosen—that influenced his thinking about offensive strategy and 'panacea bombing,' theories that he continued to develop post-war (Hastings 1979: 38; Boyne 2003: 129–30; Orange 2011; Black 2016: 28; Biddle 2002: 128).

As the war neared conclusion, British officials, like their counterparts in other countries, debated the proper use of bombers. Focusing on long-range bombers, some wondered if resources were better allocated elsewhere. Air power advocates fell into two camps: those who favoured air power for ground support and those who promoted bombing cities and enemy industry. Although he would later be 'an arch-proponent of strategic bombing,' in 1918 Trenchard supported the former position. The German bombing of London, however, persuaded him to consider, then embrace, a more independent and more aggressive approach (Buckley 1999: 61).

During the interwar period, Trenchard, Douhet and Mitchell solidified their theories that resulted from the idea of using bombers to deliver a war ending blow to the enemy. These three great air power theorists incorporated a defining aspect of this as they discussed what air forces in their respective countries would look like and how they could—or perhaps should—be utilised in time of war (Buckley 1999: 72). James Corum argued that in the 1920s and 1930s the RAF and the US Army Air Corps (USAAC) developed a 'coherent theory and doctrine of strategic' bombing as a result of First World War experiences and that the modern US Air Force (USAF) built on the foundation crafted during the interwar period (Corum 1997: 4).

It was an uphill battle to get countries to dedicate funds for the creation of air forces, particularly after the start of the Great Depression. Postwar backlash and disillusionment made governments reluctant to fund any military force (Fussell 1975). Rivalries with the established services—army and navy—made it hard for air forces to convince lawmakers to allocate sufficient funds, but some progress was made. In many cases, however, the air forces were 'subordinate to' the dominant services—the army and the navy. This was certainly the case in Italy. Lack of confidence in the Italian Army, which was predicted to fail on the battlefield, and the navy, which favoured 'convoy duties' over battles against opposing navies, gave *faute de mieux* credibility to Douhet's principles. Nevertheless, financial constraints, and the failure of powerful advocates to support it, resulted in the Italian air force never reaching the size or capability as envisioned by Douhet (Boyne 2003: 139–40; Buckley 1999: 81).

Similarly, on the other side of the world, competition between the Imperial Japanese Army (IJA) and the Imperial Japanese Navy (IJN) for financial and operational support prevented the Japanese from developing 'a unified air power theory.' Japan did, however, have talented tacticians who gave their country an advantage—even if it did not last—in 1941 (Boyne 2003: 165–8). Rather than focus on large bomber squadrons, the Japanese copied the USN's model. This was partially the result of the creation of an independent naval headquarters in Tokyo. The IJN developed the Mitsubishi G3M Type 96 Attack Bomber, called by its opponents in the Second World War the Nell. It was a land-based torpedo and high-level bomber. By 1937, however, the Nell was deemed lacking in speed, range and performance, and the IJN wanted to replace it. The new twin-engine Mitsubishi G4M Type 1 Attack Bomber received the codename 'Betty.' Although designed for speed and range, the Betty lacked armour, which reduced its survivability against modern fighters (Dwyer 2014).

By the late 1920s, the IJN also possessed an aircraft carrier fleet, which it viewed as crucial for the defence of Japan in a war against the United States and as a counter to the US fleet. Admiral Isoroku Yamamoto concurred with that assertion although it was contrary to the established doctrine. In the hierarchical ranking of vessels, aircraft carriers would play a secondary role to the most important ships in the IJN—the battleship. By the late 1930s, Yamamoto had succeeded in shifting the doctrine to 'include a knockout air strike against the US fleet.' By tactically paralysing the US fleet, the Japanese would have free strategic reign to continue the quest for resources in Southeast Asia. The 'surface fleet' would engage the USN in any other engagements. Carrier aircraft, like bomber airplanes of the First World War, were tactical plans playing a strategic role. Under the revised doctrine, Japanese air power would be crucial during the 'knockout' phase. Consequently, Japan developed an air force and doctrine to meet that specific goal (Buckley 1999: 94–7; Till 1998: 211–13).

While the Italians and the Japanese did not have the resources or enthusiasm to construct big fleets of bombers, the Germans were restricted in their interwar development of aircraft and air power doctrine by the terms of the Versailles Treaty. They nevertheless, predictably, found ways to get around the treaty's constraints. During the period 1920–1926, Generaloberst Hans von Seeckt supported 'experiments of a wartime aviation doctrine.' As the Germans devised a military doctrine, it was predicated on a 'combined-arms concept' that allowed a commander to utilise all military resources—'infantry, artillery, tanks, cavalry, and aircraft.' The Germans developed one air force with two missions—ground support and bombing—that embodied German air doctrine which had a strategic dimension—the interdiction of enemy transport and 'industrial and commercial centers' (Corum and Muller 1998: 1–10; Buckley 1999: 33–4). As the German military devised air doctrine, they had to determine the extent to which they could adapt the theories of Douhet, Mitchell or Trenchard to their situation. The Nazis particularly found value in the theory of strategic bombing. A limitation of resources meant, however, that air power was best utilised to support operations on the ground. Many in the military doubted that advocates for air power accurately predicted bombing outcomes. Despite that lack of consensus, the *Luftwaffe* was armed with 'the most effective short-to-medium range bombing fleet in the world' when the war started. What it lacked was 'a heavy-bomber fleet.' Furthermore, because Adolf Hitler was more focused on numbers than quality of aircraft, the *Luftwaffe* became tasked with implementing strategic air power with operational aircraft (Buckley 1999: 80–1; Murray 1998: 108–11).

Italy, Japan, and Germany all developed air forces to varying degrees. In each case, the existing government invested in air power as part of an agenda and not necessarily because it embraced a particular strategic air power doctrine (Boyne 2003: 169). The United States and Britain, on the other hand, had less freedom and fewer financial resources to pursue the creation of air forces to the extent that was desired—or promoted. Particularly after 1929, elected officials were increasingly reluctant to vote for an increase in defence spending that would be funded by increased taxation. In addition, after 1934, as Germany tested the waters with violations of the Versailles Treaty in support of European expansion, US President Franklin D. Roosevelt walked a fine line between trying to provide diplomatic support to Britain and France and placating Congressional leaders who opposed any commitments that might drag the United States into another war (Boyne 2003: 169).

Despite the financial constraints, the United States, like Great Britain, increasingly recognised, as the 1930s progressed, the importance of maintaining, if not expanding, their military forces, including their fledgling air forces. That did not mean, however, that air forces in these countries did not experience problems. Unlike Britain, the United States did not have an independent air force, despite General Mitchell's recommendation for the creation of one. Under the Army Reorganization Act of 1940, the Air Service was attached to the Army and became the US Army Air Forces (USAAF). This was a step forward from the case during and after the First World War. Because of its limited engagement in the First World War, US airmen had not benefited from their experiences in the same way that their British counterparts had. That did not make the fledgling service immune to debates about strategic bombing. By the late 1930s, bomber advocates such as Mitchell influenced airmen while innovations advanced bomber technology. New designs and engines for bombers improved range and capability and resulted in performance similar to that of the existing fighters—a potential tactical advantage (Murray 1998: 122–7).

During this period as well, there was a mutual sharing of technological advancements in bombers between the United States and Great Britain, where Trenchard continued to promote 'the bomber offensive concept.' Trenchard presented persuasive arguments to his government by suggesting that the RAF could perform the same functions as the more 'expensive standing garrisons of troops.' All it would take would be occasional air attacks, and troublesome indigenous populations could be brought back in line (Hastings 1979: 38–9). Tami Davis Biddle argued that the British felt more pressure to develop substantive bomber forces, because unlike the United States, Britain was not isolated from its enemies by oceans. In the 1920s, France was their major theoretical target. In addition, unlike his American counterparts, Trenchard was less concerned about the accuracy of the bombers.

He considered bombs dropping in 'the general vicinity of the target' acceptable, while the Americans were more concerned about accuracy (Biddle 2002: 128–30).

This issue would influence both policy and strategy in Anglo-American relations during the war. Strategic bombing was Britain's only way of striking Germany directly for most of the conflict. For strategic bombing to work, however, the RAF had to bring the war to the enemy. The lack of resources, including financial ones, delayed the development of technologies, including navigation equipment, needed for improved bombers to take the conflict successfully to the enemy's industry, infrastructure and population. The situation was exacerbated in 1937 when the Chamberlain government removed resources from the bomber programme and allocated them to Fighter command and an 'air defense system' that could protect the nation from an enemy air attack. While Fighter Command 'pushed the limits of technological development in aircraft, radar support, and communication links needed to protect the British Isles,' Bomber Command did not advance at the same pace. Problems with bombing accuracy did not help. When the war broke out in 1939, according to Williamson Murray, Bomber Command, the branch to implement strategic bombing was not where it needed to be (Murray 1998: 118–21). Arguably, it never quite got there.

During the Interwar Period, the allocation of resources for the construction and testing of aircraft lagged behind what air power advocates recommended. While air power theorists tried to shape what constituted a good air force, they did not always succeed in convincing government leadership, the military or other airmen to get on board. As a result, air forces remained in flux on the eve of the Second World War. While Douhet, Mitchell and Trenchard developed the theory and were influential globally, the Second World War became the testing ground for strategic air power.

The Second World War

While scholars might disagree, Bernard Brodie argued, in *Strategic Air Power in World War II*, that the strategic bombing campaigns that the Allies launched against the Germans and the Japanese were complex, large and 'among the most brilliantly illuminated campaigns of all time' (Brodie 1957: 2). Only after 1943, however, did the combined Allied bombing campaigns result in a global 'air war on the apocalyptic scale envisioned by Douhet, Trenchard, and Mitchell' (Boyne 2003: 194–5).

European Theatre

The 'big three' air power theorists increasingly advocated using aircraft to take the war to the enemy's industry and to civilian populations to bring a conflict to conclusion. In *The Luftwaffe: Creating the Operational Air*, James S. Corum contended, however, that in September 1939, the only air force capable of 'conduct[ing] strategic bombing' was the *Luftwaffe*. It was the only air force that possessed even rudimentary equipment, training and navigation aids necessary for strategic bombing (Corum 1997: 6). Furthermore, he argued that the *Luftwaffe*'s air power had a comprehensive structure: a strategic bombing doctrine; the implementation of joint air–ground operations; civil defence, including anti-aircraft artillery; and a transport air fleet that could convey paratroopers and supplies (Corum 1997: 284).

Although these advantages contributed to the *Luftwaffe*'s early successes in 1939 and 1940, they were not sustainable, as the Battle of Britain demonstrated. The doctrine was sound, but the *Luftwaffe*—structure, numerical strength, personnel and equipment—was not. It was suited to operations, not strategy, both in terms of command and doctrine. The Battle of Britain was a case study in overreach. The decline in the *Luftwaffe*'s air superiority was increasingly apparent by early 1943. Around the clock, raids by Allied bomber groups indicated an air power doctrine that was essentially different from and 'developing at a faster rate than' what the Germans could implement (Corum and Muller 1998: 12–14). Furthermore, Allied bombing raids adversely affected German industry, and by March 1944 increasingly targeted German oil-production facilities as well. Brodie maintained that Germany is generally understood as not living up to its full war potential. Even if doing so would have been possible, the destruction of Germany's liquid fuel supplies and chemical industries by Allied bombs created a stumbling block that the *Luftwaffe* could not overcome. The loss of fuel supplies resulted in shorter *Luftwaffe* sorties. The *Luftwaffe* became less effective and could no longer give German ground forces an edge (Brodie 1957: 5–7).

By 1943, Allied air forces launched a major, consistent strategic bombing offensive against Germany. From January until the war's end, Allied air forces made a concerted effort to disrupt all aspects of German life—military, industrial and economic resources and civilian morale. Targets included submarine construction yards, Germany's aircraft industry, transportation nodes in Germany and German-occupied Europe, Germany's oil industry and general war-related industry. Another goal, especially for Britain, was to undermine the enemy's civilian population's support for the war. The theory was that if civilians no longer supported the war effort, the conflict would end. A debate on the ethical and moral issues of targeting civilians began during the war—and has continued ever since. There are no easy answers (Buckley 1999: 152–5, 164–5; Brodie 1957: 12–13).

Although they ultimately approved it, it is possible that air officers and political leaders had not envisioned what strategic bombing would mean. Coordinated attacks by US (bombing by day) and British (bombing at night) air forces created firestorms in Hamburg, Dresden and Berlin. The firestorms resulted from the use of incendiary bombs, in addition to the usual ordinance dropped from Allied bombers. In July 1943 the Hamburg firestorms resulted in 45,000 dead, 'the vast majority from the devastating RAF firestorm of the night of 27 July,' which was the 'first such conflagration caused by bombing.' In November 1943, Bomber Command—without the help of the USAAF—launched an assault on Berlin, which was not the war ending attack that Sir Arthur 'Bomber' Harris envisioned. In mid-February 1945, the Allies launched a round-the-clock attack on Dresden, Germany. Americans attacked during the day while the British executed nighttime bombardments. Poor visibility, caused by smoke from the fires, obscured the targets. The exact number of the killed is unclear, although estimates place the figure between 25,000 and 35,000. The majority 'died from inhaling hot gases or carbon monoxide in the tornado of the firestorm caused by RAF incendiaries,' although the inaccuracy of American bombing of marshalling yards is to be noted. The cost rose even higher when the RAF 'mounted a follow-up attack with mostly high-explosive bombs just as the firestorm burnt itself out and before citizens could flee from their cellars.' Post-war German and Soviet propaganda claimed that approximately 135,000 people died in the Dresden firestorms. As tragic as Dresden was, there is no evidence to substantiate this claim (Crane 1993: 102, 114–15; Buckley 1999: 164; Messenger 1984: 142–4; Hastings 1979: 256–69). The bombing of Dresden is best understood as an aftermath to the strategic air war. What was the point? The war was basically over. German resistance was failing. At the end of the day, the bombing of Dresden was a paradigm of what strategic air power advocates hoped strategic bombing would be. The price may have been high—but lost wars are seldom cheap.

Asia-Pacific Theatre

Unlike their German and Italian allies, the Japanese did not invest in fleets of bombers to destroy factories and cities. Instead, they focused on a different mission. Their principal targets were generally enemy ships and aircraft. Not only did Japan want to project its power far into Southeast Asia and to protect the lands it conquered, but its leaders also planned to establish a defensive perimeter designed to protect the homeland. In order to accomplish these goals, the Japanese allocated resources to the construction of aircraft carriers. Japan's carrier fleet faced other obstacles in addition to dwindling financial resources in the 1920s. Initially, there was not much support in the IJN for investment in naval aviation. When he was vice navy minister, Yamamoto was one of a small group of naval aviation advocates who were eventually able to shift the IJN's position.

In 'Adopting the Air Craft Carrier: The British, American, and Japanese Case Studies,' Geoffrey Till argued that because their air forces were not independent services, like the RAF, the Americans and Japanese focused more of their 'air effort' on an operational force used for strategic means. The construction of carrier fleets by the Americans and Japanese was, to some extent, fueled by the belief—on both sides—that they would be in a war against each other (Till 1998: 191–203). Just as there had been a race to construct dreadnaughts prior to the First World War, the Americans and Japanese engaged in a carrier race as tensions between the two nations increased in Southeast Asia in the 1930s. On the eve of war, both the Japanese and American navies possessed sufficient numbers of naval aircraft to pose high-level threats to the surface vessels of other navies. Expected attack results on enemy battleships and carriers was exponentially high. Some of the more memorable naval battles were ones between carriers that were conducted by planes fighting each other in the skies or dropping bombs on enemy ships (Till 1998: 212, 220–1, 225–6). Till argued that naval aviation took centre stage in the Pacific War (Till 1998: 225). It also became another component in the strategic air power playbook.

The aerial war in the Pacific was not limited to battles between aircraft carriers or against other naval vessels. While, for the most part, the Japanese did not take the fight to the continental United States, the USAAF, once in range, did not hesitate to take the war to Japan. In the 1942 Doolittle raid, US bombers took flight from an aircraft carrier and bombed Tokyo. This raid would have been impossible just a decade earlier. To some extent, the bombing sorties against Japanese territory were similar to air force operations conducted against the Germans. In many respects, however, the Americans took off the gloves and moved toward a 'no-holds-barred' approach with the use of incendiary bombs. On 9–10 March 1945, US-AAF targeted the Japanese capital—Tokyo—an attack that shocked and horrified air crews, 'who had observed from low-level the carnage they caused.' By the time the two-day raid had ended, between 90,000 and 100,000 had not survived. The firestorm created an intense heat that 'consumed the oxygen, boiled water in canals, and sent liquid glass rolling down the streets. Thousands suffocated in shelters or parks. Crowds trying to escape the flames panicked and trampled those who fell in the middle of the fleeing group' (Crane 1993: 132–4).

Not all air operations involved the creation of firestorms. Prior to the firebombing of Tokyo, heavy bomber raids by the new B-29s, which began in the fall of 1943, caused damage, but had little strategic value. In March 1945, however, General Curtis LeMay ordered new tactics—'maximum effort' low-level night attacks against the Japanese. The culmination of this 'maximum effort' to destroy the morale of the Japanese people and their will to fight were the atomic bomb attacks on Hiroshima (6 August 1945), which caused 80,000–100,000 deaths, and Nagasaki (9 August 1945), which killed another 35,000 Japanese. These figures do not reflect the number of people who died later from their burns or radiation poisoning (Brodie 1957: 26–9; Buckley 1999: 194–5).

Conclusion

The air forces of Britain, Germany, the United States and Japan looked very different when the Second World War ended from how they had looked 30 years earlier, when the idea that aircraft could be used for something other than aerial reconnaissance had begun to take hold. The evolution of these air forces—of strategic air power—is more than a story of research, development and testing of airplanes. It is a story about the development of the theory behind the new power projection that air fleets promoted. While historians trace the beginning of strategic air power to the First World War, one could argue that the crucial period in its development was the interwar years during which three men—Giulio Douhet, Billy Mitchell and Hugh Trenchard—emerged as influential developers of strategic air power theory; theories that evolved because of their applications during the Second World War. Strategic air power theory continues to have an impact on the air forces globally.

Bibliography

Biddle, Tami Davis, *Rhetoric and Reality in Air Warfare: The Evolution of British and American Ideas about Strategic Bombing, 1914–1945* (Princeton: Princeton University Press, 2002).

Black, Jeremy, *Air Power: A Global History* (New York: Rowman & Littlefield Publishers, 2016).

Bourque, Stephen Alan, *Beyond the Beach: The Allied War Against France* (Annapolis: Naval Institute Press, 2018).

Boyne, Walter J., *The Influence of Air Power Upon History* (Gretna, LA: Pelican Publishing, 2003).

Brodie, Bernard, *Strategic Air Power in World War II* (Santa Monica, CA: Rand Corp., 1957).

Buckley, John, *Air Power in the Age of Total War* (Bloomington, IN: Indiana University Press, 1999).

Corum, James S., *The Luftwaffe: Creating the Operational Air* (Lawrence, KS: University Press of Kansas, 1997).

Corum, James S., and Muller, Richard R., *The Luftwaffe's Way of War: German Air Force Doctrine, 1911–1945* (Mt. Pleasant, SC: Nautical & Aviation Publishing Company of America, 1998).

Crane, Conrad C., *Bombs, Cities, and Civilians: American Airpower Strategy in World War II* (Lawrence, KS: University Press of Kansas, 1993).

Dwyer, Larry, 'Mitsubishi G4M Betty.' The Aviation History Online Museum (2014), accessed 16 July 2019. http://www.aviation-history.com/mitsubishi/g4m.html

Fussell, Paul, *The Great War and Modern Memory* (Oxford: Oxford University Press, 1975).

Gray, Peter, *Air Warfare: History, Theory, and Practice* (London: Bloomsbury Academic, 2015).

Hastings, Max, *Bomber Command: The Myths and Reality of the Strategic Bombing Offensive, 1939–45* (New York: Dial Press/James Wade, 1979).

Hippler, Thomas, *Bombing the People: Giulio Douhet and the Foundations of Air-Power Strategy, 1884–1939* (Cambridge: Cambridge University Press, 2013).

Jones, Minnie L., 'William "Billy" Mitchell—"The Father of the United States Air Force,"' US Army website, 28 January 2010; accessed 4 July 2019. https://www.army.mil/article/33680/william_billy_mitchell_the_father_of_the_united_states_air_force

Messenger, Charles, *'Bomber' Harris and the Strategic Bombing Offensive, 1939–1945* (New York: St. Martin's Press, 1984).

Millett, Allan R., 'Patterns of Military Innovation in the Interwar Period' in Williamson Murray and Allan R. Millett (eds.), *Military Innovation in the Interwar Period* (Cambridge: Cambridge University Press, 1998), pp. 329–68.

Mitchell, William, *Winged Defense: The Development and Possibilities of Modern Air Power—Economic and Military* (Port Washington, NY: Kennikat Press, 1925).

Muller, Richard R. 'Close Air Support: The German, British, and American experiences, 1918–1941,' in Williamson Murray and Allan R. Millett (eds.), *Military Innovation in the Interwar Period* (Cambridge: Cambridge University Press, 1998), pp. 144–90.

Murray, Williamson, 'Strategic Bombing: The British, American, and German Experiences,' in Williamson Murray and Allan R. Millett (eds.), *Military Innovation in the Interwar Period* (Cambridge: Cambridge University Press, 1998), pp. 96–143.

Orange, Vincent, 'Trenchard, Hugh Montague, first Viscount Trenchard,' *Oxford Dictionary of National Biography* (online version), 6 January 2011, accessed 6 July 2019. https://www.oxforddnb.com/view/10.1093/ref:odnb/9780198614128.001.0001/odnb-9780198614128-e-36552.

Till, Geoffrey, 'Adopting the Aircraft Carrier: The British, American, and Japanese Case Studies,' in Williamson Murray and Allan R. Millett (eds.), *Military Innovation in the Interwar Period* (Cambridge: Cambridge University Press, 1998), pp. 191–226.

PART VII

Wars of Decolonisation and Cold War

29

WARFARE IN THE MIDDLE EAST SINCE 1945

Rob Geist Pinfold

In his enduring epic, *On War*, the Prussian military strategist Carl von Clausewitz concluded that 'war is the pursuit of politics by other means' (Clausewitz 1993: 87). What Clausewitz sought to convey was that war is not simply an overload of passions, but is instead a perpetuation of the struggle for power and influence between two or more competing actors. Framed through the prism of Clausewitz's realpolitik, war is an instrument of statecraft that can secure a state's goals, alongside diplomacy, economics or various 'soft power' implements. Though *On War* was published in the nineteenth century, governments before and after have long employed war as a strategic tool to advance the national interests of a given state.

Nevertheless, the conflicts that have plagued the Middle East since 1945 challenge this rationalist understanding of war's utility. Instead, in a region rife with sectarian, ethnic and socio-economic tensions, wars are frequent and unpredictable, rarely achieving the instigator's desired ends. Supposedly defensive wars have created prolonged occupations with no end in sight; wars to entrench liberal democracy have birthed extremist movements that threaten the world order; stateless groups purporting to represent the oppressed have launched brutal terrorist tactics against innocent civilians. Despite this grim history, 'war has profoundly shaped the Middle East regional system' (Hinnebusch 2015: 175). Conflict has been driven by actors within and without the Middle East and influenced by ideology, territory, values and national interests alike.

Accordingly, this chapter provides a history of the key conflicts that have shaped the Middle East since 1945. Though focusing on the Cold War era, it also identifies how historical wars affect the Middle East today. Much of this chapter focuses on the Arab–Israeli conflict, which has 'dominated' regional events since 1945 (Goldschmidt and Davidson 2016: 326). Concurrently, it also scrutinises conflicts that were equally brutal, but often forgotten, such as the Iran–Iraq War and the Lebanese Civil War. It divides regional conflict along three themes: (i) wars of external intervention; (ii) interstate regional wars and; (iii) intrastate and asymmetric wars. In each chase, this chapter unpacks the conflict's origins, outcomes and consequences. Simultaneously, it demonstrates that each of these four types of wars frequently wrought more unintended conflict in the Middle East and failed to achieve an actor's goals.

DOI: 10.4324/9780429437915-37

Historical and Case Background

What territorial expanse constitutes 'the Middle East'? Some definitions expand conventional conceptions of the region, to include North African states such as Libya and Tunisia; others look eastwards and include Afghanistan. By contrast, this chapter adopts the least-contested definition of the 'classic' Middle East that extends from Egypt in the west, to Turkey in the north, Iran in the east and Yemen in the south. Though more limited in geographical scope than some conceptions of the region, this understanding of the Middle East features multiple states and peoples, affected by competing and different interests, beliefs and perceptions. As such, it provides a thorough framework for charting and understanding the key conflicts that have characterised the region since 1945.

The principal power brokers within the Middle East have fluctuated throughout history. Now-defunct great powers, such as Rome, Persia and finally the Ottomans colonised vast swathes of the region. Following the end of the First World War, the Ottoman Empire collapsed, and the region's territory was divided up by Great Britain and France, as envisioned by the Sykes–Picot Agreement, a secret pact signed by both these powers in 1916. Accordingly, Britain gained control of Iraq, Jordan and Palestine, whereas Syria and Lebanon fell under French authority.

Problematically, these states had little prior history as homogenous sovereign entities. Furthermore, borders were frequently delineated by planners in London and Paris and failed to take the interests of indigenous peoples into account. As a result, these new states were often ethnically and religiously divided. In Palestine, for instance, British officials often struggled to control and court both local Arabs and the increasing number of Zionist Jews. Concurrently, traditional regional power structures were increasingly challenged by ideologies such as republicanism, secularism and pan-Arabism.

These trends and realities precipitated the plethora of conflicts that defined regional history after 1945. Following the Second World War, Britain and France increasingly withdrew from the region, leaving weak regimes with little public legitimacy in their wake. The foundation of the State of Israel in 1948 fundamentally altered regional politics and generated multiple, bloody conflicts. Meanwhile, Arab royalists and republicans aligned with either the United States or the Soviet Union, as the Cold War facilitated further extra-regional involvement in the Middle East's affairs. Today, the Middle East remains a hotspot for global conflict, affected by the local rise of Islamism and non-state actors who pioneered deadly terrorist violence, most notably Al-Qaeda and the Islamic State of Iraq and Syria (ISIS). In sum, the region has long suffered from violence, wrought by external intervention and internal strife alike.

Wars of External Intervention

The first category of wars examined here is those driven by extra-regional actors. As illustrated above, major powers not native to the region have long shaped conflict within the Middle East. The post-1945 era was no exception to this rule and saw three major wars, in which extra-regional actors played a key, causal role: the 'Suez Crisis' of 1956–1957, the 'Gulf War' of 1991 and the 2003 'Iraq War.'

The Suez Crisis

In 1952, a collective of Egyptian army officers overthrew the pro-British King Farouq and declared an anti-colonial republican regime, which tentatively courted Soviet support. In the short term, this move backfired badly: the United States and Britain withdrew funding for the Aswan Dam, a mega-project to block the Nile backed by the new Egyptian President, Gamal Abdel Nasser. Incensed, the year 1956 saw Nasser announce the nationalisation of the Suez Canal, a man-made waterway intersecting Egypt's mainland and the Sinai Peninsula that permitted the passage of goods between Europe, East Africa, Australasia and South Asia. This move directly threatened British and French interests, since Britain and France jointly controlled the conglomerate operating the Suez Canal.

The new Egyptian regime also found itself in conflict with Israel. The 1950s saw both sides frequently engage in violent border skirmishes, a process of escalation that culminated in an Egyptian multi-million dollar deal to purchase arms from Soviet-aligned Czechoslovakia. Egypt also closed the Straits of Tiran, a waterway adjoining eastern Sinai, which completely blocked shipping access to Israel's Red Sea port of Eilat. Israel, in turn, purchased arms from, and grew increasingly close to, France. France's Prime Minister, Guy Mollet, persuaded Israel to join France and Britain in a tripartite alliance against Nasser's Egypt (Kyle 2011).

The stage was therefore set for war. Israel's goals were to forcefully open the Straits of Tiran and curb raids from the Egyptian-controlled Gaza Strip, whilst preventing Egypt from absorbing its Czechoslovakian arms. Britain and France, in turn, sought to precipitate the downfall of Nasser's republican regime. The three parties agreed a ruse: Israel would attack Egypt. In response, Britain and France would demand that Israel and Egypt accept ceasefire. Anticipating an Egyptian refusal, British and French troops would then land and seize control of the Suez Canal.

The Israeli attack began on 29 October 1956. The Israeli Defence Forces (IDF) emphasised rapid attack and manoeuvrability and rapidly routed the more static Egyptian forces within the Sinai and the Gaza Strip. On 31 October, Britain and France joined the war effort, storming Port Said and capturing the Suez Canal. In total, Israel lost fewer than 200 soldiers, while British and French losses were fewer than 20 men each. Egyptian casualties, on the other hand, numbered as many as 3,000 soldiers and civilians.

Though an unquestionable tactical victory, the Suez Crisis soon degenerated into a strategic failure for the tripartite alliance. The United States and the USSR were equally furious and demanded an unconditional withdrawal of Israeli, French and British forces; both the British and French publics also vigorously protested their countries' involvement. Facing unprecedented public and global pressure, the European powers conducted a humiliating and rapid exit. Similarly, despite Prime Minister David Ben-Gurion proclaiming that parts of the Sinai would be annexed, Israel withdrew from the entire Sinai Peninsula and Gaza Strip by March 1957. The alliance's only victory was that the UN guaranteed the Straits of Tiran would remain open to Israeli shipping. In contrast, Nasser's Egypt, which was seen throughout the Arab world as having resisted an unjust colonial incursion, received an unprecedented popularity boost.

The long-term consequences of the Suez Crisis were diametrically opposed to the war aims of Britain and France. First, Britain and France's regional reputations nosedived, ensuring that former local allies turned either to the USSR or the United States. Second, the Suez Crisis precipitated the export of Nasserite republicanism, with subsequent attempted coups against pro-Western regimes in Syria, Lebanon, Iraq and Yemen. The conflict therefore precipitated the 'Arab Cold War' between republican and monarchist forces across the region. Finally, the rivalry between Israel and Nasser's Egypt would culminate in the 'Six Day War' of June 1967 (Halliday 2012).

The Gulf War

Compared with the milieu of the Suez Crisis, the Middle East's balance of power had changed drastically by the early 1990s. Israel and Egypt had long since signed a peace treaty, whilst the Soviet Union teetered on the brink of collapse. But in August 1990, Iraq—led by Saddam Hussein—sent shockwaves through the international system by invading neighbouring Kuwait. Kuwaiti forces put up little resistance, with the invasion ending in an Iraqi victory in less than 12 hours of fighting.

Though Hussein was an uncompromising dictator, the seizure of Kuwait was a rational policy. Previous Iraqi administrations had attempted to annex oil-rich Kuwait in the 1930s and the 1960s. A bankrupt Iraq also owed Kuwait $15 billion in debts. Hussein also sought to cement his position as a contemporary Arab hero in Nasser's mould, by challenging US and Saudi Arabian local hegemony. Conversely, Hussein did not expect an overwhelming response. Prior to the invasion, the American ambassador to Iraq had told the Iraqi leader that the United States takes 'no opinion on the Arab-Arab conflicts, like your border disagreement with Kuwait' (Walt 2011).

Iraq rejected 12 UN Security Council resolutions mandating a withdrawal, precipitating a multi-national coalition to declare war in January 1991. The United States led this largest assembly of nations since the Second World War against Hussein's Iraq. The UN Security Council, the USSR and most of the often-divided Middle East, spanning Iran, Egypt, Syria and Saudi Arabia, all backed the coalition. Iraq attempted to splinter the alliance by firing missiles on Tel Aviv, calculating that a forceful Israeli response would bolster his anti-Western credentials and embarrass the United States' local partners. Yet, the United States convinced Israel not to respond, aided by the fact that the attacks themselves caused very few casualties.

For their part, the coalition unleashed a show of force unseen since the Vietnam War. The Gulf War was the first conflict that showcased the 'Revolution in Military Affairs' (RMA), an umbrella term for recent technological advances and tactical innovations in warfare. Rather than send in ground forces at the beginning of the conflict, which would inevitably generate casualties, the coalition bombarded Iraq through new weapons such as cruise missiles and stealth bombers, aided by pinpoint, Special Forces operations. In the desert expanses of Kuwait and Southern Iraq, the Iraqi Army was particularly exposed. Only in the final stages of 'Operation Desert Storm,' when ordnance targets were depleted, did coalition ground forces enter Iraq en masse. The invasion lasted less than 100 hours before all parties accepted a cease-fire on 28 February 1991. Overall, coalition losses totalled fewer than 300, while as many as 50,000 Iraqis were killed (Stansfield 2007).

Thus, mimicking the Suez Crisis, the war's results diverged drastically from Iraqi goals. Iraq was forcibly expelled from Kuwait and became a pariah state, isolated by crippling sanctions. Contrarily, despite promises by some US officials, liberal democracy did not expand in the region, or in Iraq. Hussein's resilient dictatorship crushed US-inspired Kurdish and Shiite uprisings immediately after the war, leaving up to 100,000 Iraqis dead. Rather than constitute the decisive end to a conflict, then, the Gulf War would effectively be replayed in 2003, when a US-led coalition launched another offensive against Hussein's Iraq.

The Iraq War

The inconclusive results of the Gulf War, alongside the 9/11 Attacks and the subsequent 'War on Terror,' precipitated the 2003 Iraq War. Officials within the George W. Bush administration highlighted Saddam Hussein's brutality, while alleging that Iraq supported Al-Qaeda. Similarly, US officials claimed that Iraq still sought to disrupt the international system and was covertly building weapons of mass destruction (WMD). Thus, a post-9/11, more interventionist United States again intervened in the Middle East, to promote regional stability.

The invasion, termed 'Operational Desert Fox,' began on 20 March 2003 and, as US planners had hoped, bore many similarities to the Gulf War. The fighting lasted just over a month and caused fewer than 300 coalition losses, while as many as 30,000 Iraqi troops were killed. Practising 'shock and awe,' coalition forces employed their technological superiority, monopoly over Iraq's airspace and long-range ordnance to eliminate most targets before ground forces were deployed. By 1 May, Hussein's regime no longer effectively controlled Iraq (Fawn and Hinnebusch 2006).

However, crucial aspects of the Iraq War differed significantly from the Gulf War. First, the US-led coalition was depleted compared with 1993: global audiences were sceptical of the Bush administration's claims and the United States failed to secure UN support. Most critically, US-led forces did not rapidly withdraw and instead occupied Iraq until 2011. Though President Bush declared 'mission accomplished' in May 2003, the occupation heralded a vicious Iraqi insurgency and civil war. Insurgents caused multiple coalition casualties through Improvised Explosive Devices (IEDs), which were relatively primitive roadside bombs. Conversely, the advances wrought by the RMA were ineffective when fighting irregular forces in dense urban environments. The coalition soon became bogged down operationally, while administrators struggled to traverse the labyrinthine sectarian allegiances that characterised Iraqi politics.

Overall, though, the Iraq War of 2003 failed to transform Iraq into a peaceful, liberal democracy, with subsequent violence killing hundreds of thousands of Iraqis and nearly 5,000 coalition troops. Iraqi politics remains fractured and unstable, while the United States' most prolific regional state and non-state adversaries—Iran and Al Qaeda—exploited the power vacuum to establish significant local presences. Eventually, the ISIS would form from the remnants of Al-Qaeda in Iraq, the consequences of which continue to unfold today. The United States' regional clout and image has, by contrast, been substantially damaged by the violent consequences of the Iraq War. As such, from Suez to Iraq, Middle Eastern interventions by extra-regional actors have demonstrated scant strategic utility; even the relatively successful Gulf War's gains were undone, from 2003 onwards.

Interstate Regional Wars

Equally, the region has witnessed devastating conflicts between local powers, which rarely fulfilled the initiator's political goals. The most prominent milieu for interstate regional wars is the Israel–Arab conflict, which is divided here into three episodes: the First Israeli–Arab War of 1947–1949, the Six Day War of June 1967 and the Yom Kippur War of October 1973. Concurrently, this section scrutinises a war that—though often forgotten in the West—caused a greater loss of life than any of the above Arab–Israeli conflicts: the Iran–Iraq War of 1980–1988.

The First Israeli–Arab War

Though it evolved into a regional war involving an unprecedented number of Middle Eastern states, the Arab–Israeli conflict began as an internecine struggle within one territory: Palestine. Between 1918 and 1948, Palestine was one contiguous territorial unit, ruled by the British. However, British officials increasingly struggled to appease the territory's Palestinian Arab and Jewish residents, both of whom sought their own ethnically homogenous state in the same territory. Depleted resources after the Second World War and Arab and Jewish terrorism alike precipitated Britain handing over the problem to the UN and declaring its intent to withdraw. Consequently, the UN General Assembly voted to partition the territory into separate Arab and Jewish states on 29 November 1947. Palestine's Jews reluctantly accepted partition, while the Arabs rejected it, leading to a civil war that broke out immediately.

Once British forces left Palestine in May 1948, the conflict became regional as multiple Arab armies from Egypt, Jordan, Syria, Iraq and other local states entered the fray. The Arab forces fought to restrict the nascent Israeli state's territory, or destroy it entirely if possible. Individually, the Arab states had more specific goals: Egypt, for instance, sought to seize Southern Palestine (including the Gaza Strip), while Jordanian forces annexed eastern Palestine (later known as the West Bank). Conversely, Israel fought to expand its control as much as possible, including beyond the territory the UN allocated for Jewish settlement (Morris 2009).

Ostensibly, the Arabs were a united front spanning Islamists, republicans and royalists, who far outgunned and outnumbered Israeli forces. Nevertheless, individual Palestinian villages and Arab armies in fact rarely collaborated, allowing Israeli forces to defeat each in turn. The Arab states also committed only a fraction of their strength, fielding 23,500 troops collectively at the start of the conflict. Simultaneously, Israeli policymakers purchased foreign arms and trained more fighters. By the end of the conflict, Israeli troops outnumbered all Arab forces by a two-to-one margin. At the war's outset, Israel requested truces that the Arabs rejected; by the war's end, it was the Arabs who sought to end hostilities. Nonetheless, both sides suffered dearly, with Israel experiencing approximately 6,000 casualties, or 1 per cent of its total population, while around 20,000 Arabs were killed (Bregman 2015).

Thus, when the war ended in July 1949, Israel had won a stunning victory, while the Arab's rejection of the UN-sponsored partition plan was a dire strategic mistake. The war's consequences were particularly severe for the Palestinians: up to 800,000 fled the country, or were expelled by Israel. Though they had sought a contiguous state, some Palestinians fell under Israeli rule, others—in the Gaza Strip and West Bank—became subject to Egyptian and Jordanian administration. By contrast, Israel annexed almost 80 per cent of Palestine and consolidated the plethora of pre-state Jewish militias and political organisation into a unitary regime and army. But the Arab states only signed armistice agreements with Israel, rather than peace treaties. As such, the First Arab–Israeli War had ended, but the Israeli–Arab conflict had only just begun to shape the region's politics.

The Six Day War

Almost 20 years later, in June 1967, another severe outbreak of Arab–Israeli violence engulfed the Middle East. In the early–mid-1960s, the Israel–Egypt border was quiet, with the Straits of Tiran open and a UN buffer force positioned in the Sinai following the Suez Crisis. On other fronts, though, Israel was regularly exchanging bouts of low-intensity fire with Jordan and Syria. For reasons that remain obscure, the USSR warned Egypt in May 1967 that Israeli troops were poised to invade Syria. The report was entirely false, but it led Egypt's

President Nasser to close the Straits of Tiran, expel the Sinai's UN buffer force and announce that he was preparing for war with Israel. Jordan and Syria agreed to join Egypt in a future conflict, with countries as far-flung as Morocco, Kuwait, Saudi Arabia and Iraq also pledging troops. Israel and the Arabs alike knew war was imminent (Oren 2003).

Though the Arab states precipitated the conflict, Israel struck first. On the morning of 5 June, a massive, surprise Israeli airstrike wiped out Egypt's air force. Backed by total air supremacy, the IDF quickly routed Egypt's forces in the Sinai and stormed the Gaza Strip. Israel then captured the West Bank from Jordan and the Golan Heights from Syria. Hence, in six days of fighting, until an armistice was reached on 10 June, Israel expanded its territorial holdings more than four-fold. Israeli policymakers never envisioned seizing the entire West Bank, Gaza Strip or Golan Heights; the conquest was purely a result of the IDF's battlefield successes. Combined, the Arab states lost up to 20,000 personnel, while Israeli losses were fewer than 1,000. As in 1948, the IDF out-classed the Arab forces, which rarely co-ordinated and even fed each other false information to hide their own failures (Bregman 2015).

The war's regional consequences were profound. Arab leaders had promised to capture Tel Aviv, while Israelis had feared a second Holocaust. Instead, Nasser's gamble backfired spectacularly. Within Israel, the capture of all of Jerusalem and renewed access to Judaism's holiest sites provoked a renewed national self-confidence. Operationally, Israel received a territorial buffer zone, providing 'strategic depth' to protect the narrow country. On the other hand, Israel could now trade territory for political concessions from the Arab states. This logic inspired UN Security Council Resolution 242, which advocates 'land for peace' and remains the consensus formula for resolving the conflict today. Adopting this approach, on 13 June, Israeli Defence Minister Moshe Dayan enthusiastically claimed he was waiting for a phone call from the Arab states to discuss peace terms.

But the call never came. Despite Israel's military triumph, it was deprived of a political victory as the Arab states adopted a collective policy of 'no peace, no negotiations and no recognition' of Israel (Oren 2003). Simultaneously, Israelis disagreed amongst themselves as to what territory, if any, should be conceded. Israeli governments increasingly cooperated with a religious, right-wing lobby that sought to prevent 'land for peace,' by pre-emptively filling the occupied territory with Jewish settlements. Finally, Israel now ruled over hundreds of thousands of Palestinians, which would precipitate further traumatic and divisive wars—the First and Second *Intifadas*. As such, in retrospect, the Six Day War precipitated further long-term troubles for Israel, the Palestinians and the Arab world alike.

The Yom Kippur War

Though the Arab states rejected bargaining with Israel, Egyptian President Anwar Sadat had other ideas. Inheriting the presidency in 1970 after Nasser's death, Sadat furtively advocated Israel's withdrawal from the Sinai, in exchange for Egyptian non-belligerency guarantees. Yet, Israeli policymakers, overconfident following their 1967 victory, flatly rejected Sadat's offer. While Egypt sought US mediation, American officials were equally unenthusiastic, with National Security Advisor Henry Kissinger advising Sadat that 'I cannot deal with your problem unless it is a crisis' (Bregman 2015: 108). Egypt duly created a crisis, pursuing war to advance its political goals. Sadat sought Arab partners in Jordan and Syria; the latter agreed to join the war effort, whereas Jordan—unlike in 1948 and 1967—opted for neutrality.

Launching their attack on 6 October 1973, Syria and Egypt institutionalised many lessons from 1967. First, the Arab armies seized the initiative by opening hostilities on Yom Kippur, Judaism's holiest day, when most IDF troops were away from the front. Second, rather than proclaim Palestine's imminent liberation, the Arab armies fought a limited war within the territories they lost in 1967: the Golan Heights and the Sinai. Third, Egyptian and Syrian forces initially avoided penetrating Israeli-held territory too deeply and stayed within range of their own surface-to-air missile (SAM) defences, denying Israel air supremacy. The IDF, in turn, found itself statically defending existing fortifications, a paradigm shift for an army that emphasised manoeuvrability and offence. Hence, Israel was a victim of its own successes, changing its tactics to reflect the post-1967 local balance of power.

Unlike the Six Day War, the Yom Kippur War ended inconclusively. At first, Israel's fortifications along the Suez Canal—the 'Bar Lev Line'—fell to the Egyptians, whilst the IDF evacuated civilians from the Golan Heights. However, both Arab armies then advanced beyond the cover of their SAM batteries, allowing the IDF to re-assert its air superiority. As the war dragged on, Israel also mobilised significant numbers of reservists. When a cease-fire was finally agreed on 24 October, the IDF were a scant 101 kilometres from Cairo and 45 kilometres from Damascus. Both sides suffered devastating losses: Israeli and Arab fatalities exceeded 2,500 and 15,000 respectively (Bregman 2015).

In time, the Yom Kippur War facilitated dramatic political breakthroughs in the Israel–Arab conflict. The conflict precipitated unprecedented superpower pressure on both parties to make concessions and cemented the United States' role as a mediator. Additionally, many Israelis questioned the relationship between security and territory; occupation had, apparently, precipitated the war, rather than make Israel safer. The result was that Israel and Egypt implemented 'land for peace' for the first time: beginning in 1974, Israel gradually left the Sinai, while Egypt took incremental steps to end the state of war. This process culminated the 'Camp David Accords' of 1979, Israel's final withdrawal from the entire Sinai in 1982 and a comprehensive Israel–Egypt peace treaty. The Yom Kippur War is therefore the exception to the rule for the Middle East's wars, in that the instigator—Egypt—achieved its goals.

The Iran–Iraq War

Away from the Arab–Israeli milieu, a war was brewing that would dwarf all of the above conflicts in duration and human costs. The Iranian Revolution of 1979, when Islamists overthrew the monarch Mohammed Reza Pahlavi, rekindled long-running disputes between Iran and its Arab neighbour, Iraq. A treaty signed in 1975 apparently resolved these states' territorial disputes in Iran's favour. However, Iran's new government rekindled tensions by exporting its revolutionary Islamist message to Iraqi Shiites, incensing Iraq's totalitarian Sunni leader, Saddam Hussein. Concurrently, Hussein still coveted the territory Iraq had conceded to Iran in 1975.

Employing the element of surprise, Iraq declared war on 22 September 1980. Iraqi planners anticipated a quick resolution to what they dubbed the 'whirlwind war,' since the Iranian Army was decimated by post-revolutionary purges. The new regime was also globally isolated; both the United States and the USSR alike refused to support Iran's Islamist government. In the conflict's first months, Iraqi forces rapidly seized several key Iranian border provinces, such as Kurdistan and Kuzistan. Ostensibly, Hussein's calculations were vindicated, as oil-rich regions fell into Iraqi hands and commentators predicted the collapse of Iran's regime (Black 2010).

Yet, Iran was resilient; the war actually consolidated domestic support for the regime. By 1982, Iran was able to counterattack and re-take all of the occupied provinces. Iraq now sought a ceasefire, but Iranian leaders demanded compensation and the removal of Hussein from power. When this was not forthcoming, Iran took the war into Iraqi territory in 1983. Iran's leaders emulated Hussein's pre-war optimistic hubris and projected an easy victory, since local Iraqi Shiites might rebel en masse. Instead, the war became a quagmire for five more years. Exhausted, both sides agreed to a truce in August 1988 and returned to the *status quo ante bellum*.

The Iran–Iraq War is unique in that it was a Cold War conflict between two non-aligned states that both rejected the existing world order. Monarchist and pro-Western Arab states such as Egypt and Saudi Arabia calculated that a revolutionary Iran was more dangerous than a republican Iraq and channelled funds to Hussein. Israel made the opposite calculation and assisted Iran. Consequently, the United States and the USSR frequently switched their support to the weaker party, turning the war into a prolonged stalemate with a regionally unprecedented death toll of around 1,000,000 Iranians and up to 500,000 Iraqis. Iran and Iraq alike deliberately targeted enemy civilians, as well as their own citizens who were suspected of disloyalty. Iraq employed poison gas, whereas Iran sent thousands of child soldiers to their deaths (Murray and Woods 2014).

The Iran–Iraq War therefore saw shocking brutality, with neither side achieving their goals of neutralising the opposing regime. Despite the stalemate, the war's effects continue to resonate. A mere three years after the conclusion of hostilities, Iraq's need to acquire oil to offset its vast debts triggered the Gulf War. Similarly, Hussein's brutal repression of the Kurds and Shiites who took advantage of the Iran–Iraq war to fight for autonomy precipitated the enduring sectarian violence that characterises Iraq today. The fall of the Hussein regime in 2003 notwithstanding, accusations of irredentism continue to define Iraqi–Iranian relations.

Intra-State and Asymmetric Wars

Though the above conflicts reshaped Middle Eastern politics, state-level actors lacked a monopoly over regional violence, particularly after 1945. The Cold War era particularly saw the proliferation of intrastate civil wars and asymmetric warfare. Correspondingly, the conflicts where non-state actors played a pivotal role examined here are: the Lebanese Civil War of 1975–1990, the Palestinian *Intifadas* of 1987–1993 and 2000–2005, and the 'Kurdish Struggle for Statehood' of 1978 to the present.

The Lebanese Civil War

Lebanon, a state harbouring a volatile demographic mix of Shiites, Sunnis, Druze and Christians, has long witnessed violent internecine struggles for supremacy. The bloodiest and longest of these, the Lebanese Civil War, erupted on 13 April 1975 as a conflict between Christian and Palestinian militias. The conflict quickly escalated, with all of Lebanon's militias soon fighting to increase their power in that divided country. Violence begot more violence: rather than solely fight other ethno-religious groups, Shiite, Sunni and Christian militias began warring amongst themselves. Conflict only ceased in 1990, after the signing of a multilateral peace accord in 1989. By then, over 150,000 Lebanese had died and over 1 million emigrated, though the official divisions of influence between each ethnic group changed little.

Because the civil war exacerbated the Lebanese state's inherent weakness, the fighting primarily involved non-state militias and interventions by other powers. From 1976, Syria occupied Lebanon and backed and fought divergent militias, at different times. Israel invaded in 1982, seeking to expel the Syrians and the Palestine Liberation Organization (PLO), a powerful Palestinian non-state group. Though the PLO left Lebanon, Israel's occupation culminated in the Sabra and Shatila massacre of September 1982, where Israeli-backed Christian–Lebanese forces massacred up to 3,500 Palestinian civilians. Facing domestic and international criticism alike, Israel withdrew in 1985. The United States also intervened to force a ceasefire, an effort that ended after an October 1983 bombing by Hezbollah, an Iranian-backed non-state actor, which killed 241 American troops. These and other interventions prolonged the civil war: cumulatively, Lebanon's militias received around $30 billion from other states, throughout the conflict (Choueiri 2007).

The impact of non-state groups in Lebanon's Civil War resonates to this day. Hezbollah continues to wield significant clout over Lebanon's government and possesses an arsenal that both out-classes and outnumbers the Lebanese Army. Hezbollah's confrontations with Israel regularly drag in the entire Lebanese polity and cause widespread destruction. Concurrently, Hezbollah's recent re-settling of 1 million Syrian refugees fleeing that country's civil war threatens to disrupt Lebanon's delicate demographic balance. Finally, non-state actors' innovations during the Lebanese Civil War spread worldwide: tactics such as suicide bombing, roadside bombs, hostage taking and the hijacking of airliners were pioneered and augmented during the period 1975–1989, often to devastating effects.

The Palestinian Intifadas

The Israel–Arab conflict has also seen a shift away from state-level violence towards asymmetric warfare. One unintended consequence of Israel's occupation of the Gaza Strip and West Bank was the re-emergence of Palestinian nationalism. Held in the limbo of occupation and abandoned by the Arab states, Palestinians formed their own militant and political organisations to fight for statehood. In the 1970s and 1980s, groups such as the PLO often employed terrorism; Israel, in response, utilised its overwhelming military advantage to suppress Palestinian nationalism.

However, the First Intifada of 1987–1993 proved an unprecedented challenge for the IDF. Turning to riots, strikes and protests, the Palestinians succeeded where the Arab states had failed, neutralising Israel's military edge. The IDF's tanks, technology, artillery and planes were little use against crowds and civil disobedience. Palestinian persistence and the ineffectiveness of hard power convinced many Israelis that there was no military solution to the conflict. Thus, though costing the lives of around 1,000 Palestinians and 150 Israelis, the First *Intifada* ended in an unprecedented development: the Oslo Accords. This agreement, which attempted to provide Israel with security and the Palestinians with autonomy, was the first negotiated treaty between Israel and the PLO (Bregman 2015)

Despite the initial optimism, negotiations soon became deadlocked, which precipitated the Second *Intifada* of 2000–2005. Unlike the First *Intifada*, which was often defined by spontaneous eruptions of frustration, the Second *Intifada* was propelled by organised cells of militants affiliated to the PLO or Islamist groups such as Hamas and Islamic *Jihad*. The First *Intifada* was concentrated in the Occupied Territories and directed mainly against Israeli soldiers and settlers. By contrast, the Second *Intifada* primarily targeted civilians inside Israel through terrorism, particularly suicide bombings. Accordingly, the Second *Intifada* led to significantly more fatalities: up to 1,000 Israelis and over 3,000 Palestinians (Catignani 2008).

Though exacting a heavy toll, the Palestinian employment of unprecedented, civilian-focused violence was a strategic mistake. Each increasingly brutal attack damaged Israel's 'peace camp' and legitimised the use of force. Thus, re-employing its qualitative military edge, the IDF re-took swathes of the West Bank and targeted Palestinian leaders in surgical airstrikes. By 2005, Israel had ground down its Palestinian opponents. As such, the Second *Intifada*'s repercussions were the exact opposite of what PLO leaders such as Yasser Arafat sought. Rather than induce Israeli concessions, the conflict solidified a political deadlock that continues to this day. Concurrently, the regular, ongoing violent flare-ups between Israelis and Palestinians serve as a reminder that there is no purely military solution to this long-running, asymmetric conflict between two peoples.

The Kurdish Struggle for Statehood

Alongside the Palestinians, the Kurds are the Middle East's most prominent stateless ethnic group. A distinctive, non-Arab people, the Kurds saw their nationalist aspirations dashed after the First World War, when Kurdish-majority, former Ottoman provinces were divided up between Syria, Iran, Turkey and Iraq. Thus, the Kurds today are separated by nation, religion, culture and language. Though these divisions meant the Kurds briefly 'slipped off the pages of history,' Kurdish nationalism soon re-emerged after 1945 and precipitated multiple intrastate conflicts (Fuller 1993: 108).

The recent Kurdish struggle for statehood has been concentrated in Turkey, where Kurds comprise around 20 per cent of the population. The authoritarian Turkish state banned the Kurdish language and denied that the Kurds constituted a people with national rights. In turn, the lack of a democratic outlet empowered the PKK, a Marxist-Leninist militant group, who employed violence to fight the Turkish state in a bloody campaign from 1978. The PKK targeted Turkish civilians and security forces alike, often employing terrorism. Turkey responded with ethnic cleansing, torture and extra-judicial killings against its own Kurdish citizens. The civil war apparently ended in 1999, when a ceasefire was agreed after Turkey captured the PKK's leader, Abdullah Ocalan. Subsequently, Turkey cautiously liberalised its policies, allowing Kurds greater cultural freedoms. Yet, these reforms were retrenched after 2013, when a lack of progress in negotiations spurred a return to violence. Overall, the conflict has killed around 50,000 Kurds and Turks (Gunter 2018).

The governments of Iraq, Syria and Iran employed equally authoritarian measures to repress Kurdish nationalism. Iraq was particularly brutal, unleashing a campaign of genocide against the Kurds known as al-Anfar. Between 1986 and 1989, Hussein's security forces killed around 100,000 Kurds; over 100,000 Kurds were later killed in a subsequent uprising in 1991. Subsequently, though, Iraq's Kurds have enjoyed unprecedented autonomy following the Iraq War of 2003, while every Iraqi President post-invasion has been an ethnic Kurd. Similarly, the outbreak of the Syrian Civil War in 2011 allowed local Kurds the opportunity to establish an autonomous region in the country's north-east. Syria's Kurds have since been on the front-line of the war against ISIS, while seeking to defend their autonomy from Turkey. At the time of writing, the Kurdish struggle for statehood and/or autonomy is ongoing, with no clear end in sight.

Conclusions: The Future of Regional Conflicts

The contemporary Middle East remains a region defined by conflict. Today, Lebanon is rocked by anti-corruption protests, deadlock characterises the Israeli–Palestinian conflict, while extremist transnational groups such as ISIS and Al Qaeda retain regional presences. The scourge of civil war blights Yemen and Syria and may re-emerge in Iraq. Meanwhile, the Israel–Iran conflict has been limited to proxy struggles, but threatens to elevate to a state-level war. This chapter illustrated that this contemporary reality is not an exception to the rule. Instead, warfare between and within states, involving regional actors, extra-regional powers and non-state actors, has shaped the Middle East since 1945.

Thus, the region's projected future, at least in the short-term, does not seem optimistic. Whereas state-level warfare once dominated the Middle East, contemporary threats are not just the strength of states, but also their weakness; violent, non-state actors have frequently exploited regional power vacuums, particularly in Iraq and Syria. These non-state actors could extend their violent campaigns worldwide, as evidenced by previous atrocities masterminded by Middle Eastern groups, most notably the 9/11 Attacks and the November 2015 Paris Attacks. Concurrently, powerful revisionist states that seek to re-make the world order remain entrenched within the region. Contemporary Syria, for instance, represents a 'strange hybrid of a Russian and Iranian-dominated country,' where these powers coexist with and fight against a proliferation of local and foreign non-state actors (Freilich 2019: 186).

Yet, as this chapter has illustrated, regional wars rarely end in a linear fashion. Instead, the Middle East's divergent conflicts have been often underlined by a commonality of unintended consequences and quagmire. Despite the hopes and expectations wrought by the 2011 'Arab Spring,' the region has returned to familiar patterns of authoritarian dictatorship and brutal repression that often feeds further grievances and culminates in sectarian and state-level violence. Clausewitz argued that 'the result in war is never final' (Clausewitz 1993: 75). Similarly, until solutions to the Middle East's political problems—of which violence is only a symptom—can be found, the region is unlikely to see an end to war.

Bibliography

Black, Ian, 'Iran and Iraq Remember War that Cost Over a Million Lives,' 2010, *The Guardian*: https://www.theguardian.com/world/2010/sep/23/iran-iraq-war-anniversary
Bregman, Ahron, *Israel's Wars: A History Since 1947* (London: Routledge, 2015).
Catignani, Sergio, *Israeli Counter-Insurgency and the Intifadas: Dilemmas of a Conventional Army* (London: Routledge, 2008).
Choueiri, Y.M. (ed.), *Breaking the Cycle: Civil Wars in Lebanon* (London: Centre for Lebanese Studies/Stacey International, 2007).
Clausewitz, Carl von, *On War*, trans. and eds. Michael Howard and Peter Paret (New York: Knopf, 1993).
Fawn, Rick and Hinnebusch, Raymond, *The Iraq War: Causes and Consequences* (Boulder, CO: Lynne Rienner Publishers, 2006).

Freilich, Charles, 'Security Challenges and Opportunities in the Twenty-First Century' in Joel Peters and Rob Geist Pinfold (eds.), *Understanding Israel: Political, Societal and Security Challenges* (London: Routledge, 2019), pp. 185–8.

Fuller, Graham E., 'The Fate of the Kurds,' *Foreign Affairs*, vol. 72, no. 2 (1993), pp. 108–21.

Goldschmidt, Arthur and Davidson, Lawrence, *A Concise History of the Middle East* (Boulder, CA: Westview Press, 2016).

Gunter, Michael M (ed.), *The Routledge Handbook of the Kurds* (London: Routledge, 2018).

Halliday, Fred, *The Middle East in International Relations: Politics, Power and Ideology* (Cambridge: Cambridge University Press, 2012).

Hinnebusch, Raymond, *The International Politics of the Middle East* (Manchester: Manchester University Press, 2015).

Kyle, Keith, *Suez: Britain's End of Empire in the Middle East* (London: IB Tauris, 2011).

Morris, Benny, *1948: A History of the First Arab-Israeli War* (New Haven, NY: Yale University Press, 2009).

Murray, Williamson and Woods, Kevin M., *The Iran-Iraq War: A Military and Strategic History* (Cambridge: Cambridge University Press, 2014).

Oren, Michael, *Six Days of War: June 1967 and the Making of the Modern Middle East* (Toronto, CA: Presidio Press, 2003).

Stansfield, Gareth, *Iraq: People, History, Politics* (Cambridge: Cambridge University Press, 2007).

Walt, Stephen M., 'Wikileaks, April Glaspie and Saddam Hussein,' 2011, *Foreign Policy.* https://foreignpolicy.com/2011/01/09/wikileaks-april-glaspie-and-saddam-hussein/

30

WARS OF DECOLONISATION IN SOUTH ASIA

The Indian Army, 1945–1947

Alan Jeffreys

The Indian Army and its predecessor, the Honourable East India Company armies, propped up the British Empire in India (modern day India, Pakistan and Bangladesh). After the Rebellion in 1857–1858, the Company armies of the three presidencies were eventually combined to form the basis of today's Indian Army. The nineteenth-century Indian Army was instrumental in suppressing insurgency on the North-West and North-East Frontiers, civil unrest in India, as well as fighting further afield in Southeast Asia, China and even Africa. As the imperial strategic reserve, the Indian Army was capable of projecting military power around the world. In the First World War, almost 1.4 million men fought on an imperial and global scale. By the end of the Second World War, over 2 million personnel served in the Indian Armed Forces and India was the base for supplies for the Middle East and Southeast Asian theatres.

Historians Raymond Callahan and Daniel Marston have stated in a recent chapter on the Indian Army, aptly named 'Neglected Soldiers,' that 'The Memory of the Indian Army fell into a curious limbo after 1947' (Callahan and Marston 2018: 17). Indeed, until recently, the cut-off date was more like 1945. Since the independence of India, writing the history of the Indian Army has relied heavily on narrative histories, largely written by retired Indian Army officers and civil servants. For example, Philip Mason published *A Matter of Honour: An Account of the Indian Army, Its Officers and Men* (1974) that concentrates on British officers through the prism of secondary sources written by these officers. He finishes his history of the army at the end of the war with only brief mention of the Indian National Army (INA) trials in 1945, the Punjab Boundary Force and dividing the army between India and Pakistan (Mason 1974: 521–6). As a product of the Indian Civil Service, Mason views the army very much in paternalistic terms and upholds the institution on the whole. The book 'is heavily anecdotal and drenched in nostalgia' (Callahan and Marston 2018: 18). Peter Stanley goes as far as classifying this work as the 'romantic strand' in Indian military history (Stanley 2006: 226). In contrast, Victor Longer, as a post-Independence Ministry of Defence official, was wary of the colonial version, commenting:

> Needless to say that much of this [literature] was draped in the imperial ermine and the 'Indian' part of the Indian Army found little reference; much less attention was paid

DOI: 10.4324/9780429437915-38

to it than was warranted by the prefix. If there was some praise for the Indian Sepoys it was tardy and patronizing. In this wilderness of coloured landscape one has to tread carefully avoiding many pitfalls and proceed with caution and reserve, so that the Sahib's story is not swallowed hook, line and sinker.

(Longer 1974: viii)

However, Longer also notes that 'whereas some British source material is tainted and has to be avoided the Indian source material which is purely chauvinistic and flamboyant in its braggartism has also to be discarded' (Longer 1974: viii); but he concludes with the conundrum that Indian sources are not so easily available. Like Mason, Longer skims over the period after the war (Longer 1974: 246–71).

David Omissi has suggested that the rich archives, mainly in the India Office Library held in the British Library in London are due to the army being the instrument of control in India. As a result, the Indian soldiers were more scrutinised than other members of Indian society by the bureaucracy of the time (Omissi 1994: xvii). The first two academic studies of the Indian Army stopped short of the Second World War (Heathcote 1974; Omissi 1994; Heathcote 2013: xv; Cohen 1990 is the exception). Thus it is only in comparatively recent years that serious academic study of the Indian Army in the twentieth century has been attempted. For example, with the anniversaries of the First World War, a welcome raft of both academic and more popular books have been published on the Indian contribution (Morton-Jack 2014, 2018; Stanley 2015; Roy 2018; Jeffreys 2018). Similarly, the Indian Army during the Second World War has attracted much academic assessment since the beginning of the twenty-first century (Marston 2003; Moreman 2005; Callahan 2007; Raghavan 2016; Roy 2016; Jeffreys 2017; Barkawi 2017). By contrast, the army from the end of the Second World War until the 1947 Partition and its role in South, Southeast and East Asia has been largely forgotten until very recently (Marston 2014; Bayly and Harper 2007; Wilkinson 2015; Deshpande 2016; Roy 2013; Jeffreys and Rose 2012; White-Spunner 2017).

There has been a huge amount of work on the 1947 Partition and the period after the end of the Second World War by historians (Sarkar 2013; Bandyopadhyay 2015; Moore 1999; Bates 2007; Guha 2007; Brown 1985; Mansergh and Moon 1970–1983; Wolpert 2006; Khan 2007; Talbot and Singh 2009). The important studies of Christopher Bayly and Tim Harper have made the history of South and Southeast Asia in the 1940s and decolonisation in the region more widely accessible (Bayly and Harper 2004, 2007). Similarly, there is a growing amount of work on academic studies of decolonisation (Darwin 1988: 79–97; Hyam 2006: 105–16; Thomas 2014: 62–6, 97–115; Kennedy 2018; Chatterji 2018).

This chapter will look at the far-reaching role of the Indian Army in the two years between the end of the Second World War and the formation of the independent states of India and Pakistan in August 1947. At the end of the Second World War, all Indian forces, including those who had cooperated with the British Empire, those who chose non-cooperation and those who openly opposed the regime such as the INA, all combined in ending British rule in India as Britain could no longer sustain the Raj in India. Thus, India's war experience paved the way for independence (Khan 2015: 320; Brown 1985: 310–15; Bates 2007: 155–66; Moore 1979; Mukherjee 2015: 2; Siegel 2018).

After the disastrous fall of Singapore on 15 February 1942, the prestige of the British Empire plummeted across Southeast and South Asia (Buchanan 2011; Kamtekar 2002). The INA had two incarnations; initially what the wartime intelligence officer Hugh Toye has termed 'The First Indian National Army' was established in February 1942 as a branch of the Indian Independence League. It comprised about 20,000 Indian soldiers under the leadership of Captain Mohan Singh, largely recruited from prisoners of war. However, Mohan Singh became increasingly unpopular with the Indian Independence League, and the Japanese paymasters. He was arrested in December 1942 and the INA dissolved on 1 January 1943, after which he spent the remainder of the war in isolation (Toye 1984; Marston 2014: 119–22).

The leadership of the better-known INA was undertaken by the Bengali nationalist politician, Subhas Chandra Bose, which he combined with the leadership of the Indian Independence League (Bose 2013; Fay 1997; Sundaram 2007; Noles 2014, 2017). This incarnation of the INA was created into three divisions. The 1st INA Division fought at Imphal and the 2nd INA Division, commanded by Colonel G.S. Dhillon, was on the Irrawaddy during Operation Extended Capital where it faced the Indian soldiers of 7th Indian Division. However, as a military organisation it was flawed. This formation was overrun by 7th Indian Division and in retreat after two days of fighting, with the majority of the troops deserting or surrendering to the British. GHQ India categorised the surrendering INA into three groups— 'whites,' 'greys' and 'blacks' according to their loyalty to the crown with white being loyal and black being disloyal and eligible for prosecution. Indeed, there was strong feeling against them in the Indian Army as well as those prisoners of war who remained loyal (Mansergh and Moon vol. VI 1976: 110).

Despite being a flawed military army, the INA were much more of a political threat to the Raj as they were seen as freedom fighters. Gajendra Singh has described the INA as the second of the two largest rebellions in British military history (Singh 2014: 157); the first being the 1857 Rebellion followed by the Jallianwala Bagh massacre in 1919 as the early portents of decolonisation. The trials of the three leading INA officers: Captain Shah Nawaz, Captain P.K. Saghal and Lieutenant G.S. Dhillon, held at the Red Fort in Delhi in November 1945, attracted huge negative publicity for the Indian Army and the Raj. The INA was increasingly supported and feted by the Indian National Congress Party. The three officers were found guilty of waging war against the King and sentenced to transportation for life, which General Claude Auchinleck, Commander-in-Chief India, commuted to prevent further unrest, and many of the 'blacks' were released as they had not carried out 'brutal attacks,' with further trials cancelled. This in effect was a step down for the Indian Army and portrayed as a victory for the Indian National Congress, even uniting the Congress Party and the Muslim League over the INA issue (Marston 2014: 119–42, 145–50; Kuracina 2010). Many of the 'greys' and 'blacks' who had to leave the army became 'military advisors' to various political paramilitary groups such as the Hindu nationalist Rashtriya Swayamsevak Sangh (RSS), the Muslim League National Guards and the Sikh *Jathas* and implicated in the violence during Partition (Marston 2014: 119).

There were problems in both the RAF and Royal Indian Air Force (RIAF) in India in 1946 due to demands for early demobilisation and better pay. Although these issues were dealt with peaceably by British and Indian Army units, the Royal Indian Navy Mutiny of 1946, concurrent with the INA trials, started at the signal school HMIS Talwar in Bombay on 8 February when several sailors were court-martialled for insubordination after racial abuse by the commanding officer, Commander Freddie King. It spread to over 10,000 men, 56 ships and ten naval barracks as well as rioting in Bombay, Calcutta, and later in Karachi and other cities. The mutineers had a list of initial demands for improvement within the service due to a combination of disinterested officers and dissatisfaction with food, pay and living conditions, but later extended it to more political demands including the release of all INA prisoners. According to David Omissi, mutinies in the Indian Army in the period 1886–1930 had one thing in common: 'In each case, a British action broke the tacit bargain between officers and men which informed the discipline of the Indian Army' (Omissi 1994: 138). This divide between the perennial problem of mutinies due to bad conditions or more political reasons continues in the historiography with many historians viewing the mutiny through the nationalist narrative whilst others see it as a service issue (Deshpande 2016; Spector 1981; Davies 2013; Meyer 2017). However, Daniel Spence has demonstrated that a range of factors were responsible for the uprising. There was little common cause between the sailors, let alone the civilians, and indeed the sailors were often alienated from Indian society themselves (Spence 2015: 492, 495–99). Demobilisation and the fear of lack of employment was predominant amongst the sailors, just like the RAF and RIAF mutinies in Karachi, the mutiny was put down by British and Indian soldiers but 1,000 people were arrested, 1,000 were injured and 200 killed. The mutiny had little effect on the Indian soldiers involved, many of whom had little sympathy for their naval counterparts (Marston 2014: 144–5). Anirudh Deshpande has pointed out the response of the Indian National Congress Party, under Jawaharlal Nehru, which was to disown the uprising as it had not been instigated by the party and he was busy 'supervising the transformation of the Congress from a party of political agitation to a party of governance' (Deshpande 2016: 5). Following independence, the naval mutineers were largely ignored, unlike the INA who were feted as freedom fighters.

At the end of the Second World War, Indian Army formations helped restore colonial empires in Saigon in French Indo-China (Vietnam) and on Java and Sumatra in the Dutch East Indies (Indonesia), as well as contributing to the occupation forces in Japan and Southeast Asia. Auchinleck's appraisal of the situation in November 1945 stated that the Indian armed forces were currently capable of dealing with communal and anti-Government disturbances but this situation would not necessarily last until the following year due to demobilisation, the INA trials, the nationalism of most Indian officers and the Congress campaign against Indian soldiers propping up European empires in French Indo-China and the Netherlands East Indies (NEI). In his appraisal of the situation, he concluded:

> Our action in Java and French Indo-China is already being represented as European repression of national risings of Eastern peoples. If this is made a major political issue as is likely, it may have a serious effect upon the loyalty of the Indian Armed Forces. It is certainly very undesirable that any further Indian troops should be sent to these or other similar countries.
>
> *(L/WS/1/1008, India Office Record, British Library)*

Nevertheless, Indian troops remained in Saigon, Java and Sumatra until 1946. They had to deal with a multitude of issues, ranging from returning prisoners of war and civilian internees, disarming Japanese soldiers and COIN operations against nationalist guerrillas (Marston 2014: 153). Major-General Douglas Gracey and 20th Indian Division arrived in Saigon in early September 1945 with the role of establishing control, providing support for the French authorities and disarming the IJA. Although Japanese troops continued to often act in a defensive role due to the small numbers of Allied troops available, a much smaller number even joined the Viet Minh as military advisors (Spector 2005: 1131). Gracey was much criticised, even at the time, for imposing martial law but it was an impossible situation that

20th Indian Division had been sent to sort out with the French, viewing them as 'more or less convivial peace enforcers' (Thomas 2014: 121; Bayly and Harper 2007: 147–8; Smith 2014: 2–6, 44–7; Goscha 2016: 216–17). There was serious fighting between the division and the Viet Minh which the historian of the 9th/14th Punjab Regiment described as 'an unsatisfactory sort of fighting. The enemy wore no uniform and usually did not carry arms visibly' (*Ninth Battalion Fourteenth Punjab Regiment* 1948: 97). The troops had been previously well trained and thus were capable of adapting to the situation (Jeffreys 2017: 209–10; Smith 2014: 31). Between October 1945 and January 1946, the division suffered more than 100 casualties; 40 soldiers had died in the period and 54,000 Japanese troops had been disarmed (Marston 2014: 158–74; Bayly and Harper 2007: 140–58; Smith 2014: 42–56; Dennis 1987; Springhall 2005; Hughes 2006). It left Saigon by the end of March 1946. As Daniel Marston has concluded, the British government was unclear what role British and Indian Armies should undertake in the restoration of the colonial empires of their European allies. Indeed, as he states,

> The other irony is that the new Labour Government, in haste to end the British presence in India, nevertheless saw the Indian Army as an imperial reserve with unlimited abilities to project British power and support for allies, with the added benefit of few casualties to trouble the British electorate.
>
> *(Marston 2012: 177; Smith 2014: 54–5)*

There were similarities with the situation in the NEI with a lack of clear direction from both South East Asia Command under Lord Louis Mountbatten and the British government. The role included maintaining control of the main cities until the return of the Dutch, even though Dr Sukarno had declared independence on Java on 17 August 1945. Ultimately three Indian divisions were deployed under the command of General Sir Philip Christison as Commander-in-Chief, NEI (Marston 2014: 174–99; Bayly and Harper 2007: 158–89; McMillan 2005). Christison met Jack Lawson, the Secretary of State for War before taking up his post, who told him:

> Mr Bevin (Foreign Secretary) has asked me to make it quite clear to you that HM Government are determined that nothing should be done to suggest your troops are going to re-impose Dutch Colonial rule. You must not take sides. Carry out your role; it may take up to 6 months before Dutch troops can be trained and sent out from Holland.
>
> *('Life and Times of General Sir Philip Christison': 176)*

However, the operations involved some violent fighting, particularly at Sourabaya on 28 October 1945 where 49th Indian Infantry Brigade (ca. 4,000 largely Indian troops) faced about 20,000 Tentara Keamana Rakyat (Indonesian Republican Army—trained and equipped by the Japanese) and 100,000 armed civilians resulting in the capture, torture and killing of several British and Indian officers—16 officers and 217 other ranks died, including the commander Brigadier Maltby ('Report of Operations of 49th Indian Infantry Brigade'; Marston 2014: 188; Bayly and Harper 2007: 177; Springhall 1996; Jordan 2000). The 5th Indian Division, commanded by Major-General Robert Mansergh, was drafted in to quell the situation, which was achieved by the end of November. By the withdrawal of the Indian divisions at the end of November 1946, over 600 men and officers had been killed on Java and Suma-tra. Even with the impending independence of India and Pakistan, the Indian Army was still considered by the British government and Mountbatten as the strategic reserve in Southeast Asia. Only four of the 30 battalions were British (Bayly and Harper 2007: 172). Auchinleck suggested that 2nd British Division be sent to the NEI but was overridden by Mountbatten with 5th Indian Division backing up 23rd and 26th Indian Divisions (Dennis 1987: 121). Indeed, when Christison had to get his plan approved for the clearing up of Batavia, he wrote in his memoir:

> Mountbatten and Alanbrook then approved the plan and Mountbatten said to see Indian troops were used; he did not want British troops widowed at this time so long after the war. This angered Bill [Lieutenant Colonel Bill Ridley] who bravely said: 'Sir, do you really think it is different if Mrs Poop Singh is made a widow?' 'Tell your commander his plan is approved' said Alanbrook [Alanbrooke].
>
> *('Life and Times of General Sir Philip Christison': 188)*

Indian Army troops were also deployed as a brigade of the occupation forces in Japan, in Burma (Myanmar), Hong Kong, Siam (Thailand), Borneo and Malaya (Malaysia) (Singh 2012; Charney 2009: 58–71; McEnery 1990). The army continued to adapt to their new roles, for instance William Arthur has argued that Indian troops became 'soldier-administrators' who were instrumental for nation-building in Malaya (Arthur 2013); although the corruption and behaviour of British and Indian troops in Malaya and Singapore under-mined public confidence even further in 1945, and at least as much as in 1942 (Springhall 2001: 639; Spector 2007: 85). Between 1 June and 30 November 1946, 180,000 British and Indian troops returned to India including: HQ 15 Indian Corps, 26th Indian Division, 50th Indian Tank Brigade, 80th Indian Infantry Brigade (all from NEI), 7th Indian Division from Malaya and 32nd Indian Infantry Bridge from Borneo. In addition, 23rd Indian Division was moved from NEI to Malaya and 17th Indian Division remained in Burma ('Draft Post-War Despatch': 38).

The Indian Army was a thoroughly professional army by 1945 (Callahan 2007; Jeffreys 2017; Marston 2003; Moreman 2005). It was also a politically neutral national army, but Indian soldiers were much more aware of political issues through their experience of the war and the use of discussion or 'Josh' groups, underpinned by such educational pamphlets as *Current Affairs* (Jeffreys 2017: 211; Wheeler 1944, e.g., 'Current Affairs No. 6'). It was deemed essential to keep Indian troops well informed regarding the political situation, and junior officers were forbidden to denounce M.K. Gandhi, Nehru or other members of the Congress Party or Muslim League being regarded as national leaders by both the Indian officers and soldiers. For instance, Major-General A.A. Rudra consulted with Gandhi on whether he should remain in the Indian Army (Palit 1997: 51–2, 71–2). The Indian officers in particular, who had only numbered 344 at the beginning of the Second World War, had increased to 15,540 by 1945. This meant experienced Indian officers were in a position to take over the army once British officers had left after independence and there were more opportunities for promotion (Jeffreys 2017: 37, 48, 202; Khanduri 1995: 132). There was a shortfall of officers in the Pakistan Army, which was made up with about 500 British offic-ers (Talbot 2015: 58). General Sir Frank Messervy initially commanded The Pakistan Army, only to be succeeded by General Sir Douglas Gracey in February 1948. British officers held most of the senior positions, although officers in both armies were quickly promoted, sometimes beyond their capabilities (Nawaz 2017: 32; 'Recommendations of the Armed Forces' 1947: 4).

The Calcutta Riots of August 1946, between the Muslim and Hindu communities, resulted in 3,000 dead and 17,000 wounded. They spread to East Bengal, Bihar and the United Provinces. The British government decreed in 1946 that British Army units could only be deployed in India in 'aid to civil power' if British lives were at risk. This meant that virtually all peacekeeping duties in 1946–1947 fell to the Indian Army. In addition, the deepening breakdown of order in the Punjab, the main recruiting ground of the Indian Army, further affected morale in the Indian Army. Much of the violence was orchestrated by demobilised soldiers (Marston 2014: 215–32, 290–5; Tuker 1950; Wainright 1970). At the same time, the army underwent demobilisation and nationalisation, and went about the colossal task of dividing between the new states of Pakistan and India whilst simultaneously using it to try to curtail the spreading of violence (Marston 2014: 239–80; Prasad 2012: 195–248; Deshpande 2005: 145–84; 'Recommendations of the Armed Forces' 1947; Nath 2012; Omissi 2012; Bayly and Harper 2007: 287; Khan 2007: 114–16). Christine Fair has concluded that 'Pakistan did get the short end of the stick in terms of the division of fixed assets, because the bulk of the infrastructure was located in India' (Fair 2014: 56–9).

The Labour government under Prime Minister Clement Attlee announced on 20 February 1947 that the British would transfer power no later than June 1948. Lord Mountbatten was made the last Viceroy of India to oversee the process. Indian leaders accepted his 3 June Partition plan that included a Partition Council and Boundary Commission. The date for Independence was brought forward to 14 August for Pakistan and 15 August 1947 for India. Sir Cyril Radcliffe, Chairman of the Punjab and Bengal Boundary Commission, quickly drew up the boundary awards that led to mass migration and massacres in the Punjab (Chester 2009). Many demobilised soldiers in the Punjab joined militias to carry out 'ethnic cleansing campaigns' (Chester 2009: 132–3; Savory 1947). The Punjab Boundary Force (PBF) was established in July 1947, commanded by Major-General Thomas 'Pete' Wynford Rees, who had the impossible task of policing this transfer of power. It initially comprised five brigades to cover 12 districts numbering a population of 14.5 million. The PBF had to deal with the eruption of violence alone as the police and civil administration collapsed, which meant that the PBF acting as an 'aid to civil power' was redundant. Daniel Marston has concluded that it tested the Indian Army to the extreme and that 'if the army had collapsed in the Punjab, there would have been nothing left but anarchy' (Marston 2014: 337); although Penderel Moon, in his memoir of the period, commented that 'I had written off the Boundary Force completely' (Moon 1998: 95).

The Military Evacuation Organisation (MEO) in India and Pakistan organised the movements of populations, resulting in the population exchanges of East Punjab by Muslims and of West Punjab by Hindus and Sikhs. For example, from August 1947 until May 1948, 2,341,040 were evacuated by the MEO Pakistan from East to West Punjab by foot, 218,405 by military transport and 1,156,474 by rail (Talbot and Singh 2009: 96; Marston 2014: 339). By contrast, in East and West Bengal, migrants continued for decades after the 1947 Partition (Chatterji 2007). Many urban refugees travelled for years before finally settling down. The trauma of separation was reflected in the 'Refugees Forum' in the newspapers of the Punjab.

There was a sense of inevitability of decolonisation at the end of the Second World War as Britain was exhausted by the war and the empire had lost any remaining legitimacy in South

Asia, together with the drive of nationalist support from within. Partition was a consequence of both the need to divide the empire in South Asia and the divisions between the Congress Party and the Muslim League. The bringing forward of the timetable rushed the process further in the vain hope of curtailing growing communal violence. The numbers of deaths during the 1947 Partition are unknown, but range from 200,000 to 1 million; the violence was bloody, with the legacy lasting for decades.

Bibliography

Anon, *Ninth Battalion Fourteenth Punjab Regiment* (Cardiff: Western Mail and Echo, 1948).

Bandyopadhyay, Sekhar, *From Plassey to Partition and after: A History of Modern India*, 2nd ed. (New Delhi: Orient Blackswan, 2015).

Barkawi, Tarak, *Soldiers of Empire: Indian and British Armies in World War II* (Cambridge: Cambridge University Press, 2017).

Bates, Crispin, *Subalterns and Raj: South Asia since 1600* (London: Routledge, 2007).

Bayly, Christopher and Harper, Tim, *Forgotten Armies: The Fall of British Asia, 1941–1945* (London: Allen Lane, 2004).

Bayly, Christopher and Harper, Tim, *Forgotten Wars: The End of Britain's Asian Empire* (London: Allen Lane, 2007).

Bose, Sugata, *His Majesty's Opponent: Subhas Chandra Bose and India's Struggle against Empire* (Gurgaon: Penguin, 2013).

Brown, Judith, *Modern India: The Origins of an Asian Democracy* (Oxford: Oxford University Press, 1985).

Buchanan, Andrew, 'The War Crisis and the Decolonization of India, December 1941–September 1942: A Political and Military Dilemma,' *Global War Studies*, vol. 8, no. 2 (2011), pp. 5–31.

Callahan, Raymond, *Churchill and his Generals* (Lawrence, KS: University Press of Kansas, 2007).

Callahan, Raymond and Marston, Daniel, 'Neglected Soldiers,' in Alan Jeffreys (ed.), *The Indian Army in the First World War: New Perspectives* (Solihull: Helion, 2018), pp. 17–39.

Charney, Michael, *A History of Modern Burma* (Cambridge: Cambridge University Press, 2009).

Chatterji, Joya, *The Spoils of Partition: Bengal and India, 1947–1967* (Cambridge: Cambridge University Press, 2007).

Chatterji, Joya, 'Decolonization in South Asia: The Long View,' in Martin Thomas and Andrew Thompson (eds.), *The Oxford Handbook of the Ends of Empire* (Oxford: Oxford University Press, 2018), doi:10.1093/oxfordhb/9780198713197.013.15.

Chester, Lucy, *Borders and Conflict in South Asia: The Radcliffe Boundary Commission and the Partition of Punjab* (Manchester: Manchester University Press, 2009).

Cohen, Stephen, *The Indian Army: Its Contribution to the Development of a Nation* (Delhi: Oxford University Press, 1990).

'Current Affairs No. 6 (Indian Troops), December 1944—Release from Service: The Official Plan of Release Explained and Illustrated,' Imperial War Museum (IWM), London.

Darwin, John, *Britain and Decolonisation: The Retreat from Empire in the Post-War World* (Basingstoke: Palgrave Macmillan, 1988).

Davies, Andrew D., 'From "Landsman" to Seaman'? Colonial Discipline, Organisation and Resistance in the Royal Indian Navy, 1946,' *Social & Cultural Geography*, vol. 14, no. 8 (2013), pp. 868–87.

Dennis, Peter, *Troubled Days of Peace: Mountbatten and South East Asia Command, 1945–46* (Manchester: Manchester University Press, 1987).

Deshpande, Anirudh, *British Military Policy in India, 1900–1945* (New Delhi: Manohar, 2005).

Deshpande, Anirudh, *Hope and Despair: Mutiny, Rebellion and Death in India, 1946* (New Delhi: Primus, 2016).

'Draft Post-War Despatch of General Sir Montagu Stopford Acting Supreme Allied Commander South East Asia, 1 Jun 1946 to 30 Nov 1946,' Documents 6697, Private Papers of General Sir Robert Mansergh (IWM).

Fair, C. Christine, *Fighting to the End: The Pakistan Army's Way of War* (New Delhi: Oxford University Press, 2014).

Fay, Peter Ward, *The Forgotten Army: India's Armed Struggle for Independence 1942–1945* (New Delhi: Rupa, 1997).

Goscha, Christopher, *The Penguin History of Modern Vietnam* (London: Penguin, 2016).

Guha, Ramachandra, *India after Gandhi* (London: Macmillan, 2007).

Heathcote, T.A., *The Indian Army: The Garrison of British Imperial India 1822–1922* (Newton Abbot: David & Charles, 1974).

Heathcote, T.A., *The Military in British: The Development of British Land Forces in South Asia 1600–1947* (1995, reprint, Barnsley: Praetorian Press, 2013).

Hughes, Geraint, 'A "Post-war" War: The British Occupation of French-Indochina, September 1945–March 1946,' *Small Wars & Insurgencies*, vol. 17, no. 2 (2006), pp. 263–86.

Hyam, Ronald, *Britain's Declining Empire: The Road to Decolonisation 1918–1968* (Cambridge: Cambridge University Press, 2006).

'Internal Situation in India, Appreciation by the Commander-in-Chief,' 24 November 1945, L/WS/1/1008, India Office Records, British Library, London.

Jeffreys, Alan, *Approach to Battle: Training the Indian Army during the Second World War* (Solihull: Helion, 2017).

Jeffreys, Alan (ed.), *The Indian Army in the First World War: New Perspectives* (Solihull: Helion, 2018).

Jeffreys, Alan and Rose, Patrick (eds.), *The Indian Army, 1939–47: Experience and Development* (Farnham: Ashgate, 2012).

Jordan, David, '"A Particularly Exacting Operation": British Forces and the Battle of Surabaya, November 1945,' *Small Wars & Insurgencies*, vol. 11, no. 3 (2000), pp. 89–114.

Kamtekar, Indivar, 'The Shiver of 1942,' *Studies in History*, vol. 18, no. 1 (2002), pp. 81–102.

Kennedy, Dane, *The Imperial History Wars: Debating the British Empire* (London: Bloomsbury, 2018).

Khan, Yasmin, *The Great Partition: The Making of India and Pakistan* (New Haven, CT: Yale University Press, 2007).

Khan, Yasmin, *The Raj at War: A People's History of India's Second World War* (London: Bodley Head, 2015).

Khanduri, Brigadier C.B., *Field Marshal K. M. Cariappa: His Life and Times* (New Delhi: Lancer, 1995).

Kuracina, William, 'Sentiments and Patriotism: The Indian National Army, General Elections and the Congress's Appropriation of the INA Legacy,' *Modern Asia Studies*, vol. 44, no. 4 (2010), pp. 817–56.

'Life and Times of General Sir Philip Christison,' Documents 4370, Private Papers of General Sir Philip Christison, IWM.

Longer, V., *Red Coats to Olive Green: A History of the Indian Army 1600–1974* (Bombay: Allied Publishers, 1974).

Mansergh, Nicholas and Moon, Penderel (eds.), *The Transfer of Power 1942–7* (London: HMSO, 1970–83).

Marston, Daniel, *Phoenix from the Ashes: The Indian Army in the Burma Campaign* (Westport, CT: Praeger, 2003).

Marston, Daniel, 'The 20th Indian Division in French Indo-China,' in Alan Jeffreys and Patrick Rose (eds.), *Indian Army, 1939–47: Experience and Development* (Farnham: Ashgate, 2012), pp. 157–78.

Marston, Daniel, *The Indian Army and the End of the Raj* (Cambridge: Cambridge University Press, 2014).

Mason, Philip, *A Matter of Honour: An Account of the Indian Army, Its Officers and Men* (London: Jonathan Cape, 1974).

McEnery, John, *Epilogue in Burma 1945–48* (Tunbridge Wells: Spellmount, 1990).

McMillan, Richard, *The British Occupation of Indonesia, 1945–1946* (London: Routledge, 2005).

Meyer, John M., 'The Royal Indian Navy Mutiny of 1946: Nationalist Competition and Civil-Military Relations in Postwar India,' *Journal of Imperial and Commonwealth History*, vol. 45, no. 1 (2017), pp. 46–69.

Moon, Penderel, *Divide and Quit: An Eyewitness Account of the Partition of India* (Delhi: Oxford University Press, 1998).

Moore, R.J., *Churchill, Cripps, and India: 1939–1945* (Oxford: Clarendon Press, 1979).

Moore, Robin J., 'India in the 1940s,' in Robin W. Winks (ed.), *The Oxford History of the British Empire: Historiography* (Oxford: Oxford University Press, 1999), pp. 231–42.

Moreman, Tim, *The Jungle, the Japanese and the British Commonwealth Armies at War 1941–45: Fighting Methods, Doctrine and Training for Jungle Warfare* (London: Frank Cass, 2005).

Morton-Jack, George, *The Indian Army on the Western Front: India's Expeditionary Force to France and Belgium in the First World War* (Cambridge: Cambridge University Press, 2014).

Morton-Jack, George, *The Indian Empire at War: From Jihad to Victory: The Untold Story of the Indian Army in the First World War* (London: Little, Brown, 2018).

Mukherjee, Janam, *Hungry Bengal: War, Famine and the End of Empire* (London: Hurst, 2015).

Nath, Ashok, 'Re-forging the Damascus Blade: Partition of the Indian Armoured Corps, 1947,' Alan Jeffreys and Patrick Rose (eds.), *The Indian Army: Experience and Development* (Farnham: Ashgate, 2012), pp. 179–93.

Nawaz, Shuja, *Crossed Swords: Pakistan, its Army, and the Wars Within* (Karachi: Oxford University Press, 2017).

Noles, Kevin, '"Waging War against the King": Recruitment and Motivation of the Indian National Army, 1942–1945,' *British Empire at War Research Paper*, no. 6 (2014), pp. 1–48.

Noles, Kevin, 'Renegades in Malaya: Indian Volunteers of the Japanese, F. Kikan,' *British Journal for Military History*, vol. 3, no. 2 (2017), pp. 100–18.

Omissi, David, *The Sepoy and the Raj: The Indian Army, 1860–1940* (Basingstoke: Macmillan, 1994).

Omissi, David, 'A Dismal Story? Britain, the Gurkhas and the Partition of India, 1945–1948,' in Alan Jeffreys and Patrick Rose (eds.), *The Indian Army, 1939–47: Experience and Development* (Farnham: Ashgate, 2012), pp. 195–214.

Palit, Major-General D.K., *Major General A.A. Rudra: His Service in Three Armies and Two World Wars* (New Delhi: Reliance, 1997).

Prasad, Sir Nandan, *Expansion of the Armed Forces and Defence Organisation 1939–45: Official History of the Indian Armed Forces in the Second World War 1939–45* (New Delhi: Pentagon Press, 2012).

Raghavan, Srinath, *India's War: The Making of Modern South Asia 1939–1945* (London: Allen Lane, 2016).

'Recommendations of the Armed Forces Nationalisation Committee 1947,' subject files II, File no. 1, Private Papers of General K.S. Thimayya, New Delhi, Nehru Memorial Museum and Library (NMML).

'Report of Operations of 49th Indian Infantry Brigade 25 October–8 November 1946,' Documents 6697, Private Papers of General Sir Robert Mansergh, IWM.

Roy, Kaushik, *The Army in British India: From Colonial Warfare to Total War 1857–1947* (London: Bloomsbury, 2013).

Roy, Kaushik, *India and World War II: War, Armed Forces and Society, 1939–45* (New Delhi: Oxford University Press, 2016).

Roy, Kaushik, *Indian Army and the First World War 1914–1918* (New Delhi: Oxford University Press, 2018).

Sarkar, Sumit, *Modern India: 1885–1947* (Delhi: Macmillan, 2013).

Savory, Reginald, 'The Sikhs' by Reginald Savory written in September 1947, Private Papers of Lieutenant General Sir Reginald Savory, London, National Army Museum, 7603-93-92.

Siegel, Benjamin, *Hungry Nation: Food, Famine, and the Making of Modern India* (Cambridge: Cambridge University Press, 2018).

Singh, Gajendra, *The Testimonies of Indian Soldiers and the Two World Wars* (London: Bloomsbury, 2014).

Singh, Brigadier Rajendra, *Post-War Occupation Forces: Japan and South-East Asia* (New Delhi: Pentagon Press, 2012) reprint of the Official History of the Indian Armed Force in the Second World War 1939–45, published in 1958.

Smith, T.O., *Vietnam and the Unravelling of Empire: General Gracey in Asia 1942–1951* (Basingstoke: Palgrave Macmillan, 2014).

Spector, Ronald, 'The Royal Indian Navy Strike of 1946: A Study of Cohesion and Disintegration in Colonial Armed Forces,' *Armed Forces and Society*, vol. 7, no. 2 (1981), pp. 271–84.

Spector, Ronald H., 'After Hiroshima: Allied Military Operations and the Fate of Japan's Empire, 1945–1947,' *Journal of Military History*, vol. 69 (2005), pp. 1121–36.

Spector, Ronald H., *In the Ruins of Empire: The Japanese Surrender and the Battle for Postwar Asia* (New York: Random House, 2007).

Spence, Daniel Owen, 'Beyond *Talwar*: A Cultural Reappraisal of the 1946 Royal Indian Navy Mutiny,' *Journal of Imperial and Commonwealth History*, vol. 43, no. 3 (2015), pp. 489–508.

Springhall, John, '"Disaster in Surabaya": The Death of Brigadier Maltby during the British Occupation of Java, 1945–46,' *Journal of Imperial and Commonwealth History*, vol. 24, no. 3 (1996), pp. 422–43.

Springhall, John, 'Mountbatten versus the Generals: British Military Rule of Singapore, 1945–46,' *Journal of Contemporary History*, vol. 36, no. 4 (2001), pp. 635–52.

Springhall, John, '"Kicking out the Vietminh": How Britain Allowed France to Reoccupy South Indochina, 1945–46,' *Journal of Contemporary History*, vol. 40, no. 1 (2005), pp. 115–30.

Stanley, Peter, 'Imperial Military History,' in Matthew Hughes and William J. Philpott (eds.), *Modern Military History* (Basingstoke: Palgrave, 2006), pp. 214–30.

Stanley, Peter, *Die in Battle, Do Not Despair: The Indians on Gallipoli, 1915* (Solihull: Helion, 2015).

Sundaram, Chandar S., 'The Indian National Army, 1942–1946: A Circumstantial Force,' in Daniel Marston and Chandar Sundaram (eds.), *A Military History of India and South Asia* (Westport, CT: Praeger, 2007), pp. 123–30.

Talbot, Ian, *Pakistan: A New History* (London: Hurst, 2015).

Talbot, Ian and Singh, Gurharpal, *The Partition of India* (Cambridge: Cambridge University Press, 2009).

Thomas, Martin, *Fight or Flight: Britain, France, and their Roads from Empire* (Oxford: Oxford University Press, 2014).

Toye, Hugh, 'The First Indian National Army, 1941–42,' *Journal of Southeast Asian Studies*, vol. 15, no. 2 (1984), pp. 365–81.

Tuker, Lieutenant-General Sir Francis, *While Memory Serves* (London: Cassell, 1950).

Wainright, Mary Doreen, 'Keeping the Peace in India, 1946–1947: The Role of Lieutenant-General Sir Francis Tuker in Eastern Command,' in C.H. Philips and M.D. Wainright (eds.), *The Partition of India: Policies and Perspective, 1935–1947* (London: Allen & Unwin, 1970), pp. 127–47.

Wheeler, Geoffrey, *Swords and Plowshares: The Indian Army as a Social Force* (Washington: Government of India Information Services, 1944).

White-Spunner, Barney, *Partition: The Story of Indian Independence and the Creation of Pakistan in 1947* (London: Simon & Schuster, 2017).

Wilkinson, Steven, *Army and Nation: The Military and Indian Democracy since Independence* (Cambridge, MA: Harvard University Press, 2015).

William, Arthur, 'The Padang, the Sahib and the Sepoy: The Role of the Indian Army in Malaya, 1945–1946' (Oxford University, DPhil thesis, 2013).

Wolpert, Stanley, *Shameful Flight: The Last Years of the British Empire in India* (Oxford: Oxford University Press, 2006).

31

THE FALKLANDS CONFLICT

The Last British Imperial War?

Rob McLaughlin

Introduction

The Falklands War is often described as a punctuation mark, or a plateau, in the hitherto apparently dominant (at the time) narrative of post-imperial British decline (Ward 2001: 1–20; Srinivasan 2006: 257; McCormick 2013: 102). Yes, Britain suffered 240 military and civilian casualties, and lost ships and aircraft (UK Government Falklands Official Statistics). Argentina also suffered significant losses of life (655) and assets, and more than 10,000 were, for a short time, prisoners of war (POWs) at the conclusion of hostilities (Middlebrook 1989: 282; Rattner 1982). But at the end of the war, the Falkland Islands and their inhabitants remained under the British flag, although the diplomatic tensions over sovereignty and possession of the Islands continue to this day (Ince 2012).

In the near aftermath, a Conservative government in Britain reaped the benefits of victory through re-election (Sanders et al. 1987: 313; Franklin and Freedman 1987: 29; Clarke et al. 1990: 80–1; UN Special Committee 2017), whilst the defeated military junta in Argentina soon fell (Pion-Berlin 1985: 73). Yet victory brought introspection for Britain—the focus of this chapter—as well; lessons identified or re-affirmed from the war. The lessons of the war ranged across military preparedness and equipping from the flammability of fittings and materials in warships (Hanrahan and Fox 1982: 24–8; Richards 1983: 791; Winnefeld 1983: 8); the utility of aircraft carriers in expeditionary operations; the need for improved anti-air warfare and airborne early warning capabilities (Speller 2002), through to executive and political communication processes (not least with the Falkland Islanders themselves) (Franks Report 1983: paras 83, 88–9, 91, 106; Lebow 1983: 6; Seymore-Ure 1984: 181). Journalists mused (and still do) on their role in the war (Humphries 1983: 60–4; McLaughlin 2016: 118–23; Rowley 2015), diplomats discussed the fitness for purpose of the international system in the face of such a crisis (Parsons 1983: 169; Windsor 1983: 88), and historians and strategists looked for continuities with the past, indicators for the future (Secretary of State for Defence 1982; Grove 2002: 312), and considered the revitalisation of Just War theory that the Falklands conflict wrought (Bluth 1987: 9–17; Bellot 2018: 79)—at least in Britain, anyway.

For Britain, consequently, the Falklands conflict marked, and continues to mark, a counter-intuitive, remedial moment in the national narrative. It was a short war, a limited war,

DOI: 10.4324/9780429437915-39

but it was (on most measures) a successful war,[1] even though some successes caused disquiet away from the battlefield (New Statesman 1984: 3; PM Letter 1984; PM Letter 1985; PM Principal Private Secretary Letter 1984; Leak Investigation 1984). The sinking of the light cruiser *General Belgrano*, for example—'the engagement with the greatest casualties' in the war—'came right at the start of the actual fighting. It was an important military victory for Britain, yet it turned into a political defeat because of the premium the international community put on the appearance of avoiding escalation' (Freedman 1988: 81). As Bluth has noted,

> It will remain a matter of controversy as to whether there ever was a chance of a peaceful settlement during this week-end while the so-called Peruvian peace proposals were being negotiated. It is beyond doubt, however, that whatever possibilities there may have been were destroyed by the sinking of the Belgrano on 2 May.
>
> *(Bluth 1987: 15)*

But the Falklands conflict was not the only conflict Britain has been involved in over the last 40 years that has links to the history and consequences of imperialism. Operations in Sierra Leone, for example, held some imperial resonance as a place that was once under the British flag. But the Falklands War almost uniquely speaks to the notion of a late twentieth-century British 'imperial war'? This is the focus of this chapter.

In order to explore this question, I will focus on five lenses of 'imperial war,' asking if and whether they are relevant to the case study of the Falklands conflict. These five lenses are by no means the only ways by which to frame the question, and none of them are entirely correlate; but each provides a degree of insight as to the ways in which the Falklands War might, or might not, be described as an 'imperial war.' The five lenses are: (1) Commonwealth participation; (2) imperial policing; (3) end of empire defiance; (4) post-imperial wars of unwinding; and (5) imperial attitudes. Prior to this, however, a short recapitulation of the context, and identification of the key potential 'imperial' concept at play—sovereignty—is useful.

Context

The Conflict

The sudden escalation of tensions between Britain and Argentina was a product of a situation, as Freedman observes, 'when a relatively low-priority issue in one country [had] a high priority in another' (Freedman 1988: 2). Simmering grievance, exacerbated and empowered by—and in part designed to redirect—domestic frustrations and challenges in Argentina (Cawkell 1983: 86; Dabat and Lorenzano 1984: 60–77), prompted an unexpected (from the UK view) attempt at a fait accompli. But this was based on miscalculations born of the fact that Britain had managed to convey the impression of intransigence in negotiations on the principle of sovereignty but no real interest in holding on to the islands (Boyce 2005: 27–34; Freedman 2005: 17–41, 99–133). As one commentary—drawing parallels with the South China Sea context—has observed:

> States will fight for pointless rocks, especially when they're not pointless: Unlike many of the rocks currently at dispute in East Asia, the Falklands support a small

but enthusiastic population of British citizens. The Argentines fooled themselves into thinking that, despite this population, the British would not fight.

(The Diplomat *2013*)

However, miscalculation dogged both camps—the UK was, in parallel, 'insensitive to the Junta's fast waning freedom of action with regard to the Falklands' (Lebow 1983: 12). Faced with both opportunity and need, Argentina calculated that the combined effect of relative political disinterest, limited practical investment in the Islands and pragmatic realism, would be that Britain would fume, and then negotiate, but not go to war (Arquilla and Rasmussen 2001: 750–4); however the risk was acknowledged and the potential cost of miscalculation or setback recognised (US Embassy 1982). It is equally clear that the Argentinian invasion, whilst contemplated, was not imminently anticipated by Britain, and the fact that Cabinet did not meet in urgent session to consider the situation until the day before gives some credence to this view. Perhaps the strategic idea that—as one writer has argued—'Islands . . . are particularly tempting as a territorial anomaly more easily subject to both low-cost seizure and containment of the conflict' (Kinney 1989: 16), also played a role in these calculations. Ultimately, however, the Argentine miscalculation was compounded by British electoral politics, and the challenge posed was one the government of the day felt it had to answer with force, rather than diplomacy alone, as a matter of electoral survival—although, prior to combat operations commencing, there was a lingering sense of 'bluff' and 'show of force alone' that still pervaded British public views (Hastings and Jenkins 1983: 161). But miscalculation—as one narrative points out—can be fertile, and the precipitate Argentinian act of invasion which commenced on 1 April 1982 came to look like (from Westminster, and in the immediate aftermath of initial shock) 'less a problem for the Conservative Government than a golden opportunity to repackage its ideological project (of a nation moving simultaneously into the past and the future) into a more acceptable form' (Monaghan 1998: 7, 14–24).

The subsequent tactical and operational course of the conflict is equally well known and—whilst relevant—is of secondary importance in terms of the question posed for this chapter. In short, however, the main points are that the UN Security Council condemned the invasion on 4 April 1982 (UNSCR 502), although two of the P5 (PRC and USSR) abstained, and US Secretary of State, Haig—deeply engaged in seeking a peaceful solution (Haig cable 1982)—advised the US administration to consider supporting Argentina rather than the UK, if Argentina proved more flexible diplomatically (O'Sullivan 2012). The British task force encountered Argentinian forces on the journey south (ROE 20 April 1982; ROE 6 May 1982)—including the loss to Argentinian air attack of *Atlantic Conveyor on* 25 May (Atlantic Conveyor BOI 1982: A-3)—and by 22 April were in the vicinity of the Falklands. On 27 April, South Georgia Island was retaken, Haig's proposal for a political and diplomatic solution was rejected by the junta on 29 April, and British forces attacked the Port Stanley runway on 1 May (Ward 1992: 205–7; Woodward 1992: 132–5). On 2 May, the Argentinian cruiser *General Belgrano* was sunk by *HMS Conqueror*, and *HMS Sheffield* was sunk on 4 May (Hanrahan and Fox 1982: 22–3; Hastings and Jenkins 1983: 178–83). These were not the first Argentinian losses of the war, nor the last British losses of the war. British troop landings commenced in force on 21 May. SAS and SBS reconnaissance teams had been ashore since much earlier in the campaign and UNSCR 505, calling for a ceasefire, was passed on 26 May. The subsequent British land campaign achieved tactical and operational success, and Argentinian forces in the Falklands formally surrendered on 15 June (Prime Minister Thatcher 1982: 350–2).

The dominant narrative generated immediately in the post-war period portrayed the war as a short, limited, successful war. Nevertheless, there were challenges to this initial post-conflict discourse of a just fight fairly fought, a cabinet system that functioned coherently and cohesively to deliver the outcome, and a seamless continuity between the political, strategic, operational and tactical levels of endeavour: The circumstances of the attack on *Belgrano* and the nature of *HMS Conqueror*'s orders (and the significance—if any—of a missing Control Room log) (Ponting 1985; Rossiter 2007; Prebble 2013), the rationale and processes adopted for rules of engagement changes designed for force protection as the task force sailed south, and inquiries into the release of classified material into the public domain are indicative examples. Longer term, the legitimacy of the war, its resonating effects, and its role in a touted revival of British fortunes were subject to deeper and more eclectic analyses, and the future of the Islands and the Islanders given greater attention. However, a key—if not the dominant—theme in these discussions, from the affront at invasion to the analysis of aftermath, was sovereignty.

Sovereignty in Assessments of 'Imperial-ness'

The proximate trigger for the Falklands conflict was—as always—politics. To this end, it matters not whether the Islands were or are of strategic, economic or other significance—indeed, as Lawrence Freedman asserted in 1988, whether the Falklands 'represent an important economic and strategic asset is at best unproved' (Freedman 1988: 111–12). Rather, what mattered in terms of the 'spark' were domestic conditions in both Britain and Argentina, different perspectives on likely international tolerance of robust military action, and the current state of negotiations between the two states. When thinking about empire and imperial-ness in relation to the Falklands War, the contextual mercuriality of politics is important, but perhaps not as an anchor point for analysis. To this end, the most useful and tangible indicator is probably 'sovereignty'—the long, slow-burning fuse for the conflict on most accounts. This is because framing the degree to which 'empire' and 'imperial obligation' shaped the British response must in part be informed by the multifarious notion of post-imperial British 'sovereignty.'

In terms of the Falklands conflict, the legal and historical 'sovereignty' argument is well traversed and (unsurprisingly) generally partisan (FCO 1982: 2–9; Rubin 1985: 9; Beck 1988: parts 2 and 3; Dillon 1989: 1–8; Hoffman and Hoffman 1984: chs 6–12; Keeling 2013: 159); and it requires no rehearsal in this chapter. Nevertheless, some have asked what it was about claimed British sovereignty over the Falkland Islands that engaged the British media and public in a different way from other discussions about (and assertions of) sovereignty—such as during 'The Troubles in Northern Ireland' (Calvert 2016: 1; Boyce 2005: 2–5; Washington 1992: 95). Indeed, Britain had in fact been willing to discuss sovereignty (albeit at a glacial pace) since 1968, and even considered—and started shaping the Islanders for—a potential recognition of Argentine sovereignty followed by an immediate lease-back by the UK (Lockhart 2015; Boyce 2005: 16–18). By early 1982, Britain was guilty of identifiable long-term neglect of the islands and Islanders (Shackleton Report 1982: paras 2.2.1–2, 2.7.1–2, 2.9.1–3), and had recently refused to extend 'the full rights of British citizens' to the Falkland Islanders.[2] Clearly, sovereignty was not as intransigent an issue as the military response to the invasion might appear to suggest.

Why and how, then, did 'sovereignty' become *the* issue, and the UK's strongest argument, in April 1982? This is the really interesting question in terms of assessing the 'imperial' dog-whistle (or blaring horn, depending upon one's view) of the discourse on sovereignty, for whilst it turns in part on sovereignty as a political and legal artefact in and of itself, it also engages sovereignty as a cipher. That is, if sovereignty over the Islands was not really the indispensable interest at stake—evidenced by a pre-conflict willingness to start negotiating with Argentina on this issue—then why was it so centrally invoked? It has been argued, for example, that emphasising the threat to sovereignty was one way of putting a respectable, definable visage on achieving an imperially resonant revitalisation of 'spirit' (more on this below). Both manifestations of sovereignty played a role. As Freedman (for example) has observed:

> The conflict demonstrated that people are prepared to fight for goals related to the essence of nationhood; that is, in the face of a direct threat to British territory and British people and British values. An interesting question is whether the government and the task force would have commanded the same level of public support if the conflict had been over something not so directly British but with implications far more substantial for the security and well-being of the country. Suppose, for instance, the crisis had involved a direct threat to the Omani regime, which Britain has done much to sustain.
>
> *(Freedman 1988: 91)*

On a political and legal level, a state might indicate a willingness to discuss sovereignty, but this does not mean it should then be expected to sit back and allow a precipitous assault on that sovereignty to go unanswered; consequently, sovereignty as a political and legal artefact was certainly in play. But, as the short, framed assessments below will hopefully illustrate, sovereignty as cipher—invoking other less tangible, attitudinal things—was also at play in the British Falklands War experience.

Five Lenses of 'Imperial War'

There is no doubt that the British Falklands campaign was understood to be, and was presented as, an imperial war in some quarters (Connell-Smith 1982: 343–6). In some cases, this was doctrinal and polemic:

> There could not be a more brazen expression of imperial chauvinism. Though it echoes 19th century British gunboat diplomacy, we must not forget that the parallel is very limited. Whereas in the 19th century, imperialism was very much on the offensive, today imperialism is just as much on the defensive. How different forces around the world, and particularly in Latin America, have responded to the Malvinas conflict clearly illustrates this fact.
>
> (Economic and Political Weekly *1982: 1139*)

In other cases, reference to British imperial designs and purposes was more political and nuanced; for example, there was a clear thread of 'anti-imperialism' through (in particular) Central and Latin American assessments of the crisis, although the predominant themes expressed by diplomats and governments still tended to coalesce around validating Argentina's claim to sovereignty over the islands, whilst advocating negotiations and peaceful settlement (US State Department 1982). But this chapter is concerned with indicia from a British perspective of an imperial context to the war. The following five lenses offer one set (but by no means the only set) of ways to ask whether the Falklands War was Britain's last imperial war.

Commonwealth Participation in, and Adoption of, the War?

Did the 'empire' come to Britain's aid in the Falklands War, making it an 'imperial war'? In one sense, it clearly did not—the post-imperial Commonwealth gave no military support—indeed, for example, Australian officers on exchange in deploying British units were specifically withdrawn from those units so that they would not take part in the hostilities. However, some Commonwealth states certainly gave political, logistics, intelligence and other support to Britain throughout the crisis (Yorke 2005: 179–88; West 1997: 37–8, 59). But this was not the Second Anglo–Boer War, nor the Malayan Emergency, nor the Indonesian Confrontation, where Commonwealth forces operated as components of—and in visible commitment to—'the Empire.' In this war, in terms of fighting, the British had to go it alone, so it was certainly not an 'imperial' war in that sense.

A War of Imperial Policing?

Was the Falklands War a final manifestation of the twentieth-century form of imperial gendarmerie, as (for example) described by Moreman (1996; distinct from the concept as described in Sinclair 2010)—that is, a post-colonial colonial war aimed at retaining a far-flung vestige of empire. This is in part true—Britain went to war to defend and police the Islanders and their right to be linked to Britain against all comers. The Falklands were (and remain) a British Overseas Territory governed by an appointed Governor advised by local legislative and executive bodies in a mode reflective of colonial administration, and secured by British military forces. In this sense, the Falkland arrangements did hearken back to those set in place for small imperial outposts governed, defended, and policed from London.

This said, however, the idea of the Falklands campaign as one of imperial policing is inadequate as a typology for this war because the 'policing' was neither internally focused nor prompted (as it was in Northern Ireland, or Palestine, or Kenya, or Malaya), but rather a response to external actions and actors. The campaign was not designed to keep a colony within the empire against the will of its peoples—it was designed to stop another state from doing that. Furthermore, it is hard to identify this as an incidence of imperial policing when—as was identified at the time—there was no discernible coherence or strategy for determining which vestiges of empire would be policed and which not—in itself a problem in terms of communicating resolve and signalling the wax and wane of sovereign interest:

> No one supposes that we would ever offer more than token resistance in the defence of Hong Kong. No one imagines that we would engage in major military operations in defence of Gibraltar. But what degree of commitment do ministers accept in relation to Belize, to give one example? If Belize is invaded, do we repel the invaders? Is that what the noble Baroness had in mind? If so, do we have the necessary capacity? Do we

know what reactions there would be among the other Latin American countries? If we are not committed to the defence of Belize, do the people of Belize realise this, so that they can adjust their policies and attitudes?

(Lord Mayhew 1983)

In this regard, the symbiosis between colonial policing and counterinsurgency (COIN) should be noted. There is no doubt that these two things, at the tactical and operational levels, can be markedly similar or indeed the same—one recent analysis of British operations in Afghanistan, for example is entitled 'Colonial Warfare in a Post-Colonial State: British Military Operations in Helmand Province, Afghanistan' (Chin 2010: 215). That is, most post-1945 incidences of the imperial gendarmerie function were characterised by COIN; but the Falklands War was not a counterinsurgency campaign. Thus, whilst there were elements of imperial policing evident in the Falklands War, they were present in the form of 'imperial' legacy and context, rather than any 'policing' indicia, strategy or purpose.

A Defiant End of Empire War?

Another link to empire found in some analyses of British conflicts since 1945 is the idea of an end of empire 'dirty war,' such as those between 1945 and 1957 as described by Benjamin Grob-Fitzgibbon (2011). Other end of empire dirty wars stretched post-Suez—such as against the Mau Mau in Kenya (Berman 1976: 143). Was the Falklands War a delayed, but successful, incidence of such an end of empire war? Again, this conception is ill-fitted in this case, not least because the Falkland Islanders wanted to remain within Britain's political and sovereignty orbit, and because the conflict was initiated by a foreign incursion rather than domestic rebellion. Not all incidents of British decolonisation were dirty and violently resisted—the negotiated decolonisation of Hong Kong being a case in point. Nor, it must be said, is the process of decolonisation complete, as a recent opinion of the International Court of Justice in relation to the British Indian Ocean Territory (BIOT) has pointed out (ICJ 2019: para 183[3]). The Falklands conflict was certainly related to the disappearing empire, but it was not an imperial war of the dirty, defiant, resisted, internal security-focused end of empire type.

A Post-Imperial War of Unwinding?

The opposite conception to that noted above is to think of the Falklands War as Britain's last imperial war in that it was fought to resolve what Dillon describes as 'a post-imperial problem' (Dillon 1989: 1–8). This reading would privilege the challenge faced by Britain in the lead up to, and the decision to be made upon, the Argentine invasion as being how to extricate itself, with some consideration for the Islanders and their interests, from this seemingly intractable, lingering and distracting—but far from existential—unwinding of empire problem. Cawkell has argued that, in fact, it was the rush to decolonisation that

> failed to see—or cared not to—that this intransience [of the Islanders against the proposal for transfer of sovereignty followed by leaseback] came out of the islanders' lack of basic liberties. The colonial system, under which they were forced to live, denying them natural self-expression, had bred in them a deep frustration which was

compounded when an un-asked for Argentine connection was foisted on them, followed by their never being allowed to know what was supposedly being done on their behalf. . . . Because the islander numbers were small, British Governments tended to assume that their requirements were equally diminutive, not least in their need to be decolonised. . . . It did not occur to them that people held in colonial bondage could be induced into a relationship with another country only when they were freed, and that attempting to do so the other way round could result only in superimposing on top of their existing bondage one of a much worse kind, with a country whose record in human rights was appalling.

(Cawkell 1983: 85)

This view—partisan, but nevertheless telling—asserts that it was Britain's overbearing rush to decolonisation, to dismantling this—by 1982 rare—vestige of empire, rather than any rear-guard action aimed at preservation of these last territorial indicia of empire, that created the simultaneously permissive yet intransigent policy backdrop to the Argentine calculus. There is clearly something to this idea that the Falklands War was a consequence of trying to unwind empire, as opposed to a successful attempt to maintain it. It has long been part of the Falklands War narrative that Britain fought to defend the Islanders' right to self-determination, including explicit reaffirmation in the 1985 and 2008 Constitution Orders (The Falkland Islands Constitution Order 2008: s:1[1]). But even when this is taken into account, this idea of a war of imperial unwinding is still insufficient as an explanation, if only because the end result was in fact the opposite—the islands and the Islanders were bound more tightly to Britain.

A War of Imperial Attitude?

There are (at least) two ways in which the idea of the Falklands War as a war of imperial attitude might be canvassed. The first is, as Boyce has outlined (although it is not necessarily his view) (Boyce 2005: 208), that the conflict could be described as 'a re-enactment of the colonial wars that should have disappeared with the dissolution of the British Empire' (Boyce 2005: 221); an attempt to halt, or dispense with, the post-Suez narrative of 'imperial retreat'; a re-positioning in which 'the war [was] a kind of last gasp of British imperial policing; only this time the result was not a piece of imperial retreating, but of imperial reassertion, since the Falklands Islands were, as a result of the war, not given away, but retained' (Boyce 2005: 3). This is perhaps part of the subtext beneath Prime Minister Margaret Thatcher's statement at Cheltenham on 3 July 1982 that:

we have ceased to be a nation in retreat. . . . Britain has rekindled the spirit which has fired her for generations past and which truly has begun to burn as brightly as before. Britain found herself again in the South Atlantic and will not look back from the victory she has won.

(Thatcher 1993: 235)

This revealing statement serves many purposes. In one sense, Thatcher was indeed revelling in a finalised, successful act of imperial defence—an immediate threat on the outskirts of British sovereign interest was dealt with and sovereignty (and sovereign reputation) restored. But Maltby's (and others') reading of this and similar statements tend to highlight more particularly the 'resonance' with 'a particular construction of Britain's colonial, imperial history . . . [a timely] evocation of "the spirit of Britain at its best"' that was 'critical to a (re) generation of national pride, confidence and self-esteem among the UK population' (Maltby 2016: 21). In this more enduring, sentimental sense, perhaps Thatcher's triumphalist speech revealed a deeper truth about the 'imperial' nature of the Falklands War—that it was not really about sovereignty or empire at all, but rather about the perceived, recalled, assumed *attendants* of empire—the can-do attitudes, resilient fortitudes and cleverly employed capabilities that had facilitated empire. That is, there was no longing for a renewed empire, or resurgence in imperial spirit in and of itself; rather, the want was for a renewal of the attitudes believed central to a cleansed, moderated, modern version of the 'spirit' that underwrote empire. In this sense, the Falklands War could be described as an artefact evidencing imperial sentiment, whilst not being an imperial war.

The second—and more pragmatic—way in which the Falklands War might be described as a war of imperial attitude is in the much narrower sense of proof of expeditionary capabilities. Expeditionary capacity underpinned the empire (Fry 2005: 60), and in this sense, the Falklands conflict simply re-proved the lesson of empire—to defend sovereignty a long way from the homeland, military forces capable of and equipped for expeditionary warfare are required. Speller points out how:

> The Suez Crisis provided added impetus to moves that were already under way to improve Britain's expeditionary capabilities. The defense review conducted by Duncan Sandys in 1957 coincided with NATO's adoption of the strategy of Massive Retaliation. The review announced the end of conscription, with a new emphasis on nuclear weapons to maintain the peace in Europe and more mobile and flexible conventional forces to protect British interests overseas.
>
> *(Speller 2002: 364)*

Thus, as Grove has observed, the Falklands War provided an almost unparalleled (post-imperial) opportunity for proof of concept of the renewed, or enduring, capacity of British forces to fulfil 'out of area commitments' as part of NATO (Grove 2002: 309). In this sense, the Falklands War could be described as being as much about proving and indicating sovereign capability and sustaining expeditionary resolve as it was about sovereignty itself.

Conclusion

In the final assessment, the Falklands conflict is probably best described as a non-imperial war within a legacy imperial context, which was animated to some extent by selected post-imperial, but imperial-ish, sensibilities. It is inescapable that the campaign was an immediate and politically rational response, in the form of a limited war, to an act of aggression against a sovereign territorial component of the UK—a war to defend British overseas territorial (rather than imperial) integrity. This is perhaps indicated by the fact that analyses of Britain's role in the Falklands War have commonly focused around (inter alia) the question of whether it was a 'just war' rather than an imperial war—perhaps even to a much more significant degree than the idea of 'just war' has animated analyses of other British operations until the invasion of Iraq in 2003 (Bluth 2004: 885; Fisher and Biggar 2011: 687). Furthermore, the government was actively considering trading sovereignty to Argentina.

But the historical context of empire is not silent here—the Falklands originally came within British sovereignty (on the British argument) during the era of imperialism. The governance arrangements for the territory, and the status of the Islanders were colonial. Equally, there is no doubt that the challenge posed by the Argentinian invasion was more than just immediately political—as the discourse around sovereignty and the Islanders' rights to choose their future indicated, the invasion also properly engaged a 'protective' response to a distant population forcibly severed from the post-imperial body politic. Empire was an inescapable element of contextual legacy. It is also fair to say that resonances of imperial attitude—ethereal sentiment or spirit, but also the capacity to engage in independent, self-contained, expeditionary warfare—clearly coloured and influenced the framing and perception of the conflict. Thus, it is perhaps only in these more interstitial senses—of contextual legacy, and of imperial sentiment and expeditionary capability—that it is right to speak of the Falklands War as Britain's last imperial war.

Notes

1 An Ipsos-MORI poll on 14 April 1982: 'Q: How would you vote if there were a general election tomorrow? Conservative 33%; by 21–23 June 1982—Conservative 51%'; similarly, 'Q: Are you satisfied or dissatisfied with the way the Government are now handling the situation in the Falkland Islands?' 14 April 1982—satisfied 60%; 21–23 June 1982—satisfied 84%', https://www.ipsos.com/ipsos-mori/en-uk/falklands-war-panel-survey
2 *British Nationality Act* 1981, ss—under Schedule 6 to the Act, the Falkland Islands and Dependencies were classified as 'British Dependent Territories,' and thus held 'British Dependent Territories Citizenship' in accordance with Part II of the Act, and did not hold automatic rights to live and work in the UK; this changed in the *British Overseas Territories Act* 2002, under which British Dependent Territories became British Overseas Territories [s:1(1)].

Bibliography

Anon, 'Argentina: Colonialism and the Malvinas Conflict,' *Economic and Political Weekly*, vol. 17, nos. 28–9 (1982), pp. 1139–40.
Anon, *The Falkland Islands: The Facts* (London: FCO, 1982).
Arquilla, J. and Rasmussen, M.M., 'The Origins of the South Atlantic War,' *Journal of Latin American Studies*, vol. 33, no. 4 (2001), pp. 739–75.
Beck, P., *The Falkland Islands as an International Problem* (London: Routledge, 1988).
Bellot, A., 'The Faces of the Enemy: The Representation of the "Other" in the Media Discourse of the Falklands War Anniversary,' *Journal of War and Culture Studies*, vol. 11, no. 1 (2018), pp. 79–97.
Berman, B.J., 'Bureaucracy and Incumbent Violence: Colonial Administration and the Origins of the "Mau Mau" Emergency in Kenya,' *British Journal of Political Science*, vol. 6, no. 2 (1976), pp. 143–75.
Bluth, C., 'The British Resort to Force in the Falklands/Malvinas Conflict 1982: International Law and Just War Theory,' *Journal of Peace Research*, vol. 24, no. 1 (1987), pp. 5–20.
Bluth, C., 'The British Road to War: Blair, Bush and the Decision to Invade Iraq,' *International Affairs*, vol. 80, no. 5 (2004), pp. 871–92.
Board of Inquiry Report: Loss of SS Atlantic Conveyor, 21 July 1982, https://webarchive.national archives.gov.uk/20121109064631/http://www.mod.uk/NR/rdonlyres/EC14467A-DFAF-4030-BDFB-9E1AAF00205E/0/boi_atlanticconveyorpt1.pdf
Boyce, D.G., *The Falklands War* (Basingstoke: Palgrave Macmillan, 2005).
Calvert, P., *The Falklands War: The Rights and Wrongs* (London: Bloomsbury, 2016).
Cawkell, M., *The Falkland Story 1592–1982* (Oswestry: Nelson, 1983).
Chin, W., 'Colonial Warfare in a Post-Colonial State: British Military Operations in Helmand Province, Afghanistan,' *Defence Studies*, vol. 10, nos. 1–2 (2010), pp. 215–47.

Clarke, H.D., Mishler, W. and Whiteley, P., 'Recapturing the Falklands: Models of Conservative Popularity, 1979–83,' *British Journal of Political Science*, vol. 20, no. 1 (1990), pp. 63–81.

Connell-Smith, G., 'The OAS and the Falklands Conflict,' *The World Today*, vol. 38, no. 9 (1982), pp. 340–7.

Dabat, A. and L. Lorenzano, *Argentina: The Malvinas and the End of Military Rule* (London: Verso, 1984).

Dillon, G.M., *The Falklands, Politics and War* (Basingstoke: Macmillan, 1989).

Fisher, D. and Biggar, N., 'Was Iraq an Unjust War? A Debate on the Iraq War and Reflections on Libya,' *International Affairs*, vol. 87, no. 3 (2011), pp. 687–707.

'Four Lessons of the Falklands War,' *The Diplomat*, 10 April 2013, https://thediplomat.com/2013/04/four-enduring-lessons-of-the-falklands-war/

Franklin, M.N. and Freedman, L., 'The Falklands Factor,' *Contemporary Record*, vol. 1, no. 3 (1987), pp. 27–9.

'Franks Report' Falkland Islands Review: Report of a Committee of Privy Counsellors presented to Parliament by the Prime Minister by Command of Her Majesty, January 1983, https://api.parliament.uk › commons › 1983 › jan › fal.

Freedman, L., *Britain and the Falklands War* (Oxford: Basil Blackwell, 1988).

Freedman, L., *The Official History of the Falklands Campaign*, vol. 1, *The Origins of the Falklands War* (London: Routledge, 2005).

Fry, R., 'Expeditionary Operations in the Modern Era,' *The RUSI Journal*, vol. 150, no. 6 (2005), pp. 60–3.

Grob-Fitzgibbon, B., *Imperial Endgame: Britain's Dirty Wars and the End of Empire* (London: Palgrave Macmillan, 2011).

Grove, E., 'The Falklands War and British Defense Policy,' *Defence and Security Analysis*, vol. 18, no. 4 (2002), pp. 307–17.

Hanrahan, B. and Fox R., *I Counted Them All Out and I Counted Them All Back: The Battle for the Falklands* (London: BBC, 1982).

Hastings, M. and Jenkins S., *The Battle for the Falklands* (London: Pan, 1983).

Hoffmann, F.L. and Hoffmann, O.M., *Sovereignty in Dispute: The Falklands/Malvinas, 1493–1982* (Boulder: Westview, 1984).

Humphries, A.A., 'Two Routes to the Wrong Destination: Public Affairs in the South Atlantic War,' *Naval War College Review*, vol. 36, no. 3 (1983), pp. 56–71.

Ince, M., '30 Years On: The Diplomatic Fight for the Falkland Islands,' *Small Wars Journal*, (2012), https://smallwarsjournal.com/jrnl/art/30-years-on-the-diplomatic-fight-for-the-falkland-islands

Keeling, D.J., 'A Geopolitical Perspective on Argentina's Malvinas/ Falkland claims,' *Global Discourse*, vol. 3, no. 1 (2013), pp. 158–65.

Kinney, D., *National Interest/National Honor: The Diplomacy of the Falklands Crisis* (New York: Praeger, 1989).

Lebow, R.N., 'Miscalculation in the South Atlantic: The Origins of the Falkland War,' *Journal of Strategic Studies*, vol. 6, no. 1 (1983), pp. 5–35.

Legal Consequences of the Separation of the Chagos Archipelago from Mauritius in 1965 (2019) Advisory Opinion of 25 February 2019, ICJ, https://www.icj-cij.org/files/case-related/169/169-2019 0225-01-00-EN.pdf

Lockhart, J., 'New Revelations from Argentina's Falklands Campaign,' *War on the Rocks*, 25 September 2015, https://warontherocks.com/2015/09/a-tortured-war-on-the-south-atlantic-rocks-new-revelations-from-argentinas-falklands-campaign/

Lord Mayhew, Debate on 'The Falklands Campaign: The Lessons' (Cmnd. 8758) House of Lords Debate, 17 January 1983, vol. 437, cc1205–68.

Maltby, S., *Remembering the Falklands War: Media, Memory and Identity* (London: Palgrave Macmillan, 2016).

McCormick, C., 'From Post-Imperial Britain to Post-British Imperialism,' *Global Discourse*, vol. 3, no. 1 (2013), pp. 100–14.

McLaughlin, G., *The War Correspondent*, 2nd ed. (London: Pluto Press, 2016).

Middlebrook, M., *The Fight for the 'Malvinas': The Argentine Forces in the Falklands War* (New York: Viking Press, 1989).

Monaghan, D., *The Falklands War: Myth and Counter-Myth* (Basingstoke: Macmillan, 1998).

Moreman, T.R., '"Small Wars" and "Imperial Policing": The British Army and the Theory and Practice of Colonial Warfare in the British Empire, 1919–1939,' *Journal of Strategic Studies*, vol. 19, no 4 (1996), pp. 105–31.

O'Sullivan, J., 'How the US Almost Betrayed Britain: Alexander Haig Wanted Reagan to Side with the Argentines over the Falklands, Newly Released Papers Show,' *Wall Street Journal*, 2 April 2012, https://www.wsj.com/articles/SB10001424052702303816504577313852502105454

Parsons, A., 'The Falklands Crisis in the United Nations, 31 March–14 June 1982,' *International Affairs*, vol. 59, no. 2 (1983), pp. 169–78.

Pion-Berlin, D., 'The Fall of Military Rule in Argentina: 1976–1983,' *Journal of Interamerican Studies and World Affairs*, vol. 27, no. 2 (1985), pp. 55–76.

Ponting, C.S., *The Right to Know, The Inside Story of the Belgrano Affair* (London: Sphere Books, 1985).

Prebble, S., *Secrets of the Conqueror* (London: Allen and Unwin, 2013).

Prime Minister Thatcher, House of Commons, 15 June 1982, in *The Falklands Campaign: A Digest of Debates in the House of Commons 2 April to 15 June 1982* (London: HMSO, July 1982).

Rattner, S., 'Britain and Argentina Agree on Return of Most POWs,' *New York Times*, 19 June 1982, https://www.nytimes.com/1982/06/19/world/britain-and-argentina-agree-on-return-of-most-pow-s.html

Richards, T., 'Medical Lessons from the Falklands,' *British Medical Journal*, vol. 286 (1983), pp. 790–2.

Rossiter, M., *Sink the Belgrano* (Reading: Corgi Books, 2007).

Rowley, T., 'Margaret Thatcher Papers—BBC "Assisted the Enemy during the Falklands War": Baroness Thatcher "Very Angry" that BBC Broadcasted "the Next Likely Steps" in the Falklands War, Newly Published Documents Disclose,' *The Telegraph*, 18 June 2015, https://www.telegraph.co.uk/news/politics/margaret-thatcher/11684868/Margaret-Thatcher-papers-BBC-assisted-the-enemy-during-the-Falklands-War.html

Rubin, A.P., 'Historical and Legal Background of the Falkland/Malvinas Dispute,' in A.R. Coll and A.C. Arend (eds.), *The Falklands War: Lessons for Strategy, Diplomacy, and International Law* (Boston: Allen and Unwin, 1985), pp. 9–21.

Sanders, D., Ward, H., Marsh, D., and Fletcher, T., 'Government Popularity and the Falklands War: A Reassessment,' *British Journal of Political Science*, vol. 17, no. 3 (1987), pp. 281–313.

Secretary of State for Defence *The Falklands Campaign: The Lessons*, Presented to Parliament by the Secretary of State for Defence by Command of Her Majesty, December 1982, Cmnd 8758.

Seymore-Ure, C., 'British War Cabinets in Limited Wars: Korea, Suez and the Falklands,' *Public Administration*, vol. 62, no. 2 (1984), pp. 181–200.

'Shackleton Report,' *Falkland Islands: Economic Study 1982* (Cmnd. 8653) (London: HMSO, September 1982).

Sinclair, G., *At the End of the Line: Colonial Policing and the Imperial Endgame 1945–80* (Manchester: Manchester University Press, 2010).

Speller, I., 'Delayed Reaction: UK Maritime Expeditionary Capabilities and the Lessons of the Falklands Conflict,' *Defense and Security Analysis*, vol. 18, no. 4 (2002), pp. 363–78.

Srinivasan, K., 'Nobody's Commonwealth? The Commonwealth in Britain's Post-Imperial Adjustment,' *Commonwealth and Comparative Politics*, vol. 44, no. 2 (2006), pp. 257–69.

Thatcher, M., *The Downing Street Years* (London: Harper Collins, 1993).

The Falkland Islands Constitution Order 2008, https://www.legislation.gov.uk/uksi/2008/2846/contents/made

UK Government, *Falklands Official Statistics* https://www.gov.uk/government/news/falklands-official-statistics-released

Ward, S. (ed.), *British Culture and the End of Empire* (Manchester: Manchester University Press, 2001).

Ward, Sharkey, *Sea Harrier over the Falklands* (London: Cassell, 1992).

Washington, L., *Ten Years On: The British Army in the Falklands War* (London: National Army Museum, 1992).

West, N., *The Secret War for the Falklands* (London: Little, Brown and Co, 1997).

'Why the Belgrano Papers Matter,' *New Statesman*, 24 August 1984, p. 3.

Windsor, P., 'Diplomatic Dimensions of the Falklands Crisis,' *Millennium*, vol. 12, no. 1 (1983), pp. 88–96.

Winnefeld, J.A., 'Surface Ship Survivability: An Enduring Issue,' *Naval War College Review*, vol. 36, no. 3 (1983), pp. 4–13.

Woodward, S., *One Hundred Days* (London: Fontana, 1992).

Yorke, E., '"The Empire Strikes Back"? The Commonwealth Response to the Falklands Conflict,' in S. Badsey et al. (eds), *The Falklands Conflict Twenty Years On: Lessons for the Future* (Abingdon: Frank Cass, 2005), pp. 170–92.

Official Documents

Operation Corporate Rules of Engagement dated 30 April 1982 (D/DS11/10/6/7) (UK National Archives DEFE 69/787)

Operation Corporate Rules of Engagement change brief dated 06 May 1982 (UK National Archives DEFE 69/787)

PM Thatcher to Denzil Davies MP, dated 4 April 1984 https://www.margaretthatcher.org/document/105651

United Nations Special Committee on the Situation with regard to the Implementation of the Declaration on the Granting of Independence to Colonial Countries and Peoples: Question of the Falkland Islands (Malvinas) (A/AC.109/2017/L.29), 20 June 2017, https://undocs.org/en/A/AC.109/2017/L.26

US Secretary of State Haig to President Regan (DTG: 090131Z Apr 82), https://nsarchive2.gwu.edu/dc.html?doc=329522-19820409-memo-to-the-president-discussions-in

US Embassy report to the Secretary of State (DTG: 222103Z Apr 82), https://nsarchive2.gwu.edu/dc.html?doc=329550-19820422-a-considered-argentine-view-of-the

US State Department cable, 'Latin American Reaction to South Atlantic Crisis' (DTG: 070246Z May 82), https://nsarchive2.gwu.edu//dc.html?doc=329558-19820507-latin-american-reaction-to-south

PART VIII

Postmodern/New Wars

32

THE GLOBAL WAR ON TERROR

The American Operations in Afghanistan and Iraq

Robert Johnson

Al Qaeda and 9/11

The Global War on Terror (GWOT) was announced by the United States government in response to the '9/11' terrorist attack of 2001, a profoundly shocking incident in which over 3,000 were killed in a series of attacks by suicidal extremists of the Al Qaeda movement. The attackers had hijacked civil airliners and crashed them into iconic sites in New York and Washington DC, with all their passengers on board. Al Qaeda had declared war on the United States and the Western world in 1993, but they had been unsuccessful until this incident in 2001. The attacks on the World Trade Center, the Pentagon and (in an unfulfilled attack against the Whitehouse) a flight that crashed in Pennsylvania, marked a shift in the West's reactions, from a defensive posture of counter-terrorism to a proactive, counter-offensive and global 'pursuit' strategy.

It was relatively easy to ascertain that the attackers were all members of Al Qaeda, but it was less clear why they had made their attack. At first it was thought they sought to expunge an affront to Islam's holiest sites, because Western troops had been based in Saudi Arabia during the First Gulf War (1990–1991) to liberate Kuwait from Iraqi occupation. Later it emerged that Al Qaeda wanted to tear down the entire Western world, to enslave and kill Westerners, destroy their economies and wreak revenge for being ahead of the Muslim world in economic, political, military and social power (Bobbit 2008). The extremists of the austere Salafi Wahabi branch of Sunni Islam believed they possessed the final revelation of God and therefore they, not the United States, should govern the world.

Al Qaeda were led by Osama bin Laden, a Saudi dissident who, despite Al Qaeda propaganda claims, had played almost no part in the Muslim resistance to the Soviet occupation of Afghanistan (1979–1989). Nevertheless, there was little doubt that, in the 1990s, he was determined to make Islam the dominant idea of the world. To lend legitimacy to his grandiose plans he made use of the language of *jihad*. He had been coached in this radical thinking by obscure extremists, each of whom drew their inspiration from the Egyptian writer Sayyid Qutb (1906–1966), recognised as the architect of modern *jihadism*. Qutb had been influenced by the Pakistani ideologue, Abdul Ala Maududi (1903–1979) who had advocated a

DOI: 10.4324/9780429437915-41

jihad in politics, but Qutb took these ideas further, rejecting Western 'reason' as unspiritual, and advocating instead radicalised politics, using Islam. Qutb argued that the West should be destroyed and he suggested using airliners to attack their cities and kill their peoples. To paralyse the West with fear, he posited that the Western media should be manipulated to terrorise Western audiences.

Jihadists tried to link together all conflicts between Muslims and secular authorities, even where no such link existed, over a period of many decades, and claimed that a spearhead of fighters were protecting the entire Muslim people, the *umma*. Osama bin Laden was unsuccessful in leading or sustaining the sentiments of *Jihadism* in the 1990s. Although the Soviet Union was compelled to withdraw from Afghanistan in 1989, he was disappointed by the outbreak of a civil war in the country between former *Jihadist* groups. He was forced to leave, and, in 1993, Al Qaeda (the foundation) was established in Sudan, immediately declaring 'war' on the West. His followers made an unsuccessful attack on the World Trade Center in the United States with a truck bomb, and failed to sustain their terrorist campaign in Bosnia. But in 1998 Al Qaeda made a more significant attack on the US embassies in Nairobi and Dar el Salam, which killed many innocent bystanders. This was the first indication of the signature of Al Qaeda attacks, mass-casualty terrorism: their intention was to gain notoriety and provoke reactions against all Muslims by inflicting the highest number of deaths possible. In the years that followed, Al Qaeda operatives were instructed to acquire expertise in weapons of mass destruction or to find ways to kill the maximum number in each attack. Al Qaeda failed to take over Egypt, but the victory of the Pakistan-backed Taliban in Afghanistan in 1998 meant Osama bin Laden quickly established himself there.

America responded to the first attacks with an airstrike against Al Qaeda's base in Sudan, and then, after the 1998 bombings, a limited missile strike against the new Al Qaeda bases in Afghanistan. Bin Laden was undeterred, perhaps even encouraged, that his attacks were getting the attention of the United States. Al Qaeda attempted to sink a US warship in Aden in 1999, but failed, then made another attempt the following year against the *USS Cole*, killing a number of sailors on board. They also assassinated Ahmad Shah Masood, America's Afghan ally, in a suicide bomb attack. The main plot was 9/11. This was to be a spectacular suicide attack by Al Qaeda, aiming to kill the maximum number of people possible. As they expected, there was global media coverage, and a horrified, mesmerised public watched not only the airliners' impacts, and the scene of workers throwing themselves from the burning buildings, but also the collapse of the enormous twin towers of the World Trade Center with the injured and emergency services personnel inside. The reaction was a mixture of shock, anguish, anxiety and anger. It seemed an assault not only on the people of America, but on decency itself. However deeply felt a cause might be, to murder so many, in such a callous way, simply could not be justified.

The new US administration of George W. Bush announced that, with or without allies, it would take swift action against Al Qaeda and its affiliates. The objective was to eliminate Osama bin Laden and to destroy the Al Qaeda organisation. The first step in that mission was to neutralise Al Qaeda's base in Afghanistan, and to secure vital intelligence on its extent, membership and planned operations. Within days, the United States commenced its intervention in Afghanistan, with airstrikes and more sustained bombing of Al Qaeda bases and those of the Taliban, their allies. To avoid becoming embroiled in a ground war against the Afghan people, which was regarded as the mistake made by the Soviets, the Americans and their closest partners deployed elite Special Operations Forces (SOF) and CIA officers to work in conjunction with the anti-Taliban Afghan Northern Alliance. In just three weeks, Al Qaeda and the Taliban were routed (Yousafzai and Ron Moreau 2009).

The Afghan Northern Alliance consisted of many groups and communities that had been subjected to brutal repression by the Taliban, and they were eager for revenge. To contain their desire for retribution, and to re-establish Afghan governance, an interim force of Western ground troops was deployed to Kabul. Meanwhile, American SOF continued their pursuit of Al Qaeda into the mountainous eastern border in a mission known as Operation Enduring Freedom (OEF). The *Jihadists* were attempting to escape into Pakistan, or to conceal themselves in the mountains where they had established themselves during the war against the Soviets. One complex of tunnels, known as Tora Bora, was of particular interest to the United States, because there were tip offs that bin Laden was based there. For a variety of reasons, bin Laden escaped the closing net and disappeared into Waziristan in Pakistan.

A brief attempt to break out of Kunduz Prison by defiant Al Qaeda captives in November 2001 was rapidly suppressed. But after the incident the United States decided not to take any chances with Al Qaeda personnel it held. Prisoners were removed from the region by 'extraordinary rendition,' interrogated and incarcerated at Guantanamo Bay Prison in Cuba. Characterised by their bright orange prison suits, these men were soon the subject of intense debates in the United States because their legal status was unclear. If they were prisoners, they were subject to the protection of the law and entitled to a trial. The Bush administration argued that these were special wartime conditions, so the prisoners were held as military detainees and had to be questioned to derive critical intelligence material. There were hints that waterboarding might be used to force prisoners to disclose information, but this generated protests.

American forces in Afghanistan were unable to pursue Al Qaeda and their Taliban allies across the border into Pakistan and turned to drone strikes instead. Effectively, Pakistan's frontier areas provide a 'safe haven,' but it was not simply that the *Jihadists* concealed themselves. There were reports that, in October 2006 for example, Western Coalition troops had to fight not only entrenched Afghan Taliban south of Kandahar, but their Taliban reinforcements who had arrived in pickup trucks 'waved on by Pakistani border guards' (Rashid 2006). The Pakistan border area certainly gave the Taliban a crucial respite, but more active operations by them, and by the Haqqani network, against American and Western interests, suggested more than Pakistani passivity (Schippert 2006).

Between 2003 and 2006, Al Qaeda and the Taliban attempted to regenerate forces in western Pakistan, but they also made efforts to strike elsewhere. Their hope was to rouse the Muslim world with bombings, assassinations and propaganda. In 2002, they murdered dozens in a suicide attack on a nightclub in Bali using Indonesian partners called *Jemaah Islamiyah*. In 2003, they detonated bombs in Istanbul, but their biggest opportunity to expand their operations came with the US-led invasion of Iraq.

The Iraq War

After 9/11 and the successful destruction of Al Qaeda in Afghanistan, the Bush administration considered how to take more proactive measures to prevent weapons of mass destruction (WMD) falling into the hands of terrorist groups and how to change the political conditions that gave rise to the extremist views of Al Qaeda. There was considerable faith in the idea that tackling rogue states, disrupting terrorists in 'ungoverned spaces' of the world, and fostering democracy and the rule of law were the solution. British Prime Minister Tony Blair, a close ally of Bush, offered to bring various antagonistic Middle Eastern leaders into the new way of thinking, encouraging them to abandon their programmes of nuclear or chemical weapons production, starting with Bashar al Assad of Syria and Colonel Muamarr Gadhafi of Libya. There was growing concern that Iran, North Korea and Iraq were likely to be resistant to diplomatic entreaties, but while Iran and North Korea might be contained, the Iraqi dictator, Saddam Hussein al Tikriti, had a reputation for aggression and defiance that was less likely to be managed peacefully.

In 1980 Saddam had invaded Iran to seize oil fields while the Iranians were distracted by the Islamic Revolution. His policy backfired, and he became embroiled in a protracted and costly war. To acquire compensation, he invaded Kuwait in 1990, arguing that the country was no more than a province of ancient Mesopotamia, but the real reason was to acquire its oil reserves. He hoped to intimidate Saudi Arabia into compliance but found himself confronted by a US-led Coalition that not only liberated Kuwait but defeated the Iraqi military forces. Saddam survived in power and took measures to crush internal resistance. Eager to contain the dictator, the United States acquired UN backing to assert a 'No-Fly Zone' and demanded that UN weapons inspectors have access to, and destroy, all of Saddam's WMD stockpiles. For some years, Iraqi officials played a cat-and-mouse game with the inspectors, and eventually stopped co-operating altogether. Although sanctions kept Iraq weakened, by the late 1990s it was unclear whether Saddam was secretly developing new WMD, including nuclear weapons. Pakistan's nuclear scientist, Dr A.Q. Khan, had been selling nuclear secrets to North Korea and it was not known whether Iraq had accepted the same support.

The United States argued that it had to prevent another 9/11-style attack by invoking the right of self-defence. It presented evidence to the UN that Iraq was developing a WMD programme and probably had stockpiles of chemical weapons. The UN already had a Security Council Resolution that sanctioned some form of action, and while it insisted that weapons inspections be resumed under Hans Blix and General Baradei of the IAEA (which Iraq interfered with), it was not asked for a second resolution that sanctioned specific military operations (UNSCR 2002). Herein lay the controversy.

Bush and his ally Blair were later accused of having acted beyond the remit of the UN. They were also accused of pursuing 'regime change' in Iraq, an action that is illegal under international law outside of war. Bush and Blair both defended their actions on the grounds that the threat was imminent, existential and serious enough to warrant immediate action. They were supported by a 'coalition of the willing' of Poland, Italy, Australia, Denmark, Japan, and Spain. Critics of the invasion plan argued that it was 'a war for oil' and that it was illegal. Both Congress in the United States and the UK parliament supported the decision, but mass protests indicated public scepticism. Saddam also bore responsibility for the war for refusing to co-operate with the UN, but he feared that, having established his power on the basis of being a strong Arab nationalist leader, in emulation of Nasser of Egypt, failure to defy the West would result in civil unrest.

The American plan against Saddam (US Operation Iraqi Freedom; UK Operation Telic; Australia Operation Falconer) was to launch a devastating series of air strikes, nick-named 'shock and awe,' with precision, to cut off the ability of Saddam and commanders from communicating with their forces. This was successful: many Iraqi units were simply unable to get direction or develop any situational awareness of the American offensive. Simultaneously, an American ground campaign was launched on two axes, one in the north and the other in the south. Some 248,000 personnel from the United States, 45,000 British, 2,000 Australian and 194 Polish troops took part, accompanied by 70,000 Kurdish militia.

Turkish objections prevented a large force attacking from the north, so this wing consisted of SOF and airmobile units, operating closely with local anti-Saddam militias (mainly Kurds), and supported with air power. This model, which had been so successful in Afghanistan in 2001, was equally effective in Iraq. Even Iraqi armoured brigades found themselves engaged by fast-moving SOF teams armed with accurate anti-tank missiles and subjected to precision strikes by American F16 fighters, A10 ground attack aircraft, and AH64D 'Apache' and AH1 Super Cobra attack helicopters. At Sargat, Kurdish and American forces overran a chemical weapons facility and neutralised *Jihadist* militants, which vindicated at least some of the claims of the American administration, although, even today, many maintain that Bush and Blair lied about the existence of WMD, or *Jihadists*, in Iraq.

In the south, the main Coalition offensive, led by the US Army 3rd Infantry Division and 1 Marine Expeditionary Force (MEF), crossed from Kuwait heading towards Baghdad via desert roads to the west of the Tigris. At the same time, US and Coalition forces landed at Al Faw to secure the port, refining facilities and oil fields to prevent the Iraqis repeating the environment catastrophe they had inflicted in 1991 when they had released tons of oil into the Gulf and set fire to hundreds of wells.

Despite conventional resistance, and pre-planned irregular resistance behind the US front line (with troops in civilian clothes), the 1 MEF fought its way to the Nasiriyah crossroads, and the 3rd Infantry Division secured the Talil airfield. The British 1st Armoured Division advanced into the eastern marshlands, close to Basra. Meanwhile the 101st Airborne Division seized vital crossing points on the Tigris and Euphrates, through which the US armoured forces drove towards Baghdad. 1 MEF pushed on too, reaching the eastern outskirts of Baghdad. In every encounter, Iraqi forces, both regular and irregular, were defeated decisively.

To link the two axes and penetrate into Baghdad itself, orders were issued for a 'Thunder Run,' an armoured thrust that would first connect the forces to the south of the city and the US forces at the international airport. Although bitterly contested, often in close quarter battles that prevented the use of American airpower, the armoured units broke through repeatedly. Saddam's media spokesman, Ali, tried to persuade the world's press that the Americans were being defeated outside of the city and that no Americans had reached the capital. He was silenced by the appearance of American Bradleys and M1 Abrams driving into the city behind him.

On 9 April, Baghdad therefore fell into American hands, and there were public demonstrations of relief that Saddam, the hated dictator, had been ousted. His statue was torn down, with the assistance of a nearby American engineer vehicle. Saddam was later discovered by SOF in December that year hiding in a 'spider hole,' an ignominious end for a man accustomed to a lavish lifestyle amid his many palaces. His sons, who had an equally brutal reputation, were killed in shootouts. Saddam was later tried and executed by his countrymen for a record of murder, torture, extortion and expropriation. Many Baghdad citizens meanwhile embarked on a spree of looting, creating chaos.

Nevertheless, in a matter of days, the US military had defeated the Iraqi armed forces, the largest in the Middle East, with fewer numbers. US air supremacy was total. They had taken the capital and Saddam was overthrown. On 1 May, Bush announced, in front of a banner entitled 'mission accomplished,' that Iraq's conventional resistance was at an end. On the ground, however, bands of insurgents were still operating. A mixture of Ba'ath party loyalists, military personnel and extremists were taking advantage of the abundance of munitions in the country to make hit and run attacks, detonate bombs and conduct sniping.

The United States was now officially the occupying power and it quickly moved to establish a Coalition Provisional Authority (CPA) under UNSCR 1483 to administer Iraq. The main figure to oversee this administration was Paul Bremer and it was for ideological reasons that he wanted to expunge the Ba'ath Party from the country, dismantle the armed forces which might be loyal to Saddam, and construct an entirely new security and civil apparatus prior to democratic elections under a new constitution and legal system. This proved to be a catastrophic mistake (Ricks 2006). Iraqi officers and civilian leaders, who had abandoned Saddam, found themselves redundant. They turned their anger against the American occupation and took up arms. As a result, the insurgency suddenly increased at an exponential rate. There were multiple bomb attacks, ambushes and shootings. Criminal gangs took advantage of the disorder to kidnap middle-class Iraqis for ransoms. Sectarian violence also broke out as old jealousies were enflamed. Insurgents attacked not only American and international personnel; they destroyed infrastructure too, delaying any chance of establishing a better system.

It soon emerged that little consideration had been given to the post-invasion phase. The military saw their duty as defeating the Iraqi armed forces and insurgents. Many officials in the Bush administration did not want to get involved in 'nation building' and saw their role as establishing a responsible Iraqi government. The United Kingdom had failed to provide a reconstruction plan and this caused significant antagonism for Iraqis.

The number of insurgent attacks increased dramatically through 2004–2005. They took a variety of forms, but sniping and improvised explosive devices were the most difficult to counter. Small arms, rocket-propelled grenades (RPGs), and ambushes were more readily defeated. There was a noticeable increase in the number of suicide bomb attacks, delivered either on foot or in vehicles. The size of the ordnance involved was enough to destroy light-skinned and medium-sized armoured vehicles.

The American efforts to counter the insurgents was undermined by ongoing criticism of the decision to invade, but also by stories of abuse by frustrated military personnel. The nadir of this phenomenon was the Abu Ghraib Prison revelations in 2004, where it emerged that unsupervised guards had abused Iraqi prisoners (McNamee 2008). The dereliction of standards and the broadcast of images of the abuse undermined the moral claims made for the invasion. The incident enflamed Iraqi opinion and added recruits to the insurgency.

Sectarian insurgent groups fought the Coalition and each other. In April 2004, there were uprisings by the Shia 'Mahdi army' in the south. In the 'Sunni triangle' at the centre of the country, Al Qaeda Iraq (AQI), under Abu Zarqawi, established itself in Fallujah. In the so-called First Battle of Fallujah, the city fell under the control of Al Qaeda and its insurgent partners. They imposed a reign of terror, executing any opposition with beheadings and intimidating the rest of the population into compliance. They set about fortifying the city with a warren of tunnels and accumulated munitions ready for a confrontation with the Americans.

The re-taking of Fallujah (November–December 2004) exemplified the nature and the challenges of high-intensity urban combat fought with twenty-first-century tools. The first concern of the American forces was the 'Information Operations Threshold.' This was the point at which media reporting would focus on negative effects of the fighting, and, through political pressures, render further operations impossible. Support could be lost, despite operational success, if there were significant civilian casualties, excessive destruction of property or damage to mosques and other sites of cultural importance (unless it could be proven that, under the Law of Armed Conflict and Fourth Geneva Convention, they were being deliberately used as fighting positions). Any question of breaching the law of armed conflict with regard to civilians, surrendered enemy personnel or the wounded, could also cost the legitimacy of the operation. General William Casey Jr., commanding American forces, imposed strict rules of engagement and insisted on the highest standards of conduct by his troops. Taking photographs and video footage of misuse of hospitals and mosques by insurgents was also encouraged, so as to create a body of empirical evidence.

There was never any doubt about the Americans' ability to win the operation or the tactical actions. Nevertheless, Colonel Tucker, whose RCT-7 unit suffered some losses (one platoon suffered casualties of 50 per cent), described this phase in the following terms: 'The real story—in terms of skill, discipline, and courage—was in the clearing phase' (Lowry 2010: 212). From a strategic point of view, when the Iraqi government declared the operation had concluded successfully, most media teams left (Lowry 2010: 216). Fallujah had therefore underscored some enduring realities of urban warfare, but it had also highlighted the sensitivity of reporting and how crucial political and public support could be to the outcome of a conflict. But Fallujah was cleared, and in June 2006 an airstrike killed Abu al-Zarqawi, the leader of Al Qaeda in Iraq.

Counterinsurgency in Iraq

The Coalition realised that, to end the conflict, it would need to reduce the level of violence through a more sophisticated campaign of counterinsurgency. In essence this consisted of four phases (shape, clear, hold and build) from a secured area, spreading outwards like an oil spot. In the cleared and held areas, the Coalition would build both governance and economic activity, restoring normal life to the population and demonstrating that, while the insurgents' offered violence, the people could enjoy security and prosperity under the government. In January 2005, Iraqi elections were held and an Iraqi Transitional Government was established. One of its first tasks was to draft a new constitution. This was ratified by a national referendum that October and a new constitutional assembly was convened the following summer. In 2005, fresh security operations were conducted in Baghdad, and the Syrian border was sealed off to prevent munitions crossing the frontier. But, despite these efforts, the security situation continued to deteriorate (Camp 2005). Suicide attacks increased and Baghdad experienced daily bombings, shootings and sectarian murders by 'death squads' following the bombing of al-Askari (Shia) mosque by Al Qaeda in February 2006. In April 2007, the Shia holy city of Karbala was also bombed, generating more Sunni–Shia reprisals.

In August 2004, there were riots in Basra, a city that had been the responsibility of the British. This unrest came as a surprise to the British Army, which had adopted a conciliatory approach towards the Iraqis based on its experience of counter-terrorism in Northern Ireland. It took some time for them to realise that the Iraqi insurgents were using the British system as a useful cover to develop their networks and influence. Their campaign of violence developed rapidly and unsettled the British government. The UK Foreign Office was instructed to open secret talks with the insurgent leadership of the *Jaysh al-Mahdi* militia, offering a withdrawal in return for the restoration of order in the city. In 2007, the British forces were pulled out of the city, but, far from producing peace, the insurgents imposed a reign of terror against those it accused of collaboration. The British commanders were eager to go back in, but the government refused, fearing public disapproval at home. Civil–military relations deteriorated, especially when, in January 2007, after the Americans announced they would be sending more forces to suppress the rising violence, the so-called 'surge,' Blair announced his plans for UK withdrawal. It was a betrayal of the United States' closest ally and undermined the justification for joining the Americans in the invasion in the first place, to strengthen the 'Special Relationship' between the two countries. In March 2009, the British withdrew from Basra, much to the disgust of the army.

In December 2006, the American Iraq Study Group published their report with recommendations for ending the insurgency. This consisted of additional troops and spending, with an acceleration of the handover of responsibilities to the Iraqi government. Concurrently, the Iraqi Security Forces were to be expanded and training developed. The 'surge' reduced levels of violence significantly in mid-2007, but so too did the increased targeted strikes and night raids by Special Forces and the US Air Force. Each mission yielded valuable intelligence which was quickly processed and disseminated, ready to facilitate the next mission. A significant breakthrough also occurred with the so-called 'Anbar Awakening' on the Euphrates (Green 2010). Locals, angered and sickened by Al Qaeda's brutal and abusive control of the province, looked to partner with the US forces after several unsuccessful attempts to oust the terrorists themselves. The COIN approach adopted by the Americans meant a willing attitude was easily converted into physical protection of the population. Anbaris readily provided human intelligence to US forces, who followed up and defeated Al Qaeda. In Basra, the Americans also supported an Iraqi security forces operation, 'Charge of the Knights,' between March and June 2008, which defeated Jaysh al-Mahdi and recovered control of the city. In May 2008, the Iraqi Army captured Mosul from Al Qaeda.

That December, the US and Iraqi governments concluded a State of Forces Agreement (SOFA), which clarified the respective authorities and jurisdictions of the Coalition and Iraqi security forces. It also cleared the way for the United States to commence its withdrawal from the country (Johnson and Clack 2014). This was completed in December 2011 when the SOFA was not renewed. In the three years that followed, the governance of Nouri-al Maliki, the Prime Minister (2006–2014), unfairly favoured Shias, which increased disaffection amongst Sunni Iraqi majority. Excluded from jobs and discriminated against, some Sunnis welcomed the subsequent arrival of the extremist movement calling itself 'Islamic State.' That movement, which burst into Iraq from Syria, murdered Shia and minorities in large numbers, but was defeated by Iranian and American intervention between 2015 and 2018.

Counterinsurgency in Afghanistan

While the fighting was underway in Iraq, the United States and its NATO allies had established a UN-sanctioned International Security and Assistance Force (ISAF), and brokered a settlement between Afghan factions to form a provisional government (Emergency *Loya Jirga*) under interim President, Hamid Karzai. The new administration, which was consolidated with elections, lacked revenue and presided over a country in ruins. Thirty years of civil war had left it bereft of even basic infrastructure. Aid agencies estimated it would take decades to get Afghanistan onto a footing where it could sustain itself. Until then it would be dependent on aid.

The weak central authority in Kabul meant that, in the provinces, influential leaders, many of whom had been prominent warlords in the civil war of the 1990s, established fiefdoms in different regions. Their 'auxiliary police' units were militiamen who tended to extract resources from the local population arbitrarily. This attracted opposition.

Meanwhile, the Taliban, encouraged by Al Qaeda, had been steadily rebuilding their forces. In 2003, there were insurgent attacks which increased in frequency so that, by 2005, they could no longer be ignored. ISAF leaders decided that they should extend the security envelope they provided from the area around Kabul to the entire country. National contingents would be deployed to different provinces and regions. The Germans opted for the benign north around Mazar-i Sharif; the Italians established themselves in Herat. The Canadians would create an ink spot at Kandahar and extend outward, the British would do the same in the Helmand Valley, while the Americans took on the extensive area from Nangrahar in the north-east, the eastern region and the south-east. Their air operations would cover the entire country, and their SOF would continue their own OEF mission in the hunt for Al Qaeda and bin Laden.

In April 2006, the extension of ISAF military assistance to the provinces of Afghanistan got underway, but in Helmand and Kandahar province the British and Canadians ran directly into large Taliban forces which were preparing for an assault on Kandahar. The presence of the British evoked Afghan memories of past colonial wars, but more importantly, their stated intention to back certain local leaders, most of whom, unknown to the British, were involved in abusive or corrupt practices, fuelled local anger. As the Taliban commenced fighting the outnumbered British and Canadian detachments, southern Afghans volunteered to fight the foreign 'invaders.' Local casualties mounted and more Afghans came forward to join the insurgency.

The Taliban did not enjoy full support. The majority of the population knew that the presence of the Taliban was just as destabilising as the Coalition. Local leaders tried to end the fighting, but the Taliban response was to execute those Afghans who opposed them. This disciplining of the population, often with rudimentary 'courts,' created fear, but Western military forces rarely had access to the tribal politics or public sentiment of the population.

One of the chief areas of local contestation, to which the Western forces seemed almost oblivious, was narcotics. Opium production soared when the fighting developed, largely because it was easier to store and guaranteed higher profits than wheat or other crops. The Taliban also offered down payments, in advance of the harvest, to guarantee support from the farmers, but they made huge profits by exporting the refined drugs, revenue which fuelled the insurgency. In the east of Afghanistan, American forces had to contend with raiding forces that crossed the Pakistan border to make their attacks, which then withdrew to relative immunity. Local communities exploited the situation, offering to support the Americans to get aid money, but then turning to the Taliban to get protection.

Violence increased throughout the period 2006–2009, prompting the United States to orchestrate an Iraq-style surge of military forces which would create a shield behind which an Afghan security force could be recruited, trained, tested and deployed. Initial efforts to generate a police force, led by the Europeans, was too slow, small-scale and expensive. The Americans therefore put the entire effort on an industrial scale. Thousands were attracted by the salaries, brought into large training establishments, and formed into regional corps commands as light infantry with an air corps. As this army matured, it developed a mobile capability, light artillery, a heliborne commando force and an air force.

COIN measures were also employed. ISAF units were collocated with companies of the Afghan National Army, with police units attached (Johnson 2011). Forward Operating Bases (FOBs) were constructed, and from these, patrols were mounted, while more offensive operations were launched to drive the Taliban out of the populated areas. Provincial Reconstruction Teams (PRTs) replaced the Taliban, consisting of civil and military personnel working to encourage licit agriculture, adherence to Afghan law, and infrastructural development (such as road building, canal repairs, irrigation projects and electricity generation).

Nevertheless, there were significant problems. Afghan military personnel were initially unreliable, while police, with tribal allegiances, did not always obey the law or apply it consistently. Opium production was still far more attractive then PRT offerings, and dysfunctional policies created contradictions that made the situation worse. Many Afghans saw opportunities to take what they could from the foreigners while still supporting the Taliban. The Taliban also developed their own rival systems of intelligence-gathering, administration ('shadow governance') and smuggling. Many locals found it easier to appeal to the nearest Mahez ('front') commander than any foreign and government figures to get things done. The announcement by President Barak Obama that the United States would withdraw from Afghanistan by 2014 was regarded as an error by specialists who believed the Taliban would marshal their strength until the US forces left.

The biggest challenge for the US and ISAF forces was the prevalence of improvised explosive devices (IEDs) and periodic attacks of suicide bombers, snipers and insurgent team assaults. In 2006–2007, the insurgents were defeated in all their swarming attacks, and, although there were episodic concentrations where the terrain could conceal them, as in the mountains of the east or the highly vegetated 'green zones' of the south, it was far more common for there to be a low-intensity posture. Days could pass without incidents, but then a patrol might be struck by IEDs, be subject to a complex ambush or be surprised by a suicide bombing delivered by a child (Gates and Roy 2014: 148–58).

In May 2011, the West enjoyed a significant success when Osama bin Laden was killed in a US Special Forces raid in Abbottabad in Pakistan. To the embarrassment of Islamabad, bin Laden was located in a compound in a military town, at the heart of the Pakistani establishment. The US attack was resisted by Al Qaeda guards, but bin Laden was killed in the firefight. His body was extracted and buried at sea.

There was also positive development in the Afghan security situation. Despite ongoing violence, the Afghan Army and police forces continued to expand, develop and gain experience. The US commanders started to transfer tranches of territory to Afghan officers during the transition, and in December 2014, as planned, they brought to an end the combat phase of the mission to Afghanistan. The drawing down of US and Coalition personnel was completed but a contingent remained, over 10,000 strong, to conduct a 'train, advise and assist' programme. By 2018, some interim Afghan organisations, like the Afghan Local Police, were reduced in size, while paramilitary police were transferred to the Afghan Ministry of Defence.

Most importantly, the Afghan government continued to function. Elections were held, at provincial and national level, which resulted in the presidency of Ashraf Ghani. The economy continued to thrive, albeit in an uneven way. In the years of Western aid injections, health care and life expectancy improved, education and literacy levels increased, more land was brought under cultivation and new industries flourished, especially in telecommunications. If the purpose of the Western COIN had been to get Afghanistan to a position where it could sustain itself, it was well on track by 2014.

There were still severe problems. In 2015 and 2016, the Taliban made important gains over parts of the country, conducted spectacular attacks on urban areas (especially Kabul), and established influence over some local political leaders. What they did not anticipate was divisions amongst their own leaders, and the fragmentation became serious in 2016–2017. A succession of Taliban leaders was killed, but disputes over the way forward, especially when the withdraw of international forces threw into doubt the *raison d'etre* of the whole insurgency, were not resolved. Some elements were attracted to the establishment of a new group calling itself *Vilayet* Khorassan (the Province of Khorassan, sometimes known as IS-K) which affiliated itself to Islamic State in Syria. There was even fighting between the Taliban and this new, more ruthless, movement. What was clear was that, faced with American airpower, a more resilient Afghan government and security force, and its own internal divisions, the Taliban opted for talks.

Conclusion

The phrase Global War on Terror was used less frequently after the administration of George W. Bush, and disappeared by the end of the Obama administration in 2017. It began with the objective of destroying Al Qaeda's ability to threaten the United States, to hunt down bin Laden, and to neutralise his partners. In that, the purpose was largely achieved. Bush had warned that this would be 'a long war' where there would not be a sudden victory but a patient accumulation of successes.

Al Qaeda's ability to operate and its coherence as a movement was dealt a serious blow by America's campaigns and the death of bin Laden, but there have been severe, even visceral, criticisms of Bush's decision to invade Iraq, which seemed irrelevant to a counter-terrorism agenda. Indeed, critics argued that the US occupation of Iraq created terrorists and insurgents, caused unnecessary and avoidable suffering, and distracted the West from its successful hunt for Al Qaeda operatives.

Terrorist attacks on the West have not ceased because of the GWOT. In 2011, Al Qaeda re-established itself in Yemen, and, in 2012, Al Shabaab in Somalia declared its affiliation with Al Qaeda. In 2013, Sahara militants did the same, under the title Al Qaeda in the Maghreb (AQIM), while Libyan factions joined Al Qaeda after the overthrow of Gadhafi. The Syrian Civil War fostered extremists too, although a dispute between Al Qaeda and Al Nusra, a rival organisation, spawned *Jihadist* factions, and opened the way for a new unifying catalyst to emerge in the form of ISIS, or 'Daesh.'

Critics argue that these movements and ongoing attacks were the result of Bush's misjudgements, but that is an accusation which ignores the fact that Al Qaeda intended to attack the West over the long term, regardless of the Iraq War. Few have advocated trying to negotiate with Al Qaeda because its intent is to destroy the entire global system on which ways of life, international commerce and political organisations depend. Al Qaeda's views on women are demeaning. It exhibits a gross, murderous intolerance to those who do not share its religious beliefs. The current Western posture is active containment, and the continuous neutralisation of its bases of operation. The West has supported anti-ISIS operations alongside the Iraqi Army, primarily with air power, and seeks to limit Al Qaeda in Yemen. The United States sustains its vigilance against Al Qaeda and its successors.

Bibliography

Bobbitt, Philip, *Terror and Consent: The Wars of the Twenty-First Century* (London: Allen Lane, 2008).

Camp, Richard D., *Battle for the City of the Dead: In the Shadow of the Golden Dome, Najaf, August 2004* (Grand Rapids, MI: Zenith, 2005).

Gates, S. and Roy, K., *Unconventional Warfare in South Asia: Shadow Warriors and Counterinsurgency* (Farnham: Ashgate, 2014).

Green, Daniel R., 'Fallujah Awakening: A Case Study in Counter-insurgency,' *Small Wars & Insurgencies*, vol. 21, no. 4 (2010), pp. 591–609.

Johnson, Rob, *The Afghan Way of War* (Oxford: Oxford University Press, 2011).

Johnson, Rob, and Clack, Timothy (eds.), *At the End of Military Intervention: Historical, Theoretical and Applied Approaches to Transition, Handover and Withdrawal* (Oxford: Oxford University Press, 2014).

Lowry, Richard, *New Dawn* (New York: Savas Beattie, 2010).

McNamee, T., *War Without Consequences* (London: RUSI, 2008).

Rashid, Ahmed, 'NATO's Top Brass Accuse Pakistan over Taliban Aid,' *The Daily Telegraph*, 6 October 2006.

Ricks, T., *Fiasco: The American Military Adventure in Iraq* (New York: Penguin, 2006).

Schippert, Steve, 'Pakistan Cedes North Waziristan to Taliban,' 6 September 2006, http://inbrief.ThreatsWatch.org/2006/09/pakistan-cedes-north-waziristan/

UNSCR 1441 (2002), https://www.un.org

Yousafzai, Sami, and Moreau, Ron, 'The Taliban in Their Own Words,' *Newsweek*, 26 September 2009.

33

PRIVATE MILITARY CONTRACTORS AND CONFLICTS IN THE NEW MILLENNIUM

Christopher Kinsey

Introduction

When US and UK troops crossed into Iraq in March 2003, they were accompanied by one of the largest assembled groups of military contractors since Vietnam. According to the 'Commission on Wartime Contracting in Iraq and Afghanistan Interim Report' published in June 2009, the ratio of military contractors to soldiers that make up in-theatre support personnel is now 1:1, while during the Vietnam war it was 1:6 (one contractor for six soldiers) (Gansler and Lucyshyn 2012: 279). Furthermore, this trend is set to continue for the foreseeable future. This chapter examines the role of military contractors[1] in conflict since the start of the new millennium. While part of the examination will focus on the present period, the chapter will also emphasise the future, raising important questions about the future character of conflict. Whatever this looks like, what is certain is that it will include more military contractors; nor is the idea new, with some arguing that military contractors are as old as conflict itself (see Parrott 2012), while their exploits can be traced as far back as ancient times.[2] However, with the French Revolution, and the rise of the *levée en masse*, the role of military contractors gradually declined as state militaries internalised the roles they had previously performed. Even so, they did not totally disappear, remaining an important, if under-reported, actor in conflict until the US interventions in Afghanistan and Iraq in 2001 and 2003, when their presence on the battlefield grew significantly.

Military contracting is undeniably a global phenomenon. One only need look at the number of state militaries employing them, where they are employed and where they come from for evidence to support this observation. Military contractors from the Global North and South are now employed by state militaries in conflict zones around the world. They perform routine tasks from logistics, construction and maintenance to medical and technical support. More controversial is their role in guarding military bases in operational theatres and providing direct combat support[3] on the battlefield, as was the case with the South African Private Military Company (PMC) STTEP International in 2015. The company's personnel worked alongside the Nigerian military to support its counterinsurgency operation against Boko Haram in the north of the country (Freeman 2015).

DOI: 10.4324/9780429437915-42

More recently, the Russian PMC Wagner has undertaken combat operations in Eastern Ukraine, operating alongside Russian military units. However, after the company's leader, Yevgeny Prigozhin, launched what some in the Kremlin described as a military coup, its military contractors were told by Russia's defence ministry that they could either sign regular army contracts, go home or join Prigozhin in Belarus where he went into exile before his death (BBC 2023a). The company has also provided direct combat support to the Syria military in its fight against local rebel militias in the country. Indeed, its mercenaries were reported to have engaged US forces in a firefight that resulted in the company suffering dozens of casualties (Isenberg 2018). Nor is it alone in wanting to undertake combat operations for states. Erik Prince, the founder of Blackwater, wanted to substitute the majority of US and coalition forces in Afghanistan with private military contractors. Prince believed a small footprint of US special operators and military contractors could achieve what thousands of conventional forces had failed to accomplish over two decades (Copp 2018).

But it is not only the military that now employ contractors; though this chapter is solely concerned with the military use of contractors in conflict zones. They are also utilised by other government departments, multinational corporations, and non-government organisations (NGOs). The reasons for this are twofold. First, staff/employees need protecting from insurgencies and this protection is usually provided by security contractors. Second, successfully intervening in a country to stop a conflict and then help to rebuild it requires special skill-sets that are often absent amongst government and military personnel, but are available through the market. Conflict zones are complex socio-political and economic environments, where actors that intervene face multiple and diverse challenges, usually centred on the provision of security and support for development (Duffield 2001). Without the market to provide the special skills needed for reconstruction and development, a country devastated by conflict may struggle to recover and may subsequently plunge back into conflict.

This chapter sets out to answer a series of interrelated questions (see below) on military contractors and conflict. It does this through a broad understanding of conflict, including insurgencies where the level of violence is low but persistent, to conventional state-on-state war where the level of violence is high. These questions also constitute the sections in the chapter. The first question explores where military contractors are used. This question is primarily concerned with the presence of contractors in ongoing conflicts being fought today. To put it differently, are there any ongoing conflicts where military contractors are not used? Question two is related to what military contractors do in conflict; while question three asks why state militaries have come to rely on them so much, and what this relationships look like. The last question considers the role of contractors in future conflicts and the challenges this will present governments and state militaries.

Contractors and their Presence in Conflict Zones

Wherever there is conflict, you will find military contractors. Whether it is a counter-insurgency operation, a peacekeeping or peace-enforcement intervention or, as in the case of the invasion of Iraq by the United States and its coalition of the willing in 2003, a state-on-state conflict, one only has to scratch below the surface to find contractors of every type supporting military action. Military contractors are now part of a transnational network of state and non-state actors that seeks to enmesh government agencies, international financial institutes, private military and security companies, the business sector, NGOs and so on in a cooperative and competing relationship with one another. Whereas Duffield refers to this arrangement as a network of strategic complexes, whereby the relationship within the complexes is increasingly privatised and militarised (Duffield 2001), and Abrahamsen and Williams talk about global assemblages of 'transnational structures and networks in which a range of different actors and normativities interact, cooperate and compete' (Abrahamsen and Williams 2011: 90), both are concerned with explaining the increasing reliance by states in general on non-state actors, including military contractors, to help them resolve the conflicts they have become entangled in. Nowhere is this more obvious than in Afghanistan and Iraq, where the United States and the UK have relied on thousands of military contractors to perform the most basic of tasks such as food preparation, to supplying technical support to some of the most advanced weapon systems in the world. But while these two conflicts are the obvious focus when discussing military contractors and conflict, they are by no means unique with regards to their use of military contractors. More recently, Russian military contractors have been providing military support, including combat capacity according to some press reports (Gibbons-Neff 2018), to Syrian President Bashar al-Assad (BBC 2018). Neither is Syria the only country in the region to rely on military contractors to help fight their wars. Both the United Arab Emirates (UAE) and Saudi Arabia have employed contractors to augment their military forces fighting in Yemen. In the case of UAE, for example, Erik Prince, the founder of Blackwater, was hired by the Crown Prince of Abu Dhabi to put together a security force of 800 foreign troops to conduct military operations inside and outside the country, while some of those troops have fought in Yemen's Civil War (Mazzetti and Hager 2011). At the same time, Saudi Arabia has also reportedly recruited mercenaries from Chad to help them fight the Houthi rebels in Yemen.

Then again, it is not only in the Middle East where military contractors are found. Africa is experiencing the phenomenon as well. What is more, it is so widespread that it is hard to imagine a conflict in Africa where military contractors are not present. Since 2000, military contractors have been involved in fighting or providing military supplies and training in Libya, Nigeria, Côte d'Ivoire, Sierra Leone, Liberia and Somalia. This list does not include African countries that rely on US military contractors to train their military (McFate 2014). Importantly, the phenomenon has reached such proportions in the last five years, with military contractors undertaking combat, military training, the provision of technical skill-sets and equipment and logistical support, that it has become truly internationalised and commodified, reaching every area where there is a conflict. But what exactly is it that military contractors do? This is discussed in the next section.

Defining Military Contractor Support

This section discusses the wide range of military support services military contractors provide in conflict zones. These services can be divided into four broad categories. The first category is troop support services. These include providing food services, the provision of messing, laundry, waste management facilities, transportation, performing engineering and construction tasks and maintaining vehicles. Most of the tasks that fall under troop support services are performed by third-country and local nationals. What the category does not include is support to weapon systems and the provision of security. An example of a service support contract is the US Department of Defence (DoD) LOGCAP contract (Stanger 2012: 191) and the UK Ministry of defence (MoD) Operational Support Capability Contract (OSCC). Both are enabling contracts that help sustain expeditionary operations. Furthermore, the provision of these services to the military has seen a noticeable increase over the past two decades, but especially since the start of the Iraq War in 2003.

The second category is system support services. This category covers maintenance of advanced weapon systems and information technology (IT)-based systems. It includes, for example, maintenance of fighter aircraft, such as the Typhoon and F-35, military satellites and IT-communication systems. This type of service is usually provided by the original equipment manufacturer. There are two types of contracts usually associated with system support services. The first type is for capability. This is where a supplier is responsible for providing a capability and outputs to agreed performance standards. An example of this is the UK Royal Air Force (RAF) air-to-air refuelling contract (See RAF 2018). The other type of contract is for availability. This is where the supplier is responsible for supplying a weapons platform or other equipment to agreed performance and outputs standard. This is the case with the Typhoon Availability Service contract, awarded on 4 March 2009 (BAE Systems 2018).

The third category is security protection services. This category provides armed security services, which can include heavily armed security contractors, for base security in operational spaces, convoy protection for normally nonessential supplies and close protection for individual persons and groups working in the operational space. This last group is dominated by government officials and private employees working on stabilisation projects. Most of these types of contracts involve routine protection of buildings and other facilities. They are performed by low-skilled local and foreign nationals employed as guards, and often managed by Europeans. The exception is close protection, which is usually outsourced to former Special Forces soldiers. The most publicised companies running these contracts include Academi (previously known as Blackwater), Control Risks, Olive Group Security and GardaWorld.[4] This type of contract includes the US Department of State Worldwide Protective Services (WPS) programme, which provides multi-billion-dollar funding for the protection of life, property and information for the US Department of State. The programme employs upwards of 30,000 private security contractors to provide diplomatic security in high-threat regions such as Libya and Afghanistan (Cullen 2019: 23–4).

The final category is direct combat support services. This is the most controversial of the four categories as it usually involves fighting alongside those being supported. Contractors that belong to this group are frequently called mercenaries, since there is often very little difference in the types of activities they both perform. They are also distinct from other types of private security actors, in that their intention is to have a strategic impact on a country's security and political environment. This is usually achieved by operating alongside the client's military forces, as part of a private military company (PMC) and acting as a force multiplier during an operation (Kinsey 2006: 14). This is what STTEP did in Nigeria, when it supported the army's COIN operation against Boko Haram in the north of the country.[5] Where this group of contractors differ from mercenaries is in how they are employed. They are often part of a permanent structure that extends beyond the requirement to service a single contract as is normally the case with mercenaries. Instead, they are employed by a registered corporate body (a PMC) with a legal personality and subject to legislation. As a corporation they also adopt business practices that might include the use of advertising material, a corporate ethos and doctrine, and a staff vetting system (Kinsey 2006: 14). PMCs are the official military transformed into the private sector in a business guise (Kinsey 2006: 14–15). Finally, a subset of the PMC is the proxy military company. These companies work solely for their own government, supporting its defence and foreign policies by providing strategic/ operational advice and military capabilities. An example of such a company is Military Professional Resources Incorporated (MPRI).[6] The company trains foreign military and police enforcement agencies under licence from the US government.

As this section has shown, contractors now provide many of the services and capabilities the military rely on to conduct operations. While some of these services and capabilities were once considered the sole preserve of the military, it now seems no service or capability is beyond the reach of the market. Even so, to suggest that a single driving force (for example economic efficiency) is responsible for pushing the military down the path of military contracting is simplistic and ultimately wrong. As the next section explains, the contracting out of military functions is a multidimensional and multi-layered response to the range of diverse challenges the military face today and into the future.

Explaining the Need to Employ Military Contractors

There are numerous theories that seek to explain why state militaries have turned to contractors to support their expeditionary operations since the start of the new millennium and why this reliance is set to continue for the foreseeable future. None of these theories, however, are able to explain entirely this phenomenon. Even so, some do appear to have greater explanatory powers than others in helping us to understand it. This section will focus on four of these theories. They are functionalism, ideationalism, political instrumentalism and organisational theory. Each theory brings a different perspective to the question of why contractors will remain a part of conflict in the twenty-first century.

International relations scholars who embrace functionalist explanations to elucidate the role of contractors in conflict do so by accepting a rationalist epistemology that seeks to explain state behaviour as an outcome of means–ends rationality (problem solving), the purpose of which is to effectively target limited resources on achieving the state's policies (Waltz 1987). In the context of defence policy, this means defence planners employing contractors as material and intellectual solutions to the problems of utilising military power so that it is effectively deployed in conflicts and is therefore able to meet the objectives set out in defence policy. This is no easy task as it often involves defence bureaucrats working within domestic, social and political constraints that they may have no authority over to remove, alter or influence. Examples of these problem-solving approaches include hiring contractors with specific skill-sets to perform technical roles, such as repairing and maintenance of technically advanced communications equipment, which the military can no longer perform. Another example of functionalism within defence is the use of contractors for low-skilled jobs. This is because they are assumed to be cheaper than using uniformed personnel, but also because they do not impose long-term budgetary costs through state pensions or other state entitlements and therefore seen as a financial benefit to the taxpayer (Kinsey 2009). Finally, outsourcing non-combat roles to contractors is seen as part of a wider solution to the problem of recruiting enough manpower into the ranks of the military (Kinsey 2014: 494–509).

Moving the focus on to ideational explanations, ideas have always played a very important role in determining how military forces should be organised. To understand this point, one only has to examine the impact on the military of the social changes that occurred during the nineteenth and twentieth centuries, which were the result of the creative process in generating, developing and communicating new ideas. Nowhere is this clearer than with the military's use of contractors to support its operations. What is more, as I explain below, some ideas actively seek to encourage the phenomenon (Krieg 2016).

For example, since the 1980s, neo-Liberal economic theory has had an important influence on government attitude towards outsourcing. The theory posits that outsourcing functions to the market (in our case the military functions) can deliver greater material efficiency, which in turn leads to economic benefits in the shape of financial savings for the taxpayer. The practice, moreover, is seen as a way of making the military financially sustainable (Stanger 2009: 84–108). But ideational explanations are not just concerned with financial efficiency. Ideas have also influenced how democratic states have shaped their military organisations, while first finding expression in the French Revolution. Since then, two different models of civil–military relations have emerged, one grounded in Republicanism and the other in liberalism. Advocates of the Republican model believe in a self-sustainable and centralised military force, organised around a conscript citizen soldier, while supporters of the liberal model favour a smaller, professional and politically neutral military force working alongside military contractors, one that strives to fragment and limit government power. It is unclear, however, which of the two ideas will dominate our thinking about how we should manage military force in the future and thus whether more or fewer military contractors will find themselves supporting military operations.

A particularly powerful explanation for why military contracting has grown, and will continue to grow, has its foundation in political instrumentalism. The argument here is that the increase in military contracting is due to domestic political constraints and calculations. What is more, the explanation does appear to adhere to the actions of some Western states since the interventions in Afghanistan and Iraq. In both cases, military contractors have been used to circumvent domestic political objections to mobilising military forces to fight unpopular wars, while at the same time meeting national security objectives (Avant and Sigelman 2010: 230–65). Enthusiasm for humanitarian intervention has declined over the past decade, with fewer states prepared to put their soldiers in harm's way for abstract notions about protecting human rights in conflicts where they have no or very limited interests. In the case of Afghanistan, for example, the UK government capped the number of military personnel, replacing them with contractors, to make the war palatable to the public, while also heading off criticism that it was throwing away soldiers' lives in a conflict it cannot win. Nor is the UK unique in taking this approach. As mentioned in the introduction, the Syrian Army is receiving support from Russian mercenaries in its fight against Syrian rebel forces and ISIS (Isenberg 2018). The use of Russian mercenaries is a worrying sign as it suggests the trend towards using contractors to avoid domestic political constraints is not confined to the West only. Consequently, one can easily envisage a situation where contractors become proxy forces for countries unable to deploy their own military, but still determined to protect their overseas interests.

The final explanation used to explain the growth of military contracting is grounded in organisational theory. Military contracting not only alters the relationship between different government branches; it redistributes capabilities and power across, as well as within, branches of government. The theory advances the notion that contractors provide bureaucrats with possibilities to perform functions free from interference by other government branches. What this means in the case of military contracting is their use by separate government branches to achieve their own objectives independently from each other. Importantly, by taking this approach, each government branch is able to conduct its activities according to its preferred operating procedures (Cusumano and Kinsey 2014). An example of this is where the UK's Department for International Development (DIFD)[7] uses private security contractors to protect its staff in conflict zones instead of relying on the military. By doing it this way, DIFD's staff are free to perform activities how they want to and not how the military might want them to if it were providing security. The military, on the other hand, might also feel they benefit from this situation. This is because it does not have to divert precious resources from core tasks to what may seem peripheral activity. Such a move, though, could lead to a decrease in the morale and motivation of military staff who are responsible for performing core tasks, but now with fewer resources. Finally, as conflicts become more complex in nature, governments will inevitably turn to contractors because only they will have the required skill-sets. This, in turn, is likely to result in a redistribution of power between different government branches, and, as is already happening, with branches other than the military taking the lead in resolving conflicts.

Military Contractors and Future Conflict

While the previous section attempted to explain the socio-political, economic, technological and bureaucratic forces driving military contracting, the last section explores what the future prospects are for the phenomenon. Predicting the future is, of course, notoriously difficult and there is no guarantee the picture this chapter offers will resemble what it will eventually look like. Even so, it's important to undertake this exercise so that governments and their militaries can plan for the future. This is particularly important in the field of international security, where lead-in times to organise against threats may take months or even years, while the threat might only take days or weeks to materialise. As McFate reminds us, the international system appears to be moving towards a new post-Westphalian era (McFate 2014: 72–100) where threats are harder to identify because they are more nebulous in nature, faster to appear and emerge from the most unlikely sources. If this is so, then it is essential governments are ready for it. But this will entail a clearer understanding of what the future security environment is likely to resemble and the challenges it will pose.

In the case of military contracting and its impact on conflict, a number of security trends have already started to emerge. What is more, they are set to continue for the foreseeable future. In addition, these trends are accompanied by a set of challenges that governments will need to resolve, centred around the regulation, accountability and transparency of military and security contracting. Military contracting also raises legal and ethical challenges, as seen in Russia after Prigozhin appeared to stage a military coup, as well as the potential to create a democratic deficit for Western governments in particular. The chapter will return to these issues later.

Following Executive Outcome's operations in Angola and Sierra Leone in the mid-1990s, recent events in Syria, Ukraine, Yemen and Nigeria point to the willingness of some governments to continue the trend of outsourcing combat operations. What is unclear is the extent that this trend will reach. In all the operations mentioned above, for example, contractors have played a very important, but still only a supporting role, to local forces. This role has usually entailed boosting operational capacity through better training and providing additional capabilities, such as air support. It has also involved direct combat support whereby contractors fight alongside local forces. It did not involve contractors replacing local forces and taking over sole responsibility for combat operations until Russia invaded the Ukraine in Feburary 2022. Until this time, the experience of Wagner in Syria and

Ukraine (2014) had suggested this was an opportunity too far for the industry (Gibbons-Neff 2018). Even Erik Prince, a strong advocate of contracting out military operations, did not suggest replacing the whole international force in Afghanistan with contractors, but only the conventional component of the forces (Copp 2018). Military contractors would then work alongside local security forces and US Special Forces to defeat the Taliban. To be sure, Wagner's combat operations in Ukraine since February 2022 may be an anomaly, driven by certain conditions that are unique to Ukraine and Russia, for example their location, they are neigbours, and the Kremilin's fear of NATO's eastward expansion. If this is the case, then the rest of the industry may have already reached its political and operational limits, thereby restricting the future of this particular trend.[8] It would also support the second trend, which is the continuing outsourcing of non-combat functions until the point is reached whereby the military only performs its core combat functions, with all other functions being undertaken by military contractors.

The industry is already some way down this route. As Kinsey points out, the military structure is fast approaching a core-competency model as governments, especially in the West, continue to outsource all but fighting functions (Kinsey 2009: 95–9). This trend is less dramatic than the first. Even so, it still points to an important and continued role for military contractors in international security. To support this point, Erbel and Kinsey point to the fact that military contractors are becoming integral to many armies' ability to operate (Erbel and Kinsey 2018: 302–12). The employment of contractors allows the military to focus on what characterises it as a profession: combat. As Kinsey points out, in the future, militaries, but particularly Western militaries, will concentrate their efforts on only a handful of functions performed by teeth arm and specialist units and will comprise warfighting, peace enforcement and peacekeeping, COIN operations, counter-terrorist operations and providing military capabilities to civilian authorities (Kinsey 2014).

Consequently, Kinsey differs from McFate in that he believes states will continue to be the dominant actor in international security for the foreseeable future (Erbel and Kinsey 2018: 311). Furthermore, rather than erode state military power, contractors are helping to strengthen it. This last point leads into the next trend, the wider use of military contractors by supranational organisations, for example the EU, NATO and the UN, and countries without a history of military contracting, such as China.

Supranational organisations have already begun using military contractors to support their operations. This, in part, has been driven by its members, who are often unable, or simply cannot afford, to generate certain capabilities by themselves. Strategic air and sea lift are two such capabilities. In the case of Afghanistan, for example, many of NATO's members relied on the organisation to supply their operation in the country. NATO, on the other hand, was only able to do this by hiring military contractors, while the cost was shared among member countries. Ultimately, adopting this approach meant members that could not afford their own strategic air/sea lift could still support their military participation in the operation by utilising NATO contractors. What is more, sharing resources in this way is likely to be a feature of future conflict. Other supranational organisations, however, may use contractors to help improve the standard of military training of their members. It is not beyond imagining military contractors working alongside African Union peacekeeping troops to develop new skills or expand/enhance existing ones. In fact, this is already happening in the case of the UN, which now relies on the market for a range of services it is unable or unwilling to provide itself. Finally, we know that the US and Russia rely on military contractors, but what about China, the other major power? The country is not new to the provision of private security. Chinese corporations are already using private security companies to protect their infrastructure projects in Africa. Furthermore, there does not appear to be any objections to this move from the Chinese government. The fact that China seems willing to allow its corporations to employ the services of security contractors may suggest it is also willing to employ contractors to support its military activities on the African Continent. If it is willing to use contractors in this way, it will make it very difficult to argue that military contracting is the sole result of neo-liberal economic policies. Rather, it is the outcome of other objectives/ motives.

Finally then, what are the challenges associated with these future trends? The most obvious is how to regulate military contractors, make them accountable for their behaviour and ensure their activities are transparent. Military contractors also raise legal and ethical challenges, as well as how to address any democratic deficit their employment might cause. Ultimately, it may be necessary to introduce an international convention that is able to strictly control what military activities should and shouldn't be outsourced, and that is also able to hold governments and military contractors to account for their behaviour if something goes wrong. Introducing such a convention will not be easy given the lack of support for it from the international community, but particularly from key states such as the United States. A previous attempt to introduce an international convention to control the actions of mercenaries took many years to implement. The International Convention against the Recruitment, Use, Financing and Training of Mercenaries was a response to the mercenary activity in Angola in 1976. The convention was concluded in 1989. However, it only entered into force on 20 October 2001 after enough states had ratified it.[9] In all probability, the same is likely to happen with a new convention aimed at controlling military contractors. But, this is not a reason to abandon the idea. After all, the Laws of Armed Conflict (LOAC) have not been particularly successful in achieving its aims. Yet, no government has suggested doing away with them. The fact is, the LOAC also act as a normative signpost by reminding us what acceptable military behaviour in conflict is, and what militaries should aspire to if they have not already. A similar case can be made for an international convention on military contractors operating in conflict zones. The convention would set the normative boundary on what roles they can perform and what is acceptable and unacceptable behaviour. There is also a need for national and international oversight mechanisms to ensure governments and military contractors do not erode our democratic principles (this is not an issue for non-democratic states that employ contractors). After all, we do not want to repeat the past which saw governments turn to the market to conduct covert military action using mercenaries (Jones 2004). There is every reason to believe this will happen again unless we have oversight strong enough to prevent it. What these mechanisms might look and how they might work is beyond the scope of this chapter. What is clear is the present reliance on self-regulation[10] is not enough to control the actions of military contractors.

Conclusion

This chapter set out to answer three interrelated questions on the extent and roles of military contractors participating in conflicts. Military contractors were shown to operate on a global scale. In truth, it is hard to imagine a conflict where they are not operating. This is, in part, because they are now integral to the structure of many armed forces. For example, in the case of the United States it is now inconceivable to think of its military operating without them. But, it is not only states that employ the services of contractors. As the chapter has shown, supranational organisations such as NATO, the UN and the EU have also turned to them, employing their services across a range of activities. This is a trend set to continue. The chapter also stressed the range of services performed by military contractors and the rationales behind the move to employ them. Other than combat, the chapter suggests everything is on the table. This does not mean that in the future military we will see contractors replace, wholesale, military personnel. This is because there are numerous roles the military need to hold on to for operational and strategic reasons.

As the chapter mentions above, predicting the future is fraught with difficulties. Even so, the chapter emphasised three likely trends and the challenges associated with them. The first trend pointed to the outsourcing of combat, but it was thought the market for combat may have already been reached, in that military contractors might prefer to limit their role to direct combat support for state militaries instead of acting as an alternative force as 5 Commando did in the Congo from 1964–1965, or Wagner did Ukraine between 2022–2023.[11] The second trend pointed to more roles being outsourced to military contractors. As mentioned above, other than combat, everything else is on the table to be negotiated. Moreover, as state militaries becomes more reliant on technically sophisticated weapons and communication systems, so there will be a need for more technical contractors. Many of these may find themselves working alongside military personnel in the conflict zone, operating and maintaining equipment. The last trend drew attention to international organisations and their use of military contractors. It was explained that, like the second trend, this one was also set to grow in the future. The primary reasons for this were the high cost associated with sustaining expeditionary operations and the professionalism military contractors bring to an international organisation that its member states may lack. Lastly, the chapter touched on some of the challenges associated with these trends. In so doing, it pointed out that a considerable amount of political effort will be needed to overcome them. It also noted that this will take time, but was vital to ensuring military contractors maintain the same high standards of state military personnel, and that if not, they can be held legally accountable. That said, whatever conflict looks like in the future the genie is now truly out of the bottle. Moving forward, we should therefore expect military contractors to continue being a part of it.

Notes

1 The term private contractor is used interchangeably in this chapter to refer to those supplying logistics and technical services and those who engage in actual fighting. The latter group is often referred to as mercenaries.

2 One example is Xenophon's army of Greek mercenaries, known as the Ten Thousand (401–399).

3 Direct combat support involves fighting alongside troops you are giving operational support to. It is different from giving logistical/technical support that involves generating capabilities soldiers need to be able to fight.

4 See the website for the International Code of Conduct Association for PSC that provides security services in conflict zones https://www.icoca.ch/en/membership (accessed 7 December 2018).

5 See Eeben Barlow's presentation to the Symposium on Leadership, UNG, 14–15 November 2018. https://drive.google.com/file/d/1F1MQI8z7RUlIDfQL1TX6XsxzzKSV18Q-/view?usp=drivesdk (accessed 10 December 2018).

6 Mohlin, for example, argues that such organisations can be used as an alternative clandestine instrument of state power instead of using the state's own military force (Mohlin 2016: 109–16).

7 The Department for International Development (DIFD) was replaced by the Foreign, Commonwealth & Development Office in September 2020.

8 After Wagner's attempted military coup and the ease by which Wagner contractors took control of the city of Rostov-on-Don and then drove north to Moscow the Kremlin may decide to restrict the future activities of private military companies in the country. It will certainly not want to expose further major weaknesses in the Kremlin's control of security in Russia as revealed by Wagner's actions (BBC 2023b).

9 See, Hin-Yan Liu 2015: 172–8, for a detailed explanation on how the International Convention against the Recruitment, Use, Financing and Training of Mercenaries came into force.

10 To see what self-regulation involves, see https://www.icoca.ch (accessed 17 June 2019).

11 At the time of writing this chapter it is not clear if Wagner will continue operating in Ukraine after its attempted military coup against Putin.

Bibliography

Abrahamsen, Rita and Williams, Michael C., *Security Beyond the State: Private Security in International Politics* (Cambridge: Cambridge University Press, 2011).

Avant, Deborah and Sigelman, Lee, 'Private Security Contractors and Democracy: Lessons from the US in Iraq,' *Security Studies*, vol. 19, no. 2 (2010), pp. 230–65.

BAE Systems, 2018, 'Supporting Typhoon for the Frontline', https://www.baesystems.com/en-uk/product/supporting-typhoon-for-the-frontline (accessed 13 December 2018).

BBC, 2018, 'Syria Conflict: "Russians Killed" in US Air Strikes,' 13 February 2018, https://www.bbc.co.uk/news/world-middle-east-43051333 (accessed 6 December 2018).

BBC, 2023a, 'Belarus Leader Welcomes Wagner Boss Prigozhin into Exile,' https://www.bbc.co.uk/news/world-europe-66029636, 28 June 2023 (accessed 29 June 2023).

BBC, 2023b, 'Wagner Chief Vows to Topple Russian Military Leaders,' 24 June 2023, https://www.bbc.co.uk/news/world-europe-66005256# (accessed 29 June 2023).

Copp, Tara, 'Here's the Blueprint for Erik Prince's $5 Billion Plan to Privatize the Afghanistan War,' 5 September 2018, https://www.militarytimes.com/news/your-military/2018/09/05/heres-the-blueprint-for-erik-princes-5-billion-plan-to-privatize-the-afghanistan-war/#.W5EJczSaDxQ.twitter (accessed 4 October 2018).

Cullen, Patrick, 'A Century of US Diplomatic Security,' in Eugenio Cusumano and Christopher Kinsey, *Diplomatic Security: A Comparative Analysis* (Stanford: Stanford University Press, 2019), pp. 11–36.

Cusumano, Eugenio and Kinsey, Christopher, 'Bureaucratic Interests and the Outsourcing of Security,' *Armed Forces and Society*, vol. 41, Issue 4 (2014), pp. 591–615.

Duffield, Mark, *Global Governance and the New Wars* (London: Zed Books, 2001).

Erbel, Mark and Kinsey, Christopher, 'The Role of Private Military Corporations in Defence,' in David Galbreath and John R. Deni (eds.), *Routledge Handbook of Defence Studies* (Oxford: Routledge, 2018), pp. 302–14.

Freeman, Colin, 'South African Mercenaries' Secret War on Boko Haram,' *The Telegraph*, 10 May 2015.

Gansler, Jacques S. and Lucyshyn, William, 'Contractors Supporting Military Operations: Many Challenges Remain,' in Christopher Kinsey and Malcolm Hugh Patterson (eds.), *Contractors and War: The Transformation of US Expeditionary Operations* (Stanford: Stanford University Press, 2012), pp. 278–96.

Gibbons-Neff, Thomas, 'How a 4-Hour Battle Between Russian Mercenaries and US Commandos Unfolded in Syria,' *The New York Times*, 24 May 2018.

Hin-Yan, Liu, *Law's Impunity* (Oxford: Hart Publishing, 2015).

Isenberg, David, 'Wagner vs. the Russian Media,' 13 August 2018, https://lobelog.com/wagner-vs-the-russian-media/ (accessed 4 October 2018).

Jones, Clive, *Britain and the Yemen Civil War: 1962–1965* (Sussex: Sussex Academic Press, 2004).

Kinsey, Christopher, *Corporate Soldiers and International Security* (Oxford: Routledge, 2006).

Kinsey, Christopher, *Private Contractors and the Reconstruction of Iraq* (Oxford: Routledge, 2009).

Kinsey, Christopher, 'Transforming War Supply,' *International Journal*, vol. 69, no. 4 (2014), pp. 494–509.

Kinsey, Christopher and Patterson, Malcolm Hugh (eds.), *Contractors and War: The Transformation of US Expeditionary Operations* (Stanford: Stanford University Press, 2012).

Krieg, Andreas, *Commercialising Cosmopolitan Security* (London: Palgrave Macmillan, 2016).

Mazzetti, Mark and Hager, Emily B., 'Secrete Desert Force Set Up by Blackwater's Founder,' *The New York* Times, 14 May 2011.

McFate, Sean, *The Modern Mercenary* (Oxford: Oxford University Press, 2014).

Mohlin, Marcus, 'Merchants of Security,' in Joakim Berndtsson and Christopher Kinsey (eds.), *The Routledge Research Companion to Security Outsourcing* (Oxford: Routledge, 2016), pp. 109–16.

Parrott, David, *The Business of War* (Cambridge: Cambridge University Press, 2012).

RAF, 2018, 'Voyager', https://www.raf.mod.uk/aircraft/voyager/ (accessed 11 December 2018).

Stanger, Allison, *One Nation Under Contract* (New Haven: Yale University Press, 2009).

Stanger, Allison, 'Contractors' Wars and the Commission on Wartime Contracting,' in Christopher Kinsey and Malcolm Hugh Patterson (eds.), Christopher Kinsey and Malcolm Hugh Patterson (eds.), *Contractors and War: The Transformation of US Expeditionary Operations* (Stanford: Stanford University Press, 2012), pp. 184–204.

Waltz, Kenneth, *Theory of International Politics* (New York: Random House, 1987).

34

THE ARMS RACE IN CONTEMPORARY EAST ASIA

Andrew T.H. Tan

The Arms Build-up in East Asia

Despite the end of the Cold War in 1989, Asia as a whole has since increased defence spending as states in the region expanded and modernised their armed forces. According to the authoritative Stockholm International Peace Research Institute (SIPRI), global defence spending measured in constant US dollars in 2016 rose from $1,424 billion in 1988 to US$ 1,686 in 2017 (SIPRI 2018a). While this is a modest increase over a 20-year period, defence spending in Asia (including Oceania) rose from $132 billion in 1988 to $469 billion in 2017. Over this period, East Asia's defence spending rose from $80 billion to $322 billion, which means that not only does it constitute the bulk of Asia's defence spending, the increase in its spending has also been significantly higher than the global norm (SIPRI 2018b).

According to the authoritative International Institute for Strategic Studies (IISS), Asia's total defence spending exceeded Europe in 2012 (*Military Balance* 2013: 33). In 2018, Asia and Australasia as a whole accounted for 24.3 per cent of global military expenditure, compared with 39.2 per cent for North America and 16.9 per cent for Europe (*Military Balance* 2019: 21). In 2017, China had the highest defence expenditure in Asia, spending US$ 228 billion. This was followed by India at US$ 63.9 billion, Japan at US$ 45.4 billion and South Korea at US$ 39.1 billion (SIPRI n.d.). The predominance of East Asian states amongst the highest defence spenders in Asia indicates that it is this sub-region that is leading Asia's defence expansion and modernisation.

It was thus no surprise that, in 2010, Desmond Ball warned of an impending arms race in Asia, given the increased defence budgets and capabilities, as well as evidence of an action–reaction dynamic. As well, these developments, coupled with the lack of effective regional institutions that could manage and contain regional conflicts, meant that the possibility of interstate conflict had increased (Ball 2010: 30–51). This chapter will investigate the phenomenon of East Asia's arms build-up, which some have feared is in grave danger of becoming a full-blown arms race (Tan 2014a). What are the trends in this build-up, and is there evidence of an arms race? What explains it? What are its implications?

DOI: 10.4324/9780429437915-43

For the purpose of this chapter, East Asia is defined as consisting of China, Taiwan, North Korea, South Korea and Japan. This is consistent with most authoritative definitions which differentiate North-East and South-East Asia, usually equating East Asia with North-East Asia. Moreover, the East Asian states (excluding North Korea) spent US$ 323 billion on defence in 2017, compared with US$ 40.1 billion for all Southeast Asian countries combined (SIPRI n.d.). Given the constraints of space, and the relative importance of East Asia/North-East Asia, this chapter will focus on this sub-region as defined here.

While SIPRI includes Mongolia and the IMF leaves out North Korea in their definition of East Asia, this chapter will exclude Mongolia as its security dynamics relate more to Central Asia, and will include North Korea as SIPRI has done, since its security dynamics relate to the other states in this region (SIPRI 2018a, SIPRI 2018b, IMF 2018). However, given the lack of reliable data on North Korea, the analysis of North Korea will be based on a qualitative assessment of its military development.

Arms Build-ups and Arms Races

Before examining in detail the key issues in the arms build-up in East Asia, it is useful to assess if there is indeed an 'arms race' in the region. A useful framework for defining an arms race was developed by Colin Gray in 1971. Gray suggested that an arms race has four fundamental attributes: two or more parties who are conscious of their antagonism; the development of their armed forces with other arms race participants as referents; competition in terms of quantity and quality; and, rapid increases in quantity as well as improvement in quality. In addition, Gray asserted that for an arms race to take place, some kind of interactive action–reaction dynamic has to be present (Gray 1971: 41–77).

In evaluating the question of an arms race in Asia, Desmond Ball argued in 2010 that there was clear evidence of the emergence of increased military capabilities, citing the rapid increase in China's defence budget as well as the proliferation of nuclear weapons and ballistic missiles. This, coupled with evidence of an interactive action–reaction dynamic, meant that there existed 'arms racing behaviour.' Ball thus warned that 'in this environment, with many parties and many levels and directions of interactions, the possibilities for calamity are high' (Ball 2010: 49). What is thus happening in East Asia is therefore at least a nascent arms race, given evidence of arms racing behaviour characterised by increased defence spending and increased military capabilities, as well as evidence of an action–reaction phenomenon.

Trends in the Arms Race in East Asia

A survey of defence spending trends over the decade from 2008 to 2017 demonstrates a clear pattern of increased defence spending as well as enhanced capabilities. Interestingly, Table 34.1 shows that despite the increases in defence spending by China and South Korea, overall, defence spending as a percentage of GDP in East Asia has been low, in the 0.9–2.6 per cent range. In 2017, for instance, China spent only 1.9 per cent of its GDP on defence, despite the huge outlay of US $228 billion, which is the second largest in the world after the United States. Similarly, Japan has consistently spent 1 per cent or below of its GDP on defence.

Table 34.1 East Asia's Defence Expenditure as Percentage of GDP, 2008–2017

Country	2008	2009	2010	2011	2012	2013	2014	2015	2016	2017
China	1.9%	2.1%	1.9%	1.8%	1.8%	1.9%	1.9%	1.9%	1.9%	1.9%
Japan	0.9%	1.0%	1.0%	1.0%	1.0%	1.0%	1.0%	1.0%	0.9%	0.9%
South Korea	2.6%	2.7%	2.6%	2.6%	2.6%	2.6%	2.7%	2.6%	2.6%	2.6%
Taiwan	2.1%	2.3%	2.0%	2.1%	2.1%	1.9%	1.8%	1.9%	1.9%	1.8%
North Korea	n.a.	n.a.	n.a.	n.a.	n.a.	n.a.	n.a.	n.a.	n.a.	n.a.

Source: SIPRI 2018c.

Table 34.2 East Asia's Defence Expenditure, 2008–2017

Country	2008	2009	2010	2011	2012	2013	2014	2015	2016	2017
China	86362	105644	115712	137967	157390	179881	200772	214093	216031	228231
Japan	46362	51465	54656	60762	60012	49024	46881	42106	46471	45387
South Korea	26072	24575	28175	30992	31952	34137	37552	36571	36934	39153
Taiwan	8960	9123	9092	9998	10497	9964	9802	9803	9924	10569
North Korea	n.a.	n.a.	n.a.	n.a.	n.a.	n.a.	n.a.	n.a.	n.a.	n.a.

Note: Figures in current US$ m.

Source: SIPRI n.d., Data for All Countries, 1949–2017, Military Expenditure Database, https://www.sipri.org/databases

However, Table 34.2 provides a different picture. It demonstrates that defence spending in East Asia (excluding North Korea), rose from US$ 167.7 billion in 2008 to US$ 323.3 billion in 2017. China is also the leading region in terms of defence spending, which has grown exponentially in tandem with its rapid economic rise. Thus, China's defence spending has risen dramatically, from US$ 86.3 billion in 2008 to US$ 228 billion in 2017. South Korea has also raised its defence spending, from US$ 26 billion in 2008 to US$ 39 billion in 2017. However, despite the existential threat from China, Taiwan has only moderately increased its defence spending, raising it from US$ 8.9 billion in 2008 to US$ 10.5 billion in 2017. Japan too has not significantly raised its defence spending, spending about US$ 45.4 billion in 2017, compared with around US$ 46.3 billion in 2008.

A comparison with Europe illustrates the tremendous strides East Asia has taken during the 20 years from 1998 to 2017. In 1998, Europe's defence spending, measured in constant US dollars in 2016, totalled US$ 267 billion, of which Western Europe accounted for US$ 235 billion. In 2017, this had increased to US$ 327 billion, with Western Europe accounting for US$ 243 billion. Over the same period, East Asia alone had increased from US$ 80 billion to $ 322 billion (SIPRI 2018b). Thus, East Asia in 2017 is well ahead of Western Europe in terms of defence spending.

A survey of weapons systems deployed by the armed forces of East Asia provides proof of the impressive strides they have made in expanding their range of capabilities by acquiring the most sophisticated weapons systems available. There is also clearly an interactive action–reaction element at play.

Tables 34.3 and 34.4 compare the numerical holdings of key weapons systems of the East Asian states between 1998 and 2018. In some cases, they show a decrease in the numerical holdings of key weapons systems. However, it is clear that new capabilities have been acquired, such as Airborne Early Warning and Control (AEWC) aircraft, and aircraft/helicopter carriers. The two tables also show that China is the dominant conventional military power in East Asia, as well as the fact that it is building a powerful navy by substantially increasing the number of principal surface warships.

A qualitative analysis provides a better picture of the sustained military development in East Asia. In 2018, China's armed forces consisted of around 2 million personnel (excluding reserves), with 5,800 main battle tanks and 8,950 armoured personnel carriers in its army, 2,798 combat aircraft in its air force and naval aviation force, as well as 87 principal surface warships (including an aircraft carrier) and 62 submarines in its navy (Military Balance 2019: 256–65).

More significantly, China has, in the past decade, steadily introduced new weapons systems, technologies and capabilities that it did not previously possess. In particular, its air force and navy have made huge strides, enabling China to project military power regionally as well as further afield. China's J-10 jet-fighter is a fourth-generation combat aircraft and is described as 'the most capable single engine fighter in service' anywhere in the world, particularly after it received sophisticated upgrades after entering into service in 2006. Today,

Table 34.3 East Asia's Military Capabilities, 1999

	Main battle tanks	Armoured personnel carriers	AEWC	Combat aircraft	Aircraft/ helicopter carriers	Principal surface warships	Submarines	Major landing ships
China	7,060	4,800	0	3,000	0	60	65	59
Japan	1,070	850	14	331	0	55	16	9
South Korea	1,000	2,520	0	555	0	39	19	14
Taiwan	739	1,175	0	570	0	33	4	18
North Korea	3,500	2,500	0	621	0	3	26	10

Source: *The Military Balance 2000* (London: International Institute for Strategic Studies, 2000).

Table 34.4 East Asia's Military Capabilities, 2018

	Main battle tanks	Armoured personnel carriers	AEWC	Combat aircraft	Aircraft/ helicopter carriers	Principal surface warships	Submarines	Major landing ships
China	5,800	8,950	13	2,798	1	86	62	54
Japan	667	863	17	625	4	45	20	3
South Korea	2,514	3,330	4	604	0	26	22	9
Taiwan	565	1,445	6	479	0	26	4	9
North Korea	3,500	2,532	0	545	0	2	73	10

Source: *The Military Balance 2019* (London: International Institute for Strategic Studies, 2019).

it is equipped with an advanced active electronically scanned array (AESA) fire control radarandinfrared search and track systems, making it more than the equal of the US F-16 (Military Watch n.d.).

In 2015, China also upgraded its copy of the Russian Su-27 jet-fighter, the J-11, by incorporating AESA, which would give it similar capabilities to current front-line US combat aircraft (Keck 2015). More significantly, China announced in 2018 that its new fifth-generation J-20 stealth combat aircraft, considered the rough equal of the US F-22 Raptor, had become operational, making China only the second country in the world to field stealth combat aircraft (Siedel 2018a). Reportedly, China is also developing a stealth long-range bomber, the H-20, which will be similar in capability to the US Air Force's B-2 stealth 'flying wing' bomber (Martin 2018).

In 2017, China's navy launched the *Liaoning*, a refurbished ex-Russian aircraft carrier, complete with J-15 combat aircraft. In 2018, China began sea-trails of a second aircraft carrier, which was locally built and based on the *Liaoning* design. Reportedly, construction on a third aircraft carrier has begun, and there are plans for a fourth to be built. Significantly, the third aircraft carrier will have an advanced electromagnetic aircraft launch system, which the USN has only just deployed on its latest aircraft carriers (Wang 2018). This third carrier will be around 85,000 tonnes in size, making it comparable to current US aircraft carriers, and will be operational in 2024 (Siedel 2018b). China is also developing the large and capable Type 055 cruisers, which are armed with the advanced active phased array radar system that can detect air targets from afar, and has the ability to launch 112 anti-air, anti-ship and land-attack cruise missiles. The Type 055 is considered to be more than a match for the current fleet of ageing Ticonderoga-class cruisers in the USN (Huang 2018). China is thus steadily acquiring the same aircraft carrier battle group capabilities of the USN, which will enable it to carry out significant power projection, both regionally and further afield.

Significantly, China's nuclear deterrent capabilities have improved, beginning with the deployment of its first sea-based, nuclear-armed ballistic missile submarine in 2014 (Babiarz 2014). In 2018, China deployed four Jin class Type 094 nuclear ballistic missile submarines, armed with JL-2 strategic nuclear missiles (Military Balance 2019: 258). This is a new capability which it did not previously possess. Once the new H-20 stealth bomber is developed, China will possess a very sophisticated nuclear triad of air, sea and land-launched nuclear missiles, a capability that only the United States and Russia have. Thus, China has been developing capabilities designed to match its peer competitor, the United States. In particular, China is attempting to build A2/AD (anti-access/area denial) capabilities that would be able to counter US military freedom of action in East Asia in the event of conflict, thus preventing the United States from coming to the aid of its allies, namely, South Korea, Japan and Taiwan.

China's power projection capabilities pose an obvious threat to US allies in East Asia, especially Japan and Taiwan, both of which have had sometimes tense and conflictual relations with China. Japan has followed developments closely and introduced new weapons systems and capabilities in order to counter China's military development. According to Japan's defence white paper in 2017, Japan aims to boost its amphibious warfare, power projection and ballistic missile defence capabilities, particularly on account of China's capabilities and aggressive military posture as well as North Korea's ballistic missile tests and nuclear weapons development (Defence of Japan 2017).

Japan thus established its first amphibious rapid deployment brigade in 2017 (Defence of Japan 2017: 326). While its three Osumi class amphibious ships were procured between

1998 and 2003, Japan added to its amphibious capabilities by purchasing 17 V-22 Osprey tilt-rotor transport aircraft from the United States in 2015 (LaGrone 2015). There are also plans to convert Japan's four helicopter carriers into small aircraft carriers using F-35B stealth combat aircraft (Panda 2018). Japan's air force has already acquired 42 F-35A stealth combat aircraft, which are being manufactured locally. In 2018, an additional 20 F-35As were purchased directly from the United States (Reuters 2018a). Japan has also been exploring the possibility of developing its own fifth-generation F-3 stealth combat aircraft, albeit in collaboration with international partners, that will have capabilities similar to the F-22 Raptor which the United States has declined to sell (Gady 2018). Japan's deep interest in stealth combat capabilities and aircraft carriers can be understood in the context of China's emerging capabilities in these areas, as well as its assertive and aggressive stance towards territorial disputes with Japan, such as its frequent violations of Japan's territorial waters (Defence of Japan 2018: 319–20). Japan is also actively building a ballistic missile defence system centred on new PAC-3 MSE Patriot missiles and sea-based SM-3 anti-ballistic missiles to be deployed on its Aegis-equipped destroyers, primarily to counter the missile threat from North Korea (Reuters 2018b).

Despite numerous constraints, Taiwan has also acquired new capabilities. Diplomatically isolated due to China's demand that all countries adhere to the 'One-China' stance under which Taiwan is part of China, Taiwan has not been able to procure new weapons systems from abroad, except from its close ally, the United States. Even the United States, however, has been reluctant to sell sophisticated weapons systems, limiting its sales to Taiwan to defensive items. Within these considerable constraints, Taiwan has still managed to acquire some new capabilities in order to counter the existential threat from China. For instance, it has developed RT 2000 Thunder artillery multiple-launch rocket systems (AMLRS) for coastal defence, and the Hsiung Feng 2E land-attack cruise, missile which is similar to the US Tomahawk missile, providing Taiwan with counter-strike capabilities against China's growing array of ballistic missiles. Taiwan has also acquired Patriot PAC-3 air defence systems from the United States to counter the missile threat from China. In addition, Taiwan's 146 ageing F-16A/B combat aircraft are being upgraded with AESA radars. Taiwan has also acquired a squadron of US-built E2C Hawkeye AWAC aircraft, giving it early warning capabilities (Tan 2014b: 48–9). In 2017, Taiwan announced that it would manufacture eight conventional submarines, although the first boat would not enter sea-trials until 2024, provided Taiwan is able to overcome technological hurdles and also obtain the necessary systems and other components to build them (Gady 2017).

North Korea has been isolated and has few economic resources to import the latest and most sophisticated weapons systems. Out of necessity, it has emphasised autarky in its defence production, resulting in the mass production of basic conventional weapons systems which are outdated. However, it has made up for its own military deficiencies with a 'military first' policy of emphasising military defence as a national priority. While no reliable data is available, it is estimated that North Korea spends around 25 per cent of its GDP on defence (CFR 2018). North Korea has opted to develop asymmetric capabilities including weapons of mass destruction. While it first tested a nuclear bomb in 2006, the pace of nuclear weapons and ballistic missile development accelerated markedly after Kim Jong-un came to power in 2011. In 2017, it tested a hydrogen bomb with a yield exceeding 100 kilotons. In addition, it has developed a large stockpile of chemical weapons. North Korea has also developed ballistic missiles of increasing range (CFR 2018). This has enabled it to target not just South Korea but also Japan and US territory and bases in the Pacific. Its conventional forces are huge, with almost 1.3 million troops, over 21,000 pieces of artillery, some 6,000 tanks and armoured vehicles, as well as 73 tactical submarines (Military Balance 2019: 280–3). By positioning its conventional forces, especially artillery, close to the border with South Korea, it is in a position to wreck huge damage to Seoul, the capital of the South, in the event of conflict (CFR 2018).

South Korea, which faces an existential threat from North Korea, has made substantial efforts at upgrading its land, air and naval forces with new capabilities. Its navy currently has one helicopter-carrying amphibious warfare vessel, the *Dokdo*, which was commissioned in 2007. However it is building another, with plans to turn these ships into small aircraft carriers deploying F-35B stealth combat aircraft (Keck 2018). Its navy also boasts three powerful Sejong-class, Aegis-equipped cruisers, which were acquired between 2008 and 2014. Each vessel carries 128 missiles, and they are reputedly some of the most advanced and heavily armed warships in the world (Military Today n.d.). In 2011, South Korea's air force took delivery of four Boeing 737–700IGW airborne warning and control system (AWACS) aircraft (Carey 2011). In 2018, it took delivery of the first of its 40 F-35 stealth combat aircraft (Straits Times 2018). Like Japan, it plans to acquire the PAC-3 MSE Patriot anti-ballistic missile, placing orders for it in 2018 (Defence News 2018). South Korea's military development can thus be understood in the context of the existential threat to it from North Korea's ballistic, nuclear and vast land conventional forces. In sum, the increased defence spending, the acquisition of new capabilities and the interactive element in East Asia's military development all point to at least a nascent arms race in the region.

Explaining the Arms Race in East Asia

What explains the arms race in East Asia? Three key explanations underlie the phenomenon: the availability of economic resources, the increasing importance of maritime security and the presence of interstate tensions.

Availability of Economic Resources

The availability of economic resources is one key reason for East Asia's dramatic military development over the past three decades. Table 34.1 demonstrated that despite the increases in defence spending, in particular by China and South Korea, overall, defence spending as a percentage of GDP in East Asia has been low. Indeed, East Asia's military spending as a percentage of GDP actually declined over 1988–2017. In 1988, defence spending was in the 0.9–5.1 per cent GDP range, but by 2017, this had declined to 0.9–2.6 per cent (SIPRI 2018c). In 2017, for instance, China spent only 1.9 per cent of its GDP on defence, despite the huge outlay of US$ 228 billion, the second largest in the world after the United States (Tables 34.1 and 34.2).

This is indicative of the strong economic growth of this region, which has provided the necessary resources to fund military modernisation and expansion. Table 34.5 demonstrates the rapid economic development of China over the ten-year period from 2008–2017. China posted high economic growth rates through the Global Financial Crisis in 2008, helping to sustain the global economy at a crucial time. According to Trading Economics, the size of China's economy has grown tremendously, from US$ 4,598 billion in 2008 to US$ 11,199 billion in 2016 (Trading Economics n.d.). In 2018, China's economy was the second largest in the world after the United States, with a GDP of about US$ 14,000 billion, while Japan

Table 34.5 East Asia's Economic Growth (Percentage of GDP), 2008–2017

Country	2008	2009	2010	2011	2012	2013	2014	2015	2016	2017
China	9.6	9.2	10.6	9.5	7.9	7.8	7.3	6.9	6.6	6.2
Japan	−1.0	−5.5	4.7	−0.5	1.7	1.4	0	0.5	0.5	0.6
South Korea	2.8	0.7	6.5	3.7	2.3	2.9	3.3	2.6	2.7	3
Taiwan	0.7	−1.6	10.6	3.8	2.1	2.2	3.9	0.7	1	1.7
North Korea	n.a.	n.a.	n.a.	n.a.	n.a.	n.a.	n.a.	n.a.	n.a.	n.a.

Source: Global Finance 2018, Country Data, https://www.gfmag.com/global-data/country-data/

remained the world's third largest economy despite its relative economic decline, with a gross GDP of about US$ 5,100 billion. South Korea, with a GDP of about US$ 1,700 billion, was ranked eleventh in the world in 2018 (IMF 2018).

China's dramatic economic rise and relatively steady economic growth in South Korea have provided the necessary economic resources for military modernisation, resulting in large increases in defence spending. Thus, even without the presence of other factors, the availability of economic resources would have ensured continued military development.

Increasing Importance of Maritime Security

One key feature of East Asia's military development has been the build-up of naval forces due to the increasing importance of maritime security. This can be attributed to the growth of the global interlinked economy in an age of globalisation, which has resulted in 90 per cent of global trade being carried out by sea. This has meant that sea lines of communications (SLOCs) and strategic waterways have become strategically important to export-oriented economies which are also dependent on foreign sources of energy, particularly those in East Asia. In addition, the United Nations Convention of the Law of the Sea (UNCLOS III) in 1982 granted coastal states 200 nautical mile (322 km) Exclusive Economic Zones (EEZs), which meant that most coastal states now had to defend larger maritime territories (United Nations 1982). Thus, even in the absence of other factors, such as interstate tensions, the need for increased naval capability to ensure maritime security would be present.

A more salient reason for the increased emphasis on naval power has been the presence of maritime territorial disputes. China's expansive claims to maritime territory in the East and South China Seas has been a major driver of its naval build-up. After 2008, when China's economy grew rapidly even as the West faced the Global Financial Crisis, China began to perceive that the balance of power was shifting in its favour. Given the increasing importance of maritime security due to its rapid economic growth, China has been focusing on developing its maritime power (ICG 2013: 15). This explains China's aggressive and assertive moves over the disputed Senkaku islands, which are potentially rich in oil and gas resources. China has since carried out a number of intrusions into Japanese air and maritime space, raising tensions and fears over an accidental war (Defence of Japan 2017: 319–20).

Similarly, China has also been very aggressive in asserting its claims to disputed maritime territory in the potentially oil-rich South China Sea, where it has been engaged in reclamation works and the building of artificial islands as well as military facilities in its quest to consolidate control over the area (Siedel 2018c). In March 2018, China carried out its largest naval exercise to date in the South China Sea, involving some 40 warships including the aircraft carrier, the *Liaoning*, in a show of force to underline its claims to the area (Clover 2018).

Interstate Tensions

Finally, there are serious interstate tensions in East Asia, which have been a key factor in the arms race in the region. The most significant are tensions between China and the United States. Described as 'the most consequential bilateral relationship of our time,' China, the rising power, is challenging the dominant position of the United States in the region (Steinberg and O'Hanlon 2014: 1). The tensions were epitomised by the *Impeccable* incident in March 2009, when Chinese ships harassed a USN surveillance vessel that was in South China Sea, on the grounds that it was trespassing Chinese territorial waters. China accused the US vessel of trespassing on Chinese waters, but the United States protested that the South China Sea constituted international waters (CNN 2009). The United States however, has refused to recognise China's expansive claims in the South China and East China Seas and has carried out FONOPs (freedom of navigation operations) to challenge China's claims. In 2011, in his seminal 'Asia Pivot' speech, then-US President Obama promised that the United States would 'allocate the resources necessary to maintain our strong military presence in this region,' and that 'we will preserve our unique ability to project power and deter threats to peace' (Sydney Morning Herald 2011).

Another serious set of interstate tensions is that between China and Japan. China's use of anti-Japanese nationalism to sustain the ruling communist party's legitimacy has led to sustained anti-Japanese sentiments in China which has affected bilateral relations (Jacques 2009: 310). Violent anti-Japanese protests took place in China in 2012 following the nationalisation of the disputed Senkaku Islands by the government of Japan (Ogura and Mullen 2012). In early 2013, Chinese warships locked fire-control radars on a Japanese destroyer, a dangerous move that could have led to an accidental war (South China Morning Post 2013). In late 2013, China's 'air-defence identification zone' around the Senkaku Islands led to the United States flying B-52 bombers over the island to challenge this (BBC 2013). By then, tensions between China and Japan were described as having reached their highest levels since the end of the Second World War in 1945 (Hughes 2013).

A third set of interstate tensions exist on the Korean Peninsula. Since the end of the Korean War in 1953, North Korea has used provocative brinkmanship as part of its strategy to force the United States and its allies, namely, South Korea and Japan, to bargain with it. North Korea has 21,000 pieces of artillery aimed at South Korea, most of it directly threatening its capital, Seoul, which is 40 km from the border between the two countries (Military Balance 2019: 281). North Korea also possesses nuclear weapons of increasing yield, as well as ballistic missiles which are steadily increasing in range, threatening not just South Korea but also Japan and US bases and territory in the West Pacific. The tensions between the two Koreas have led to an arms race between the two. While the South has increased defence spending over the years and developed new conventional capabilities to counter the threat from the North, the North has focused on asymmetric warfare capabilities centred around weapons of mass destruction, vast artillery capabilities, tactical submarines and ballistic missiles. While the two countries held a summit in 2018, and both President Trump and North

Korean leader Kim Jong-un signed an agreement on denuclearisation in Singapore in June 2018, there is still a long road to travel before genuine peace can replace the decades of hostility and tension in the Korean Peninsula (CNBC 2018).

Finally, tensions also exist along the Taiwan Strait. The reunification of Taiwan with the mainland is an emotive nationalist issue for China, and there are signs that China is increasingly impatient for this to occur. China's military modernisation has focused on building the necessary anti-access capabilities to prevent US forces from coming to the assistance of Taiwan in a conflict, as well as conventional capabilities to either coerce or forcibly reunify Taiwan if necessary (Blumenthal 2010: 17). This has led to an increasingly stark military imbalance on the Taiwan Strait, as Taiwan's relative decline in defence capabilities in the context of China's dramatic rise has increased the possibility of conflict since it has given China a military option it did not previously possess.

Conclusions: Implications of East Asia's Arms Race

It is clear that the East Asian states, namely, China, South Korea, North Korea, Japan and Taiwan, have been increasing their defence spending and capabilities in recent years, acquiring new capabilities that had not been present before. In particular, this phenomenon has been driven in large part by China's rise as a global economic power. These new and sophisticated capabilities include AWACs, stealthy combat aircraft, aircraft/helicopter carriers, missile cruisers and ballistic missiles. Existing weapons systems, such as main battle tanks, armoured personnel carriers, submarines and amphibious warfare vessels, have also been upgraded with newer, more capable models. In addition, military developments in East Asia demonstrate the presence of an action–reaction dynamic, where the states involved have attempted to acquire similar weapons systems to counter other referent states in the region. The rapid build-up of arms and the presence of an interactive element are the essence of arms racing behaviour, and can be described as at least a nascent arms race.

The problem is that this emerging arms race in East Asia is taking place in the context of heightened regional tensions arising from the increasing intense strategic competition between China and the United States, as well as high tensions between China and Japan, on the Korean Peninsula (at least until the recent possibly temporary thaw arising from the Trump–Kim Summit in 2018), and on the Taiwan Strait. This leads to heightened risks arising from the security dilemma, such as misperceptions and conflict spirals that could lead to war (Jervis 1976). The danger is accentuated by the absence of effective regional institutions, norms and regimes that could ameliorate interstate tensions and conflicts amongst the states in the region.

More significantly, any outbreak of conflict in East Asia could rapidly escalate, given the presence of substantial conventional and nuclear capabilities in the region. After all, three of the key players, namely, China, North Korea and the United States—which maintains security guarantees over South Korea and Japan—possess nuclear weapons. Any open conflict in East Asia would also have immense global consequences as it would also invariably involve the three largest economies of the world, namely, the United States, China and Japan.

The solution to the arms race in East Asia might lie with the experience of Europe during the Cold War, where confidence and security building measures (CSBMs) helped to ease tensions and contain the Cold War arms race. It is imperative that East Asia begin the process of confidence-building which can ease tensions and lead to greater stability.

In this respect, the Trump-Kim summit in Singapore in 2018 and subsequently in Vietnam in 2019 (despite its failure) are important initial steps. Given the spiralling tensions in East Asia, it is precisely such bold moves that are required.

Bibliography

Babiarz, Renny, 'China's Nuclear Submarine Force,' *China Brief*, vol. 17, no. 10 (2014), 21 July, Jamestown Foundation, https://jamestown.org/program/chinas-nuclear-submarine-force/, accessed 15 March 2019.

Ball, Desmond, 'Arms Modernization in Asia: An Emerging Complex Arms Race,' in Andrew T.H. Tan (ed.), *The Global Arms Trade* (London: Routledge, 2010), pp. 30–51.

BBC News, 'US B-52 Bombers Challenge Disputed China Air Zone,' 26 November 2013, http://www.bbc.com/news/world-asia-25110011, accessed 15 March 2019.

Blumenthal, Dan, *Sino-US Competition and US Security: How Do We Assess the Military Balance?* (Seattle, Washington: NBR Analysis, National Bureau of Asian Research, December, 2010).

Buzan, Barry, *An Introduction to Strategic Studies: Military Technology and International Relations* (London: St. Martin's Press, 1987).

Carey, Bill, 'First 737 AWACS Arrives in South Korea,' *AIN Online*, 8 August 2011, https://www.ainonline.com/aviation-news/defense/2011-08-08/first-737-awacs-arrives-south-korea, accessed 15 March 2019.

CFR (Council on Foreign Relations), *Backgrounder: North Korea's Military Capabilities*, 6 June 2018, https://www.cfr.org/backgrounder/north-koreas-military-capabilities, accessed 15 March 2019.

Clover, Charles, 'China Offers Show of Naval Force in South China Sea,' *Financial Times*, 29 March 2018, https://www.ft.com/content/ff945850-333c-11e8-b5bf-23cb17fd1498, accessed 20 September 2023.

CNBC, *Full Text of the Trump-Kim Agreement*, 12 June 2018, https://www.cnbc.com/2018/06/12/full-text-of-the-trump-kim-summit-agreement.html, accessed 15 March 2019.

CNN, 'Pentagon Says Chinese Vessels Harassed U.S. Ship,' 9 March 2009, http://edition.cnn.com/2009/POLITICS/03/09/us.navy.china/index.html, accessed 15 March 2019.

Defence News, 'Seoul to Order New Pac-3 Interceptors to Counter North Korea,' 20 February 2018, https://www.defensenews.com/global/asia-pacific/2018/02/19/seoul-to-order-new-pac-3-interceptors-to-counter-north-korea/, accessed 20 September 2023.

Defence of Japan 2017, Ministry of Defence, Tokyo, http://www.mod.go.jp/e/publ/w_paper/pdf/2017/DOJ2017_3-1-2_web.pdf, accessed 15 March 2019.

Financial Adviser, 'The World's Biggest Economies in 2018,' 1 August 2018, https://www.fa-mag.com/news/the-world-s-biggest-economies-in-2018-39645.html?section=3&page=1, accessed 15 March 2019.

Gady, Franz-Stefan, 'Japan Denies Scrapping 5th Generation Stealth Fighter Program,' *The Diplomat*, 7 March 2018, https://thediplomat.com/2018/03/japan-denies-scrapping-5th-generation-stealth-fighter-program/, accessed 15 March 2019.

Gady, Franz-Stefan, 'Taiwan to Build Own Attack Submarines Within Next 10 Years,' *The Diplomat*, 23 March 2017, https://thediplomat.com/2017/03/taiwan-to-build-own-attack-submarines-within-next-10-years/, accessed 15 March 2019.

Global Finance 2018, Country Data, https://www.gfmag.com/global-data/country-data/

Gray, Colin S., 'The Arms Race Phenomenon,' *World Politics*, vol. 24, no. 1 (1971), pp. 39–79.

Huang, Paul, 'Chinese Regime Races for Naval Supremacy, Building 8 Cruisers While US Builds None,' 27 January 2018, *Epoch Times*, https://www.theepochtimes.com/chinese-regime-races-for-naval-supremacy-building-8-cruisers-while-us-builds-none_2416921.html, accessed 15 March 2019.

Hughes, Christopher, 'Viewpoints: How Serious are China-Japan Tensions?' *BBC News*, 8 February 2013, www.bbc.co.uk/news/world-asia-21290349, accessed 15 March 2019.

ICG (International Crisis Group), 'Dangerous Waters: China-Japan Relations on the Rocks,' Asia Report No. 245, 8 April 2013, International Crisis Group, https://d2071andvip0wj.cloudfront.net/dangerous-waters-china-japan-relations-on-the-rocks.pdf, accessed 15 March 2019.

IMF (International Monetary Fund), World Economic Outlook Database, 2018, https://www.imf.org.

Jacques, Martin, *When China Rules the World: The Rise of the Middle Kingdom and the End of the Western World* (London: Allen Lane, 2009).

Jervis, Robert, *Perception and Misperception in International Politics* (Princeton, New Jersey: Princeton University Press, 1976).

Keck, Zachary, 'The Chinese Air Force's Super Weapon: Beware the J-11D Fighter,' *The National Interest*, 30 April 2015, http://nationalinterest.org/blog/the-buzz/the-chinese-air-forces-super-weapon-beware-the-j-11d-fighter-12777, accessed 15 March 2019.

Keck, Zachary, 'South Korea May Turn Its Assault Ships into F-35 Armed Aircraft Carriers,' *The National Interest*, 5 January 2018, http://nationalinterest.org/blog/the-buzz/south-korea-may-turn-its-assault-ships-f-35-armed-aircraft-23960, accessed 15 March 2019.

LaGrone, Sam, 'Pentagon Notifies Congress of Potential $3 Billion V-22 Osprey Sale to Japan,' *USNI News*, 5 May 2015, https://news.usni.org/2015/05/05/u-s-notifies-congress-of-potential-3-billion-v-22-osprey-sale-to-japan, accessed 15 March 2019.

Martin, Sean, 'China Teases Glimpse of Devastating Stealth Bomber Which Could Reach the US,' *Sunday Express*, 16 May 2018, https://www.express.co.uk/news/science/960733/china-news-world-war-3-stealth-bomber-nuclear-war-H-20, accessed 15 March 2019.

Military Today, *Sejong the Great Class*, n.d., http://www.military-today.com/navy/sejong_the_great_class.htm, accessed 15 March 2019.

Military Watch (n.d.), 'Comparing the World's Best Light Multirole Fighters, Part Five: J-10 Firebird,' http://militarywatchmagazine.com/read.php?my_data=70457, accessed 15 March 2019.

Ogura, Junko and Mullen, Jethro, 'Fresh Anti-Japanese Protests in China on Symbolic Anniversary,' *CNN.com*, 19 September 2012, http://edition.cnn.com/2012/09/18/world/asia/china-japan-islands-dispute/, accessed 15 March 2019.

Panda, Ankit, 'Japan Plans to Convert the Izumo-class Into a True Aircraft Carrier?' *The Diplomat*, 7 January 2018, https://thediplomat.com/2018/01/japan-plans-to-convert-the-izumo-class-into-a-true-aircraft-carrier/, accessed 15 March 2019.

Reuters, 'Exclusive: Japan to Buy at Least 20 More F-35A Stealth Fighters—Sources,' 21 February 2018a, https://www.reuters.com/article/us-japan-defence-f35-exclusive/exclusive-japan-to-buy-at-least-20-more-f-35a-stealth-fighters-sources-idUSKCN1G507W, accessed 15 March 2019.

Reuters, 'Exclusive: Japan to Upgrade Patriot Batteries for Olympics as North Korean Missile Threat Grows: Sources,' 29 July 2018b, https://www.reuters.com/article/us-japan-northkorea-patriot-exclusive-idUSKCN1082W1, accessed 15 March 2019.

Siedel, Jamie, 'China's J-20 Stealth Fighter Designer Promises His Aircraft Will Be a Battlefield 'Trump Card,' 23 March 2018a, *News.com.au*, https://www.news.com.au/technology/innovation/military/chinas-j20-stealth-fighter-designer-promises-his-aircraft-will-be-a-battlefield-trump-card/news-story/88a65a9e151f26825eb472a6fad753d3, accessed 15 March 2019.

Siedel, Jamie, 'Here's What We Know About China's Newest Aircraft Carriers,' *News.com.au*, 23 April 2018b, https://www.news.com.au/technology/innovation/military/heres-what-we-know-about-chinas-newest-aircraft-carriers/news-story/7460539e54fe026976276dd4cbfaa2e1, accessed 15 March 2019.

Siedel, Jamie, 'Photos Reveal China's South China Sea Island Fortresses are Complete,' *News.com.au*, 7 February 2018c, https://www.news.com.au/technology/innovation/photos-reveal-chinas-south-china-sea-island-fortresses-are-complete/news-story/776e1a695fb41ccb7e47a436594c1530, accessed 15 March 2019.

SIPRI, Military Expenditure by Country, in Constant (2016) US$ M., 1988–2017 (2018a), https://www.sipri.org/sites/default/files/1_Data%20for%20all%20countries%20from%201988%E2%80%932017%20in%20constant%20%282016%29%20USD.pdf, accessed 15 March 2019.

SIPRI, Military Expenditure by Region in Constant US Dollars, 1988–2017 (2018b), https://www.sipri.org/sites/default/files/4_Data%20for%20world%20regions%20from%201988%E2%80%932017.pdf, accessed 15 March 2019.

SIPRI, Military Expenditure by Country as Percentage of Gross Domestic Product, 1988–2017 (2018c), https://www.sipri.org/sites/default/files/3_Data%20for%20all%20countries%20from%201988%E2%80%932017%20as%20a%20share%20of%20GDP.pdf, accessed 15 March 2019.

SIPRI (n.d.), SIPRI Arms Transfers Database, https://www.sipri.org/databases/armstransfers, accessed 15 March 2019.

South China Morning Post, 'China Military Officials Admit Radar Lock on Japanese Ship, Says Report,' 29 August 2013, http://www.scmp.com/news/china/article/1193600/china-military-officials-admit-radar-lock-japanese-ship-says-report, accessed 15 March 2019.

Steinberg, James and O'Hanlon, Michael E., *Strategic Reassurance and Resolve: U.S.-China Relations in the Twenty-First Century* (Princeton/Oxford: Princeton University Press, 2014).

Straits Times, 'South Korea's First F-35 Stealth Fighter Unveiled at a Sensitive Time for Inter-Korean Diplomacy,' 29 March 2018, https://www.straitstimes.com/asia/east-asia/south-koreas-first-f-35-stealth-fighter-unveiled-at-a-sensitive-time-for-inter-korean, accessed 15 March 2019.

Sydney Morning Herald, Text of Obama's Speech to Parliament, 17 November 2011, http://www.smh.com.au/national/text-of-obamas-speech-to-parliament-20111117-1nkcw.html, accessed 15 March 2019.

Tan, Andrew T.H., *The Arms Race in Asia: Trends, Causes and Implications* (London: Routledge, 2014a).

Tan, Andrew T.H., 'The Implications of Taiwan's Declining Defense,' *Asia-Pacific Review*, vol. 21, no. 1 (2014b), pp. 41–62.

The Military Balance (London: International Institute for Strategic Studies, 2000).

The Military Balance (London: International Institute for Strategic Studies, 2013).

The Military Balance (London: International Institute for Strategic Studies, 2014).

The Military Balance (London: International Institute for Strategic Studies, 2019).

Trading Economics, *China: Economic Indicators*, https://tradingeconomics.com/china/indicators, accessed 15 March 2019.

United Nations, *Convention on the Law of the Sea*, 1982, http://www.un.org/Depts/los/convention_agreements/texts/unclos/unclos_e.pdf, accessed 15 March 2019.

Wang, Brian (2018), 'Military Confirms China's High Tech Third Aircraft Carrier Construction Details,' *Next Big Future*, 5 January 2018, https://www.nextbigfuture.com/2018/01/military-confirms-chinas-high-tech-third-aircraft-carrier-construction-details.html, accessed 15 March 2019.

35

EMPLOYMENT OF AIR POWER IN FOURTH GENERATION WARFARE

An Indian Perspective

Air Vice Marshal Arjun Subramaniam

Introduction

In the prevailing global conflict milieu that is characterised by diverse forms at the lower end of the conflict spectrum, the flexibility, reach and precision targeting offered by air power make it an increasingly attractive instrument of force in the hands of the state in its fight against the non-state actor. In such an environment, it is critical to examine the various opportunities and challenges that air power strategists will face in the twenty-first century in this most complex domain that comprises Fourth Generation Warfare (4GW) and several similar genres of conflict.

Several terms attempting to define and explain the lower end of the spectrum of conflict have emerged with great rapidity over the last few decades, with 4GW being one such honest early endeavour. Since then, however, the constantly changing nature of warfare has given birth to terms like Irregular Warfare, sub-conventional warfare, proxy war, hybrid war, Fifth Generation Warfare (5GW) and even Sixth Generation Warfare. While sticking to the employment of air power in 4GW in the narrative, readers would do well to extrapolate and consider the relevance of what is argued in the chapter across the domain of what can be broadly called today the 'lower end of the spectrum of conflict.'

In the early hours of the morning of 26 February 2019, an Indian Air Force (IAF) strike on a suspected Jaish-e-Mohammad training camp in Balakot in the Khyber Pakhtunwa region of Pakistan changed the paradigms of the employment of air power as a military instrument at the lower end of the conflict spectrum in South Asia (Subramaniam 2019: 1). Although offensive air power has been a preferred option for Western powers, including Russia, as a first responder in the Global War on Terror (GWOT) and in several instances of conflict at the lower end of the spectrum of warfare, the Balakot Strike only reinforced the emerging appeal of air power in this genre of conflict. After examining the characteristics of 4GW, this chapter will examine the impact, consequences and challenges of employing air power in 4GW. Drawing on two case studies from South Asia wherein offensive air power was successfully employed at the lower end of the spectrum of conflict, the chapter will argue that air power remains a powerful instrument of statecraft in the fight against the

DOI: 10.4324/9780429437915-44

non-state actor. In doing so, the chapter will also reinforce the full-spectrum warfighting capabilities of air forces across the world.

Understanding 4GW: Fitting Air Power In

William Lind, an ex-US Marine Corps Officer, was among the early proponents of 4GW. He argued, in the early 1990s, that it is not novel but a return to the way war worked before the rise of the state. Now, as then, many different entities, not just governments of states, will wage war. They will do so for many different reasons, not just 'the extension of politics by other means.' And they will use many different tools to fight war, not restricting themselves to what we recognise as military forces (Lind 2004).

Lind was among those who challenged the myth of modern conventional conflict in the post-Cold-War era and warned established Western nation states of the perils of sticking to the strategy of only fighting big conventional battles with technology and firepower. His warning was prescient as the world since then has been plagued by continually evolving small conflicts, driven by a host of issues ranging from the awakening of ethnic and religious aspirations at the end of the Cold War, to globalisation and a clash of ideologies, aspirations and cultures (Huntington 1997: 13). There is increasing competition for resources and recognition, even while there is rising pressure to avoid conflict. Additionally, even small conflicts can have an extended impact in an interconnected world. Earlier, 4GW was localised in its effect, whereas now a seemingly insignificant insurgency can disrupt economies and political systems on the other side of the planet (Beckett 2001: 23–39). Thus, while major wars are becoming infrequent, 4GW, with limited objectives and scope, is becoming more significant and proliferating across continents. It only makes sense that powered flight, one of the central enabling technologies of globalisation, is also central to effectively prosecuting the lead counter to 4GW by the more powerful protagonist: the state, and in some cases as will be highlighted later, by the non-state actor too. The unique capabilities of air power lend themselves readily to many of the requirements of combating proxy/non-state practitioners of 4GW. However, a great capability misused is often more destructive than helpful, and the same holds true of air power. Air power employed foolishly and in isolation in 4GW is just as likely to contribute to the defeat of the nation that wields it. To further an understanding of the employment of air power in 4GW it is critical to identify the tools of air power that fit the model.

The two Gulf Wars of recent times (1991 and 2003) epitomised what third generation warfare was all about: speed, surprise, focused firepower, physical dislocation and non-linear operations that seek to bypass and collapse the enemy (USAF Doctrine on Irregular Warfare 2007: 3). However, third generation warfare had no inbuilt mechanisms for reconstruction, or 'winning the hearts and minds of the defeated enemy.' Suddenly, in the twenty-first century, there was a distinct shift whereby the state lost its monopoly on the warfare of mass, mobility, speed or firepower (Fuller 1961: 16–22). Instead, it found itself staring down the barrel at culturally distinct, militarily well-trained, and sometimes fanatic, non-state actors who had perfected a 'potpourri' of guerrilla tactics combined with conventional firepower, terrorism and the extremism of Islamist groups like the Al Qaeda, LeT, Hezbollah and Taliban. Suddenly, even powerful states like the United States and India found themselves on the back foot while dealing with such adversaries who practised what was initially termed as 4GW, but later morphed into the more acceptable genre of hybrid warfare (USAF Doctrine 2007: 4). Overnight, twenty-first-century warfare was transformed with military theorists at a loss and debating to fix a name for a brand of warfare that loomed on the horizon. The increasing threat of non-state and proxy entities which are driven by religious fundamentalism arising from the spread of Islamist fundamentalism and extremism in Iran, Iraq, Yemen, Syria, Pakistan and Afghanistan has only made it more difficult for states to evolve suitable strategies to militarily counter them.

Simply put, 4GW is truly a broad genre of modern warfare with diverse tools, props and feeders. Terrorism, insurgencies, guerrilla movements and wars of liberation are but the various tools for waging 4GW. The main props and feeders for this are: religious fundamentalism and desire for religious domination; ethnic diversity; proxy ambitions; internal economic disparity; developmental bankruptcy; and the desire for the revival of historical legacies and geographical boundaries (Rashid 2012: 27). Every 4GW conflict that is raging today would have one or more of the above-mentioned props and feeders. As we look back on the first two decades of the twenty-first century, even 4GW seems to be giving way to a new kind of warfare that relies overwhelmingly on information highways and networks for reducing decision time cycles. Some analysts and writers even go to the extent of describing this as the emergence of the 5GW (Berkowitz 2003: preface and ch. 1).

Seeking to create its own space, the United States Air Force (USAF) first articulated its doctrine on this genre of warfare through its *Air Force Document Doctrine 2–3: Irregular Warfare* published in August 2007. Writing in the foreword to the document, the USAF Chief of Staff, General Michael T. Mosely argued:

> Employed properly, airpower (to include air, space, and cyberspace capabilities) produces asymmetric advantages that can be effectively leveraged by joint force commanders in virtually every aspect of irregular warfare. Irregular warfare is sufficiently different from traditional conflict to warrant a separate keystone doctrine document. While the fighting experiences in Iraq and Afghanistan should weigh heavily in the development of our doctrine, we intend this doctrine document to be broad, enduring, and forward-looking.
>
> *(USAF Doctrine on Irregular Warfare 2007: Foreword)*

Around the same time, the French Air Force too found it necessary to articulate its views on the employment of air power at the lower end of the spectrum of conflict after its experience over the Balkans and its inherent objection to the overwhelming use of firepower against poorly defined targets. Colonel Regis Chamagne, in his book, *The Art of Air War,* firmly positions air power as a tool that must strive to shorten the OODA (observe, orient, decide, act) loop and work towards a 'Zero Dead and Zero Collateral Damage' strategy (Chamagne 2007: 207–11). The fourth edition of the *British Air and Space Power Doctrine* (AP 3000; British Ministry of Defence 2009) clearly enunciates the importance of hybrid and irregular warfare in the RAF mind space in its chapter 'Air Power in the Contemporary Operating Environment.' The doctrine accepts the constancy of violent conflict, but recognises the change in its character as 'adversaries have come to understand the extent of Western warfighting dominance and have developed irregular strategies to counter it' (British Ministry of Defence 2009). Consequently, it goes on to highlight the blurring of 'boundaries between the strategic, operational and tactical levels' and the ability of air power to offer 'unique capability to provide strategic influence almost independently,' thereby 'remaining a very attractive option politically' to deliver effects without commitment of surface forces. The IAF was not far behind to articulate its understanding of sub-conventional warfare in its *Basic Doctrine of the Indian Air Force* that was published in 2012. Demonstrating a clarity and shift from its traditional focus on conventional operations to what it calls Full Spectrum Operations, it recognises the growing proliferation of sub-conventional warfare and showcases the diverse and effective tools that air power possesses to make a significant contribution in such operations (IAF Headquarters 2012: ch. 8).

Tenets, Roles and Missions

Colonel Phillip Meilinger, an accomplished practitioner-scholar and former Dean of the School for Advanced Air and Space Power Studies (SAASS) at the USAF's Air University, laid out 'Ten Propositions Regarding Airpower' in 1996 (Meilinger 1996: 2–3). Essentially meant to simplify an understanding of roles and missions in conventional conflict, most propositions remain relevant across the lower end of the conflict spectrum and merit some examination below:

- Whoever controls the air generally controls the surface.
- Air power is an inherently strategic force.
- Air power is primarily an offensive weapon.
- In essence, air power is targeting, targeting is intelligence: and intelligence is analysing the effect of air operations.
- Air power produces physical and psychological shock by dominating the fourth dimension—time.
- Air power can simultaneously conduct parallel operations at all levels of war.
- Precision weapons have redefined the meaning of mass.
- Air power's unique characteristics require centralised control by airmen.
- Technology and airpower are synergistically related.
- Air power not only includes military assets but aerospace industry and commercial aviation.

The first proposition does not really hold well because, although controlling the air may enable the state to retain situational awareness with respect to the non-state actor, it does not afford dominance unless complemented by swift surface action. The second proposition holds true as air power retains the ability to create strategic effects even in 4GW by its characteristics of precision, speed, vertical envelopment, surprise and shock effect. Some examples could be given. Take for instance, the use of air power to target the LTTE (Liberation Tigers for Tamil Eelam) leadership in 2007–2008 during the closing stages of the Sri Lanka–LTTE conflict (Subramaniam 2008: 30–2). Then, the air strikes over Tora Bora in the mountains of Afghanistan during the early days of *Operation Enduring Freedom* in January 2002 that resulted in the displacement of the top Al Qaeda leadership or the persistent targeting of Taliban/Al Qaeda leadership by drones in AF-Pak over the past 15 years (Lambeth 2010: 255–77). All these cases show that air power that has continued to shape the battle space.

Are the offensive characteristics of air power the primary drivers of air power employment in 4GW? While they surely are important, it must be understood that several non-kinetic characteristics of air power are also of significance in the fight against the non-state actor. Of the main options for the elimination of Osama Bin Laden during *Operation Neptune Spear*, precision attacks from the air and vertical envelopment from the air by special forces were the two main options that were finally considered (Subramaniam 2011: 47–50). One was a direct kinetic option, and the other, a non-kinetic application of air power. Therefore, it is not only offensive action, but a wide range of non-kinetic applications of air power that also have the potential to defeat the non-state actor. Air power retains the capability of creating physical and psychological shock across the spectrum of conflict. This shock is caused by the application of firepower without warning. Another dimension of shock that merits serious study is the disorientation caused by the stealthy insertion of combat ready special forces from the air into the battle space where the adversary is least expected by the enemy. Nothing typifies this proposition better than *Operation Neptune Spear* and the ongoing drone strikes against Taliban and Al Qaeda terrorists in the AF-Pak region. The ability of air power to simultaneously conduct operations across the spectrum of conflict can be questionable in the context of the long, drawn out war in Iraq and the need to redefine air power strategies when conventional combat operations came to an end, both in Iraq and Afghanistan. It took a while for the coalition forces to see value in shifting from fixed wing air strikes to covert CIA controlled drone strikes (Bergen and Tiedman 2011: 12–15).

It was not a seamless shift, lending strength to the argument that while air power can conduct highly effective parallel operations across nuclear to conventional conflicts, a significant shift in focus is required for 4GW/conflict at the lower end of the spectrum of warfare. Air-delivered precision capability is all pervasive and equally effective against a camouflaged communications node in the Tactical Battle Area (TBA) and a terrorist leadership node. In fact, the need to reduce collateral damage in 4GW scenarios has spurred efforts for increasing the accuracy of precision weapons, contrary to popular belief that small team operations in LICO or 4GW are decentralised to a very large extent; and that air power too must be decentralised in support of such operations. One must look no further than the highly centralised CIA-controlled drone attacks in the AF-Pak region that are closely orchestrated from Langley by drone specialists. Even Operation Neptune Spear, a classic anti-terrorist operation that used air power as a critical tool to facilitate the elimination of Osama bin Laden, was a highly centralised operation. In the final analysis, barring two of the propositions, the rest of Colonel Meilinger's propositions are equally applicable to 4GW. They are truly enduring propositions considering that they were laid down at a time when air power strategists were reluctant to look beyond the application of air power in conventional warfare.

Intelligence, Surveillance and Reconnaissance (ISR) remains the most important aerospace mission in 4GW. To locate, fix, identify, track, target and engage the non- state actor is significantly different from doing the same for conventional targets in a conventional scenario. Unmanned aerial vehicles (UAV) and satellites would form the primary tier of sensors that need to be synergised as part of a 'persistent stare mosaic' that needs to be created to continuously monitor the activities of non-state adversaries. Aggressive ISR immediately puts the terrorist or insurgent on the back foot and demands that all sensors be linked to a centralised intelligence dissemination centre that can be immediately accessed by the apex decision-making or crisis management group. Put very simply, decision-makers should be able to 'see and hear the non-state actor' on a real-time basis and transfer relevant information to the soldier on the ground or pilot in the air who is being tasked with targeting. The only way that the 'fog of uncertainty' in such warfare is penetrated, in the requisite time frame, is if there is synergy between all intelligence agencies. This is something that is an absolute imperative if we want to stay ahead of the non-state actor.

Use of offensive air power in 4GW has always been a debatable and contentious issue. It has been indiscriminately used by the US and Russia in Chechnya, Iraq, Afghanistan and Syria with mixed results; effective if one looks at the target systems that were engaged that mainly comprised non-state and terrorist leadership/feeder systems, and ineffective if one looks at the collateral damage and inability to control the alienation of the local population. However, advances in modern air-delivered weaponry with longer ranges, precision with discrimination and proportionate destructive capability, create circumstances that encourage the use of air power to avoid direct engagement by surface forces until sufficient attrition has been caused to the non-state actor.

There is no doubt about the importance of all the non-kinetic and supporting roles of air power, ranging from casualty evacuation and joint operations with special forces, to humanitarian relief and logistic support missions at the lower end of the spectrum of warfare. Apart from leveraging such capabilities in integrated operations, such missions are vital in showing the 'human face' of the state and accelerating the process of 'Winning the Hearts and Minds' of citizens affected by 4GW.

The more specific roles of air power that lend themselves very easily to 4GW are listed below (Subramaniam 2009: 162–3).

- Surveillance and reconnaissance by all available platforms (satellites, Unmanned Aerial Vehicles and manned aircraft) are vital to find, fix, target and engage the mobile and highly elusive non-state actor as compared with static targets that are associated with conventional conflict.
- Targeting terrorist and non-state leadership along with the destruction of terrorist infrastructure, launch pads and training camps has emerged as a key role for offensive air power and has yielded rich dividends in various conflicts in the recent past. Systematic elimination of the Al-Qaeda leadership in Iraq, Yemen and Afghanistan and the LTTE leadership in Sri Lanka has contributed significantly to the path that the conflicts have taken.
- Integrated operations in conjunction with Special Forces using medium lift transport aircraft and utility helicopters for insertion and extraction of forces into and out of the combat area. Complementing them would be Attack/armed helicopters for sanitising the area and providing fire support and, lastly, fighter aircraft to ensure air defence cover so that the 'proxy' country that is providing support to the non-state actor is not able to provide air support to him.
- Air mobility and logistics support to ground forces that are engaged in COIN, counter terrorism and proxy war operations using fixed wing aircraft and helicopters remains a critical role.
- Other key roles for air power in 4GW involve casualty evacuation and search and rescue missions, humanitarian relief missions, 'show of force' and peacekeeping missions under the UN flag.

Some Case Studies

Most analyses and studies of the employment of air power in this genre of warfare over the past two decades have involved Western powers and revolved around conflicts in the Balkans, Lebanon, the AF-Pak region, Iraq and Afghanistan. This chapter, however, will examine the employment of air power during the closing stages of the Sri Lankan Armed Forces' campaign against the LTTE and the coercive and preventive strikes by the Indian Air Force against a suspected terrorist camp of the Jaish-e-Mohammad at Balakot in the North-West Frontier Province of Pakistan.

Sri Lanka: Air Power Against the LTTE

The ethnic conflict between the LTTE and the Sri Lankan government, which came to a bloody end in early 2009, was one of the longest ethnic conflicts of modern times (Wickremesekara 2016: ch. 1). The growth of the LTTE from a rag-tag bunch of guerrilla fighters into the world's most dreaded terrorist organisation with superb hybrid fighting skills commenced in the late 1980s during the Indian military intervention in Sri Lanka, which lasted from mid-1987 to early 1990. Following the withdrawal of the Indian Peacekeeping Force (IPKF) from the island after it failed to execute the mandate of the India–Sri Lanka Accord that was meant to resolve the ethnic crisis and stop the vicious fighting between the LTTE and the Sri Lankan state forces, the LTTE emerged as a force to reckon with. As the country lurched from one ceasefire to another, the LTTE morphed into a multi-dimensional fighting force that became adept at switching between classical guerrilla tactics and perpetrating ruthless terror attacks against urban targets and innocent civilians. An interesting feature of the closing years of the conflict, also called the Eelam Wars, was the psychological impact of the use of air power by a non-state actor (Subramaniam 2008: 33–5).

Faced with a feeble military response from the state through much of the 1990s, an emboldened LTTE even acquired a few slow-moving Czech Zlin turbo-prop aircraft and launched aerial attacks against a major Sri Lanka Air Force Base. Till very recently, air power assets with non-state actors mainly comprised a variety of ground-based air defence weapons like anti-aircraft guns and Man Portable Air Defence Systems (MANPADS) of the Stinger and SAM-7 class. Taking a leaf out of the Al Qaeda attacks on the twin towers, the LTTE made remarkable progress in putting together an air wing of its own. Apart from training aircrews in Europe, it acquired a few light aircraft with limited armament carrying capability, supposedly of Czech origin, and built a few airstrips in Northern Sri Lanka in the area northwest of Mullaitivu. More than any long-term impact, the Tamil Eelam Air Force (TAF) was meant to be used as a psychological weapon against Colombo. In an air strike on Katunayake air base on the outskirts of Colombo on the night of 26 March 2007, two unidentified TAF aircraft bombed a brightly lit hangar facility of the Sri Lanka Air Force (SLAF), inflicting significant but unconfirmed damage to equipment and aircraft. What was surprising in this stealthy attack was the inability of air defence sensors or weapons to intercept the aircraft. It is also believed that locally manufactured and fused 250 kg bombs were fitted on crude bomb racks and dropped with the help of GPS assistance for reasonable accuracy. In a second coordinated strike on 22 October 2007, a suicide commando group of the Black Tigers in conjunction with two aircraft of the TAF carried out a precision attack on the Anuradhapura air base destroying many aircraft that are said to have included three helicopters (Mi-24/17), two fixed wing aircraft, three drones/UAVs and an expensive Beechcraft surveillance plane. The raid was executed with unbelievable courage and precision, sending shocks through the world community at use of such tactics by non-state actors. The attack was well planned with the commandoes neutralising a radar station and an anti-aircraft position before calling for the air strike. The series of air attacks by the TAF galvanised the SLAF into action and what followed was a series of strikes by Mi-24s Kfirs and MiG-27s against LTTE camps and patrol boats, the LTTE airstrip complex at Irramadu and, most importantly, the LTTE leadership (Subramaniam 2008: 30–1; Raman 2007: 107).

Post-2004, the Sri Lankan Armed Forces embarked on a modernisation spree and more importantly displayed an increasing willingness to use offensive air power and target the LTTE leadership and destroy LTTE camps even if they were embedded in urban areas. While the Al Qaeda and Taliban leadership remain highly mobile and had safe havens in multiple countries, the LTTE leadership was confined to the Jaffna Peninsula and Northern Sri Lanka. Additionally, Prabhakaran, the LTTE supremo, was known to prefer bunker-based refuges and is said to have moved around with a large entourage, making him relatively vulnerable to air strikes, something that the SLAF took advantage of. The air strike by SLAF jets on Thiruvairu, a suburb of Kilinochchi, on 2 November 2007 resulted in the elimination of the LTTE Political Commissar, SP Tamilselvan (Subramaniam 2008: 31–2).

Considering that Kilinochchi was considered as the capital of the LTTE-controlled northern province of Wanni, the SLAF exploited the coercive and deterrent capability of air power without any ground action. This scenario was re-created on 26 and 28 November when the SLAF hit two 'high value' targets in Jeyanthi Nagar in Kilinochchi. The targets comprised bunkers in which Velupillai Prabhakaran had taken refuge and was reported to have been injured during the strike. The extent of the injuries to Prabhakaran is not as important as the fact that Prabhakaran was made to look extremely vulnerable in his limited area of operation. Eelam War-4, as the undeclared final phase of the war was termed by the LTTE, proved to be extremely critical as it saw a comprehensive defeat of the LTTE by the resurgent Sri Lankan armed forces backed by effective utilisation of air power that targeted the LTTE's centres of gravity.

India Strikes Terror Camps

As the Central Reserve Police Force (CRPF) convoy crawled along the Jammu-Srinagar highway in the Indian state of Jammu and Kashmir on 14 February 2019, it was waylaid at Pulwama by a Jaish-e-Mohammad suicide bomber. Ramming into the convoy with his explosive-laden SUV, 22-year-old Adil Ahmad Dar triggered off a chain of events that not only killed over 40 security personnel, but also changed the trajectory of India's response mechanisms following terrorist attacks. Aided and abetted for decades by the Pakistani 'Deep State,' comprising elements of the Pakistan Army, its highly effective intelligence agency, the Inter-Services Intelligence Agency (ISI), and components of the *jihadi* complex, the broad aims of this conglomerate have been the secession of Jammu and Kashmir, and the erosion of Indian power by covert means. The Indian strategic establishment, which in 2016 had responded to an earlier terrorist attack with shallow cross-border strikes by special forces, re-alised that Pakistan would be ready for any kind of surface retaliation following the Pulwama attack. This meant that an air strike against non-military targets emerged as the only other viable coercive response (Subramaniam 2019: 8–9).

Air Force Station Gwalior hummed like a hornets' nest in the early hours of the morning of 26 February 2019, even as the rest of the country slept. While the ground crew were busy strapping on Israeli-made Spice and Crystal Maze bombs and air-to-air missiles on a combination of both older and upgraded Mirage-2000s, all the commanding officers of the three Mirage squadrons in Gwalior from the Tigers (1 Squadron), Battle Axes (7 Squadron) and Wolf Pack (9 Squadron) were busy conducting their individual element briefings and going over contingency plans and escape drills. It was made clear to the air crews that there was no possibility of a planned rescue, in case they ejected in enemy territory. The target was Juba Top and the strike element comprised 12 Mirage-2000 THs, mostly armed with a single Spice 2000 bomb and a few with the older Crystal Maze bombs. Escorting them were several upgraded Mirage-2000s with MICA Beyond Visual Range (BVR) Missiles—the closest IAF counter to the potent AIM-120 AMRAAM that were carried by PAF F-16s. There were no surprises though in Gwalior, as the roar of 20 Mirages becoming airborne from the two runways clearly indicated that something indeed was cooking. Joining them en route were four SU-30 MKIs from 15 Squadron for additional air defence protection, while a diversionary package of Jaguars and SU-30s headed towards the Jaish-e-Mohammad's headquarters at Bahawalpur to the south, to draw away PAF air defence fighters and AWACS, which had been seen orbiting near the PAF air base of Murid. Demonstrating reach and precision and maintaining complete radio silence, the fuel-efficient Mirages flew over 1,300 km one way with an aerial refuelling; used the mountains to mask their entry into Pakistan-Occupied Kashmir (POK) airspace; and delivered five Spice 2000 weapons with unerring accuracy onto selected buildings that housed terrorists. Unfortunately, the Crystal Maze bombs could not be dropped due to some last-minute technical glitches—which proved costly in the post-strike narrative as these stand-off weapons are video-linked with the targeting pod till impact, unlike the more accurate Spice weapons that are autonomous after launch. The cloud cover over the target area had prompted a rethink and possible delay, but the IAF leadership took a call that was vindicated by the ability of the Spice weapon to gather adequate target discrimination details and strike the target with unerring accuracy.

Launching their weapons from well inside POK onto targets across the International Border, barely 160 km north of the Pakistani capital, Islamabad, the coercive signaling to induce a change in behaviour was clear and palpable. Contrary to reports that the PAF had been caught napping, a post-strike IAF review claims that several pairs of PAF interceptors under Airborne Early Warning and Control (AEWC) had been scrambled, but had failed to intercept the Mirages, or were drawn away by the IAF deception strikes, seemingly headed towards Bahawalpur, the headquarters of the Jaish-e-Mohammad. The decoys turned back short of the International Border, allowing a free run to the Mirage package to the north.

The Balakot strikes were planned in utmost secrecy and executed with finesse and surprise, but there were also a few technical glitches that diluted their overall impact and denied the Indians real-time information on target destruction. The synergy between the three services was evident as the Indian Army and Indian Navy ensured that no unnecessary troop or fleet movement was evident prior to the air strike and army formations moved to their forward locations only after the strike mission returned. At the politico-strategic level, proactive deterrence remains the consistent intent of the Modi government with prevention and pre-emption emerging as likely operational strategies against possible terrorist attacks. The escalation ladder was tested further when, for the first time since 1971, Indian forces attacked targets in mainland Pakistan. Notwithstanding the political rhetoric and hyperbole that followed the attacks, and the chaotic information war that ensued, the Pakistani Deep State got the message that India would not only resort to diplomacy to expose Pakistani machinations, but accompany it with suitable punitive action. Not a single country, including China, criticised the Indian air strikes, and its broader strategy of preventive military action against non-military terror targets inside Pakistan.

Challenges

The single biggest dilemma facing the state in general, and democracies specifically, relates to the ethical dimension of civilian casualties and collateral damage whenever offensive or kinetic air power is used against non-state actors, especially if they are embedded in local population or constitutionally belong to the same stock of people who comprise the nation state. The United States or Israel have used offensive air power against non-state actors, but it has been on foreign soil against non-state actors who, presumably, were a danger to their sovereignty, way of life and national interest. In that context, as far as these two countries are concerned, they are at war and are willing to pay the price for protecting their sovereignty even if it evokes international condemnation and censure. The bottom line as far as they are concerned is that any action that saves lives of American or Israeli soldiers, and results in the destruction of the centres of gravity of the non-state actor is justified. This strategy is not without merit, particularly when the non-state actor of recent times has become ruthless and has adopted unethical and unscrupulous war-fighting techniques that follow no rules or conventions. India, however, is a fascinating case study when it comes to analysing national strategy to fight 4GW.

Unlike the United States and Israel, for whom the source and pedigree of terror are reasonably well defined, India faces a combination of indigenous, proxy and transnational terror threats from both within its own territory and outside, which are increasingly being referred to in the West as proxy and hybrid threats. It is this 'hybridity' that poses a significant challenge to the Indian strategic community. Even though India has fought insurgencies and a proxy war for over four decades, been the victim of numerous terrorist attacks that have killed its leadership and hundreds of citizens, and has the experience of fighting a ruthless guerrilla turned terrorist force (LTTE) in Sri Lanka, there is no comprehensive national strategy to fight a 4GW either on home or foreign soil. The whole dilemma of using offensive or kinetic air power in 4GW boils down to the 'attrition threshold' of a state. Attrition threshold is a term I coined to indicate the extent of the punishment a state can allow a non-state actor to inflict upon it. The higher the threshold, the greater will be the desire of the non-state actor to expand his capabilities and erode the sovereignty of the state. The use of offensive air power by the Indian state against the non-state actor clearly indicates the lowering of India's attrition threshold and marks a new phase in 4GW in South Asia.

Conclusion

The underlying essence of the chapter aims at reinforcing the argument that air power as an instrument of stated national policy can be used effectively even at the lower end of the spectrum of conflict in general, and in 4GW. The proliferation of terrorism and myriad aspirations of non-state actors, along with their capability to target the soft underbelly of established democracies and nations, has created an 'asymmetry' that is cause for some concern. This asymmetry has led to the emergence of new genres of warfare that have placed fresh challenges on militaries the world over. In such a scenario, it is important to realign the roles, missions and capabilities of air power to tackle sub-conventional threats from guerrillas, insurgents, terrorists and religious fundamentalists who threaten national security.

Air power resources must be effectively networked with intelligence sources, command and control elements and other parallel lines of operations, both military and civilian, to respond rapidly and precisely. Air power has the unique capability of exercising the 'carrot and stick' policy over non-state actors and the environment they operate in by operating simultaneously over multiple lines of operation (LOOs). The 'carrot' is offered in the form of non-kinetic missions that include humanitarian and relief missions, casualty evacuation and supply drops in hostile territory, while the 'stick' can be effectively employed, only when necessary, in the form of well-orchestrated precision strikes against non-state leadership and infrastructure and fire support to ground forces. Most importantly, weighing the pay-offs and the risks of employing air power in a calibrated manner would be the hallmark of a mature democracy.

Bibliography

Beckett, Ian, *Modern Insurgencies and Counter Insurgencies: Guerrillas and Their Opponents since 1750* (New York: Routledge, 2001).

Bergen, Peter and Tiedman, Katherine, 'Washington's Phantom War: The Effects of the U.S. Drone Program in Pakistan,' *Foreign Affairs*, vol. 90, no. 4 (July/August 2011), pp. 12–18.

Berkowitz, Bruce, *The New Face of War* (New York: The Free Press, 2003).

British Ministry of Defence, AP 3000: *British Air and Space Power Doctrine Fourth Edition*, 2009, accessed online on 18 December 2019 at http://www.defencesynergia.co.uk/wp-content/uploads/2015/05/RAF-AP3000-Air-Power-Doctrine-4th-edition-2009.pdf

Chamagne, Regis, *The Art of Air War* (Paris: L'Espirit du Livre, 2007).

Fuller, J.F.C., *The Conduct of War* (New Brunswick: Rutgers University Press, 1961).

Huntington, Samuel P., *The Clash of Civilizations and the Remaking of the World Order* (New Delhi: Penguin, 1997).

IAF Headquarters, *Basic Doctrine of the Indian Air Force: IAP 2000–12* (New Delhi: IAF Press, 2012).

Johnson, Ed and Ondaatje, Anusha, 'Sri Lankan Economy feels the heat of LTTE's air wing raids,' *Live Mint*, 11 May 2007, accessed online 20 December 2019 at https://www.livemint.com/Politics/OAwjAm7Bm3rTTUHZThw9wO/Sri-Lankan-economy-feels-the-heat-of-LTTEs-air-wing-raids.html

Kainikara, Sanu (ed.), *Friends in High Places: Air Power in Irregular Warfare* (Canberra: Air Power Development Centre, 2009).

Lambeth, Benjamin S, 'Operation Enduring Freedom,' in John Andreas Olsen (ed.), *A History of Air Warfare* (Washington DC: Potomac Books, 2010), pp. 255–77.

Lind, William S., 'Understanding Fourth Generation War,' 15 January 2004, accessed online 16 November 2019 at https://original.antiwar.com/lind/2004/01/15/understanding-fourth-generation-war/

Meilinger, Phillip S., 'Ten Propositions Regarding Airpower,' *Air Power Journal*, vol. 10, no. 1 (Spring 1996), pp. 50, 52–72.

Olsen, John Andreas (ed.), *Airpower Applied* (Annapolis: Naval Institute Press, 2017).

Raman, B., 'LTTE's Anuradhapura Raid,' *Indian Defence Review*, October–December (2007), p. 107.

Rashid, Ahmed, *Pakistan on the Brink* (London, Allen Lane, 2012).

Stein, Sam, 'Obama had Multiple Options But Did Not Hesitate,' *Huffington Post*, 5 February 2011, accessed online on 9 August 2012 at http://www.huffingtonpost.com/2011/05/02/obama-had-multiple-option_n_856280.html

Subramaniam, Arjun, 'The Use of Air Power in Sri Lanka: Operation Pawan and Beyond,' *Air Power Journal*, vol. 3, no. 3 (July–September 2008), pp. 15–35.

Subramaniam, Arjun, 'Air Dominance in 4th Generation Warfare,' *Air Power Journal*, vol. 4, no 2 (Summer 2009), pp. 149–65.

Subramaniam, Arjun, 'Strategies to Tackle Fourth Generation Warfare (4GW): An Aerial Perspective,' *Strategic Analysis*, vol. 34, no. 5 (September 2010), pp. 756–65.

Subramaniam, Arjun, 'Air Power in Stability and Anti-Terrorist Operations,' *Air Power Journal*, vol. 6, no. 3 (2011), pp. 25–51.

Subramaniam, Arjun, 'Doctrinal Evolution in the Indian Air Force: Towards a Strategic Future,' in Harsh Pant (ed.), *Handbook of Indian Defence Policy: Themes Structures and Doctrines* (New Delhi: Routledge, 2016), pp. 219–32.

Subramaniam, Arjun, 'The Indian Air Force, Sub-Conventional Operations and Balakot: A Practitioner's Perspective,' *ORF Issue Brief: National Security*, Issue No. 294, May 2019, accessed online 1 June 2019 at https://www.orfonline.org/wp-content/uploads/2019/05/ORF_Issue_Brief_294_IAF-Balakot.pdf

USAF HQ AFDDEC, *Irregular Warfare: Air Force Document Doctrine 2–3*, 1 August 2007, accessed online 19 December 2019 at https://fas.org/irp/doddir/usaf/afdd2-3.pdf

Wickremesekara, Channa, *The Tamil Separatist War in Sri Lanka* (Oxon: Routledge, 2016).

INDEX

Note: Locators in *italic* indicate figures, in **bold** tables and in ***italic-bold*** boxes.

Reid, Richard 132–134, 135, 138
Renault, Louis 378
Rice, James D. 323
The Rise of the Carthaginian Empire (Turner) 400
Robert M. Utley 324
Rodger, N.A.M. 270, 271
Roman Gaul to Merovingian France warfare transition 8–9, 97–106; civil population militia levies 102; *civitas* 99, 101, 102–103, 104; Clovis, Aquitaine conquest 103–105; *comitatenses* 99, 100, 101; field army to *civitas* power transfer 102, 103; geographical, social-historical context 97–98; *limitanei* 100, 101, 102; military background 9, 98–99; military effectiveness 101–105; *Notitia Dignitatum* 99, 100; recruitment system failure 100–102; Roman army structure (*comitatenses, limitanei*) 99–100; Roman fortification systems 8, 98–99; Roman fortification systems / fortress cities (*urbes*) 105–106; *see also* pre-modern warfare, polities and armed forces
Rostam 202, 204
Royal Navy 355; British Army (1870–1918) 367, 368; British naval power rise, Indian Ocean 262, 265, 269, 270, 271, 272, 273; Modern Naval Warfare (1815–2000) 404, 408, 410
Royster, Charles 361
'Rules for the Use of Aircraft in War' (Douhet) 416
Russo–Japanese War (1904–1905) 405

sābu 28, 39n1
samurai xvii, 195–196, 199, 243
Scammell, G.V. 262, 266–267
Second World War (1939–1945): armoured technology and warfare, tank developments 16, 383–387 (*see also* tank development); Atlantic naval war (1939–1945) 407; Indian Army (1945–1947) 441–443; Italian Ethiopia invasion (1936) as precursor 381; naval warfare 17, 405–407; race and conflict, Asia–Pacific Region (1941–1945) 17 (*see also under own heading*); strategic air power 17, 420–423
Servan, Joseph 346, 348
Shahnameh (Ferdowsi) 202, 204, 207, 208
shaikh xvii, 131
shaman xvii, 322
Shannon, Timothy J.
shogunate xvii, 247, 333, 338
Silverman, David J. 322

Singh, Mohan 442
Sino–Japanese War (1894–1895) 404
sipahi xvii, 127, 208
Six Day War (1967) 432–433, 434
Small Wars, as colonial/imperial counterinsurgency 4, 16, 17, 18, 369; *see also* British Army (1870–1918), small to trench war
Small Wars: Their Principles and Practice (Callwell) 17, 369
smallpox: inoculation as immunity protection 15, 308–309, 310–312; previous exposure as immunity protection 308, 309; smallpox characteristics/symptoms and immunity 307–308, 309
smallpox, American War of Independence (1775–1783) 311–312, 313
smallpox, Amerindian society devastation 327
smallpox, Spanish–American War (1898) 314
smallpox and war, European vs Native American conquests 305–315; Columbian Exchange model 306, 307, 315; conquest via disease, early contacts 15, 306–307; fear, effect on Amerindian alliance decisions 310, 313, 315; immunity by inoculation, European armed forces 310–312; immunity by previous exposure, European armed forces 308, 309; indigenous narratives and rumours 312–313; smallpox spread strategy as biological warfare 305, 313, 315; war action and smallpox epidemic outbreaks 309–310, 312; *see also* gunpowder, influence on warfare and economy
smallpox as biological weapon 305, 313; American War of Independence (1775–1783) 313; Ottawa smallpox narrative (1757) 313; smallpox blankets (Fort Pitt / Pontiac War, 1763) 305; smallpox blankets, handkerchiefs (Fort Pitt / Pontiac War) 313–314; US vaccination campaign (2002–2003) 313
smallpox outbreaks, warfare related 309–310
Sneferu 217–218
Sokoto *Caliphate* 131
South African War / Second Boer War (1899–1902) 370–371
South Asia: decolonisation wars 440–447
South Korea, contemporary defence spending 491, 492, 493, **493, 494**, 497, 498, **498**, 499–500; *see also* East Asia, contemporary arms race
Southeast Asia: Americans-Japanese aircraft carrier race in (Second World War) 422; Burma (Myanmar) 198, 333, 335,

For Product Safety Concerns and Information please contact our EU
representative GPSR@taylorandfrancis.com
Taylor & Francis Verlag GmbH, Kaufingerstraße 24, 80331 München, Germany

www.ingramcontent.com/pod-product-compliance
Lightning Source LLC
Chambersburg PA
CBHW081216220326
41598CB00037B/6801